Textbook of Energy Balance, Neuropeptide Hormones, and Neuroendocrine Function

Eduardo A. Nillni

Editor

Textbook of Energy Balance, Neuropeptide Hormones, and Neuroendocrine Function

Springer

Editor
Eduardo A. Nillni
Emeritus Professor of Medicine, Molecular Biology
Cell Biology & Biochemistry
Department of Medicine, Molecular Biology
Cell Biology & Biochemistry
The Warren Alpert Medical School of Brown University
Providence, RI, USA

ISBN 978-3-030-07787-7   ISBN 978-3-319-89506-2   (eBook)
https://doi.org/10.1007/978-3-319-89506-2

This Springer imprint is published by the registered company Springer International Publishing AG part of Springer Nature.
The registered company address is: Gewerbestrasse 11, 6330 Cham, Switzerland

To my wife Marina and my two daughters Yael and Anna. Without the constant support of my family over the years to go through the treacherous scientific career, my life in science would not have been possible.

Preface

Nutrition is a fundamental mechanism for survival and eating is one of the greatest human pleasures, but it can also lead to under-nutrition, over-nutrition, and eating disorders. In modern human society, excess eating for susceptible individuals is leading to a growing obesity epidemic in the United States and other developed countries. Obesity or overweight state is the most dangerous nutritional disorder of the twenty-first century affecting 2.1 or more billion individuals worldwide. There are more obese and overweight people on the planet than people suffering from malnutrition. Recent data suggest that the prevalence of obesity in the United States is 35% for men and 40% for women. According to the Center for Disease Control, by 2030, medical costs associated with treating preventable obesity-related diseases are estimated to increase from $48 billion to $66 billion per year in the United States, and the loss in economic productivity could be between $390 billion and $580 billion annually by 2030. There have been many interventions developed to treat obesity, including dietary changes, increased activity and exercise, behavior modification, prescription weight-loss medication, and weight-loss surgery; however, most of the interventions have failed. Specifically, most diets offered by different outlets have a failure rate of 75–95% after 1 or 2 years of intervention. Therefore, despite all of these tremendous efforts, obesity not only remains an unsolved problem, but obesity continues to rise.

The survival of all species depends on their ability to acquire energy for its daily use and storage. From an evolutionary standpoint, feeding or intake of calories from meal to meal is necessary to (a) satisfy nutritional and metabolic requirements and (b) prepare for periods of food shortage during seasonal changes. Energy balance is the relationship that exists between energy intake (i.e., calories taken from food and drink) and energy expenditure (i.e., calories used for our daily energy requirements). Hence, the maintenance of this balance is achieved by the integration of (a) environmental signals (i.e., environmental cues steer individual's decisions concerning food intake and food choice); (b) physiological and metabolic signals (i.e., hormones, nutrient sensors, and critical organs); (c) genetic makeup (i.e., multiple genetic interactions); and (d) social and hedonic influences (i.e., the drive to eat to obtain pleasure without an energy deficit). When this balance is disrupted, we witness either a weight gain or a weight loss. In the case of weight gain, energy imbalance is caused by a higher calorie intake versus the number of calories burned. Therefore, we can define obesity as a disorder of energy balance. This longstanding imbalance between energy intake and energy expenditure is influenced by a very complex set of biological pathway systems regulating appetite. In summary, obesity results from multiple genetic and environmental factors that likely interact with each other. Specifically, genes operate additively and through gene-gene interactions to influence body weight. A longstanding obesity state will eventually cause type 2 diabetes, dyslipidemia, mood disorders, heart disease, liver disease, hypertension, reproductive disorders, and cancer risk. The aim of this textbook, *Energy Balance, Neuropeptide Hormones, and Neuroendocrine Function*, is to provide a comprehensive description and discussion of the latest knowledge in peptide hormones and their action on energy balance, as well as a cutting-edge analysis by leading experts in the field. Peptide hormones play a central role in many physiological and metabolic processes acting in concert with other molecules to regulate an array of molecular mechanisms. Research from the last 30 years has shown the importance of peptide hormones in fields such as neuroscience, immunology, pharmacology, cell biology, and energy metabolism.

This book is divided into 5 parts. ***The first part*** begins by examining the evolutionary history of the human society from a thin phenotype to the obese phenotype. Within that context, this part analyzes how modern society, social habits, and the development of industrial food production did not respect the evolutionary traits resulting in changes in the energy balance weight set point. It defines obesity in the context of the hedonic influence in the modern world, the importance of thermogenesis in the conquest of new environments, and the thermogenic adaptation during human migration across the globe. The heat production of uncoupling proteins in brown adipocyte tissue mitochondria is believed to be a key driver behind the conquest of a variety of environments in mammals 65 million years ago. This ability to produce and maintain heat contributed to the evolution of mammals to explore and settle in uninhabitable territories throughout the planet by adjusting the thermoregulatory response to sharply different environments. Several early evolutionary hypotheses to explain the development of obesity and metabolic syndrome are also discussed in this part. ***The second*** part emphasizes the analysis of a particular region of the forebrain below the thalamus, the hypothalamus. The hypothalamus coordinates both the autonomic nervous system and the activity of the pituitary gland, which controls body temperature, thirst, other homeostatic systems, and hunger. It is also involved in sleep and emotion activity. The emphasis in this part of the book is on the biosynthesis of neuropeptides and their role in energy balance regulation, the transcriptional regulation of genes expressed within the hypothalamus, and discusses the questions that remain for us to understand the hypothalamic transcriptome and resulting proteome. It examines the interaction between neuropeptides and some essential peripheral hormones toward controlling feeding behavior. It also covers the impact of chronic hypothalamic inflammation and endoplasmic reticulum stress developed in chronic obesity. A comprehensive understanding of the causes of the central nervous system inflammation and endoplasmic reticulum stress, and how these processes interact with the metabolic regulatory routes will help us devise novel therapeutic opportunities for the treatment of metabolic diseases such as obesity. This part also describes the cell biology of pro-hormones and their conversion to the biologically active peptide through a post-translational mechanism. To fully understand the biology of neuropeptide hormones controlling energy balance, it is essential to uncover the mechanisms by which specific pro-hormones are post-translationally modified to its active form under normal and pathological conditions, a process that happens in a tissue-specific manner. Finally, this part explains the biology of nutrient sensors; we summarize the findings in the literature on the role of nutrient sensors in the hypothalamus and their role in the regulation of energy homeostasis.

The third part describes the role of external signals contributing to energy homeostasis. It describes the gastrointestinal hormones, which constitute a group of hormones secreted by enteroendocrine cells located in the stomach, pancreas, and small intestine. Gut hormones control various functions of the digestive organs, and peptides such as secretin, cholecystokinin, or substance P act as neurotransmitters or neuromodulators in both the central and peripheral systems. It also describes the complexity of the adipose tissue and how various adipose tissues characteristics are altered in the lipodystrophic state and the obese state. It also describes the physiological and molecular functions of adipokines in the obesity-induced inflammation and insulin resistance. *The fourth part* deals with the neuroendocrine axes and obesity. It defines the hypothalamic-pituitary-thyroid axis and its role in body weight regulation, weight loss, and obesity. It shows that thyroid hormone action in peripheral tissues can be just as significant to energy expenditure as the central regulation of thyroid hormone. Obesity and stress are also covered, describing the interaction with the melanocortin system. The melanocortin system can coordinate a wide variety of behavioral and physiological responses to internal and environmental cues ranging from the control of adrenal function, pain, and inflammation to surprising behavioral outputs such as grooming. Finally, the last portion of this part focuses mainly on growth hormone, the clinical conditions, and mouse lines with alterations in this axis along with their adiposity phenotype and concludes by considering the role and therapeutic use of growth hormone in obesity. *The fifth part* deals with nutrition. It focuses on how different internal and external factors influence central control of appetite, energy intake, and weight status, as well as how eating patterns may alter the brain's response to food stimuli.

An important aspect described in this book is related to the role of leptin and nutrient sensors on the biosynthesis of neuropeptide hormones through the action of pro-hormone convertases, which strongly suggests that chronic high-fat feeding induces brain inflammation and endoplasmic reticulum stress, both of which obstruct the normal biosynthesis and post-translational processing of peptides. This new evidence represents an understudied and underappreciated novel mechanism of altered pro-hormone processing caused by these stress mechanisms in the obese state. These new observations could constitute an additional, yet significant, explanation regarding why an anorectic activity is reduced in this state, and cannot compensate for excessive calorie intake. The changes observed in neuropeptide production in the obese state are correlated with their impact on the neuroendocrine axes. Recent data show that abnormalities in pro-hormone processing cause profound obesity, and members of the family of pro-hormone convertases and other processing enzymes are crucial partners in the production of orectic and anorectic peptides. It examines why the obese state impedes the production of anorectic neuropeptides in neuronal cells of the hypothalamus that curbs appetite and inspires calorie burning. One of the causes, among many, appears to be a breakdown in a protein-processing mechanism essential for the production of biologically active hormones, which are altered in the obese state. Therefore, we could say that obesity can sustain itself by impeding hormones that would curb appetite or increase the burn rate for calories. This is a novel concept of a vicious cycle involving a breakdown of key processes in brain cells that allows obesity to beget further obesity.

The contributors are well-recognized academic experts in their respective medical and basic science fields. I would also like to thank each of the participating authors for their insightful and rigorous contributions. This book will be a great source of consultation about the current understanding of hypothalamic neuropeptide hormone action in the endocrine axes in obesity. It is aimed towards researchers in biochemistry, cell biology, and molecular biology; neuroscientists; physician endocrinologists; and nutritionists. In addition, this book is an excellent resource for teaching graduate and medical school courses. It provides a novel conceptualization of the obesity problem by considering the biochemistry of peptide hormones and entertaining novel ideas on multiple approaches to the issues of energy balance. It demonstrates and explains why changes in pro-hormone processing are paramount to understand the metabolic disease. It stimulates the reader to find greater scientific insight and motivation in the understanding the pathogenesis of obesity/metabolic diseases and aid in the development of effective therapies. Finally, I would like to express my enormous gratitude to the National Institutes of Health and National Science Foundation for supporting my research endeavors for more than twenty-six consecutive years at Brown University.

Providence, RI, USA   Eduardo A. Nillni

Contents

Part I Evolution and Origens of Obesity

**1 The Evolution from Lean to Obese State
and the Influence of Modern Human Society** 3
Eduardo A. Nillni

Part II The Hypothalamus

2 Neuropeptides Controlling Our Behavior 29
Eduardo A. Nillni

**3 Transcriptional Regulation of Hypothalamic
Energy Balance Genes** 55
Deborah J. Good

**4 Brain Inflammation and Endoplasmic
Reticulum Stress** 75
Isin Cakir and Eduardo A. Nillni

5 The Cell Biology Neuropeptide Hormones 109
Eduardo A. Nillni

6 Nutrient Sensors Regulating Peptides 141
Isin Cakir and Eduardo A. Nillni

**Part III Peripheral Contributors Participating
in Energy Homeostasis and Obesity**

**7 Gastrointestinal Hormones Controlling Energy
Homeostasis and Their Potential Role in Obesity** 183
María F. Andreoli, Pablo N. De Francesco,
and Mario Perello

8 The Complexity of Adipose Tissue 205
Katie M. Troike, Kevin Y. Lee, Edward O. List,
and Darlene E. Berryman

**9 Adipokines, Inflammation, and Insulin
Resistance in Obesity** 225
Hyokjoon Kwon and Jeffrey E. Pessin

Part IV   Neuroendocrine Axes and Obesity

10   **The Thyroid Hormone Axis: Its Roles
     in Body Weight Regulation, Obesity, and Weight Loss**   255
     Kristen Rachel Vella

11   **Obesity and Stress: The Melanocortin Connection**   271
     Sara Singhal and Jennifer W. Hill

12   **Obesity and the Growth Hormone Axis**   321
     Brooke Henry, Elizabeth A. Jensen, Edward O. List,
     and Darlene E. Berryman

Part V   Nutrition

13   **Brain, Environment, Hormone-Based Appetite,
     Ingestive Behavior, and Body Weight**   347
     Kyle S. Burger, Grace E. Shearrer, and Jennifer R. Gilbert

Index   371

Contributors

María F. Andreoli School of Biochemistry and Biological Sciences, National University of Litoral (UNL) and Institute of Environmental Health [ISAL, Argentine Research Council (CONICET)- (UNL)], Santa Fe, Argentina

Darlene E. Berryman The Diabetes Institute, Department of Biomedical Sciences, Heritage College of Osteopathic Medicine, Ohio University, Athens, OH, USA

The Diabetes Institute, Konneker Research Labs, Ohio University, Athens, OH, USA

Edison Biotechnology Institute, Konneker Research Labs, Ohio University, Athens, OH, USA

Department of Biomedical Sciences, Heritage College of Osteopathic Medicine, Ohio University, Athens, OH, USA

Kyle S. Burger Department of Nutrition, University of North Carolina at Chapel Hill, Chapel Hill, NC, USA

Isin Cakir Life Sciences Institute and Department of Molecular and Integrative Physiology, University of Michigan, Ann Arbor, MI, USA

Pablo N. De Francesco Laboratory of Neurophysiology, Multidisciplinary Institute of Cell Biology [IMBICE, Argentine Research Council (CONICET), National University of La Plata and Scientific Research Commission, Province of Buenos Aires (CIC-PBA)], La Plata, Buenos Aires, Argentina

Jennifer R. Gilbert Department of Nutrition, University of North Carolina at Chapel Hill, Chapel Hill, NC, USA

Deborah J. Good Department of Human Nutrition, Foods, and Exercise, Virginia Tech, Blacksburg, VA, USA

Brooke Henry The Diabetes Institute, Konneker Research Labs, Ohio University, Athens, OH, USA

Edison Biotechnology Institute, Konneker Research Labs, Ohio University, Athens, OH, USA

Jennifer W. Hill University of Toledo, Toledo, OH, USA

Elizabeth A. Jensen Translational Biomedical Sciences, Graduate College, Ohio University, Athens, OH, USA

Edison Biotechnology Institute, Konneker Research Labs, Ohio University, Athens, OH, USA

Hyokjoon Kwon Department of Medicine and Molecular Pharmacology, Albert Einstein College of Medicine, Bronx, NY, USA

Kevin Y. Lee Department of Biomedical Sciences, The Diabetes Institute, Heritage College of Osteopathic Medicine, Athens, OH, USA

Edward O. List Edison Biotechnology Institute, Konneker Research Labs, Ohio University, Athens, OH, USA

The Diabetes Institute, Konneker Research Labs, Ohio University, Athens, OH, USA

Eduardo A. Nillni Emeritus Professor of Medicine, Molecular Biology, Cell Biology & Biochemistry, Department of Medicine, Molecular Biology, Cell Biology & Biochemistry, The Warren Alpert Medical School of Brown University, Providence, RI, USA

Mario Perello Laboratory of Neurophysiology, Multidisciplinary Institute of Cell Biology [IMBICE, Argentine Research Council (CONICET), National University of La Plata and Scientific Research Commission, Province of Buenos Aires (CIC-PBA)], La Plata, Buenos Aires, Argentina

Jeffrey E. Pessin Department of Medicine and Molecular Pharmacology, Albert Einstein College of Medicine, Bronx, NY, USA

Grace E. Shearrer Department of Nutrition, University of North Carolina at Chapel Hill, Chapel Hill, NC, USA

Sara Singhal University of Toledo, Toledo, OH, USA

Katie M. Troike The Diabetes Institute, Konneker Research Labs, Ohio University, Athens, OH, USA

Kristen Rachel Vella Weill Cornell Medical College, New York, NY, USA

Part I
Evolution and Origens of Obesity

The Evolution from Lean to Obese State and the Influence of Modern Human Society

Eduardo A. Nillni

1.1 Introduction

One of the paramount changes in human evolution was the development of large brain size with a tremendous impact on the nutritional behavior of our species. A larger brain demands more food intake to keep up with the need of the overall energy budget. This high demand for energy to maintain the brain metabolism forced early humans to move from a strictly vegetarian diet to more energy-rich diet. Changes in nutrient-rich energy diets evolving from an exclusively vegetarian to an omnivorous diet were among many evolutionary factors developed to maintain the high cost of a large human brain. Paleontological data indicates that fast brain evolution occurred with the appearance of *Homo erectus* 1.8 million years ago, which was related to critical changes in diet, body size, and foraging behavior. Then, the survival of more advanced humans depended on their ability to acquire energy for its daily use and storage, which was well balanced before the advent of modern humans. The energy balance of early humans is disrupted today by an excessive food intake, processed food, and an increase in sedentary life. Obesity pandemic has undoubtedly coincided with not only an increase in unhealthy eating habits but also with migratory movements of different ethnic communities to dissimilar environmental pressures. The heat producing of uncoupling proteins in mitochondria brown adipocyte tissue is believed to be a key driver behind the conquest of a variety of environments in mammals 65 million years ago. This ability to produce and maintain heat contributed to the evolution of mammals to explore and settle in uninhabitable territories throughout the planet by adjusting the thermoregulatory response to sharply different environments. It is also discussed in this chapter several early evolutionary hypotheses to explain the development of obesity and metabolic syndrome, the evolutionary changes from hominoids 20 million years ago to industrialized humans, and the effects on traits causing profound changes in the evolution of human nutritional requirements.

1.2 Brain Evolution and Changes in Human Nutrition

One of the chief characteristics of humans is that they are holding big brains, and the evolution of this large brain size has had significant implications for the nutritional biology of our species. On average, our brain size as part of the

E. A. Nillni (✉)
Emeritus Professor of Medicine, Molecular Biology, Cell Biology & Biochemistry,
Department of Medicine, Molecular Biology, Cell Biology & Biochemistry,
The Warren Alpert Medical School of Brown University, Providence, RI, USA
e-mail: Eduardo_Nillni@Brown.edu

© Springer International Publishing AG, part of Springer Nature 2018
E. A. Nillni (ed.), *Textbook of Energy Balance, Neuropeptide Hormones, and Neuroendocrine Function*, https://doi.org/10.1007/978-3-319-89506-2_1

Table 1.1 Evolution from early anthropoid primates to industrialized humans, and their journey through food intake, stress and less physical activity transitions through time. Industrial development has altered seven essential nutritional traits of ancestral hominin diets: (a) glycemic load, (b) fatty acid composition, (c) micronutrient density, (d) macronutrient composition, (e) sodium-potassium ratio, (f) acid-base balance, and (g) fiber content. From this point forward, the changes introduced by modern humans to industrially processed foods associated with less physical activity represented the springboard to a deviation of our natural nutritional environment unfamiliar to our genetic repertoire

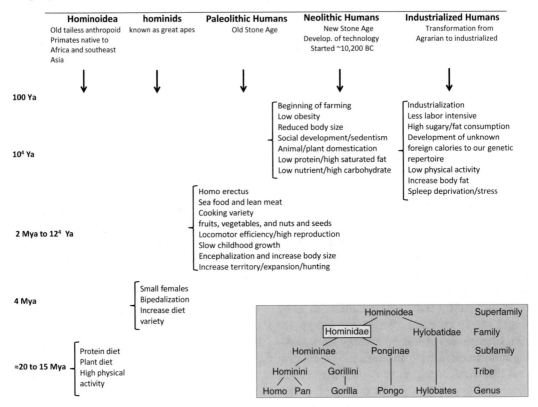

primates group is nearly double from mammals of the same body size. Throughout nearly 7 million years of evolution, the human brain has tripled in size, and the past 2 million years represents the most significant brain growth. The large human brain is energetically expensive and uses a more significant proportion of its energy budget on brain metabolism as compared with other primates with a lower energy budget. For example, humans consume 400 more calories than chimpanzees and 635 more calories than gorillas and 820 more calories than orangutans. Paleontological data pointed out that a rapid brain evolution occurred with the emergence of *Homo erectus* 1.8 million years ago. These posture changes are related to significant changes in diet, body size, and hunting behavior

(Table 1.1). These alterations are dramatic enough that separated us from our close relatives, including the great apes or hominids, which are a taxonomic family of primates. They include seven species in four genera, *Pongo*, the Bornean and Sumatran orangutan; *Gorilla*, the Eastern and Western gorilla; *Pan*, the common chimpanzee and the bonobo; and *Homo*, the human and the near-human ancestors and relatives (e.g., the Neanderthal) (Chatterjee et al. 2009). The combination of larger brains, high reproductive output with slow childhood growth, changes in history of social behavior, and an extraordinary longevity separated us far apart from other apes (Isler and van Schaik 2012; Schuppli et al. 2012; van Schaik et al. 2012; van Woerden et al. 2012). Consistent with these observations, analysis from African human fossils indicates that significant changes in both brain size and diet were linked with the rise of early members of the genus *Homo* between 2.0 and 1.7 millions of years ago. The question that arises from these observations was then how much energy supply would be needed to maintain metabolism in these brains in a steady-state status? Changes in nutrient-rich energy diets evolving from an exclusively vegetarian to an omnivorous diet were among many evolutionary factors developed to maintain the high cost of a large human brain.

The fact that early genus *Homo* consumed more animal foods was a turning point in providing elevated levels of essential long-chain polyunsaturated fatty acids (docosahexaenoic acid and arachidonic acid) highly necessary for brain growth. This evolutionary adaptation forced humans to use a larger proportion of their resting energy budget on brain metabolism separating them again from other primates or non-primate mammal. One of the key features developed by humans was learning to share food, making social groups of early humans more resilient, spreading their diets, and obtaining more energy from rich foods such as meat. Humans also developed much larger deposits of body fat, which can be used to sustain them during periods of food scarcity. Also, compare with other primates, humans have a relatively small gastrointestinal tract and reduced colon allowing for greater energy allocation to reproduction and limiting the increase in basic metabolic rate. This type of adaptation is consistent with a high in energy and nutrient intake quickly to digest. Another important feature was the improved walking efficiency in the evolution of early *H. erectus* 1.8 million of years ago. Evidence also supports an important evolutionary adaptive change involving hunting and hoarding, which resulted in greater consumption of animal foods and sharing the prey within other social groups. Dietary changes to energy-dense foods and the discovery of cooking (of thermal and nonthermal food processing) in modern humans successfully increased the net energy gained. These were significant contributors to the evolutionary expansion of the hominin

energy budget (humans and great apes together form a superfamily called hominoids). The genus *H. erectus* appears to develop a rapid rate of a bigger part of the central nervous system contained within the cranium and comprising of the forebrain (prosencephalon), midbrain (mesencephalon), and hindbrain (rhombencephalon). Finally, humans have smaller muscle mass and more fat tissue as compared with other primates helping to offset the high-energy demands of our brains. These high levels of adiposity in humans are especially prominent in infants to accommodate the growth of their large brains with enough supply of stored energy (Pontzer et al. 2016a, b). In summary, these evolutionary traits caused profound changes in the evolution of human nutritional requirements (Anton et al. 2014; Cordain et al. 2005; Eaton 2006; Garn and Leonard 1989; Leonard and Robertson 1992, 1994) separating us further apart from other primates in terms of distinctive nutritional needs (Leonard 2002; Leonard et al. 2007).

Through evolution time, our ancestors ate poorly, particularly during climate disadvantages, and they often had vitamin deficiencies, food-borne diseases, and neurotoxins. Dirt, grit, and fiber constituted a significant part of most early diets. With the advent of modern technologies, these food components diminished. The profound changes in the environmental components including diet and lifestyle circumstances introduced by agriculture and animal husbandry approximately 10,000 years ago are too recent on an evolutionary time scale for the human genome to change. The lack of agreement between our ancient genes adjusted to the earlier way of nutritional behaviors compared to the cultural patterns of contemporary Western life created the so-called diseases of modern civilizations. The evolutionary clash of our ancient genome with the nutritional variants of recently introduced processed foods may be the cause of the established chronic diseases of Western civilization. Food-processing procedures introduced during the Neolithic period was considered the last part of the Stone Age. The New Stone Age is a time limit in the development of human technology starting about 10,200 BC and ending between 4500 and 2000 BC. In addition, industrial development has essentially altered seven essential nutritional traits of ancestral hominin diets: (a) glycemic load, (b) fatty acid composition, (c) micronutrient density, (d) macronutrient composition, (e) sodium-potassium ratio, (f) acid-base balance, and (g) fiber content. From this point forward, the changes introduced by modern humans to industrially processed foods associated with less physical activity represented the springboard to a deviation of our natural nutritional environment unfamiliar to our genetic repertoire.

1.3   Definition of Obesity

As described above, the human genome has hardly changed since the emergence of modern humans leaving East Africa 70,000 years ago. Genetically, humans remain adapted for the foods consumed at that time in history. From the records, the proposal is that human ancestors obtained about 35% of their dietary energy from fats, 35% from carbohydrates, and 30% from protein. Saturated fats contributed approximately 7.5% total energy. Polyunsaturated fat intake was high, cholesterol consumption was significant, and carbohydrate came from uncultivated fruits and vegetables. The latter represented 50% energy consumption as compared with the 16% energy intake consumed today by Americans. While high levels of fruits and vegetables and minimal grain consumption constituted the ancestral diet, today's diet is far apart from that regime. Honey included 2–3% energy intake as compared with the 15% added sugars which contribute to the present time. Also, fiber consumption in ancient humans was high. Although a significant progress was made in understanding ancient human diet, nutritionists are still searching for a unifying hypothesis on which to build a dietary strategy for prevention. Therefore, a better understanding of human evolutionary nutritional habits and its impact on contemporary nutritional requirements could help us with strategies to better define obesity and combat this malady.

Over the last five decades, there has been a major widespread of obesity, which is associated with many comorbidities or metabolic syndrome, mostly in the Western world but reaching now a global dimension. The development of these chronic diseases in the culture of the west is related to high genetic components within different populations. As obesity rates soared between 1980 and 2017, the number of Americans who are obese has doubled. In the United States alone, 70% of the adult population is overweight, and 36% are obese. There are ~700 million obese people worldwide, and another ~2.1 or more billion who are overweight, according to the World Health Organization. There is today a financial burden in the United States with obesity-related healthcare costs. In 2005 and 2006 alone, 150–190 billion dollars was spent on obesity-related diseases. In 2010, no state had a prevalence of obesity less than 20%.

Thirty-six states had occurrence equal to or greater than 25%; 12 of these states (Alabama, Arkansas, Kentucky, Louisiana, Michigan, Mississippi, Missouri, Oklahoma, South Carolina, Tennessee, Texas, and West Virginia) had prevalence equal to or greater than 30%. The estimation is that by the year 2030, healthcare costs will increase by more than 50 billion dollars annually. There are more obese and overweight people on the planet than people suffering from malnutrition. In spite of the great progress made in the field of energy balance, our understanding of some basic mechanisms to combat this malady remains unclear. Obesity and its associated medical complications including type 2 diabetes, cardiovascular disease, dyslipidemia, mood disorders, reproductive disorders, hypertension, asthma, and potential for cancer development account for more than 300,000 deaths per year in the United States. Obesity treatment strategies often do not result in adequate, sustained weight loss, and the prevalence and severity of obesity in the United States and many other countries are progressively increasing (Ahima 2005). Current treatments include dietary changes, increased physical activity, prescription medications, weight loss surgery, and behavior modification.

Most surgical and pharmacological treatments require lifestyle changes to achieve sustained weight loss fully. However, dangerous side effects may accompany these treatment strategies. In addition, some pharmacotherapies may not work in certain individuals. The complexity of the obese condition results from the interaction between environmental and predisposing genetic factors interacting with each other. Specifically, genes operate additively and through gene-gene interactions to influence body weight (Clement 2005). A more thorough understanding of the molecular mechanisms underlying the pathogenesis of obesity and regulation of energy metabolism is essential for the development of effective therapies. Therefore, it is necessary to characterize the molecular and behavioral mechanisms governing body weight to identify the abnormality or impairment in the regulation of the metabolic, physiological, and psychological mechanisms causing obesity.

One of the major obstacles encountered in the United States to combat the obesity prevalence is related to the fact that junk foods are the largest source of calories in the American diet. They include grain-based desserts like cookies, doughnuts, granola bars, sugary soda, and fruit juices, an excess of pasta and pizza, and pieces of bread with high sugar content to name some. What all these foods have in common, different from the same meals made in the 1950s, is that they are mainly the products of seven crops and farm foods. They are corn, soybeans, wheat, rice, sorghum, milk, and meat heavily subsidized through decades by the federal government, ensuring that junk foods are cheap and plentiful. As a matter of fact, between 1995 and 2010, the government contributed with $170 billion in agricultural subsidies to finance these foods. While many of these foods are not innately unhealthy, only a small percentage of them are eaten as is (New York Times, How the Government Supports Your Junk Food Habit, by Anahad O'Connor, July 19, 2016). All these products are converted to cheap foods and additives like corn sweeteners, industrial oils, processed meats, and refined carbohydrates. It is quite ironic that on the one hand, the government promotes healthy diets (organic fruits and vegetables) while at the same time has the complicity in supporting the industrial production of junk food leaving a very small fraction of its subsidies to support the production of fresh produce. The result is that taxpayers are paying for the privilege of making our country sick (Anahad O'Connor, New York Times). The subsidies program was started decades ago in part to support struggling farmers and to secure America's food supply. Since 1995 the government has provided farmers with close to $300 billion in agricultural subsidies overall; today the grants program no longer helps its original purpose because it continues to give subsidies to large producers of grains, corn, sorghum, and oilseeds like soybeans instead of small farmers who grow fruits, nuts, and vegetables. In summary, we created an evolution to self-destruction (Fig. 1.1).

Having described our ancient and modern diets, we can now define obesity as a state of

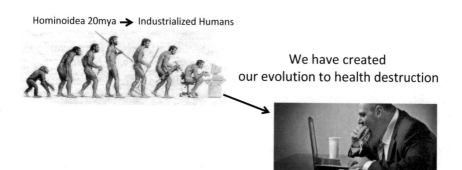

Hominoidea 20mya ➔ Industrialized Humans

We have created
our evolution to health destruction

Fig. 1.1 Evolution to self-destruction: The recent substantial increase in the prevalence of obesity in susceptible individuals has been mostly caused by our modern urban societies in which demand for physical activity is extremely reduced, and highly palatable and relatively cheap food is ubiquitously available. Geographic migrations with an adaptive thermogenesis added an additional variable to the confounding problem of obesity

excess adipose tissue mass, which translates into excessive body weight and an alteration of energy balance (this condition may not be confused with a body builder that can develop a remarkable overweight state without excessive body fatness). Energy balance is defined by the number of calories consumed versus the amount of energy used either via exercise, physical activity, or resting metabolism. Energy derived from food intake enters the plasma from the intestine and then to cells involved in energy consumption. Under normal conditions, any minor excess is dealt with by cells that function in energy storage. When the number of calories expended is the same with the number of calories consumed, the energy caloric balance is neutral, and no change in weight occurs. On the other hand when the number of calorie intake is greater than a number of calories expended, an energy balance disruption occurs (positive energy balance), and obesity develops. Therefore, obesity occurs as a result of a long-standing imbalance between energy intake and energy expenditure, which is influenced by a very complex set of biological pathway systems regulating appetite. We then can say that obesity is a "disorder of energy balance." According to the World Health Organization (WHO), obesity is classified as class I for a BMI between 30 and 34.9 kg/m2, class II for a BMI between 35 and 39.9 kg/m2, and class III for a BMI ≥ 40 kg/m2 (Obesity: preventing and managing the global epidemic 2000). Class I obesity is associated with a moderate risk, class II with a high risk, and class III with a very high risk of mortality (Gonzalez et al. 2007). Anatomically obesity can be classified for the prevalence of visceral or subcutaneous deposition of fat. The ratio of waist circumference to hip circumference (WHR) is used to serve the purpose of defining the degree of central (i.e., visceral) vs. peripheral (i.e., subcutaneous) obesity. Visceral adiposity is a major risk factor for metabolic syndrome, while subcutaneous fat seems to be much more benign and in some cases even protective against the development of metabolic complications (Jensen 2008). The metabolic syndrome can be defined as a group of risk factors that increases blood pressure, high blood sugar, excess body fat around the waist, abnormal cholesterol, or triglyceride levels causing to increase the risk of heart disease, stroke, and diabetes. To understand the imbalance in obesity from the calorie in and calorie out is a complicated matter. While calories from food are easy to control, the way calories burn represents a different undertaking. It consists mainly of the energy required for the basal metabolism of the body, at rest, in the absence of external work. That's called resting energy expenditure, which represents 60–70% of the total energy expenditure. However, it is highly variable from individual to individual. The second component is the physical activity that is the sum of basal activities of daily living and voluntary exercise. The third part of total energy expenditure, although small, is diet-induced thermogenesis, which is the energy associated with a postprandial rise in metabolic rate to process food during digestion, usually amounting about 10% of calories.

Obesity can also be caused by treating diseases with pharmacological treatments including steroids, antipsychotics, some antidepressants, and some anti-epileptics but could also be a consequence of some diseases or conditions, including polycystic ovary syndrome (PCOS), Cushing's syndrome, hypothyroidism, hypothalamic defects, and growth hormone deficiency. Obesity is frequently associated with low androgen levels causing hypogonadism. Hyperinsulinemia is believed to be the primary etiological factor for the development of PCOS, but there are other factors involved as well such as obesity-induced hyperestrogenism and a male pattern adipokine gene expression observed in these women. Obesity also causes a reduction in growth hormone secretion in the pituitary gland. The decrease in growth hormone does not appear to translate into a similar reduction in IGF-1. Therefore it is unlikely that obesity represents a condition of growth hormone deficiency reflected at the tissue level. Both growth hormone deficiency and growth hormone excess are associated with increase in fat mass.

1.4 Nutritional Balance, Metabolic and Hedonic Set Point

The survival of all species depends on their ability to acquire energy for its daily use and storage. From an evolutionary standpoint, feeding or intake of calories from meal to meal is necessary to (a) satisfy nutritional and metabolic requirements and (b) prepare for periods of food shortage during seasonal changes. Energy balance is the relationship that exists between energy intake (i.e., calories taken from food and drink) and energy expenditure (i.e., calories being used for our daily energy requirements). Hence, the maintenance of this balance is achieved by the integration of (a) environmental signals (i.e., environmental cues steer individuals' decisions concerning food intake and food choice), (b) physiological and metabolic signals (i.e., neurohormone, peripheral hormones, nutrient sensors, and key organs), (c) genetic makeup (i.e., multiple genetic interactions, epigenetic actions), and (d) social and hedonic influences (i.e., the drive to eat to obtain pleasure without an energy deficit). When this balance is disrupted, we witness either a weight gain or a weight loss. In the case of weight gain or positive energy balance, energy imbalance is caused by a higher calorie intake versus the number of calories burned.

However, defining when energy balance is disrupted represents a complex undertaking. For example, excessive food intake is likely to be the primary cause of positive energy balance (obese phenotype) driven by both nonconscious (homeostatic) and conscious (perceptual, emotional, and cognitive) phenomena processed in the brain. Functional neuroimaging in a few studies has provided evidence of functional differences between obese and lean individuals in the brain's response to energy intake (DelParigi et al. 2005a). Connecting hyperphagia to actual weight gain has proved remarkably difficult (Stunkard et al. 1999; Tataranni et al. 2003). The experimental evidence connecting the relative contribution among people who have differences in energy intake, expenditure, and resting metabolic rate or due to physical activity to weight gain is limited.

In addition, food intake and development of obesity involve diet composition (Astrup 1999; Astrup et al. 1997), energy density of food (Bell and Rolls 2001) (Drewnowski 2003), rate of meal consumption, taste preferences (Cooling and Blundell 2001), eating behavioral style (Keski-Rahkonen et al. 2003), and subphenotypes (DelParigi et al. 2005a, b), all of them contribute but complicate matters with some contradictory results. Within the United States, a significant decline in the percentage of energy from fat foods during the last two decades has paralleled with a massive increase in obesity. Therefore, diets high in fat do not seem to be the cause of high prevalence in excess body fat in our society, suggesting that decreases in fat content will not be the answer (Willett and Leibel 2002). The genetics of obesity is also partially understood because it is not clear whether obesity is caused by a single genetic mutation, by multiple allelic defects, and which one of those determines susceptibility to environmental factors including epigenetic contribution. Epigenetics is defined as a stable heritable traits or phenotypes that cannot be explained by changes in DNA sequence or changes to the genome that do not involve a change in the nucleotide sequence. This genetic change means features that are "on top of" or "in addition to" the traditional genetic basis for inheritance (those are shifts in a chromosome that affect gene activity and expression) (Pomp and Mohlke 2008). It's hard to predict who will or will not develop obesity in an obesogenic environment. It depends on an individual combination of alleles in gene-gene interaction and how it reacts with the environment in a particular way. People who carry only one or some of these alleles may still not develop obesity because they either lack another allele in gene-gene interaction needed or are not exposed to the stimulating environment causing gene-environment interaction. Further clarification will be necessary to resolve the controversy that exists among genotypes and lifestyle (Holzapfel et al. 2010) or anatomical phenotype of obesity (Bauer et al. 2009; de Krom et al. 2009).

For quite some time, it has been suggested that there are two systems controlling eating behavior (Saper et al. 2002). The metabolic system is regulated by mediators such as leptin and ghrelin, neuropeptide Y (NPY), agouti-related peptide (AgRP), melanocortins, orexins, and melanin-concentrating hormone, among the important ones. The second one is the hedonic behavior that is regulated by taste and reward systems to certain foods (Fig. 1.2). Hence, one of the common questions regarding the epidemic of obesity in modern society is whether hedonic feeding overcomes metabolic feeding. It is well accepted that body weight is determined via both mechanisms, and recurrence of food consumption above the minimum energy requirements to fulfill basic metabolism is the hallmark of obesity. Metabolic needs drive food intake in response to changes in body energy status, which is dictated by the brain at the hypothalamic level and responsibly in determining the "body weight set point" to maintain energy homeostasis at a constant level. This programmed set point is regulated by a circuitry of neuronal cells in the hypothalamus and other specific brain regions. It is controlled mostly by leptin, insulin, and ghrelin acting on melanocortin anorectic neurons pro-opiomelanocortin/cocaine- and amphetamine-regulated transcript (POMC/CART) and orectic AgRP/NPY (Kim et al. 2014; Koch et al. 2015; Waterson and Horvath 2015). Another second order of neurons located in the lateral hypothalamus (LH) contains orexins and melanocyte-concentrating hormone (MCH); both peptides are potent stimulators of food intake (Ludwig et al. 2001; Flier 2004). The ventromedial hypothalamus (VMH), which is controlled by leptin through the regulation of brain-derived neurotrophic factor (BDNF), also regulates energy balance. The paraventricular nucleus (PVN) of the hypothalamus contains several groups of neurons all involved in energy balance regulation of corticotropin-releasing hormone (CRH) signaling in the PVN which increases leptin signaling in the VMH (Gotoh et al. 2005). Hypophysiotropic thyrotropin-releasing hormone (TRH) neurons expressing the leptin receptor (ObRb) are considered as one of the primary hypothalamic centers controlling food intake and energy homeostasis (Nillni 2010; Nillni and Sevarino 1999; Perello et al. 2006; Sanchez et al. 2004; Elmquist et al. 1998; Elias

et al. 1999). Oxytocin produced in the PVN and vasopressin generated in the PVN and supraoptic nucleus together with the brain stem interacting with glucagon-like peptide-1 (GLP-1), cholecystokinin (CCK), serotonin, and melanocortin (see full description in Chap. 2) are all involved in the energy balance regulation (Fig. 1.3, Chap. 2). The metabolic body weight set point is genetically regulated, but exposure to a constant obesogenic environment may provoke allostatic adaptation and upward drift of the set point, leading to a new body weight set point of higher maintained body weight. However, an elevated body weight set point may also be achieved without changes in the metabolic homeostasis, but rather a sustained hedonic overeating is driven by the rewarding property of palatable foods, which is primarily controlled by the mesolimbic reward system and dopamine signaling (Rui 2013). The amygdala, prefrontal cortex, and ventral striatum (including the core and shell of the nucleus accumbens) are the networks linking hedonic effect. The reward system of the brain can override homeostatic metabolic signals. These two different response systems are heavily entangled (Murray et al. 2014). Therefore, research laboratories continue to search the contributive factors involved in homeostatic and hedonic mechanisms related to eating behavior. The peripheral hormones leptin, ghrelin, and insulin play a major role in food reward as demonstrated in studies done in laboratory animals and humans, which show relationships between hyperphagia and neural pathways involved in reward. These results have provoked questions pertaining the possibility of addictive-like features in food consumption.

From the hedonic side, dopamine depletion drastically impairs feeding and causes starvation in animals. Metabolic hormones including ghrelin and leptin can stimulate on ventral tegmental area midbrain dopamine neurons to affect feeding. This area is the origin of mesolimbic dopamine neurons that project to the nucleus accumbens (NAc) that in turn influences behavior. Therefore, peripheral hormones affecting the hypothalamus can also affect feeding behavior via action on the midbrain circuits (Narayanan

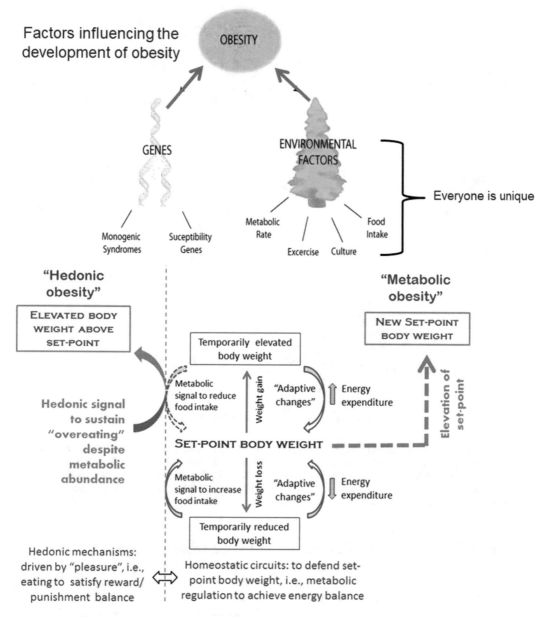

Fig. 1.2 Metabolic and hedonic obesity altering the body weight set point: The hypothalamus and brain stem interacting with peripheral organs and tissues are the homeostatic regulators of the body weight set point. Chronic deviation of body weight from its original set point provokes a compensatory increase or decrease in food intake (cumulative over an extended period) and energy expenditure (both resting and non-resting) in opposite direction to restore the original body weight set point. However, metabolic obesity, as defined by Yu and colleagues (Yu et al. 2015) results from an elevation of the metabolic set point characterized by an elevated body weight which is metabolically protected as a new normal body weight set point. Hedonic consumption, on the other hand, is ruled by the reward dopamine system to gratify the need of pleasure independent of the metabolic set point. Deviation of the reward system may lead to hedonic overeating in susceptible individuals (drifty genotype) leading to continued weight gain above the metabolic set point weight (hedonic obesity). (Figure reproduced with permission from Yu et al. (2015))

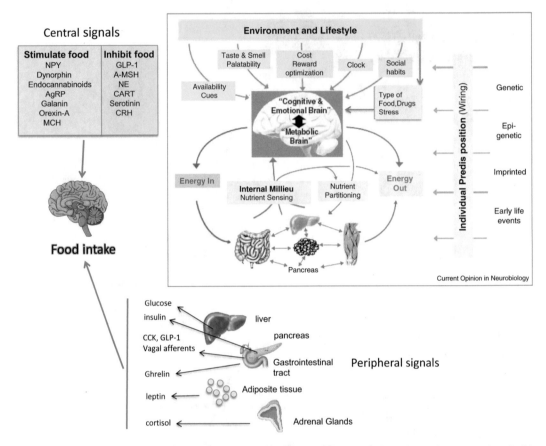

Fig. 1.3 Neuronal components of body weight regulation: Body weight in adulthood is influenced by changes in the environment of subsequent generations affecting the genetic and epigenetic propensity for weight gain and modern lifestyle that promotes sedentary behaviors and provides an oversupply of energy-dense foods. Figure reproduced with permission from Berthoud (2011). The brain maintains the homeostasis through a number of hormonal and neural nutrient-sensing inputs and from the environment and lifestyle through the cognitive and emotional brain

et al. 2010). One potential hub for the hedonic behavior is the lateral hypothalamus LH, also called "feeding center." Lesion in this region suppresses eating and causes weight loss (Delgado and Anand 1953), while electrical stimulation causes insatiable feeding (Delgado and Anand 1953). The LH stimulation is rewarding and leads to self-stimulation in animals (Olds and Milner 1954). The LH is one of the evolutionary oldest parts of the brain, and groundbreaking work of physiologists and psychologists from the middle of the last century demonstrated that the hypothalamus is essential for the control of motivated behaviors. The LH integrates large amounts of information and arranges adaptive responses including energy homeostasis by receiving meta-

bolic state information through both neural and humoral routes and having direct access to behavioral, autonomic, and endocrine effector pathways.

Optogenetic (the combination of genetics and optics to control well-defined events within specific cells of living tissue) experiments further demonstrated the critical roles of the LH in behavior (Stuber and Wise 2016) (Berthoud and Munzberg 2011). The LH receives numerous inputs from reward-processing centers such as medial prefrontal cortex (mPFC), nucleus accumbens (NAc), bed nucleus of the stria terminalis (BNST), and dorsal raphe (DR). The LH likely integrates the hedonic and metabolic signals for eating. From the metabolic side, the anorexigenic hormone leptin can suppress the reward value of palatable food (Hommel et al. 2006). The presence of ObRB in ventral tegmental area (dopamine neurons) is critical for feeding behavior providing direct evidence of peripheral metabolic signals affecting dopamine activity. On the other hand, the hunger hormone ghrelin potentiates the hedonic response (Malik et al. 2008) by favoring food consumption and enhancing the hedonic and incentive responses to food-related cues. When an excess of weight is due to elevation of the metabolic set point, energy expenditure is supposed to fall onto the standard energy mass regression line. In contrast, when a steady-state weight is above the metabolic set point due to hedonic overeating, a persistent compensatory increase in energy expenditure per unit metabolic mass may be demonstrable (Fig. 1.2). Recognition of the two types of obesity origin may trigger to more effective treatment and prevention of obesity. In humans, the hedonic consumption of high-calorie food is a major driver for obesity (Volkow and Wise 2005). Similar to drugs food activates a common dopamine brain reward circuitry. Addiction and obesity result in habits that persist and strengthen despite the threat of catastrophic consequences. Feeding above the metabolic needs and drug use habits are imprinted behavioral preferences that reinforce properties of great and repetitive rewards. Palatable sugary foods raise glucose concentration in the blood and brain, and drugs with pharmacological agents activate the same brain reward circuitry. The magnitude and duration of increases in dopamine induced by either excess of food or drugs in the nucleus accumbens to maintain the level demand are an intensive target of an investigation. While hedonic and metabolic mechanisms are working together at any given time or any given individual, derangement in either or both may lead to obesity. Therefore, it should be taken into account for the management of obesity and treatment modalities whether the target is behavioral changes in the case of hedonic obesity or those related to changes in body weight set point because of metabolic obesity. Identification of the neural bases separating these two systems at the molecular, cellular, and neuron-neuron inter-

action is key to understand how they are coordinated, and dysregulated, under healthy and obesogenic conditions (Yu et al. 2015).

We could argue that an energy-dense diet, high in saturated fat and sugar, should cause weight gain and increased adiposity but can be easily reversed by a more natural regimen of foods and lower calorie intake. However, it appears that these high-calorie diets, maintained long term, cause a profound change in the energy balance set point not so easy to reverse, particularly for those individuals who pass the 30 BMI mark. Diets containing long-chain saturated fats result in metabolic dysfunction with increased adiposity and body weight that are protected, so any subsequent weight loss through calorie restriction is difficult to maintain. The profound changes observed in the energy balance controlled by the hypothalamus result in the loss of central leptin and insulin sensitivity, which perpetuates the development of both obesity and peripheral insulin insensitivity. This hypothalamic dysfunction causes changes in the set point between energy intake and energy expenditure, which is protected by the brain at any cost. Continuous ingestion of an excess of high-calorie diet induces hypothalamic dysfunction, which includes an increase of oxidative stress; chronic atypical neuronal inflammation; endoplasmic reticulum (ER) stress; lipid metabolism; changes in neuronal cellular death, called apoptosis; neuronal rewiring; or neuronal and synaptic plasticity (see Chap. 4). All these hypothalamic changes induced by a high-calorie diet linked to inflammation increase the development of obesity. Although obesity is a consequence of the modern lifestyle society and other evolutionary and genetic factors, in my view, it is not a disease per se, a condition that over time leads to severe side effects including a range of metabolic diseases as depicted above. These diseases have increased in gigantic proportions in the United States and lesser degree in other countries, with no reversal despite educational programs and treatment options. These unsuccessful strategies are a consequence of a lack of knowledge about the precise pathology and etiology of metabolic disorders. Different independent studies had demonstrated that obesity has a strong genetic component when predisposed individuals are living in an obesogenic environment (Sorensen et al. 1989), signifying a potential gene-environment interaction (Speakman 2006). The most accepted model by different scientists is that obesity and its consequences are a result of a gene-environment interplay, an ancient genetic evolutionary selection to store fat efficiently that is poorly adapted to modern times. Interestingly, certain human populations are susceptible to obesity and metabolic syndrome (Caballero 2007), whereas others appear resistant to the forces inducing obesity (Beck-Nielsen 1999) (Neel 1962). Much emphasis has been placed on individuals and geographic populations suggesting that evolutionary traits play a key role in obesity and metabolic syndrome.

1.5   Evolutionary Traits

The high prevalence of obesity is seemingly a detrimental condition inconsistent with the evolutionary progress of all species including humans in their adaptation journey to new environments. Several early evolutionary hypotheses have been proposed to explain the development of obesity and metabolic syndrome. In 1962, James Neel introduced the first evolutionary explanation for the modern obesity epidemic that is founded on the notion that the development of diabetes or obesity is an adaptive trait incompatible with modern lifestyles. Neel's "thrifty gene" hypothesis proposes that genes enable humans to efficiently collect and process food to store fat during periods of food abundance to save for times of food shortage. It would be advantageous for hunter-gatherer populations and childbearing women. Therefore, more obese individuals carrying the thrifty genes will better survive times of food shortage. Contrarily to this paradigm, in modern societies where the abundance of food is the norm, this genotype resulted on an incongruity between the environment in which the brain evolved and today's environment with widespread of chronic obesity and diabetes. In that sense, this hypothesis represents a regression and

inadequate in modern times as compared to our ancestors who undergone positive selection for genes that favored energy storage, a consequence of the cyclical episodes of famine and surplus after the advent of farming 10,000 years ago (Neel 1962, 1999). This hypothesis is based on the assumption that during human evolution, humans were always subjected to periods of feast and famine, favoring individuals who had more capacity for energy stores. This evolutionary trait allowed people more likely to survive and produce more offspring. In other words, evolution acted to select those genes in individuals who possessed high efficiency at storing fat during times of plenty.

However, in the modern environment, this genetic predisposition, which prepares us for a famine stage that never comes, an epidemic of obesity and diabetes with their induced maladies, made our society ill. We now know the genes determine the propensity for obesity or lack thereof. The dominance of obesity in modern human societies has two contributory components: a) an environmental change in the industrialized society that has happened around hundred years ago and a genetic predisposition that has its origins in our evolutionary history 2 million years ago. Around 70 percent of the variation among people in their amount of body fat is justified by inherited differences constructed into our genetic makeup and passed to next consecutive generations. According to the thrifty gene hypothesis, it was advantageous for early humans letting them store fat in times of plenty and survive in times of food scarcity. What could this hypothesis not explain in modern times is why isn't everyone fat? John Speakman from Aberdeen University showed evidence that supporting the famine hypothesis has fundamental flaws, and he has come up with an alternative theory, nicknamed the "drifty gene" hypothesis or "predation release hypothesis." To start, Speakman argues that famines weren't a real threat before the advent of farming around 15,000 years ago. He suggests that there was not sufficient differential impact on survival of the lean and obese to cause such a powerful selective effect (most human populations have only experienced at most 100 famine events in their evolutionary history). Also, famines involve increases in total mortality that only rarely exceeded 10% of the population, and people in famines die of disease rather than starvation. He proposes that modern human distribution of obesity stems from a genetic drift in genes encoding the system that regulates metabolism controlling our body fatness. This drift may have started around 2 million years ago during the Paleolithic stage or Old Stone Age where our ancestors developed newer abilities to avoid being preyed upon, developing of cooking, encephalization, increase body size, expansion of new territory, and hunting (Table 1.1). In other words, the "drifty genotype" hypothesis argues that the prevalence of thrifty genes is not a result of positive selection for energy storage genes but, in reality, a genetic drift caused by the removal of predatory selection pressures.

To further counterbalance the long-held acceptance of the thrifty genotype hypothesis proposed by Neel as the most reliable model for the genetic basis of obesity, John Speakman in 2008 introduced the "drifty phenotype" or predation release hypothesis. For this interpretation in opposition to being selected for, obesogenic energy-efficient genes favoring fat storage are present in Western populations because early hominids removed the selection pressure previously exerted on them by predation. The concept of Speakman is that around 2 million years ago, the ancient ancestors, *Homo habilis* and *Homo erectus*, evolved to acquire the capability of using fire and stone tools, building weapons, and organizing social communities. For the first time in evolutionary history, an animal that was not the top predator in its ecosystem was able to remove the threat of predatory danger (Speakman 2008). This hypothesis then suggests that vital genes involved in the evasion of predators that include athletic fitness, speed, agility, stamina, and leanness were no longer needed in the life of modern humans but continue to be present for all other animals (Speiser et al. 2013; Spence et al. 2013). In other words, in the absence of predation selection pressure, genes that promote energy storage and obesity were not eliminated by natural selection. In fact, they were allowed to drift in the

genetic journey of human evolution explaining why the obesity pandemic in modern Western societies has developed. Both theories, thrifty and drifty genotypes, assume that the selection pressures that ancestors of modern humans living in Western societies faced were the same. However, neither theory sufficiently explained the influence of globalization and population demographic changes that started 70,000 years ago from Africa. In the face of clear evidence, ethnic variation in obesity susceptibility and related metabolic syndrome demographics also plays a role. Having said that, although both the thrifty and drifty genotype hypotheses have considerable merit and may be responsible for the genetic susceptibility to obesity, in a particular group of individuals, neither theory can conclusively explain for the contemporary obesity pandemic in industrialized countries. The additional point is that obesity is not adaptive and may never even have existed in our evolutionary past, but it is evident today as a maladaptive by-product of positive selection on some other trait. For example, obesity may result from variation in brown adipose tissue (BAT) thermogenesis (see next topic). Another view is that most mutations in the genes that predispose us to obesity are neutral or not exacerbated, but they were drifted over evolutionary time leading some individuals to be obese while others resistant to obesity.

The general concept of Neel's hypothesis is attractive as pointed out by Andrew Prentice (Prentice 2001) where "the genetic influences on body weight are the product of natural selection from lean times" suggesting that there is no advantage of fatness that had much to do with mortality. In his example with women from Africa, he found that food scarcity influences fertility by a cessation to ovulate, while higher body weight individuals have a greater reproductive achievement. With the advent of the agricultural society, periods of plenty increased. The advantage was that while thinner individuals are more likely to die, bear fewer children, and pass their genes to the next generation, the fatter or well-fed individuals instead were able to be more successful in generating offspring. The other problem with the Neel hypothesis was that if through our evolution it was advantageous to be fat conferring a survival trait, then why the highest percentage of people in society is not obese? Is this fact suggesting that not all of us inherit the thrifty genes? To summarize these concepts, we could say that the thrifty hypothesis is based on feast or famine events giving humans the advantage of being exceptionally efficient at storing fat which were more likely to survive. On the contrary, the drifty gene hypothesis claims that fatness was not a survival advantage but rather being a disadvantage when humans no longer had to run from predators; consequently, obesity drifted into the population. Other alternative hypotheses came along to contribute or complement the thrifty and drifty genotype hypotheses. The "thrifty phenotype hypothesis," or Barker hypothesis, addresses the insufficiencies of thrifty gene hypothesis and also explains that newborns with low birth weight and poor nutrition in the uterus are especially prone to diabetes, obesity, heart disease, and other metabolic disorders later in life even when food is abundant in adulthood (Hales and Barker 1992). Barker proposes that the developing undernourished fetus suffering from energy shortage will allocate energy away from the pancreas in favor of other tissues such as the brain. There are additional hypotheses related to the same principle: "weather forecast model" where the fetal setting predicts the quality of the childhood environment, "maternal fitness model" where fetal environment uses nutritional signals to support its metabolism with the mother's, "intergenerational phenotypic inertia model" where intrauterine nutritional signals are related to the history of the mother, and recent ancestors through epigenetic mechanisms: "predictive adaptive response model" where fetal environment predicts adult environment (Hales and Barker 1992; Bateson 2001).

Interestingly, as we recognize that obesity is a result of gene-environment interactions and that predisposition to obesity lies predominantly in our evolutionary past, the concept that human metabolism runs on old unmodified genes and unprepared for modern eating habits is actively debated. A diet based on foraging (collecting wild plants and pursuing wild animals), which represents a diet high in proteins and low in carbohydrates, should make us of a lean phenotype; however, that premise is more complicated than a simple hunter-gatherer's diet. Hunting and gathering were the most successful of human adaptation, occupying at least 90 percent of human history. Following the development of agriculture, which relies on agricultural societies, and domesticated species, hunter-gatherers were displaced or conquered by farming or pastoralist groups in most parts of the world. Table 1.1 summarizes dynamic transitions through human evolution (Bellisari 2008). The deleterious changes seen in our modern society are the result of the interaction between the evolutionary human biology and development of culture over the long period of human evolution. The encephalization (the tendency for a species to evolve larger brains through time involving a change of function from noncortical parts of the brain to the cortex) of humans evolved in complex genetic and physiological systems to protect against starvation and defend stored body fat. Besides, the advantage of technological development providing access to significant quantities of mass-produced high-calorie food caused an increase in consumption. The latter event associated with reduced physical effort, the decrease in physical labor, transportation devices abolishing starvation, and heavy manual work all contributed to the current state of obesity in our society.

With the arrival of the industrial and agricultural revolution, maximizing energy intake and minimizing physical effort and energy expenditure became the norm causing a dramatic decline in nutritional health. Combined with the high genetic predisposition (O'Rahilly and Farooqi 2006) and efficient metabolic system for energy accumulation, storage, and protection (Woods and Seeley 2000), the high rates of obesity became a new trend in modern society. The factors contributing to obesity in a community with unnatural access to calories and processed food are multiple and in great part due to an exacerbation of our evolutionary genes to promote survival. Genes enhancing obesity bring up an interesting observation because obesity seems to cause with time a host of negative consequences.

During evolution by natural selection, all species, including humans, develop genes throughout the natural selection, which favors advantages in dealing with the environment, not disadvantages. Therefore, how is it possible for us to become an obese species if obesity is a negative trait that will threaten our survival and should have eliminated us as species? But, in modern society only 36% of the people are obese, and the rest have average weight or slightly overweight. It is important to point out that we cannot entirely compare us with a certain group of animals that accumulate body fat in amounts that would be considered obese in humans. The most typical examples are hibernating animals, which deposit large fat stores before entering hibernation, and migratory birds, which store similar stores before starting on migratory journeys. It is clear that these situations of temporary obesity, as a mechanism of survival in anticipation of a future shortfall of energy, are well established and do not cause future obesity in those animals. It will be catastrophic for hibernating animals to be unable to feed in winter and for migratory birds to be unable to feed enough before flying over oceans. A lack of genes favoring fat accumulation will exterminate these species. Therefore, primitive humans had a more complex set of evolutionary genes to contemplate survival during periods of starvation (see below).

Why understanding these evolutionary factors is important to grasp the meaning of obesity today? In part because medical research is focusing on the contribution that nutritional programming (a process through which a stimulus during a critical window of time lastingly effects following structure, function or developmental schedule of the organism) has to disease in later life. The idea of the thrifty phenotype, first proposed by Hales and Barker (1992), used in medical research today goes in opposition to the thrifty genotype model, to interpret associations between early-life experience and adult health status. However, one of the caveats in the thrifty phenotype hypothesis is that it fails to explain why plasticity is lost so early in development in species with extensive growth, maybe because developing animals cannot maintain phenotypic

plasticity during growth. Allowing the preservation of maternal strategy in offspring phenotype buffered against environmental fluctuations during the most sensitive period of development ensures a logical adaptation of growth to the state of the environment. Therefore, strategies in public health oriented for improving birth weight may be more effective if they target maternal development rather than nutrition during pregnancy. In addition, based on the thrifty phenotype hypothesis, several evolutionary models proposed include (1) the weather forecast model of Bateson, (2) the maternal fitness model of Wells, (3) the intergenerational phenotypic inertia model of Kuzawa, and (4) the predictive adaptive response model (Gluckman and Hanson 2006) (Wells 2007). From all these models, the weather forecast model is widely accepted because it proposes that developing organisms respond to cues of environmental quality and that mismatches between this forecast and subsequent reality generate significant adverse effects on adult phenotype. For more reading see Gluckman's work (2006). One of the recent hypotheses, a consequence of the progress in molecular biology, is the thrifty epigenome hypothesis that claims that there are epigenetic modifications in response to environmental conditions (Stoger 2008) susceptibly to epigenetic variations corresponding epigenotypes with the potential to be inherited across generations. Furthermore, recent evidence suggests that early prenatal or postnatal environmental changes cause permanent metabolic modifications that are in part due to epigenetic changes in essential genes and areas of the central nervous system involved in the control of energy balance. This interaction between genetic and environmental factors including nutrition, maternal health, unknown chemicals, and lifestyle during the prenatal or perinatal period has influenced the development of energy balance causing unwanted changes. In studies done in both humans and animal models, prenatal or perinatal nutritional manipulations lead to chronic metabolic alteration affecting leptin sensitivity, glucose metabolism, and in turn energy expenditure and feeding behavior. These metabolic flaws may be a result of abnormal development of appetite-regulating neuronal circuits due to peri-natal programming (Contreras et al. 2013).

"Genetically unknown foods hypothesis" pro-poses that obesity and diabetes occur when popu-lations are introduced to new foods that they haven't adapted to (Baschetti 1998). That is the case when certain "new-world populations that kept to traditional dietary habits were virtually free from diabetes"; then, after they began eating some foods that are common in Europe, the dis-ease reached epidemic proportions. This hypoth-esis certainly has a lot of merits, especially as in modern times, processed fatty and sugary (natu-ral and synthetic) foods were introduced and heavily consumed in society today, a diet that does not match with our homeostatic gene reper-toire for energy balance. Another group proposes that insulin resistance is believed to have evolved as an adaptation to periodic starvation, and there-fore they propose a hypothesis that insulin resis-tance is a socio-ecological adaptation that mediates two phenotypic transitions. A reproduc-tive strategy deals from a large number of off-spring with little investment in each to a smaller number of offspring with more investment in each (Watve and Yajnik 2007).

Multiple and intricate mechanisms have evolved to control energy balance to maintain body weight. Energy intake has to match energy expenditure to keep body weight at a constant level, but also macronutrient intake must balance macronutrient oxidation. This situation of equili-brated balance seems to be predominantly diffi-cult to achieve in individuals with low-fat oxidation, low energy expenditure, low sympa-thetic activity, or low levels of spontaneous phys-ical activity. All of these factors, among many, explain the tendency of some people to gain weight. Since there is a considerable variability in weight change in different individuals as observed when energy surplus is imposed experi-mentally or spontaneously, recent data suggest a strong genetic influence on body weight regula-tion when normal physiology is subjected to an "obesogenic" environment. In the modern world, we no longer eat only when metabolically hun-gry; on the contrary, we frequently eat in the complete absence of appetite and in spite of hav-ing large fat reserves in our bodies. Therefore, hedonic eating that refers to the participation of cognitive, reward, and emotional factors disrupts the homeostatic model for the regulation of energy balance. Although substantial progress has been made in recognizing the metabolic sig-nals and neural circuitry between the brain stem and hypothalamus representing the homeostatic metabolic regulator (Berthoud 2011) (Galgani and Ravussin 2008), the neural pathways located in cortico-limbic structures responsible for hedonic behavior are much less understood. Figure 1.3 depicts a brief integration of major components of body weight regulation in an obe-sogenic environment as described by Berthoud (2011).

1.6   Thermogenesis and Human Migration

Among the various physiological mechanisms, homeotherms (animals that maintain body tem-perature generally above of the environment at a constant level through metabolic activity) utilize the heat production to maintain body temperature in their adaptation to different environments. BAT is responsible for the thermogenic mecha-nisms involved in energy expenditure. BAT in mitochondria uniquely express uncoupled pro-tein 1 (UCP1), an inner mitochondrial membrane protein that uncouples ATP synthesis from oxida-tive phosphorylation, liberating energy in the form of heat (Lowell and Spiegelman 2000). It is important during cold stress by producing heat using lipids and glucose as metabolic fuels. Additionally, white adipose tissue (WAT) or beige cells have also been found to exhibit a thermogenic action similar to BAT. The heat pro-ducing of uncoupling proteins in BAT mitochon-dria is believed to be a key driver behind the conquest of a variety of environments in mam-mals 65 million years ago (Oelkrug et al. 2013; Saito et al. 2008). This ability to produce and maintain heat contributed to the evolution of mammals to explore and settle in uninhabitable territories throughout the planet (Saito et al. 2008) by adjusting the thermoregulatory response to sharply different environments. The impact of BAT thermogenesis to survival was critical in a way that it probably drove mammalian placental radiation at the end of the Cretaceous. It was a global event that led to mammals displacing the dinosaurs as the dominant class of animal on earth (Oelkrug et al. 2013).

One of the latest hypotheses to explain obesity (Sellayah et al. 2014) proposes that the current obesity pandemic in industrialized countries is also a result of the differential exposure of human ancestors to environmental factors that began when humans left Africa around 70,000 years ago and then migrated through the globe by settling in varied climates. They noted that diabetes and obesity are unequally distributed among populations from different parts of the world. This striking finding is related to the fact that survival in colder parts of the world amplified genes that help preserve body temperature. A higher metabolic rate that keeps the body warm would confer some resistance to obesity. Genes adapted for warmer climates would lower the metabolic rate, burn calories at a slower pace, and make the body more inclined to accumulate fat. It is therefore proposed that genetic factors played a role in ancestral environmental exposures in a way that affected energy expenditure even in groups of peoples from heterogeneous populations. These environmental pressures caused a great selection giving an advantage of cold-adapted genes. The high basic metabolic rate was seen in arctic people (Leonard et al. 2002), intermediate in white Europeans, and lowest in African-Americans (Weyer et al. 1999; Wong et al. 1999). The obesity rates in white Europeans with similar lifestyles and caloric intake, but in different regions of Europe, were seen to have a significant disparity. For example, among Scandinavian countries, whose population's ancestry has generations of genetic adaptation to extreme cold, some have much lower rates of obesity than the rest of Europe, despite having similar lifestyles and consuming similar calorie foods.

A summary of their demographic hypothesis is depicted in Fig. 1.4, which shows the historic human migration out of Africa 70,000 years ago (Sellayah et al. 2014). By 60,000 years ago,

humans populated Central Asia, and from that location, they migrated to northeast into Siberia and Northeast Asia. In this new environment, human acquired genes for cold adaptation with a higher resting metabolic rates and thus more resistance to obesity. The second group of migrants from Central Asia moved north and west into Europe, which also forced them to acquire genes for cold adaptation, displacing the resident Neanderthals. A third group migrated into Australia and maintained genes for heat adaptation. The Aborigines in Australia from that migration then develop a low resting metabolic rate and an increased propensity for obesity and type 2 diabetes. From the Northeast of Asia, a group crossed the Bering Strait 20,000 years ago into Alaska. Some of their descendants still live in the Canadian Arctic and are highly resistant to cold with an exceptionally high resting metabolic rate. Migration through the Pacific coast to North America and Mexico encountered hotter climates and reacquired genes for heat adaptation. The Pima Indians, which are the descendants of these groups, have the highest rates of obesity and cardiovascular disease in the world. Their evolutionary cousins, the Yaghan from Tierra del Fuego whose ancestors continued the southern migration toward the Antarctic South American Cone, probably recuperated their high BAT capability producing high resting metabolic rates and a thinner phenotype. The overall concept of this hypothesis is that ethnic differences, which resulted from different migrations to cold, mild, and hot environments 70,000 years ago, produced different genetic adaptations and susceptibilities to obesity and metabolic syndrome particularly in those individuals with a low basic metabolic rate.

1.7   Conclusions and Further Thoughts

To put all these hypotheses in perspective, the contributing factors causing the global obesity pandemic we are witnessing in today's human society reside in three distinct attributes. The first one is the environmental changes that occurred

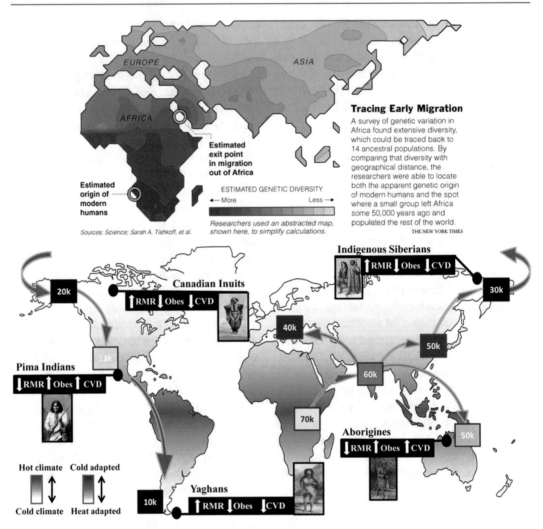

Fig. 1.4 Historical human migration and the impact of thermogenesis: This figure depicts the impact that ancient human migration 70,000 years ago from Africa has on selection of genes for heat and cold adaptation and the consequences in the prevalence of obesity in modern society. Individuals who migrated to cold regions acquired genes for cold adaptation, conferring them higher resting metabolic rates and thus more resistance to obesity. On the contrary, individuals who migrated to warm climates have low resting metabolic rates and an increased propensity for obesity and type 2 diabetes. (Figure reproduced with permission from Sellayah et al. (2014))

through the industrial revolution and beyond. The second one is a genetic tendency that has its origins in our evolutionary traits (genetically determined characteristics). The third is lifestyle changes introducing an excess of sedentary life in the way daily work is performed in most settings. To untangle these aspects is rather complicated because of the multiple factors involved. Understanding the evolutionary mechanisms that allow for obesity to take place in human society is highly relevant to clinical and public health management of the epidemic. The thrifty genotype hypothesis posits that although the modern environment is different from the evolutionary environment, the body is still adapted to the past where it was advantageous to store fat against future food insecurity. These genes enable individuals to efficiently collect and process food to deposit fat during periods of food abundance to provide energy for periods of famine.

From the above description, we can conclude that there are several competing hypotheses for the evolutionary origins of the widespread obesity from gene to life conditions, ethnic groups, and migrations to different climates. Although in appearance some of the hypotheses appear incompatible, complementary features among them do exist. For example, the thrifty epigenome hypothesis is believed to be a link between the thrifty gene and thrifty phenotype hypotheses (Genne-Bacon 2014). The called behavioral switch hypothesis is also in some regards compatible with the thrifty family of hypotheses (Watve and Yajnik 2007). It proposes an integration of both social and physiological mechanisms into a combined theory for the evolutionary origins of insulin resistance and obesity. It argues that metabolic diseases are by-products of a socio-ecological adaptation that switches between both reproductive and socio-behavioral strategies. This hypothesis justifies the modern pandemic of metabolic diseases as based on extreme environmental incentives: population density, urbanization, social competition, caloric access, and sedentary lifestyles exaggerated broadly to a degree never before seen in human evolutionary history (Watve and Yajnik 2007). The lack of food available is an important factor in mediating the switch between reproductive and lifestyle strategies. Changes in energy balance set point are still an important evolutionary component in the behavioral switch hypothesis. Selection for thrifty genes could have been the hallmark of a predation release/freedom from selective group providing metabolic thriftiness and weight control to avoid predation. Once predator threat was eliminated because of human social progress, there was no more selection for leanness. This means that there is room for more than one hypothesis to be corrected depending upon the nature of the natural pressures of the environment. Although, in general, the thrifty gene hypothesis has been accepted as the central hypothesis, it has impacted the way research and clinical management of obesity and diabetes are conducted.

According to Sellayah and colleagues (Sellayah et al. 2014) that propose current obesity pandemic is a result of the differential exposure of the ancestors to environmental factors that began when they left Africa around 70,000 years ago and migrated through the globe, the thrifty and drifty genotype hypotheses do not answer all questions. They claim that a lack of full understanding of the genetic basis for ethnic variability could also represent an obstacle in the interpretation of susceptibility to obesity in the developed world that caused an obesity pandemic. According to this hypothesis, obesity pandemic has certainly coincided with not only an increase in unhealthy eating habits but also a bulk of immigration of various ethnicities with different BAT energy requirements of basic metabolic rate. They argue that the thrifty and drifty genotype hypotheses assume that the selection pressures faced by the ancestors living today in all countries are the same, while they argue that this is not an accurate statement. For example, the descendants of early humans who remained in Africa and those who migrated to similar environments such as Black Americans and Pacific Islanders maintained heat adaption genes. On the other hand, those groups who migrated to colder regions including Europe and Siberia such as Caucasians and Chinese acquired genes for cold adaptation. Siberians who migrated to the American continent and established in subtropical and tropical regions in North, Central, and South America lost their cold-adaptive genes and developed genes for heat adaptation. They propose that positive selection for cold adaptation provided Caucasians and East Asians such as Chinese, Japanese, and Koreans with efficient BAT and UCP1 function with higher metabolic rate and resistance to obesity. On the opposite side, Africans and South Asians, whose ancestors did not need to evolve efficient BAT and UCP1 function, have a major propensity for obesity because of their more sedentary and hypercaloric Western lifestyle. The evolutionary origins of obesity as briefly described here to explain the global obesity epidemic are still at odds with ways how scientists struggle to understand the biological, cultural, and evolutionary basis of this condition. Furthermore, a better understanding of the interaction between physical activity and the endocrine system regulating metabolism in lean and obese could potentially help the evolutionary pathways that our ancestors took and develop tools to combat this condition. Any one theory could not explain an evolutionary tendency to become obese since humans in different parts of the world experienced different evolutionary pressures, so what's true in one population of migrants might not apply to another living in a different climate. It is important to consider that evolutionary changes can be evident in a single generation when one or more alleles from genetic variants could change. The genetics of obesity has many influences over time and not just starvation or plenty or cold or warm. Another important negative consequence of the obesity state is the endocrine disarray seen in the overall metabolism of obese individuals. Modern obese humans have many endocrine changes in the hypothalamic-pituitary-endocrine axis homeostasis including low androgen levels (hypogonadism), polycystic ovary syndrome (PCOS), reduction in GH secretion by the pituitary gland, changes in cortisol levels, and thyroid dysfunction that is frequently associated with changes in body weight and composition, body temperature, energy expenditure, adipose tissue, food intake, and glucose and lipid metabolism. All these topics are discussed in detail throughout the different chapters of this book. Besides from being a fascinating academic pursuit, understanding human evolution is exceedingly important to comprehend the health of modern humans.

Questions

1. Define the changes seen in the modern human brain.
2. How is obesity defined, and what does energy balance regulation mean?
3. What is the difference between metabolic and hedonic behavior in the control of body weight?
4. Describe the role of thermogenesis in human migration.
5. Which one is the best genetic hypothesis to define obesity in the modern world?

References

Ahima, R. S. (2005). Central actions of adipocyte hormones. *Trends in Endocrinology and Metabolism: TEM*, 307–313.

Anton, S. C., Potts, R., & Aiello, L. C. (2014). Human evolution. Evolution of early Homo: An integrated biological perspective. *Science, 1236828*. https://doi.org/10.1126/science.1236828.

Astrup, A. (1999). Macronutrient balances and obesity: The role of diet and physical activity. *Public Health Nutrition*, 341–347.

Astrup, A., Toubro, S., Raben, A., & Skov, A. R. (1997). The role of low-fat diets and fat substitutes in body weight management: What have we learned from clinical studies? *Journal of the American Dietetic Association*, S82–S87.

Baschetti, R. (1998). Diabetes epidemic in newly westernized populations: Is it due to thrifty genes or to genetically unknown foods? *Journal of the Royal Society of Medicine*, 622–625.

Bateson, P. (2001). Fetal experience and good adult design. *International Journal of Epidemiology*, 928–934.

Bauer, F., Elbers, C. C., Adan, R. A., Loos, R. J., Onland-Moret, N. C., Grobbee, D. E., van Vliet-Ostaptchouk, J. V., Wijmenga, C., & van der Schouw, Y. T. (2009). Obesity genes identified in genome-wide association studies are associated with adiposity measures and potentially with nutrient-specific food preference. *The American Journal of Clinical Nutrition*, 951–959. https://doi.org/10.3945/ajcn.2009.27781.

Beck-Nielsen, H. (1999). General characteristics of the insulin resistance syndrome: Prevalence and heritability. *European Group for the study of Insulin Resistance (EGIR) Drugs, 7–10*, 75–82.

Bell, E. A., & Rolls, B. J. (2001). Energy density of foods affects energy intake across multiple levels of fat content in lean and obese women. *The American Journal of Clinical Nutrition*, 1010–1018.

Bellisari, A. (2008). Evolutionary origins of obesity. *Obesity Reviews: An Official Journal of the International Association for the Study of Obesity*, 165–180. https://doi.org/10.1111/j.1467-789X.2007.00392.x.

Berthoud, H. R. (2011). Metabolic and hedonic drives in the neural control of appetite: Who is the boss? *Current Opinion in Neurobiology*, 888–896. https://doi.org/10.1016/j.conb.2011.09.004.

Berthoud, H. R., & Munzberg, H. (2011). The lateral hypothalamus as integrator of metabolic and environmental needs: From electrical self-stimulation to optogenetics. *Physiology & Behavior*, 29–39. https://doi.org/10.1016/j.physbeh.2011.04.051.

Caballero, B. (2007). The global epidemic of obesity: An overview. *Epidemiologic Reviews*, 1–5. https://doi.org/10.1093/epirev/mxm012.

Chatterjee, H. J., Ho, S. Y., Barnes, I., & Groves, C. (2009). Estimating the phylogeny and divergence times of primates using a supermatrix approach.

BMC Evolutionary Biology, 259. https://doi.org/10.1186/1471-2148-9-259.

Clement, K. (2005). Genetics of human obesity. *The Proceedings of the Nutrition Society*, 133–142.

Contreras, C., Novelle, M. G., Leis, R., Dieguez, C., Skrede, S., & Lopez, M. (2013). Effects of neonatal programming on hypothalamic mechanisms controlling energy balance. *Hormone and Metabolic Research*, 935–944. https://doi.org/10.1055/s-0033-1351281.

Cooling, J., & Blundell, J. E. (2001). High-fat and low-fat phenotypes: Habitual eating of high- and low-fat foods not related to taste preference for fat. *European Journal of Clinical Nutrition*, 1016–1021. https://doi.org/10.1038/sj.ejcn.1601262.

Cordain, L., Eaton, S. B., Sebastian, A., Mann, N., Lindeberg, S., Watkins, B. A., O'Keefe, J. H., & Brand-Miller, J. (2005). Origins and evolution of the western diet: Health implications for the 21st century. *The American Journal of Clinical Nutrition*, 341–354.

de Krom, M., Bauer, F., Collier, D., Adan, R. A., & la Fleur, S. E. (2009). Genetic variation and effects on human eating behavior. *Annual Review of Nutrition*, 283–304. https://doi.org/10.1146/annurev-nutr-080508-141124.

Delgado, J. M., & Anand, B. K. (1953). Increase of food intake induced by electrical stimulation of the lateral hypothalamus. *The American Journal of Physiology*, 162–168.

DelParigi, A., Pannacciulli, N., Le, D. N., & Tataranni, P. A. (2005a). In pursuit of neural risk factors for weight gain in humans. *Neurobiology of Aging*, 50–55. https://doi.org/10.1016/j.neurobiolaging.2005.09.008.

DelParigi, A., Chen, K., Salbe, A. D., Reiman, E. M., & Tataranni, P. A. (2005b). Sensory experience of food and obesity: A positron emission tomography study of the brain regions affected by tasting a liquid meal after a prolonged fast. *NeuroImage*, 436–443. https://doi.org/10.1016/j.neuroimage.2004.08.035.

Drewnowski, A. (2003). The role of energy density. *Lipids*, 109–115.

Eaton, S. B. (2006). The ancestral human diet: What was it and should it be a paradigm for contemporary nutrition? *The Proceedings of the Nutrition Society*, 1–6.

Elias, C. F., Aschkenasi, C., Lee, C., Kelly, J., Ahima, R. S., Bjorbaek, C., Flier, J. S., Saper, C. B., & Elmquist, J. K. (1999). Leptin differentially regulates NPY and POMC neurons projecting to the lateral hypothalamic area. *Neuron*, 775–786.

Elmquist, J. K., Bjorbaek, C., Ahima, R. S., Flier, J. S., & Saper, C. B. (1998). Distributions of leptin receptor mRNA isoforms in the rat brain. *The Journal of Comparative Neurology*, 535–547.

Flier, J. S. (2004). Obesity wars: Molecular progress confronts an expanding epidemic. *Cell*, 337–350.

Galgani, J., & Ravussin, E. (2008). Energy metabolism, fuel selection and body weight regulation. *International Journal of Obesity*, (2005), S109–S119. https://doi.org/10.1038/ijo.2008.246.

Garn, S. M., & Leonard, W. R. (1989). What did our ancestors eat? *Nutrition Reviews*, 337–345.

Genne-Bacon, E. A. (2014). Thinking evolutionarily about obesity. *The Yale Journal of Biology and Medicine*, 99–112.

Gluckman, P. D., & Hanson, M. A. (2006). Evolution, development and timing of puberty. *Trends in endocrinology and metabolism: TEM*, 7–12. https://doi.org/10.1016/j.tem.2005.11.006.

Gonzalez, R., Sarr, M. G., Smith, C. D., Baghai, M., Kendrick, M., Szomstein, S., Rosenthal, R., & Murr, M. M. (2007). Diagnosis and contemporary management of anastomotic leaks after gastric bypass for obesity. *Journal of the American College of Surgeons*, 47–55. https://doi.org/10.1016/j.jamcollsurg.2006.09.023.

Gotoh, K., Fukagawa, K., Fukagawa, T., Noguchi, H., Kakuma, T., Sakata, T., & Yoshimatsu, H. (2005). Glucagon-like peptide-1, corticotropin-releasing hormone, and hypothalamic neuronal histamine interact in the leptin-signaling pathway to regulate feeding behavior. *The FASEB Journal*, 1131–1133. https://doi.org/10.1096/fj.04-2384fje.

Hales, C. N., & Barker, D. J. (1992). Type 2 (non-insulin-dependent) diabetes mellitus: The thrifty phenotype hypothesis. *Diabetologia*, 595–601.

Holzapfel, C., Grallert, H., Huth, C., Wahl, S., Fischer, B., Doring, A., Ruckert, I. M., Hinney, A., Hebebrand, J., Wichmann, H. E., Hauner, H., Illig, T., & Heid, I. M. (2010). Genes and lifestyle factors in obesity: Results from 12,462 subjects from MONICA/KORA. *International Journal of Obesity*, (2005), 1538–1545. https://doi.org/10.1038/ijo.2010.79.

Hommel, J. D., Trinko, R., Sears, R. M., Georgescu, D., Liu, Z. W., Gao, X. B., Thurmon, J. J., Marinelli, M., & DiLeone, R. J. (2006). Leptin receptor signaling in midbrain dopamine neurons regulates feeding. *Neuron*, 801–810. https://doi.org/10.1016/j.neuron.2006.08.023.

Isler, K., & van Schaik, C. P. (2012). Allomaternal care, life history and brain size evolution in mammals. *Journal of Human Evolution, 63*, 52. https://doi.org/10.1016/j.jhevol.2012.03.009.

Jensen, M. D. (2008). Role of body fat distribution and the metabolic complications of obesity. *The Journal of Clinical Endocrinology and Metabolism*, S57–S63. https://doi.org/10.1210/jc.2008-1585.

Keski-Rahkonen, A., Kaprio, J., Rissanen, A., Virkkunen, M., & Rose, R. J. (2003). Breakfast skipping and health-compromising behaviors in adolescents and adults. *European Journal of Clinical Nutrition*, 842–853. https://doi.org/10.1038/sj.ejcn.1601618.

Kim, J. G., Suyama, S., Koch, M., Jin, S., Argente-Arizon, P., Argente, J., Liu, Z. W., Zimmer, M. R., Jeong, J. K., Szigeti-Buck, K., Gao, Y., Garcia-Caceres, C., Yi, C. X., Salmaso, N., Vaccarino, F. M., Chowen, J., Diano, S., Dietrich, M. O., Tschop, M. H., & Horvath, T. L. (2014). Leptin signaling in astrocytes regulates hypothalamic neuronal circuits and feeding. *Nature Neuroscience*, 908–910. https://doi.org/10.1038/nn.3725.

Koch, M., Varela, L., Kim, J. G., Kim, J. D., Hernandez-Nuno, F., Simonds, S. E., Castorena, C. M., Vianna, C. R., Elmquist, J. K., Morozov, Y. M., Rakic, P., Bechmann, I., Cowley, M. A., Szigeti-Buck, K., Dietrich, M. O., Gao, X. B., Diano, S., & Horvath, T. L. (2015). Hypothalamic POMC neurons promote cannabinoid-induced feeding. *Nature*, 45–50. https://doi.org/10.1038/nature14260.

Leonard, W. R. (2002). Food for thought. Dietary change was a driving force in human evolution. *Scientific American*, 106–115.

Leonard, W. R., & Robertson, M. L. (1992). Nutritional requirements and human evolution: A bioenergetics model. *American Journal of Human Biology: Official Journal of the Human Biology Council*, 179–195. https://doi.org/10.1002/ajhb.1310040204.

Leonard, W. R., & Robertson, M. L. (1994). Evolutionary perspectives on human nutrition: The influence of brain and body size on diet and metabolism. *American Journal of Human Biology: Official Journal of the Human Biology Council*, 77–88. https://doi.org/10.1002/ajhb.1310060111.

Leonard, W. R., Sorensen, M. V., Galloway, V. A., Spencer, G. J., Mosher, M. J., Osipova, L., & Spitsyn, V. A. (2002). Climatic influences on basal metabolic rates among circumpolar populations. *American Journal of Human Biology: The Official Journal of the Human Biology Council*, 609–620. https://doi.org/10.1002/ajhb.10072.

Leonard, W. R., Snodgrass, J. J., & Robertson, M. L. (2007). Effects of brain evolution on human nutrition and metabolism. *Annual Review of Nutrition, 27*, 311. https://doi.org/10.1146/annurev.nutr.27.061406.093659.

Lowell, B. B., & Spiegelman, B. M. (2000). Towards a molecular understanding of adaptive thermogenesis. *Nature*, 652–660. https://doi.org/10.1038/35007527.

Ludwig, D. S., Tritos, N. A., Mastaitis, J. W., Kulkarni, R., Kokkotou, E., Elmquist, J., Lowell, B., Flier, J. S., & Maratos-Flier, E. (2001). Melanin-concentrating hormone overexpression in transgenic mice leads to obesity and insulin resistance. *The Journal of Clinical Investigation*, 379–386. https://doi.org/10.1172/JCI10660.

Malik, S., McGlone, F., Bedrossian, D., & Dagher, A. (2008). Ghrelin modulates brain activity in areas that control appetitive behavior. *Cell Metabolism*, 400–409. https://doi.org/10.1016/j.cmet.2008.03.007.

Murray, S., Tulloch, A., Gold, M. S., & Avena, N. M. (2014). Hormonal and neural mechanisms of food reward, eating behaviour and obesity. *Nature Reviews Endocrinology*, 540–552. https://doi.org/10.1038/nrendo.2014.91.

Narayanan, N. S., Guarnieri, D. J., & DiLeone, R. J. (2010). Metabolic hormones, dopamine circuits, and feeding. *Frontiers in Neuroendocrinology*, 104–112. https://doi.org/10.1016/j.yfrne.2009.10.004.

Neel, J. V. (1962). Diabetes mellitus: A "thrifty" genotype rendered detrimental by "progress"? *American Journal of Human Genetics*, 353–362.

Neel, J. V. (1999). The "thrifty genotype" in 1998. *Nutrition Reviews*, S2–S9.

Nillni, E. A. (2010). Regulation of the hypothalamic thyrotropin releasing hormone (TRH) neuron by neuronal and peripheral inputs. *Frontiers in Neuroendocrinology*, 134–156. doi: S0091-3022(10)00002-6 [pii]. https://doi.org/10.1016/j.yfrne.2010.01.001.

Nillni, E. A., & Sevarino, K. A. (1999). The biology of pro-thyrotropin-releasing hormone-derived peptides. *Endocrine Reviews*, 599–648.

Obesity: preventing and managing the global epidemic. (2000). Report of a WHO consultation. World Health Organization technical report series:i–xii, 1–253.

Oelkrug, R., Goetze, N., Exner, C., Lee, Y., Ganjam, G. K., Kutschke, M., Muller, S., Stohr, S., Tschop, M. H., Crichton, P. G., Heldmaier, G., Jastroch, M., & Meyer, C. W. (2013). Brown fat in a protoendothermic mammal fuels eutherian evolution. *Nature Communications*, 2140. https://doi.org/10.1038/ncomms3140.

Olds, J., & Milner, P. (1954). Positive reinforcement produced by electrical stimulation of septal area and other regions of rat brain. *Journal of Comparative and Physiological Psychology*, 419–427.

O'Rahilly, S., & Farooqi, I. S. (2006). Genetics of obesity. *Philosophical Transactions of the Royal Society of London. Series B, Biological Sciences*, 1095–1105. https://doi.org/10.1098/rstb.2006.1850.

Perello, M., Stuart, R. C., & Nillni, E. A. (2006). The role of intracerebroventricular administration of leptin in the stimulation of prothyrotropin releasing hormone neurons in the hypothalamic paraventricular nucleus. *Endocrinology*, 3296–3306.

Pomp, D., & Mohlke, K. L. (2008). Obesity genes: So close and yet so far. *Journal of Biology*, 36. https://doi.org/10.1186/jbiol93.

Pontzer, H., Brown, M. H., Raichlen, D. A., Dunsworth, H., Hare, B., Walker, K., Luke, A., Dugas, L. R., Durazo-Arvizu, R., Schoeller, D., Plange-Rhule, J., Bovet, P., Forrester, T. E., Lambert, E. V., Thompson, M. E., Shumaker, R. W., & Ross, S. R. (2016a). Metabolic acceleration and the evolution of human brain size and life history. *Nature*, 390–392. https://doi.org/10.1038/nature17654.

Pontzer, H., Durazo-Arvizu, R., Dugas, L. R., Plange-Rhule, J., Bovet, P., Forrester, T. E., Lambert, E. V., Cooper, R. S., Schoeller, D. A., & Luke, A. (2016b). Constrained total energy expenditure and metabolic adaptation to physical activity in adult humans. *Current Biology*, 410–417. https://doi.org/10.1016/j.cub.2015.12.046.

Prentice, A. M. (2001). Obesity and its potential mechanistic basis. *British Medical Bulletin*, 51–67.

Rui, L. (2013). Brain regulation of energy balance and body weight. *Reviews in Endocrine & Metabolic Disorders*, 387–407. https://doi.org/10.1007/s11154-013-9261-9.

Saito, S., Saito, C. T., & Shingai, R. (2008). Adaptive evolution of the uncoupling protein 1 gene contributed to the acquisition of novel nonshivering thermogenesis in ancestral eutherian mammals. *Gene*, 37–44. https://doi.org/10.1016/j.gene.2007.10.018.

Sanchez, V. C., Goldstein, J., Stuart, R. C., Hovanesian, V., Huo, L., Munzberg, H., Friedman, T. C., Bjorbaek, C., & Nillni, E. A. (2004). Regulation of hypothalamic prohormone convertases 1 and 2 and effects on processing of prothyrotropin-releasing hormone. *The Journal of Clinical Investigation*, 357–369.

Saper, C. B., Chou, T. C., & Elmquist, J. K. (2002). The need to feed: Homeostatic and hedonic control of eating. *Neuron*, 199–211.

Schuppli, C., Isler, K., & van Schaik, C. P. (2012). How to explain the unusually late age at skill competence among humans. *Journal of Human Evolution*, 843–850. https://doi.org/10.1016/j.jhevol.2012.08.009.

Sellayah, D., Cagampang, F. R., & Cox, R. D. (2014). On the evolutionary origins of obesity: A new hypothesis. *Endocrinology*, 1573–1588. https://doi.org/10.1210/en.2013-2103.

Sorensen, T. I., Price, R. A., Stunkard, A. J., & Schulsinger, F. (1989). Genetics of obesity in adult adoptees and their biological siblings. *BMJ*, 87–90.

Speakman, J. R. (2006). Thrifty genes for obesity and the metabolic syndrome--time to call off the search? *Diabetes & Vascular Disease Research*, 7–11. https://doi.org/10.3132/dvdr.2006.010.

Speakman, J. R. (2008). Thrifty genes for obesity, an attractive but flawed idea, and an alternative perspective: The 'drifty gene' hypothesis. *International Journal of Obesity*, (2005), 1611–1617. https://doi.org/10.1038/ijo.2008.161.

Speiser, D. I., Lampe, R. I., Lovdahl, V. R., Carrillo-Zazueta, B., Rivera, A. S., & Oakley, T. H. (2013). Evasion of predators contributes to the maintenance of male eyes in sexually dimorphic Euphilomedes ostracods (Crustacea). *Integrative and Comparative Biology*, 78–88. https://doi.org/10.1093/icb/ict025.

Spence, R., Wootton, R. J., Barber, I., Przybylski, M., & Smith, C. (2013). Ecological causes of morphological evolution in the three-spined stickleback. *Ecology and Evolution*, 1717–1726. https://doi.org/10.1002/ece3.581.

Stoger, R. (2008). The thrifty epigenotype: An acquired and heritable predisposition for obesity and diabetes? *BioEssays: News and Reviews in Molecular, Cellular and Developmental Biology*, 156–166. https://doi.org/10.1002/bies.20700.

Stuber, G. D., & Wise, R. A. (2016). Lateral hypothalamic circuits for feeding and reward. *Nature Neuroscience*, 198–205. https://doi.org/10.1038/nn.4220.

Stunkard, A. J., Berkowitz, R. I., Stallings, V. A., & Schoeller, D. A. (1999). Energy intake, not energy output, is a determinant of body size in infants. *The American Journal of Clinical Nutrition*, 524–530.

Tataranni, P. A., Harper, I. T., Snitker, S., Del Parigi, A., Vozarova, B., Bunt, J., Bogardus, C., & Ravussin, E. (2003). Body weight gain in free-living pima Indians: Effect of energy intake vs expenditure. *International Journal of Obesity and Related Metabolic Disorders: Journal of the International Association for the Study of Obesity*, 1578–1583. https://doi.org/10.1038/sj.ijo.0802469.

van Schaik, C. P., Isler, K., & Burkart, J. M. (2012). Explaining brain size variation: From social to cultural brain. *Trends in Cognitive Sciences*, 277–284. https://doi.org/10.1016/j.tics.2012.04.004.

van Woerden, J. T., Willems, E. P., van Schaik, C. P., & Isler, K. (2012). Large brains buffer energetic effects of seasonal habitats in catarrhine primates. *Evolution: International Journal of Organic Evolution*, 191–199. https://doi.org/10.1111/j.1558-5646.2011.01434.x.

Volkow, N. D., & Wise, R. A. (2005). How can drug addiction help us understand obesity? *Nature Neuroscience*, 555–560. https://doi.org/10.1038/nn1452.

Waterson, M. J., & Horvath, T. L. (2015). Neuronal regulation of energy homeostasis: Beyond the hypothalamus and feeding. *Cell Metabolism*, 962–970. https://doi.org/10.1016/j.cmet.2015.09.026.

Watve, M. G., & Yajnik, C. S. (2007). Evolutionary origins of insulin resistance: A behavioral switch hypothesis. *BMC Evolutionary Biology*, 61. https://doi.org/10.1186/1471-2148-7-61.

Wells, J. C. (2007). Environmental quality, developmental plasticity and the thrifty phenotype: A review of evolutionary models. *Evolutionary Bioinformatics Online*, 109–120.

Weyer, C., Snitker, S., Bogardus, C., & Ravussin, E. (1999). Energy metabolism in African Americans: Potential risk factors for obesity. *The American Journal of Clinical Nutrition*, 13–20.

Willett, W. C., & Leibel, R. L. (2002). Dietary fat is not a major determinant of body fat. *The American Journal of Medicine*, 47S–59S.

Wong, W. W., Butte, N. F., Ellis, K. J., Hergenroeder, A. C., Hill, R. B., Stuff, J. E., & Smith, E. O. (1999). Pubertal African-American girls expend less energy at rest and during physical activity than Caucasian girls. *The Journal of Clinical Endocrinology and Metabolism*, 906–911. https://doi.org/10.1210/jcem.84.3.5517.

Woods, S. C., & Seeley, R. J. (2000). Adiposity signals and the control of energy homeostasis. *Nutrition*, 894–902.

Yu, Y. H., Vasselli, J. R., Zhang, Y., Mechanick, J. I., Korner, J., & Peterli, R. (2015). Metabolic vs. hedonic obesity: A conceptual distinction and its clinical implications. *Obesity Reviews : An Official Journal of the International Association for the Study of Obesity*, 234–247. https://doi.org/10.1111/obr.12246.

Part II

The Hypothalamus

Neuropeptides Controlling Our Behavior

2

Eduardo A. Nillni

2.1 Introduction

Major contributors in the control of food intake include behavioral response to the environment, hedonic behavior, and metabolism: nutrient sensors, neuropeptide hormones, and peripheral hormones. All combined contribute to three primary stimuli to eat: hunger, reward, and stress. The brain plays a critical role in relaying information about the neuroendocrine and endocrine system regulating the energy network. The mechanisms for controlling food intake require interaction among three major components: the gut, brain, and adipose tissue. The parasympathetic, sympathetic, and other systems are also needed for communication between the brain satiety center, gut, and adipose tissue. These neuronal circuits include many neuropeptide hormones and peptide hormones coming from the periphery, all acting in concert in the regulation of food intake and energy homeostasis. This chapter describes mostly brain peptide hormones and some coming from the periphery. The interested reader could find a full description of all peptide hormones involved in energy balance regulation in a recent review by Crespo et al. (2014). Although regulation of energy homeostasis engages several brain regions including the brainstem, cortex, amygdala, and the limbic system, it is the hypothalamus that is responsible for integrating neuronal and humoral signals in the control feeding behavior. In response to sensory and social stimulation including visual, smell, taste, stress, reward, culture, and exercise, a feeding center in the hypothalamus initiates food uptake that is then terminated by a satiety center. Both the gastrointestinal tract and the white fat cells are responsible for releasing hormone signals that are integrated with the hypothalamus and nucleus of the solitary tract (NTS) to control feeding neural circuits. In spite of the sophistication of these interconnected systems, which are tightly regulated to control energy demand with energy expenditure, the recent sizeable increase in the prevalence of obesity seen in modern urban societies represents a deviation of this evolutionary control of weight homeostasis. Among the major contributors to this disarrangement of the feeding control system are highly palatable and relatively cheap foods ubiquitously available and unfamiliar to our genetic repertoire. The massive reduction in physical activity also contributed to this condition. These current life factors combined with the evolutionary genetic predisposition are the leading causes of the common obesity. This

E. A. Nillni (✉)
Emeritus Professor of Medicine, Molecular Biology, Cell Biology & Biochemistry,
Department of Medicine, Molecular Biology, Cell Biology & Biochemistry,
The Warren Alpert Medical School of Brown University, Providence, RI, USA
e-mail: Eduardo_Nillni@Brown.edu

© Springer International Publishing AG, part of Springer Nature 2018
E. A. Nillni (ed.), *Textbook of Energy Balance, Neuropeptide Hormones, and Neuroendocrine Function*, https://doi.org/10.1007/978-3-319-89506-2_2

chapter describes the role of neuropeptide and some essential peripheral hormones interacting in the hypothalamus toward controlling feeding behavior. The role of hypothalamic nutrient sensors, important in metabolic sensing, is discussed in Chap. 7.

2.2   The Hypothalamus

From its Greek derivation, the hypothalamus (hypo = below, thalamus = bed) is the portion of the diencephalon in all vertebrates that lies inferior to the thalamus. The main function of the hypothalamus is the integration of endocrine, autonomic, and behavioral responses. Even though the hypothalamus comprises only 2% of the total brain volume, it is the primary regulator of pituitary function, homeostatic balance, and neuroendocrine downstream stability. The hypothalamus lies directly above the pituitary gland and is composed of some neuronal cell groups as well as fiber tracts that are symmetric around the third ventricle. The median eminence, which is a midline structure located in the basal hypothalamus ventral to the third ventricle and adjacent to the arcuate nucleus, plays an essential role in communicating the hypothalamus with the pituitary gland. The median eminence contains three zones: the ependymal zone, the internal zone (or zona interna), and the external zone (or zona externa). The portal capillaries in the median eminence that lie outside of the blood-brain barrier functions of the ependymal zone are to create a barrier to the brain, preventing substances released into the periportal capillary spaces from entering the cerebrospinal fluid. The hypothalamus arranges both the autonomic nervous system and the activity of the pituitary, managing energy balance, body temperature, thirst, and hunger and involved in sleep and emotional activity. Figure 2.1 illustrates the hypothalamic-pituitary axis that is considered the coordinating center of the endocrine system. It consolidates signals derived from upper cortical inputs, autonomic function, environmental cues such as light and temperature, and peripheral endocrine feedback. The hypothalamus delivers neuropeptide hormones to the pituitary gland that

in turn releases hormones affecting most endocrine systems including the thyroid gland, the adrenal gland, and the gonads, as well as influencing growth, milk production, and water balance.

The hypothalamus integrates local (neurotransmitters and neuropeptides), circulating metabolic signals to promote appropriate neuroendocrine and autonomic function, peripheral hormones, nutrients, and hedonic signals. Three different components are regulating the hypothalamus, input (afferent), central processing unit, and output (efferent). The afferent component conveys peripheral information on hunger and peripheral metabolism (in the form of hormonal and neural inputs) to the hypothalamus. They include nutrients such as glucose, amino acids, and free fatty acids, peripheral hormones such as leptin and insulin, and hedonic signals such as odor, visual, taste, stress, culture, and reward circuitry (Figs. 2.2 and 2.3; see also Fig. 2.2, Chap. 1). The vagus nerve is the primary neural connection between the brain and the gut and conveys information regarding mechanical stretch of the stomach and duodenum and feelings of gastric fullness to the nucleus tractus solitarius (NTS). The brainstem neurons control energy balance by processing energy status information by sensing circulating metabolites and hormones released from peripheral organs, receiving vagal inputs from the gastrointestinal tract, receiving neuronal inputs from midbrain and forebrain nuclei, and integrating circuits and other neuronal regions of the brain. Ghrelin, an octanoylated 28-amino acid peptide, increases food intake (see Chap. 8). Efferent signals are responsible for hunger/satiety, which include feeding and energy expenditure. They are the ventromedial hypothalamus (VMH), consisting of the ventromedial (VMN) and arcuate (ARC) nuclei that integrate afferent peripheral signals necessary in feeding behavior as well as other central stimuli, and the paraventricular nuclei (PVN) and lateral hypothalamic area (LHA) with neurotransmitter and hormonal system to regulate neural signals for changes in feeding and energy expenditure (Fig. 2.3). Other brain areas serve as neuromodulators for this system. Efferent signals from the hypothalamus to the pituitary axes are considered the coordinating

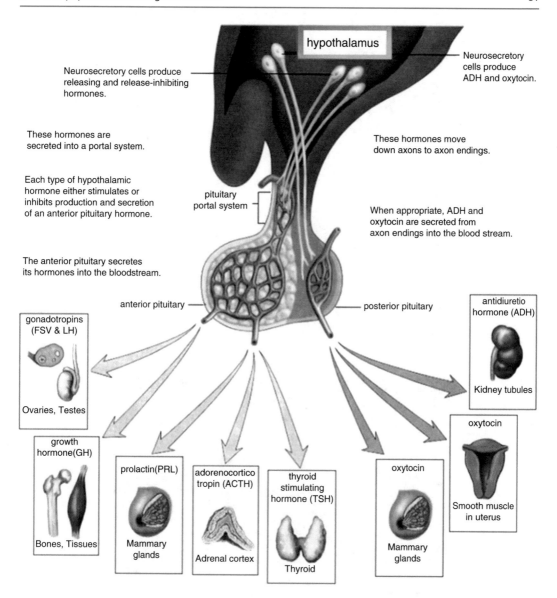

Fig. 2.1 The hypothalamic-pituitary axis and its affected organs located in the periphery. The hypothalamus checks many aspects of the state of the body systems integrating a large amount of information from many sensory pathways coming from the periphery and the brain. The neuroendocrine system combines the hypothalamus with the pituitary, a small gland hanging from underneath the hypothalamus. The anterior pituitary or the adenohypophysis secretes the adrenocorticotrophic hormone (ACTH), thyroid-stimulating hormone (TSH), growth hormone (GH), follicle-stimulating hormone (FSH), luteinizing hormone (LH), and prolactin (PRL). These hormones are released in response to stimulation by the appropriate ligand or secretagogue. For growth hormone and prolactin, there are also hypothalamic inhibitory hormones such as somatostatin, which stop their release, providing a control mechanism. The posterior pituitary or neurohypophysis releases oxytocin and vasopressin (also known as the antidiuretic hormone, ADH). Nerves originating in the supraoptic nuclei and the paraventricular nuclei of the hypothalamus produce oxytocin and vasopressin. These hormones, different from the anterior pituitary, are transported down the nerve cells to the posterior pituitary where they are then released into the bloodstream to affect their target organs

The hypothalamus integrates local (neurotransmitters and neuropeptides) and circulating metabolic signals to promote appropriate neuroendocrine and autonomic function

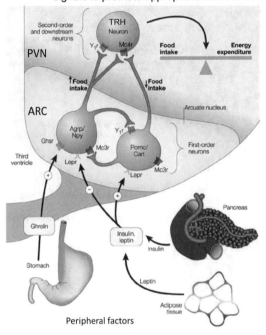

1. Appetite control
2. Energy expenditure
3. Carbohydrate and lipid metabolism
4. Nutrient sensing

Efferent Signals
1. Hunger/satiety
2. Feeding
3. Energy expenditure
4. Reproduction
5. Growth
6. Lifespan

Afferent Signals
1. Nutrients (glucose, amino acids, FFA)
2. Hormones (leptin, insulin)
3. Hedonic signals (odor, reward circuitry)

Nature Reviews | Genetics

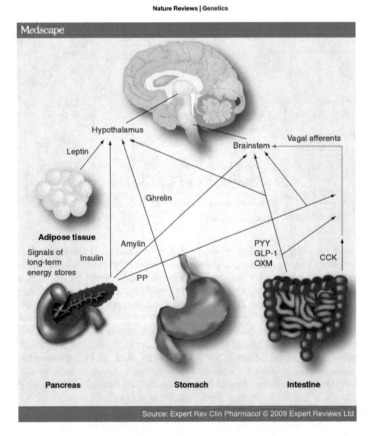

Fig. 2.2 The hypothalamic-feeding network. The hypothalamus integrates local neurotransmitters, neuropeptide hormones, peripheral hormones, nutrient/metabolic signals, hedonic signals, neuronal transmission in the brainstem, and activating vagal afferents to promote appropriate neuroendocrine and autonomic function. Circulating

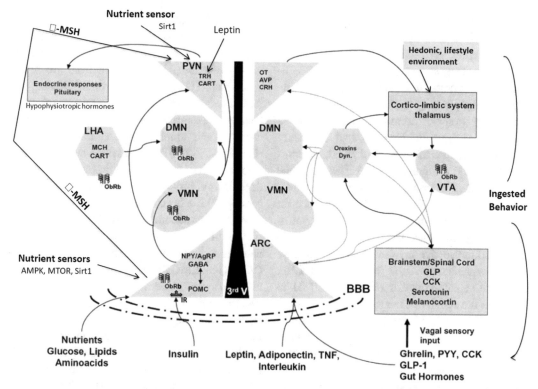

Fig. 2.3 Partial overview of the hypothalamic network involved in the regulation of energy balance. This figure shows the integration of different hypothalamic nuclei within the hypothalamus. The ARC receives and integrates external signals of nutritional status and controls adipose metabolism. The PVN is also a region implicated in controlling food intake and body weight by regulating feeding and satiety. Interaction with other nuclei including the LA and hedonic signals, DMN, and VMN interacting with the ARC and LHA is also shown. NPY/AgRP and POMC neurons are responsive to nutrients and leptin sending their projections to other neurons in- and outside of the hypo-thalamus. Leptin also directly acts on TRH neurons in the PVN to increase basic metabolic rate (see Fig. 2.4, Chap. 5). The areas of the CNS that contain the leptin receptor (ObRb) are known to be involved in the leptin-mediated control of food intake. AVP, arginine vasopressin; BBB, blood-brain barrier; CART, cocaine- and amphetamine-regulated transcript; CCK, cholecystokinin; DMN, dorso-medial nucleus; Dyn dynein; GABA, g-aminobutyric acid; GLP, glucagon-like peptide; LHA, lateral hypothalamus; OT, oxytocin; PVN, paraventricular nucleus; PYY, peptide YY; 3rd V, third ventricle; VMN, ventromedial nucleus. (Modified illustration from Obici 2009)

Fig. 2.2 (continued) hormones leptin and insulin regulate the AgRP/NPY and POMC/CART neurons in the ARC. AgRP and NPY stimulate food intake and decrease energy expenditure, whereas the POMC-derived peptide α-MSH inhibits food intake and increases energy expenditure. Leptin and insulin that circulate in proportion to body adipose stores are signals considered of long-term energy stores. They inhibit AgRP/NPY neurons and stimulate adjacent POMC/CART neurons. Lower insulin and leptin levels activate AgRP/NPY neurons while inhibiting POMC/CART neurons. Stomach-produced ghrelin is the only gut hormone released during fasting and is a meal initiator. Ghrelin is a circulating peptide hormone that can activate AgRP/NPY neurons, thereby stimulating food intake. There are also gastrointestinal peptide hormone signals of short-term nutrient availability on a meal-to-meal basis. They are released postprandially: pancreatic polypeptide and amylin from the pancreas and cholecystokinin, peptide YY, and the incretins glucagon-like peptide-1 and oxynto-modulin from the intestine. CCK, cholecystokinin; GLP, glucagon-like peptide; OXM, oxyntomodulin; PP, pancreatic polypeptide; PYY, peptide YY. GHSR, growth hormone secretagogue receptor; LEPR, leptin receptor; MC3R/MC4R, melanocortin 3/melanocortin 4 receptor; Y1r, neuropeptide Y1 receptor. (Illustrations from Salem and Bloom 2010; Barsh and Schwartz 2002)

center of the endocrine system. It consolidates signals derived from upper cortical inputs, autonomic function, environmental cues such as light and temperature, and peripheral endocrine feedback. The hypothalamus delivers peptide hormones to the pituitary gland that in turn releases the second group of hormones that affect most endocrine systems in the body and especially affect the functions of the thyroid gland, the adrenal gland, and the gonads, as well as influence growth, milk production, water balance, and lifespan (see Chaps. 10, 11, and 12). The melanocortin receptors present in the PVN containing thyrotropin-releasing hormone (TRH) neurons are essential to stimulate the thyroid axis and increase energy expenditure. Also, the PVN and LHA transduce signals emanating from the VMH to modulate the activity of the sympathetic nervous system (SNS), which promotes energy expenditure, and the efferent vagus, which promotes energy storage. Another component is the autonomic effectors with origins in the locus coeruleus (LC) and dorsal motor nucleus of the vagus (DMV), which regulate energy intake, expenditure, and storage (Druce et al. 2004; Morton et al. 2006). Neuronal disruptions, genetic or metabolic alterations of either the afferent, central processing, or efferent components, can modify energy intake or expenditure, leading to either obesity or cachexia or wasting syndrome (loss of body mass, muscle atrophy, less fatty tissue accumulation, fatigue, weakness, and loss of appetite).

2.3 The Arcuate Nucleus (ARC) of the Hypothalamus

The ARC involved in the control of energy homeostasis is located below the VMN, on both sides of the third ventricle, and immediately adjacent to the median eminence (ME). This area has a semipermeable blood-brain barrier system (Broadwell and Brightman 1976), and thus it is advantageously located to sense hormonal and nutrient fluctuations in the bloodstream. The ARC is composed of two major populations of neurons controlling appetite and energy expenditure by releasing specific neuropeptide hormones from

the distinct subset of neurons. The first subset of neurons co-express the orectic neuropeptide Y (NPY) derived from pro-NPY and the agouti-related peptide (AgRP) derived from pro-AgRP processing. These hormones promote feeding and inhibit energy expenditure. The second subset of neurons co-express the anorectic neuropeptide cocaine- and amphetamine-regulated transcript (CART (CARTPT)) derived from pro-CART and α-melanocyte-stimulating hormone (α-MSH), derived from pro-opiomelanocortin (POMC) processing. These hormones reduce food intake and increase the catabolic process. These two populations of neurons, together with downstream target neurons expressing the melanocortin receptor 4 (MC4R) and MC3R, comprise the central melanocortin system. The melanocortin system is crucial for sensing and integrating some peripheral signals allowing for a precise control of food intake, energy expenditure, and glucose homeostasis. The α- and b-MSH act on the MC3R and MC4R to activate anorectic responses (Biebermann et al. 2006; Lee et al. 2006), while AgRP is a MC3R/MC4R competitive antagonist of a-MSH and thereby reduces MSH signaling to promote food intake (Ollmann et al. 1997). Experiments done in vitro and in vivo showed that AgRP acts as an inverse agonist, modulating MC3R/MC4R independently of the presence of a-MSH (Haskell-Luevano and Monck 2001; Tolle and Low 2008).

ARC neurons containing POMC is central to the maintenance of energy homeostasis, and the central melanocortin system is one of the best-characterized neuronal pathways involved in the regulation of energy balance. The POMC-derived peptide α-MSH is well known for inhibiting food intake and increasing energy expenditure (Ellacott and Cone 2006; Fan et al. 1997; Nillni et al. 2000; Perello et al. 2006; Fekete et al. 2000a; Kim et al. 2014) by upregulating, for example, hypophysiotropic pro-TRH neurons in the PVN (Nillni et al. 2000; Perello et al. 2006; Fekete et al. 2000a). Since antagonists of melanocortin 3/melanocortin 4 receptors (MCRs) completely block the autonomic, satiety, and metabolic effects of leptin, it is believed that the melanocortin system mediates several central actions of leptin (Ellacott and Cone 2006; da Silva et al. 2004). This system is particularly important since mutations in MCRs, POMC, or pro-converting enzymes (PCs; see Chap. 5) have been associated with obesity in humans and rodents (Coll et al. 2004). In humans with POMC, gene mutations display early-onset obesity (Krude et al. 1998; Krude and Gruters 2000), and a similar obese phenotype also occurs in POMC-deficient mice. Though acute ablation of POMC neurons in adult mice results in an obese phenotype with hyperphagia (Gropp et al. 2005), postnatal ablation of POMC neurons causes an obese phenotype with reduced energy expenditure but no hyperphagia (Greenman et al. 2013). Interestingly, reactivation of hypothalamic POMC at different stages of development in neural-specific POMC-deficient mice reduces food intake and weight gain and weakens comorbidities including hyperglycemia and hyperinsulinemia (Bumaschny et al. 2012). Different to POMC mutations, neuropeptide Y (NPY) and agouti-related peptide (AgRP) gene mutations produce a negligible change in phenotype and no changes in food intake and body weight, suggesting that compensatory mechanisms may occur during development, a phenomena seen in other metabolic systems (Erickson et al. 1996a; Erickson et al. 1996b; Mathews et al. 2002; Gropp et al. 2005; Ste Marie et al. 2005; Wu and Palmiter 2011).

In the ARC, POMC neurons express the leptin receptor ObRb (Mercer et al. 1998). Low leptin signaling (e.g., fasting) directly inhibits POMC gene expression, resulting in a decrease in all POMC-derived peptides, including desacetyl-α-MSH, and administration of leptin can attenuate this response (Schwartz et al. 1996a; Cheung et al. 1997; Thornton et al. 1997; Perello et al. 2007). The commissural part of the NTS also produces POMC (Palkovits et al. 1987; Bronstein et al. 1992). The NTS contains mainly POMC products within the brainstem since it excludes the parabrachial nucleus, locus coeruleus, and most of the paragigantocellular reticular nucleus, which are recipients of fibers from both ARC and NTS (Joseph and Michael 1988; Pilcher and Joseph 1986). The lateral reticular nucleus and

the rostral NTS itself also receive dual innervations; however, the contribution from the ARC seems to be minimal, as established by neuronal fiber-tracing studies (Joseph and Michael 1988; Zheng et al. 2005).

Although the expression of NPY throughout the CNS is very diverse, it is, however, densely localized in the ARC (Gehlert et al. 1987). NPY peptide release in the ARC responds to changes in energy status, being reduced under feeding conditions and increased under fasting conditions (Beck et al. 1990; Kalra 1997) (Kalra et al. 1991). Pharmacological administration of NPY results in hyperphagia and reduced thermogenesis of brown adipose tissue (BAT), linked with diminished thyroid axis activity (Clark et al. 1984; Stanley et al. 1986; Egawa et al. 1991). Also, NPY acting on NPYY1 and Y5 receptors mediate the beneficial effects on energy balance (Nguyen et al. 2012; Sohn et al. 2013). Recent studies done in our laboratory demonstrated that NPY inhibited the posttranslational processing of POMC by decreasing the prohormone convertase 2 (PC2; see Chap. 5) (Cyr et al. 2013). We also found that early growth response protein 1 (Egr-1) and NPY-Y1 receptors mediated the NPY-induced decrease in PC2. NPY given intra-PVN also decreased PC2 in PVN samples suggesting a reduction in PC2-mediated pro-TRH processing. Also, NPY attenuated the α-MSH-induced increase in TRH production by two mechanisms. First, NPY decreased α-MSH-induced CREB phosphorylation, which typically enhances TRH transcription. Second, NPY decreased the amount of α-MSH in the PVN. Collectively, these results underscore the significance of the interaction between NPY and α-MSH in the central regulation of energy balance and indicate that posttranslational processing is a mechanism that plays a particular role in this interaction (Cyr et al. 2013) (Fig. 2.4).

AgRP is only expressed in the ARC, and it co-localizes with NPY neurons and the neurotransmitter g-aminobutyric acid (GABA) (Broberger et al. 1998; Cowley et al. 2001). Conclusive studies showed that AgRP or its genetic overexpression stimulates food intake, reduces energy expenditure, and produces obesity (Small et al.

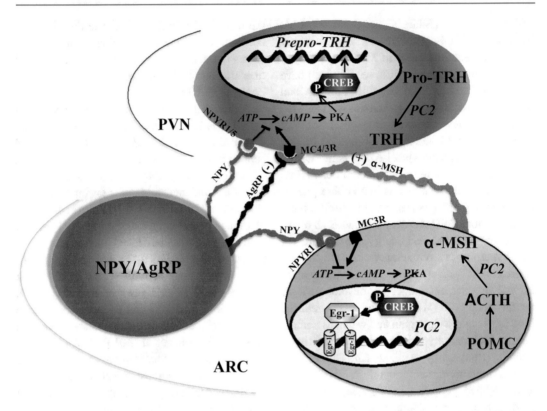

Fig. 2.4 NPY regulation of POMC (α-MSH) and TRH neurons. This figure summarizes, in a very simplified manner, the various mechanisms suggested to mediate the effects of NPY in two key anorexigenic neuropeptide hormones. Our data support the hypothesis that NPY regulates TRH by both direct and indirect pathways. In the direct pathway, NPY regulates TRH directly by binding to Y1/Y5 receptor(s) on TRH neurons in the PVN and decreasing pCREB levels, which decreases pro-TRH. In the indirect pathway, NPY reduces α-MSH peptide content by altering its maturation as well as its release into the PVN. Both pathways by which NPY regulates TRH function by reducing α-MSH's actions on TRH production. (Figure from Cyr et al. 2013)

2003; Graham et al. 1997). AgRP neurons are regulated by insulin (Marks et al. 1990), leptin (Elmquist et al. 1998; Elmquist et al. 1999), and ghrelin (Willesen et al. 1999). These neurons project their axons mainly into the PVN, dorsomedial nucleus (DMN), and lateral hypothalamus (LHA). Despite the well-known effects of NPY and AgRP as positive modulators of energy balance, genetic studies have yielded conflicting results. Contrarily to the expected role for NPY and AgRP in energy homeostasis, deletion of NPY and AgRP genes causes only mild effects in energy balance in mice (Palmiter et al. 1998; Qian et al. 2002; Corander et al. 2011). These findings suggest that other pathways may compensate for the loss of both AgRP and NPY. To directly determine the functional role of these neurons avoiding

compensatory responses, experiments with selective removal of AgRP/NPY neurons by toxin-mediated ablation proved this hypothesis, but not during the neonatal period. With a similar approach, POMC neuron removal caused hyperphagia, obesity, and hypocortisolism.

The α-MSH peptide derived from POMC exerts a potent anorectic effect by binding to MC3R and MC4R (Mercer et al. 2013). The levels of POMC expression and α-MSH peptide are increased during feeding and decrease under fasting conditions (Seeley et al. 1997). The intracerebroventricular infusion α-MSH or its direct delivery into the PVN suppresses food intake and reduces body weight (Poggioli et al. 1986; Wirth et al. 2001). Overexpression of POMC gene has been shown to cause anti-obesity effects in genetic and diet-induced obesity (DIO) models (Mizuno et al. 2003; Savontaus et al. 2004; Lee et al. 2007; Lee and Wardlaw 2007). POMC and AgRP neurons express leptin and insulin receptors and are targeted by these hormones to rise POMC mRNA expression and decrease NPY and AgRP mRNA levels (Benoit et al. 2002; Kitamura et al. 2006). Both neuronal populations express the inhibitory neurotransmitter GABA (Hentges et al. 2004). AgRP neurons inhibit neighboring POMC neurons through GABA release from synaptic terminals (Cowley et al. 2001). POMC and AgRP neurons deliver an important first interface in the communication of peripheral organs and the CNS and thus have the capability of responding to various food-related cues coming from the periphery. Therefore, a key role for POMC in whole-body energy homeostasis is evident, since studies in mice lacking POMC-derived peptides develop obesity (Yaswen et al. 1999; Xu et al. 2005). Consistent with this notion, mutations in the POMC gene have been reported to be associated with morbid obesity in humans (Krude et al. 1998; Lee et al. 2006). In addition, GABAergic and glutamatergic subpopulations of POMC neurons in the hypothalamus have been described, even though their physiological actions remain unclear (Mercer et al. 2013).

2.4 Paraventricular Nucleus (PVN)

The PVN is located in the anterior hypothalamus, just above the third ventricle, and expresses high levels of MC3R/MC4R. This brain nucleus receives axon terminals not only from the AgRP and POMC neurons of the ARC but also from extra-hypothalamic regions including the nucleus of the tractus of the solitarius (NTS) and leptin coming from the adipose tissue (Nillni 2010). It is composed of two major parts, the magnocellular and the parvocellular divisions. The magnocellular hypothalamic neurohypophysial tract carries vasopressin and oxytocin to the posterior pituitary. The parvocellular section is composed of some subcompartments including anterior, medial, periventricular, ventral, dorsal, and lat-eral parvocellular subdivisions. A large population of TRH neurons is located in medial and periventricular parvocellular subdivisions, organized in a triangular configuration, symmetric to the dorsal aspect of the third ventricle, whereas neurons expressing TRH in the anterior parvocellular subdivision are more dispersed (Nillni 2010; Nillni and Sevarino 1999). However, not all TRH-containing neurons in the PVN project to the ME (Ishikawa et al. 1988). Only TRH neurons in the medial and periventricular parvocellular subdivisions project to the ME. Most likely the function of these neurons differs from those of the anterior parvocellular subdivision. Non-hypophysiotropic TRH neurons also present in the PVN have no known projections to the ME. It is presumed that they do not serve a direct hypophysiotropic function but rather play a different role. Among other functions, they might act on the autonomic nervous system to induce thermoregulation. For example, after ICV administration of α-MSH (Sarkar et al. 2002), the peptide stimulated both hypophysiotropic neurons and pro-TRH from the anterior and ventral parvocellular subdivisions of the PVN, a region associated with the stimulation of the autonomic regulation. This region sends neuronal projections to the brainstem and spinal cord targets (Roeling et al. 1993; Swanson et al. 1980; Swanson and Sawchenko 1980). Given those facts, it is possible that non-hypophysiotropic TRH neurons might have an action in autonomic regulation by activating uncoupling protein-1 (UCP-1) in the brown adipose tissue (Haynes et al. 1999) in concert with a simultaneous stimulation of the HPT axis by hypophysiotropic TRH neurons.

In the human PVN, the magnocellular and parvocellular neurons are not restricted to any particular subdivision of the nucleus. Although the literature on TRH neurons in the human PVN is limited, it was reported that most TRH neurons are present in its dorsomedial part (Fliers et al. 1994). Using in situ hybridization in expressing TRH mRNA neurons was reported not only in the human PVN but also in the suprachiasmatic nucleus, perifornical area, and lateral hypothalamus (Guldenaar et al. 1996). Therefore, the neuroanatomy of the TRH neuron is certainly not identical in all species, and the rat hypothalamus cannot just be taken as universal. On the other hand, NPY, AgRP, and α-MSH-containing axons are frequently present in close juxtaposition to TRH neurons in the human PVN, which is very similar to the rat PVN (Mihaly et al. 2000). It is, therefore, important to point out that most data generated to date are derived from rat studies, and its extrapolation to humans may have its limitations in interpretation.

The four different neuropeptides produced in the ARC as described above are involved in regulating TRH by projecting axon terminals in synaptic contact with the hypophysiotropic TRH neurons (Nillni 2010). These neuronal groups expressing the leptin receptor (ObRb) are considered as one of the main hypothalamic centers controlling food intake and energy homeostasis (Elmquist et al. 1998; Elias et al. 1999). Alpha-MSH and CART (Cheung et al. 1997; Schwartz et al. 1996b; Schwartz et al. 2000) are potent and signals in the hypothalamus through both the MC3R and MC4R. TRH neurons in the PVN express MC4R and are innervated by α-MSH nerve terminals (Kielar et al. 1998; Kishi et al. 2003) (Fig. 2.3). ICV administration of α-MSH can prevent the fasting-induced suppression of the thyroid axis (Fekete et al. 2000b, a). TRH neurons are also innervated by other ARC neurons involved in energy balance expressing ObRb receptor. In the rat, α-MSH and CART are expressed in the same neurons situated in lateral portions of the ARC, while NPY and AgRP are co-expressed in a distinct population of neurons located in more medial parts of the ARC. This neuronal organization is somewhat similar in the human brain suggesting an evolutionary significance of this neuroendocrine system. However, unlike in rodents, CART in humans is absent from the perikarya and axons of α-MSH-synthesizing neurons but is expressed in approximately one-third of NPY/AgRP neurons in the human infundibular nucleus (Menyhert et al. 2007). This anatomic distribution for CART seen in humans raises the question of whether CART is associated with an orexigenic role in the human brain. However, CART and α-MSH co-localize

in the human lateral hypothalamus, which is similar to anatomy described in rodents. AgRP and NPY are downregulated by leptin and upregulated during fasting (Schwartz et al. 1996b; Schwartz et al. 2000; Stephens et al. 1995; Mizuno et al. 1999; Mizuno and Mobbs 1999). When AgRP, an MC4R antagonist, and NPY are administered centrally, they cause central hypothyroidism by downregulating prepro-TRH mRNA expression in the PVN (Mihaly et al. 2000; Fekete et al. 2001; Fekete et al. 2002). Anatomically, NPY appears most prominent in its inputs to the periventricular and medial parvocellular divisions of the PVN (Sawchenko and Pfeiffer 1988). NPY cell bodies principally reside in the medulla, often coexisting with NE and E (Everitt et al. 1984).

Corticotrophin-releasing hormone (CRH) neurons are also relevant in the PVN and are directly involved in the control of energy balance through AgRP innervation or indirectly through the regulation of adrenal glucocorticoids controlling the expression of POMC (Richard and Baraboi 2004). The major role of CRH neurons present in the PVN is to regulate the adrenal axis (Toorie et al. 2016; Toorie and Nillni 2014). Stimulation of the adrenal axis by CRH produces in an increase of circulating glucocorticoids (GC) from the adrenal gland affecting energy metabolism. Chronic increases of basal GC are associated with increased food drive and enhanced abdominal adiposity. Early studies by Vale and others showed a role for CRH (Vale et al. 1981) in mediating the stress response (Kovacs 2013). However, CRH also regulates metabolic, immunologic, and homeostatic changes under normal and pathologic conditions (Sominsky and Spencer 2014; Chrousos 1995; Seimon et al. 2013). Although CRH is expressing in mast cells (Cao et al. 2005), amygdala, locus, and hippocampus, the one expressed in the PVN has a significant role in metabolism (Ziegler et al. 2007; Raadsheer et al. 1993; Korosi and Baram 2008). It is produced in the medial parvocellular division of the PVN and functions as the central regulator of the HPA axis (Aguilera et al. 2008). The levels of bioactive CRH secreted to the circulation are dependent on different peripheral and brain inputs and on the ability of the cell in performing an effective posttranslational processing from its precursor pro-CRH by the PCs.

Like most hypophysiotropic neurons, CRH is released from nerve terminals anteriorly juxtapose to the median eminence where fenestrated capillaries in the hypophyseal portal system facilitate a rapid exchange between the hypothalamus and the pituitary (Rho and Swanson 1987). Upon binding to its corticotropin-releasing hormone receptor 1 (CRHR1) (Lovejoy et al. 2014) in corticotropic cells, CRH stimulates the synthesis and secretion of adrenocorticotropic hormone (ACTH) derived from POMC as well as other bioactive molecules such as β-endorphin (Solomon 1999). ACTH engages the melanocortin 2 receptor expressed by cells of the adrenal cortex and stimulates the production and secretion of steroid hormones such as cortisol (Veo et al. 2011). Both ACTH and GC function to regulate HPA axis activity via long and short negative feedback loops that signal at the level of the hypothalamus, extra-hypothalamic brain sites, and the adenohypophysis. The complete description of the HPA axis, the stress response, and body weight is depicted in Chap. 11.

Among the brain circuitries involved in the control of food intake and energy expenditure capable of integrating peripheral signals is the CRH system, which has many clusters of brain neurons closely related to peptide urocortin. The CRH system showed to have some level of plasticity in obesity and starvation. Based on those observations, it is possible that obesity can block or activate the expression of the CRH type 2alpha receptor in the ventromedial hypothalamic nucleus and induce the expression of the CRH-binding protein in brain areas involved in the anorectic and thermogenic actions of CRH (Richard et al. 2000; Mastorakos and Zapanti 2004; Toriya et al. 2010). On the other hand, CRH acting in the adrenal axis stimulated the production of GC that promotes positive energy balance partly by affecting glucose metabolism and lipid homeostasis and increasing appetite drive (Tataranni et al. 1996). Although a consensus on the exact role of adrenal activity in relation to energy dysfunction has yet to be reached,

increased and sustained basal GC is implicated in the development of visceral obesity, insulin resistance, and metabolic disease (Laryea et al. 2013; Kong et al. 2014; Spencer and Tilbrook 2011). In summary, the PVN is as important as the ARC in integrating mechanisms involved in whole-body energy homeostasis, as shown by the diverse afferent inputs and its high sensitivity to the administration of endogenous neuropeptides involved in the regulation of food intake such as NPY, AGRP, a-MSH, and norepinephrine (Stanley et al. 1986; Nillni 2010; Kim et al. 2000b). Part of these effects is mediated by a subset of neurons expressing prepro-TRH and prepro-CRH (Nillni 2010; Toorie and Nillni 2014).

2.5 Lateral Hypothalamus Area

The LHA, also known as the hunger center, plays an essential role in the mediation of orexigenic responses. This function is attributed to orexins and melanin-concentrating hormone (MCH, a 17-amino acid peptide expressed in the zona incerta and LHA) neurons. MCH neurons synapse on neurons in the forebrain and the locus coeruleus. The LHA was early identified as a critical neuroanatomical brain nucleus for motivated behavior. For example, electrical stimulation of the LHA induces insatiable feeding behavior even in well-fed animals (Stuber and Wise 2016). In the lateral hypothalamus and perifornical area, a cluster of neurons produce orexin-A/hypocretin-1 and orexin-B/hypocretin-2 made of 33- and 28-amino acid sequence neuropeptides. Orexin neurons receive many signals related to environmental, physiological, and emotional stimuli, and their neuronal projections are broadly distributed to the entire CNS. The multitasking role of orexin neurons includes vital body functions such as sleep/wake states, feeding behavior, energy homeostasis, reward systems, cognition, and mood. Orexin knockout mice develop narcolepsy, hypophagia, and obesity (Chieffi et al. 2017), suggesting an important role in the energy balance system and in maintaining wakefulness. Orexin deficiency in animal models showed to induce obesity even if they consume fewer calories than the wild-type counterpart. Reduced physical activity seems to be one of the main reasons of weight gain in these models. On the contrary, stimulated orexin signaling promotes obesity resistance via enhanced spontaneous physical activity and increase energy expenditure. The fact that orexinergic neurons have connections to areas responsible for cognition and mood regulation including hippocampus makes matters more complicated. Orexin also enhances hippocampal neurogenesis and improves learning, memory, and mood, while orexin deficiency results in depression associated with learning and memory deficit.

The orexins or hypocretins are a pair of neuropeptides with extensive CNS projections. This system is strongly associated with feeding, arousal, and the maintenance of waking to show a dichotomy in orexin function. Orexin produced in the lateral hypothalamus regulates reward processing for food and abused drugs. They stimulate NPY and the increase of CRH and sympathetic nervous system (SNS) output to promote wakefulness, energy expenditure, learning and memory, and the hedonic reward system. On the other hand, orexin produced in the perifornical and dorsomedial hypothalamus regulates arousal and response to stress. Orexin gene expression is increased under fasting conditions (Sakurai et al. 1998). Intra-cerebroventricular administration of orexins increases food intake (Sakurai et al. 1998; Dube et al. 1999) and promotes behavioral responses to food reward and increases arousal (Cason et al. 2010). This multiple functions can be explained by the fact that orexin neurons project not only within the LHA, ARC, PVN, and NTS but also into other regions implicated in functions such as body temperature and wakefulness control (de Lecea et al. 1998; Peyron et al. 1998). Likewise, fasting augments the gene expression of MCH, and ICV administration of MCH peptide or genetic overexpression causes an orexigenic action (Hu et al. 2001). On the contrary, mice with reduced MCH tone or disruption of its receptor MCH1 receptor caused the animals to be lean (Marsh et al. 2002).

2.6 Dorsomedial Nucleus

The dorsomedial hypothalamic nucleus (DMN) is involved in feeding, drinking, body-weight regulation, and circadian rhythm. It receives information from neurons and humors involved in feeding regulation, body weight, and energy consumption and then passes this information onto brain regions involved in sleep and wakefulness regulation, body temperature, and corticosteroid secretion. The DMN receives projections from the ARC and sends neuronal connections into the PVN and LHA. In addition, NPY, CRH, and receptors for peptides responsible in the control of appetite and energy balance are expressed within the DMN. It has been reported in several rodent models of obesity increase the expression of NPY at the level of the DMN (Guan et al. 1998; Bi et al. 2001). A thermogenic role had also been assigned for NPY in this region during the development of diet-induced obesity (Chao et al. 2011).

2.7 Ventromedial Nucleus

The ventromedial nucleus of the hypothalamus (VMN, also referred VMH) is a distinct morphological nucleus involved in feeding, fear, thermoregulation, and sexual activity, and it is associated with satiety. Early studies indicated that VMN lesions caused overeating and an increase in body weight in rats. Arcuate AgRP and POMC neurons are connected to the VMN by sending their fibers, and in sequence VMN neurons project into hypothalamic and the brainstem (Cheung et al. 2013). Studies using laser identified VMN-enriched genes (Segal et al. 2005), including steroidogenic factor 1 (Sf1 (Nr5a1)) that are directly associated with the development of the VMN (Davis et al. 2004). The second abundantly expressed peptide in the VMN is the brain-derived neurotrophic factor (BDNF). Both Sf1 (Bingham et al. 2008; Zhang et al. 2008; Kim et al. 2011) and BDNF are involved in energy balance. For example, the absence of BDNF or its receptor leads to hyperphagia and obesity in humans and mice (Yeo et al. 2004). On the contrary, central or peripheral administration of BDNF leads to loss of body weight and reduction in food intake through MC4R signaling (Xu et al. 2003). Also, the VMN heavily express pituitary adenylate cyclase-activating polypeptide (PACAP) type I receptors (PAC1R), and stimulation of the hypothalamic ventromedial nuclei by pituitary adenylate cyclase-activating polypeptide induces hypophagia and thermogenesis (Resch et al. 2011; Lopez et al. 2010; Martinez de Morentin et al. 2012; Whittle and Vidal-Puig 2012).

2.8 Brainstem

The dorsal vagal complex (DVC) within the brainstem is a critical region for the integration of energy-related signals by depending on peripheral cues through vagal afferents that project into the hypothalamus. Morphologically the DVC is divided into the dorsal motor nucleus of the vagus, the NTS, and the area postrema (AP) that has an incomplete blood-brain barrier system and accessible to peripheral signals. Many of the brainstem neurons express POMC, tyrosine hydroxylase (TH), proglucagon, CART, GABA, NPY, and BDNF important in the regulation of appetite, and receptors affecting secretion. Receptors for the circulating hormones leptin, ghrelin, glucagon-like peptide 1 (GLP1), and cholecystokinin (CCK) are present in brainstem neurons or vagal afferent projections to brainstem areas (Figs. 2.2 and 2.3). Information from the GI tract about luminal distension, nutritional content, and locally produced peptides via glutamate neurotransmission is delivered to the NTS (Travagli et al. 2006) that in turn send projections into the hypothalamus and other basal forebrain areas. The demonstration of the nervous vagus's role in energy balance was shown through a series of experiments to eliminate or enhance vagus activity. Chronic or acute vagus nerve stimulation led to a decrease in body weight and food intake, strongly suggesting direct vagal afferent interventions that in turn influence feeding behavior (Gil et al. 2011). Through the action of the hormone CCK, vagal signaling to the

hypothalamus also plays important roles in the regulation of meal size and duration (Schwartz et al. 1999). The NTS also receives inputs from the ARC POMC neurons through the action of the POMC-derived peptide a-MSH where there is a high expression level of MC4R (Kishi et al. 2003). The NTS also receives melanocortin agonist signals from NTS POMC neurons (Palkovits and Eskay 1987). MC4R agonist administration to the hindbrain (the lower part of the brainstem, comprising the cerebellum, pons, and medulla oblongata) leads to a reduction in food intake and an increase in energy expenditure, while MC4R antagonism causes the opposite effect (Williams et al. 2000; Skibicka and Grill 2009). MC4Rs present in the NTS facilitate satiation effects of CCK, and also the anorexigenic effects of hypothalamic and brainstem leptin signaling (Skibicka and Grill 2009; Zheng et al. 2010). Additionally, orexin A and MCH produced in the LHA send projections to the NTS, and orexin-A into the hindbrain increases food intake (Ciriello et al. 2003; Parise et al. 2011). These findings suggest that orexigenic peptides of the LHA with their anatomical connection with the NTS could function as a mechanism to control the satiety signals from the GI tract. It has been demonstrated long ago by Sawchenko that hypothalamic paraventricular nucleus also sends projections into the NTS (Swanson and Sawchenko 1980; Sawchenko and Swanson 1982; Luiten et al. 1985). The TRH neurons contained in neural terminals innervating brainstem vagal motor neurons have been shown to enhance vagal outflow to modify multisystemic visceral functions and food intake. More recently it has been shown that TRH as part of the brainstem regulation system takes part in the endocrine and vagal-sympathetic responses (Ao et al. 2006; Zhao et al. 2013).

2.9 Peripheral Signals Leptin and Insulin Involved in Energy Homeostasis

At the central level, the hypothalamus is the primary component of the nervous system in interpreting adiposity or nutrient-related inputs; it delivers hormonal and behavioral responses with the ultimate purpose of regulating energy intake and consumption. The hormone leptin (16kDa), the product of the Ob gene, is expressed predominantly in adipose cells, and the lack thereof, as in the ob/ob mouse, produces severe obesity (Zhang et al. 1994). Leptin is an important anorexigenic peptide hormone that circulates in proportion to fat mass initially considered as a hormone to prevent obesity. It was later shown that the major role of leptin is to signal the switch from the fed to the starved state at the hypothalamic level (Ahima et al. 1996; Flier and Maratos-Flier 1998; Flier 1998). The fall in circulating levels of leptin is perceived in the hypothalamus to increase appetite, decrease energy expenditure, and change neuroendocrine function in a direction that favors survival (Fig. 2.2). The major consequences of falling leptin include activation of the stress axis and suppression of reproduction, linear growth, and thyroid axis (Ahima et al. 1996). Leptin has a central physiologic role in providing information on energy stores and energy balance to brain centers that regulate appetite, energy expenditure, and neuroendocrine function (Zhang et al. 1994; Ahima et al. 1996; Campfield et al. 1995; Halaas et al. 1995; Pelleymounter et al. 1995). When leptin signaling is deficient, due either to mutation of leptin peptide or leptin receptor genes, severe obesity results in both rodents and humans (Zhang et al. 1994; Chen et al. 1996; Clement et al. 1998; Lee et al. 1996; Montague et al. 1997), underscoring the fundamental role of leptin in the physiology of energy balance. There is multiple leptin receptor (LEPR) isoforms, and the long form b (ObRb) showed to be essential for the effects of leptin. The lack of leptin or the long form ObRb in both rodents and humans causes hyperphagia, reduced energy expenditure, and severe obesity (Halaas et al. 1995; Chen et al. 1996; Clement et al. 1998; Montague et al. 1997; Clement et al. 1998). Interestingly, most obese patients are leptin resistant, while at the same time, they exhibit high levels of circulating leptin unable to exert its central anorexigenic function, which completely precludes the use of leptin as a therapeutical approach as initially thought. Therefore, obesity

is associated with hypothalamic leptin resistance, which contributes to maintaining obesity despite hyperleptinemia. The importance of leptin regulation of POMC neurons was demonstrated in earlier studies where deletion of the leptin receptor specifically in POMC neurons (POMC-Cre: LepRloxP/loxP) resulted in increased body weight and adiposity, hyperleptinemia, and altered hypothalamic neuropeptide expression (Balthasar et al. 2004). However, the leptin receptor is also highly expressed in different hypothalamic areas of the CNS, besides POMC neurons, implicated in the control of energy balance (Elmquist et al. 1998). AgRP neurons in the ARC are also the direct targets of leptin (Cheung et al. 1997; Cowley et al. 2001; Elias et al. 1999). The ablation of ObRb in POMC and AgRP neurons causes increased body weight. However, the scale of these changes is not as significant as compared to globally knockout mice for ObRb. These findings suggest that additional neurons mediating the effects of leptin on food intake and body weight do exist as demonstrated in our previous studies. We suggested that the partial leptin resistance, and not full resistance, was due in part to the ability of leptin to activate TRH neurons in the PVN through a direct pathway instead of leptin stimulating POMC and POMC-derived a-MSH peptide stimulating TRH neurons in the PVN (see Fig. 2.4, Chap. 5) (Perello et al. 2006; Nillni 2010).

Leptin binding to its receptor ObRb induces receptor-associated Janus tyrosine kinases (primarily JAK2) to phosphorylate both itself and the leptin receptor on specific tyrosine residues (Ghilardi et al. 1996; Baumann et al. 1996; Bjørbæk et al. 1997). Intracellular signal transducer and activator of transcription 3 (STAT) isoforms then bind to these receptor phosphotyrosines, which allow STAT to be phosphorylated by JAK on tyrosine residue Y705. Phosphorylated STATs dimerize and enter the nucleus, where STAT regulates gene transcription by binding to STAT-responsive DNA elements (Ihle 1995). In the hypothalamus, STAT3, but not STAT1, STAT5, or STAT6, was shown to be activated after in vivo leptin administration (Vaisse et al. 1996). The short leptin receptor isoform, ObRa, does not contain intracellular tyrosine residues and cannot activate STAT3 signaling (Bjørbæk et al. 1997). Although STATs are important for leptin signaling through ObRb, like other members of this receptor class, ObRb can modulate other intracellular proteins and pathways, including insulin receptor substrate proteins and the RAS-MAPK pathway (Bjørbæk et al. 1997). It is critical to emphasize that, for the biologically relevant actions of leptin regulation of appetite, energy expenditure, and neuroendocrine function, the relative roles of signaling via STATs and these or other pathways are unknown. Our laboratory in collaboration with C. Bjorbaek from Harvard Medical School was the first to demonstrate that STAT3 activation is critical for regulation of hypothalamic POMC and prepro-TRH gene expression by leptin and that this occurs by leptin acting directly via leptin receptors expressed on POMC and TRH neurons (Munzberg et al. 2003; Huo et al. 2004). Consistent with a key role of this transcription factor in leptin action, recent data demonstrate that removal of the STAT3-binding site (Y1138) of the leptin receptor in mice results in leptin resistance and extreme obesity (Bates et al. 2003). Combined, these findings underscore the importance of transcriptional events in leptin action and imply that STAT3 is likely to be critical for regulation of the melanocortin pathway to maintain energy balance by leptin. Leptin also activates the phosphatidylinositol-3-kinase (PI3K) pathway, and leptin is required for its mediated regulation of energy balance. PI3K generates phosphatidylinositol-3,4,5-triphosphate (PIP3) and activates downstream targets including phosphoinositide-dependent kinase 1 (PDK1) and AKT (also known as protein kinase B) and in turn phosphorylates the transcription factor forkhead box protein O1 (FOXO1). Phosphorylated FOXO1 is excluded from the nucleus, allowing STAT3 to bind to POMC and AgRP promoters, thus stimulating and inhibiting, respectively, the expression of these neuropeptides (Kitamura et al. 2006). Leptin action on melanocortin peptides can be summarized as follows: leptin stimulates POMC transcription, depolarizes POMC neurons, and

increases the biosynthesis of POMC and further processing to produce α-MSH peptide for release as we showed earlier (Cowley et al. 2001; Munzberg et al. 2003; Guo et al. 2004), while at the same time there is an attenuation in the expression and release of orexigenic NPY and AgRP neuropeptides (Stephens et al. 1995; Mizuno and Mobbs 1999). Suppressor of cytokine signaling-3 (SOCS3) and protein-tyrosine phosphatase 1B (PTP1B) negatively regulate leptin signaling (Ernst et al. 2009). Genetic deletion in high-fat diet animals of either SOCS3 or PTP1B in POMC neurons leads to reduced adiposity, improved leptin sensitivity, and increased energy expenditure. In addition to its effects on food intake and energy expenditure, leptin also regulates thermogenesis and locomotor activity via central circuits. Mice lacking the leptin gene (ob/ob mice) have decreased body temperature and are cold-intolerant (Trayhurn et al. 1977), yet these mice can survive at low temperatures (40 C) when properly acclimated (Coleman 1982). Leptin delivery to ob/ob mice induces thermogenesis via increased sympathetic activity to brown adipose tissue (BAT) and induction of UCP-1 expression (Commins et al. 2000; Commins et al. 1999).

Insulin, a peptide hormone produced by pancreatic β-cells, also acts as an anorectic signal inside the central nervous system. Physiologically, glucose induces insulin secretion into the bloodstream in proportion to fat stores (Bagdade et al. 1967). Then, insulin goes into the brain through a rate-limiting transport mechanism (Baura et al. 1993). The proof of the anorexigenic role of insulin was demonstrated by intro-cerebral ventricular or intra-hypothalamic administration to primates and rodents' brain which showed reduced food intake (Woods et al. 1979; McGowan et al. 1993; Air et al. 2002). Similar to leptin, the insulin receptor and its downstream signaling machinery are expressed in hypothalamic areas involved in the control of hunger (Havrankova et al. 1978a, b) and co-localize with AgRP and POMC neurons (Benoit et al. 2002). Interestingly, the loss of insulin receptor in either POMC or AgRP neurons does not cause alterations in energy balance, although hepatic glucose production deficiency has been observed in mice missing insulin receptor in AgRP neurons (Konner et al. 2007), demonstrating that insulin action on AgRP neurons is required for withholding hepatic glucose production. Therefore, insulin signaling in AgRP and POMC neurons is involved in glucose metabolism and energy expenditure. A more detailed description of insulin signaling and its biological role is described in Chap. 9.

2.10　Conclusions

This chapter presented an overview and recent progress in understanding the role of the hypothalamus and its interaction with external inputs to maintain the energy balance in a steady-state status. This metabolic homeostatic state is achieved by the tight communication between different hypothalamic neuronal centers releasing specific neuropeptide hormones, neurotransmitters, and metabolic signals. Neuronal circuits and metabolic pathways are affected by long-term signals, including leptin and insulin, and short-term signals such as gastrointestinal hormones and vagal inputs (Figs. 2.2 and 2.3). Additionally, hedonic, rewarding, and motivational aspects of eating behavior significantly contribute to control energy balance. The circulating signals leptin and ghrelin also target hedonic networks to modulate appetite causing an override in homeostatic control and produce energy imbalance (Berthoud 2011). Similar to drug addiction, food activates a common dopamine brain reward system (DiLeone et al. 2012) suggesting that these complex interactions between the homeostatic and non-homeostatic systems coordinate appetite and energy balance regulation through the modulation of endocrine, autonomic, and behavioral outputs (Fig. 2.2). The exact mechanisms by which these different levels of control are integrated represent among the central unsolved paradoxes of the central regulation of energy balance. Recently, although not discussed in the chapter, it has been suggested that non-neuronal glial cells and tanycytes play a role in energy balance.

An example of an integrated regulation of one of the hypophysiotropic neurons, TRH, is depicted in Fig. 2.5. TRH neurons in the PVN are regulated by thyroid hormone 3 (T3) through a feedback mechanism. However, other modulators are affecting the TRH neuron including leptin, α-MSH, AgRP, and NPY. The TRH neuron integrates these inputs to determine the set point of prepro-TRH expression and biosynthesis of TRH. These mechanisms of gene regulation are acting in concert with tight control of pro-TRH processing by PC1, PC2, CPE/D, and PAM. For more details on the pro-TRH biology, see Chap. 5 (Nillni 2010).

The disruption of the sensing and integration of these neural signals might be a crucial factor that leads to the development of obesity. Of importance is the fact that understanding these neuronal and metabolic pathways represents a likely target for therapeutic intervention. The incidence of obesity is still rising at an alarming rate, which highlights the importance of the search for new and novel effective therapies to combat this weight disorder. It is worth to repeat that unlimited availability of high-calorie processed food and decreased physical activity with population carrying responsive genes for obesity are the primary factors responsible for the modern human obesity epidemic.

Although it is clear that decreasing food intake and increasing energy expenditure are both essential to treat obesity, success with this approach has been limited, and only useful in the short term. It was originally thought that excess of fat consumption was the real problem of obesity. Today we know that it is more complicated than that, and the excess of sugar in the diet is a real killer. "Fat, though, has so far resisted every chemical assault, with the result that more and more people are reduced to mutilating their stomachs in gastric bypass surgery". "Mammals," says Dr. Nillni from Brown University, "are very complex animals." The most practical solution, for now, he says, is not to fight the basic biology of the fat cell. It is to "eat less and exercise more" (Anna Underwood and Jerry Adler, *Fat Cells: What You Do Not Know* Sep 19, 2004, pp 41–47, Newsweek Magazine). Thus, to effectively fight the current obesity epidemic, it is vital to understand better

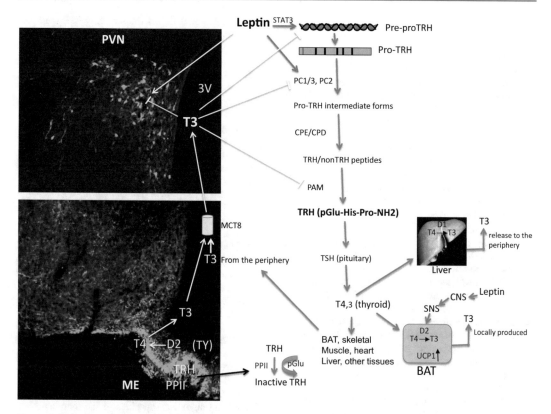

Fig. 2.5 This figure depicts the mechanisms regulating the biosynthesis of pro-TRH its posttranslational processing and downstream action on thyroid hormone as well as the inactivation of TRH peptide. Leptin and T3 control the output of TRH. Tanycytes incorporate T4 and converted it to T3 by D2, and together with T3 coming from the periphery reach TRH neurons via specific transporters such as the monocarboxylate transporter (MCT8) inhibiting gene expression by affecting the TRH promoter through binding to the THRb2 isoform leading to the recruitment of cofactors such as SRC-1. Prepro-TRH, PC1, PC2, and PAM are inversely regulated by T3. Post-translational processing of pro-TRH occurs while the pro-TRH is transported through the axons reaching the median eminence. The degrading tanycyte-bound enzyme PPII, which is positively regulated by thyroid hormone, controls TRH released from the axon terminals located in the median eminence to the defenestrated capillaries. The potential effect of leptin via activation of the sympathetic nervous system has been suggested, and in turn, increases the expression of BAT-UCP1 mRNA by the conversion of T4 to T3 facilitated by D2 increased activity. The staining observed in the PVN and ME was done using an antibody against some of the epitopes of the pro- TRH sequence

the neuronal and molecular mechanisms involved and the evolutionary impact of our ancestral genes (see in Chap. 1 on genes and evolution).

Questions

1. What is the primary role of the hypothalamus, and in which way the afferent signals control the orexigenic and anorexigenic hormones?

2. What is the physiological role of the brainstem?

3. Is the paraventricular nucleus of the hypothalamus a significant center in controlling energy balance?

4. In which way (mechanism) hedonic signals control food intake?

5. Does neuropeptide Y inhibit TRH neurons or only POMC neurons?

References

Aguilera, G., Subburaju, S., Young, S., & Chen, J. (2008). The parvocellular vasopressinergic system and responsiveness of the hypothalamic pituitary adrenal axis during chronic stress. *Progress in Brain Research*, 29–39. https://doi.org/10.1016/S0079-6123(08)00403-2.

Ahima, R. S., Prabakaran, D., Mantzoros, C., Qu, D., Lowell, B., Maratos-Flier, E., & Flier, J. S. (1996). Role of leptin in the neuroendocrine response to fasting. *Nature, 382*, 250–252.

Air, E. L., Benoit, S. C., Blake Smith, K. A., Clegg, D. J., & Woods, S. C. (2002). Acute third ventricular administration of insulin decreases food intake in two paradigms. *Pharmacology, Biochemistry, and Behavior, 72*, 423–429.

Ao, Y., Go, V. L., Toy, N., Li, T., Wang, Y., Song, M. K., Reeve, J. R., Jr., Liu, Y., & Yang, H. (2006). Brainstem thyrotropin-releasing hormone regulates food intake through vagal-dependent cholinergic stimulation of ghrelin secretion. *Endocrinology*, 6004–6010. https://doi.org/10.1210/en.2006-0820.

Bagdade, J. D., Bierman, E. L., & Porte, D., Jr. (1967). The significance of basal insulin levels in the evaluation of the insulin response to glucose in diabetic and nondiabetic subjects. *The Journal of Clinical Ivestigation*, 1549–1557. https://doi.org/10.1172/JCI105646.

Balthasar, N., Coppari, R., McMinn, J., Liu, S. M., Lee, C. E., Tang, V., Kenny, C. D., McGovern, R. A., Chua, S. C., Jr., Elmquist, J. K., & Lowell, B. B. (2004). Leptin receptor signaling in POMC neurons is required for normal body weight homeostasis. *Neuron, 42*, 983–991.

Barsh, G. S., & Schwartz, M. W. (2002). Genetic approaches to studying energy balance: Perception and integration. *Nature Reviews Genetics, 3*(8), 589–600.

Bates, S. H., Stearns, W. H., Dundon, T. A., Schubert, M., Tso, A. W., Wang, Y., Banks, A. S., Lavery, H. J., Haq, A. K., Maratos-Flier, E., Neel, B. G., Schwartz, M. W., & Myers, M. G., Jr. (2003). STAT3 signalling is required for leptin regulation of energy balance but not reproduction. *Nature, 421*, 856–859.

Baumann, H., Morella, K. K., White, D. W., Dembski, M., Bailon, P. S., Kim, H., Lai, C.-F., & Tartaglia, L. A. (1996). The full-length leptin receptor has signaling capabilities of interleukin 6-type cytokine receptors. *Proceedings of the National Academy of Sciences, 93*, 8374–8378.

Baura, G. D., Foster, D. M., Porte, D., Jr., Kahn, S. E., Bergman, R. N., Cobelli, C., & Schwartz, M. W. (1993). Saturable transport of insulin from plasma into the central nervous system of dogs in vivo. A mechanism for regulated insulin delivery to the brain. *The Journal of Clinical Ivestigation, 92*, 1824–1830. https://doi.org/10.1172/JCI116773.

Beck, B., Jhanwar-Uniyal, M., Burlet, A., Chapleur-Chateau, M., Leibowitz, S. F., & Burlet, C. (1990). Rapid and localized alterations of neuropeptide Y in discrete hypothalamic nuclei with feeding status. *Brain Research, 528*, 245–249.

Benoit, S. C., Air, E. L., Coolen, L. M., Strauss, R., Jackman, A., Clegg, D. J., Seeley, R. J., & Woods, S. C. (2002). The catabolic action of insulin in the brain is mediated by melanocortins. *Journal of Neuroscience, 22*, 9048–9052.

Berthoud, H. R. (2011). Metabolic and hedonic drives in the neural control of appetite: Who is the boss? *Current Opinion in Neurobiology*, 888–896. https://doi.org/10.1016/j.conb.2011.09.004.

Bi, S., Ladenheim, E. E., Schwartz, G. J., & Moran, T. H. (2001). A role for NPY overexpression in the dorsomedial hypothalamus in hyperphagia and obesity of OLETF rats. *American Journal of Physiology-Regulatory, Integrative and Comparative Physiology, 281*, R254–R260.

Biebermann, H., Castaneda, T. R., van Landeghem, F., von Deimling, A., Escher, F., Brabant, G., Hebebrand, J., Hinney, A., Tschop, M. H., Gruters, A., & Krude, H. (2006). A role for beta-melanocyte-stimulating hormone in human body-weight regulation. *Cell Metabolism*, 141–146. https://doi.org/10.1016/j.cmet.2006.01.007.

Bingham, N. C., Anderson, K. K., Reuter, A. L., Stallings, N. R., & Parker, K. L. (2008). Selective loss of leptin receptors in the ventromedial hypothalamic nucleus results in increased adiposity and a metabolic syndrome. *Endocrinology*, 2138–2148. https://doi.org/10.1210/en.2007-1200.

Bjørbæk, C., Uotani, S., da Silva, B., & Flier, J. S. (1997). Divergent signaling capacities of the long and short isoforms of the leptin receptor. *Journal of Biological Chemistry, 272*, 32686–32695.

Broadwell, R. D., & Brightman, M. W. (1976). Entry of peroxidase into neurons of the central and peripheral nervous systems from extracerebral and cerebral blood. *The Journal of Comparative Neurology*, 257–283. https://doi.org/10.1002/cne.901660302.

Broberger, C., Johansen, J., Johansson, C., Schalling, M., & Hokfelt, T. (1998). The neuropeptide Y/agouti gene-related protein (AGRP) brain circuitry in normal, anorectic, and monosodium glutamate-treated mice. *Proceedings of the National Academy of Sciences, 95*, 15043–15048.

Bronstein, D. M., Schafer, M. K., Watson, S. J., & Akil, H. (1992). Evidence that beta-endorphin is synthesized in cells in the nucleus tractus solitarius: Detection of POMC mRNA. *Brain Research, 587*, 269–275.

Bumaschny, V. F., Yamashita, M., Casas-Cordero, R., Otero-Corchon, V., de Souza, F. S., Rubinstein, M., & Low, M. J. (2012). Obesity-programmed mice are rescued by early genetic intervention. *The Journal of Clinical Investigation*, 4203–4212. https://doi.org/10.1172/JCI62543.

Campfield, L. A., Smith, F. J., Guisez, Y., Devos, R., & Burn, P. (1995). Recombinant mouse OB protein: Evidence for peripheral signal linking adiposity and central neural networks. *Science, 269*, 546–549.

Cao, J., Papadopoulou, N., Kempuraj, D., Boucher, W. S., Sugimoto, K., Cetrulo, C. L., & Theoharides, T. C. (2005). Human mast cells express corticotropin-releasing hormone (CRH) receptors and CRH leads to selective secretion of vascular endothelial growth factor. *The Journal of Immunology, 174*, 7665–7675.

Cason, A. M., Smith, R. J., Tahsili-Fahadan, P., Moorman, D. E., Sartor, G. C., & Aston-Jones, G. (2010). Role of orexin/hypocretin in reward-seeking and addiction: Implications for obesity. *Physiology & Behavior*, 419–428. https://doi.org/10.1016/j.physbeh.2010.03.009.

Chao, P. T., Yang, L., Aja, S., Moran, T. H., & Bi, S. (2011). Knockdown of NPY expression in the dorsomedial hypothalamus promotes development of brown adipocytes and prevents diet-induced obesity. *Cell Metabolism*, 573–583. https://doi.org/10.1016/j.cmet.2011.02.019.

Chen, H., Chatlat, O., Tartaglia, L. A., Woolf, E. A., Weng, X., Ellis, S. J., Lakey, N. D., Culpepper, J., Moore, K. J., Breitbart, R. E., Duyk, G. M., Tepper, R. I., & Morgenstern, J. P. (1996). Evidence that the diabetes gene encodes the leptin receptor: Identification of a mutation in the leptin receptor gene in db/db mice. *Cell, 84*, 491–495.

Cheung, C. C., Clifton, D. K., & Steiner, R. A. (1997). Proopiomelanocortin neurons are direct targets for leptin in the hypothalamus. *Endocrinology, 138*, 4489–4492.

Cheung, C. C., Kurrasch, D. M., Liang, J. K., & Ingraham, H. A. (2013). Genetic labeling of steroidogenic factor-1 (SF-1) neurons in mice reveals ventromedial nucleus of the hypothalamus (VMH) circuitry beginning at neurogenesis and development of a separate non-SF-1 neuronal cluster in the ventrolateral VMH. *The Journal of Comparative Neurology*, 1268–1288. https://doi.org/10.1002/cne.23226.

Chieffi, S., Carotenuto, M., Monda, V., Valenzano, A., Villano, I., Precenzano, F., Tafuri, D., Salerno, M., Filippi, N., Nuccio, F., Ruberto, M., De Luca, V., Cipolloni, L., Cibelli, G., Mollica, M. P., Iacono, D., Nigro, E., Monda, M., Messina, G., & Messina, A. (2017). Orexin system: The key for a healthy life. *Frontiers in Physiology, 357*. https://doi.org/10.3389/fphys.2017.00357.

Chrousos, G. P. (1995). The hypothalamic-pituitary-adrenal axis and immune-mediated inflammation. *The New England Journal of Medicine*, 1351–1362. https://doi.org/10.1056/NEJM199505183322008.

Ciriello, J., McMurray, J. C., Babic, T., & de Oliveira, C. V. (2003). Collateral axonal projections from hypothalamic hypocretin neurons to cardiovascular sites in nucleus ambiguus and nucleus tractus solitarius. *Brain Research, 991*, 133–141.

Clark, J. T., Kalra, P. S., Crowley, W. R., & Kalra, S. P. (1984). Neuropeptide Y and human pancreatic polypeptide stimulate feeding behavior in rats. *Endocrinology*, 427–429. https://doi.org/10.1210/endo-115-1-427.

Clement, K., Vaisse, C., Lahlou, N., Cabrol, S., Pelloux, V., Cassuto, D., Gourmelen, M., Dina, C., Chambaz, J., Lacorte, J. M., Basdevant, A., Bougneres, P., Lebouc, Y., Froguel, P., & Guy-Grand, B. (1998). A mutation in the human leptin receptor gene causes obesity and pituitary dysfunction. *Nature, 392*, 398–401.

Coleman, D. L. (1982). Thermogenesis in diabetes-obesity syndromes in mutant mice. *Diabetologia, 22*, 205–211.

Coll, A. P., Farooqi, I. S., Challis, B. G., Yeo, G. S., & O'Rahilly, S. (2004). Proopiomelanocortin and energy balance: Insights from human and murine genetics. *The Journal of Clinical Endocrinology & Metabolism, 89*, 2557–2562.

Commins, S. P., Watson, P. M., Padgett, M. A., Dudley, A., Argyropoulos, G., & Gettys, T. W. (1999). Induction of uncoupling protein expression in brown and white adipose tissue by leptin. *Endocrinology, 140*, 292–300.

Commins, S. P., Watson, P. M., Levin, N., Beiler, R. J., & Gettys, T. W. (2000). Central leptin regulates the UCP1 and ob genes in brown and white adipose tissue via different beta-adrenoceptor subtypes. *Journal of Biological Chemistry*, 33059–33067. https://doi.org/10.1074/jbc.M006328200. M006328200 [pii].

Corander, M. P., Rimmington, D., Challis, B. G., O'Rahilly, S., & Coll, A. P. (2011). Loss of agouti-related peptide does not significantly impact the phenotype of murine POMC deficiency. *Endocrinology*, 1819–1828. https://doi.org/10.1210/en.2010-1450.

Cowley, M. A., Smart, J. L., Rubinstein, M., Cerdan, M. G., Diano, S., Horvath, T. L., Cone, R. D., & Low, M. J. (2001). Leptin activates anorexigenic POMC neurons through a neural network in the arcuate nucleus. *Nature, 411*, 480–484.

Cyr, N. E., Toorie, A. M., Steger, J. S., Sochat, M. M., Hyner, S., Perello, M., Stuart, R., & Nillni, E. A. (2013). Mechanisms by which the orexigen NPY regulates anorexigenic alpha-MSH and TRH. *American Journal of Physiology-Endocrinology and Metabolism, 67*, E640–E650. https://doi.org/10.1152/ajpendo.00448.2012.

da Silva, A. A., Kuo, J. J., & Hall, J. E. (2004). Role of hypothalamic melanocortin 3/4-receptors in mediating chronic cardiovascular, renal, and metabolic actions of leptin. *Hypertension, 43*, 1312–1317.

Davis, A. M., Seney, M. L., Stallings, N. R., Zhao, L., Parker, K. L., & Tobet, S. A. (2004). Loss of steroidogenic factor 1 alters cellular topography in the mouse ventromedial nucleus of the hypothalamus. *Journal of Neurobiology*, 424–436. https://doi.org/10.1002/neu.20030.

de Lecea, L., Kilduff, T. S., Peyron, C., Gao, X., Foye, P. E., Danielson, P. E., Fukuhara, C., Battenberg, E. L., Gautvik, V. T., Bartlett, F. S., 2nd, Frankel, W. N., van den Pol, A. N., Bloom, F. E., Gautvik, K. M., & Sutcliffe, J. G. (1998). The hypocretins: Hypothalamus-specific peptides with neuroexcitatory activity. *Proceedings of the National Academy of Sciences, 95*, 322–327.

DiLeone, R. J., Taylor, J. R., & Picciotto, M. R. (2012). The drive to eat: Comparisons and distinctions between mechanisms of food reward and drug addiction. *Nature Neuroscience*, 1330–1335. https://doi.org/10.1038/nn.3202.

Druce, M. R., Small, C. J., & Bloom, S. R. (2004). Minireview: Gut peptides regulating satiety.

Endocrinology, 2660–2665. https://doi.org/10.1210/en.2004-0089.

Dube, M. G., Kalra, S. P., & Kalra, P. S. (1999). Food intake elicited by central administration of orexins/hypocretins: Identification of hypothalamic sites of action. *Brain Research, 842,* 473–477.

Egawa, M., Yoshimatsu, H., & Bray, G. A. (1991). Neuropeptide Y suppresses sympathetic activity to interscapular brown adipose tissue in rats. *American Journal of Physiology-Regulatory, Integrative and Comparative Physiology, 260,* R328–R334.

Elias, C. F., Aschkenasi, C., Lee, C., Kelly, J., Ahima, R. S., Bjorbaek, C., Flier, J. S., Saper, C. B., & Elmquist, J. K. (1999). Leptin differentially regulates NPY and POMC neurons projecting to the lateral hypothalamic area. *Neuron, 23,* 775–786.

Ellacott, K. L., & Cone, R. D. (2006). The role of the central melanocortin system in the regulation of food intake and energy homeostasis: Lessons from mouse models. *Philosophical Transactions of the Royal Society of London B: Biological Sciences, 361,* 1265–1274.

Elmquist, J. K., Bjorbaek, C., Ahima, R. S., Flier, J. S., & Saper, C. B. (1998). Distributions of leptin receptor mRNA isoforms in the rat brain. *Journal of Comparative Neurology, 395,* 535–547.

Elmquist, J. K., Elias, C. F., & Saper, C. B. (1999). From lesions to leptin: Hypothalamic control of food intake and body weight. *Neuron,* 221–232. https://doi.org/S0896-6273(00)81084-3 [pii].

Erickson, J. C., Clegg, K. E., & Palmiter, R. D. (1996a). Sensitivity to leptin and susceptibility to seizures of mice lacking neuropeptide Y. *Nature,* 415–421. https://doi.org/10.1038/381415a0.

Erickson, J. C., Hollopeter, G., & Palmiter, R. D. (1996b). Attenuation of the obesity syndrome of ob/ob mice by the loss of neuropeptide Y. *Science, 274,* 1704–1707.

Ernst, M. B., Wunderlich, C. M., Hess, S., Paehler, M., Mesaros, A., Koralov, S. B., Kleinridders, A., Husch, A., Munzberg, H., Hampel, B., Alber, J., Kloppenburg, P., Bruning, J. C., & Wunderlich, F. T. (2009). Enhanced Stat3 activation in POMC neurons provokes negative feedback inhibition of leptin and insulin signaling in obesity. *Journal of Neuroscience,* 11582–11593. 29/37/11582 [pii]. https://doi.org/10.1523/JNEUROSCI.5712-08.2009.

Everitt, B. J., Hokfelt, T., Terenius, L., Tatemoto, T., Mutt, V., & Goldstein, M. (1984). Differential co-existence of neuropeptide Y (NPY)-like immunoreactivity with catecholamines in the central nervous system of the rat. *Neuroscience, 11,* 443.

Fan, W., Boston, B. A., Kesterson, R. A., Hruby, V. J., & Cone, R. D. (1997). Role of melanocortinergic neurons in feeding and the agouti obesity syndrome. *Nature, 385,* 165–168.

Fekete, C., Legradi, G., Mihaly, E., Huang, Q. H., Tatro, J. B., Rand, W. M., Emerson, C. H., & Lechan, R. M. (2000a). alpha-Melanocyte-stimulating hormone is contained in nerve terminals innervating thyrotropin-releasing hormone-synthesizing neurons in the hypothalamic paraventricular nucleus and prevents fasting-induced suppression of prothyrotropin-releasing hormone gene expression. *Journal of Neuroscience, 20,* 1550–1558.

Fekete, C., Mihaly, E., Luo, L. G., Kelly, J., Clausen, J. T., Mao, Q., Rand, W. M., Moss, L. G., Kuhar, M., Emerson, C. H., Jackson, I. M., & Lechan, R. M. (2000b). Association of cocaine- and amphetamine-regulated transcript-immunoreactive elements with thyrotropin-releasing hormone-synthesizing neurons in the hypothalamic paraventricular nucleus and its role in the regulation of the hypothalamic-pituitary-thyroid axis during fasting. *Journal of Neuroscience, 20,* 9224–9234.

Fekete, C., Kelly, J., Mihaly, E., Sarkar, S., Rand, W. M., Legradi, G., Emerson, C. H., & Lechan, R. M. (2001). Neuropeptide Y has a central inhibitory action on the hypothalamic-pituitary-thyroid axis. *Endocrinology, 142,* 2606–2613.

Fekete, C., Sarkar, S., Rand, W. M., Harney, J. W., Emerson, C. H., Bianco, A. C., Beck-Sickinger, A., & Lechan, R. M. (2002). Neuropeptide Y1 and Y5 receptors mediate the effects of neuropeptide Y on the hypothalamic-pituitary-thyroid axis. *Endocrinology, 143,* 4513–4519.

Flier, J. S. (1998). Clinical review 94: What's in a name? In search of leptin's physiologic role. *The Journal of Clinical Endocrinology and Metabolism, 83,* 1407–1413.

Flier, J. S., & Maratos-Flier, E. (1998). Obesity and the hypothalamus: Novel peptides for new pathways. *Cell, 92,* 437–440.

Fliers, E., Noppen, N. W., Wiersinga, W. M., Visser, T. J., & Swaab, D. F. (1994). Distribution of thyrotropin-releasing hormone (TRH)-containing cells and fibers in the human hypothalamus. *Journal of Comparative Neurology, 350,* 311–323.

Gehlert, D. R., Chronwall, B. M., Schafer, M. P., & O'Donohue, T. L. (1987). Localization of neuropeptide Y messenger ribonucleic acid in rat and mouse brain by in situ hybridization. *Synapse,* 25–31. https://doi.org/10.1002/syn.890010106.

Ghilardi, N., Ziegler, S., Wiestner, A., Stoffel, R., Heim, M. H., & Skoda, R. C. (1996). Defective STAT signaling by the leptin receptor in diabetic mice. *Proceedings of the National Academy of Sciences, 93,* 6231–6235.

Gil, K., Bugajski, A., & Thor, P. (2011). Electrical vagus nerve stimulation decreases food consumption and weight gain in rats fed a high-fat diet. *Journal of Physiology and Pharmacology, 62,* 637–646.

Graham, M., Shutter, J. R., Sarmiento, U., Sarosi, I., & Stark, K. L. (1997). Overexpression of Agrp leads to obesity in transgenic mice. *Nature Genetics,* 273–274. https://doi.org/10.1038/ng1197-273.

Greenman, Y., Kuperman, Y., Drori, Y., Asa, S. L., Navon, I., Forkosh, O., Gil, S., Stern, N., & Chen, A. (2013). Postnatal ablation of POMC neurons induces an obese phenotype characterized by decreased food intake and enhanced anxiety-like behavior. *Molecular Endocrinology*, 1091–1102. https://doi.org/10.1210/me.2012-1344.

Gropp, E., Shanabrough, M., E Borok, A. W. X., Janoschek, R., Buch, T., Plum, L., Balthasar, N., Hampel, B., Waisman, A., Barsh, G. S., Horvath, T. L., & Bruning, J. C. (2005). Agouti-related peptide-expressing neurons are mandatory for feeding. *Nature Neuroscience*, 1289–1291. https://doi.org/10.1038/nn1548.

Guan, X. M., Yu, H., Trumbauer, M., Frazier, E., Van der Ploeg, L. H., & Chen, H. (1998). Induction of neuropeptide Y expression in dorsomedial hypothalamus of diet-induced obese mice. *Neuroreport, 9*, 3415–3419.

Guldenaar, S. E., Veldkamp, B., Bakker, O., Wiersinga, W. M., Swaab, D. F., & Fliers, E. (1996). Thyrotropin-releasing hormone gene expression in the human hypothalamus. *Brain Research, 743*, 93–101.

Guo, L., Munzberg, H., Stuart, R. C., Nillni, E. A., & Bjorbaek, C. (2004). N-acetylation of hypothalamic alpha-melanocyte-stimulating hormone and regulation by leptin. *Proceedings of the National Academy of Sciences of the United States of America, 101*, 11797–11802.

Halaas, J. L., Gajiwala, K. S., Maffei, M., Cohen, S. L., Chait, B. T., Rabinowitz, D., Lallone, R. L., Burley, S. K., & Friedman, J. M. (1995). Weight-reducing effects of the plasma protein encoded by the obese gene. *Science, 269*, 543–546.

Haskell-Luevano, C., & Monck, E. K. (2001). Agouti-related protein functions as an inverse agonist at a constitutively active brain melanocortin-4 receptor. *Regulatory Peptides, 99*, 1–7.

Havrankova, J., Roth, J., & Brownstein, M. (1978a). Insulin receptors are widely distributed in the central nervous system of the rat. *Nature, 272*, 827–829.

Havrankova, J., Schmechel, D., Roth, J., & Brownstein, M. (1978b). Identification of insulin in rat brain. *Proceedings of the National Academy of Sciences, 75*, 5737–5741.

Haynes, W. G., Morgan, D. A., Djalali, A., Sivitz, W. I., & Mark, A. L. (1999). Interactions between the melanocortin system and leptin in control of sympathetic nerve traffic. *Hypertension, 33*, 542–547.

Hentges, S. T., Nishiyama, M., Overstreet, L. S., Stenzel-Poore, M., Williams, J. T., & Low, M. J. (2004). GABA release from proopiomelanocortin neurons. *Journal of Neuroscience*, 1578–1583. https://doi.org/10.1523/JNEUROSCI.3952-03.2004.

Hu, J., Ludwig, T. E., Salli, U., Stormshak, F., & Mirando, M. A. (2001). Autocrine/paracrine action of oxytocin in pig endometrium. *Biology of Reproduction, 64*, 1682–1688.

Huo, L., Munzberg, H., Nillni, E. A., & Bjorbaek, C. (2004). Role of signal transducer and activator of transcription 3 in regulation of hypothalamic trh gene expression by leptin. *Endocrinology, 145*, 2516–2523.

Ihle, J. N. (1995). Cytokine receptor signalling. *Nature, 377*, 591–594.

Ishikawa, K., Taniguchi, Y., Inoue, K., Kurosumi, K., & Suzuki, M. (1988). Immunocytochemical delineation of thyrotropic area: Origin of thyrotropin-releasing hormone in the median eminence. *Neuroendocrinology, 47*, 384.

Joseph, S. A., & Michael, G. J. (1988). Efferent ACTH-IR opiocortin projections from nucleus tractus solitarius: A hypothalamic deafferentation study. *Peptides, 9*, 193–201.

Kalra, S. P. (1997). Appetite and body weight regulation: is it all in the brain. *Neuron, 19*, 227–230.

Kalra, S. P., Dube, M. G., Sahu, A., Phelps, C. P., & Kalra, P. S. (1991). Neuropeptide Y secretion increases in the paraventricular nucleus in association with increased appetite for food. *Proceedings of the National Academy of Sciences, 88*, 10931–10935.

Kielar, D., Clark, J. S., Ciechanowicz, A., Kurzawski, G., Sulikowski, T., & Naruszewicz, M. (1998). Leptin receptor isoforms expressed in human adipose tissue. *Metabolism, 47*, 844–847.

Kim, M. S., Small, C. J., Stanley, S. A., Morgan, D. G., Seal, L. J., Kong, W. M., Edwards, C. M., Abusnana, S., Sunter, D., Ghatei, M. A., & Bloom, S. R. (2000a). The central melanocortin system affects the hypothalamo-pituitary thyroid axis and may mediate the effect of leptin. *The Journal of Clinical Investigation, 105*, 1005–1011.

Kim, M. S., Rossi, M., Abusnana, S., Sunter, D., Morgan, D. G., Small, C. J., Edwards, C. M., Heath, M. M., Stanley, S. A., Seal, L. J., Bhatti, J. R., Smith, D. M., Ghatei, M. A., & Bloom, S. R. (2000b). Hypothalamic localization of the feeding effect of agouti-related peptide and alpha-melanocyte-stimulating hormone. *Diabetes, 49*, 177–182.

Kim, K. W., Zhao, L., Donato, J., Jr., Kohno, D., Xu, Y., Elias, C. F., Lee, C., Parker, K. L., & Elmquist, J. K. (2011). Steroidogenic factor 1 directs programs regulating diet-induced thermogenesis and leptin action in the ventral medial hypothalamic nucleus. *Proceedings of the National Academy of Sciences of the United States of America*, 10673–10678. https://doi.org/10.1073/pnas.1102364108.

Kim, J. D., Leyva, S., & Diano, S. (2014). Hormonal regulation of the hypothalamic melanocortin system. *Frontiers in Physiology, 480*. https://doi.org/10.3389/fphys.2014.00480.

Kishi, T., Aschkenasi, C. J., Lee, C. E., Mountjoy, K. G., Saper, C. B., & Elmquist, J. K. (2003). Expression of melanocortin 4 receptor mRNA in the central nervous system of the rat. *Journal of Comparative Neurology, 457*, 213–235.

Kitamura, T., Feng, Y., Kitamura, Y. I., Chua, S. C., Jr., Xu, A. W., Barsh, G. S., Rossetti, L., & Accili, D. (2006). Forkhead protein FoxO1 mediates Agrp-dependent effects of leptin on food intake. *Nature Medicine, 12*, 534–540.

Kong, X., Yu, J., Bi, J., Qi, H., W Di, L. W., Wang, L., Zha, J., Lv, S., Zhang, F., Y Li, F. H., Liu, F., Zhou, H., Liu, J., & Ding, G. (2014). Glucocorticoids transcriptionally regulate miR-27b expression promoting body fat accumulation via suppressing the browning of white adipose tissue. *Diabetes*. https://doi.org/10.2337/db14-0395.

Konner, A. C., Janoschek, R., Plum, L., Jordan, S. D., Rother, E., X Ma, C. X., Enriori, P., Hampel, B., Barsh, G. S., Kahn, C. R., Cowley, M. A., Ashcroft, F. M., & Bruning, J. C. (2007). Insulin action in AgRP-expressing neurons is required for suppression of hepatic glucose production. *Cell Metabolism*, 438–449. https://doi.org/10.1016/j.cmet.2007.05.004.

Korosi, A., & Baram, T. Z. (2008). The central corticotropin releasing factor system during development and adulthood. *European Journal of Pharmacology*, 204–214. https://doi.org/10.1016/j.ejphar.2007.11.066.

Kovacs, K. J. (2013). CRH: The link between hormonal-, metabolic- and behavioral responses to stress. *Journal of Chemical Neuroanatomy*, 25–33. https://doi.org/10.1016/j.jchemneu.2013.05.003.

Krude, H., & Gruters, A. (2000). Implications of proopiomelanocortin (POMC) mutations in humans: The POMC deficiency syndrome. *Trends in Endocrinology and Metabolism: TEM, 11*, 15–22.

Krude, H., Biebermann, H., Luck, W., Horn, R., Brabant, G., & Gruters, A. (1998). Severe early-onset obesity, adrenal insufficiency and red hair pigmentation caused by POMC mutations in humans. *Nature Genetics*, 155–157. https://doi.org/10.1038/509.

Laryea, G., Schutz, G., & Muglia, L. J. (2013). Disrupting hypothalamic glucocorticoid receptors causes HPA axis hyperactivity and excess adiposity. *Molecular Endocrinology*, 1655–1665. https://doi.org/10.1210/me.2013-1187.

Lee, M., & Wardlaw, S. L. (2007). The central melanocortin system and the regulation of energy balance. *Frontiers in Bioscience, 12*, 3994–4010.

Lee, G. H., Proenca, R., Montez, J. M., Carroll, K. M., Darvishzadeh, J. G., Lee, J. I., & Friedman, J. M. (1996). Abnormal splicing of the leptin receptor in diabetic mice. *Nature, 379*, 632–635.

Lee, Y. S., Challis, B. G., Thompson, D. A., Yeo, G. S., Keogh, J. M., Madonna, M. E., Wraight, V., Sims, M., Vatin, V., Meyre, D., Shield, J., Burren, C., Ibrahim, Z., Cheetham, T., Swift, P., Blackwood, A., Hung, C. C., Wareham, N. J., Froguel, P., Millhauser, G. L., O'Rahilly, S., & Farooqi, I. S. (2006). A POMC variant implicates beta-melanocyte-stimulating hormone in the control of human energy balance. *Cell Metabolism*, 135–140. https://doi.org/10.1016/j.cmet.2006.01.006.

Lee, M., Kim, A., Chua, S. C., Jr., Obici, S., & Wardlaw, S. L. (2007). Transgenic MSH overexpression attenuates the metabolic effects of a high-fat diet. *American Journal of Physiology-Endocrinology and Metabolism*, E121–E131. https://doi.org/10.1152/ajpendo.00555.2006.

Lopez, M., Varela, L., Vazquez, M. J., Rodriguez-Cuenca, S., Gonzalez, C. R., Velagapudi, V. R., Morgan, D. A., Schoenmakers, E., Agassandian, K., Lage, R., Martinez de Morentin, P. B., Tovar, S., Nogueiras, R., Carling, D., Lelliott, C., Gallego, R., Oresic, M.,

Chatterjee, K., Saha, A. K., Rahmouni, K., Dieguez, C., & Vidal-Puig, A. (2010). Hypothalamic AMPK and fatty acid metabolism mediate thyroid regulation of energy balance. *Nature Medicine*, 1001–1008. https://doi.org/10.1038/nm.2207.

Lovejoy, D. A., Chang, B. S., Lovejoy, N. R., & del Castillo, J. (2014). Molecular evolution of GPCRs: CRH/CRH receptors. *Journal of Molecular Endocrinology*, T43–T60. https://doi.org/10.1530/JME-13-0238.

Luiten, P. G. M., Horst, G. Z., Karst, H., & Steffans, A. B. (1985). The course of paraventricular hypothalamic efferents to autonomic structures in medulla and spinal cord. *Brain Research, 329*, 374–378.

Marks, J. L., Porte, D., Jr., Stahl, W. L., & Baskin, D. G. (1990). Localization of insulin receptor mRNA in rat brain by in situ hybridization. *Endocrinology*, 3234–3236. https://doi.org/10.1210/endo-127-6-3234.

Marsh, D. J., Weingarth, D. T., Novi, D. E., Chen, H. Y., Trumbauer, M. E., Chen, A. S., Guan, X. M., Jiang, M. M., Feng, Y., Camacho, R. E., Shen, Z., EG Frazier, H. Y., Metzger, J. M., Kuca, S. J., Shearman, L. P., Gopal-Truter, S., MacNeil, D. J., Strack, A. M., MacIntyre, D. E., Van der Ploeg, L. H., & Qian, S. (2002). Melanin-concentrating hormone 1 receptor-deficient mice are lean, hyperactive, and hyperphagic and have altered metabolism. *Proceedings of the National Academy of Sciences of the United States of America*, (5), 3240. https://doi.org/10.1073/pnas.052706899.

Martinez de Morentin, P. B., Whittle, A. J., Ferno, J., Nogueiras, R., Dieguez, C., Vidal-Puig, A., & Lopez, M. (2012). Nicotine induces negative energy balance through hypothalamic AMP-activated protein kinase. *Diabetes*, 807–817. https://doi.org/10.2337/db11-1079.

Mastorakos, G., & Zapanti, E. (2004). The hypothalamic-pituitary-adrenal axis in the neuroendocrine regulation of food intake and obesity: The role of corticotropin releasing hormone. *Nutritional Neuroscience*, 271–280. https://doi.org/10.1080/10284150400020516.

Mathews, S. T., Singh, G. P., Ranalletta, M., Cintron, V. J., Qiang, X., Goustin, A. S., Jen, K. L., Charron, M. J., Jahnen-Dechent, W., & Grunberger, G. (2002). Improved insulin sensitivity and resistance to weight gain in mice null for the Ahsg gene. *Diabetes, 51*, 2450–2458.

McGowan, M. K., Andrews, K. M., Fenner, D., & Grossman, S. P. (1993). Chronic intrahypothalamic insulin infusion in the rat: Behavioral specificity. *Physiology & Behavior, 54*, 1031–1034.

Menyhert, J., Wittmann, G., Lechan, R. M., Keller, E., Liposits, Z., & Fekete, C. (2007). Cocaine- and amphetamine-regulated transcript (CART) is colocalized with the orexigenic neuropeptide Y and agouti-related protein and absent from the anorexigenic alpha-melanocyte-stimulating hormone neurons in the infundibular nucleus of the human hypothalamus. *Endocrinology, 148*, 4276–4281.

Mercer, J. G., Moar, K. M., & Hoggard, N. (1998). Localization of leptin receptor (Ob-R) messenger ribonucleic acid in the rodent hindbrain. *Endocrinology, 139,* 29–34.

Mercer, A. J., Hentges, S. T., Meshul, C. K., & Low, M. J. (2013). Unraveling the central proopiomelanocortin neural circuits. *Frontiers in Neuroscience, 19.* https://doi.org/10.3389/fnins.2013.00019.

Mihaly, E., Fekete, C., Tatro, J. B., Liposits, Z., Stopa, E. G., & Lechan, R. M. (2000). Hypophysiotropic thyrotropin-releasing hormone-synthesizing neurons in the human hypothalamus are innervated by neuropeptide Y, agouti-related protein, and alpha-melanocyte-stimulating hormone. *The Journal of Clinical Endocrinology & Metabolism, 85,* 2596–2603.

Mizuno, T. M., & Mobbs, C. V. (1999). Hypothalamic agouti-related protein messenger ribonucleic acid is inhibited by leptin and stimulated by fasting. *Endocrinology, 140,* 814–817.

Mizuno, T. M., Makimura, H., Silverstein, J., Roberts, J. L., Lopingco, T., & Mobbs, C. V. (1999). Fasting regulates hypothalamic neuropeptide Y, agouti-related peptide, and proopiomelanocortin in diabetic mice independent of changes in leptin or insulin. *Endocrinology, 140,* 4551–4557.

Mizuno, T. M., Kelley, K. A., Pasinetti, G. M., Roberts, J. L., & Mobbs, C. V. (2003). Transgenic neuronal expression of proopiomelanocortin attenuates hyperphagic response to fasting and reverses metabolic impairments in leptin-deficient obese mice. *Diabetes, 52,* 2675–2683.

Montague, C. T., Farooqi, I. S., Whitehead, J. P., Soos, M. A., Rau, H., Wareham, N. J., Sewter, C. P., Digby, J. E., Mohammed, S. N., Hurst, J. A., Cheetham, C. H., Earley, A. R., Barnett, A. H., Prins, J. B., & O'Rahilly, S. (1997). Congenital leptin deficiency is associated with severe early-onset obesity in humans. *Nature, 387,* 903–908.

Morton, G. J., Cummings, D. E., Baskin, D. G., Barsh, G. S., & Schwartz, M. W. (2006). Central nervous system control of food intake and body weight. *Nature, 443,* 289–295.

Munzberg, H., Huo, L., Nillni, E. A., Hollenberg, A. N., & Bjorbaek, C. (2003). Role of signal transducer and activator of transcription 3 in regulation of hypothalamic proopiomelanocortin gene expression by leptin. *Endocrinology, 144,* 2121–2131.

Nguyen, A. D., Mitchell, N. F., Lin, S., Macia, L., Yulyaningsih, E., Baldock, P. A., Enriquez, R. F., Zhang, L., Shi, Y. C., Zolotukhin, S., Herzog, H., & Sainsbury, A. (2012). Y1 and Y5 receptors are both required for the regulation of food intake and energy homeostasis in mice. *PLoS ONE,* e40191. https://doi.org/10.1371/journal.pone.0040191.

Nillni, E. A. (2010). Regulation of the hypothalamic thyrotropin releasing hormone (TRH) neuron by neuronal and peripheral inputs. *Frontiers in Neuroendocrinology,* 134–156 .doi: S0091-3022(10)00002-6 [pii]. https://doi.org/10.1016/j.yfrne.2010.01.001.

Nillni, E. A., & Sevarino, K. A. (1999). The biology of pro-thyrotropin-releasing hormone-derived peptides. *Endocrine Reviews, 20,* 599–648.

Nillni, E. A., Vaslet, C., Harris, M., Hollenberg, A., Bjorbak, C., & Flier, J. S. (2000). Leptin regulates prothyrotropin-releasing hormone biosynthesis. Evidence for direct and indirect pathways. *Journal of Biological Chemistry, 275,* 36124–36133.

Obici, S. (2009). Molecular targets for obesity therapy in the brain. *Endocrinology, 150*(6), 2512–2517.

Ollmann, M. M., Wilson, B. D., Yang, Y. K., Kerns, J. A., Chen, Y., Gantz, I., & Barsh, G. S. (1997). Antagonism of central melanocortin receptors in vitro and in vivo by agouti-related protein. *Science, 278,* 135–138.

Palkovits, M., & Eskay, R. L. (1987). Distribution and possible origin of beta-endorphin and ACTH in discrete brainstem nuclei of rats. *Neuropeptides, 9,* 123–137.

Palkovits, M., Mezey, E., & Eskay, R. L. (1987). Pro-opiomelanocortin-derived peptides (ACTH/beta-endorphin/alpha-MSH) in brainstem baroreceptor areas of the rat. *Brain Research, 436,* 323–338.

Palmiter, R. D., Erickson, J. C., Hollopeter, G., Baraban, S. C., & Schwartz, M. W. (1998). Life without neuropeptide Y. *Recent Progress in Hormone Research, 53,* 163–199.

Parise, E. M., Lilly, N., Kay, K., Dossat, A. M., Seth, R., Overton, J. M., & Williams, D. L. (2011). Evidence for the role of hindbrain orexin-1 receptors in the control of meal size. *American Journal of Physiology-Regulatory, Integrative and Comparative Physiology,* R1692–R1699. https://doi.org/10.1152/ajpregu.00044.2011.

Pelleymounter, M. A., Cullen, M. J., Baker, M. B., Hecht, R., Winters, D., Boone, T., & Collins, F. (1995). Effects of the obese gene product on body weight regulation in ob/ob mice. *Science, 269,* 540–543.

Perello, M., Stuart, R. C., & Nillni, E. A. (2006). The role of intracerebroventricular administration of leptin in the stimulation of prothyrotropin releasing hormone neurons in the hypothalamic paraventricular nucleus. *Endocrinology, 147,* 3296–3306.

Perello, M., Stuart, R. C., & Nillni, E. A. (2007). Differential Effects of Fasting and Leptin on Pro-Opiomelanocortin Peptides in the Arcuate Nucleus and in the Nucleus of the Solitary Tract. *American Journal of Physiology Endocrinology and Metabolism.* Am J Physiol Endocrinol Metab. 2007 May;292(5):E1348–57. Epub 2007 Jan 16.

Peyron, C., Tighe, D. K., van den Pol, A. N., de Lecea, L., Heller, H. C., Sutcliffe, J. G., & Kilduff, T. S. (1998). Neurons containing hypocretin (orexin) project to multiple neuronal systems. *The Journal of Neuroscience, 18,* 9996–10015.

Pilcher, W. H., & Joseph, S. A. (1986). Differential sensitivity of hypothalamic and medullary opiocortin and tyrosine hydroxylase neurons to the neurotoxic effects of monosodium glutamate (MSG). *Peptides, 7,* 783–789.

Poggioli, R., Vergoni, A. V., & Bertolini, A. (1986). ACTH-(1-24) and alpha-MSH antagonize feeding behavior stimulated by kappa opiate agonists. *Peptides, 7,* 843–848.

Qian, S., Chen, H., Weingarth, D., Trumbauer, M. E., Novi, D. E., X Guan, H. Y., Shen, Z., Feng, Y., Frazier, E., Chen, A., Camacho, R. E., Shearman, L. P.,

Gopal-Truter, S., MacNeil, D. J., Van der Ploeg, L. H., & Marsh, D. J. (2002). Neither agouti-related protein nor neuropeptide Y is critically required for the regulation of energy homeostasis in mice. *Molecular and Cellular Biology, 22*, 5027–5035.

Raadsheer, F. C., Sluiter, A. A., Ravid, R., Tilders, F. J., & Swaab, D. F. (1993). Localization of corticotropin-releasing hormone (CRH) neurons in the paraventricular nucleus of the human hypothalamus; age-dependent colocalization with vasopressin. *Brain Research, 615*, 50–62.

Resch, J. M., Boisvert, J. P., Hourigan, A. E., Mueller, C. R., Yi, S. S., & Choi, S. (2011). Stimulation of the hypothalamic ventromedial nuclei by pituitary adenylate cyclase-activating polypeptide induces hypophagia and thermogenesis. *American Journal of Physiology. Regulatory, Integrative and Comparative Physiology*, R1625–R1634. https://doi.org/10.1152/ajpregu.00334.2011.

Rho, J. H., & Swanson, L. W. (1987). Neuroendocrine CRF motoneurons: Intrahypothalamic axon terminals shown with a new retrograde-Lucifer-immuno method. *Brain Research, 436*, 143–147.

Richard, D., & Baraboi, D. (2004). Circuitries involved in the control of energy homeostasis and the hypothalamic-pituitary-adrenal axis activity. *Treatments in Endocrinology, 3*, 269–277.

Richard, D., Huang, Q., & Timofeeva, E. (2000). The corticotropin-releasing hormone system in the regulation of energy balance in obesity. *International Journal of Obesity and Related Metabolic Disorders: Journal of the International Association for the Study of Obesity, 24*, S36–S39.

Roeling, T. A., Veening, J. G., Peters, J. P., Vermelis, M. E., & Nieuwenhuys, R. (1993). Efferent connections of the hypothalamic "grooming area" in the rat. *Neuroscience, 56*, 199–225.

Sakurai, T., Amemiya, A., Ishii, M., Matsuzaki, I., Chemelli, R. M., Tanaka, H., Williams, S. C., Richardson, J. A., Kozlowski, G. P., Wilson, S., Arch, J. R., Buckingham, R. E., Haynes, A. C., Carr, S. A., Annan, R. S., McNulty, D. E., Liu, W. S., Terrett, J. A., Elshourbagy, N. A., Bergsma, D. J., & Yanagisawa, M. (1998). Orexins and orexin receptors: A family of hypothalamic neuropeptides and G protein-coupled receptors that regulate feeding behavior. *Cell, 92*, 573–585.

Salem, V., & Bloom, S. R. (2010). Approaches to the pharmacological treatment of obesity. *Expert Review of Clinical Pharmacology, 3*(1), 73–88.

Sarkar, S., Legradi, G., & Lechan, R. M. (2002). Intracerebroventricular administration of alpha-melanocyte stimulating hormone increases phosphorylation of CREB in TRH- and CRH-producing neurons of the hypothalamic paraventricular nucleus. *Brain Research, 945*, 50–59.

Savontaus, E., Breen, T. L., Kim, A., Yang, L. M., Chua, S. C., Jr., & Wardlaw, S. L. (2004). Metabolic effects of transgenic melanocyte-stimulating hormone overexpression in lean and obese mice. *Endocrinology*, 3881–3891. https://doi.org/10.1210/en.2004-0263.

Sawchenko, P. E., & Pfeiffer, S. W. (1988). Ultrastructural localization of neuropeptide Y and galanin immunoreactivity in the paraventricular nucleus of the hypothalamus in the rat. *Brain Research, 474*, 231.

Sawchenko, P. E., & Swanson, L. W. (1982). Immunohistochemical identification of neurons in the paraventricular nucleus of the hypothalamus that project to the medulla or to the spinal cord in the rat. *Journal of Comparative Neurology*, 260–272. https://doi.org/10.1002/cne.902050306.

Schwartz, M. W., Baskin, D. G., Bukowski, T. R., Kuijper, J. L., Foster, D., Lasser, G., Prunkard, D. E., Porte, D. J., Woods, S. C., Seeley, R. J., & Weigle, D. S. (1996a). Specificity of leptin action on elevated blood glucose levels and hypothalamic neuropeptide Y gene expression in ob/ob mice. *Diabetes, 45*, 531–535.

Schwartz, M. W., Seeley, R. J., Campfield, L. A., Burn, P., & Baskin, D. G. (1996b). Identification of targets of leptin action in rat hypothalamus. *The Journal of Clinical Investigation, 98*, 1101–1106.

Schwartz, G. J., Whitney, A., Skoglund, C., Castonguay, T. W., & Moran, T. H. (1999). Decreased responsiveness to dietary fat in Otsuka Long-Evans Tokushima fatty rats lacking CCK-A receptors. *The American Journal of Physiology, 277*, R1144–R1151.

Schwartz, M. W., Woods, S. C., Porte, D., Jr., Seeley, R. J., & Baskin, D. G. (2000). Central nervous system control of food intake. *Nature, 404*, 661–671.

Seeley, R. J., Yagaloff, K. A., Fisher, S. L., Burn, P., Thiele, T. E., van Dijk, G., Baskin, D. G., & Schwartz, M. W. (1997). Melanocortin receptors in leptin effects. *Nature, 390*, 349.

Segal, J. P., Stallings, N. R., Lee, C. E., Zhao, L., Socci, N., Viale, A., Harris, T. M., Soares, M. B., Childs, G., Elmquist, J. K., Parker, K. L., & Friedman, J. M. (2005). Use of laser-capture microdissection for the identification of marker genes for the ventromedial hypothalamic nucleus. *Journal of Neuroscience*, 4181–4188. https://doi.org/10.1523/JNEUROSCI.0158-05.2005.

Seimon, R. V., Hostland, N., Silveira, S. L., Gibson, A. A., & Sainsbury, A. (2013). Effects of energy restriction on activity of the hypothalamo-pituitary-adrenal axis in obese humans and rodents: Implications for diet-induced changes in body composition. *Hormone Molecular Biology and Clinical Investigation*, 71–80. https://doi.org/10.1515/hmbci-2013-0038.

Skibicka, K. P., & Grill, H. J. (2009). Hindbrain leptin stimulation induces anorexia and hyperthermia mediated by hindbrain melanocortin receptors. *Endocrinology*, 1705–1711. https://doi.org/10.1210/en.2008-1316.

Small, C. J., Liu, Y. L., Stanley, S. A., Connoley, I. P., Kennedy, A., Stock, M. J., & Bloom, S. R. (2003). Chronic CNS administration of Agouti-related protein (Agrp) reduces energy expenditure. *International Journal of Obesity and Related Metabolic Disorders: Journal of the International Association for the Study of Obesity*, 530–533. https://doi.org/10.1038/sj.ijo.0802253.

Sobrino Crespo, C., Perianes Cachero, A., Puebla Jimenez, L., Barrios, V., & Arilla Ferreiro, E. (2014).

Peptides and food intake. *Frontiers in Endocrinology, 58.* https://doi.org/10.3389/fendo.2014.00058.

Sohn, J. W., Elmquist, J. K., & Williams, K. W. (2013). Neuronal circuits that regulate feeding behavior and metabolism. *Trends in Neurosciences, 504–512.* https://doi.org/10.1016/j.tins.2013.05.003.

Solomon, S. (1999). POMC-derived peptides and their biological action. *Annals of the New York Academy of Sciences, 885,* 22–40.

Sominsky, L., & Spencer, S. J. (2014). Eating behavior and stress: A pathway to obesity. *Frontiers in Psychology, 434.* https://doi.org/10.3389/fpsyg.2014.00434.

Spencer, S. J., & Tilbrook, A. (2011). The glucocorticoid contribution to obesity. *Stress, 233–246.* https://doi.org/10.3109/10253890.2010.534831.

Stanley, B. G., Kyrkouli, S. E., Lampert, S., & Leibowitz, S. F. (1986). Neuropeptide Y chronically injected into the hypothalamus: A powerful neurochemical inducer of hyperphagia and obesity. *Peptides, 7,* 1189–1192.

Ste Marie, L., Luquet, S., Curtis, W., & Palmiter, R. D. (2005). Norepinephrine- and epinephrine-deficient mice gain weight normally on a high-fat diet. *Obesity Research,* 1518–1522. https://doi.org/10.1038/oby.2005.185.

Stephens, T. W., Basinski, M., Bristow, P. K., Bue-Valleskey, J. M., Burgett, S. G., Craft, L., Hale, J., Hoffmann, J., Hsiung, H. M., Kriauciunas, A., et al. (1995). The role of neuropeptide Y in the antiobesity action of the obese gene product. *Nature, 377,* 530–532.

Stuber, G. D., & Wise, R. A. (2016). Lateral hypothalamic circuits for feeding and reward. *Nature Neuroscience,* 198–205. https://doi.org/10.1038/nn.4220.

Swanson, L. W., & Sawchenko, P. E. (1980). Paraventricular nucleus: A site for the integration of neuroendocrine and autonomic mechanisms. *Neuroendocrinology, 31,* 410–417.

Swanson, L. W., Sawchenko, P. E., Wiegand, S. J., & Price, J. L. (1980). Separate neurons in the paraventricular nucleus project to the median eminence and to the medulla or spinal cord. *Brain Research, 198,* 190–195.

Tataranni, P. A., Larson, D. E., Snitker, S., Young, J. B., Flatt, J. P., & Ravussin, E. (1996). Effects of glucocorticoids on energy metabolism and food intake in humans. *The American Journal of Physiology, 271,* E317–E325.

Thornton, J. E., Cheung, C. C., Clifton, D. K., & Steiner, R. A. (1997). Regulation of hypothalamic proopiomelanocortin mRNA by leptin in ob/ob mice. *Endocrinology, 138,* 5063–5066.

Tolle, V., & Low, M. J. (2008). In vivo evidence for inverse agonism of Agouti-related peptide in the central nervous system of proopiomelanocortin-deficient mice. *Diabetes,* 86–94. https://doi.org/10.2337/db07-0733.

Toorie, A. M., & Nillni, E. A. (2014). Minireview: Central Sirt1 regulates energy balance via the melanocortin system and alternate pathways. *Molecular Endocrinology,* 1423–1434. https://doi.org/10.1210/me.2014-1115.

Toorie, A. M., Cyr, N. E., Steger, J. S., Beckman, R., Farah, G., & Nillni, E. A. (2016). The nutrient and energy sensor Sirt1 regulates the hypothalamic-pituitary-adrenal (HPA) axis by altering the production of the prohormone convertase 2 (PC2) essential in the maturation of corticotropin releasing hormone (CRH) from its prohormone in male rats. *The Journal of Biological Chemistry.* https://doi.org/10.1074/jbc.M115.675264.

Toriya, M., Maekawa, F., Maejima, Y., Onaka, T., Fujiwara, K., Nakagawa, T., Nakata, M., & Yada, T. (2010). Long-term infusion of brain-derived neurotrophic factor reduces food intake and body weight via a corticotrophin-releasing hormone pathway in the paraventricular nucleus of the hypothalamus. *Journal of Neuroendocrinology,* 987–995. https://doi.org/10.1111/j.1365-2826.2010.02039.x.

Travagli, R. A., Hermann, G. E., Browning, K. N., & Rogers, R. C. (2006). Brainstem circuits regulating gastric function. *Annual Review of Physiology,* 279–305. https://doi.org/10.1146/annurev.physiol.68.040504.094635.

Trayhurn, P., Thurlby, P. L., & James, W. P. (1977). Thermogenic defect in pre-obese ob/ob mice. *Nature, 266,* 60–62.

Vaisse, C., Halaas, J. L., Horvath, C. M., Darnell, J. J. E., Stoffel, M., & Friedman, J. M. (1996). Leptin activation of Stat3 in the hypothalamus of wild-type and ob/ob mice but not db/db mice. *Nature Genetics, 14,* 95–97.

Vale, W., Spiess, J., Rivier, C., & Rivier, J. (1981). Characterization of a 41-residue ovine hypothalamic peptide that stimulates secretion of corticotropin and beta-endorphin. *Science,* 18;213(4514):1394–1397.

Veo, K., Reinick, C., Liang, L., Moser, E., Angleson, J. K., & Dores, R. M. (2011). Observations on the ligand selectivity of the melanocortin 2 receptor. *General and Comparative Endocrinology,* 3–9. https://doi.org/10.1016/j.ygcen.2011.04.006.

Whittle, A. J., & Vidal-Puig, A. (2012). NPs – heart hormones that regulate brown fat? *The Journal of Clinical Investigation,* 804–807. https://doi.org/10.1172/JCI62595.

Willesen, M. G., Kristensen, P., & Romer, J. (1999). Co-localization of growth hormone secretagogue receptor and NPY mRNA in the arcuate nucleus of the rat. *Neuroendocrinology, 70,* 306–316.

Williams, D. L., Kaplan, J. M., & Grill, H. J. (2000). The role of the dorsal vagal complex and the vagus nerve in feeding effects of melanocortin-3/4 receptor stimulation. *Endocrinology, 141,* 1332–1337.

Wirth, M. M., PK Olszewski, C. Y., Levine, A. S., & Giraudo, S. Q. (2001). Paraventricular hypothalamic alpha-melanocyte-stimulating hormone and MTII reduce feeding without causing aversive effects. *Peptides, 22,* 129–134.

Woods, S. C., Lotter, E. C., McKay, L. D., & Porte, D., Jr. (1979). Chronic intracerebroventricular infusion of insulin reduces food intake and body weight of baboons. *Nature, 282,* 503–505.

Wu, Q., & Palmiter, R. D. (2011). GABAergic signaling by AgRP neurons prevents anorexia via a melanocortin-independent mechanism. *European Journal of Pharmacology*, 21–27. https://doi.org/10.1016/j.ejphar.2010.10.110.

Xu, B., Goulding, E. H., Zang, K., Cepoi, D., Cone, R. D., Jones, K. R., Tecott, L. H., & Reichardt, L. F. (2003). Brain-derived neurotrophic factor regulates energy balance downstream of melanocortin-4 receptor. *Nature Neuroscience*, 736–742. https://doi.org/10.1038/nn1073.

Xu, A. W., Kaelin, C. B., Morton, G. J., Ogimoto, K., Stanhope, K., Graham, J., Baskin, D. G., Havel, P., Schwartz, M. W., & Barsh, G. S. (2005). Effects of hypothalamic neurodegeneration on energy balance. *PLoS Biology*, e415. https://doi.org/10.1371/journal.pbio.0030415.

Yaswen, L., Diehl, N., Brennan, M. B., & Hochgeschwender, U. (1999). Obesity in the mouse model of pro-opiomelanocortin deficiency responds to peripheral melanocortin. *Nature Medicine*, 1066–1070. https://doi.org/10.1038/12506.

Yeo, G. S., Connie Hung, C. C., Rochford, J., Keogh, J., Gray, J., Sivaramakrishnan, S., O'Rahilly, S., & Farooqi, I. S. (2004). A de novo mutation affecting human TrkB associated with severe obesity and developmental delay. *Nature Neuroscience*, 1187–1189. https://doi.org/10.1038/nn1336.

Zhang, Y., Proenca, R., Maffei, M., Barone, M., Leopold, L., & Friedman, J. M. (1994). Positional cloning of the mouse obese gene and its human homologue. *Nature, 372*, 425–432.

Zhang, R., Dhillon, H., Yin, H., Yoshimura, A., Lowell, B. B., Maratos-Flier, E., & Flier, J. S. (2008). Selective inactivation of Socs3 in SF1 neurons improves glucose homeostasis without affecting body weight. *Endocrinology*, 5654–5661. https://doi.org/10.1210/en.2008-0805.

Zhao, K., Ao, Y., Harper, R. M., Go, V. L., & Yang, H. (2013). Food-intake dysregulation in type 2 diabetic Goto-Kakizaki rats: Hypothesized role of dysfunctional brainstem thyrotropin-releasing hormone and impaired vagal output. *Neuroscience*, 43, 54. https://doi.org/10.1016/j.neuroscience.2013.05.017.

Zheng, H., Patterson, L. M., Phifer, C. B., & Berthoud, H. R. (2005). Brain stem melanocortinergic modulation of meal size and identification of hypothalamic POMC projections. *American Journal of Physiology-Regulatory, Integrative and Comparative Physiology, 289*, R247–R258.

Zheng, H., Patterson, L. M., Rhodes, C. J., Louis, G. W., Skibicka, K. P., Grill, H. J., Myers, M. G., Jr., & Berthoud, H. R. (2010). A potential role for hypothalamomedullary POMC projections in leptin-induced suppression of food intake. *American Journal of Physiology-Regulatory, Integrative and Comparative Physiology*, R720–R728. https://doi.org/10.1152/ajpregu.00619.2009.

Ziegler, C. G., Krug, A. W., Zouboulis, C. C., & Bornstein, S. R. (2007). Corticotropin releasing hormone and its function in the skin. *Hormone and metabolic research = Hormon- und Stoffwechselforschung = Hormones et metabolisme*, 106–109. https://doi.org/10.1055/s-2007-961809.

Transcriptional Regulation of Hypothalamic Energy Balance Genes

3

Deborah J. Good

Within the postnatal human brain, there are at least 4627 expressed transcription factors (Hawrylycz et al. 2012). These transcription factors are differentially expressed within the various lobes and structures of our brain, including the hypothalamus and its distinct set of nuclei. Data from my own laboratory indicates that there are as many as 2089 mRNAs coding for transcription factors within the hypothalamus alone (Jiang and Good, unpublished, referring to dataset within (Jiang et al. 2015)). When one begins to contemplate the enormity of having over 2000 different transcription factors to specify the thousands of proteins that make up the hypothalamus, it is clear that we are still a long way from completely figuring out how each of these are coordinated in response to energy balance signals. In addition to the transcription factors, thousands of noncoding RNAs mediate posttranscriptional regulatory signals, and both proteins and noncoding RNAs direct differential translation of mRNAs which specifies the hypothalamic proteome. This chapter provides an overview of what we know about transcriptional regulation of genes expressed within the hypothalamus, and

discusses the questions that remain in order for us to fully understand the hypothalamic transcriptome and resulting proteome.

3.1 Transcriptional Mechanisms

In its simplest model, transcription of a gene involves RNA polymerase binding to or near the transcriptional start site of a gene and starting the process of making a new mRNA. It is, unfortunately, not that simple. Tissue-specific transcription is guided by the chromatin state (methylated/acetelyated/biotinylated, etc. DNA and histones), specific transcription factor binding sites within the promoter region of a gene, and the complement of transcription factors active within the nucleus at any given moment. While some researchers look at multiple transcription factors or transcribed genes at once, others focus on resolving individual gene-transcription factor interactions. Still others examine the delineation of regulatory regions—in other words, what constitutes a hypothalamic promoter? In this section, we will examine some of the ways that hypothalamic transcriptional mechanisms can be studied, using specific examples from the literature, and then piece together what is known about how changes in energy balance, including leptin or glucose signals, food intake, and food deprivation, change the transcriptional landscape.

D. J. Good (✉)
Department of Human Nutrition, Foods, and
Exercise, Virginia Tech, Blacksburg, VA, USA
e-mail: goodd@vt.edu

© Springer International Publishing AG, part of Springer Nature 2018
E. A. Nillni (ed.), *Textbook of Energy Balance, Neuropeptide Hormones,
and Neuroendocrine Function*, https://doi.org/10.1007/978-3-319-89506-2_3

3.1.1 Models and Cell Lines for Studying Hypothalamic-Specific Transcription

3.1.1.1 Gene Expression Localization

Some of the first studies of gene expression in the hypothalamus utilized antibodies and immunohistochemical localization of hypothalamic proteins, Northern blots of isolated hypothalamic RNA, and in situ hybridization probes to hypothalamic tissue sections to show localized expression of genes in response to stimuli. For example, Northern analysis of vasopressin mRNA from the hypothalamus of the Brattleboro rat was used to show that vasopressin was expressed but at a lower level and at a slightly different size than the Long-Evans rat (Majzoub et al. 1984). We now know that the Brattleboro rat carries a point mutation within the protein-coding region of vasopressin, which affects the open reading frame of the mRNA and leads to diabetes insipidus (Schmale and Richter 1984) The technique of in situ hybridization was used to show c-fos expression in the paraventricular nucleus of the hypothalamus in response to water deprivation of rats (Sagar et al. 1988). As stated by the authors of this historic paper, in situ hybridization allowed for cellular-level resolution of expression within these neurons, as opposed to antibody labeling, which could be more diffuse. Dual-label in situ hybridization techniques enabled co-localization of gene expression, while some RNA probe-based methods allowed researchers to look at specific transcripts of genes (i.e., differentially spliced genes), but none of these could get at the actual mechanisms of gene regulation—only the fact that mRNA and protein were expressed in hypothalamic locations and in response to various stimuli. For understanding mechanisms involved in hypothalamic gene expression, other model organisms and cell lines were developed.

3.1.1.2 Hypothalamic Cell Lines

One of the first established hypothalamic cell line was the HT9 cell line, established via SV-40 viral infection of primary embryonic hypothalamic cells in 1974 (De Vitry et al. 1974). This cell line expressed both vasopressin and neurophysin,

suggesting that the cell line originated from neurosecretory cells of hypothalamic magnocellular neurons. However, few published studies used these lines, as indicated by only four citations to the original paper made since 1974. Several other lines were established in the 1980s and 1990s, including the GT1–7, gonadotropin-releasing hormone (GnRH) secreting cell line, which originated from the scattered GnRH neurons in the rostral hypothalamus (Mellon et al. 1990). The developers of this line used a transgenic mouse expressing SV-40 under the control of the GnRH promoter to direct the transforming oncogene, Large T, just to the neurons of interest—namely, the scattered GnRH neurons of the hypothalamus—and it worked! The GT1–7 line continues to be in use to this day, with a 2016 publication on enhancers and noncoding RNA involvement in GNRH promoter regulation (more about his later) (Huang et al. 2016). Each of these previous examples is of one specific hypothalamic cell type, but in 2004, Dr. Denise Belsham and her group published an article detailing 38 embryonic mouse hypothalamic cell lines, each representing different neuronal cell types of the hypothalamus (Belsham et al. 2004). These cell lines allowed individuals working in hypothalamic gene regulation to tailor the gene regulation studies to the specific hypothalamic neuron type that they were interested in, meaning that the gene expression and promoter analysis studies could be hypothalamic-specific, with regard to transcription factors, DNA promoter methylation and acetylation marks, and noncoding RNAs—all of which will be discussed in the next sections. The 2004 paper has been cited more than 128 times, indicating the utility of those defined hypothalamic cell lines for gene expression and other studies. Dr. Belsham's group has also developed rat hypothalamic lines—both adult and embryonic (Gingerich et al. 2009)—and mouse adult hypothalamic cell lines (Belsham et al. 2009) which allow researchers to do even more specific hypothalamic gene expression studies.

Why is the establishment of hypothalamic cell lines so important to studying hypothalamic gene regulation? Simply put, having cell lines that match the tissue being studied allows one to do molecular "promoter-bashing" studies that provide quick and informative data on regions of the promoter needed for tissue-specific gene expression, and transcription factors involved in regulation of hypothalamic genes. Too many times the inappropriate cell line is used and later found not to represent the tissue that the researcher claimed to be studying. For example, while HeLa cells, derived from a human cervical carcinoma, have been used in many different gene expression studies, and certainly have helped establish some basic "rules" about gene expression (Landry et al. 2013), they would not be a good model for hypothalamic-specific gene expression.

Promoter-bashing refers to the use of cell lines transfected with fragments of DNA containing a promoter region of interest linked to a reporter gene such as luciferase to determine regions of DNA, and transcription factors that confer hypothalamic-specific expression patterns. We have used the Belsham hypothalamic cell lines in studies to examine melanocortin-4-receptor expression in paraventricular nucleus-like lines and the leptin responsiveness of the nescient helix-loop-helix 2 and prohormone convertase 1 promoters in arcuate nucleus-like lines (Al Rayyan et al. 2014; Fox and Good 2008; Wankhade and Good 2011). Without the cell lines, the work would have been done either using transgenic animals (see below), which are much more time-consuming, or would have had to be done in non-hypothalamic neuronal lines, which might not have been representative of hypothalamic gene expression.

3.1.1.3 In Vivo Methods and Models

Even with the best circumstances, cell lines do not give the most accurate profile of gene expression, as they are usually grown without glial cells, and on plastic in defined media, rather than on the extracellular matrix supplied by capillary blood containing growth factors, hormones, and other signals. Thus, the best mechanistic gene expression studies are those done using in vivo methods, such as developing promoter transgenes or examining DNA binding of transcrip-

tion factors using chromatin pull-down assays from tissues.

There are many ways to use transgenes, such as directing expression of a gene to a specific target tissue, but in this section, we will focus on using transgenes composed of a promoter linked to a reporter to examine hypothalamic-specific transcription and to identify minimal promoters and enhancer regions needed for this specificity. There are multiple examples of using this method in the literature, but studies using the minimal promoter and enhancer region from the GnRH promoter provide an excellent example. In these studies, cell lines had been previously used to identify both rat and mouse GnRH promoter regions that directed expression in the GnRH neuron cell line GT1–7 cells (Whyte et al. 1995). These data showed that a minimal promoter existed within the first ~173 base pairs prior to the start of transcription, and a hypothalamic-specific enhancer region could be found between −1571 and − 1863 (Lawson et al. 2002). Based on that knowledge, a minimally sized transgene was created, which contained both the minimal promoter and the enhancer within just a few base pairs of each other, and this transgene recapitulated hypothalamic expression in multiple lines of animals (Lawson et al. 2002). The use of transgenic promoters to study expression appears to have fallen out of favor in recent articles, likely due to the time and expenses needed to generate the constructs and the animals. However, the new methodology of CRISPR/Cas9 gene editing may provide a way to mutate a promoter or enhancer element in situ (i.e., within the genome, not as a transgene per se) and analyze the resulting effects on expression. While this procedure has not yet been done in vivo in hypothalamic tissues, a very recent article was able to successfully modify a gene in mouse retinal cells using CRISPR/Cas9 technology (Latella et al. 2016). A similar strategy could be employed for hypothalamic promoters, combining this in vivo method with current CRISPR/Cas9 promoter analysis methods (i.e., such as those used by Fulco et al. 2016, among others).

DNA binding of transcriptional elements and analysis of histone modifications in vivo can be done using tissue chromatin immunohistochemistry methods. In this procedure, protein-bound chromatin are isolated from a tissue of interest and then immunoprecipitated with antibodies to the transcription factor or histone modification domain of interest. The protein-DNA complexes pulled down in the immunoprecipitation are further analyzed by PCR for the region of interest within the DNA. Quantitative PCR can be used to analyze relative binding as well. My laboratory has used this method to show Stat3 interactions on the hypothalamic Nhlh2 promoter occurring 2 h following leptin injection of mice (Al Rayyan et al. 2014). This method allows one to look at a snapshot of protein-DNA interactions in vivo but with the caveat that the DNA fragments in this method are about 200 base pairs in length and thus usually contain multiple possible sites for protein binding or histone modifications. Thus, the results must be confirmed with in vitro or cell-based studies.

3.1.1.4 Genomic Transcriptional Analysis

A number of databases exist online, which allow researchers to analyze data in silico before going to the bench. First, one can look directly at gene expression in online datasets in the NCBI GEO Profiles site: https://www.ncbi.nlm.nih.gov/geo-profiles/. By entering a gene name or tissue in the search box, previously submitted transcriptome analysis data will be displayed. For example, by entering the term "hypothalamus" in the search box, more than 804,000 entries are displayed. Adding the term "leptin" narrows the search down to a more reasonable 998 entries, and further narrowing the search by putting in the gene name STAT3 results in 91 entries. This type of data may be a place to start looking to design new experimental directions.

Another worthwhile analysis using completely online tools involves promoter analysis and comparison between species. One possible way to do this, even if one doesn't know the exact location or length of the promoter is through the gVISTA tool (http://genome.lbl.gov/cgi-bin/GenomeVista). An mRNA or genomic sequence for the gene of interest can be put into the query

box, and the tool will align the sequence with human or any other species in the databases. One can zoom out from the initial comparison to see the upstream and downstream regions of homology. An example of this is shown on Fig. 3.1, where the gene for NHLH2 was originally screened and then the genome sequence surrounding NHLH2 compared between several species. As one can see, homologous noncoding regions (shown in pink) appear in both the proximal promoter, next to the transcription start site (light blue box, 5′ untranslated region), and in a region approximately 10,000 base pairs upstream which may represent an as yet uncharacterized enhancer region for the NHLH2 gene. One can use the sequences in this region to analyze putative transcription factor binding sites with the PROMO website: http://alggen.lsi.upc.es/cgi-bin/promo_v3/promo/promoinit.cgi?dirDB=TF_8.3 . This type of in silico analysis can be a starting point for cell- or animal-based studies. Other sites of use for promoter/transcription factor analysis include some of the sites listed at Epigenie: http://epigenie.com/epigenetic-tools-and-databases/. These online analysis tools provide information about histone methylation and acetylation and DNA methylation based on studies already done for multiple tissues and whole genomes. As methylation and acetylation of histones and methylation of GC dinucleotide in DNA can specify "active" and "inactive" areas of the genome, the in silico analysis can be a very good starting point for transcriptional analysis of a hypothalamic promoter.

3.1.2 Hypothalamic Promoter Regions

According to one paper that has functionally mapped human promoters, there are at least 400 different cell types in the human body, with essentially 1 genome to specify them all, and this specificity must occur through gene regulation (Consortium et al. 2014). Gene promoters have evolved to specify cell-specific gene expression, and from the work of the FANTOM consortium, neural promoters form a subgroup of these

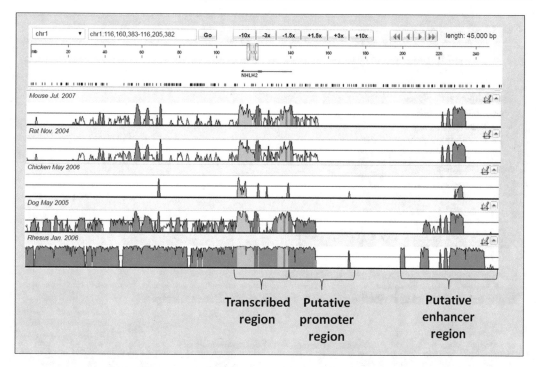

Fig. 3.1 Promoter analysis for NHLH2, using the online GenomeVISTA tool. The search term "NHLH2" was entered into the search bar (http://genome.lbl.gov/cgi-bin/GenomeVista), resulting in the following output. Pink regions are areas of homology between species, while light blue and dark blue regions are transcribed (light blue is untranslated, while dark blue is translated). The peak height indicates the strength of homology. Note that the NHLH2 gene is transcribed from the opposite strand of the genomic DNA, and runs from right to left

expression patterns (Consortium et al. 2014). In addition, there are multiple enhancer regions for each gene, located within 500 kilobases (kb) of the start of transcription (Andersson et al. 2014). Further use of the FANTOM site at http://fantom.gsc.riken.jp/5/ allows one to dig deeper into hypothalamic promoter analysis, finding a list of 1000 transcription factors that are enriched in adult diencephalon (Fig. 3.2, discussed more below). One can use a site like this to explore an individual promoter, or transcription factors, but the bulk of the confirmation work still needs to be done in the laboratory.

Some of the earliest studies on hypothalamic promoters used the pro-opiomelanocortin (POMC) gene as a model (Young et al. 1998). POMC's expression is limited to approximately 1000 POMC-specific neurons within the arcuate nucleus of the hypothalamus, as well as within hindbrain neurons and corticotrophs of the pituitary. Initial studies of the POMC promoter in pituitary cell lines found a very small ~500 base pair (bp) region directed expression in pituitary cells, and in the pituitary of transgenic mice, but not in the hypothalamus of those same animals (Rubinstein et al. 1993). Instead, a 15 kb region, containing ~13 kb of upstream sequence, was shown to direct hypothalamic-specific expression (Young et al. 1998). Further studies identified an enhancer region at ~12 kb upstream as the key regulatory element, and this region had high sequence conservation between human, mouse, dog, and bovine sequences, even though the latter two genomes were barely even started at that time (de Souza et al. 2005). Ten years later, two POMC gene enhancers, nPE1 and nPE2 located ~2 kb from each other and ~10 kb from the transcription start site, direct transcription of POMC within those 1000 POMC arcuate nucleus neurons (Lam et al. 2015). Interestingly, although deletion of just one of the enhancers still resulted in hypothalamic-specific expression, POMC

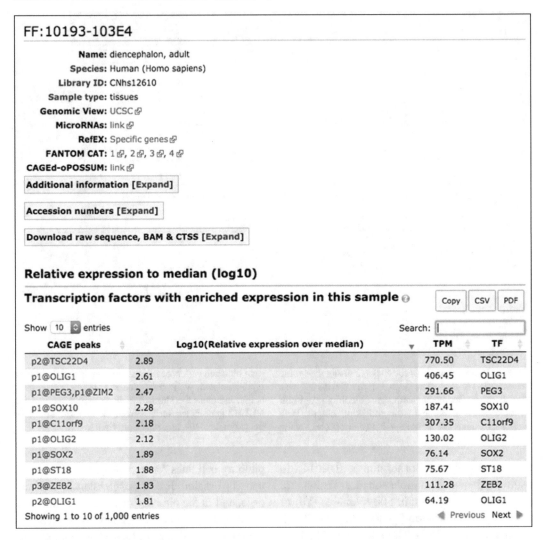

Fig. 3.2 Use of the FANTOM site (http://fantom.gsc. riken.jp/5/) to find hypothalamic transcription factors. As shown, search tools to narrow the tissue to adult dienceph-alon from adult tissues were used. The first 10, in a list of 1000, are shown, presented from highest expression to lowest expression

expression was decreased overall, suggesting the two enhancers have an additive effect on overall transcription. Furthermore, deletion of the nPE1 alone, which reduced expression of POMC and POMC cell numbers to 30% of normal, led to a significant increase in body weight and body fat in the mutant mice (Lam et al. 2015). This example alone makes a case for the relative importance of hypothalamic promoter function in maintaining POMC and other hypothalamic-specific genes needed for energy balance pathways. Expression levels and proper expression patterns are necessary for physiological maintenance of body weight and energy sensing. In a study examining single nucleotide polymorphisms within or near CpG islands, which can specify promoter activity via DNA methylation, 28 genes (including POMC) were found to have SNPs that affected promoter methylation level (Voisin et al. 2015). Studies that specifically examine differential methylation of the POMC promoter have shown a link between POMC methylation and offspring obesity in animals fed a high-fat/high-sucrose diet (Zheng et al. 2015). In a very recent study, variable methylation of the POMC gene was shown to be established early in human fetal development within a POMC variable methylation region located within intron 2, where hypermethylation equals reduced POMC expression and increased body weight (Kuhnen et al. 2016). Thus one needs to consider not only the promoter but also enhancers and downstream regions within the gene for overall transcriptional regulation.

Polymorphisms in the promoter of the melanocortin-4-receptor (MC4R) gene, whose protein product binds to the POMC peptide alpha-melanocortin-stimulating hormone, have been linked to obesity in several studies (Muller et al. 2014; Tan et al. 2014; Valli-Jaakola et al. 2006; van den Berg et al. 2011). In one of these studies, the two variations studied were shown to reduce MC4R gene transcription (Tan et al. 2014). A 2 bp deletion in the MC4R promoter, found by another study (Valli-Jaakola et al. 2006), was used by our laboratory to show that the deletion affected Nhlh2 transcription factor binding and transcription of the MC4R gene (Wankhade and Good 2011).

It is likely that many more promoter-associated polymorphisms will be identified as more and more genomes are sequenced and more and more association studies are completed. At this time, the numbers of studies of this type are meager—just 39 published articles are found in PubMed using the search terms "hypothalamus, promoter, obesity, transcription." More work is needed both to confirm the transcriptional regulatory regions of obesity- and energy balance-associated genes and to determine if newly detected variants within the promoters or enhancer regions may affect an individual's ability to respond to energy availability.

3.1.3 Hypothalamic Transcription Factors

According to the FANTOM database, a total of 1762 human and 1516 mouse transcription factors have been identified to date (Consortium et al. 2014). Note that this number is lower than we found using exon arrays (unpublished data from (Jiang et al. 2015)). However, using the FANTOM database tools http://fantom.gsc.riken.jp/5/, the relative expression level of transcription factors expressed within the adult diencephalon can be analyzed. As shown on Fig. 3.2, of 1000 transcription factors with expression of at least $\log(10) = 0.33$ (~twofold) of the median expression over all of their studies, the top 10 in the list form a less than well-known list of transcription factors—at least as far as hypothalamic transcription factors go. So what does this mean for analyzing transcription within the hypothalamus? Is it possible that the highest expressed factors do not necessarily control hypothalamic specificity? Yes, it is possible. For example, the first factor on the list shown in Fig. 3.2 is TSC22D4, a leucine zipper transcription factor family member, which according to http://www.genecards.org (Stelzer et al. 2016) is at least as highly enriched in lymphocytes and liver and gland cells as it is in nervous system. Factors such as Nhlh2 (~twofold), Stat3 (not even listed), or FoxO1 (~3.3-fold)—all of which are transcription factors that whose "hypothalamic importance" has been shown—are expressed at much lower levels, at least within the snapshot of expression data used for the FANTOM studies.

So what is known about transcription factors within the hypothalamus—which are the "important" ones and which are more general and likely just performing housekeeping duties? One of the first transcription factors identified in the paraventricular and supraoptic hypothalamus was c-fos (Sagar et al. 1988). C-fos is part of the activator protein-1 (AP1) complex, and its expression and nuclear localization appear to be stimulated when nerves are activated. In the FANTOM database, basal expression is only about twofold above the median, but it is likely that stimulated neurons show much higher levels. This point is key to the discussion—the hypothalamus sits at the base of the brain, with a median eminence region that "sips" capillary blood and can integrate external signals—likely doing this through differential activation of transcription factors and their target promoters. C-fos is likely one of those transcription factors that responds to external signals, such as water deprivation as was shown in that early study (Sagar et al. 1988). However, c-fos is expressed throughout the brain and in many different tissues, so while it is important for integrating signals, it is likely not key in mediating hypothalamic-specific responses.

Conversely, SOX2, which did make the top 10 list from the FANTOM expression analysis, is enriched in adult hypothalamic tissues at approximately 76-fold (log10 2.28) above the median gene expression levels (Fig. 3.2). SOX2 is normally thought of as a neuronal developmental transcription factor but, as shown by Hoefflin and Carter, is highly localized to the suprachiasmatic nucleus (SCN) and the periventricular nucleus (PeN), adjacent to the median eminence in adult rats (Hoefflin and Carter 2014). While the target genes for SOX2 within these hypothalamic nuclei have not been identified, the related transcription factors, SOX4 and SOX11, direct GnRH gene transcription within the hypothalamic-specific enhancer which is actually downstream of the transcription start site and within the first intron of the gene (Kim et al. 2011). Of note, this study used several of the cell lines and procedures described in the preceding section, including the GT1–7 cell line, two neuronal hypothalamic cell lines developed by the Belsham group, and ChIP analysis from hypothalamic lysates.

In summary, hypothalamic transcription factors, both those that are highly expressed and those that show a more targeted distribution with lower overall expression in the hypothalamus, require more analysis. Researchers certainly have not characterized the roles of many, if not most of the hypothalamic transcription factors that are expressed, nor understand what the hypothalamic nuclei-specific distribution of each of these might be. Further studies to identify gene targets of these transcription factors, and methods by which their expression is induced or repressed, as well as secondary modifications that might increase or decrease their activity or DNA binding, need to be elucidated.

3.1.4   DNA and Histone Modification

DNA methylation occurs via the addition of a methyl group to the 5′ cytosine of a dinucleotide CpG. DNA methyltransferase enzymes (Dnmt) perform this function—with Dnmt1, 3a, and 3b all being expressed in hypothalamic tissue (Benite-Ribeiro et al. 2016). While this simplified version makes sense from a mechanistic standpoint, the how, why, where, and when issues remain unanswered. What we know is that DNA is differentially methylated on different genes, in different conditions, and at different times, but what dictates these differences is still under intense investigation. Furthermore, in addition to DNA methylation, histones wrapping the DNA are also modified, with methylation, acetylation, and other less frequent marks like biotinylation (Benite-Ribeiro et al. 2016). The epigenetic marks of DNA methylation and histone modification can ultimately affect the level of gene expression at any point in time, irrespective of positively or negatively acting transcription factors.

Early studies using rats as model animals identified a CpG island with DNA methylation marks in the promoter region of the POMC gene and showed that the methylation of this area is inversely correlated to POMC mRNA expression (Plagemann et al. 2009). More interesting was the fact that rats that were overfed during the neonatal period showed increased body weight, accompanied by reduced POMC mRNA levels and higher POMC CpG methylation. This type of finding has been repeated in other models and appears to be consistent with the fact that overnutrition results in increased POMC promoter methylation, while undernutrition results in decreased POMC promoter methylation (Benite-Ribeiro et al. 2016). One study that has examined POMC promoter-associated histones in sheep that experienced maternal undernutrition found that there was a reduction in two different histone H3 methylation marks and an increase in histone H3 acetylation. However, these changes did not correlate into a difference in POMC mRNA expression (Begum et al. 2012). In a separate study, fasting of adult mice led to increased expression in histone deacetylases −3 and − 4 and decreased expression of histone deacetylases −10 and − 11. The remaining seven histone deacetylases tested did not change with fasting (Funato et al. 2011). Just between these two studies, there is discordance in how undernutrition and fasting might affect overall acetylation—or is there? We don't quite understand how the 11 different deacetylases contribute to gene expression changes, nor how they are induced or repressed in response to energy availability. This last study did not examine gene expression for POMC—but did look at expression of histone deacetylases in POMC neurons—and was able to detect acetylated histones in POMC neurons both in the fed and fasted state (Funato et al. 2011). That was expected, as different promoters will be differentially regulated, some through histone modifications. However, the results do not tell us whether POMC expression is modulated by histone acetylation. Further studies discussed below will follow up on this question.

3.1.5 Transcriptional Up- and Downregulation Through Changes in Energy Availability

Undoubtedly, one of the key signals to energy availability regulation by the hypothalamus is the sensing of the level of circulating leptin. The response to serum leptin levels is also likely the most characterized hypothalamic energy balance-signaling pathway. Multiple studies have shown that increased serum leptin results in up- and downregulation of hypothalamic genes, including increasing the neuropeptide genes such as POMC (Munzberg et al. 2003), and thyrotropic releasing hormone (TRH) (Huo et al. 2004), increasing transcription factors such as Nhlh2 (Al Rayyan et al. 2014), and decreasing expression of the neuropeptide genes AgRP (Toorie and Nillni 2014) and NPY (Morrison et al. 2005). The overall effect is that an increase in energy availability, as signaled by increased circulating leptin, results in increased levels of anorexigenic and decreased levels of orexigenic neuropep-

tides. In this section, the responses of genes to energy availability will be discussed, specifically transcriptional modulation of gene expression through leptin-mediated and fasting-induced signaling in the hypothalamus. Some other hypothalamic regions have been investigated, but the POMC neurons have the best mix of information on them and thus will be the focus on this section.

3.1.6 POMC Neurons in Response to Increased Energy Availability

The POMC neurons lie within the arcuate nucleus of the hypothalamus and there can directly sense leptin signal from the median eminence. This sensing occurs through the long form leptin receptor (LepRb) through a Jak-Stat signaling pathway (Flak and Myers 2016). Of note, Stat3 is required for leptin signaling (Flak and Myers 2016), even though, as mentioned earlier, it is not even found on the top 1000 list of hypothalamic transcription factors in the FANTOM database for hypothalamus (diencephalon) (Consortium et al. 2014). As we have already discussed in general, Stat3 acts as a transcription factor, regulating the POMC gene, the NHLH2 gene, and the PCSK1 gene (Fig. 3.3) (Al Rayyan et al. 2014; Fox and Good 2008; Munzberg et al. 2003). Jak2, a tyrosine kinase, first phosphorylates LepRb following leptin binding, making the activated and phosphorylated LepRb complex more "attractive" to Stat transcription factors. Stat3 and Stat5 are recruited to the activated LepRb/Jak complex and phosphorylated, resulting in translocation of Stat3 and Stat5 to the nucleus (Flak and Myers 2016). Stat transcription factors form homo- and heterodimers with each other and other transcription factors (Reich 2007).

Stat5 is dispensable for leptin signaling (Flak and Myers 2016), so we'll focus on Stat3 transcriptional regulation. What are its target genes? The first leptin-responsive target gene identified was POMC (Munzberg et al. 2003). In this study, leptin was shown to stimulate POMC promoter expression in a kidney and in a pituitary cell line

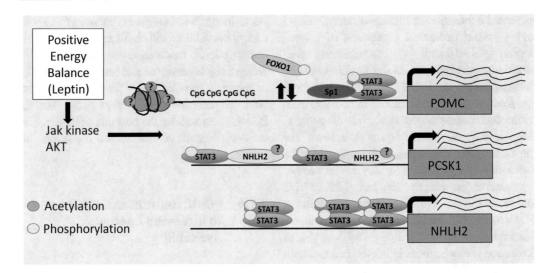

Fig. 3.3 Leptin-mediated induction of gene expression in POMC neurons. Three cartoons of gene promoters are shown with the possible set of transcription factors that regulate these genes in response to positive energy balance, in particular leptin. The posttranscriptional modification status of the transcription factors is shown or hypothesized based on references within the text

(at the time, there we no good hypothalamic lines), and a 30 bp region of the POMC promoter between −91 and − 61 from the start of transcription was shown to be sufficient for leptin-mediated expression. However, while the study showed that Stat3 was activated (phosphorylated and translocated to the nucleus), Munzberg and colleagues were not able to detect Stat3 binding to that 30 bp element using electrophoretic mobility (gel shift) assays (Munzberg et al. 2003). The researchers speculated that this might be because Stat3 was interacting with another protein/transcription factor. Only later did another group show that this was indeed the case. Yang and colleagues narrowed down the region needed for Stat3 binding to a slightly larger 50 bp fragment located between −138 and − 88 of the POMC promoter. This region contained a specificity protein 1 (Sp1) transcription factor binding site, but not a Stat3 binding site. However, immunoprecipitation of Sp1 in leptin-activated cells could bring down phosphorylate Stat3 protein, indicating that Stat3 and Sp1 interacted (Yang et al. 2009). In the Munzberg paper, the original 30 bp element which was located just upstream of the Sp1 site did contain a putative Stat site (Munzberg et al. 2003), and it is likely (although not yet directly shown) that these two elements act together to mediate leptin responsiveness of the POMC promoter (depicted on Fig. 3.3). Interestingly, the Yang et al. article also showed that the transcription factor FoxO1 was involved in POMC regulation as well, namely, that FoxO1 appeared to sequester Stat3 from interacting with Sp1, and subsequent studies revealed that the FoxO1:Stat3 interaction domain was separate from the FoxO1 DNA binding domain and that the POMC promoter also had a FoxO1 binding element in its proximal promoter region (Ma et al. 2015).

FoxO1, which is found in the FANTOM list of hypothalamic transcription factors (Consortium et al. 2014), is a member of the forkhead transcription factor family and worth mentioning in the context of hypothalamic gene regulation. FoxO1 is a.

target of the AKT kinase, which itself is regulated by phosphoinositol-3-phosphate (PIP_3) levels. PIP_3 levels are elevated by LepR (both long and short form) signaling. Overall then, activation of the leptin receptor through leptin results in phosphorylated FoxO1 protein, and this form of FoxO1 is not able to interact with Stat3 (Kwon et al. 2016) (Fig. 3.3).

Moving to the POMC enhancer regions discussed previously, these regions appear to specify hypothalamic tissue-specific transcription through homeodomain transcription factor binding motifs, and PE1 is needed for adult body weight regulation but only due to the fact that it allows a high level of hypothalamic POMC expression to occur (Lam et al. 2015). To date, no one has shown that the homeobox transcription factors that may bind in this region are leptin regulated. However, a region more proximal (at approximately −330 bp) appears to mediate POMC expression via differential DNA methylation and appears to be induced by excessive energy availability (Zheng et al. 2015). Additionally, changes in the levels of histone acetylases occur with fasting or high-fat feeding (Funato et al. 2011), suggesting that a histone-level gene regulation is occurring—but to date, this activity has not been specifically localized to the POMC promoter.

Slightly after leptin-mediated POMC gene regulation began to be characterized, regulation of the PCSK1 gene was characterized. The protein product of PCSK1 is prohormone convertase 1/3—the main neuropeptide processing enzyme for POMC. Thus, activation of PCSK1 gene by leptin makes sense, as the arcuate neurons need to not only increase POMC mRNA but make sure that the neuropeptide alpha-melanocyte-stimulating hormone is produced from the translated mRNA. In a similar fashion to POMC, Stat3 was shown to mediate PCSK1 (and the related gene PCSK2) expression following leptin stimulation (Sanchez et al. 2004). Only one promoter region containing a 971 bp fragment was used in the initial promoter analysis for this study, and although it contained Stat3 sites, there was no further specific characterization of the PCSK1 promoter at that time. However, during the same year, my laboratory showed that mice lacking the hypothalamic transcription factor Nhlh2 had low levels of PCSK1 mRNA, accompanied by low POMC peptide levels, with normal POMC RNA levels (Jing et al. 2004). This suggested that Nhlh2 might be a transcriptional regulator of the PCSK1 gene, which our laboratory was able to confirm in 2008 (Fox and Good 2008). As a

member of the basic helix-loop-helix (bHLH) transcription factor family, Nhlh2 can homodimerize (Fox and Good 2008) as well as form heterodimers with transcription factors. For PCSK1 leptin-mediated regulation, Nhlh2 appears to heterodimerize with Stat3 at two regions of the PCSK1 promoter, which contain adjacent E-box (bHLH binding) and Stat binding motifs (Fig. 3.3) (Fox and Good 2008). In a heterologous cell culture experiment, both Stat3 and Nhlh2 needed to be added to the cells for leptin-stimulated expression of PCSK1 to occur. In addition, in the absence of Stat3, Nhlh2 appeared to homodimerize, but addition of Stat3 led to heterodimerization of Stat3 and Nhlh2 in an in vitro pull-down assay (Fox and Good 2008). While Stat3 also homodimerizes, it is not clear at this time whether the heterodimer complex between a single Stat3 and a single Nhlh2 is sufficient or whether other factors, including other Stat3 molecules, are involved.

The NHLH2 gene, which is expressed within POMC neurons and throughout many other nuclei of the hypothalamus, is also induced by leptin (Vella et al. 2007), and this again makes sense if Nhlh2 is required for PCSK1 leptin-induced expression. The protein also appears to be regulated posttranscriptionally by the SIRT1 deacetylase (Libert et al. 2011). Indeed, the human and mouse Nhlh2 proteins contain seven lysine amino acids which could be the target of acetylases. Lysine #49 (K49) on Nhlh2 appears to be the deacetylation target of Sirt1, and deacetylation leads to loss of Nhlh2 promoter binding (Libert et al. 2011). While the studies by Libert and colleagues were on a negatively regulated Nhlh2 target which can only be expressed once, Nhlh2 is removed from the promoter; whether or not Nhlh2 is acetylated while binding to the PCSK1 promoter (Fig. 3.3) and whether or not acetylation of Nhlh2 is required for Stat3 interaction are still under investigation.

As mentioned above, the NHLH2 gene is also a known leptin-stimulated gene target in the hypothalamus. In order to further characterize elements in the NHLH2 promoter which might facilitate leptin-mediated regulation, we scanned the proximal 950 bps of the mouse and human

NHLH2 promoter and found multiple putative Stat3 binding sites (Al Rayyan et al. 2014). Two nuclear factor, NFκB sites were also identified, but do not appear to mediate leptin-stimulated gene expression—in fact, mutagenesis of the most distal NFκB resulted in even better leptin-mediated expression, suggesting that these sites could be negative regulators. Three of the five putative Stat3 sites were all required for leptin-induced expression of an NHLH2 heterologous promoter (Al Rayyan et al. 2014) (Fig. 3.3). In this set of experiments, the N29/2 cell line from the Belsham group was used, as this cell line best replicated a POMC neuronal cell line (Belsham et al. 2004). In this study we were also able to confirm that Stat3 interacted on the endogenous Nhlh2 promoter, by performing chromatin immunoprecipitation (ChIP) on hypothalamic nuclear extracts following leptin stimulation of mice (Al Rayyan et al. 2014). Quantification of the ChIP analysis by QPCR revealed that leptin treatment leads to a ~threefold increase in Stat3 occupation on the NHLH2 promoter (Al Rayyan et al. 2014). It is not clear if Sp1:Stat3 interactions, as seen on the POMC promoter, might also be involved in Nhlh2 regulation, as several putative Sp1 sites exist on the Nhlh2 promoter but have not been further investigated at this time.

For all of the genes discussed in this section, there are likely other transcription factors and DNA methylation/acetylation dynamics at play in leptin-stimulated expression. For now, the simplified version how the POMC, PSCK1, and NHLH2 promoters all are stimulated by leptin signals will have to suffice until more work is done to reveal any additional mechanisms.

3.1.7 Responses to Fasting and Reduced Energy Availability

According to studies completed by our laboratory (Jiang et al. 2015), serum leptin levels fall approximately 50% and glucose levels about 60% following a 24 h food deprivation, relative to ad-lib fed mice. In this whole hypothalamic analysis of total RNA, approximately equal numbers of mRNAs were upregulated or downregulated by fasting (Jiang et al. 2015). Interestingly, although 16 microRNAs were upregulated and 6 microRNAs were downregulated with the microarray data, of 6 tested further, none of these could be confirmed via QPCR. Overall, we were able to detect 536 microRNA species within the hypothalamus, which is more than the 105 out of 717 reported for mouse midbrain by http://www.microrna.org/microrna/home.do. Long noncoding RNA (lncRNA) appears to make up more of the noncoding RNA species and has more of those that are differentially regulated. In this same study, and a subsequent one, we were able to identify 421 lncRNA species that were upregulated in response to fasting, and 201 lncRNA species that were downregulated in response to fasting, and were able to detect over 12,000 lncRNA species that were expressed in hypothalamic tissues (Jiang et al. 2015). Two of three lncRNA tested by QPCR were confirmed to be differentially regulated in response to fasting. In addition, using data from an older generation microarray, we were also able to identify 60 lncRNAs that were differentially expressed with fasting (Jiang and Good 2016). However, our knowledge of the role of lncRNAs in the hypothalamus and especially in response to energy balance is still evolving. We don't yet understand how many of these lncRNAs function, nor how they are differentially expressed (Jiang and Good 2016). However, it is important to mention that within the context of differential gene expression, one should not forget the noncoding RNAs.

So what about the "more common" mRNAs—as noted from our study—there were significant changes in the mRNA transcriptome following fasting (Jiang et al. 2015). Of these, gene ontology analysis revealed a number of genes in the cell cycle category were changed—many of them surprisingly upregulated in response to fasting. One of these was Cdkn1a, which codes for the p21 protein (Jiang et al. 2015). Of note, Cdkn1a had already been shown to be upregulated in specific hypothalamic nuclei with fasting, and one study showed that the FoxO1 transcription factor played a role in its upregulation (Tinkum et al. 2013), although the transcriptional analyses were only done with the liver where p21 protein is also increased with fasting. However, as noted by these authors, Cdkn1a has FoxO1 binding sites in its proximal promoter, so it is likely that FoxO1, which is expressed in hypothalamus, mediates this gene expression upregulation. As noted previously, POMC promoter also has a FoxO1 binding element in its proximal promoter region (Ma et al. 2015), and overexpression of FoxO1 in POMC neurons reduces POMC expression and overall POMC neuron numbers (Plum et al. 2012). So, FoxO1, acting on different neurons, perhaps in different conditions, can either up- or downregulate target gene expression.

In considering gene regulation during reduced energy availability, it is important to look at how orexigenic genes, such as neuropeptide Y (NPY) and agouti-related protein (AgRP), are induced. The human and mouse promoter regions for AgRP were characterized in 2001 (Mayfield et al. 2001) and found to contain at least on Stat site, a CCAAT box, which can bind to factors of the CREB family of transcription factors, and an E-box motif (for bHLH transcription factors) which was also the site of a human DNA variant that reduced gene expression (Mayfield et al. 2001). In this study, Mayfield and colleagues were able to show, using GT1–7 hypothalamic cells, a significant reduction in promoter activity with the variant and an association of the variant genotype TT with obesity (Mayfield et al. 2001). Subsequent studies from this same laboratory did not follow up on the BHLH transcription factor in AgRP regulation, but rather showed that a Krüpple-like transcription factor, Klf4, directed expression of AgRP (Ilnytska et al. 2011). In these studies, the Belsham hypothalamic cell line N38 (Belsham et al. 2004) was used to show induction of both Klf4 and AgRP with fasting as well as direct binding of Klf4 to the AgRP promoter under these conditions (Ilnytska et al. 2011). AgRP is also regulated by FoxO1, which appears to be assisted by the activating transcription factor ATF3 (Lee et al. 2013) and antagonized by Stat3 (Kitamura et al. 2006). All of this is summarized in cartoon form in Fig. 3.4. Of note, while we know that Stat3 is likely phosphorylated, FoxO1 is normally acetylated when inactive and deacetylated by Sirt1 when it is in the nucleus and regulating AgRP (Cakir et al. 2009). Phosphorylation of ATF3 and acetylation of KLF4 have been shown in other systems (e.g., Mayer et al. 2008; Meng et al. 2009), but not confirmed in hypothalamic AgRP neurons.

Despite the fact that one of the main human polymorphism in the NPY gene which is linked to metabolic phenotypes resides in the NPY promoter region at −399 (e.g., de Luis et al. 2016), relatively little is known about the transcriptional regulation of NPY in response to fasting. It has been known for more than 25 years that NPY gene expression is induced with fasting (Sanacora et al. 1990), but the mechanisms of this remain elusive. The polymorphism at −399 appears to affect an SP1 site, which is conserved in the bovine NPY gene as well (Alam et al. 2012). According to a very large study of the response to starvation in mice in which a transcriptional network analysis was performed, the "interactome" of transcriptional responses to starvation, SP1 is listed in 4 of the 5 tissues examined, notably absent from brain (Hakvoort et al. 2011). Thus it could be that SP1 did not reach the level of significance in this study but may in fact be a fasting−/starvation-induced transcription factor responsible for NPY expression (Fig. 3.4). The NPY promoter also contains two distal Stat3 sites, but the role of these in leptin-inhibited versus fasting-induced expression is not yet clear. Notably, in a review that this author wrote 17 years ago, there was still little known about NPY promoter regulation, but several studies suggested that its regulation was not transcriptional but posttranscriptional, with increased mRNA stability of the NPY mRNA contributing to its change in expression in response to fasting (Good 2000). With more knowledge about microRNA and lncRNA expression in the hypothalamus, as well as the known function of some noncoding RNA in mRNA stability, more work should be done on NPY mRNA stability versus transcriptional activation in response to fasting.

As mentioned above, several of the transcription factors mediating fasting response in the hypothalamus undergo posttranslational modifications such as phosphorylation and acetylation

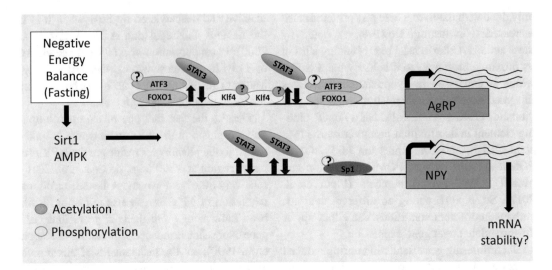

Fig. 3.4 Fasting-induced induction of gene expression in NPY neurons. AgRP and NPY genes, with their putative promoter regions, are shown as both of these genes are induced with fasting. The posttranscriptional modification status of the transcription factors is shown or hypothesized based on references within the text

(and removal of these modification) in order to either move to the nucleus or bind DNA. For example, AMP kinase and Sirt1 deacetylase are both active under conditions of low leptin, such as fasting, and both are responsive to metabolic signals (low ATP/high AMP for AMP kinase and low NADH/high NAD for Sirt1) which would be present during fasting conditions, and both target FoxO1 transcription factor (Minokoshi et al. 2008; Toorie and Nillni 2014). Thus, we are starting to build transcriptional signaling pathways to understand the molecular changes that occur in response to fasting. However, this area likely needs more attention to characterize more gene targets of fasting and how these gene targets are regulated. The roles of lncRNA should also be considered as several lines of evidence suggest that they may play a major role in energy balance control (Jiang and Good 2016).

3.2 Summary and Key Questions that Remain Unanswered

The tools are in place for researchers to build comprehensive knowledge on gene expression mechanisms in response to changes in energy balance, but much more needs to be done. It is clear, for example, that many more genes undergo transcriptional regulation by those transcription factors that we know are induced (FoxO1, Stat3, Nhlh2, etc.), and these genes need to be characterized to fully understand the response to feeding and fasting. Furthermore, there are likely many other transcription factors that have not been fully characterized; yet still many likely play important roles in energy balance regulation. One of the downsides of working with transcription factors is that they generally are not highly expressed and may be difficult to detect or missed with some previous methods used. Antibodies can sometimes detect the protein, but posttranslational modifications also interfere with immunodetection, depending on whether conformational changes occur with the modification. Many transcription factors, like Stat3, are not necessarily transcriptionally regulated but posttranscriptionally regulated through protein modifications, so studying their transcriptional regulation may be important, but may not tell us much about energy balance signaling. However, posttranslational modifications need to be more rigorously investigated for each neuron type within the hypothalamus and under conditions of both fasting and ad-lib feeding to fully understand how the transcription factors mediate differential gene expression.

It is also clear that sexual dimorphism exists in hypothalamic transcriptional and responses (Loganathan and Belsham 2016), and yet few laboratories study this area. One should consider the "sex" or gender of the cell lines and then use specific and perhaps matched cell lines to establish and study sex-specific gene expression differences for many different hypothalamic promoters. In addition, once we begin to understand these mechanisms in cell lines, we need to extend it to humans, and specifically characterize transcription factor binding sites within human promoters, and then utilize online genomic databases to identify polymorphisms that might affect transcriptional regulation, as possible mechanism for genetic mechanisms of body weight control. Genome-wide association studies often identify associated polymorphisms that are not within gene coding regions and, thus, may instead be associated with gene regulatory changes.

No chapter on transcriptional mechanisms of gene regulation would be complete without at least mentioning posttranscriptional mechanisms, which have been done with NPY and mRNA stability, posttranslational modifications, and analysis of noncoding RNA expression with fasting in this chapter. However, many other posttranscriptional mechanisms exist than were discussed above, and given the number of different outcomes for an mRNA once it is transcribed, this topic really deserves a chapter of its own.

We finish his chapter by again contemplating the enormity of having over 2000 different transcription factors to regulate the thousands of genes that are needed for energy balance control by the hypothalamus. While researchers have made some headway into this topic, it is also clear that there is still a lot to be done to understand the complexity of transcription (and posttranscriptional) control of the hypothalamic transcriptome and ultimately its resulting proteome.

Questions to Consider

1. Compare and contrast studying gene expression in hepatocytes to gene expression within the hypothalamic arcuate POMC neurons. What are the challenges and benefits for each tissue type?

2. Take a look at the NCBI gene expression profiles database (https://www-ncbi-nlm-nih-gov.ezproxy.lib.vt.edu/geoprofiles/), and put a search term into the box (such as hypothalamus, a nucleus of the hypothalamus, or a gene name). What kind of data is returned and how might one use this information to develop a research question?

3. Analyze the actual data from one or two of the papers referenced in this chapter in which transcriptional regulation of a hypothalamic gene was characterized. What would be the next step in this type of research? What tools would you need to conduct this type of research?

4. Is it better to study gene expression in cell lines or in animals? Compare and contrast these two models.

5. Can knowledge about gene expression be used in personalized medicine? Are there any known treatments to date that take gene expression patterns into account?

References

Al Rayyan, N., Zhang, J., Burnside, A. S., & Good, D. J. (2014). Leptin signaling regulates hypothalamic expression of nescient helix-loop-helix 2 (Nhlh2) through signal transducer and activator 3 (Stat3). *Molecular and Cellular Endocrinology*, 134–142. https://doi.org/10.1016/j.mce.2014.01.017.

Alam, T., Bahar, B., Waters, S. M., McGee, M., O'Doherty, J. V., & Sweeney, T. (2012). Functional characterisation of the bovine neuropeptide Y gene promoter and evaluation of the transcriptional activities of promoter haplotypes. *Molecular Biology Reports*, 919–928. https://doi.org/10.1007/s11033-011-0817-z.

Andersson, R., Gebhard, C., Miguel-Escalada, I., Hoof, I., Bornholdt, J., Boyd, M., Chen, Y., Zhao, X., Schmidl,

C., Suzuki, T., Ntini, E., Arner, E., Valen, E., Li, K., Schwarzfischer, L., Glatz, D., Raithel, J., Lilje, B., Rapin, N., Bagger, F. O., Jorgensen, M., Andersen, P. R., Bertin, N., Rackham, O., Burroughs, A. M., Baillie, J. K., Ishizu, Y., Shimizu, Y., Furuhata, E., Maeda, S., Negishi, Y., Mungall, C. J., Meehan, T. F., Lassmann, T., Itoh, M., Kawaji, H., Kondo, N., Kawai, J., Lennartsson, A., Daub, C. O., Heutink, P., Hume, D. A., Jensen, T. H., Suzuki, H., Hayashizaki, Y., Muller, F., Consortium, F., Forrest, A. R., Carninci, P., Rehli, M., & Sandelin, A. (2014). An atlas of active enhancers across human cell types and tissues. *Nature*, 455–461. https://doi.org/10.1038/nature12787.

Begum, G., Stevens, A., Smith, E. B., Connor, K., Challis, J. R., Bloomfield, F., & White, A. (2012). Epigenetic changes in fetal hypothalamic energy regulating pathways are associated with maternal undernutrition and twinning. *The FASEB Journal*, 1694–1703. https://doi.org/10.1096/fj.11-198762.

Belsham, D. D., Cai, F., Cui, H., Smukler, S. R., Salapatek, A. M., & Shkreta, L. (2004). Generation of a phenotypic array of hypothalamic neuronal cell models to study complex neuroendocrine disorders. *Endocrinology*, 393–400. https://doi.org/10.1210/en.2003-0946.

Belsham, D. D., Fick, L. J., Dalvi, P. S., Centeno, M. L., Chalmers, J. A., Lee, P. K., Wang, Y., Drucker, D. J., & Koletar, M. M. (2009). Ciliary neurotrophic factor recruitment of glucagon-like peptide-1 mediates neurogenesis, allowing immortalization of adult murine hypothalamic neurons. *The FASEB Journal*, 4256–4265. https://doi.org/10.1096/fj.09-133454.

Benite-Ribeiro, S. A., Putt, D. A., Soares-Filho, M. C., & Santos, J. M. (2016). The link between hypothalamic epigenetic modifications and long-term feeding control. *Appetite*, 445–453. https://doi.org/10.1016/j.appet.2016.08.111.

Cakir, I., Perello, M., Lansari, O., Messier, N. J., Vaslet, C. A., & Nillni, E. A. (2009). Hypothalamic Sirt1 regulates food intake in a rodent model system. *PLoS One*, e8322. https://doi.org/10.1371/journal.pone.0008322.

Consortium, F., Forrest, A. R., Kawaji, H., Rehli, M., Baillie, J. K., de Hoon, M. J., Haberle, V., Lassmann, T., Kulakovskiy, I. V., Lizio, M., Itoh, M., Andersson, R., Mungall, C. J., Meehan, T. F., Schmeier, S., Bertin, N., Jorgensen, M., Dimont, E., Arner, E., Schmidl, C., Schaefer, U., Medvedeva, Y. A., Plessy, C., Vitezic, M., Severin, J., Semple, C., Ishizu, Y., Young, R. S., Francescatto, M., Alam, I., Albanese, D., Altschuler, G. M., Arakawa, T., Archer, J. A., Arner, P., Babina, M., Rennie, S., Balwierz, P. J., Beckhouse, A. G., Pradhan-Bhatt, S., Blake, J. A., Blumenthal, A., Bodega, B., Bonetti, A., Briggs, J., Brombacher, F., Burroughs, A. M., Califano, A., Cannistraci, C. V., Carbajo, D., Chen, Y., Chierici, M., Ciani, Y., Clevers, H. C., Dalla, E., Davis, C. A., Detmar, M., Diehl, A. D., Dohi, T., Drablos, F., Edge, A. S., Edinger, M., Ekwall, K., Endoh, M., Enomoto, H., Fagiolini, M., Fairbairn, L., Fang, H., Farach-Carson, M. C., Faulkner, G. J., Favorov, A. V., Fisher, M. E., Frith,

M. C., Fujita, R., Fukuda, S., Furlanello, C., Furino, M., Furusawa, J., Geijtenbeek, T. B., Gibson, A. P., Gingeras, T., Goldowitz, D., Gough, J., Guhl, S., Guler, R., Gustincich, S., Ha, T. J., Hamaguchi, M., Hara, M., Harbers, M., Harshbarger, J., Hasegawa, A., Hasegawa, Y., Hashimoto, T., Herlyn, M., Hitchens, K. J., Ho Sui, S. J., Hofmann, O. M., Hoof, I., Hori, F., Huminiecki, L., Iida, K., Ikawa, T., Jankovic, B. R., Jia, H., Joshi, A., Jurman, G., Kaczkowski, B., Kai, C., Kaida, K., Kaiho, A., Kajiyama, K., Kanamori-Katayama, M., Kasianov, A. S., Kasukawa, T., Katayama, S., Kato, S., Kawaguchi, S., Kawamoto, H., Kawamura, Y. I., Kawashima, T., Kempfle, J. S., Kenna, T. J., Kere, J., Khachigian, L. M., Kitamura, T., Klinken, S. P., Knox, A. J., Kojima, M., Kojima, S., Kondo, N., Koseki, H., Koyasu, S., Krampitz, S., Kubosaki, A., Kwon, A. T., Laros, J. F., Lee, W., Lennartsson, A., Li, K., Lilje, B., Lipovich, L., Mackay-Sim, A., Manabe, R., Mar, J. C., Marchand, B., Mathelier, A., Mejhert, N., Meynert, A., Mizuno, Y., de Lima Morais, D. A., Morikawa, H., Morimoto, M., Moro, K., Motakis, E., Motohashi, H., Mummery, C. L., Murata, M., Nagao-Sato, S., Nakachi, Y., Nakahara, F., Nakamura, T., Nakamura, Y., Nakazato, K., van Nimwegen, E., Ninomiya, N., Nishiyori, H., Noma, S., Noma, S., Noazaki, T., Ogishima, S., Ohkura, N., Ohimiya, H., Ohno, H., Ohshima, M., Okada-Hatakeyama, M., Okazaki, Y., Orlando, V., Ovchinnikov, D. A., Pain, A., Passier, R., Patrikakis, M., Persson, H., Piazza, S., Prendergast, J. G., Rackham, O. J., Ramilowski, J. A., Rashid, M., Ravasi, T., Rizzu, P., Roncador, M., Roy, S., Rye, M. B., Saijyo, E., Sajantila, A., Saka, A., Sakaguchi, S., Sakai, M., Sato, H., Savvi, S., Saxena, A., Schneider, C., Schultes, E. A., Schulze-Tanzil, G. G., Schwegmann, A., Sengstag, T., Sheng, G., Shimoji, H., Shimoni, Y., Shin, J. W., Simon, C., Sugiyama, D., Sugiyama, T., Suzuki, M., Suzuki, N., Swoboda, R. K., t Hoen, P. A., Tagami, M., Takahashi, N., Takai, J., Tanaka, H., Tatsukawa, H., Tatum, Z., Thompson, M., Toyodo, H., Toyoda, T., Valen, E., van de Wetering, M., van den Berg, L. M., Verado, R., Vijayan, D., Vorontsov, I. E., Wasserman, W. W., Watanabe, S., Wells, C. A., Winteringham, L. N., Wolvetang, E., Wood, E. J., Yamaguchi, Y., Yamamoto, M., Yoneda, M., Yonekura, Y., Yoshida, S., Zabierowski, S. E., Zhang, P. G., Zhao, X., Zucchelli, S., Summers, K. M., Suzuki, H., Daub, C. O., Kawai, J., Heutink, P., Hide, W., Freeman, T. C., Lenhard, B., Bajic, V. B., Taylor, M. S., Makeev, V. J., Sandelin, A., Hume, D. A., Carninci, P., & Hayashizaki, Y. (2014). A promoter-level mammalian expression atlas. *Nature*, 462–470. https://doi.org/10.1038/nature13182.

de Luis, D. A., Izaola, O., de la Fuente, B., Primo, D., & Aller, R. (2016). Polymorphism of neuropeptide Y gene rs16147 modifies the response to a hypocaloric diet on cardiovascular risk biomarkers and adipokines. *Journal of Human Nutrition and Dietetics*. https://doi.org/10.1111/jhn.12406.

de Souza, F. S., Santangelo, A. M., Bumaschny, V., Avale, M. E., Smart, J. L., Low, M. J., & Rubinstein,

M. (2005). Identification of neuronal enhancers of the proopiomelanocortin gene by transgenic mouse analysis and phylogenetic footprinting. *Molecular and Cellular Biology*, 3076–3086. https://doi.org/10.1128/MCB.25.8.3076-3086.2005.

De Vitry, F., Camier, M., Czernichow, P., Benda, P., Cohen, P., & Tixier-Vidal, A. (1974). Establishment of a clone of mouse hypothalamic neurosecretory cells synthesizing neurophysin and vasopressin. *Proceedings of the National Academy of Sciences of the United States of America*, (9), 3575.

Flak, J. N., & Myers, M. G., Jr. (2016). Minireview: CNS Mechanisms of Leptin Action. *Molecular Endocrinology*, 3–12. https://doi.org/10.1210/me.2015-1232.

Fox, D. L., & Good, D. J. (2008). Nescient helix-loop-helix 2 interacts with signal transducer and activator of transcription 3 to regulate transcription of prohormone convertase 1/3. *Molecular Endocrinology*, 1438–1448. https://doi.org/10.1210/me.2008-0010.

Fulco, C. P., Munschauer, M., Anyoha, R., Munson, G., Grossman, S. R., Perez, E. M., Kane, M., Cleary, B., Lander, E. S., & Engreitz, J. M. (2016). Systematic mapping of functional enhancer-promoter connections with CRISPR interference. *Science*, 769–773. https://doi.org/10.1126/science.aag2445.

Funato, H., Oda, S., Yokofujita, J., Igarashi, H., & Kuroda, M. (2011). Fasting and high-fat diet alter histone deacetylase expression in the medial hypothalamus. *PLoS One*, e18950. https://doi.org/10.1371/journal.pone.0018950.

Gingerich, S., Wang, X., Lee, P. K., Dhillon, S. S., Chalmers, J. A., Koletar, M. M., & Belsham, D. D. (2009). The generation of an array of clonal, immortalized cell models from the rat hypothalamus: analysis of melatonin effects on kisspeptin and gonadotropin-inhibitory hormone neurons. *Neuroscience*, 1134–1140. https://doi.org/10.1016/j.neuroscience.2009.05.026.

Good, D. J. (2000). How tight are your genes? Transcriptional and posttranscriptional regulation of the leptin receptor, NPY, and POMC genes. *Hormones and Behavior*, 284–298. https://doi.org/10.1006/hbeh.2000.1587.

Hakvoort, T. B., Moerland, P. D., Frijters, R., Sokolovic, A., Labruyere, W. T., Vermeulen, J. L., Ver Loren van Themaat, E., Breit, T. M., Wittink, F. R., van Kampen, A. H., Verhoeven, A. J., Lamers, W. H., & Sokolovic, M. (2011). Interorgan coordination of the murine adaptive response to fasting. *The Journal of Biological Chemistry*, 16332–16343. https://doi.org/10.1074/jbc.M110.216986.

Hawrylycz, M. J., Lein, E. S., Guillozet-Bongaarts, A. L., Shen, E. H., Ng, L., Miller, J. A., van de Lagemaat, L. N., Smith, K. A., Ebbert, A., Riley, Z. L., Abajian, C., Beckmann, C. F., Bernard, A., Bertagnolli, D., Boe, A. F., Cartagena, P. M., Chakravarty, M. M., Chapin, M., Chong, J., Dalley, R. A., Daly, B. D., Dang, C., Datta, S., Dee, N., Dolbeare, T. A., Faber, V., Feng, D., Fowler, D. R., Goldy, J., Gregor, B. W., Haradon, Z.,

Haynor, D. R., Hohmann, J. G., Horvath, S., Howard, R. E., Jeromin, A., Jochim, J. M., Kinnunen, M., Lau, C., Lazarz, E. T., Lee, C., Lemon, T. A., Li, L., Li, Y., Morris, J. A., Overly, C. C., Parker, P. D., Parry, S. E., Reding, M., Royall, J. J., Schulkin, J., Sequeira, P. A., Slaughterbeck, C. R., Smith, S. C., Sodt, A. J., Sunkin, S. M., Swanson, B. E., Vawter, M. P., Williams, D., Wohnoutka, P., Zielke, H. R., Geschwind, D. H., Hof, P. R., Smith, S. M., Koch, C., Grant, S. G., & Jones, A. R. (2012). An anatomically comprehensive atlas of the adult human brain transcriptome. *Nature*, 391–399. https://doi.org/10.1038/nature11405.

Hoefflin, S., & Carter, D. A. (2014). Neuronal expression of SOX2 is enriched in specific hypothalamic cell groups. *Journal of Chemical Neuroanatomy*, 153–160. https://doi.org/10.1016/j.jchemneu.2014.09.003.

Huang, P. P., Brusman, L. E., Iyer, A. K., Webster, N. J., & Mellon, P. L. (2016). A Novel Gonadotropin-Releasing Hormone 1 (Gnrh1) Enhancer-Derived Noncoding RNA Regulates Gnrh1 Gene Expression in GnRH Neuronal Cell Models. *PLoS One*, e0158597. https://doi.org/10.1371/journal.pone.0158597.

Huo, L., Munzberg, H., Nillni, E. A., & Bjorbaek, C. (2004). Role of signal transducer and activator of transcription 3 in regulation of hypothalamic trh gene expression by leptin. *Endocrinology*, 2516–2523. https://doi.org/10.1210/en.2003-1242.

Ilnytska, O., Stutz, A. M., Park-York, M., York, D. A., Ribnicky, D. M., Zuberi, A., Cefalu, W. T., & Argyropoulos, G. (2011). Molecular mechanisms for activation of the agouti-related protein and stimulation of appetite. *Diabetes*, 97–106. https://doi.org/10.2337/db10-0172.

Jiang, H., & Good, D. J. (2016). A molecular conundrum involving hypothalamic responses to and roles of long non-coding RNAs following food deprivation. *Molecular and Cellular Endocrinology*, 52–60. https://doi.org/10.1016/j.mce.2016.08.028.

Jiang, H., Modise, T., Helm, R., Jensen, R. V., & Good, D. J. (2015). Characterization of the hypothalamic transcriptome in response to food deprivation reveals global changes in long noncoding RNA, and cell cycle response genes. *Genes & Nutrition, 48*. https://doi.org/10.1007/s12263-015-0496-9.

Jing, E., Nillni, E. A., Sanchez, V. C., Stuart, R. C., & Good, D. J. (2004). Deletion of the Nhlh2 transcription factor decreases the levels of the anorexigenic peptides alpha melanocyte-stimulating hormone and thyrotropin-releasing hormone and implicates prohormone convertases I and II in obesity. *Endocrinology*, 1503–1513. https://doi.org/10.1210/en.2003-0834.

Kim, H. D., Choe, H. K., Chung, S., Kim, M., Seong, J. Y., Son, G. H., & Kim, K. (2011). Class-C SOX transcription factors control GnRH gene expression via the intronic transcriptional enhancer. *Molecular Endocrinology*, 1184–1196. https://doi.org/10.1210/me.2010-0332.

Kitamura, T., Feng, Y., Kitamura, Y. I., Chua, S. C., Jr., Xu, A. W., Barsh, G. S., Rossetti, L., & Accili, D. (2006). Forkhead protein FoxO1 mediates Agrp-dependent effects of leptin on food intake. *Nature Medicine*, 534–540. https://doi.org/10.1038/nm1392.

Kuhnen, P., Handke, D., Waterland, R. A., Hennig, B. J., Silver, M., Fulford, A. J., Dominguez-Salas, P., Moore, S. E., Prentice, A. M., Spranger, J., Hinney, A., Hebebrand, J., Heppner, F. L., Walzer, L., Grotzinger, C., Gromoll, J., Wiegand, S., Gruters, A., & Krude, H. (2016). Interindividual variation in DNA methylation at a putative POMC metastable epiallele is associated with obesity. *Cell Metabolism*, 502–509. https://doi.org/10.1016/j.cmet.2016.08.001.

Kwon, O., Kim, K. W., & Kim, M. S. (2016). Leptin signalling pathways in hypothalamic neurons. *Cellular and Molecular Life Sciences*, 1457–1477. https://doi.org/10.1007/s00018-016-2133-1.

Lam, D. D., de Souza, F. S., Nasif, S., Yamashita, M., Lopez-Leal, R., Otero-Corchon, V., Meece, K., Sampath, H., Mercer, A. J., Wardlaw, S. L., Rubinstein, M., & Low, M. J. (2015). Partially redundant enhancers cooperatively maintain Mammalian pomc expression above a critical functional threshold. *PLoS Genetics*, e1004935. https://doi.org/10.1371/journal.pgen.1004935.

Landry, J. J., Pyl, P. T., Rausch, T., Zichner, T., Tekkedil, M. M., Stutz, A. M., Jauch, A., Aiyar, R. S., Pau, G., Delhomme, N., Gagneur, J., Korbel, J. O., Huber, W., & Steinmetz, L. M. (2013). The genomic and transcriptomic landscape of a HeLa cell line. *G3 (Bethesda)*, 1213–1224. https://doi.org/10.1534/g3.113.005777.

Latella, M. C., Di Salvo, M. T., Cocchiarella, F., Benati, D., Grisendi, G., Comitato, A., Marigo, V., & Recchia, A. (2016). In vivo Editing of the Human Mutant Rhodopsin Gene by Electroporation of Plasmid-based CRISPR/Cas9 in the Mouse Retina. *Molecular Therapy – Nucleic Acids, e389*. https://doi.org/10.1038/mtna.2016.92.

Lawson, M. A., Macconell, L. A., Kim, J., Powl, B. T., Nelson, S. B., & Mellon, P. L. (2002). Neuron-specific expression in vivo by defined transcription regulatory elements of the GnRH gene. *Endocrinology*, 1404–1412. https://doi.org/10.1210/endo.143.4.8751.

Lee, Y. S., Sasaki, T., Kobayashi, M., Kikuchi, O., Kim, H. J., Yokota-Hashimoto, H., Shimpuku, M., Susanti, V. Y., Ido-Kitamura, Y., Kimura, K., Inoue, H., Tanaka-Okamoto, M., Ishizaki, H., Miyoshi, J., Ohya, S., Tanaka, Y., Kitajima, S., & Kitamura, T. (2013). Hypothalamic ATF3 is involved in regulating glucose and energy metabolism in mice. *Diabetologia*, 1383–1393. https://doi.org/10.1007/s00125-013-2879-z.

Libert, S., Pointer, K., Bell, E. L., Das, A., Cohen, D. E., Asara, J. M., Kapur, K., Bergmann, S., Preisig, M., Otowa, T., Kendler, K. S., Chen, X., Hettema, J. M., van den Oord, E. J., Rubio, J. P., & Guarente, L. (2011). SIRT1 activates MAO-A in the brain to mediate anxiety and exploratory drive. *Cell*, 1459–1472. https://doi.org/10.1016/j.cell.2011.10.054.

Loganathan, N., & Belsham, D. D. (2016). Nutrient-sensing mechanisms in hypothalamic cell models: neuropeptide regulation and neuroinflammation in male- and female-derived cell lines. *American Journal of Physiology. Regulatory, Integrative and Comparative Physiology*, R217–R221. https://doi.org/10.1152/ajpregu.00168.2016.

Ma, W., Fuentes, G., Shi, X., Verma, C., Radda, G. K., & Han, W. (2015). FoxO1 negatively regulates leptin-induced POMC transcription through its direct interaction with STAT3. *Biochemical Journal*, 291–298. https://doi.org/10.1042/BJ20141109.

Majzoub, J. A., Pappey, A., Burg, R., & Habener, J. F. (1984). Vasopressin gene is expressed at low levels in the hypothalamus of the Brattleboro rat. *Proceedings of the National Academy of Sciences of the United States of America*, 5296–5299.

Mayer, S. I., Dexheimer, V., Nishida, E., Kitajima, S., & Thiel, G. (2008). Expression of the transcriptional repressor ATF3 in gonadotrophs is regulated by Egr-1, CREB, and ATF2 after gonadotropin-releasing hormone receptor stimulation. *Endocrinology*, 6311–6325. https://doi.org/10.1210/en.2008-0251.

Mayfield, D. K., Brown, A. M., Page, G. P., Garvey, W. T., Shriver, M. D., & Argyropoulos, G. (2001). A role for the Agouti-Related Protein promoter in obesity and type 2 diabetes. *Biochemical and Biophysical Research Communications*, 568–573. https://doi.org/10.1006/bbrc.2001.5600.

Mellon, P. L., Windle, J. J., Goldsmith, P. C., Padula, C. A., Roberts, J. L., & Weiner, R. I. (1990). Immortalization of hypothalamic GnRH neurons by genetically targeted tumorigenesis. *Neuron, 5*, 1–10.

Meng, F., Han, M., Zheng, B., Wang, C., Zhang, R., Zhang, X. H., & Wen, J. K. (2009). All-trans retinoic acid increases KLF4 acetylation by inducing HDAC2 phosphorylation and its dissociation from KLF4 in vascular smooth muscle cells. *Biochemical and Biophysical Research Communications*, 13–18. https://doi.org/10.1016/j.bbrc.2009.05.112.

Minokoshi, Y., Shiuchi, T., Lee, S., Suzuki, A., & Okamoto, S. (2008). Role of hypothalamic AMP-kinase in food intake regulation. *Nutrition*, 786–790. https://doi.org/10.1016/j.nut.2008.06.002.

Morrison, C. D., Morton, G. J., Niswender, K. D., Gelling, R. W., & Schwartz, M. W. (2005). Leptin inhibits hypothalamic Npy and Agrp gene expression via a mechanism that requires phosphatidylinositol 3-OH-kinase signaling. *American Journal of Physiology. Endocrinology and Metabolism*, E1051–E1057. https://doi.org/10.1152/ajpendo.00094.2005.

Muller, Y. L., Thearle, M. S., Piaggi, P., Hanson, R. L., Hoffman, D., Gene, B., Mahkee, D., Huang, K., Kobes, S., Votruba, S., Knowler, W. C., Bogardus, C., & Baier, L. J. (2014). Common genetic variation in and near the melanocortin 4 receptor gene (MC4R) is associated with body mass index in American Indian adults and children. *Human Genetics*, 1431–1441. https://doi.org/10.1007/s00439-014-1477-6.

Munzberg, H., Huo, L., Nillni, E. A., Hollenberg, A. N., & Bjorbaek, C. (2003). Role of signal transducer and activator of transcription 3 in regulation of hypothalamic proopiomelanocortin gene expression by leptin. *Endocrinology*, 2121–2131. https://doi.org/10.1210/en.2002-221,037.

Plagemann, A., Harder, T., Brunn, M., Harder, A., Roepke, K., Wittrock-Staar, M., Ziska, T., Schellong, K., Rodekamp, E., Melchior, K., & Dudenhausen, J. W. (2009). Hypothalamic proopiomelanocortin promoter methylation becomes altered by early overfeeding: an epigenetic model of obesity and the metabolic syndrome. *The Journal of Physiology*, 4963–4976. https://doi.org/10.1113/jphysiol.2009.176156.

Plum, L., Lin, H. V., Aizawa, K. S., Liu, Y., & Accili, D. (2012). InsR/FoxO1 signaling curtails hypothalamic POMC neuron number. *PLoS One*, e31487. https://doi.org/10.1371/journal.pone.0031487.

Reich, N. C. (2007). STAT dynamics. *Cytokine & Growth Factor Reviews*, 511–518. https://doi.org/10.1016/j.cytogfr.2007.06.021.

Rubinstein, M., Mortrud, M., Liu, B., & Low, M. J. (1993). Rat and mouse proopiomelanocortin gene sequences target tissue-specific expression to the pituitary gland but not to the hypothalamus of transgenic mice. *Neuroendocrinology, 58*, 373–380.

Sagar, S. M., Sharp, F. R., & Curran, T. (1988). Expression of c-fos protein in brain: metabolic mapping at the cellular level. *Science, 240*, 1328–1331.

Sanacora, G., Kershaw, M., Finkelstein, J. A., & White, J. D. (1990). Increased hypothalamic content of preproneuropeptide Y messenger ribonucleic acid in genetically obese Zucker rats and its regulation by food deprivation. *Endocrinology*, 730–737. https://doi.org/10.1210/endo-127-2-730.

Sanchez, V. C., Goldstein, J., Stuart, R. C., Hovanesian, V., Huo, L., Munzberg, H., Friedman, T. C., Bjorbaek, C., & Nillni, E. A. (2004). Regulation of hypothalamic prohormone convertases 1 and 2 and effects on processing of prothyrotropin-releasing hormone. *The Journal of Clinical Investigation*, 357–369. https://doi.org/10.1172/JCI21620.

Schmale, H., & Richter, D. (1984). Single base deletion in the vasopressin gene is the cause of diabetes insipidus in Brattleboro rats. *Nature, 308*, 705–709.

Stelzer, G., Rosen, N., Plaschkes, I., Zimmerman, S., Twik, M., Fishilevich, S., Stein, T. I., Nudel, R., Lieder, I., Mazor, Y., Kaplan, S., Dahary, D., Warshawsky, D., Guan-Golan, Y., Kohn, A., Rappaport, N., Safran, M., & Lancet, D. (2016). The GeneCards suite: from gene data mining to disease genome sequence analyses. *Current Protocols in Bioinformatics*, 1 30 1–1 30 33. https://doi.org/10.1002/cpbi.5.

Tan, K. M., Ooi, S. Q., Ong, S. G., Kwan, C. S., Chan, R. M., Seng Poh, L. K., Mendoza, J., Heng, C. K., Loke, K. Y., & Lee, Y. S. (2014). Functional characterization of variants in MC4R gene promoter region found in obese children. *The Journal of Clinical Endocrinology and Metabolism*, (5), E931. https://doi.org/10.1210/jc.2013-3711.

Tinkum, K. L., White, L. S., Marpegan, L., Herzog, E., Piwnica-Worms, D., & Piwnica-Worms, H. (2013). Forkhead box O1 (FOXO1) protein, but not p53, contributes to robust induction of p21 expression in fasted mice. *The Journal of Biological Chemistry*, 27999–28008. https://doi.org/10.1074/jbc.M113.494328.

Toorie, A. M., & Nillni, E. A. (2014). Minireview: Central Sirt1 regulates energy balance via the melanocortin system and alternate pathways. *Molecular Endocrinology*, 1423–1434. https://doi.org/10.1210/me.2014-1115.

Valli-Jaakola, K., Palvimo, J. J., Lipsanen-Nyman, M., Salomaa, V., Peltonen, L., Kontula, K., & Schalin-Jantti, C. (2006). A two-base deletion -439delGC in the melanocortin-4 receptor promoter associated with early-onset obesity. *Hormone Research*, 61–69. https://doi.org/10.1159/000093469.

van den Berg, L., van Beekum, O., Heutink, P., Felius, B. A., van de Heijning, M. P., Strijbis, S., van Spaendonk, R., Piancatelli, D., Garner, K. M., El Aouad, R., Sistermans, E., Adan, R. A., & Delemarre-van de Waal, H. A. (2011). Melanocortin-4 receptor gene mutations in a Dutch cohort of obese children. *Obesity (Silver Spring)*, 604–611. https://doi.org/10.1038/oby.2010.259.

Vella, K. R., Burnside, A. S., Brennan, K. M., & Good, D. J. (2007). Expression of the hypothalamic transcription factor Nhlh2 is dependent on energy availability. *Journal of Neuroendocrinology*, 499–510. https://doi.org/10.1111/j.1365-2826.2007.01556.x.

Voisin, S., Almen, M. S., Zheleznyakova, G. Y., Lundberg, L., Zarei, S., Castillo, S., Eriksson, F. E., Nilsson, E. K., Bluher, M., Bottcher, Y., Kovacs, P., Klovins, J., Rask-Andersen, M., & Schioth, H. B. (2015). Many obesity-associated SNPs strongly associate with DNA methylation changes at proximal promoters and enhancers. *Genome Medicine, 103*. https://doi.org/10.1186/s13073-015-0225-4.

Wankhade, U. D., & Good, D. J. (2011). Melanocortin 4 receptor is a transcriptional target of nescient helix-loop-helix-2. *Molecular and Cellular Endocrinology*, 39–47. https://doi.org/10.1016/j.mce.2011.05.022.

Whyte, D. B., Lawson, M. A., Belsham, D. D., Eraly, S. A., Bond, C. T., Adelman, J. P., & Mellon, P. L. (1995). A neuron-specific enhancer targets expression of the gonadotropin-releasing hormone gene to hypothalamic neurosecretory neurons. *Molecular Endocrinology*, 467–477. https://doi.org/10.1210/mend.9.4.7659090.

Yang, G., Lim, C. Y., Li, C., Xiao, X., Radda, G. K., Li, C., Cao, X., & Han, W. (2009). FoxO1 inhibits leptin regulation of pro-opiomelanocortin promoter activity by blocking STAT3 interaction with specificity protein 1. *The Journal of Biological Chemistry*, 3719–3727. https://doi.org/10.1074/jbc.M804965200.

Young, J. I., Otero, V., Cerdan, M. G., Falzone, T. L., Chan, E. C., Low, M. J., & Rubinstein, M. (1998). Authentic cell-specific and developmentally regulated expression of pro-opiomelanocortin genomic fragments in hypothalamic and hindbrain neurons of transgenic mice. *The Journal of Neuroscience, 18*, 6631–6640.

Zheng, J., Xiao, X., Zhang, Q., Yu, M., Xu, J., Wang, Z., Qi, C., & Wang, T. (2015). Maternal and post-weaning high-fat, high-sucrose diet modulates glucose homeostasis and hypothalamic POMC promoter methylation in mouse offspring. *Metabolic Brain Disease*, 1129–1137. https://doi.org/10.1007/s11011-015-9678-9.

Brain Inflammation and Endoplasmic Reticulum Stress

Isin Cakir and Eduardo A. Nillni

4.1 Inflammation

In his medical treatise *De Medicina* dating back two millenniums, Aulus Cornelius Celsus refers to the signs of inflammation as "redness and swelling with heat and pain." Inflammation is a reaction of the body to foreign stimuli and has presumably evolved to restore homeostasis in response to infections, tissue damage, or toxins. In conditions when the basal homeostatic state cannot be restored, persistent inflammatory signals usually lead to a maladaptive state as we observe in diet-induced obesity. Inflammatory process typically involves an inducing factor, such as bacterial infection, which is recognized by sensory molecules, e.g., as toll-like receptors, which then leads to secretion of inflammatory mediators including cytokines, chemokines, and a subclass of eicosanoids called prostaglandins.

I. Cakir
Life Sciences Institute and Department of Molecular and Integrative Physiology, University of Michigan, Ann Arbor, MI, USA

E. A. Nillni (✉)
Emeritus Professor of Medicine, Molecular Biology, Cell Biology & Biochemistry,
Department of Medicine, Molecular Biology, Cell Biology & Biochemistry,
The Warren Alpert Medical School of Brown University, Providence, RI, USA
e-mail: Eduardo_Nillni@Brown.edu

Final stage of the acute inflammatory cycle involves a resolution stage aimed to return to the pre-inflammatory homeostatic boundaries (Serhan et al. 2007), and failure to do so might lead to chronic inflammation. As opposed to acute inflammatory response commonly observed, following tissue injury or bacterial or viral infections, metabolic syndrome, and obesity per se manifest in the form of a chronic low-grade inflammatory state also called para−/meta-inflammation (Medzhitov 2008; Hotamisligil 2017). The inducing factor(s) of this chronic low-grade inflammation is still not entirely clear; however common mediators are involved in the acute and chronic inflammation. In contrast to many other chronic inflammatory conditions where the response is localized to the site of action of the inducing factor (e.g., site of infection), obesity-associated inflammation is manifested at a systemic level incorporating the peripheral as well as central tissues. Low-grade chronic inflammation has been demonstrated in a variety of metabolic tissues, such as the white adipose (Xu et al. 2003), liver (Cai et al. 2005), skeletal muscle (8, 9), and pancreas (Ehses et al. 2008; Donath et al. 2010), and appears to play a causative role in metabolic dysregulations including insulin resistance during obesity. For example, treatment with an anti-inflammatory agent amlexanox, an inhibitor of the NF-κB kinases IKKε and TBK-1, reduces obesity in rodents and improves glucose homeostasis in mice (Reilly et al. 2013) and a subset of diabetics (Oral et al. 2017).

Inflammation in the central nervous system (CNS) also parallels the excess calorie intake and obesity, and this chapter will focus on the key mediators, signaling components, and the various cell types involved in the inflammatory process in the CNS. We will describe genetic and pharmacological approaches utilized to target the inflammatory mediators or signaling pathways involved. A significant number of studies now suggest that obesity and the associated inflammatory state form a positive feedback loop where induction of inflammation exacerbates various abnormalities observed in obesity such as the impaired glucose metabolism (Okin and Medzhitov 2016). We further address the central inflammation in regard to neurons and glial cell populations with specific emphasis on the hypothalamus and regulation of energy homeostasis.

4.1.1 Hallmarks of Inflammation

Classical activators of inflammation, such as lipopolysaccharide (LPS or endotoxin), tumor necrosis factor alpha (TNFα), and interleukin-1 (IL-1), induce a series of physiological outcomes involving fever and anorexia through their central action. Fever is mediated through activation of cyclooxygenase expression, which is targeted – either directly or indirectly – by most commonly used antipyretic (fever reducing) agents including common nonsteroidal anti-inflammatory drugs (NSAID), such as salicylic acid (aspirin) and ibuprofen. Some inflammatory signals derived from the periphery act through their receptors in the brain; however their action is further amplified through local production of other inflammatory mediators. For example, peripheral administration of LPS (endotoxin) triggers hypothalamic expression of pro-inflammatory cytokines including TNFα, IL-6, and IL-1 (Hillhouse and Mosley 1993; Layé et al. 1994; Breder et al. 1994; Wong et al. 1997), and the LPS-induced anorexia is presumably an integrated response to all these factors, which activate an overlapping set of signaling cascades: These include the stress-activated kinase pathways including, but not limited

to, c-Jun N-terminal kinase (JNK) and p38 kinase, canonical and noncanonical nuclear factor kappa B (NF-κB) pathways, and protein kinase R (PKR). Some of these pathways are also utilized by metabolic hormones and important for the regulation of energy balance. For example, sickness and leptin-induced anorexia are mediated in part by acute hypothalamic inflammation and require central NF-κB signaling (Jang et al. 2010). Although LPS also results in a systemic increase in TNFα concentrations, LPS-induced upregulation of central TNFα seems to follow a different mechanism and precedes the rise in circulating TNFα concentration (Sacoccio et al. 1998). Furthermore, there appears to be a functional redundancy on certain phenotypes induced by the central action of the pro-inflammatory cytokines. For example, most effects of systemic LPS administration, including anorexia, hypoglycemia, and activation of the hypothalamus-pituitary-adrenal axis, were intact in IL-1β knockout mice. However, the anorexic response to central administration of LPS still required central IL-1β production (Yao et al. 1999).

A classical hallmark of inflammation is upregulation of the hypothalamic-pituitary-adrenal (HPA) axis. Upon inflammatory challenge, plasma ACTH concentrations rise, which act on the adrenal glands and result in systemic elevated glucocorticoid levels. Glucocorticoids are anti-inflammatory agents and therefore act as a protective negative feedback loop to block persistent HPA activation (Besedovsky and del Rey 1996). Classical pro-inflammatory cytokines such as TNFα, IL-1, and IL-6 as well as LPS stimulate transcription of hypothalamic corticotropin-releasing hormone/factor (CRH/F), which acts on pituitary to induce ACTH release into the circulation. There is a certain degree of overlap in the physiology as well as the mechanism of action of leptin and the pro-inflammatory cytokines. For example, central IL-1 infusion into the VMH results in acute decreases in food intake and body weight in rats (Kent et al. 1994) and mice (Kent et al. 1996), and LPS administration upregulates the immediate early gene c-fos, which has been widely used as a marker for neuronal activation (Sagar and Sharp 1993), in the hypothalamus, the amygdala, and the nucleus tractus solitarius (NTS) (Elmquist et al. 1993). LPS, like leptin, induces the hypothalamic STAT3 phosphorylation (Hosoi et al. 2004), although this is a secondary response to LPS-induced IL-6 upregulation. In addition, LPS-induced anorexia and fever depend in part on circulating leptin (Sachot et al. 2004). Rats lacking leptin receptor signaling are defective in activating the HPA axis upon LPS exposure (Steiner et al. 2004). However, central action of leptin counteracts the fasting-induced activation of HPA axis, such that restoration of fasting-induced decrease in plasma leptin concentrations back to the fed levels blocks the rise in plasma corticosterone and ACTH levels observed upon food deprivation (Ahima et al. 1996).

As discussed below, NF-κB activation is a common hallmark of diet-induced obesity in the hypothalamus and peripheral metabolic tissues (Arkan et al. 2005; Wunderlich et al. 2008; Chiang et al. 2009). The canonical (classical) NF-κB signaling is comprised of a family of five transcription factors (p65 (RelA), RelB, c-Rel, p105/p50 (NF-κB1), and p100/52 (NF-κB2)), which are under normal conditions sequestered in the cytosol by inhibitor of κB (IκB). Phosphorylation of IκB by upstream IκB kinases (IKK) results in its proteasome-mediated degradation. Free NF-κB proteins then translocate to the nucleus to activate the transcription of target genes. Besides classical IKKs (IKKα, IKKβ), there are IKK-related kinases (IKKε, TBK1) that play a critical role in the obesity-induced inflammation. In the adipose tissue and liver of diet-induced obese mice, the activity of these noncanonical (alternate) IKKs increases (Chiang et al. 2009). IKKε knockout mice are protected from diet-induced inflammation and obesity mainly due to elevated energy expenditure (Chiang et al. 2009). Furthermore, treatment of diet-induced knockout mice with an IKKε/TBK1 inhibitor reverses diet-induced obesity and associated complications including glucose metabolism (Reilly et al. 2013). Obesity activates IKKε in the hypothalamus as well (Weissmann et al. 2014). Pharmacological or genetic inhibition of hypothalamic IKKε signaling in obese rodents leads to decreased NF-κB activity and ameliorates central insulin and leptin resistance (Weissmann et al. 2014).

IKKβ/NF-κB pathway plays a key role in neuronal stem cells (NSCs). For example, CNTF, a key factor involved in the maintenance of NSCs (Shimazaki et al. 2001), can suppress food intake and body weight even in leptin-resistant mice through a mechanism that involves neurogenesis in the hypothalamus (Kokoeva et al. 2005). Obese rodents have decreased numbers of hypothalamic NSCs, and inhibition of IKKβ/NF-κB increases their survival (Li et al. 2012). NSC-specific activation of IKKβ/NF-κB in adult mice results in decreased NSC number and increased food intake and weight gain (Li et al. 2012). One of the hallmarks of obesity is induction of a pathophysiology of aging at earlier ages in peripheral as well as central tissues. Hypothalamic markers of inflammation increase as mice get older. Most notably, ablating neuronal IKKβ expression or blocking IKKβ activity specifically in the mediobasal hypothalamus results in extended life-span accompanied by improved cognitive and motor functions, whereas activation of NF-κB induces an opposite phenotype and results in mice with shorter life-span (Zhang et al. 2013a). Most of these phenotypes could be mimicked by microglia-specific IKKβ inhibition as well, suggesting the importance of non-neuronal cells in hypothalamic inflammation and associated metabolic abnormalities.

Inflammatory ligands in the form of foreign pathogens or endogenous signals are typically recognized by pattern recognition receptors (toll-like receptors, TLRs) expressed on the cell surface or intracellular membranes such as ER. The ligands recognized by different TLRs have been characterized; for example, TLR4 acts as the receptor for LPS. TLR activation classically leads to NF-κB activation and induces the expression of interleukin-1 (IL-1) family of cytokines (IL-1 and IL-18), which are processed from their pro-interleukin precursors to mature forms by active caspase 1. The canonical pathway of caspase 1 activation involves inflammasomes (Martinon et al. 2002), large protein complexes assembled by innate immune receptors called nucleotide-binding domain leucine-rich repeat-containing receptors (NLRs), absent in melanoma

2 (AIM2) and pyrin (Stutz et al. 2009; Broz and Dixit 2016). Glial cells in the CNS including microglia and astrocytes express TLRs (Lehnardt 2010), and inflammasome components are constitutively expressed in the brain (Yin et al. 2009), where they act as major regulators of central inflammation (Song et al. 2017). For example, the microglial IL-1β production observed in Alzheimer's disease is mediated by amyloid beta recognized by the inflammasome NALP3 (also called NLRP3), which leads to caspase 1 activation and the processing of pro-IL-1β to active IL-1β (Halle et al. 2008). Inflammasome components are also detected in neurons (Kummer et al. 2007) and seem to play essential roles during conditions such as brain injury (de Rivero Vaccari et al. 2008), headache (Silverman et al. 2009; Karatas et al. 2013), and response to certain viral infections (Ramos et al. 2012). The overall clinical importance of inflammasomes extends beyond infections, and its dysregulation plays a key role in peripheral as well as central inflammation associated with metabolic syndrome including obesity. For example, NLRP3 ablation protected mice from age-associated central inflammation and astrogliosis, improved glucose homeostasis, attenuated bone loss and thymus dysfunction (Youm et al. 2013), and enhanced the obesity-associated defects in insulin signaling (Vandanmagsar et al. 2011). We next discuss in more detail the inflammation of the nervous system or "neuroinflammation."

4.1.2 Central Nervous System (CNS) Components of the Inflammatory Process

4.1.2.1 Neurons Are Not Alone in the CNS

In order to understand the CNS inflammation, it is important first to get a broad understanding of the cellular architecture of the CNS as term neuroinflammation extends beyond the neurons. Majority of the cells in the CNS are non-neuronal cells called glia. There are three glial cell population: astrocytes, microglia, and oligodendrocytes. The main function of these cells is to maintain the homeostasis in the nervous system. Astrocytes

send projections (called astrocytic end feet) to meet the blood capillaries and have traditionally been viewed as a gate between the circulation and the rest of the CNS. For example, the classical view of CNS glucose utilization has been the import of glucose from the bloodstream by the astrocytes to be first oxidized to lactic acid, which is then taken up by neurons and used as the energy source. This indirect route of glucose utilization, called the astrocyte-neuron lactate shuttle, was reported to be key to the central regulation of glucose metabolism (Lam et al. 2005). Likewise, astrocytes can oxidize fatty acids to ketones, which could then be used by neurons as energy source (Le Foll and Levin 2016). Although recent findings suggest that neurons might play a more direct role in nutrient utilization (Lundgaard et al. 2015), astrocytes play an indispensible role in the central regulation of energy homeostasis: They express the receptors for key metabolic factors including leptin, insulin, and ghrelin (Kim et al. 2014; García-Cáceres et al. 2016; Frago and Chowen 2017). Astrocytic insulin signaling is required for brain glucose uptake and systemic glucose homeostasis (García-Cáceres et al. 2016). By regulating extracellular adenosine concentration, astrocytes regulate AgRP neuronal activity and actively participate in the regulation of feeding (Yang et al. 2015). Leptin induces astrocyte proliferation in the hypothalamus (Rottkamp et al. 2015), and removal of leptin receptor from astrocytes attenuates the response to leptin-induced anorexia while potentiating ghrelin-induced food intake (Kim et al. 2014).

Microglia are resident macrophages of the CNS derived from erythromyeloid precursors in the yolk sac (Kierdorf et al. 2013). Their function includes maintenance as well as the plasticity of synaptic circuitry, synaptic transmission, neuronal surveillance, and neurogenesis (Walton et al. 2006; Paolicelli et al. 2011; Pascual et al. 2012; Wake et al. 2013). Rodents are estimated to have between 3 and 4 million microglia. Their distribution is not uniform throughout the brain, with more microglia detected in the hippocampus, olfactory telencephalon, basal ganglia, and substantia nigra than regions such as the cerebellum and brain stem (Lawson et al. 1990). Contrary to differentiated adult neurons, microglia can proliferate. Earlier studies have suggested that a source of resident microglia was monocytes that are recruited from the periphery, which in turn could differentiate into resident microglia (Lawson et al. 1992). However, recent studies show that postnatal hematopoiesis makes a marginal contribution to steady-state resident microglial cell population (Ginhoux et al. 2010). Microglia development and survival depend on a receptor tyrosine kinase called colony-stimulating factor 1 receptor (CSF1R) (Ginhoux et al. 2010; Erblich et al. 2011) whose inhibition leads to complete microglial depletion in adult mice. However, in adult mice, nestin-positive latent microglial progenitors are capable of renewing the CNS microglia pool (Elmore et al. 2014). The third glial population is comprised of the oligodendrocytes (ODs). These cells mainly function to provide electrical insulation (myelination) to neurons. Besides this fundamental role, ODs produce a variety of neurotrophic and growth factors such as BDNF and IGF-1 to promote neuronal survival (Wilkins et al. 2001, 2003; Du and Dreyfus 2002). Another glial population is NG2 glia, which are also called polydendrocytes or oligodendrocyte precursors (Nishiyama et al. 2009). These cells are characterized by their expression of a single membrane-spanning chondroitin sulfate proteoglycan, called NG2, and are able to differentiate into myelinating oligodendrocytes and engage in tissue repair (Hughes et al. 2013). Accordingly, NG2 glia gain a hypertrophic characteristic upon demyelinating insults (Keirstead et al. 1998; Di Bello et al. 1999; Levine and Reynolds 1999). In the hypothalamus, particularly in the circumventricular organ the median eminence (ME), NG2 glia play a crucial role in sensing circulating leptin (Djogo et al. 2016). Leptin receptor-positive neurons have their dendritic projections in the ME in close proximity of NG2 glia. Following depletion of NG2 glia, the ARC neurons lose their response to leptin, which in turn triggers obesity (Djogo et al. 2016). This finding has more imminent consequences for humans as NG2 depletion upon cranial X-ray irradiation was also proposed to contribute to weight gain in humans (Djogo et al. 2016).

The resident astrocytes and microglia are the innate immune cells of the CNS, and there exists a dynamic interaction between them including during central inflammation. Activated microglia can induce a subpopulation of astrocytes by secreting inflammatory mediators including IL1α, TNF, and C1q, which in turn blunts the neuroprotective capacity of the astrocytes leading to neuronal death (Liddelow et al. 2017). Contrary to the periphery, central innate immune response cannot initiate adaptive immunity: In the peripheral innate immune cells (e.g., dendritic cells), foreign antigens are typically presented by MHC class II molecules on the cell surface to T cells to trigger an adaptive immune response. However, this process is not adapted by the glia. This is consistent with the anti-inflammatory environment of the brain parenchyma as well as the physical limitations induced by the blood-brain barrier (BBB) that would block migration/communication of innate and adaptive immune cell components. However, it is worth to mention that the notion that brains are devoid of adaptive immunity is not absolute. For example, dendritic cell-like properties have been reported in meninges and choroid plexus in healthy mice (Anandasabapathy et al. 2011), and a rare expression pattern of MHCs exists in neurons (Neumann et al. 1995). Surprisingly, recent evidence suggests that inflammatory signals regulate peripheral adaptive immunity through pathways including their central actions. For example, hypothalamic responses to TNFα result in increased sympathetic tone to white adipose tissue. As a result, increased lipolysis produces long-chain fatty acids, which mediate the accumulation of the cells of adaptive immunity, the lymphocytes, in spleen and adipose tissue (Kim et al. 2015).

4.1.2.2 Hypothalamic Inflammation and Energy Homeostasis

As explained above, obesity manifests itself as a chronic low-grade inflammatory state. Increase in the adipose tissue expression and plasma levels of the pro-inflammatory cytokine TNFα in obesity was first reported in 1993 in rodents (Hotamisligil et al. 1993). TNFα inhibits insulin signaling by inducing an inhibitory phosphorylation on insulin receptor substrate 1 (IRS1) (Hotamisligil et al. 1994, 1996), an insulin receptor-interacting protein. Accordingly, TNFα knockout mice are protected from obesity-induced insulin resistance (Uysal et al. 1997). Likewise, macrophage-specific ablation of c-Jun N-terminal kinase (JNK) blocks pro-inflammatory macrophage polarization, decreases their adipose tissue infiltration, and confers protection from obesity-induced insulin resistance (Han et al. 2013). Markers of inflammation in obesity are observed in several peripheral tissues including the liver, adipose tissue, and muscle (Schenk et al. 2008). Hypothalamus is no exception; high-fat feeding results in increased activation of JNK and NF-κB pathways in the hypothalamus (De Souza et al. 2005); results in elevated expression of pro-inflammatory mediators in the hypothalamus, including IL-1β and TNFα; and attenuates insulin, leptin, and ghrelin (Naznin et al. 2015) signaling. Hypothalamic PTP1B expression, a negative regulator of insulin and leptin receptor signaling, increases in diet-induced obesity; and this response can be mimicked by TNFα administration to lean healthy mice (Bence et al. 2006; Zabolotny et al. 2008). Central injection of neutralizing antibodies targeted against TLR4 or TNFα in obese rats reduces markers of hypothalamic inflammation and improves hepatic steatosis and gluconeogenesis through parasympathetic output to the liver (Milanski et al. 2012). Furthermore, reducing hypothalamic inflammation by neuronal-, glial-, or hypothalamus-specific deletion of IKKβ in mice confers protection from diet-induced obesity (Zhang et al. 2008; Valdearcos et al. 2017).

The hypothalamic inflammation observed in obesity is temporally different than peripheral inflammation. While peripheral inflammation is thought to develop secondary to diet-induced adiposity, rapid changes in the hypothalamic inflammatory signaling cascades could be observed within only 1 day following high-fat diet consumption. For example, rodents display elevated markers of inflammation (Thaler et al. 2012) and insulin resistance in the hypothalamus-brain circuitry (Ono et al. 2008) within 24 h of exposure to high-fat diet, much earlier than accumulation of adiposity. HFD-induced proliferation and activation of glia in the hypothalamus seem to be important for the accompanying peripheral inflammation such that blocking central cell proliferation attenuates not only the central but also the peripheral inflammation (André et al. 2017). The central inflammation observed upon high-fat diet consumption, at least in part, depends on the direct effect of saturated fatty acids on the hypothalamus rather than the total calories consumed, and central infusion of palmitate, a saturated fatty acid, to lean rats mimics the hypothalamic insulin resistance and IKKβ activation observed in rats exposed to high-fat diet (Posey et al. 2009). Hypothalamic inflammation could be stimulated by administration of saturated fatty acids by intragastric gavage for 3 days; however, the same treatment does not induce systemic inflammation. In parallel with their central effects in vivo, saturated fatty acids induce endoplasmic reticulum (ER) stress in neuronal cultures (Mayer and Belsham 2010; Choi et al. 2010). However, their pro-inflammatory effects were attenuated in cultured hypothalamic neurons, suggesting a role for non-neuronal cells in the fatty acid-induced central inflammation (Choi et al. 2010). Consumption of high-calorie diet results in accumulation of saturated fatty acids in the hypothalamus (but not the cerebral cortex) and activates the microglia in the ARC (Valdearcos et al. 2014). Diet-induced hypothalamic gliosis, activation of the glial population, is also reported in humans (Thaler et al. 2012). At least in high-fat diet-fed rodents, some of the hypothalamic gliosis is also attributed to monocyte-derived non-microglial myeloid cells recruited by resident microglia (Berkseth et al. 2014). Furthermore, the detected inflammation was restricted to hypothalamic microglia, but not the astrocytes, although both cell population accumulate in the mediobasal hypothalamus following high-fat diet consumption. Accordingly, saturated fatty acids induce secretion of inflammatory mediators, including TNFα, IL-6, and monocyte chemoattractant protein-1 (also called CCL2), from primary microglia but not primary astrocytes (Valdearcos et al. 2014), and hypothalamic infiltration of astrocytes following high-fat diet does not seem to blunt leptin responsiveness (Balland and Cowley 2017). Astrocytes also express some components of the inflammasome such as NLRP2 (Minkiewicz et al. 2013) but not others (e.g., NLRP3) (Gustin et al. 2015), and microglia appear to be the main inflammatory glial population involved in CNS (Gustin et al. 2015). The hypothalamic microglial inflammation is accompanied by elevated markers of neuronal injury and decreased cognitive function, and this was reported in both rodents and humans (Thaler et al. 2012; Valdearcos et al. 2014; Puig et al. 2015). Saturated fatty acids induce hypothalamic inflammation mainly acting through TLR4 (Milanski et al. 2009), and in line with above findings, activation of microglial TLR4 by LPS induces neurodegeneration (Lehnardt et al. 2003). Upon microglial depletion, the response of the CNS to LPS challenge is compromised for certain pro-inflammatory markers including IL-1β and TNFα (Elmore et al. 2014). Microglia also express TLR2, which was proposed to acutely mediate the sickness-induced anorexia and POMC activation (Jin et al. 2016), at least in part through microglia-derived TNFα release (Yi et al. 2017). Although the inflammatory pathways might contribute to the regulation of energy balance by anorectic hormones including leptin (Jang et al. 2010), these effects seem to be acute, and blocking TNFα signaling in the mediobasal hypothalamus results in attenuation of diet-induced weight gain (Yi et al. 2017). Depleting microglia by CSF1R antagonism augmented leptin signaling and suppressed food intake, suggesting a negative role of microglial activation on central leptin action (Valdearcos et al. 2014). When mice are fed with HFD supplemented with the CSF1R antagonist, the food intake and weight gain on the mice are attenuated (Valdearcos et al. 2017). Likewise, blocking microglial activation by specific depletion of IKKβ from microglia confers resistance to high-fat diet-induced weight gain in part by reducing food intake (Valdearcos et al. 2017). When hypothalamic gliosis is induced by specifically activating the microglial NF-κB pathway, the transgenic mice had increased food intake and weight gain and reduced energy expenditure (Valdearcos et al.

2017). Considering that microglial depletion in adult mice does not result in overt behavioral phenotypes or cognitive abnormalities (Elmore et al. 2014), CSF1R antagonism might be a promising anti-obesity mechanism. Astrocyte activation is also observed upon acute HFD challenge. Surprisingly, blocking HFD-induced astrocyte activation by astrocyte-specific inhibition of NF-κB results in increased food intake, although this response is very acute and does not persist upon continuous HFD feeding (Buckman et al. 2015). In obesity as well as aging, elevated TGFβ production by the astrocytes impairs glucose homeostasis by acting on POMC neurons and inducing an inflammatory response (Yan et al. 2014). Diet-induced hypothalamic inflammation and gliosis are not permanent in rodents and could largely be reversed upon exercise (Ropelle et al. 2010; Yi et al. 2012a) or weight loss (Berkseth et al. 2014). In humans, at least one study showed that hypothalamic inflammation in obese human MBH was not reversed following gastric bypass-induced weight loss and elevated insulin sensitivity (Kreutzer et al. 2017). More research is certainly warranted to dissect the potential differences between humans and rodents in regard to the kinetics as well as the reversible nature of diet-induced hypothalamic inflammation.

Neurons are rather resistant to saturated fatty acid-induced inflammation compared to glia (Choi et al. 2010). There is relatively higher expression of IKKβ and IκBα in the hypothalamus compared to peripheral tissues including the liver, skeletal muscle, adipose tissue, and kidney (Zhang et al. 2008). In diet-induced obese mice, IKKβ-NF-κB signaling gets activated in the neurons located in the mediobasal hypothalamus (MBH) (Zhang et al. 2008). Neuronal activation of IKKβ in MBH neurons stimulates food intake and weight gain, while a dominant negative IKKβ reverses both parameters (Zhang et al. 2008). AgRP-specific IKKβ knockout mice also eat less and gain less weight on HFD (Zhang et al. 2008), and activating IKKβ specifically in AgRP neurons impairs glucose homeostasis without changing body weight (Tsaousidou et al. 2014). Activation of the inflammatory signals and ER stress are mechanistically coupled at least through NF-κB signaling (Zhang et al. 2008). Active IKKβ triggers ER stress, which in turn leads to further activation of IKKβ, resulting in a positive feedback loop to attenuate central insulin and leptin signaling leading to positive energy balance (Zhang et al. 2008). Besides overnutrition, inhibition of hypothalamic autophagy also leads to IKKβ activation and hypothalamic inflammation, increased food intake, and decreased energy expenditure (Meng and Cai 2011). Increased hypothalamic IKKβ-NFκB signaling decreases the number of hypothalamic neural stem cells, results in dysregulated neurogenesis (Li et al. 2012), inhibits gonadotropin-releasing hormone (GnRH) expression (Zhang et al. 2013a), and triggers hypertension (Purkayastha et al. 2011a). For example, increased MBH-specific IKKβ activation increases blood pressure, and POMC-specific IKKβ KO mice are protected from obesity-induced hypertension (Purkayastha et al. 2011a). Consequently, neuronal IKKβ KO mice exhibit extended life-span (Zhang et al. 2013a).

Prominent role of NF-κB signaling in the regulation of energy balance is further emphasized by studies utilizing glucocorticoids (GC). GCs are potent anti-inflammatory molecules, secreted from the adrenal glands typically in response to stress-induced activation of the HPA axis. GCs are steroids and act through the intracellular glucocorticoid receptor (GR). GR directly interacts with p65 (Rel A) (Ray and Prefontaine 1994) and inhibits NF-κB signaling by increasing the nuclear export rate of p65 (Nelson et al. 2003) and increasing IkB expression level (Deroo and Archer 2001). Targeting a GR agonist, dexamethasone (Dexa), to glucagon-like peptide-1 (GLP-1) expressing cells by using a GLP-1/Dexa conjugate results in decreased hypothalamic and systemic inflammation, improves glucose homeostasis, and decreases body weight in DIO mice (Quarta et al. 2017). GLP-1 receptor (GLP1R) is mostly expressed by neurons in human and rodent brain. It is also expressed by the immune cells in the circulation. Accordingly, the metabolic improvements observed upon GLP-1/Dexa treatment are mediated by both central and peripheral GLP1R-expressing cells (Quarta et al. 2017).

JNKs are typically activated by high-fat feeding (Prada et al. 2005), and genetic studies have indicated the ablation of peripheral JNK activity to be protective from diet-induced metabolic abnormalities. JNK1 global KO mice, but not JNK2 KO mice, are protected from diet-induced obesity (Hirosumi et al. 2002). Adipose tissue-specific ablation of JNK1 does not alter body weight or adiposity but confers significant protection from hepatic insulin resistance (Sabio et al. 2008). Similar protection from peripheral insulin resistance is observed upon deletion of JNK1 and JNK2 from macrophages (Han et al. 2013) or skeletal muscle (Sabio et al. 2010b). Although liver-specific JNK1 KO mice were reported to exhibit hepatic insulin resistance, glucose intolerance, and steatosis (Sabio et al. 2009), deletion of both JNK1 and JNK2 from hepatocytes significantly improves systemic glucose homeostasis and insulin signaling (Vernia et al. 2014).

In the CNS, JNKs are essential for proper brain development (Kuan et al. 1999; Chang et al. 2003; Amura et al. 2005). JNK2 and JNK3 deletion confers protection from various forms of neurodegeneration including cerebral ischemic hypoxia (Kuan et al. 2003; Hunot et al. 2004). In JNK1 KO mice, glucocorticoid-induced feeding was exacerbated (Unger et al. 2010); however the animals exhibit increased central insulin sensitivity (Unger et al. 2010). Accordingly, neuron-specific ablation of JNK1 results in increased central as well as peripheral insulin sensitivity, reduced hepatic steatosis, and decreased body weight in diet-induced obese mice, without altering central leptin sensitivity (Kleinridders et al. 2009; Sabio et al. 2010a; Belgardt et al. 2010). Most of these effects were due to decreased food intake and elevated HPT axis activity (Sabio et al. 2010a). Accordingly, disrupting HPT axis by blocking thyroxine production attenuated the protective effect of neuronal JNK1 depletion from diet-induced weight gain (Sabio et al. 2010a). Triiodothyronine treatment induced hypothalamic JNK1 activity, which in turn led to hepatic steatosis (Martínez-Sánchez et al. 2017).

AgRP-specific JNK1 activation blunted leptin signaling specifically in AgRP neurons, increased AgRP neuronal firing, and consequently induced weight gain (Tsaousidou et al. 2014) without significant alterations in glucose homeostasis.

Role of JNK3 in feeding and body weight regulation is different than the other isoforms. HFD activated hypothalamic JNK3 phosphorylation, and embryonic deletion of JNK3 induces food intake, significantly exacerbates weight gain, increases adipose tissue inflammation, and impairs systemic glucose homeostasis specifically on HFD (Vernia et al. 2016). These phenotypes were largely mimicked when JNK3 was ablated from leptin receptor-positive cells or specifically from AgRP neurons (Vernia et al. 2016). However, POMC-specific JNK3 KO mice had normal feeding and glucose homeostasis (Vernia et al. 2016). Despite these findings based on genetic studies, central or peripheral administration of a JNK2/3-specific inhibitor suppressed food intake and resulted in robust weight loss (Gao et al. 2017).

TLR4 acts as the endogenous receptor for LPS and saturated fatty acids and plays a critical role in the lipid-induced insulin resistance in the periphery (Shi et al. 2006). In the CNS, TLR4 is expressed predominantly in microglia (Chakravarty and Herkenham 2005) and plays a critical role for fatty acid-induced hypothalamic inflammation (Milanski et al. 2009). Adult neuronal progenitor cells also express TLR4, and its absence promotes neuronal differentiation (Rolls et al. 2007). Hypothalamic TLR4 signaling has also been implicated in the etiology of metabolic syndrome. HFD-induced obesity triggers neuronal expression of a chemokine, CX3CL1, in the hypothalamus and mediates the recruitment of peripheral monocytes and induction of hypothalamic inflammation (Morari et al. 2014). TLR4 was proposed to also interact with resistin, an adipokine that negatively regulates insulin signaling (Benomar et al. 2013), such that the central effects of resisting are blocked in TLR4 knockout mice (Benomar et al. 2016). Hypothalamic TLR4 also appears to have protective effects such that TLR4 blocks excess apoptotic neuronal death upon HFD feeding (Moraes

et al. 2009). Acting in the paraventricular nucleus of the hypothalamus, TLR4 signaling mediates the obesity-induced increase hypertension (Dange et al. 2015; Masson et al. 2015). MyD88 knockout mice are resistant to LPS- or IL-1β-induced anorexia but not weight loss (Ogimoto et al. 2006; Yamawaki et al. 2010). Neuron-specific ablation of MyD88 blocked high-fat diet-induced ARC IKKβ activation (but not JNK1 activation) and conferred partial protection from fatty acid-induced leptin resistance and weight gain (Kleinridders et al. 2009).

Diet-induced obesity and associated inflammation have been shown to alter the structure and permeability of the BBB (Mauro et al. 2014; Varatharaj and Galea 2017). Certain regions of the CNS involved in the regulation of energy homeostasis lie outside the BBB; these include, among other regions, the ME and area postrema (Maolood and Meister 2009; Riediger 2012). The barriers between the CNS-resident cells and circulation have a fundamental role in the regulation of energy homeostasis during normal physiology as well as pathophysiological conditions including inflammation. Permeability of BBB is not rigid especially at the blood-hypothalamus barrier. A specialized and modified ependymal microglial population called tanycytes line the ventricles and form a barrier between the cerebrospinal fluid and circulation (Langlet et al. 2013b). Tanycytes are critically important in the sensing of circulating factors, such as leptin by hypothalamic neurons (Balland et al. 2014), while actively participating in neurogenesis (Lee et al. 2012). During fasting, decreased blood glucose concentration triggers the release of a permeability factor (vascular endothelial growth factor-A, VEGF-A) from tanycytes (Langlet et al. 2013a). Tanycytes are glucose-sensitive cells; their stimulation by glucose evokes ATP-mediated Ca^{+2} responses, including ATP release (Frayling et al. 2011). VEGF-A results in increased access of the ARC to circulating factors during fasting compared to other regions. However, astrocytes also express VEGF-A and contribute to pathologically increased BBB permeability and lymphocyte infiltration into the brain parenchyma during neuroinflammatory diseases (Argaw et al. 2012). IL-1β administration increases BBB permeability, probably by altering the tight junction profile of the endothelium, and its effect is amplified by other inflammatory mediators such as TNFα (Quagliarello et al. 1991; Nadeau and Rivest 1999; Blamire et al. 2000; Beard et al. 2014). Upon exposure to HFD, there is elevated degeneration in the endothelium of the hypothalamic BBB vasculature, which leads to infiltration and population of the ARC by IgG-type antibodies, and a pathological increase in blood vessel length and density in the ARC (Yi et al. 2012b, c). Notably, similar findings have been observed in the hypothalami of diabetic patients (Yi et al. 2012b).

While it is not going to be extensively discussed in this chapter, it is important to note that central inflammation is not restricted to the hypothalamus and extends to other brain regions. Diet-induced weight gain increases inflammatory markers in the cortex, amygdala, cerebellum, and brain stem and is overall associated with anxiety, depression, cognitive defects, and hypertension (Russo et al. 2011; Wu et al. 2012; Speretta et al. 2016; Guillemot-Legris et al. 2016; Carlin et al. 2016; Almeida-Suhett et al. 2017; Spencer et al. 2017). Hippocampus is another site where neuroinflammation is associated with high-fat diet, where increased TLR4 expression is detected (Dutheil et al. 2016). Hippocampal inflammation acts as a causative factor for high-fat diet-induced anxiety such that blocking the activation of inflammasome attenuates the anxious phenotype in rodents. Obesity and inflammatory cytokines are associated with a decline in cognitive functions (Yehuda et al. 2005; Gunstad et al. 2006; Nguyen et al. 2014) and impaired neurogenesis in the hippocampus (Lindqvist et al. 2006; Park et al. 2010; Chesnokova et al. 2016). The observed cognitive decline appears to require microglial activation associated with decreased BDNF levels (Pistell et al. 2010). Anti-inflammatory manipulations including pharmacological interventions or exercise increase cognitive function in rodents and humans (Jeon et al. 2012; Kang et al. 2016; Veronese et al. 2017; Pérez-Domínguez et al. 2017; Wang et al. 2017). For example, treadmill exercise decreases

hippocampal microgliosis in obese mice. Furthermore, blocking hippocampal IL-1β signaling reverses the defects in synaptic plasticity and cognitive decline (Erion et al. 2014).

Genetic and pharmacological studies targeting different receptors or intracellular signaling cascades of the inflammatory molecules have revealed that, as opposed to their acute effects, inflammatory molecules in the CNS contribute to a positive energy balance, and their inhibition contributes to weight loss and improves systemic glucose metabolism. Diet-induced obesity and aging display similar signs of neuroinflammation, neurodegeneration, and cognitive decline, some of which could be blocked or in some cases reversed by noninvasive interventions in lifestyle including a healthy diet and exercise. However it remains a reality that obesity has reached epidemic proportions throughout the world, which increases the demand on pharmacological interventions to combat obesity and associated metabolic abnormalities. Studies obtained in rodents and in part in humans collectively suggest that targeting central inflammation might present promising approaches.

4.2 Endoplasmic Reticulum Stress and Unfolded Protein Response

Endoplasmic reticulum (ER) is a multifaceted membranous structure extending from the outer nuclear membrane to the cytoplasm and is involved in several biological processes including protein folding, secretion, lipid biosynthesis, calcium storage, and posttranslational modifications such as glycosylation and lipidation. About one third of the human proteome including the secreted, ER-resident, and membrane proteins go through ER. Depending on the tissue, representation of the secreted and membrane proteins in the total proteome greatly varies. For example, the majority of pancreatic proteome is composed of secretory proteins, which however constitute a small percentage of the skeletal muscle proteome. The intracellular expression profile of the chaperones has evolved to match the nature of the proteome of the corresponding tissue: While pancreas and liver, tissues with higher secreted and/or membrane protein pools, have higher ER chaperone expression, tissues such as skeletal muscle and skin, where the soluble proteins constitute the majority of the proteome, have higher expression of cytosolic chaperones. Accordingly, of more than 300 proteins in the human chaperone proteome (chaperome), 48 are ER-specific proteins (Brehme et al. 2014). Together with the ER-associated degradation (ERAD) pathways, the ER chaperones coordinate the folding capacity of the ER to regulate ER protein homeostasis. When the ER protein load exceeds the folding capacity, a condition called ER stress, a series of signaling pathways collectively called the unfolded protein response – UPR – is activated. The primary function of UPR is to restore the ER homeostasis by informing the cytosol and nucleus about the excess protein load to activate a series of protective mechanisms. Among these responses are global shutdown of the protein synthesis while specifically activating the translation of chaperone proteins, elevated membrane synthesis to increase the ER volume to remodel the secretory apparatus, and degradation of unfolded proteins through either ubiquitin-proteasome pathway (ER-associated degradation) or lysosomes (autophagy). While UPR might achieve in reinstating the ER homeostasis, persistent ER stress leads to apoptotic cell death.

We discuss below the signaling components of UPR and the results of genetic studies designed to study the role of individual UPR components in metabolism. Before that, we find it helpful to briefly summarize some of the chemical tools employed to study the biology of ER stress. Some of the commonly used chemical agents known to induce ER stress are tunicamycin, thapsigargin, and brefeldin A (BFA). Tunicamycin is an antifungal and antibiotic nucleoside, which acts as an inhibitor of N-acetylglucosamine transferases to block N-linked glycosylation (Takatsuki and Tamura 1971). Resulting defective protein folding triggers ER stress. Thapsigargin is an inhibitor of the ER Ca^{2+} importer called sarco−/endoplasmic reticulum Ca^{2+} ATPase (SERCA). Upon thapsigargin treatment, cytosolic Ca^{2+} con-

centration rises, while ER Ca^{2+} concentration gets depleted. There is a significant difference between the cytosolic and ER Ca^{2+} concentrations (300 μM in the ER vs. 5–50 μM in the cytosol) (Pozzan et al. 1994). Ca^{2+} is important for protein-protein interactions and ER chaperone function, thereby affecting protein synthesis, folding, and posttranslational modifications (Corbett et al. 1999). BFA is an antiviral lactone, which blocks ER-Golgi transport (Helms and Rothman 1992), resulting in accumulation of the polypeptides in the ER, and triggers ER stress. Another ER stress inducer, homocysteine, is a nonprotein cysteine analog and triggers the expression of ER stress-responsive genes, including GRP78, and decreases the expression of antioxidant enzymes, without inducing heat shock response (Outinen et al. 1998). There is also a set of other compounds, called chemical chaperones, which help stabilize the native conformation of proteins, which are now extensively used in research to alleviate ER stress (Welch and Brown 1996). 4-phenyl butyric acid (4PBA) and tauroursodeoxycholic acid (TUDCA) are two chemical chaperones commonly used in biomedical research. Both compounds are pleiotropic agents: While 4PBA also acts as a broad-spectrum HDAC inhibitor (Bora-Tatar et al. 2009), TUDCA is a bile acid that acts as a ligand for the G-protein-coupled receptor TGR5 (also called G-protein-coupled bile acid receptor 1), which is expressed in BAT, microglia, and other tissues (Kawamata et al. 2003; Yanguas-Casás et al. 2017), and acts as an anti-inflammatory molecule.

UPR is classically composed of three arms that are each initiated by distinct ER membrane proteins: PERK (protein kinase RNA (PKR)-like ER kinase) (Harding et al. 1999), IRE1 (inositol-requiring protein-1) (Morl et al. 1993; Cox et al. 1993), and ATF6 (activating transcription factor 6) (Haze et al. 1999). These proteins function as stress sensors that detect the ER protein load and induce adaptive responses aimed to bring the ER folding capacity back to homeostatic boundaries.

PERK's lumenal domain is a stress sensor, whereas its cytosolic domain has enzymatic activity for its substrate eIF2α (eukaryotic initiation factor 2 subunit α). The other known target of PERK is a transcription factor nuclear erythroid-related factor 2 (Nrf2), the master regulator of the antioxidant response that is involved in the PERK-mediated cell survival in stressed cells (Cullinan et al. 2003). The enzyme(s) that dephosphorylate and inactivate PERK are currently unknown. eIF2α dephosphorylation is accomplished by the phosphatase PP1 through its association with either the growth arrest and DNA damage-inducible protein GADD34 (also called PPP1R15A) (Novoa et al. 2001; Lee et al. 2009; Rojas et al. 2015) or PPP1R15B (Jousse et al. 2003). GADD34 itself is an ER stress-inducible gene acting in a negative feedback loop, whereas PPP1R15B is a constitutive repressor of eIF2α phosphorylation.

PERK is one of four eIF2α kinases (other being GNC2, PKR, HRI) that halt translation in response to different stimuli (Donnelly et al. 2013). Besides ER stress, eIF2α phosphorylation could be triggered by amino acid starvation, double-stranded RNA accumulation, oxidants, and heme depletion. Amino acid deficiency (as well as UV light) is sensed by GCN2 (general control nonderepressible 2, also called EIF2AK4). PKR (protein kinase R, also known as RNA-activated or double-stranded RNA-activated protein kinase, EIF2AK2) responds to viral infections, and HRI (heme-regulated eIF2α kinase, EIF2AK1) senses heme deficiency. Depletion of all four eIF2α kinases in mouse embryonic fibroblasts results in complete loss of eIF2α phosphorylation in response to various stress stimulations (Taniuchi et al. 2016). All these stimuli merge on eIF2α and regulate the same downstream pathways and collectively constitute the integrated stress response (ISR) (Pakos-Zebrucka et al. 2016). In unstressed cells, PERK appears as a monomer bound to the major ER chaperone GRP78 (78 kDa glucose-regulated protein also known as binding immunoglobulin protein-BiP). GRP78 is the primary chaperone in the ER lumen that interacts with the translocating nascent polypeptides. Accumulation of unfolded proteins in the ER lumen triggers the dissociation of the GRP78-PERK complex, which results in PERK homodimerization followed by trans-autophosphorylation. Therefore, GRP78 dissoci-

ation couples the ER protein load (or ER stress) to PERK dimerization and activation. PERK phosphorylates eIF2α at Ser51 (Harding et al. 1999), which results in the GDP-bound, inactive form of eIF2α. While eIF5 inactivates eIF2α by hydrolyzing GTP to GDP, eIF2B acts as the guanine nucleotide exchange factor for eIF2α. Phosphorylation status and duration of eIF2α have significant impact in mammalian physiology. An activator of eIF2B called ISRIB, e.g., acts as a chemical inhibitor of the ISR and blocks eIF2α phosphorylation, and has been implicated in increased cognitive function in rodents (Sidrauski et al. 2013). GTP-bound eIF2α interacts with the initiator methionyl-tRNA, which recognizes the start codon for initiation of translation. Therefore, eIF2α phosphorylation links ER proteostasis to the regulation of global protein synthesis. Translational attenuation, however, is not the only function of eIF2α. Induction of ER stress by an inhibitor of glycosylation (tunicamycin) in PERK knockout cells results in accumulation of endogenous peroxides (Harding et al. 2003). A list of mRNAs whose transcription is linked to eIF2α phosphorylation has been characterized and includes genes regulating amino acid sufficiency and resistance to oxidative stress (Harding et al. 2003). Accordingly, there are selective mRNAs whose translation actually increases upon eIF2α phosphorylation (Harding et al. 2000). For example, ATF4 (activating transcription factor 4) mRNA translation is elevated following eIF2α phosphorylation by PERK (Harding et al. 2000), which then induces the expression of genes involved in amino acid and cholesterol metabolism, glutathione biosynthesis, resistance to oxidative stress, and the pro-apoptotic transcription factor CHOP (Harding et al. 2000, 2003; Fusakio et al. 2016).

The mechanism of translational control of the mRNAs whose translation increases by ER stress represents a prominent example of adaptation. ATF4, e.g., has two upstream open reading frames (uORF), uORF1 and uORF2, which are followed by the ATF4 coding sequence. Under ER stress or other conditions that result in phosphorylated (thus inactive) eIF2α, ribosomes do not have sufficient time to reassemble at the uORF2, which would otherwise be translated.

uORF2 is inhibitory and translated in unstressed cells where active eIF2α is abundant (Vattem and Wek 2004; Lu et al. 2004). Therefore, uORF1 has a stimulatory effect on the reinitiation of translation at AUGs downstream of uORF2.

UPR activation also leads to activation of the NF-κB pathway. Phospho-eIF2α-induced NF-κB activation involves two possible mechanisms: Phosphorylated eIF2α triggers the release of the I-κB from NF-κB (Jiang et al. 2003); when bound, I-κB sequesters NF-κB in the cytosol and blocks its nuclear localization. An alternative mechanism is that the amount of I-κB decreases due to the translational inhibition imposed by phospho-eIF2α (Deng et al. 2004). ATF6 branch (discussed below) and the IRE1-IKK-TRAF2 complex can also induce NF-κB activity (Hu et al. 2006; Yamazaki et al. 2009). ER stress and NF-κB connection is of particular importance in the context of metabolism. For example, ER stress-induced NF-κB triggers hypothalamic leptin resistance (Zhang et al. 2008). High-fat diet feeding results in elevated ER stress in the hypothalami of obese mice. Resulting activation of NF-κB upregulates the transcription of SOCS3 (Zhang et al. 2008), which is a negative regulator of the leptin-induced STAT3 phosphorylation.

IRE1 has a lumenal domain, which interacts with GRP78 and can also recognize the unfolded proteins, and a cytosolic kinase domain (Morl et al. 1993; Cox et al. 1993). As described below, the best characterized function of IRE1 involves the splicing of the mRNA of a transcription factor (HAC1 in yeast and XBP1 in metazoans) (Calfon et al. 2002), and the IRE1-HAC1/XBP1 arm constitutes the most ancient UPR pathway. Dissociation of GRP78 triggers oligomerization of the monomeric IRE1s, but is not sufficient for IRE1 activation. IRE1 oligomers form a luminal surface which allows IRE1 to directly sense unfolded proteins (Kimata et al. 2007), which in turn is thought to trigger a conformational change in the protein allowing IRE1 dimers to trans-autophosphorylate each other. Other than itself, IRE1 does not have any other known substrates as far as its kinase activity is concerned. IRE1 activation is prone to regulation by other factors as well. For example, the cytosolic non-receptor ABL tyrosine kinases localize to the ER membrane during ER stress and potentiate IRE1's RNase activity (Morita et al. 2017).

Active IRE1 has endoribonuclease activity toward X-box-binding protein-1 (XBP1) mRNA (Yoshida et al. 2001; Calfon et al. 2002). Unprocessed XBP1 encodes a protein that represses UPR genes. Excision of a 26 nucleotide intron results in a frameshift in the XBP1 reading frame. Upon removal of this intron, cleaved XBP1 mRNA is ligated by RtcB (Lu et al. 2014), and the resulting shorter mRNA encodes an active transcription factor of UPR called XBP1 spliced (XBP1s). XBP1 transcription is upregulated by UPR (through ATF6) as well (Yoshida et al. 2001), resulting in increased unspliced form of XBP1, which serves as an inhibitor of UPR, thereby forming a negative feedback loop (Yoshida et al. 2006). XBP1s are involved in the transcriptional regulation of genes regulating ER biogenesis and ER-associated degradation. Apparently, these two key processes that have to be carried out in order to relieve ER stress reveal the vital role of XBP1 in UPR. For example, XBP1s upregulate the expression of several ER-resident chaperones and increase the biosynthesis of the predominant ER phospholipid phosphatidylcholine to enable ER expansion (Sriburi et al. 2004). XBP1 mediates the upregulation of ER degradation-enhancing α-mannosidase-like protein, which is required for the degradation of misfolded glycoprotein substrates (Yoshida et al. 2003). XBP1s play a fundamental role in overall mammalian development and physiology. Global deletion of XBP1 results in embryonic lethality from anemia (Reimold et al. 2000). In the liver, XBP1 is required for recovery from ER stress such that liver-specific XBP1 KO mice display liver injury and fibrosis following ER stress induction (Olivares and Henkel 2015). Hepatic XBP1 positively regulates the expression of lipogenic genes, and liver-specific XBP1 KO mice have decreased plasma levels of cholesterol and triglyceride (TG) and hepatic lipid biosynthesis (Lee et al. 2008). In the liver, there is a postprandial transient upregulation of UPR, which is important for the remodeling of hepatic metabolic flux in the transition from fasting to refeeding, and liver-specific XBP1s overexpression can mimic this metabolic switch (Deng et al. 2013).

Deletion of IRE1α from hepatocytes decreases very low-density lipoprotein (VLDL) assembly and secretion, without altering TG synthesis or de novo lipogenesis (Wang et al. 2012). Adipocyte-specific XBP1 KO females have defective milk production resulting in decreased litter growth (Gregor et al. 2013). Depletion of XBP1 from the pancreas and the hypothalamus by RIP-Cre (rat insulin promoter)-mediated recombination triggers glucose intolerance due to β-cell failure and decreased insulin secretion (Lee et al. 2011). XBP1 deletion results in constitutive activation of IRE1, which is capable of cleaving prohormone convertase 1 and 2, and carboxypeptidase E mRNAs in pancreatic β cells (Lee et al. 2011). Likewise, deletion of IRE1α from RIP-positive cells leads to glucose intolerance and attenuates the HFD-induced β-cell proliferation commonly observed in obesity, but does not affect body weight or food intake of lean or HFD-induced obese mice (Xu et al. 2014). In the brain XBP1 regulates memory formation (Martínez et al. 2016), regulates neurite growth (Hayashi et al. 2007), and confers neuroprotection from degenerative diseases (Valdés et al. 2014). However, neuron-specific XBP1 KO mice are not different than wild-type mice in neuronal loss or survival under conditions associated with prion protein misfolding (Hetz et al. 2008). Furthermore, the neuronal XBP1 KOs have comparable body weights to wild-type mice (Hetz et al. 2008; Martínez et al. 2016).

XBP1 plays an important role in innate and adaptive immunity: It is required for the development and survival of dendritic cells, differentiation of lymphocytes (Reimold et al. 2001; Iwakoshi et al. 2003), and TLR2/4-mediated activation of pro-inflammatory cytokines by macrophages (Martinon et al. 2010). Eosinophil differentiation also requires XBP1s activity (Bettigole et al. 2015). Despite its well-characterized function in peripheral immune system, the role of XBP1 in microglia and CNS inflammation is not uncovered.

Another important function of IRE1 – that is implicated in linking ER stress to the metabolic diseases – was revealed by David Ron and coworkers, where they showed that IRE1 can activate c-Jun N-terminal kinases (JNK) (Urano et al. 2000). Tumor necrosis factor receptor (TNFR)-associated factor 2 (TRAF2), an adaptor protein, is recruited by phospho-IRE1 so that TRAF2 associates with the JNK kinases that ultimately result in JNK phosphorylation and activation (Urano et al. 2000). Tumor necrosis factor (TNF) receptor family is known to elicit its activity partly through JNK activation (Smith et al. 1994), and this pathway depends on intact TRAF2 (Lee et al. 1997). JNK could phosphorylate insulin receptor substrate 1 (IRS1) at Ser307, which results in the blockage of insulin-mediated IRS1 activation (Aguirre et al. 2000). Upon high-fat feeding, certain peripheral tissues such as the liver develop insulin resistance, at least in part, due to the high-fat diet-induced activation of ER stress, which in turn results in inhibition of JNK-mediated insulin signaling (Ozcan et al. 2004). Ser307Ala mutation eliminates phosphorylation of IRS-1 by JNK and removes the inhibitory effect of TNFα on insulin-stimulated tyrosine phosphorylation of IRS-1 (Aguirre et al. 2000).

IRE1 is also engaged in an RNA degradation pathway called IRE1-dependent decay of mRNA (RIDD) (Maurel et al. 2014), which is not limited to mRNA cleavage and extends to pre-microRNA processing (Upton et al. 2012). Substrates for IRE1's RNase activity are comprised of nuclear, cytosolic, ER, and extracellular targets (Maurel et al. 2014). IRE1β, one of the two IRE1 isoforms, e.g., regulates selective degradation of secretory pathway protein mRNAs (Nakamura et al. 2011).

ATF6 is an ER transmembrane protein that has a stress-sensing lumenal domain and a cytoplasmic basic leucine zipper domain that acts as a transcription factor upon dissociation from the rest of the molecules (Haze et al. 1999). In unstressed cells, ATF6 exists bound to GRP78. During ER stress, GRP78 dissociates from ATF6, which triggers ATF6 to localize to Golgi, where it is processed to release its cytoplasmic DNA-binding domain (Shen et al. 2002). ATF6 is processed by Site 1 and Site 2 proteases in Golgi in a similar manner as the processing of sterol regulatory element-binding proteins (SREBPs) (Brown and Goldstein 1997; Ye et al. 2000).

ATF6 has two isoforms, ATF6α and ATF6β. Knocking out both gene results in embryonic lethality, while single knockouts develop normally (Yamamoto et al. 2007). Another ER transmembrane protein that is activated in a similar manner to ATF6 is CREBH (cyclic AMP response element-binding protein hepatocyte), which is a liver transcription factor. CREBH has a cytoplasmic domain, which upon cleavage of CREBH in Golgi transits to the nucleus to activate genes involved in acute inflammatory response (Zhang et al. 2006). Because ER stress activates cleavage of CREBH in Golgi, CREBH acts as a mediator of ER stress-induced inflammatory response (Zhang et al. 2006). Activated ATF6 increases phosphatidylcholine synthesis independently of XBP1, suggesting a redundancy between ATF6 and XBP1 (Lee et al. 2003; Yamamoto et al. 2007) for certain genes including the ones responsible for ER stress-activated lipid biosynthesis and ER expansion (Bommiasamy et al. 2009). In hypothalamic cultures obtained from ATF6α KO mice, ER stress-induced regulation of various genes including CHOP-, GRP78-, and ERAD-associated genes is defective, suggesting that hypothalamic XBP1s cannot compensate for the absence of ATF6α (Lu et al. 2016). ATF6 and XBP1s can physically interact, and a dominant negative truncated version of XBP1 can suppress the activity of XBP1s or AFT6 (Lee et al. 2003). ATF6 activates the transcription of XBP1 and ER chaperones including GRP78 (Yoshida et al. 2001). ATF6 participates in the hepatic control of glucose homeostasis. In fasting, gluconeogenesis is triggered in the liver partly through the activity of CREB-regulated transcription coactivator 2 (CRTC2). Induction of ER stress in hepatocytes stimulates the dephosphorylation and nuclear entry of CRTC2. ATF6 recruits CRTC2 to ER stress-responsive genes, which results in increased expression of ER quality control genes and decreased hepatic glucose output (Wang et al. 2009). In the brain, ATF6 activation confers neuroprotection (Naranjo et al. 2016), and astroglial activation is attenuated in ATF6 knockout mice upon ischemic brain injury (Yoshikawa et al. 2015).

As discussed above, UPR has evolved to maintain the ER homeostasis acting in a cell-autonomous manner. Additionally, there are other proteostatic signaling routes evolved to enable communication between the nucleus and the cytosolic proteome (i.e., heat shock response) (Li et al. 2017) or the mitochondria (mitochondrial UPR) (Fiorese and Haynes 2017). However, there is an evolutionarily conserved yet incompletely understood intertissue signaling mechanism, also called the transcellular chaperone signaling (van Oosten-Hawle et al. 2013), described at least in metazoans that enable non-autonomous protein homeostasis by communication between different tissues (van Oosten-Hawle and Morimoto 2014). As UPR has evolved to allow communication between ER and other parts of the cellular structures, such as the nucleus, this trans-chaperone signaling enables the individual cells/tissues to communicate with each other. While it is beyond the scope of this chapter, it is worth to note that the regulation of ER homeostasis including ER stress within individual tissues should not be evaluated independently of its effect on other tissues. While induction of ER stress and impairment of UPR at individual tissues have detrimental metabolic consequences, some tissues respond to organelle stress, including ER stress, by secreting factors that improve the overall physiology. For example, during ER stress in the liver, hepatocytes secrete FGF21 (Schaap et al. 2013), which confers broad metabolic improvements including ameliorating ER stress and inducing weight loss (Kharitonenkov et al. 2005; Coskun et al. 2008). Likewise, neuronal overexpression of XBP1 in *C. elegans* activates UPR in non-neuronal cells, confers stress resistance, and increases longevity (Taylor and Dillin 2013).

4.2.1 ER Stress and the Regulation of Energy Balance

Regulation of metabolism in mammals requires a coordination between multiple tissues that is mediated by secreted factors, hormones and others, and neuronal inputs. The past decades in the metabolism field have uncovered numerous secreted factors, most notably leptin, and the dysregulation in the secretion, processing, recognition, or signaling of these factors play a fundamental role in the etiology of metabolic disorders. The canonical secretory pathway in mammals involves the translocation of nascent polypeptides into the ER, their regulated processing, posttranslational modifications, packing, and secretion. The hypothalamus is a heterogeneous population of chemically distinct neurons, with at least 50 distinct cell types in the ARC and ME identified based on their transcriptome profile (Campbell et al. 2017). Even within defined classes of neuronal populations, such as AgRP and POMC neurons, there are various subtypes. For example, there are at least three different subsets of POMC neurons in the MBH (Campbell et al. 2017). Likewise, among the AgRP neurons, which are exclusively expressed in the ARC, the leptin receptor-positive subset does not project to intra-hypothalamic sites (Betley et al. 2013). Therefore, the secretory profile of these neuronal populations, and the relative role of UPR in the respective neuronal subsets, is likely not uniform. The wide abundance and variety of the hypothalamic neuropeptides, which are synthesized as inactive precursors that require processing, demand a well-orchestrated ER quality control machinery. The physiological relevance and requirement of UPR in hypothalamic feeding centers are evident by a study on the transcriptional profile of AgRP and POMC neurons under satiated and fasted states showing UPR activation in AgRP neurons upon fasting (Henry et al. 2015). Numerous target genes for XBP1s including ER chaperones were upregulated by fasting in AgRP neurons. Genes involved in ER protein translocation and Golgi trafficking were also upregulated by fasting (Henry et al. 2015). ATF4 and ATF6 transcriptions also increased in AgRP neurons, although ATF6 nuclear localization was not altered by fasting. AgRP-specific ERAD-related transcripts were also upregulated by food deprivation. Besides UPR, fasting activated Nrf2 oxidative stress pathway in AgRP neurons. Notably, most of these fasting-induced changes in UPR pathway were specific to AgRP neurons, and no significant regulation was detected in POMC cells (Henry et al. 2015). However, upon short-term refeeding of fasted mice, POMC-specific XBP1s, ATF4, and ATF6 expression increased (Williams et al. 2014). Changes in AgRP neurons are probably related to fasting-induced AgRP neuronal activation and increased AgRP biosynthesis and secretion. For example, hypothalamic ATF4 regulates AgRP expression by stimulating FOXO1 (Deng et al. 2017), a transcription factor that positively regulates AgRP expression (Kitamura et al. 2006; Kim et al. 2006; Ren et al. 2012). However, whether and/or to what extent UPR contributes to the orexigenic response elicited by AgRP neurons is not known.

Overexpression of ATF4 in the MBH marginally induces food intake and weight gain and results in an attenuated hepatic insulin signaling, while a dominant negative ATF4 improves insulin signaling and glucose metabolism (Zhang et al. 2013b). Adverse effects of ATF4 overexpression could be reversed by hepatic vagotomy or by knocking down hypothalamic S6K expression (Zhang et al. 2013b). As explained in more detail in Chap. 8, while acute activation of hypothalamic S6K activity appears to block hepatic insulin signaling, overexpression of constitutively active S6K in the MBH blocks weight gain and protects against adverse consequences of HFD (Ono et al. 2008; Blouet et al. 2008). AgRP ATF4 KO mice are protected from weight gain on regular chow and HFD due to decreased food intake and increased energy expenditure (Deng et al. 2017). Deletion of ATF4 from POMC neurons also result in a similar phenotype with mice resistant to HFD-induced weight gain (Xiao et al. 2017a, b). Effect of ATF4 on POMC neurons was largely dependent on the negative regulation of ATG5 by ATF4 such that mice lacking both genes in POMC neurons had reduced energy expenditure on HFD and gained more weight (Xiao et al. 2017a, b). ATG5 is required for the regulation of autophagy, and POMC-specific ATG5 KO mice have defective autophagy in POMC neurons, although their body weight is not affected (Malhotra et al. 2015). Deletion of ATF4 led to the ATG5-dependent autophagy and increased αMSH production, which in turn led to increased energy expenditure (Xiao et al. 2017a, b).

Deletion of XBP1 from POMC neurons does not significantly alter body weight on regular chow or HFD, although the mice show a tendency for elevated weight gain (https://tez.yok. gov.tr/UlusalTezMerkezi/TezGoster?key= 1zw6GvYMe-q3Hf6HR-3USykM7dHpyqsbqQ-p1MsCgPMp7KeLv7nO_Vnsn0mL6Afc). While suppressing caloric intake on lean mice, HFD-fed POMC-specific XBP1 KOs tend to eat more (https://tez.yok.gov.tr/UlusalTezMerkezi/Tez Goster?key=1zw6GvYMe-q3Hf6HR-3USykM 7dHpyqsbqQ-p1MsCgPMp7KeLv7nO_ Vnsn0mL6Afc). Increased food intake of obese mice is however compensated by increased energy expenditure, and the obese KO mice have decreased fat percentage and decreased respiratory exchange ratios, indicative of elevated fat oxidation, than their POMC-Cre control counterparts (https://tez.yok.gov.tr/UlusalTezMerkezi/ TezGoster?key=1zw6GvYMe-q3Hf6HR-3 USykM7dHpyqsbqQ-p1MsCgPMp7KeLv7nO_ Vnsn0mL6Afc). These results should, however, be evaluated taking into account the phenotype of the POMC-Cre mice itself and the significant difference in the co-localization of POMC-Cre-expressing cells and POMC neurons in adult mice (Padilla et al. 2012). Genetic depletion of IRE1α expression from POMC neurons does not alter body weight on regular diet, but induces food intake and weight gain with increased adiposity on HFD (Yao et al. 2017). Obese POMC-IRE1α KO mice also display decreased energy expenditure, decreased cold tolerance, and impaired glucose tolerance (Yao et al. 2017). Additionally, these KO mice had elevated expression of negative regulators of leptin receptor signaling including PTP1B and SOCS3, which might contribute to their hyperphagic phenotype (Yao et al. 2017). Overexpression of XBP1s in POMC neurons results in hypophagic mice with increased energy expenditure and confers resistance to obesity (Williams et al. 2014). Interestingly, POMC-specific XBP1 overexpression also led to increased hepatic XBP1s expression and increased the BAT- and beige-specific

genes in the adipose tissue. Considering that hypothalamic induction of ER stress attenuates POMC processing and αMSH production (Cakir et al. 2013), it is worth investigating whether the metabolic improvements observed in mice with POMC-specific XBP1s overexpression are secondary to improved POMC processing.

ER stress and UPR play a fundamental role in metabolic regulation. Studies conducted in rodents showed that HFD-induced obesity leads to ER stress in multiple peripheral systems including the liver and adipose tissue. Furthermore, there is a functional interaction between ER stress and inflammation. For example, ER stress-induced inflammation in the liver couples obesity to insulin resistance (Ozcan et al. 2004), and deletion of IRE1α from the cells of myeloid lineage, including the macrophages but not the lymphocytes (T and B cells), confers almost complete protection from HFD-induced obesity and other associated metabolic abnormalities (Shan et al. 2017). The role of central ER stress and its effect on energy metabolism have been an area of active research during the past decade. Findings from our laboratory and others have indicated that development of obesity is accompanied by elevated markers of ER stress in the hypothalamus, predominantly in the MBH. Elevated ER stress in turn triggers central inflammation, resulting a positive feedback loop that is ultimately detrimental to the CNS circuitry regulating food intake and energy expenditure. ER stress has a more direct, cell-autonomous effect on the neuroendocrine hypothalamic population of neurons encoding bioactive neuropeptides, such as POMC (Cakir et al. 2013).

Initial evidence on the relationship between HFD-feeding and hypothalamic ER stress came from in vitro studies as well as results obtained from rodents using pharmacological and genetic tools. Treatment of leptin-responsive cells with inducers of ER stress, such as homocysteine, tunicamycin, and thapsigargin, induces the expression of negative regulators of leptin receptor signaling including SOCS3 and PTP1B and reduces leptin-induced STAT3 phosphorylation (Hosoi et al. 2008; Cakir et al. 2013). These results are reproduced when ER stress was induced in the hypothalamus of lean rodents (Hosoi et al. 2008; Cakir et al. 2013). Homocysteine treatment induced XBP1 splicing in the brain (Hosoi et al. 2010), also blocked the leptin-induced ERK phosphorylation, and did not affect the phospho-JNK levels (Hosoi et al. 2008). Furthermore, 4PBA treatment reversed the ER stress-induced attenuation in leptin receptor signaling. Central infusion of thapsigargin also blocked leptin and insulin signaling in the hypothalamus and had an orexigenic effect.

Diet-induced obesity correlates with increased markers of hypothalamic ER stress. IRE1 and PERK phosphorylation, XBP1 splicing, and expression of CHOP and GRP78 increase in HFD-fed rodent hypothalamus (Zhang et al. 2008; Won et al. 2009; Cakir et al. 2013). It is possible that the diet composition plays a fundamental role in the induction of hypothalamic ER stress. For example, central ceramide infusion induces hypothalamic ER stress and causes weight gain (Contreras et al. 2014). Ceramide-induced ER stress and weight gain can be reversed by VMH-specific overexpression of GRP78, while a dominant negative GRP78 induces ER stress and weight gain (Contreras et al. 2014). Overexpression of GRP78 in the VMH of Zucker rats decreased ER stress, suppresses weight gain, and increases BAT-mediated thermogenesis without altering food intake (Contreras et al. 2014). Effect of GRP78 and ER stress in the VMH is likely independent of leptin signaling and involves the sympathetic output to the adipose tissue and involves browning of WAT- and UCP1-induced thermogenesis (Contreras et al. 2017). Saturated fatty acids, such as palmitate, activate ER stress in hypothalamic neurons and attenuate leptin and insulin signaling (Kleinridders et al. 2009; Mayer and Belsham 2010; Diaz et al. 2015). As discussed above, TLR4 acts as a receptor for saturated fatty acids, and deletion of the TLR adaptor protein MyD88 from neurons confers protection from diet-induced obesity (Kleinridders et al. 2009). Central infusion of palmitate induces inflammatory cytokine expression and ER stress through activation of toll-like receptor 4 (TLR4) (Milanski et al. 2009; Kanczkowski et al. 2013), and central

inhibition or genetic inactivation of TLR4 confers protection from diet-induced obesity in rodents (Milanski et al. 2009). Likewise, knockout of the TLR4 adapter protein MyD88 prevents excess weight gain in mice on high-fat diet (Kanczkowski et al. 2013). TLR4 signaling can also activate IRE1-XBP1 axis in macrophages while suppressing the ATF4 pathway (Woo et al. 2009). Distinct from the ER-related XBP1 targets, TLR4-mediated XBP1 activation triggers the production of pro-inflammatory cytokines (Martinon et al. 2010). In rodent and human CNS, XBP1 expression is highest in microglia (Zhang et al. 2014; Bennett et al. 2016), and whether a similar link exists between fatty acids and IRE1-XBP1 pathway in CNS immune cells, such as in microglia, is worth investigating.

As discussed above, diet-induced obese mice have elevated hypothalamic IKKβ-NF-κB axis, which in turn activated the inflammatory gene expression in the hypothalamus, but also act as a factor in a positive feedback loop where its constitutive activation promoted ER stress (Zhang et al. 2008): HFD-fed mice have elevated levels of phosphorylated PERK and eIF2α in the hypothalamus (Zhang et al. 2008). Induction of central ER stress in lean mice by tunicamycin led to the activation of NF-κB, suggesting a causative relationship between ER stress and central inflammation. Accordingly, overexpression of constitutively active IKKβ in the MBH of lean mice was sufficient to induce ER stress, while neuronal deletion of IKKβ decreased PERK phosphorylation. Activated IKKβ-NF-κB signaling induced the expression of SOCS3, which is a negative regulator of insulin and leptin signaling. While activation of ER stress increased NF-κB activity, relieving ER stress with chemical chaperone TUDCA alleviated the HFD-induced hypothalamic NF-κB activation (Zhang et al. 2008). In summary, hypothalamic IKKβ and ER stress can activate each other, forming a positive feedback loop leading to dysregulation of central energy balance. DIO rats also display increased hypothalamic PERK and eIF2α phosphorylation (Cakir et al. 2013), and these responses seem to be ARC specific, as DIO rats do not suffer ER stress in the PVN (Cakir et al. 2013). Accordingly, PVN retains its leptin sensitivity in DIO rodents (Perello et al. 2010). While induction of ER stress in lean animals induces markers of leptin resistance, including SOCS3 and PTP1B, and suppressed energy expenditure (Zhang et al. 2008; Cakir et al. 2013), central infusion of TUDCA in obese rats did not alter STAT3 phosphorylation but acutely suppressed food intake and triggered increased oxygen consumption (Cakir et al. 2013).

Central ER stress does not only alter energy metabolism but also adversely affect other metabolic parameters. For example, thapsigargin-induced hypothalamic ER stress results in glucose tolerance, peripheral insulin resistance, and increases blood pressure, mimicking the phenotype in DIO rodents (Purkayastha et al. 2011b; Cakir et al. 2013). These responses could be reversed by genetic inhibition of IKK signaling in the ARC or partially reversed by treatment of DIO mice with TUDCA without altering food intake or body weight (Purkayastha et al. 2011b; Horwath et al. 2017). 4PBA treatment of rats resulted in attenuation of LPS-induced heart rate thought to be mediated, at least in part, through TLR4 action in the PVN (Masson et al. 2015). In obese mice that undergo vertical sleeve gastrectomy, hypothalamic markers of ER stress were decreased, which was accompanied by a significant decrease in mean arterial pressure (McGavigan et al. 2017).

The physical interaction between the ER and mitochondria is also affected by ER stress. Mitofusin 1 and mitofusin 2 (MTF1/2) are proteins located on the ER membrane and the outer surface of the mitochondria tethering the two organelles. MTF deficiency in the liver results in ER stress and causes glucose intolerance and defective insulin signaling in the liver and muscle, a defect that can be rescued with TUDCA treatment (Sebastián et al. 2012). AgRP-specific MTF1 or MTF2 KO mice are protected from HFD-induced weight gain (Dietrich et al. 2013). On the contrary, ablation of MTF2 in POMC neurons results in elevated ER stress and promotes food intake and weight gain (Schneeberger et al. 2013), while POMC-specific MTF1 KOs have dysfunctional glucose metabolism without

altered hypothalamic ER stress (Ramírez et al. 2017). As we continue to discuss in the following section, several studies have indicated that ER stress leads to a profound defect in the processing of prohormones including POMC.

4.2.2 Hypothalamic ER Stress and Neuropeptides

ER stress-induced abnormalities in energy as well as glucose metabolism are causally related to altered hypothalamic POMC processing. POMC-specific MTF2 KO mice develop glucose intolerance prior to development of their obese phenotype largely due to increased hepatic gluconeogenesis (Schneeberger et al. 2015). These mice had reduced hypothalamic αMSH levels that could be rescued with central TUDCA treatment (Schneeberger et al. 2015). In addition, acute HFD exposure also increased the hypothalamic ER stress, decreased αMSH levels, and impaired glucose homeostasis, which could be rescued by central αMSH replenishment (Schneeberger et al. 2015). MTF2 deletion in POMC neurons was proposed to decrease ER-mitochondrial contacts in the POMC neurons of obese mice; however the observed decreased ER-mitochondrial physical interaction is likely not a direct effect of the MTF2 deletion but rather secondary to the obese phenotype of the KO mice. MTF2 depletion in mouse embryonic fibroblasts results in increased ER-mitochondrial tethering (Filadi et al. 2015). AgRP-specific MTF1 or MTF2 KOs did not have differences in mitochondria-ER contacts, compared to control mice, and did not display altered markers of ER stress (Dietrich et al. 2013). POMC-MTF2 KO mice are resistant to anorectic action of exogenous leptin (Schneeberger et al. 2013). As discussed below, these mice have a decreased expression of POMC mRNA but an accumulation in the POMC protein. Although the expression of the POMC-processing enzymes PC2, PAM, and CPE was also elevated in the KO mice, α-MSH level was significantly lower, which accounts for the obese phenotype of the mice (Schneeberger et al. 2013). However, upon chronic central infusion of 4PBA or TUDCA, their food intake decreased, αMSH levels increased, and the mice lost their excess adiposity, and their body weight returned to the range observed in their chow-fed counterparts (Schneeberger et al. 2013). Accordingly, overexpression of MTF2 in the ARC protects against HFD-induced weight gain and decreases markers of ER stress (Schneeberger et al. 2013). The adverse effect of HFD on hypothalamic MTF expression might be, at least in part, dependent on SFAs. HFD feeding or palmitate alone decreases the expression of mitofusin 2 (MTF2) the ARC (Diaz et al. 2015).

In HFD-induced obese rodents, there is a gradual decrease in the amount of POMC processing evidenced by accumulation of POMC but decreased amounts of its processing products ACTH and αMSH in the MBH (Enriori et al. 2007; Cakir et al. 2013; Schneeberger et al. 2013). Furthermore, although leptin could stimulate, albeit at a compromised level, the expression of POMC mRNA and hypothalamic ACTH levels, αMSH level, is completely insensitive to leptin administration (Cakir et al. 2013). Decreased αMSH levels in DIO rodents were not due to increased αMSH degradation (Wallingford et al. 2009; Cakir et al. 2013; Schneeberger et al. 2013). These findings suggested that there was a defect in the step leading from the POMC precursor to the production of αMSH. While there was no significant difference in the mRNA levels of POMC-processing enzymes between lean and DIO animals, there was a significant reduction in the protein levels of PC2 in DIO ARC. Reduced PC2 expression and αMSH levels could be induced in lean rat hypothalamus simply by central infusion of thapsigargin or tunicamycin, which suggested that ER stress negatively regulated PC2 protein levels (Cakir et al. 2013), which in turn leads to decreased αMSH levels. The thapsigargin- or tunicamycin-induced defects in POMC processing could be partially rescued by chemical chaperones PBA or TUDCA (Cakir et al. 2013). As explained in detail in Chap. 5, PC2 is critical in the processing of ACTH to αMSH. Decreased PC2 and αMSH levels could also be induced in different cell lines upon induction of ER stress in a way that could

be rescued by chemical chaperones. Furthermore, central TUDCA infusion rescued the defect in the POMC processing in DIO rats and increased the αMSH levels. However, the same treatment did not alter the level of hypothalamic NPY, suggesting that DIO-induced ER stress affects the processing of hypothalamic neuropeptides differently (Cakir et al. 2013). Salubrinal is an agent that selectively inhibits the dephosphorylation of p-eIF2α (Boyce et al. 2005) and enhances the FFA-induced activation of the PERK pathway, but not the ATF6 and IRE1 branches (Ladrière et al. 2010). Blocking the dephosphorylation of eIF2α with salubrinal increased PC2 and reversed the effect of thapsigargin on PC2 levels, suggesting that PERK-eIF2α branch of the UPR could protect the decrease in PC2 during ER stress.

A recent study has further revealed the overall importance of ER homeostasis on POMC processing in regard to energy balance. Proper trafficking and processing of POMC require a protein complex between suppressor enhancer of lin-like 1 (Sel1L) and hydroxymethylglutaryl reductase degradation protein (Hrd1, also called synoviolin). Sel1L-Hrd1 is an evolutionarily conserved arm of ERAD, and their genetic deletion results in embryonic lethality in mice (Yagishita et al. 2005; Francisco et al. 2010). Both genes are expressed in the hypothalamus and in POMC neurons (Kim et al. 2017). Feeding and leptin positively regulate their expression. Deletion of Sel1L from POMC neurons results in loss of response to leptin's anorectic effect, hyperphagia, and age-associated obesity. Sel1L deficiency did not induce central ER stress or inflammation (Kim et al. 2017). Although POMC mRNA level was unaltered in POMC-specific Sel1L deficiency, there was a significant accumulation of the POMC protein. Hrd1 is an E3 ligase and together with Sel1L targets POMC to proteasomal degradation. Consequently, in the absence of this quality control system, the nascent POMC polypeptide cannot be properly targeted for further processing resulting in aggregation and accumulation in the ER (Kim et al. 2017). It is possible that diet-induced obesity also results in a functional dysregulation of the ERAD system in

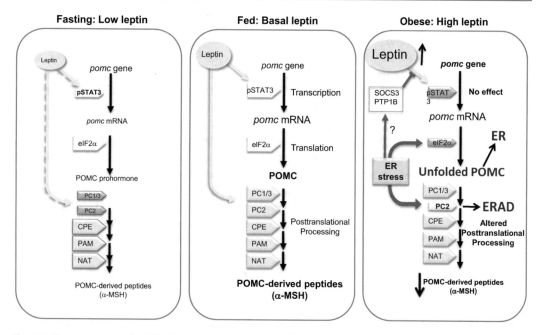

Fig. 4.1 Propose changes in POMC processing and biosynthesis of α-MSH in the obese state. While the levels of processing enzyme PC2 adjust under fasting (i.e., low leptin, first panel) and fed (i.e., basal leptin, second panel), in the DIO (i.e., high leptin, third panel), there is a decrease in PC2 and an accumulation of POMC compared to the fed control. Compared to the basal leptin condition, fasting causes diminished α-MSH by a leptin-mediated reduction in POMC mRNA along with reduced POMC processing due to lower PC1 and PC2 levels. In the DIO condition, endoplasmic reticulum stress triggers an unfolded response causing an accumulation of POMC in the ER partly caused by an unfolded response and a possible degradation of PC2 through the ERAD system, conditions that ultimately cause less α-MSH production. (Image from our prior publication (Cakir et al. 2013))

POMC neurons, which might contribute to the accumulation of POMC observed in obese rodents (Cakir et al. 2013). Finally, while it is beyond the scope of this chapter, it is worth noting that the effect of hypothalamic ER stress extends beyond regulation of energy balance. Central ER stress affects the polyA tail length of mRNA for certain neuropeptides including vasopressin and oxytocin, resulting in decreased mature mRNA levels (Morishita et al. 2011), which could be reversed by TUDCA treatment (Morishita et al. 2011). Furthermore, ERAD deficiency also leads to aggregation and accumulation of proAVP, the precursor of the precursor for the antidiuretic hormone arginine vasopressin, resulting in central diabetes insipidus (Shi et al. 2017) (Fig. 4.1).

4.3 Conclusion

Across species in all kingdoms of life, maintenance of homeostasis is key for survival. Both at the cellular and the organismal level, various signaling routes have evolved and are activated in response to internal or environmental stimuli to keep the homeostasis within certain boundaries. This chapter has focused on two of these adaptive mechanisms, inflammation and ER stress/UPR.

Infection, tissue damage, or stress could trigger inflammation, which in turn increases defense mechanisms to cope with the infection (immune response) or tissue repair and adaptation to the stress stimuli. As outlined above, the inflammatory state associated with obesity is different than classical inflammatory response; it is a chronic but rather low-grade inflammatory state, which, unlike the infection-induced inflammation, is not coupled to induction of an immune response. Furthermore, the central vs. the peripheral inflammation associated with high-calorie intake are different. Central markers of inflammation can be detected following consumption of high-calorie diets prior to any weight gain, while peripheral inflammation, tissue macrophage infiltration, and polarization proceeds gain of adiposity. The associated cell types in the periphery vs. the CNS are also different, and the anti-inflammatory nature of the brain parenchyma is not suitable for survival of immune cell components in the brain. CNS has its own macrophage pool, the microglia, and we are only beginning to understand the role of the glial cell populations in the regulation of energy balance and overall metabolism.

ER stress is classically triggered by an imbalance between the ER's folding capacity and the steady-state unfolded protein pool in the ER lumen. There are numerous physiological or pathological triggers of ER stress, and nutrient deprivation, redox status, infection, and ER calcium levels are among some of them. The dysfunction of the ER is communicated to the nucleus and the translational machinery through a set of signaling routes called the UPR. Much like other stress response pathways, UPR initially functions to restore the homeostasis, in this case, of the ER. The temporal aspect of the stress-activated pathways including the UPR is of significant importance; prolonged UPR can result in detrimental consequences as induction of apoptosis. This makes sense from an organismal point of view to eliminate overly stressed/damaged cells which cannot be rescued. However, a chronic ER stress detected in an entire tissue as observed in obesity results in a maladaptive state which triggers a shift in homeostatic set points. This, in turn, results in maladaptive states, such as insulin and leptin resistance, which contributes to the etiology of obesity and diabetes.

Both ER stress and inflammation are intimately connected for a number of reasons. First of all, mediators of inflammation and ER stress utilize, at least in part, common downstream signaling pathways (e.g., NF-κB activation). This is one of the reasons that ER stress and inflammation can trigger each other even at a cell autonomous level. Second, signaling components of the inflammatory process as well as ER stress are also shared by several metabolic routes, resulting in a causative relationship between these processes and the regulation of energy metabolism. Third, either process is activated chronically at a moderate level during obesity. Fourth, nutrients, most prominently saturated fatty acids, can induce central ER stress as well as an inflammatory response. Fifth, both obesity-associated ER stress and elevated inflammation appear to be reversible, at least in rodents.

Various studies conducted in rodents as well as results from humans suggest that obesity is accompanied by elevated ER stress and inflammation in the CNS. Our knowledge on central ER stress in regard to the possible role of UPR in glia is rather limited, and most of the metabolic studies have so far focused on ER stress in neurons. Central inflammation encompasses neurons as well as glia, and astrogliosis and microgliosis have detrimental consequences in the hypothalamic circuits regulating energy balance. What triggers hypothalamic ER stress or inflammation in the first place is not fully uncovered; however nutrient-related inputs, such as saturated fatty acids, appear to play a causative role. Induction of hypothalamic ER stress contributes to a positive energy balance, and this is mediated at least in part due to the direct impact on altered prohormone processing as in the case for POMC. Inflammatory mediators are classically accompanied by an anorexic response; however prolonged activation of central inflammatory pathways results in elevated food intake and/or increased weight gain. This is a major distinction in the effect of central inflammation when considered in an acute vs. chronic setting. Genetic and pharmacological evidence suggest that amelioration of the hypothalamic ER stress and inflammation that accompanies obesity result in weight loss, improved glucose metabolism, and improved cardiovascular functions. It is finally important to note the similarity between obesity and the aging process in regard to their association to central inflammation and ER stress. The results indicating that suppression of neuronal inflammation confers longevity in mice point out the importance of CNS inflammation on overall mammalian physiology.

It is vital to the understanding of biology that the cells or organisms are more than a collection of individual components (organelles or tissues) put together. As Aristotle said, "the whole is more than the sum of its parts." As such, there is a sophisticated communication system evolved to maintain the homeostasis in multicellular organisms. Components of the inflammatory process and the UPR should be evaluated from this perspective at the organismal level. Cell autonomous stress response pathways function to restore the homeostatic boundaries; however they have non-cell autonomous consequences (e.g., ER stress-induced hepatic FGF21 secretion) that translate into whole animal physiology. While cellular components of the UPR and inflammation are rather well-characterized, how these processes affect homeostatic brain centers including the hypothalamus and the brain stem is only beginning to be uncovered. A comprehensive understanding of the causes of the CNS inflammation and ER stress and how these processes interact with the metabolic regulatory routes will help us devise novel therapeutic opportunities for the treatment of metabolic diseases such as obesity.

Questions

1. What are the differences between central and peripheral inflammation? How does innate vs. acquired immunity play a role in CNS inflammation?
2. What are the causes and consequences of acute vs. chronic inflammation on energy balance?
3. What is the relationship between ER stress and inflammation? What are the signaling pathways that connect them?
4. What are the signaling routes that communicate between the ER and the nucleus? How conserved are they?
5. How does hypothalamic ER stress interfere with the regulation of energy homeostasis?

References

Aguirre, V., Uchida, T., Yenush, L., et al. (2000). The c-Jun NH(2)-terminal kinase promotes insulin resistance during association with insulin receptor substrate-1 and phosphorylation of Ser(307). *The Journal of Biological Chemistry, 275*, 9047–9054.

Ahima, R. S., Prabakaran, D., Mantzoros, C., et al. (1996). Role of leptin in the neuroendocrine response to fasting. *Nature, 382*, 250–252.

Almeida-Suhett, C. P., Graham, A., Chen, Y., & Deuster, P. (2017). Behavioral changes in male mice fed a high-fat diet are associated with IL-1β expression in specific brain regions. *Physiology & Behavior, 169*, 130–140.

Amura, C. R., Marek, L., Winn, R. A., & Heasley, L. E. (2005). Inhibited neurogenesis in JNK1-deficient embryonic stem cells. *Molecular and Cellular Biology, 25*, 10791–10802.

Anandasabapathy, N., Victora, G. D., Meredith, M., et al. (2011). Flt3L controls the development of radiosensitive dendritic cells in the meninges and choroid plexus of the steady-state mouse brain. *The Journal of Experimental Medicine, 208*, 1695–1705.

André, C., Guzman-Quevedo, O., Rey, C., et al. (2017). Inhibiting microglia expansion prevents diet-induced hypothalamic and peripheral inflammation. *Diabetes, 66*, 908–919.

Argaw, A. T., Asp, L., Zhang, J., et al. (2012). Astrocyte-derived VEGF-A drives blood-brain barrier disruption in CNS inflammatory disease. *The Journal of Clinical Investigation, 122*, 2454–2468.

Arkan, M. C., Hevener, A. L., Greten, F. R., et al. (2005). IKK-beta links inflammation to obesity-induced insulin resistance. *Nature Medicine, 11*, 191–198.

Balland, E., & Cowley, M. A. (2017). Short-term high fat diet increases the presence of astrocytes in the hypothalamus of C57BL6 mice without altering leptin sensitivity. *Journal of Neuroendocrinology*. https://doi.org/10.1111/jne.12504.

Balland, E., Dam, J., Langlet, F., et al. (2014). Hypothalamic tanycytes are an ERK-gated conduit for leptin into the brain. *Cell Metabolism, 19*, 293–301.

Beard, R. S., Jr., Haines, R. J., Wu, K. Y., et al. (2014). Non-muscle Mlck is required for β-catenin- and FoxO1-dependent downregulation of Cldn5 in IL-1β-mediated barrier dysfunction in brain endothelial cells. *Journal of Cell Science, 127*, 1840–1853.

Belgardt, B. F., Mauer, J., Wunderlich, F. T., et al. (2010). Hypothalamic and pituitary c-Jun N-terminal kinase 1 signaling coordinately regulates glucose metabolism. *Proceedings of the National Academy of Sciences of the United States of America, 107*, 6028–6033.

Bence, K. K., Delibegovic, M., Xue, B., et al. (2006). Neuronal PTP1B regulates body weight, adiposity and leptin action. *Nature Medicine, 12*, 917–924.

Bennett, M. L., Bennett, F. C., Liddelow, S. A., et al. (2016). New tools for studying microglia in the mouse and human CNS. *Proceedings of the National*

Academy of Sciences of the United States of America, 113, E1738–E1746.

Benomar, Y., Gertler, A., De Lacy, P., et al. (2013). Central resistin overexposure induces insulin resistance through toll-like receptor 4. *Diabetes, 62*, 102–114.

Benomar, Y., Amine, H., Crépin, D., et al. (2016). Central Resistin/TLR4 impairs adiponectin signaling, contributing to insulin and FGF21 resistance. *Diabetes, 65*, 913–926.

Berkseth, K. E., Guyenet, S. J., Melhorn, S. J., et al. (2014). Hypothalamic gliosis associated with high-fat diet feeding is reversible in mice: A combined immunohistochemical and magnetic resonance imaging study. *Endocrinology, 155*, 2858–2867.

Besedovsky, H. O., & del Rey, A. (1996). Immune-neuro-endocrine interactions: Facts and hypotheses. *Endocrine Reviews, 17*, 64–102.

Betley, J. N., Cao, Z. F. H., Ritola, K. D., & Sternson, S. M. (2013). Parallel, redundant circuit organization for homeostatic control of feeding behavior. *Cell, 155*, 1337–1350.

Bettigole, S. E., Lis, R., Adoro, S., et al. (2015). The transcription factor XBP1 is selectively required for eosinophil differentiation. *Nature Immunology, 16*, 829–837.

Blamire, A. M., Anthony, D. C., Rajagopalan, B., et al. (2000). Interleukin-1beta -induced changes in blood-brain barrier permeability, apparent diffusion coefficient, and cerebral blood volume in the rat brain: A magnetic resonance study. *The Journal of Neuroscience, 20*, 8153–8159.

Blouet, C., Ono, H., & Schwartz, G. J. (2008). Mediobasal hypothalamic p70 S6 kinase 1 modulates the control of energy homeostasis. *Cell Metabolism, 8*, 459–467.

Bommiasamy, H., Back, S. H., Fagone, P., et al. (2009). ATF6alpha induces XBP1-independent expansion of the endoplasmic reticulum. *Journal of Cell Science, 122*, 1626–1636.

Bora-Tatar, G., Dayangaç-Erden, D., Demir, A. S., et al. (2009). Molecular modifications on carboxylic acid derivatives as potent histone deacetylase inhibitors: Activity and docking studies. *Bioorganic & Medicinal Chemistry, 17*, 5219–5228.

Boyce, M., Bryant, K. F., Jousse, C., et al. (2005). A selective inhibitor of eIF2alpha dephosphorylation protects cells from ER stress. *Science, 307*, 935–939.

Breder, C. D., Hazuka, C., Ghayur, T., et al. (1994). Regional induction of tumor necrosis factor alpha expression in the mouse brain after systemic lipopolysaccharide administration. *Proceedings of the National Academy of Sciences of the United States of America, 91*, 11393–11397.

Brehme, M., Voisine, C., Rolland, T., et al. (2014). A chaperome subnetwork safeguards proteostasis in aging and neurodegenerative disease. *Cell Reports, 9*, 1135–1150.

Brown, M. S., & Goldstein, J. L. (1997). The SREBP pathway: Regulation of cholesterol metabolism by proteolysis of a membrane-bound transcription factor. *Cell, 89*, 331–340.

Broz, P., & Dixit, V. M. (2016). Inflammasomes: Mechanism of assembly, regulation and signalling. *Nature Reviews. Immunology, 16*, 407–420.

Buckman, L. B., Thompson, M. M., Lippert, R. N., et al. (2015). Evidence for a novel functional role of astrocytes in the acute homeostatic response to high-fat diet intake in mice. *Molecular Metabolism, 4*, 58–63.

Cai, D., Yuan, M., Frantz, D. F., et al. (2005). Local and systemic insulin resistance resulting from hepatic activation of IKK-beta and NF-kappaB. *Nature Medicine, 11*, 183–190.

Cakir, I., Cyr, N. E., Perello, M., et al. (2013). Obesity induces hypothalamic endoplasmic reticulum stress and impairs proopiomelanocortin (POMC) posttranslational processing. *The Journal of Biological Chemistry, 288*, 17675–17688.

Calfon, M., Zeng, H., Urano, F., et al. (2002). IRE1 couples endoplasmic reticulum load to secretory capacity by processing the XBP-1 mRNA. *Nature, 415*, 92–96.

Campbell, J. N., Macosko, E. Z., Fenselau, H., et al. (2017). A molecular census of arcuate hypothalamus and median eminence cell types. *Nature Neuroscience, 20*, 484–496.

Carlin, J. L., Grissom, N., Ying, Z., et al. (2016). Voluntary exercise blocks Western diet-induced gene expression of the chemokines CXCL10 and CCL2 in the prefrontal cortex. *Brain, Behavior, and Immunity, 58*, 82–90.

Chakravarty, S., & Herkenham, M. (2005). Toll-like receptor 4 on nonhematopoietic cells sustains CNS inflammation during endotoxemia, independent of systemic cytokines. *The Journal of Neuroscience, 25*, 1788–1796.

Chang, L., Jones, Y., Ellisman, M. H., et al. (2003). JNK1 is required for maintenance of neuronal microtubules and controls phosphorylation of microtubule-associated proteins. *Developmental Cell, 4*, 521–533.

Chesnokova, V., Pechnick, R. N., & Wawrowsky, K. (2016). Chronic peripheral inflammation, hippocampal neurogenesis, and behavior. *Brain, Behavior, and Immunity, 58*, 1–8.

Chiang, S.-H., Bazuine, M., Lumeng, C. N., et al. (2009). The protein kinase IKKepsilon regulates energy balance in obese mice. *Cell, 138*, 961–975.

Choi, S. J., Kim, F., Schwartz, M. W., & Wisse, B. E. (2010). Cultured hypothalamic neurons are resistant to inflammation and insulin resistance induced by saturated fatty acids. *American Journal of Physiology. Endocrinology and Metabolism, 298*, E1122–E1130.

Contreras, C., González-García, I., Martínez-Sánchez, N., et al. (2014). Central ceramide-induced hypothalamic lipotoxicity and ER stress regulate energy balance. *Cell Reports, 9*, 366–377.

Contreras, C., González-García, I., Seoane-Collazo, P., et al. (2017). Reduction of hypothalamic endoplasmic reticulum stress activates browning of white fat and ameliorates obesity. *Diabetes, 66*, 87–99.

Corbett, E. F., Oikawa, K., Francois, P., et al. (1999). Ca2+ regulation of interactions between endoplasmic reticulum chaperones. *The Journal of Biological Chemistry, 274*, 6203–6211.

Coskun, T., Bina, H. A., Schneider, M. A., et al. (2008). Fibroblast growth factor 21 corrects obesity in mice. *Endocrinology, 149*, 6018–6027.

Cox, J. S., Shamu, C. E., & Walter, P. (1993). Transcriptional induction of genes encoding endoplasmic reticulum resident proteins requires a transmembrane protein kinase. *Cell, 73*, 1197–1206.

Cullinan, S. B., Zhang, D., Hannink, M., et al. (2003). Nrf2 is a direct PERK substrate and effector of PERK-dependent cell survival. *Molecular and Cellular Biology, 23*, 7198–7209.

Dange, R. B., Agarwal, D., Teruyama, R., & Francis, J. (2015). Toll-like receptor 4 inhibition within the paraventricular nucleus attenuates blood pressure and inflammatory response in a genetic model of hypertension. *Journal of Neuroinflammation, 12*, 31.

de Rivero Vaccari, J. P., Lotocki, G., Marcillo, A. E., et al. (2008). A molecular platform in neurons regulates inflammation after spinal cord injury. *The Journal of Neuroscience, 28*, 3404–3414.

De Souza, C. T., Araujo, E. P., Bordin, S., et al. (2005). Consumption of a fat-rich diet activates a proinflammatory response and induces insulin resistance in the hypothalamus. *Endocrinology, 146*, 4192–4199.

Deng, J., Lu, P. D., Zhang, Y., et al. (2004). Translational repression mediates activation of nuclear factor kappa B by phosphorylated translation initiation factor 2. *Molecular and Cellular Biology, 24*, 10161–10168.

Deng, Y., Wang, Z. V., Tao, C., et al. (2013). The Xbp1s/GalE axis links ER stress to postprandial hepatic metabolism. *The Journal of Clinical Investigation, 123*, 455–468.

Deng, J., Yuan, F., Guo, Y., et al. (2017). Deletion of ATF4 in AgRP neurons promotes fat loss mainly via increasing energy expenditure. *Diabetes, 66*, 640–650.

Deroo, B. J., & Archer, T. K. (2001). Glucocorticoid receptor activation of the I kappa B alpha promoter within chromatin. *Molecular Biology of the Cell, 12*, 3365–3374.

Di Bello, I. C., Dawson, M. R., Levine, J. M., & Reynolds, R. (1999). Generation of oligodendroglial progenitors in acute inflammatory demyelinating lesions of the rat brain stem is associated with demyelination rather than inflammation. *Journal of Neurocytology, 28*, 365–381.

Diaz, B., Fuentes-Mera, L., Tovar, A., et al. (2015). Saturated lipids decrease mitofusin 2 leading to endoplasmic reticulum stress activation and insulin resistance in hypothalamic cells. *Brain Research, 1627*, 80–89.

Dietrich, M. O., Liu, Z.-W., & Horvath, T. L. (2013). Mitochondrial dynamics controlled by mitofusins regulate Agrp neuronal activity and diet-induced obesity. *Cell, 155*, 188–199.

Djogo, T., Robins, S. C., Schneider, S., et al. (2016). Adult NG2-glia are required for median eminence-mediated leptin sensing and body weight control. *Cell Metabolism, 23*, 797–810.

Donath, M. Y., Böni-Schnetzler, M., Ellingsgaard, H., et al. (2010). Cytokine production by islets in health and diabetes: Cellular origin, regulation and function. *Trends in Endocrinology and Metabolism, 21*, 261–267.

Donnelly, N., Gorman, A. M., Gupta, S., & Samali, A. (2013). The eIF2α kinases: Their structures and functions. *Cellular and Molecular Life Sciences, 70*, 3493–3511.

Du, Y., & Dreyfus, C. F. (2002). Oligodendrocytes as providers of growth factors. *Journal of Neuroscience Research, 68*, 647–654.

Dutheil, S., Ota, K. T., Wohleb, E. S., et al. (2016). High-fat diet induced anxiety and anhedonia: Impact on brain homeostasis and inflammation. *Neuropsychopharmacology, 41*, 1874–1887.

Ehses, J. A., Böni-Schnetzler, M., Faulenbach, M., & Donath, M. Y. (2008). Macrophages, cytokines and beta-cell death in type 2 diabetes. *Biochemical Society Transactions, 36*, 340–342.

Elmore, M. R. P., Najafi, A. R., Koike, M. A., et al. (2014). Colony-stimulating factor 1 receptor signaling is necessary for microglia viability, unmasking a microglia progenitor cell in the adult brain. *Neuron, 82*, 380–397.

Elmquist, J. K., Ackermann, M. R., Register, K. B., et al. (1993). Induction of Fos-like immunoreactivity in the rat brain following Pasteurella multocida endotoxin administration. *Endocrinology, 133*, 3054–3057.

Enriori, P. J., Evans, A. E., Sinnayah, P., et al. (2007). Diet-induced obesity causes severe but reversible leptin resistance in arcuate melanocortin neurons. *Cell Metabolism, 5*, 181–194.

Erblich, B., Zhu, L., Etgen, A. M., et al. (2011). Absence of colony stimulation factor-1 receptor results in loss of microglia, disrupted brain development and olfactory deficits. *PLoS One, 6*, e26317.

Erion, J. R., Wosiski-Kuhn, M., Dey, A., et al. (2014). Obesity elicits interleukin 1-mediated deficits in hippocampal synaptic plasticity. *The Journal of Neuroscience, 34*, 2618–2631.

Filadi, R., Greotti, E., Turacchio, G., et al. (2015). Mitofusin 2 ablation increases endoplasmic reticulum-mitochondria coupling. *Proceedings of the National Academy of Sciences of the United States of America, 112*, E2174–E2181.

Fiorese, C. J., & Haynes, C. M. (2017). Integrating the UPR(mt) into the mitochondrial maintenance network. *Critical Reviews in Biochemistry and Molecular Biology, 52*, 304–313.

Frago, L. M., & Chowen, J. A. (2017). Involvement of astrocytes in mediating the central effects of ghrelin. *International Journal of Molecular Sciences*. https://doi.org/10.3390/ijms18030536.

Francisco, A. B., Singh, R., Li, S., et al. (2010). Deficiency of suppressor enhancer Lin12 1 like (SEL1L) in mice leads to systemic endoplasmic reticulum stress and embryonic lethality. *The Journal of Biological Chemistry, 285*, 13694–13703.

Frayling, C., Britton, R., & Dale, N. (2011). ATP-mediated glucosensing by hypothalamic tanycytes. *The Journal of Physiology, 589*, 2275–2286.

Fusakio, M. E., Willy, J. A., Wang, Y., et al. (2016). Transcription factor ATF4 directs basal and stress-induced gene expression in the unfolded protein response and cholesterol metabolism in the liver. *Molecular Biology of the Cell, 27*, 1536–1551.

Gao, S., Howard, S., & LoGrasso, P. V. (2017). Pharmacological inhibition of c-Jun N-terminal kinase reduces food intake and sensitizes leptin's anorectic signaling actions. *Scientific Reports, 7*, 41795.

García-Cáceres, C., Quarta, C., Varela, L., et al. (2016). Astrocytic insulin signaling couples brain glucose uptake with nutrient availability. *Cell, 166*, 867–880.

Ginhoux, F., Greter, M., Leboeuf, M., et al. (2010). Fate mapping analysis reveals that adult microglia derive from primitive macrophages. *Science, 330*, 841–845.

Gregor, M. F., Misch, E. S., Yang, L., et al. (2013). The role of adipocyte XBP1 in metabolic regulation during lactation. *Cell Reports, 3*, 1430–1439.

Guillemot-Legris, O., Masquelier, J., Everard, A., et al. (2016). High-fat diet feeding differentially affects the development of inflammation in the central nervous system. *Journal of Neuroinflammation, 13*, 206.

Gunstad, J., Paul, R. H., Cohen, R. A., et al. (2006). Obesity is associated with memory deficits in young and middle-aged adults. *Eating and Weight Disorders, 11*, e15–e19.

Gustin, A., Kirchmeyer, M., Koncina, E., et al. (2015). NLRP3 inflammasome is expressed and functional in mouse brain microglia but not in astrocytes. *PLoS One, 10*, e0130624.

Halle, A., Hornung, V., Petzold, G. C., et al. (2008). The NALP3 inflammasome is involved in the innate immune response to amyloid-beta. *Nature Immunology, 9*, 857–865.

Han, M. S., Jung, D. Y., Morel, C., et al. (2013). JNK expression by macrophages promotes obesity-induced insulin resistance and inflammation. *Science, 339*, 218–222.

Harding, H. P., Zhang, Y., & Ron, D. (1999). Protein translation and folding are coupled by an endoplasmic-reticulum-resident kinase. *Nature, 397*, 271–274.

Harding, H. P., Novoa, I., Zhang, Y., et al. (2000). Regulated translation initiation controls stress-induced gene expression in mammalian cells. *Molecular Cell, 6*, 1099–1108.

Harding, H. P., Zhang, Y., Zeng, H., et al. (2003). An integrated stress response regulates amino acid metabolism and resistance to oxidative stress. *Molecular Cell, 11*, 619–633.

Hayashi, A., Kasahara, T., Iwamoto, K., et al. (2007). The role of brain-derived neurotrophic factor (BDNF)-induced XBP1 splicing during brain development. *The Journal of Biological Chemistry, 282*, 34525–34534.

Haze, K., Yoshida, H., Yanagi, H., et al. (1999). Mammalian transcription factor ATF6 is synthesized as a transmembrane protein and activated by proteolysis in response to endoplasmic reticulum stress. *Molecular Biology of the Cell, 10*, 3787–3799.

Helms, J. B., & Rothman, J. E. (1992). Inhibition by brefeldin A of a Golgi membrane enzyme that catalyses exchange of guanine nucleotide bound to ARF. *Nature, 360*, 352–354.

Henry, F. E., Sugino, K., Tozer, A., et al. (2015). Cell type-specific transcriptomics of hypothalamic energy-sensing neuron responses to weight-loss. *eLife*. https://doi.org/10.7554/eLife.09800.

Hetz, C., Lee, A.-H., Gonzalez-Romero, D., et al. (2008). Unfolded protein response transcription factor XBP-1 does not influence prion replication or pathogenesis. *Proceedings of the National Academy of Sciences of the United States of America, 105*, 757–762.

Hillhouse, E. W., & Mosley, K. (1993). Peripheral endotoxin induces hypothalamic immunoreactive interleukin-1 beta in the rat. *British Journal of Pharmacology, 109*, 289–290.

Hirosumi, J., Tuncman, G., Chang, L., et al. (2002). A central role for JNK in obesity and insulin resistance. *Nature, 420*, 333–336.

Horwath, J. A., Hurr, C., Butler, S. D., et al. (2017). Obesity-induced hepatic steatosis is mediated by endoplasmic reticulum stress in the subfornical organ of the brain. *JCI Insight*. https://doi.org/10.1172/jci.insight.90170.

Hosoi, T., Okuma, Y., Kawagishi, T., et al. (2004). Bacterial endotoxin induces STAT3 activation in the mouse brain. *Brain Research, 1023*, 48–53.

Hosoi, T., Sasaki, M., Miyahara, T., et al. (2008). Endoplasmic reticulum stress induces leptin resistance. *Molecular Pharmacology, 74*, 1610–1619.

Hosoi, T., Ogawa, K., & Ozawa, K. (2010). Homocysteine induces X-box-binding protein 1 splicing in the mice brain. *Neurochemistry International, 56*, 216–220.

Hotamisligil, G. S. (2017). Inflammation, metaflammation and immunometabolic disorders. *Nature, 542*, 177–185.

Hotamisligil, G. S., Shargill, N. S., & Spiegelman, B. M. (1993). Adipose expression of tumor necrosis factor-alpha: Direct role in obesity-linked insulin resistance. *Science, 259*, 87–91.

Hotamisligil, G. S., Budavari, A., Murray, D., & Spiegelman, B. M. (1994). Reduced tyrosine kinase activity of the insulin receptor in obesity-diabetes. Central role of tumor necrosis factor-alpha. *The Journal of Clinical Investigation, 94*, 1543–1549.

Hotamisligil, G. S., Peraldi, P., Budavari, A., et al. (1996). IRS-1-mediated inhibition of insulin receptor tyrosine kinase activity in TNF-alpha- and obesity-induced insulin resistance. *Science, 271*, 665–668.

Hu, P., Han, Z., Couvillon, A. D., et al. (2006). Autocrine tumor necrosis factor alpha links endoplasmic reticulum stress to the membrane death receptor pathway through IRE1alpha-mediated NF-kappaB activation and down-regulation of TRAF2 expression. *Molecular and Cellular Biology, 26*, 3071–3084.

Hughes, E. G., Kang, S. H., Fukaya, M., & Bergles, D. E. (2013). Oligodendrocyte progenitors balance growth with self-repulsion to achieve homeostasis in the adult brain. *Nature Neuroscience, 16*, 668–676.

Hunot, S., Vila, M., Teismann, P., et al. (2004). JNK-mediated induction of cyclooxygenase 2 is required for neurodegeneration in a mouse model of Parkinson's disease. *Proceedings of the National*

Academy of Sciences of the United States of America, 101, 665–670.

Iwakoshi, N. N., Lee, A.-H., Vallabhajosyula, P., et al. (2003). Plasma cell differentiation and the unfolded protein response intersect at the transcription factor XBP-1. *Nature Immunology, 4*, 321–329.

Jang, P.-G., Namkoong, C., Kang, G. M., et al. (2010). NF-kappaB activation in hypothalamic pro-opiomelanocortin neurons is essential in illness- and leptin-induced anorexia. *The Journal of Biological Chemistry, 285*, 9706–9715.

Jeon, B. T., Jeong, E. A., Shin, H. J., et al. (2012). Resveratrol attenuates obesity-associated peripheral and central inflammation and improves memory deficit in mice fed a high-fat diet. *Diabetes, 61*, 1444–1454.

Jiang, H.-Y., Wek, S. A., McGrath, B. C., et al. (2003). Phosphorylation of the alpha subunit of eukaryotic initiation factor 2 is required for activation of NF-kappaB in response to diverse cellular stresses. *Molecular and Cellular Biology, 23*, 5651–5663.

Jin, S., Kim, J. G., Park, J. W., et al. (2016). Hypothalamic TLR2 triggers sickness behavior via a microglia-neuronal axis. *Scientific Reports, 6*, 29424.

Jousse, C., Oyadomari, S., Novoa, I., et al. (2003). Inhibition of a constitutive translation initiation factor 2alpha phosphatase, CReP, promotes survival of stressed cells. *The Journal of Cell Biology, 163*, 767–775.

Kanczkowski, W., Alexaki, V.-I., Tran, N., et al. (2013). Hypothalamo-pituitary and immune-dependent adrenal regulation during systemic inflammation. *Proceedings of the National Academy of Sciences of the United States of America, 110*, 14801–14806.

Kang, E.-B., Koo, J.-H., Jang, Y.-C., et al. (2016). Neuroprotective effects of endurance exercise against high-fat diet-induced hippocampal neuroinflammation. *Journal of Neuroendocrinology.* https://doi.org/10.1111/jne.12385.

Karatas, H., Erdener, S. E., Gursoy-Ozdemir, Y., et al. (2013). Spreading depression triggers headache by activating neuronal Panx1 channels. *Science, 339*, 1092–1095.

Kawamata, Y., Fujii, R., Hosoya, M., et al. (2003). A G protein-coupled receptor responsive to bile acids. *The Journal of Biological Chemistry, 278*, 9435–9440.

Keirstead, H. S., Levine, J. M., & Blakemore, W. F. (1998). Response of the oligodendrocyte progenitor cell population (defined by NG2 labelling) to demyelination of the adult spinal cord. *Glia, 22*, 161–170.

Kent, S., Rodriguez, F., Kelley, K. W., & Dantzer, R. (1994). Reduction in food and water intake induced by microinjection of interleukin-1 beta in the ventromedial hypothalamus of the rat. *Physiology & Behavior, 56*, 1031–1036.

Kent, S., Bret-Dibat, J. L., Kelley, K. W., & Dantzer, R. (1996). Mechanisms of sickness-induced decreases in food-motivated behavior. *Neuroscience and Biobehavioral Reviews, 20*, 171–175.

Kharitonenkov, A., Shiyanova, T. L., Koester, A., et al. (2005). FGF-21 as a novel metabolic regulator. *The Journal of Clinical Investigation, 115*, 1627–1635.

Kierdorf, K., Erny, D., Goldmann, T., et al. (2013). Microglia emerge from erythromyeloid precursors via Pu.1- and Irf8-dependent pathways. *Nature Neuroscience, 16*, 273–280.

Kim, M.-S., Pak, Y. K., Jang, P.-G., et al. (2006). Role of hypothalamic Foxo1 in the regulation of food intake and energy homeostasis. *Nature Neuroscience, 9*, 901–906.

Kim, J. G., Suyama, S., Koch, M., et al. (2014). Leptin signaling in astrocytes regulates hypothalamic neuronal circuits and feeding. *Nature Neuroscience, 17*, 908–910.

Kim, M. S., Yan, J., Wu, W., et al. (2015). Rapid linkage of innate immunological signals to adaptive immunity by the brain-fat axis. *Nature Immunology, 16*, 525–533.

Kim, G. H., Shi, G., Somlo, D. R. et al. (2018). Hypothalamic ER-associated degradation regulates POMC maturation, feeding, and age-associated obesity. Mar 1;128(3):1125–1140. https://doi.org/10.1172/JCI96420. Epub 2018 Feb 19. PMID: 29457782 Free PMC Article.

Kimata, Y., Ishiwata-Kimata, Y., Ito, T., et al. (2007). Two regulatory steps of ER-stress sensor Ire1 involving its cluster formation and interaction with unfolded proteins. *The Journal of Cell Biology, 179*, 75–86.

Kitamura, T., Feng, Y., Kitamura, Y. I., et al. (2006). Forkhead protein FoxO1 mediates Agrp-dependent effects of leptin on food intake. *Nature Medicine, 12*, 534–540.

Kleinridders, A., Schenten, D., Könner, A. C., et al. (2009). MyD88 signaling in the CNS is required for development of fatty acid-induced leptin resistance and diet-induced obesity. *Cell Metabolism, 10*, 249–259.

Kokoeva, M. V., Yin, H., & Flier, J. S. (2005). Neurogenesis in the hypothalamus of adult mice: Potential role in energy balance. *Science, 310*, 679–683.

Kreutzer, C., Peters, S., Schulte, D. M., et al. (2017). Hypothalamic inflammation in human obesity is mediated by environmental and genetic factors. *Diabetes, 66*, 2407–2415.

Kuan, C. Y., Yang, D. D., Samanta Roy, D. R., et al. (1999). The Jnk1 and Jnk2 protein kinases are required for regional specific apoptosis during early brain development. *Neuron, 22*, 667–676.

Kuan, C.-Y., Whitmarsh, A. J., Yang, D. D., et al. (2003). A critical role of neural-specific JNK3 for ischemic apoptosis. *Proceedings of the National Academy of Sciences of the United States of America, 100*, 15184–15189.

Kummer, J. A., Broekhuizen, R., Everett, H., et al. (2007). Inflammasome components NALP 1 and 3 show distinct but separate expression profiles in human tissues suggesting a site-specific role in the inflammatory response. *The Journal of Histochemistry and Cytochemistry, 55*, 443–452.

Ladrière, L., Igoillo-Esteve, M., Cunha, D. A., et al. (2010). Enhanced signaling downstream of ribonucleic acid-activated protein kinase-like endoplasmic reticulum kinase potentiates lipotoxic endoplasmic reticulum stress in human islets. *The Journal of Clinical Endocrinology and Metabolism, 95*, 1442–1449.

Lam, T. K. T., Gutierrez-Juarez, R., Pocai, A., & Rossetti, L. (2005). Regulation of blood glucose by hypothalamic pyruvate metabolism. *Science, 309*, 943–947.

Langlet, F., Levin, B. E., Luquet, S., et al. (2013a). Tanycytic VEGF-A boosts blood-hypothalamus barrier plasticity and access of metabolic signals to the arcuate nucleus in response to fasting. *Cell Metabolism, 17*, 607–617.

Langlet, F., Mullier, A., Bouret, S. G., et al. (2013b). Tanycyte-like cells form a blood-cerebrospinal fluid barrier in the circumventricular organs of the mouse brain. *The Journal of Comparative Neurology, 521*, 3389–3405.

Lawson, L. J., Perry, V. H., Dri, P., & Gordon, S. (1990). Heterogeneity in the distribution and morphology of microglia in the normal adult mouse brain. *Neuroscience, 39*, 151–170.

Lawson, L. J., Perry, V. H., & Gordon, S. (1992). Turnover of resident microglia in the normal adult mouse brain. *Neuroscience, 48*, 405–415.

Layé, S., Parnet, P., Goujon, E., & Dantzer, R. (1994). Peripheral administration of lipopolysaccharide induces the expression of cytokine transcripts in the brain and pituitary of mice. *Brain Research. Molecular Brain Research, 27*, 157–162.

Le Foll, C., & Levin, B. E. (2016). Fatty acid-induced astrocyte ketone production and the control of food intake. *American Journal of Physiology. Regulatory, Integrative and Comparative Physiology, 310*, R1186–R1192.

Lee, S. Y., Reichlin, A., Santana, A., et al. (1997). TRAF2 is essential for JNK but not NF-kappaB activation and regulates lymphocyte proliferation and survival. *Immunity, 7*, 703–713.

Lee, A.-H., Iwakoshi, N. N., & Glimcher, L. H. (2003). XBP-1 regulates a subset of endoplasmic reticulum resident chaperone genes in the unfolded protein response. *Molecular and Cellular Biology, 23*, 7448–7459.

Lee, A.-H., Scapa, E. F., Cohen, D. E., & Glimcher, L. H. (2008). Regulation of hepatic lipogenesis by the transcription factor XBP1. *Science, 320*, 1492–1496.

Lee, Y.-Y., Cevallos, R. C., & Jan, E. (2009). An upstream open reading frame regulates translation of GADD34 during cellular stresses that induce eIF2alpha phosphorylation. *The Journal of Biological Chemistry, 284*, 6661–6673.

Lee, A.-H., Heidtman, K., Hotamisligil, G. S., & Glimcher, L. H. (2011). Dual and opposing roles of the unfolded protein response regulated by IRE1alpha and XBP1 in proinsulin processing and insulin secretion. *Proceedings of the National Academy of Sciences of the United States of America, 108*, 8885–8890.

Lee, D. A., Bedont, J. L., Pak, T., et al. (2012). Tanycytes of the hypothalamic median eminence form a diet-responsive neurogenic niche. *Nature Neuroscience, 15*, 700–702.

Lehnardt, S. (2010). Innate immunity and neuroinflammation in the CNS: The role of microglia in toll-like receptor-mediated neuronal injury. *Glia, 58*, 253–263.

Lehnardt, S., Massillon, L., Follett, P., et al. (2003). Activation of innate immunity in the CNS triggers neurodegeneration through a toll-like receptor 4-dependent pathway. *Proceedings of the National Academy of Sciences of the United States of America, 100*, 8514–8519.

Levine, J. M., & Reynolds, R. (1999). Activation and proliferation of endogenous oligodendrocyte precursor cells during ethidium bromide-induced demyelination. *Experimental Neurology, 160*, 333–347.

Li, J., Tang, Y., & Cai, D. (2012). IKKβ/NF-κB disrupts adult hypothalamic neural stem cells to mediate a neurodegenerative mechanism of dietary obesity and prediabetes. *Nature Cell Biology, 14*, 999–1012.

Li, J., Labbadia, J., & Morimoto, R. I. (2017). Rethinking HSF1 in stress, development, and organismal health. *Trends in Cell Biology*. https://doi.org/10.1016/j.tcb.2017.08.002.

Liddelow, S. A., Guttenplan, K. A., Clarke, L. E., et al. (2017). Neurotoxic reactive astrocytes are induced by activated microglia. *Nature, 541*, 481–487.

Lindqvist, A., Mohapel, P., Bouter, B., et al. (2006). High-fat diet impairs hippocampal neurogenesis in male rats. *European Journal of Neurology, 13*, 1385–1388.

Lu, P. D., Harding, H. P., & Ron, D. (2004). Translation reinitiation at alternative open reading frames regulates gene expression in an integrated stress response. *The Journal of Cell Biology, 167*, 27–33.

Lu, Y., Liang, F.-X., & Wang, X. (2014). A synthetic biology approach identifies the mammalian UPR RNA ligase RtcB. *Molecular Cell, 55*, 758–770.

Lu, W., Hagiwara, D., Morishita, Y., et al. (2016). Unfolded protein response in hypothalamic cultures of wild-type and ATF6α-knockout mice. *Neuroscience Letters, 612*, 199–203.

Lundgaard, I., Li, B., Xie, L., et al. (2015). Direct neuronal glucose uptake heralds activity-dependent increases in cerebral metabolism. *Nature Communications, 6*, 6807.

Malhotra, R., Warne, J. P., Salas, E., et al. (2015). Loss of Atg12, but not Atg5, in pro-opiomelanocortin neurons exacerbates diet-induced obesity. *Autophagy, 11*, 145–154.

Maolood, N., & Meister, B. (2009). Protein components of the blood-brain barrier (BBB) in the brainstem area postrema-nucleus tractus solitarius region. *Journal of Chemical Neuroanatomy, 37*, 182–195.

Martínez, G., Vidal, R. L., Mardones, P., et al. (2016). Regulation of memory formation by the transcription factor XBP1. *Cell Reports, 14*, 1382–1394.

Martínez-Sánchez, N., Seoane-Collazo, P., Contreras, C., et al. (2017). Hypothalamic AMPK-ER stress-JNK1 axis mediates the central actions of thyroid hormones on energy balance. *Cell Metabolism, 26*, 212–229.e12.

Martinon, F., Burns, K., & Tschopp, J. (2002). The inflammasome: A molecular platform triggering activation of inflammatory caspases and processing of proIL-beta. *Molecular Cell, 10*, 417–426.

Martinon, F., Chen, X., Lee, A.-H., & Glimcher, L. H. (2010). TLR activation of the transcription factor

XBP1 regulates innate immune responses in macrophages. *Nature Immunology, 11*, 411–418.

Masson, G. S., Nair, A. R., Dange, R. B., et al. (2015). Toll-like receptor 4 promotes autonomic dysfunction, inflammation and microglia activation in the hypothalamic paraventricular nucleus: Role of endoplasmic reticulum stress. *PLoS One, 10*, e0122850.

Maurel, M., Chevet, E., Tavernier, J., & Gerlo, S. (2014). Getting RIDD of RNA: IRE1 in cell fate regulation. *Trends in Biochemical Sciences, 39*, 245–254.

Mauro, C., De Rosa, V., Marelli-Berg, F., & Solito, E. (2014). Metabolic syndrome and the immunological affair with the blood-brain barrier. *Frontiers in Immunology, 5*, 677.

Mayer, C. M., & Belsham, D. D. (2010). Palmitate attenuates insulin signaling and induces endoplasmic reticulum stress and apoptosis in hypothalamic neurons: Rescue of resistance and apoptosis through adenosine 5′ monophosphate-activated protein kinase activation. *Endocrinology, 151*, 576–585.

McGavigan, A. K., Henseler, Z. M., Garibay, D., et al. (2017). Vertical sleeve gastrectomy reduces blood pressure and hypothalamic endoplasmic reticulum stress in mice. *Disease Models & Mechanisms, 10*, 235–243.

Medzhitov, R. (2008). Origin and physiological roles of inflammation. *Nature, 454*, 428–435.

Meng, Q., & Cai, D. (2011). Defective hypothalamic autophagy directs the central pathogenesis of obesity via the IkappaB kinase beta (IKKbeta)/NF-kappaB pathway. *The Journal of Biological Chemistry, 286*, 32324–32332.

Milanski, M., Degasperi, G., Coope, A., et al. (2009). Saturated fatty acids produce an inflammatory response predominantly through the activation of TLR4 signaling in hypothalamus: Implications for the pathogenesis of obesity. *The Journal of Neuroscience, 29*, 359–370.

Milanski, M., Arruda, A. P., Coope, A., et al. (2012). Inhibition of hypothalamic inflammation reverses diet-induced insulin resistance in the liver. *Diabetes, 61*, 1455–1462.

Minkiewicz, J., de Rivero Vaccari, J. P., & Keane, R. W. (2013). Human astrocytes express a novel NLRP2 inflammasome. *Glia, 61*, 1113–1121.

Moraes, J. C., Coope, A., Morari, J., et al. (2009). High-fat diet induces apoptosis of hypothalamic neurons. *PLoS One, 4*, e5045.

Morari, J., Anhe, G. F., Nascimento, L. F., et al. (2014). Fractalkine (CX3CL1) is involved in the early activation of hypothalamic inflammation in experimental obesity. *Diabetes, 63*, 3770–3784.

Morishita, Y., Arima, H., Hiroi, M., et al. (2011). Poly(A) tail length of neurohypophysial hormones is shortened under endoplasmic reticulum stress. *Endocrinology, 152*, 4846–4855.

Morita, S., Villalta, S. A., Feldman, H. C., et al. (2017). Targeting ABL-IRE1α signaling spares ER-stressed pancreatic β cells to reverse autoimmune diabetes. *Cell Metabolism, 25*, 883–897.e8.

Mori, K., Ma, W., Gething, M. J., et al. (1993). A transmembrane protein with a cdc2+ cdc28-related kinase activity is required for signaling from the ER to the nucleus. *Cell, 74*(4), 743–756.

Nadeau, S., & Rivest, S. (1999). Effects of circulating tumor necrosis factor on the neuronal activity and expression of the genes encoding the tumor necrosis factor receptors (p55 and p75) in the rat brain: A view from the blood-brain barrier. *Neuroscience, 93*, 1449–1464.

Nakamura, D., Tsuru, A., Ikegami, K., et al. (2011). Mammalian ER stress sensor IRE1β specifically down-regulates the synthesis of secretory pathway proteins. *FEBS Letters, 585*, 133–138.

Naranjo, J. R., Zhang, H., Villar, D., et al. (2016). Activating transcription factor 6 derepression mediates neuroprotection in Huntington disease. *The Journal of Clinical Investigation, 126*, 627–638.

Naznin, F., Toshinai, K., Waise, T. M. Z., et al. (2015). Diet-induced obesity causes peripheral and central ghrelin resistance by promoting inflammation. *The Journal of Endocrinology, 226*, 81–92.

Nelson, G., Wilde, G. J. C., Spiller, D. G., et al. (2003). NF-kappaB signalling is inhibited by glucocorticoid receptor and STAT6 via distinct mechanisms. *Journal of Cell Science, 116*, 2495–2503.

Neumann, H., Cavalié, A., Jenne, D. E., & Wekerle, H. (1995). Induction of MHC class I genes in neurons. *Science, 269*, 549–552.

Nguyen, J. C. D., Killcross, A. S., & Jenkins, T. A. (2014). Obesity and cognitive decline: Role of inflammation and vascular changes. *Frontiers in Neuroscience, 8*, 375.

Nishiyama, A., Komitova, M., Suzuki, R., & Zhu, X. (2009). Polydendrocytes (NG2 cells): Multifunctional cells with lineage plasticity. *Nature Reviews. Neuroscience, 10*, 9–22.

Novoa, I., Zeng, H., Harding, H. P., & Ron, D. (2001). Feedback inhibition of the unfolded protein response by GADD34-mediated dephosphorylation of eIF2alpha. *The Journal of Cell Biology, 153*, 1011–1022.

Ogimoto, K., Harris, M. K., Jr., & Wisse, B. E. (2006). MyD88 is a key mediator of anorexia, but not weight loss, induced by lipopolysaccharide and interleukin-1 beta. *Endocrinology, 147*, 4445–4453.

Okin, D., & Medzhitov, R. (2016). The effect of sustained inflammation on hepatic mevalonate pathway results in hyperglycemia. *Cell, 165*, 343–356.

Olivares, S., & Henkel, A. S. (2015). Hepatic Xbp1 gene deletion promotes endoplasmic reticulum stress-induced liver injury and apoptosis. *The Journal of Biological Chemistry, 290*, 30142–30151.

Ono, H., Pocai, A., Wang, Y., et al. (2008). Activation of hypothalamic S6 kinase mediates diet-induced hepatic insulin resistance in rats. *The Journal of Clinical Investigation, 118*, 2959–2968.

Oral, E. A., Reilly, S. M., Gomez, A. V., et al. (2017). Inhibition of IKKε and TBK1 improves glucose control in a subset of patients with type 2 diabetes. *Cell Metabolism, 26*, 157–170.e7.

Outinen, P. A., Sood, S. K., Liaw, P. C., et al. (1998). Characterization of the stress-inducing effects of homocysteine. *The Biochemical Journal, 332*(Pt 1), 213–221.

Ozcan, U., Cao, Q., Yilmaz, E., et al. (2004). Endoplasmic reticulum stress links obesity, insulin action, and type 2 diabetes. *Science, 306*, 457–461.

Padilla, S. L., Reef, D., & Zeltser, L. M. (2012). Defining POMC neurons using transgenic reagents: Impact of transient Pomc expression in diverse immature neuronal populations. *Endocrinology, 153*, 1219–1231.

Pakos-Zebrucka, K., Koryga, I., Mnich, K., et al. (2016). The integrated stress response. *EMBO Reports, 17*, 1374–1395.

Paolicelli, R. C., Bolasco, G., Pagani, F., et al. (2011). Synaptic pruning by microglia is necessary for normal brain development. *Science, 333*, 1456–1458.

Park, H. R., Park, M., Choi, J., et al. (2010). A high-fat diet impairs neurogenesis: Involvement of lipid peroxidation and brain-derived neurotrophic factor. *Neuroscience Letters, 482*, 235–239.

Pascual, O., Ben Achour, S., Rostaing, P., et al. (2012). Microglia activation triggers astrocyte-mediated modulation of excitatory neurotransmission. *Proceedings of the National Academy of Sciences of the United States of America, 109*, E197–E205.

Perello, M., Cakir, I., Cyr, N. E., et al. (2010). Maintenance of the thyroid axis during diet-induced obesity in rodents is controlled at the central level. *American Journal of Physiology. Endocrinology and Metabolism, 299*, E976–E989.

Pérez-Domínguez, M., Tovar-Y-Romo, L. B., & Zepeda, A. (2017). Neuroinflammation and physical exercise as modulators of adult hippocampal neural precursor cell behavior. *Reviews in the Neurosciences*. https://doi.org/10.1515/revneuro-2017-0024.

Pistell, P. J., Morrison, C. D., Gupta, S., et al. (2010). Cognitive impairment following high fat diet consumption is associated with brain inflammation. *Journal of Neuroimmunology, 219*, 25–32.

Posey, K. A., Clegg, D. J., Printz, R. L., et al. (2009). Hypothalamic proinflammatory lipid accumulation, inflammation, and insulin resistance in rats fed a high-fat diet. *American Journal of Physiology. Endocrinology and Metabolism, 296*, E1003–E1012.

Pozzan, T., Rizzuto, R., Volpe, P., & Meldolesi, J. (1994). Molecular and cellular physiology of intracellular calcium stores. *Physiological Reviews, 74*, 595–636.

Prada, P. O., Zecchin, H. G., Gasparetti, A. L., et al. (2005). Western diet modulates insulin signaling, c-Jun N-terminal kinase activity, and insulin receptor substrate-1ser307 phosphorylation in a tissue-specific fashion. *Endocrinology, 146*, 1576–1587.

Puig, J., Blasco, G., Daunis-I-Estadella, J., et al. (2015). Hypothalamic damage is associated with inflammatory markers and worse cognitive performance in obese subjects. *The Journal of Clinical Endocrinology and Metabolism, 100*, E276–E281.

Purkayastha, S., Zhang, G., & Cai, D. (2011a). Uncoupling the mechanisms of obesity and hyperten-

sion by targeting hypothalamic IKK-β and NF-κB. *Nature Medicine, 17*, 883–887.

Purkayastha, S., Zhang, H., Zhang, G., et al. (2011b). Neural dysregulation of peripheral insulin action and blood pressure by brain endoplasmic reticulum stress. *Proceedings of the National Academy of Sciences of the United States of America, 108*, 2939–2944.

Quagliarello, V. J., Wispelwey, B., Long, W. J., Jr., & Scheld, W. M. (1991). Recombinant human interleukin-1 induces meningitis and blood-brain barrier injury in the rat. Characterization and comparison with tumor necrosis factor. *The Journal of Clinical Investigation, 87*, 1360–1366.

Quarta, C., Clemmensen, C., Zhu, Z., et al. (2017). Molecular integration of incretin and glucocorticoid action reverses immunometabolic dysfunction and obesity. *Cell Metabolism, 26*, 620–632.e6.

Ramírez, S., Gómez-Valadés, A. G., Schneeberger, M., et al. (2017). Mitochondrial dynamics mediated by Mitofusin 1 is required for POMC neuron glucose-sensing and insulin release control. *Cell Metabolism, 25*, 1390–1399.e6.

Ramos, H. J., Lanteri, M. C., Blahnik, G., et al. (2012). IL-1β signaling promotes CNS-intrinsic immune control of West Nile virus infection. *PLoS Pathogens, 8*, e1003039.

Ray, A., & Prefontaine, K. E. (1994). Physical association and functional antagonism between the p65 subunit of transcription factor NF-kappa B and the glucocorticoid receptor. *Proceedings of the National Academy of Sciences of the United States of America, 91*, 752–756.

Reilly, S. M., Chiang, S.-H., Decker, S. J., et al. (2013). An inhibitor of the protein kinases TBK1 and IKK-ε improves obesity-related metabolic dysfunctions in mice. *Nature Medicine, 19*, 313–321.

Reimold, A. M., Etkin, A., Clauss, I., et al. (2000). An essential role in liver development for transcription factor XBP-1. *Genes & Development, 14*, 152–157.

Reimold, A. M., Iwakoshi, N. N., Manis, J., et al. (2001). Plasma cell differentiation requires the transcription factor XBP-1. *Nature, 412*, 300–307.

Ren, H., Orozco, I. J., Su, Y., et al. (2012). FoxO1 target Gpr17 activates AgRP neurons to regulate food intake. *Cell, 149*, 1314–1326.

Riediger, T. (2012). The receptive function of hypothalamic and brainstem centres to hormonal and nutrient signals affecting energy balance. *The Proceedings of the Nutrition Society, 71*, 463–477.

Rojas, M., Vasconcelos, G., & Dever, T. E. (2015). An eIF2α-binding motif in protein phosphatase 1 subunit GADD34 and its viral orthologs is required to promote dephosphorylation of eIF2α. *Proceedings of the National Academy of Sciences of the United States of America, 112*, E3466–E3475.

Rolls, A., Shechter, R., London, A., et al. (2007). Toll-like receptors modulate adult hippocampal neurogenesis. *Nature Cell Biology, 9*, 1081–1088.

Ropelle, E. R., Flores, M. B., Cintra, D. E., et al. (2010). IL-6 and IL-10 anti-inflammatory activity links exercise to hypothalamic insulin and leptin sensitivity through IKKbeta and ER stress inhibition. *PLoS Biology.* https://doi.org/10.1371/journal.pbio.1000465.

Rottkamp, D. M., Rudenko, I. A., Maier, M. T., et al. (2015). Leptin potentiates astrogenesis in the developing hypothalamus. *Mol Metab, 4,* 881–889.

Russo, I., Barlati, S., & Bosetti, F. (2011). Effects of neuroinflammation on the regenerative capacity of brain stem cells. *Journal of Neurochemistry, 116,* 947–956.

Sabio, G., Das, M., Mora, A., et al. (2008). A stress signaling pathway in adipose tissue regulates hepatic insulin resistance. *Science, 322,* 1539–1543.

Sabio, G., Cavanagh-Kyros, J., Ko, H. J., et al. (2009). Prevention of steatosis by hepatic JNK1. *Cell Metabolism, 10,* 491–498.

Sabio, G., Cavanagh-Kyros, J., Barrett, T., et al. (2010a). Role of the hypothalamic-pituitary-thyroid axis in metabolic regulation by JNK1. *Genes & Development, 24,* 256–264.

Sabio, G., Kennedy, N. J., Cavanagh-Kyros, J., et al. (2010b). Role of muscle c-Jun NH2-terminal kinase 1 in obesity-induced insulin resistance. *Molecular and Cellular Biology, 30,* 106–115.

Sachot, C., Poole, S., & Luheshi, G. N. (2004). Circulating leptin mediates lipopolysaccharide-induced anorexia and fever in rats. *The Journal of Physiology, 561,* 263–272.

Sacoccio, C., Dornand, J., & Barbanel, G. (1998). Differential regulation of brain and plasma TNFalpha produced after endotoxin shock. *Neuroreport, 9,* 309–313.

Sagar, S. M., & Sharp, F. R. (1993). Early response genes as markers of neuronal activity and growth factor action. *Advances in Neurology, 59,* 273–284.

Schaap, F. G., Kremer, A. E., Lamers, W. H., et al. (2013). Fibroblast growth factor 21 is induced by endoplasmic reticulum stress. *Biochimie, 95,* 692–699.

Schenk, S., Saberi, M., & Olefsky, J. M. (2008). Insulin sensitivity: Modulation by nutrients and inflammation. *The Journal of Clinical Investigation, 118,* 2992–3002.

Schneeberger, M., Dietrich, M. O., Sebastián, D., et al. (2013). Mitofusin 2 in POMC neurons connects ER stress with leptin resistance and energy imbalance. *Cell, 155,* 172–187.

Schneeberger, M., Gómez-Valadés, A. G., Altirriba, J., et al. (2015). Reduced α-MSH underlies hypothalamic ER-stress-induced hepatic gluconeogenesis. *Cell Reports, 12,* 361–370.

Sebastián, D., Hernández-Alvarez, M. I., Segalés, J., et al. (2012). Mitofusin 2 (Mfn2) links mitochondrial and endoplasmic reticulum function with insulin signaling and is essential for normal glucose homeostasis. *Proceedings of the National Academy of Sciences of the United States of America, 109,* 5523–5528.

Serhan, C. N., Brain, S. D., Buckley, C. D., et al. (2007). Resolution of inflammation: State of the art, definitions and terms. *The FASEB Journal, 21,* 325–332.

Shan, B., Wang, X., Wu, Y., et al. (2017). The metabolic ER stress sensor IRE1α suppresses alternative activation of macrophages and impairs energy expenditure in obesity. *Nature Immunology, 18,* 519–529.

Shen, J., Chen, X., Hendershot, L., & Prywes, R. (2002). ER stress regulation of ATF6 localization by dissociation of BiP/GRP78 binding and unmasking of Golgi localization signals. *Developmental Cell, 3,* 99–111.

Shi, H., Kokoeva, M. V., Inouye, K., et al. (2006). TLR4 links innate immunity and fatty acid-induced insulin resistance. *The Journal of Clinical Investigation, 116,* 3015–3025.

Shi, G., Somlo, D., Kim, G. H., et al. (2017). ER-associated degradation is required for vasopressin prohormone processing and systemic water homeostasis. *The Journal of Clinical Investigation, 127,* 3897–3912.

Shimazaki, T., Shingo, T., & Weiss, S. (2001). The ciliary neurotrophic factor/leukemia inhibitory factor/gp130 receptor complex operates in the maintenance of mammalian forebrain neural stem cells. *The Journal of Neuroscience, 21,* 7642–7653.

Sidrauski, C., Acosta-Alvear, D., Khoutorsky, A., et al. (2013). Pharmacological brake-release of mRNA translation enhances cognitive memory. *eLife, 2,* e00498.

Silverman, W. R., de Rivero Vaccari, J. P., Locovei, S., et al. (2009). The pannexin 1 channel activates the inflammasome in neurons and astrocytes. *The Journal of Biological Chemistry, 284,* 18143–18151.

Smith, C. A., Farrah, T., & Goodwin, R. G. (1994). The TNF receptor superfamily of cellular and viral proteins: Activation, costimulation, and death. *Cell, 76,* 959–962.

Song, L., Pei, L., Yao, S., et al. (2017). NLRP3 inflammasome in neurological diseases, from functions to therapies. *Frontiers in Cellular Neuroscience, 11,* 63.

Spencer, S. J., D'Angelo, H., Soch, A., et al. (2017). High-fat diet and aging interact to produce neuroinflammation and impair hippocampal- and amygdalar-dependent memory. *Neurobiology of Aging, 58,* 88–101.

Speretta, G. F., Silva, A. A., Vendramini, R. C., et al. (2016). Resistance training prevents the cardiovascular changes caused by high-fat diet. *Life Sciences, 146,* 154–162.

Sriburi, R., Jackowski, S., Mori, K., & Brewer, J. W. (2004). XBP1: A link between the unfolded protein response, lipid biosynthesis, and biogenesis of the endoplasmic reticulum. *The Journal of Cell Biology, 167,* 35–41.

Steiner, A. A., Dogan, M. D., Ivanov, A. I., et al. (2004). A new function of the leptin receptor: Mediation of the recovery from lipopolysaccharide-induced hypothermia. *The FASEB Journal, 18,* 1949–1951.

Stutz, A., Golenbock, D. T., & Latz, E. (2009). Inflammasomes: Too big to miss. *The Journal of Clinical Investigation, 119,* 3502–3511.

Takatsuki, A., & Tamura, G. (1971). Effect of tunicamycin on the synthesis of macromolecules in cultures of chick embryo fibroblasts infected with Newcastle disease virus. *The Journal of Antibiotics, 24,* 785–794.

Taniuchi, S., Miyake, M., Tsugawa, K., et al. (2016). Integrated stress response of vertebrates is regulated by four eIF2α kinases. *Scientific Reports, 6*, 32886.

Taylor, R. C., & Dillin, A. (2013). XBP-1 is a cell-nonautonomous regulator of stress resistance and longevity. *Cell, 153*, 1435–1447.

Thaler, J. P., Yi, C.-X., Schur, E. A., et al. (2012). Obesity is associated with hypothalamic injury in rodents and humans. *The Journal of Clinical Investigation, 122*, 153–162.

Tsaousidou, E., Paeger, L., Belgardt, B. F., et al. (2014). Distinct roles for JNK and IKK activation in agouti-related peptide neurons in the development of obesity and insulin resistance. *Cell Reports, 9*, 1495–1506.

Unger, E. K., Piper, M. L., Olofsson, L. E., & Xu, A. W. (2010). Functional role of c-Jun-N-terminal kinase in feeding regulation. *Endocrinology, 151*, 671–682.

Upton, J.-P., Wang, L., Han, D., et al. (2012). IRE1α cleaves select microRNAs during ER stress to derepress translation of proapoptotic Caspase-2. *Science, 338*, 818–822.

Urano, F., Wang, X., Bertolotti, A., et al. (2000). Coupling of stress in the ER to activation of JNK protein kinases by transmembrane protein kinase IRE1. *Science, 287*, 664–666.

Uysal, K. T., Wiesbrock, S. M., Marino, M. W., & Hotamisligil, G. S. (1997). Protection from obesity-induced insulin resistance in mice lacking TNF-alpha function. *Nature, 389*, 610–614.

Valdearcos, M., Robblee, M. M., Benjamin, D. I., et al. (2014). Microglia dictate the impact of saturated fat consumption on hypothalamic inflammation and neuronal function. *Cell Reports, 9*, 2124–2138.

Valdearcos, M., Douglass, J. D., Robblee, M. M., et al. (2017). Microglial inflammatory signaling orchestrates the hypothalamic immune response to dietary excess and mediates obesity susceptibility. *Cell Metabolism, 26*, 185–197.e3.

Valdés, P., Mercado, G., Vidal, R. L., et al. (2014). Control of dopaminergic neuron survival by the unfolded protein response transcription factor XBP1. *Proceedings of the National Academy of Sciences of the United States of America, 111*, 6804–6809.

van Oosten-Hawle, P., & Morimoto, R. I. (2014). Organismal proteostasis: Role of cell-nonautonomous regulation and transcellular chaperone signaling. *Genes & Development, 28*, 1533–1543.

van Oosten-Hawle, P., Porter, R. S., & Morimoto, R. I. (2013). Regulation of organismal proteostasis by transcellular chaperone signaling. *Cell, 153*, 1366–1378.

Vandanmagsar, B., Youm, Y.-H., Ravussin, A., et al. (2011). The NLRP3 inflammasome instigates obesity-induced inflammation and insulin resistance. *Nature Medicine, 17*, 179–188.

Varatharaj, A., & Galea, I. (2017). The blood-brain barrier in systemic inflammation. *Brain, Behavior, and Immunity, 60*, 1–12.

Vattem, K. M., & Wek, R. C. (2004). Reinitiation involving upstream ORFs regulates ATF4 mRNA translation in mammalian cells. *Proceedings of the National Academy of Sciences of the United States of America, 101*, 11269–11274.

Vernia, S., Cavanagh-Kyros, J., Garcia-Haro, L., et al. (2014). The PPARα-FGF21 hormone axis contributes to metabolic regulation by the hepatic JNK signaling pathway. *Cell Metabolism, 20*, 512–525.

Vernia, S., Morel, C., Madara, J. C., et al. (2016). Excitatory transmission onto AgRP neurons is regulated by cJun NH2-terminal kinase 3 in response to metabolic stress. *eLife, 5*, e10031.

Veronese, N., Facchini, S., Stubbs, B., et al. (2017). Weight loss is associated with improvements in cognitive function among overweight and obese people: A systematic review and meta-analysis. *Neuroscience and Biobehavioral Reviews, 72*, 87–94.

Wake, H., Moorhouse, A. J., Miyamoto, A., & Nabekura, J. (2013). Microglia: Actively surveying and shaping neuronal circuit structure and function. *Trends in Neurosciences, 36*, 209–217.

Wallingford, N., Perroud, B., Gao, Q., et al. (2009). Prolylcarboxypeptidase regulates food intake by inactivating alpha-MSH in rodents. *The Journal of Clinical Investigation, 119*, 2291–2303.

Walton, N. M., Sutter, B. M., Laywell, E. D., et al. (2006). Microglia instruct subventricular zone neurogenesis. *Glia, 54*, 815–825.

Wang, Y., Vera, L., Fischer, W. H., & Montminy, M. (2009). The CREB coactivator CRTC2 links hepatic ER stress and fasting gluconeogenesis. *Nature, 460*, 534–537.

Wang, S., Chen, Z., Lam, V., et al. (2012). IRE1α-XBP1s induces PDI expression to increase MTP activity for hepatic VLDL assembly and lipid homeostasis. *Cell Metabolism, 16*, 473–486.

Wang, S., Huang, X.-F., Zhang, P., et al. (2017). Dietary teasaponin ameliorates alteration of gut microbiota and cognitive decline in diet-induced obese mice. *Scientific Reports, 7*, 12203.

Weissmann, L., Quaresma, P. G. F., Santos, A. C., et al. (2014). IKKε is key to induction of insulin resistance in the hypothalamus, and its inhibition reverses obesity. *Diabetes, 63*, 3334–3345.

Welch, W. J., & Brown, C. R. (1996). Influence of molecular and chemical chaperones on protein folding. *Cell Stress & Chaperones, 1*, 109–115.

Wilkins, A., Chandran, S., & Compston, A. (2001). A role for oligodendrocyte-derived IGF-1 in trophic support of cortical neurons. *Glia, 36*, 48–57.

Wilkins, A., Majed, H., Layfield, R., et al. (2003). Oligodendrocytes promote neuronal survival and axonal length by distinct intracellular mechanisms: A novel role for oligodendrocyte-derived glial cell line-derived neurotrophic factor. *The Journal of Neuroscience, 23*, 4967–4974.

Williams, K. W., Liu, T., Kong, X., et al. (2014). Xbp1s in Pomc neurons connects ER stress with energy balance and glucose homeostasis. *Cell Metabolism, 20*, 471–482.

Won, J. C., Jang, P.-G., Namkoong, C., et al. (2009). Central administration of an endoplasmic reticulum stress inducer inhibits the anorexigenic effects of leptin and insulin. *Obesity, 17*, 1861–1865.

Wong, M. L., Bongiorno, P. B., Rettori, V., et al. (1997). Interleukin (IL) 1beta, IL-1 receptor antagonist, IL-10, and IL-13 gene expression in the central nervous system and anterior pituitary during systemic inflammation: Pathophysiological implications. *Proceedings of the National Academy of Sciences of the United States of America, 94*, 227–232.

Woo, C. W., Cui, D., Arellano, J., et al. (2009). Adaptive suppression of the ATF4-CHOP branch of the unfolded protein response by toll-like receptor signalling. *Nature Cell Biology, 11*, 1473–1480.

Wu, K. L. H., Chan, S. H. H., & Chan, J. Y. H. (2012). Neuroinflammation and oxidative stress in rostral ventrolateral medulla contribute to neurogenic hypertension induced by systemic inflammation. *Journal of Neuroinflammation, 9*, 212.

Wunderlich, F. T., Luedde, T., Singer, S., et al. (2008). Hepatic NF-kappa B essential modulator deficiency prevents obesity-induced insulin resistance but synergizes with high-fat feeding in tumorigenesis. *Proceedings of the National Academy of Sciences of the United States of America, 105*, 1297–1302.

Xiao, Y., Deng, Y., Yuan, F., et al. (2017a). ATF4/ATG5 signaling in hypothalamic proopiomelanocortin neurons regulates fat mass via affecting energy expenditure. *Diabetes, 66*, 1146–1158.

Xiao, Y., Deng, Y., Yuan, F., et al. (2017b). An ATF4-ATG5 signaling in hypothalamic POMC neurons regulates obesity. *Autophagy, 13*, 1088–1089.

Xu, H., Barnes, G. T., Yang, Q., et al. (2003). Chronic inflammation in fat plays a crucial role in the development of obesity-related insulin resistance. *The Journal of Clinical Investigation, 112*, 1821–1830.

Xu, T., Yang, L., Yan, C., et al. (2014). The IRE1α-XBP1 pathway regulates metabolic stress-induced compensatory proliferation of pancreatic β-cells. *Cell Research, 24*, 1137–1140.

Yagishita, N., Ohneda, K., Amano, T., et al. (2005). Essential role of synoviolin in embryogenesis. *The Journal of Biological Chemistry, 280*, 7909–7916.

Yamamoto, K., Sato, T., Matsui, T., et al. (2007). Transcriptional induction of mammalian ER quality control proteins is mediated by single or combined action of ATF6alpha and XBP1. *Developmental Cell, 13*, 365–376.

Yamawaki, Y., Kimura, H., Hosoi, T., & Ozawa, K. (2010). MyD88 plays a key role in LPS-induced Stat3 activation in the hypothalamus. *American Journal of Physiology. Regulatory, Integrative and Comparative Physiology, 298*, R403–R410.

Yamazaki, H., Hiramatsu, N., Hayakawa, K., et al. (2009). Activation of the Akt-NF-kappaB pathway by subtilase cytotoxin through the ATF6 branch of the unfolded protein response. *Journal of Immunology, 183*, 1480–1487.

Yan, J., Zhang, H., Yin, Y., et al. (2014). Obesity- and aging-induced excess of central transforming growth factor-β potentiates diabetic development via an RNA stress response. *Nature Medicine, 20*, 1001–1008.

Yang, L., Qi, Y., & Yang, Y. (2015). Astrocytes control food intake by inhibiting AGRP neuron activity via adenosine A1 receptors. *Cell Reports, 11*, 798–807.

Yanguas-Casás, N., Barreda-Manso, M. A., Nieto-Sampedro, M., & Romero-Ramírez, L. (2017). TUDCA: An agonist of the bile acid receptor GPBAR1/TGR5 with anti-inflammatory effects in microglial cells. *Journal of Cellular Physiology, 232*, 2231–2245.

Yao, J. H., Ye, S. M., Burgess, W., et al. (1999). Mice deficient in interleukin-1beta converting enzyme resist anorexia induced by central lipopolysaccharide. *The American Journal of Physiology, 277*, R1435–R1443.

Yao, T., Deng, Z., Gao, Y., et al. (2017). Ire1α in Pomc neurons is required for thermogenesis and Glycemia. *Diabetes, 66*, 663–673.

Ye, J., Rawson, R. B., Komuro, R., et al. (2000). ER stress induces cleavage of membrane-bound ATF6 by the same proteases that process SREBPs. *Molecular Cell, 6*, 1355–1364.

Yehuda, S., Rabinovitz, S., & Mostofsky, D. I. (2005). Mediation of cognitive function by high fat diet following stress and inflammation. *Nutritional Neuroscience, 8*, 309–315.

Yi, C.-X., Al-Massadi, O., Donelan, E., et al. (2012a). Exercise protects against high-fat diet-induced hypothalamic inflammation. *Physiology & Behavior, 106*, 485–490.

Yi, C.-X., Gericke, M., Krüger, M., et al. (2012b). High calorie diet triggers hypothalamic angiopathy. *Mol Metab, 1*, 95–100.

Yi, C.-X., Tschöp, M. H., Woods, S. C., & Hofmann, S. M. (2012c). High-fat-diet exposure induces IgG accumulation in hypothalamic microglia. *Disease Models & Mechanisms, 5*, 686–690.

Yi, C.-X., Walter, M., Gao, Y., et al. (2017). TNFα drives mitochondrial stress in POMC neurons in obesity. *Nature Communications, 8*, 15143.

Yin, Y., Yan, Y., Jiang, X., et al. (2009). Inflammasomes are differentially expressed in cardiovascular and other tissues. *International Journal of Immunopathology and Pharmacology, 22*, 311–322.

Yoshida, H., Matsui, T., Yamamoto, A., et al. (2001). XBP1 mRNA is induced by ATF6 and spliced by IRE1 in response to ER stress to produce a highly active transcription factor. *Cell, 107*, 881–891.

Yoshida, H., Matsui, T., Hosokawa, N., et al. (2003). A time-dependent phase shift in the mammalian unfolded protein response. *Developmental Cell, 4*, 265–271.

Yoshida, H., Oku, M., Suzuki, M., & Mori, K. (2006). pXBP1(U) encoded in XBP1 pre-mRNA negatively regulates unfolded protein response activator pXBP1(S) in mammalian ER stress response. *The Journal of Cell Biology, 172*, 565–575.

Yoshikawa, A., Kamide, T., Hashida, K., et al. (2015). Deletion of Atf6α impairs astroglial activation and enhances neuronal death following brain ischemia in mice. *Journal of Neurochemistry, 132*, 342–353.

Youm, Y.-H., Grant, R. W., McCabe, L. R., et al. (2013). Canonical Nlrp3 inflammasome links systemic low-grade inflammation to functional decline in aging. *Cell Metabolism, 18*, 519–532.

Zabolotny, J. M., Kim, Y.-B., Welsh, L. A., et al. (2008). Protein-tyrosine phosphatase 1B expression is induced by inflammation in vivo. *The Journal of Biological Chemistry, 283*, 14230–14241.

Zhang, K., Shen, X., Wu, J., et al. (2006). Endoplasmic reticulum stress activates cleavage of CREBH to induce a systemic inflammatory response. *Cell, 124*, 587–599.

Zhang, X., Zhang, G., Zhang, H., et al. (2008). Hypothalamic IKKbeta/NF-kappaB and ER stress link overnutrition to energy imbalance and obesity. *Cell, 135*, 61–73.

Zhang, G., Li, J., Purkayastha, S., et al. (2013a). Hypothalamic programming of systemic ageing involving IKK-β, NF-κB and GnRH. *Nature, 497*, 211–216.

Zhang, Q., Yu, J., Liu, B., et al. (2013b). Central activating transcription factor 4 (ATF4) regulates hepatic insulin resistance in mice via S6K1 signaling and the vagus nerve. *Diabetes, 62*, 2230–2239.

Zhang, Y., Chen, K., Sloan, S. A., et al. (2014). An RNA-sequencing transcriptome and splicing database of glia, neurons, and vascular cells of the cerebral cortex. *The Journal of Neuroscience, 34*, 11929–11947.

The Cell Biology Neuropeptide Hormones

5

Eduardo A. Nillni

5.1 Introduction

To understand peptide hormone biosynthesis and their action at a distant target cell, we need first to comprehend the cell biology of these molecules, their origin, and the mechanism by which they became biologically active. Today we know that all peptide hormones and many nonhormonal proteins derived from larger inactive precursor proteins, which are posttranslationally modified to produce an array of different peptides with specific biological function and secretion patterns. The biosynthesis of neuropeptide hormones from their larger inactive precursor proteins and their traffic to the regulated secretory pathway (RSP) for cellular release is one of the paramount cellular processes in hormone action. In early times of peptide discovery, the "peptidergic neuron" name was reserved for those neurosecretory cells within the hypothalamus that released oxytocin and vasopressin directly into the circulation from their nerve terminals in the posterior pituitary. The idea of neu-

rosecretion in the hypothalamus can be traced back to the work of Scharrer and Scharrer (Scharrer and Scharrer 1940) as early as the late 1920s. Later work by Harris and colleagues specified that the hypothalamic substances secreted into the portal vessels were pituitary specific and led to the concept of "releasing factors" whose purpose was to initiate a cascade of events resulting in the release of peripherally active hormones (Fink 1976). The discovery and chemical characterization of the first identified hypothalamic releasing factor, thyrotropin-releasing hormone (pyroGlu-His-ProNH$_2$, also known as thyroliberin, and herein referred to as TRH), by Guillemin and colleagues (Burgus et al. 1969) and Schally and colleagues (Boler et al. 1969) provided ultimate confirmation for the founding principles of neuroendocrinology which resulted later in the discovery of other releasing factor peptides (Guillemin 1978; Schally 1978). Recent progress over the last decades in genetics and molecular biology provided considerable information about the expression of brain neuropeptide hormone genes and their tissue-specific regulation. From multiple studies conducted in many laboratories including ours, it has become clear that neuropeptides acting as neurotransmitters or hormones play a significant modulatory roles in the control of the central nervous system and neuroendocrine function. Even more remarkable was the discovery that multiple neuropeptides derived from posttranslational processing of

E. A. Nillni (✉)
Emeritus Professor of Medicine, Molecular Biology, Cell Biology & Biochemistry,
Department of Medicine, Molecular Biology, Cell Biology & Biochemistry,
The Warren Alpert Medical School of Brown University, Providence, RI, USA
e-mail: Eduardo_Nillni@Brown.edu

© Springer International Publishing AG, part of Springer Nature 2018
E. A. Nillni (ed.), *Textbook of Energy Balance, Neuropeptide Hormones, and Neuroendocrine Function*, https://doi.org/10.1007/978-3-319-89506-2_5

its single gene- polypeptide precursor has distinct physiological functions (Nillni and Sevarino 1999; Eipper and Mains 1980; Nillni et al. 1996; Liston et al. 1984; Hall and Stewart 1983; Nillni 2007, 2010; Wardlaw 2011). Therefore, to fully understand the biology of neuropeptide hormones controlling energy balance, it is essential to uncover the mechanisms by which a specific prohormone is posttranslationally modified to its active form under normal and pathological conditions, a process that happens in a tissue-specific manner. This topic will be discussed in this chapter putting emphasis on three prohormones, pro-thyrotropin-releasing hormone (pro-TRH), pro-opiomelanocortin (POMC), and pro-corticotropin-releasing hormone (pro-CRH).

5.2   Origin of the Prohormone Theory and Biosynthesis of Inactive Prohormone Precursors

Even though the amino acid structures of vasopressin and insulin had been elucidated in the 1950s (Sanger 1959), it was not until the early 1960s that the mechanisms of protein biosynthesis began to be understood and the genetic code fully defined. In 1964, Sachs (Sachs and Takabatake 1964) provided the first evidence that the biosynthesis of vasopressin can be inhibited with puromycin, a protein synthesis inhibitor, and that newly synthesized vasopressin could not be detected in its producing tissue only after approximately 1 h following a pulse-labeling protocol. This observation suggested that a protein biosynthesis mechanism must be involved. Sachs demonstrated that before vasopressin becomes a biologically active peptide, it exists in a pro-form state. Posttranslational modification was then required to convert the pro-form into an active peptide hormone. Likewise, while the structure of insulin peptides was described early on (Sanger 1959), it was difficult to envision how the combination of A and B chains was attained in beta cells of the pancreas. The pioneering work from Donald F. Steiner in 1965 represented a landmark in the prohormone theory. He elegantly labeled the newly synthe-

sized bigger pro-form (proinsulin) with tritiated leucine and phenylalanine to demonstrate that its conversion to smaller forms gave rise to insulin and C-peptide. These findings were later seen as crucial to understand the molecular mechanisms of insulin biosynthesis and provided broad insight into the cell biology of proteins and their evolution. Using a pancreatic insulinoma derived from a patient, it was possible to determine that insulin could be derived in vitro from a single molecule that was converted to the A and B chains by trypsin treatment (Steiner and Oyer 1967). Studies done in rat islets subsequently confirmed the conversion of proinsulin to insulin as a relatively slow process taking approximately 40 min (Steiner et al. 1969; Steiner et al. 1967). During the same period, work done by Howell and Taylor on insulin biosynthesis showed that newly synthesized insulin was released several hours after its biosynthesis (Howell and Taylor 1967). The emerging view of these findings was that some sort of orderly vectorial transport occurred involving the rough endoplasmic reticulum (RER), the Golgi complex (GC) compartments, and secretory granules (SG).

These data (Steiner and Oyer 1967), along with the major contributions of other investigators who established that the biosynthesis of serum albumin, parathyroid hormone, and glucagon also originates from larger inactive precursors, formed the basis of the prohormone theory. *This theory states that synthesis of peptide hormones and neuropeptides begins with messenger RNA (mRNA) translation into a large, inactive precursor polypeptide, followed by posttranslational chemical modifications and limited proteolysis to release bioactive end products.* Chretien and Li (Chretien and Li 1967) also made an important contribution to the prohormone theory, when they determined the amino acid sequences of β-lipotropin (β-LPH), γ-LPH, and β-melanotropin (β-MSH). They observed that β-MSH was part of the β-LPH sequence, providing evidence that β-MSH was a conversion product of β-LPH. They also observed that cleavage occurred at the C-terminal side of paired basic lysine or arginine residues. More definitive evidence for a precursor/product was provided with the cloning of POMC (29 kDa), which revealed that the ACTH and β-LPH sequences were present within the N-terminus of POMC (Eipper and Mains 1980).

5.3 The Family of Prohormone Convertases (PCs) and Their Cellular Traffic and Action on Key Neuropeptide Prohormones: History of PCs on Prohormones

Which one are the cellular mechanisms involved in the generation of mature active peptide hormones? As described above, the biosynthesis of mammalian neuropeptide hormones follows the principles of the prohormone theory (Nillni 2007; Chretien and Li 1967; Steiner 1998; Chretien et al. 1979). The generation of the smaller active peptide hormones is achieved through a differential processing mechanism by the action of specific enzymes called prohormone convertases (PCs) acting in specific cellular and extracellular compartments (Fig. 5.1). *Posttranslational processing of any given hormone precursor protein is a critical mechanism by which cells increase their biological and functional diversity, such that two or more peptides with different biological functions originate from the same gene precursor protein.* It is through differential posttranslational processing mechanisms that cells selectively produce specific peptides for secretion (Nillni and Sevarino 1999; Nillni 2010; Matsuuchi and Kelly 1991; Arvan and Castle 1998; Dannies 1999; Chun et al. 1994; van Heumen and Roubos 1991). These covalent modifications are of enzymatic nature and occur while synthesized proteins by ribosomal translating mRNA into polypeptide chains or after full protein biosynthesis is completed. Modifications include addition of functional groups like phosphorylation, neddylation, glycosylation, ubiquitination, nitrosylation, methylation, acetylation, lipidation, and methylation. It is also referred to a limited endoproteolytic proteolysis of an initially folded protein in the endoplasmic reticulum to smaller biologically forms targeted to secretory granules for release. One important aspect of protein post-translational modifications is the remarkable increase in the functional diversity of the proteome. While human genome contains between 20,000 and 25,000 genes (Genomes Project et al. 2012; Ezkurdia et al. 2014), the total number of proteins in the human proteome is assessed at over one million (Jensen 2004). These observations indicate that one single gene can encode multiple proteins with multiple functions. Before a gene is translated into a protein, it is subjected to genomic recombination, transcription initiation at alternative promoters, differential transcription termination, and alternative splicing of the transcript. These mechanisms allow the generation of different mRNA transcripts from a single gene (Ayoubi and Van De Ven 1996). Characterizing posttranslational modifications in a given prohormone located in a distinct tissue is critical to understand consequences related to diseases caused by protein biosynthesis dysregulation.

The family of prohormone convertases (PCs) is composed of nine secretory serine proteases related to the bacterial subtilisin and yeast kexin (Hook et al. 1994; Seidah and Chretien 1994; Seidah 1995) (Fig. 5.1). Serine proteases, also called serine endopeptidases, are enzymes that cleave peptide bonds in small or larger forms in which serine assists in this enzymatic process as the nucleophilic amino acid located at the active side of the enzyme (Hedstrom 2002). These enzymes are found universally in eukaryotes and prokaryotes. Based on their structure, they could behave like chymotrypsin or trypsin or subtilisin (Madala et al. 2010). These enzymes cleave at the C-terminal side of single, paired, or tetra basic amino acid residue motifs (Rouille et al. 1995), followed by removal of the remaining basic residue(s) by a carboxypeptidases E and D (CPE, CPD) (Xin et al. 1997; Fricker et al. 1996; Nillni et al. 2002a). The selective expression of PC1 and PC2 in endocrine and neuroendocrine cells demonstrates that they are important in prohormone processing. Further processing of cleaved peptides is carried out by an amidating enzyme, which causes the conversion of glycine to an amide group. An example of a classical post-translational processing of a given prohormone is depicted in Fig. 5.2.

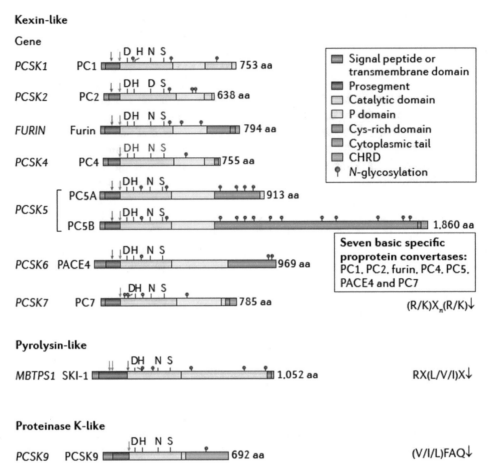

Fig. 5.1 Schematic representation of the pro-protein convertases (PCs). PC1 and PC2 are the two main enzymes involved in the processing of pro-peptides in the neuroendocrine and endocrine tissues. The membrane-bound human PCs include furin, PC4, PC5B, PC7, and SKI-1. PCs have medical significance including cholesterol synthesis. Furin and PACE4 are involved in pathological processes including viral infection, inflammation, hypercholesterolemia, and cancer and have been proposed as therapeutic targets. Similar to all secretory proteins, the PCs contain a signal peptide, a pro-segment, and a catalytic domain that exhibits the typical catalytic triad residues Asp, His, and Ser, as well as the Asn residue comprising the oxyanion hole (Asp for PC2). (Illustration from Seidah and Prat 2012)

In humans, the PCs are involved in different physiological processes, including immune response, digestion, blood coagulation, reproduction, growth, development, metabolism, endocrine and brain functions, and maturation of pro-proteins (Chretien 2011). Furin, PC1 (also known as PC3), PC2, PC4, PACE4, PC5-A (also known as PC6-A), its isoform PC5-B (also known as PC6-B), PC7 (also known as LPC), and PC8 (also known as SPC7) PC1, PC2, furin, PC4, PC5, are involved in the posttranslational processing of many prohormones and precursor proteins by producing a cleavage at single, paired, and tetra basic residues (Seidah et al. 1991; Smeekens et al. 1991; Smeekens and Steiner 1990; Seidah and Chretien 1992; Seidah et al. 1990, 1992a, 1996; Constam et al. 1996). The other two, subtilisin kexin isozyme 1 (SKI-1) and pro-protein convertase subtilisin kexin 9 (PCSK9), regulate cholesterol and/or lipid homeostasis via cleavage at nonbasic residues or through induced degradation of receptors (Seidah 2012 Nature reviews). The selective expression of PC1 and PC2 in endocrine, neuro-

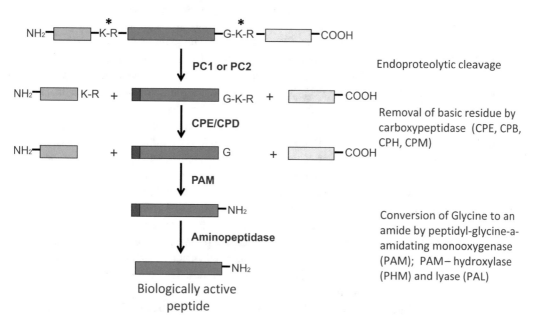

Proteolytic Processing of a Model Neuroendocrine Precursor

Fig. 5.2 Schematic representation of a typical limited proteolysis of a given prohormone by posttranslational processing mechanism. Asterisks indicate the amino acids where the PCs produce their enzymatic cleavages. Removal of the carboxyl-terminal pair of basic residues is conducted by carboxypeptidases. Amidation of peptides to confer biological activity is conducted by the PAM enzyme. (Illustration from Nillni 2016)

endocrine, and neuronal cells showed to be important in tissue-specific pro-neuropeptide hormone processing (Seidah et al. 1990, 1991, 1993, 1994; Smeekens et al. 1991; Smeekens and Steiner 1990; Seidah et al. 1990, 1996; Constam et al. 1996; Seidah et al. 1990, 1991, 1992b, 1994; Schafer et al. 1993). PC1 and PC2 have been shown to process pro-TRH (Nillni 2010; Friedman et al. 1995; Nillni et al. 1995; Pu et al. 1996; Schaner et al. 1997), proinsulin (Rouille et al. 1995; Steiner et al. 1992; Smeekens et al. 1992), pro-enkephalin (Breslin et al. 1993), prosomatostatin (Galanopoulou et al. 1993; Brakch et al. 1995), pro-growth hormone-releasing hormone (pro-GHRH) (Posner et al. 2004; Dey et al. 2004), pro-opiomelanocortin (POMC) (Benjannet et al. 1991; Thomas et al. 1991), pro-CRH (Spiess et al. 1981; Rivier et al. 1983; Perone et al. 1998; Brar et al. 1997; Castro et al. 1991), pro-NPY (Paquet et al. 1996), pro-CART (Dey et al. 2003), and pro-neurotensin (Villeneuve et al. 2000) to various intermediates and end products of processing.

Among the members of the PC family cloned thus far, PC1 and PC2 are specifically found in neuronal and endocrine cells containing secretory granules (Seidah et al. 1990, 1991, 1994). Their involvement in the processing of neuropeptide precursors has been early suggested by the finding that PC1 and PC2 transcripts and protein products are widely distributed in different areas of the brain, including the cerebral cortex, hippocampus, and hypothalamus (Schafer et al. 1993; Winsky-Sommerer et al. 2000). Within the hypothalamus, these enzymes display an extensive overlapping pattern of expression (Schafer et al. 1993; Winsky-Sommerer et al. 2000). In rodents and humans, abnormalities in prohormone processing results in pathological consequences, including metabolic dysfunctions (Nillni et al. 2002a; Lloyd et al. 2006; Naggert et al. 1995; Jing et al. 2004; Jackson et al. 1997; Challis et al. 2002).

The critical role of PC1 and PC2 in prohormone processing is underscored by studies from animals lacking the genes encoding PC1 (Zhu et al. 2002a) and PC2 (Furuta et al. 1997), as well as 7B2 (Laurent et al. 2002; Westphal et al. 1999), a neuropeptide essential for the maturation of PC2. 7B2 functions as a specific chaperone for PC2. The sequence Pro-Pro-Asn-Pro-Cys-Pro in 7B2 binds to an inactive proPC2 and facilitates its transport from the endoplasmic reticulum to the secretory granules with a pH of 5.5 where PC2 is proteolytically matured and activated. 7B2 C-terminal peptide can inhibit PC2 in vitro and may contribute to keeping the enzyme transiently inactive in vivo. The PC2-7B2 complex is a unique mechanism whereby the proteolytic activation of PC2 is delayed until it reaches later stages in the secretory pathway (Fig. 5.3). Disruption of the gene-encoding mouse PC1 results in a syndrome of severe postnatal growth impairment and multiple defects in processing for many hormone precursors, including hypothalamic pro-growth hormone-releasing hormone (pro-GHRH) to mature GHRH, pituitary pro-opiomelanocortin hormone (POMC) to adrenocorticotropic hormone, islet proinsulin to insulin, and intestinal pro-glucagon to glucagon-like peptide-1 and peptide-2. PC1$^{-/-}$ mice are normal at birth but display impaired postnatal growth and are about 60% of normal size at 10 weeks. They lack mature GHRH, have low pituitary growth hormone and hepatic insulin-like growth factor-1 mRNA levels, and phenotypically are smaller than a normal mouse. Mice with disruption of the gene-encoding PC2 appear to be normal at birth. However, they exhibit a small decrease in rate of growth. They also have chronic fasting hypoglycemia and a reduced blood glucose level during an intraperitoneal glucose tolerance test, both of which are consistent with a deficiency of circulating glucagons (Zhu et al. 2002b). The processing of pro-glucagon, prosomatostatin, and proinsulin in the alpha, delta, and beta cells of the pancreatic islets is severely impaired in PC2 null mice (Webb et al. 2004). A mouse model of PC1 deficiency generated by random mutagenesis (Lloyd et al. 2006) has a missense mutation in the PC1 catalytic domain (N222D) that leads to obesity with abnormal proinsulin processing and multiple endocrine deficiencies. Although there was defective proinsulin processing leading to glucose intolerance, neither insulin resistance nor diabetes developed despite obesity. The apparent key factor in the induction of obesity was impaired autocatalytic activation of mature PC1 causing reduced production of hypothalamic α-MSH (Lloyd et al. 2006). A patient with a compound heterozygous mutation in the PC1 gene resulting in production of nonfunctional PC1 had severe childhood obesity (Jackson et al. 1997). An analogous obese condition was found in a patient with a defect in POMC processing (Challis et al. 2002). Mice lacking PC2 (Furuta et al. 1997), as well as 7B2 (Laurent et al. 2002; Westphal et al. 1999), essential for the maturation of PC2, are also altered. As PC1 and PC2 are essential for the processing of a variety of pro-neuropeptides, alterations in the expression and protein biosynthesis of PC1 and PC2 are likely to have profound effects on neuropeptide homeostasis.

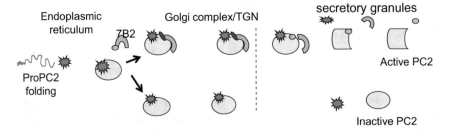

Fig. 5.3 Activation of the prohormone convertase 2 (PC2). PC2 maturation and activation requires an interaction with the neuroendocrine 7B2 peptide. The neuroendocrine peptide 7B2 binds to PC2 and exerts a profound influence on its biosynthesis. The carboxyl-terminal peptide of 7B2 is a potent inhibitor of PC2 and of pro-PC2 activation, while the amino-terminal portion of this protein of 21 kDa facilitates the intracellular transport and the activation of pro-PC2 to mature PC2 in low pH

At the physiological level, the PCs are regulated by states of hyperglycemia (Nie et al. 2000), inflammation (Li et al. 1999), suckling (Nillni et al. 2001), starvation (Sanchez et al. 2004), cold stress (Perello et al. 2007a), and morphine withdrawal (Nillni et al. 2002b). In the case of morphine withdrawal, for example, it was previously demonstrated that during opiate withdrawal, prepro-TRH mRNA increased in neurons of the midbrain periaqueductal gray matter (PAG), causing a significant change in the level of some posttranslational processing products derived from the TRH precursor, and the mature form of PC2 increased only in PAG as compared with their respective controls demonstrating a region-specific regulation of pro-TRH processing in the brain, which may engage PC2 (Nillni et al. 2002b; Legradi et al. 1996). Another interesting example was provided in alterations seen in pro-TRH processing in the PVN during lactation where PC2 was involved (Nillni et al. 2001). Other examples of the regulation of PCs include rats exposed to streptozotocin, where it was demonstrated that the diabetic state altered alpha-cell processing of pro-glucagon to give increased levels of glucagon-like peptide 1 (Nie et al. 2000). Li et al. also identified regions in PC1 and PC2 human promoters that contain putative negative thyroid hormone response elements and has shown that T_3 negatively regulates PC1 and/or PC2 expression in rat GH3 cells, rat anterior pituitary, hypothalamus, and cerebral cortex (Nillni et al. 2001; Li et al. 2000, 2001; Shen et al. 2004, 2005). In addition to the PCs, other enzymes involved in the maturation of peptides could be affected under different physiological conditions including cold exposure, fasting, or changes in thyroid status (Perello and Nillni 2007). For example, the amidating enzyme PAM activity on TRH neurons is regulated by the thyroid status. TRH and TRH-Gly levels increase under low iodine/PTU diet-induced hypothyroidism and decrease under TH-induced hyperthyroidism (Perello and Nillni 2007; Perello et al. 2006a). However, the ratio TRH/TRH-Gly, which depends on PAM enzymatic activity, is different for each condition. In hypothyroidism, the ratio TRH/TRH-Gly increases suggesting an increase in PAM activity, and in hyperthyroidism, the ratio decreases suggesting a decrease in PAM activity. Since PAM activity is responsive to changes in thyroid hormone levels, this could represent another level of control in the final production of mature TRH.

To make matters more complex in the relationship between prohormones and processing enzymes, a major breakthrough study in the biology of leptin added an extra level of complexity in the maturation of TRH, that is, in addition to regulating peptide hormone expression, leptin also controls prohormone processing by regulating PC1 and PC2 (Sanchez et al. 2004). Since leptin regulation of energy balance works through the activity of several neuropeptides, it was hypothesized that leptin might regulate processing enzymes in addition to regulating peptide production. These studies clearly supported the hypothesis that the regulation of hypophysiotropic TRH biosynthesis by leptin occurs not only at the transcriptional level but also at the posttranslational level through changes in pro-TRH processing by the action of PC1 and PC2 (Sanchez et al. 2004). Thus, the results from this study demonstrate that leptin couples the upregulation of prepro-TRH expression and its protein biosynthesis with the upregulation of the processing enzymes in a coordinated fashion. Such regulation ultimately leads to more effective processing of leptin-regulated pro-neuropeptides into mature peptides, such as TRH and α-MSH, which are critical for leptin action. Therefore, transcriptional control of PC1 and PC2 gene expression by leptin is another level at which this hormone regulates energy homeostasis. This adds a novel key checkpoint that is tightly regulated in the control of energy consumption. Since leptin-induced increase in the PC levels in vivo is partially dependent on the activation of the melanocortin system (Perello et al. 2006b) and since PC1 promoter contains two CREB response elements, which are transactivated by CREB-1 (Jansen et al. 1995), the P-CREB transcription factor could activate in a coordinated fashion the synthesis of prepro-TRH and PCs. Therefore, leptin can act on the TRH neurons of the

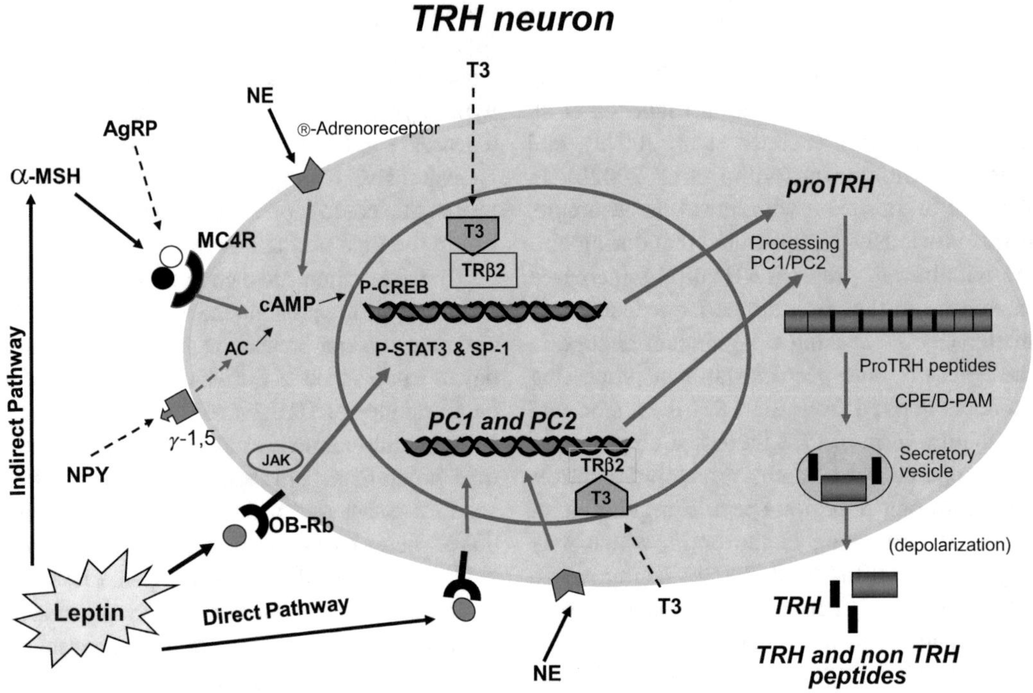

Fig. 5.4 Schematic representation of the TRH neuron with the most relevant inputs controlling its gene expression. The illustration shows the multiple signals coming from the brain and periphery affecting the TRH neuron including leptin and T3 and from the ARC, α-MSH, NPY, and AgRP. The TRH promoter integrates each of these inputs to determine the set point of the HPT axis. The different hormonal inputs regulating the PCs are also represented. (Illustration from Nillni 2010)

PVN directly (P-STAT) and indirectly via the melanocortin system (P-CREB) regulating in a synchronized manner the biosynthesis of pro-TRH and the PCs through both pathways. See an example of leptin acting on PCs and a given prohormone in pro-TRH (Fig. 5.4) (Nillni 2010).

5.4　Traffic of Prohormones and Its Derivate Peptides to the Regulated Secretory Pathway

All peptides hormones are release outside the cells through the regulated secretory pathway (RSP). This pathway is a key cellular process that involves prohormone biosynthesis, posttranslational modifications, sorting, and release to the extracellular milieu. Therefore, the regulated secretory pathway constitutes a unique character-

istic of endocrine and neuroendocrine cells (Fig. 5.5). Prohormones destined for the secretory pathway (see pro-TRH example in Fig. 5.6) begin with a ribosome bound to the rough endoplasmic reticulum (ER) followed by a co-translational translocation to the lumen of the ER. A leader signal sequence or signal recognition peptide located in the N-terminus side of the protein directs the synthesizing proteins into the lumen of the ER. Soluble proteins in this class are first localized in the ER lumen and consequently sorted to the lumen of other organelles or are secreted outside the cell. Prohormones reaching the ER lumen are incorporated into small, ≈50-nm-diameter transport vesicles, and then they either fuse with the *cis*-Golgi or with each other to form the membrane stacks known as the *cis*-Golgi reticulum. During the cisternal migration, a new *cis*-Golgi stack with its cargo of luminal prohormones physically moves from the *cis*

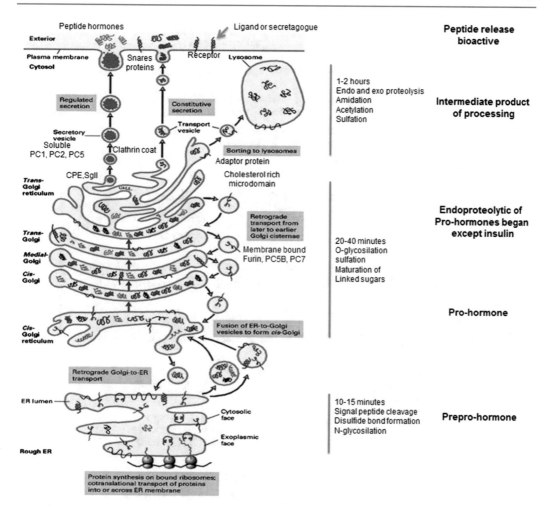

Fig. 5.5 This figure represents a typical biosynthesis, posttranslational processing, trafficking, and secretion of most prohormones. (The left side of the figure is modified from *Molecular Cell Biology* Textbook. 4th edition)

position to the *medial*-Golgi and then to the last compartment of the Golgi complex, the *trans*-Golgi network (TGN). Prohormones aimed for secretion leave the *trans* face of the Golgi complex to form immature secretory granules (ISGs).

Other types of secretory vesicles secrete proteins continuously (constitutive pathway Fig. 5.5). There are two pathways of non-stimulated release, constitutive (nongranular) secretion, and basal release from compartments that form after sorting into the regulated secretory pathway (Matsuuchi and Kelly 1991; Arvan and Castle 1998; Dannies 1999). Examples of constitutive pathway secretion include fibroblasts producing collagen and secretion of serum pro-

teins by hepatocytes. These proteins are sorted in the TGN into transport vesicles and quickly move to and fuse with the plasma membrane of the cell for immediate release of their content by exocytosis. Other vesicles released from the trans-Golgi fuse with lysosomes for protein degradation. In the case of secretory granules destined for regulated secretion, instead of being released continuously, they are stored near the plasma membrane of the cell waiting for an external stimulation for release. Upon maturation, the secretion process begins when a specific ligand binds to specific receptor triggering a signal transduction pathway that in turn causes the fusion of electron-dense-SGs containing sorted

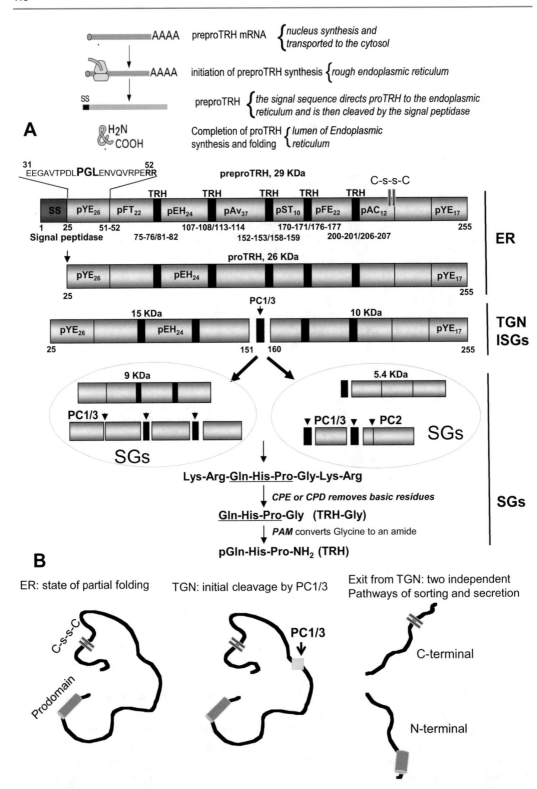

Fig. 5.6 Schematic representation of the biosynthesis and posttranslational processing of rat pro-TRH. (**a**) depicts the transcription and posttranslational modifications in the rat pro-TRH composed of 231 amino acids.

products to the plasma membrane in a calcium-dependent manner (Burgess et al. 1987; Burgess and Kelly 1987). During the vectorial transport through the GC and beyond, the newly synthesized proteins are subjected to posttranslational modifications including glycosylation, phosphorylation, amidation, acetylation, and proteolytic conversion (Seidah et al. 1992a; Nillni et al. 1993a, b). Ultimately, partially processed proteins reach the TGN where initial sorting to secretory vesicles begins (Fig. 5.6). Our laboratory was one of the first to show that prohormone endoproteolytic processing is initiated at the TGN level (Nillni et al. 1993a, b; Xu and Shields 1993; Tooze et al. 1993) and shows later that the partially processed products are differentially sorted to different immature secretory granules (ISGs) of the regulated secretory pathway (RSP) (Fig. 5.7) (Perello et al. 2008). For pro-TRH, N- and C-terminal pro-TRH intermediate forms of processing behave independently after the pro-TRH cleavage in the TGN delivering N- and C-terminal peptides to different SGs. This discovery represented the foundation for what was later the established fact that prohormones initially processed in the TGN could deliver its intermediate products of products of processing to different secretory granules of the RSP for differential release (Perello et al. 2008; Perez de la Cruz and Nillni 1996; Zhang et al. 2010). This concept of differential processing associated with

differential subcellular localization and specific cell type is further reinforced by the observation that certain regions in the brain can give rise to several different prohormone-derived peptides in addition to or instead of the well-characterized hormone.

This is a unique event involving a series of discrete and exceptional actions including protein aggregation in secretory granules/sorting, the formation of ISGs, prohormone processing, and vesicle fusion. For soluble proteins delivered for secretion in SGs, luminal pH and divalent metals are involved in hormone aggregation and interaction with enclosing membranes. Recent evidence showed that the trafficking of granule membrane proteins could be controlled by luminal and cytosolic factors. Cytosolic adaptor proteins that recognize the cytosolic domains of proteins spanning the SG membrane have been shown to play a role in the construction of functional SGs. Adaptor protein 1A (AP-1A) interacting with specific cargo proteins and clathrin heavy chain contributed to the formation of a clathrin coat. This suggests that some cargo proteins in SGs are managed by Aps (Bonnemaison et al. 2013). In addition, lipid rafts have been shown to be implicated in the sorting of peptide hormones to the TGN, and the identification of SNARE molecules (the molecular basis of exocytosis activity) in the formation of mature secretory granules ready for fusion and released of

Fig. 5.6 (continued) The signal sequence is cleaved from prepro-TRH upon delivery into the endoplasmic reticulum yielding pro-TRH. The conserved PGL sequence in the pYE26 peptide ensures the proper folding of pro-TRH within the lumen of the endoplasmic reticulum. The initial processing cleavage of pro-TRH by PC1/3 begins at the trans-Golgi network level generating an N-terminal and C-terminal intermediate forms. The intermediate forms of processed pro-TRH are then targeted to different secretory granules where processing continues by the action of the PCs, CPE/D, and PAM until TRH and non-TRH peptides are formed. The vertical bars on the right indicate in which intracellular compartment pro-TRH is cleaved and further processed. Numbers indicate the positions of paired basic residues. Non-TRH peptides are indicated in the pro-TRH molecule, and TRH is indicated by a black rectangle. Peptides are indicated as pXYZ nomenclature, where "p" means peptide, "X" is the first amino acid of each peptide, "Y" is the last one, and Z indicates the total

number of amino acids in that given peptide. The non-TRH peptides are then targeted to different secretory granules ready for secretion. (**b**) depicts the proposed model of the unfolding process for pro-TRH after its initial cleavage by PC1/3, exposing potential sorting signals responsible for the targeting to different granules. Thus far a disulfide sequence has been identified in pro-TRH as an important sorting signal for the correct targeting of peptides to secretory granules. SS signal sequence; ER endoplasmic reticulum; TGN trans-Golgi network; ISGs immature secretory granules; C-s-sC disulfide bond. The reader should note that pro-TRH-derived peptides are named by "p" for peptide followed by the single letter amino acid designation for the first and last amino acid of the peptide, along with the peptide length in subscript. Where these peptides are first mentioned, they are followed by the longer prepro-TRH name that describes their amino acid residue positions within the precursor. (From Nillni 2010)

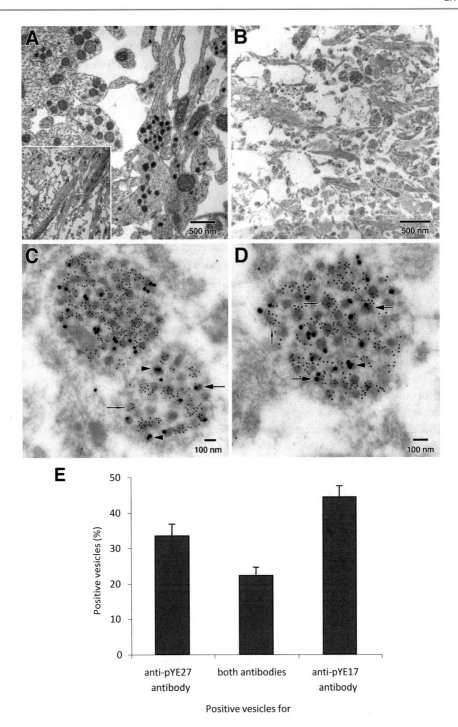

Fig. 5.7 N- and C-terminal end products of pro-TRH are partially located in different vesicles within the ME fibers. Sprague-Dawley rats were perfused with Karnovsky's fixative (**a**) inlet shows a panoramic micrograph of this region) or with 4% paraformaldehyde/0.15% glutaraldehyde in PBS (**b–d**). Panels C and D show an increased magnification of the panel B. Images show three classes of positive vesicles for either anti-pYE17 antibody alone (arrowhead), anti-pYE17 antibody alone (needles), or a combination of both antibodies (arrows). Panel E shows quantitative results obtained from sections labeled first with anti-pYE17 (10 nm gold particles) and then with anti-pYE27 (25 nm gold particles). Mean percentages represent the proportion of a particular positive vesicle compared to the total number of positive vesicles observed per micrograph. (From Perello et al. 2008)

peptide hormones broads more light in the understanding of the secretory process (Morvan and Tooze 2008; Xiong et al. 2017).

It has been arguably debated about the intracellular sorting mechanism for prohormone post-translational processing to end products, while they are sorted to the SGs (Steiner 1998; Arvan and Castle 1998; Nillni et al. 2002a; Nillni et al. 1993a, b; Perez de la Cruz and Nillni 1996; Morvan and Tooze 2008; Glombik and Gerdes 2000; Seidah and Chretien 1997; Mulcahy et al. 2006; Mulcahy et al. 2005; Cawley et al. 2016a, b) except for proinsulin that is processed when the prohormone reaches the SG (Hou et al. 2009). Several structural elements have been related to the sorting mechanism within the RSP. The N-terminal hydrophobic domain in certain prohormones could represent a sorting signal (Gorr et al. 2001). Also, a disulfide loop in the N-terminal region found in certain prohormones seems to be essential for targeting to the RSP (Chanat 1993; Glombik et al. 1999; Cool et al. 1995). Many prohormones contain pairs of basic residues, which could also act as a sorting signal (Brakch et al. 1994; Brechler et al. 1996). Additionally, the RGD (Arg-Gly-Asp) motif present in processing enzymes is also implicated in trafficking within the RSP (Rovere et al. 1999). In the pro-TRH polypeptide, there are 11 pairs of basic residues, 1 disulfide loop, and 2 RGD motifs (Fig. 5.6). While evidence showed that the RGD motif did not play a role in pro-TRH sorting, two different motifs within its sequence are involved in the trafficking and sorting to SGs. The primary sequence of pro-TRH has an N-terminal segment of 25 amino acid residues, pYE26, immediately after the signal sequence (Fig. 5.6). By expressing several deletions and point mutations in cDNA constructs within the prepro-TRH$_{31-52}$ sequence and by monitoring the steady-state production of pro-TRH in the ER, we identified a single tripeptide Pro-Gly-Leu (^{40}PGL^{42}) sequence, which is conserved in mammals, and it turns out to be important for the stability of pro-TRH, in terms of resistance to protein degradation (Romero et al. 2008). These studies revealed that PGL is primarily involved in the stability of pro-TRH in the early secretory pathway. Deletion of PGL destabilizes pro-TRH

by targeting the protein to the proteasome for degradation. This was the first evidence showing that the structural role played by a short motif located in the N-terminal region of pro-TRH takes place early in the ER and has important consequences on precursor stability rather than the sorting process to the SGs per se (Romero et al. 2008). The processing modifications occur while pro-TRH is transported from the TGN to newly formed ISGs (Perez de la Cruz and Nillni 1996). In addition, the disulfide loop present in the C-terminal side of pro-TRH was involved in the sorting of pro-TRH-derived peptides and in their retention in the SGs (Mulcahy et al. 2006).

In terms of sorting signals driving prohormones to the RSP, two models have been originally proposed to explain the trafficking to SGs. In the first hypothesis, the sorting by entry model, there is sorting signal of tertiary structure responsible for directing the prohormone and some enzymes to the budding SGs. Carboxypeptidase E (CPE) has been proposed to fulfill the role of sorting receptor (Dhanvantari et al. 2002; Cool and Loh 1998; Loh et al. 2002). The sorting signal has negatively charged amino acids at its core that bind to positively charged regions of membrane-bound proteins or receptors in the TGN and are incorporated into newly formed granules budding from the TGN. Molecular modeling analysis of CPE identified a putative sorting signal binding site in Arg255 and Lys260 (Zhang et al. 1999) that binds to N-POMC (1–26). The same group in further studies showed that CPE and secretogranin III facilitate the sorting of POMC in corticotropic AtT20 cells (Cawley et al. 2016b). The second model proposes that prohormones are sorted by retention, which means that aggregated prohormone and other proteins are packaged into ISGs. Once aggregated in the budding granule, non-regulated secretory proteins are removed from the granules by a constitutive-like secretion, leaving only the prohormones and processing enzymes in the aggregate. The idea of this hypothesis is that the prohormone must be first sorted to the ISGs and then must be retained in the same granules to be processed. There are supports for both models, but they are not mutually exclusive. New studies identified the membrane component to which the sorting signal binds such as cholesterol, a major membrane component of the TGN and the ISGs (Sun et al. 2013). They showed that deletion of cholesterol in cells caused a mis-sorting of POMC and insulin. Biochemical analyses indicated that SgII bound directly to the cholesterol-rich secretory granule membrane in a cholesterol-dependent manner and was able to retain the aggregated form of POMC supporting a sorting and retention mechanisms (Sun et al. 2013) (Fig. 5.5).

Similar to all secretory pro-proteins that undergo posttranslational modification before reaching their final bioactive form, prohormone convertases are themselves exposed to various posttranslational modifications before becoming enzymatically active. Once the ribosomes begin synthesizing these enzymes in the rough endoplasmic reticulum (ER), they enter the ER membrane co-translationally, whereby the zymogens lose their signal peptide by the action of the signal peptidase and are N-glycosylated at various sites (Fig. 5.1). The N-terminal pro-segment of these enzymes acts as an intramolecular chaperone and inhibitor resulting in folded proteins within the ER lumen into an active conformation later in the secretory pathway. The cleaved pro-segment remains bound to the mature enzymatically active protease, retaining it in an inactive state. The only pro-protein convertase that does not follow this patters is PC2, as pro-PC2 is transported as a complex with its ~30 kDa binding protein 7B2 (Fortenberry et al. 1999; Muller and Lindberg 1999; Westphal et al. 1999) to acidic immature secretory granules, where it is then autocatalytically activated (Fig. 5.3). Thus, the compartments where these enzymes are active were also determined by differential timing of their respective maturation. While some type of posttranslational modifications are reversible as a mechanism ensuring proper functioning of the cell under various physiological and/or stress conditions, posttranslational modification with limited proteolysis is irreversible and is referred to all prohormones generating multiple products. These smaller forms of the prohormone are covalently modified to generate the final bioactive hormone.

5.5 Pro-thyrotropin-Releasing Hormone

Pro-TRH biosynthesis and processing is described here in more detail as an illustrative example of prohormone maturation, a topic that was extensively studied in our laboratory for 26 years. Similar to other potent secretory molecules regulating key biological functions, the biosynthesis of TRH (pyroGlu-His-ProNH2, MW 362) and other non-TRH peptides begins with mRNA-directed ribosomal translation of a larger inactive precursor called pro-TRH (Nillni and Sevarino 1999; Nillni 2010). The maturation of TRH implicates many coordinated cellular and biochemical steps along the RSP (Fig. 5.6). First, the signal sequence, or pre-sequence, directs a co-translational translocation of prepro-TRH into the lumen of the rough ER after which the signal sequence is removed. Once in the ER, the newly synthesized pro-TRH meets a unique environment containing a number of ER-specific chaperones involved in its proper folding pathway leading to its wild-type conformation. In addition, pro-TRH, like all secretory proteins, encounter an exclusive set of posttranslational modifications by specific peptide bond cleavages provided by a network of processing peptidases. Pro-TRH is transported from the ER to the GC and then to the TGN, where an initial proteolytic processing event occurs. At the TGN, pro-TRH products are sorted to and stored into specialized SGs that undergo secretion only after appropriate stimuli. In these SGs the final processing steps take place, which involve endoproteolysis by specific prohormone convertases at pair of basic residues (Romero et al. 2008; Zhou et al. 1993; Scamuffa et al. 2006), removal of the basic residues by a carboxypeptidases (Fricker 1988; Varlamov et al. 1999; Fricker 2007), and amidation (Eipper et al. 1992; Prigge et al. 1997). If one of these regulated steps is compromised, the biosynthesis of pro-TRH-derived peptides and their secretion may be affected.

Rat prepro-TRH is a 29 kDa polypeptide composed of 255 amino acids. This precursor contains an N-terminal 25 amino acid leader sequence, 5 copies of the TRH progenitor sequence Gln-His-Pro-Gly flanked by paired basic amino acids (Lys-Arg or Arg-Arg), 4 non-TRH peptides lying between the TRH progenitor sequences, an N-terminal flanking peptide, and a C-terminal flanking peptide (Nillni 2010; Lechan et al. 1986a, b). The N-terminal flanking peptide (prepro-TRH$_{25-50}$-R-R-prepro-TRH$_{53-74}$) is further cleaved at the C-terminal side of the arginine pair site to render prepro-TRH$_{25-50}$ (pYE26) and prepro-TRH$_{53-74}$ (pFT$_{22}$), thus yielding a total of seven non-TRH peptides (Fig. 5.6). The rat prepro-TRH is approximately 88% homologous to the mouse prepro-TRH that contains 256 amino acids, also generates five copies of TRH by proteolytic processing. The human prepro-TRH contains 242 amino acids and generates 6 copies of the progenitor sequence for TRH. On the other hand, in the frog brain, there are at least 3 different TRH preprohormones ranging between 224 and 227 amino acids and containing 7 copies of the TRH progenitor sequence. Rat, mouse, and human pro-TRH contain multiple copies of the progenitor for TRH, Gln-His-Pro-Gly (Lee et al. 1988; Masanobu et al. 1990), which could represent an evolutionary trend present in all species to ensure enough production of TRH molecules in response to signals that trigger its secretion. The complete sequence of pro-TRH has been described for salmon, frog, zebra fish, chicken, rhesus monkey, rat, mouse, and human. The more conserved prohormone sequence is found among the mammalian species, but common to all of them is the presence of multiple copies of a progenitor sequence for TRH, flanked on either side by paired basic amino acids, Lys-Arg or Arg-Arg (Fig. 5.6).

An original work published in 1997 (Schaner et al. 1997) and prior supporting data provided unequivocal evidence for the role of PC1 and PC2 in the processing of pro-TRH (Nillni et al. 1996; Friedman et al. 1995; Nillni et al. 1995; Schaner et al. 1997). Initial processing of pro-TRH occurs in the TGN at the prepro$_{152-153}$-TRH-158-159 site to generate the 15 and 10 kDa peptides (Fig. 5.6a). In subsequent steps, the 15 kDa N-terminal intermediate form is process to a 9.5 kDa peptide followed by continuous processing until the end products are generated in SGs. In separate study, the data demonstrated that the initial processing of pro-TRH by PC1 in the TGN is important for the downstream sorting events that result in the storage of pro-TRH-derived peptides in mature SGs (Mulcahy et al. 2005). The same initial cleavage by PC1 produced the unfolding of a partially folded pro-TRH, and it was a key determinant mechanism for the delivering of N- and C-terminal pro-TRH-derived peptides to different SGs of the RSP (Perello et al. 2008) (Fig. 5.6b). This could be a common mechanism used by neuroendocrine cells to independently regulate the secretion of different bioactive peptides derived from the same gene product (Perello et al. 2008). The important evolutionary concept that arises from these studies is that posttranslational processing of hormone precursor proteins is a critical mechanism by which cells increase their biological and functional diversity, such that two or more peptides with different biological functions originate from the same precursor. It is through differential posttranslational processing mechanisms that cells selectively produce specific peptides for secretion (Nillni and Sevarino 1999; Nillni et al. 1993a, b). The correct sorting events of proteins and peptides to their target organelles are of fundamental importance in cell biology. However, identification of the signals or physical properties that ensure proper intracellular sorting of peptides derived from different prohormones to dense-core SGs is not yet fully established (Dikeakos and Reudelhuber 2007) which subdivided the targeting function of proteins that are sorted to SGs into three groups: membrane-associated (or traversing) tethers, tether-associated cargo, and aggregation. The enzyme PC1 is a granule-tethered protein that may act as a sorting chaperone for its substrates in addition to being a processing enzyme. In the case of pro-TRH, as described above, initial processing action of PC1 on pro-TRH, and not a cargo-receptor relationship, is important for the sorting of the pro-TRH-derived peptides. Thus, it is possible that key paired basic amino acids of pro-TRH (residues 107, 108, 113, 114, 152, 153, and 159), which constitute a cleavage site for PC1, would also function as granule sorting domains (Mulcahy et al. 2005; Dikeakos and Reudelhuber 2007).

For pro-TRH thus far, N- and C-terminal-derived fragments behave independently after the prohormone is cleaved (Fig. 5.6b). It is possible that the initial cleavage changes the folding conformation and/or exposes different motifs that allow N- and C-terminal pro-TRH fragments to aggregate independently or to interact with different components of the RSP. Interestingly, the differential sorting is independent of the cell type (Perello et al. 2008). This suggests that the N- and C-terminal ends of pro-TRH have conserved properties, which are sufficient for the sorting events. This finding also suggests that the cellular environment is not a critical factor, although the presence of RSP machinery should be necessary. Peptides derived from the same precursor, which are differentially packaged, have been shown for the egg-laying hormone (ELH) precursor in both *Aplysia californica* and *Lymnaea stagnalis* (Chun et al. 1994), but this phenomenon had never been demonstrated in mammalian prohormones, and pro-TRH represents the first mammalian case in which differential sorting of its processing products is demonstrated (Perello et al. 2008). This differential processing event serves as a mechanism to regulate the timing of production of peptides such as prepro-TRH$_{160-169}$, prepro-TRH$_{178-199}$, and prepro-TRH$_{53-75}$ and possibly TRH (Fig. 5.6).

In studies using the PC1$^{-/-}$, mice confirmed the initial in vitro finding that attributes a primary role of PC1/3 in the processing of pro-TRH (Schaner et al. 1997), while PC2 has a specific role in cleaving TRH from its extended forms (Nillni et al. 2001). PC1/3 null mice show a dramatic decrease in the biosynthesis of all pro-TRH-derived peptides analyzed, including TRH and its pro-form, TRH-Gly. However, pro-TRH is still processed to its end products suggesting, as demonstrated by us in earlier studies, that PC2 and potentially furin can compensate for the lack of PC1 in the processing of pro-TRH (Schaner et al. 1997). In the PC2 null mice, although TRH and TRH-Gly showed some decrease as compared with the wild-type animals, the concentrations of N-terminal peptides prepro-TRH$_{25-50}$ and prepro-TRH$_{83-106}$ did not change. This indicates that PC1 is the primary enzyme involved in the processing of pro-TRH that occurs at the pairs of basic residues flanking

the TRH sequence including the sequence Arg51-Arg52, which does not contain TRH. However, the real challenge for the thyroid axis occurs when animals are physiologically perturbed such as during cold stress or during changes in the nutritional status, situations in which more TRH peptide is necessary to increase the output of the thyroid axis. For example, the role of CPE in processing has been elucidated in CPE$^{fat/fat}$ mice, which lack functional CPE resulting from a naturally occurring mutation (Naggert et al. 1995). This mouse is obese, diabetic, and infertile (Nillni et al. 2002a). Using this mouse, it was found that hypothalamic TRH was depressed by at least 75% compared to wild-type controls (Nillni et al. 2002a) and that CPD was probably responsible for the generation of 21% of the TRH produced in the hypothalamus of these animals (Nillni et al. 2002a). Their body temperatures had declined by an additional 2.1 °C as compared with wild-type controls. Furthermore, these animals cannot maintain a cold challenge for 2 h at 4 °C because of alterations in the HPT axis represented by deficits in pro-TRH processing, which resulted in hypothalamic TRH depletion and a reduction in thyroid hormone levels (Nillni et al. 2002a). More insight can be obtained from other genetic models as useful tools to understand the role of leptin in the thyroid axis, but not without simultaneously raising puzzling questions. One important point to be considered is that the status of the thyroid axis for each genetic model is different, which makes the interpretation of the results obtained from different animals difficult to compare. For example, to understand the implications of TRH in the physiology of the HPT axis, previous studies generated mice that lack TRH. Although the TRH−/− mice showed signs of hypothyroidism with characteristic elevation of serum TSH level and diminished TSH biological activity, they were viable through adulthood and showed no gross anatomic abnormalities and exhibited normal development. The decrease in TSH immunopositive cells could be reversed by TRH (Yamada et al. 1997). These TRH-deficient mouse also exhibited hyperglycemia, which was accompanied by impaired insulin secretion in response to glucose, providing a good model of tertiary hypothyroidism. This hyperglycemia seen in the TRH−/− mice indicates that TRH is involved in the regulation of glucose homeostasis. Supporting this hypothesis, a recent study showed that TRH was able to reverse STZ-induced hyperglycemia by increasing pancreatic islet insulin content, preventing apoptosis, and potentially inducing islet regeneration (Luo et al. 2008). Also, perinatal TRH treatment enhanced basal insulin secretion in STZ-induced animals of both sexes and partially restored the insulin response to glucose stimulation in females (Bacova et al. 2005).

These observations suggested that, similar to other genetic models, there must be some compensatory responses during ontogeny to compensate for the lack of TRH. In the ob/ob mouse, which could be important for the understanding of the direct pathway, there is a compensatory response by other unknown mechanisms during development that keeps $T_{3/4}$ almost at normal levels. However, ob/ob mice are inefficient in their response to cold stress, indicating a deficiency in HPT axis regulation (O'Rahilly et al. 2003; Farooqi et al. 2002). A similar response was observed in CPE$^{fat/fat}$ mice where the level of $T_{3/4}$ was normal, but the animals failed to maintain their body temperature during cold stress (Nillni et al. 2002a). In both cases the thyroid axis was relatively normal under basal conditions but unable to respond to cold stress. The other interesting and highly relevant genetic model is the mouse lacking the MC4 receptor (MC4$^{-/-}$), which is essential for α-MSH activity in TRH neurons. Utilizing the MC4$^{-/-}$ mouse model, it was found that TRH peptide levels fluctuate during fasting and respond to leptin treatment in a similar fashion to wild-type controls. Additional roles for the two pathways of leptin action on TRH neurons (Perello et al. 2006b) were uncovered in studies related to the role of the melanocortin system and adaptive thermogenesis of brown adipocyte tissue (Voss-Andreae et al. 2007). In rodents, thyroid hormones act synergistically with the SNS to regulate UCP1 expression (Bianco et al. 1988). UCP1 expression relies on functional T_3 response elements and cAMP response element-binding protein motifs in the UCP1 gene upstream enhancer region (Silva 2005). Brown adipocyte tissue contains abundant

type 2 deiodinase (D2), which catalyzes the conversion of T_4 to the more biologically active and potent T_3 and which is activated by the SNS (Silva 2005). Since the melanocortin system can modulate both sympathetic outflow to the BAT (Haynes et al. 1999) and the function of the HPT axis (Fekete et al. 2002), the question was whether the MC4R-mediated upregulation of UCP1 expression in response to a high-fat diet involves the HPT axis. These studies led to the hypothesis that perhaps there may be a defect in the MC4R-HPT axis that contributes to the inability of MC4R$^{-/-}$ mice to upregulate UCP1 expression in BAT in response to a HF diet. However, MC4R$^{-/-}$- and AgRP-treated mice exhibit the same increase in total T_3 hormone levels as wild-type controls when switched from a low-fat to a high-fat diet. The total T_4 levels in high-fat-fed MC4R$^{-/-}$ and wild-type mice are decreased, which is likely related to the observed increases in T_3, because T_4 is converted to T_3. Altogether, these results suggest that the central melanocortin-mediated regulation of the HPT axis may not be involved in the diet-induced thermogenesis provoked upregulation of UCP1, although the melanocortin system may mediate part of the leptin-induced secretion of TRH from hypothalamic explants (Perello et al. 2006b).

5.6 Pro-opiomelanocortin

Similar to pro-TRH, POMC follows the intracellular trafficking of a secreted protein through the GC and ultimately the SGs where the end products of processing are stored before being secreted by exocytosis. During trafficking POMC undergoes a series of posttranslational modifications, resulting in the processing of the precursor to yield the various biologically active POMC-derived peptides (Fig. 5.8). The POMC polypeptide precursor contains eight pairs, and one quadruplet, of basic amino acids, which are the cleavage sites for PC1 and PC2 (Benjannet et al. 1991). Tissue-specific processing of POMC is one of the best known examples emphasizing the importance that the expression and activity of PC1 and PC2 have on the outcome of POMC products. For example, in the corticotroph cells of the anterior pituitary, POMC is processed predominantly to ACTH, ß-lipotropin (LPH), and a 16 kDa N-terminal fragment. ACTH is the regulator of the adrenocortical function. Only four pairs of basic residues are cleaved, which are all of the Lys-Arg type. These cleavages generate the following six peptides: amino terminal (NT), joining peptide, ACTH, ß-LPH, a small amount of γ-LPH, and small amount of ß-end (Bertagna et al. 1986; 1988). On the other hand, in the melanotroph cells of the intermediate lobe of the pituitary in rodents and the hypothalamus, nucleus of the solitary tract, and placenta in man, a different POMC processing occurs. In these tissues, most of the pairs of basic residues including the tetra basic are cleaved by the PCs. For example, the NT gives rise to the γ-MSHs, ACTH to α-MSH and CLIP (corticotropin-like intermediate lobe peptide or ACTH (18–39)), and ß-LPH to ß-MSH, ß-end (1–31), and ß-end (1–27) (Bertagna et al. 1986; Liotta et al. 1982). The key factor involved

in the processing of ACTH to α-MSH and CLIP in melanotrophs, the ARC, and the NTS is the presence of PC2. Further chemical modifications including glycosylation, amidation, phosphorylation, acetylation, and sulfation also occur in a cell-specific manner resulting in an alteration of the biological activity of the peptides (such as acetylation of ß-end and α-MSH).

In the ARC, similar to the pars intermedia of the pituitary (Barnea et al. 1982; Gramsch et al. 1980; Orwoll et al. 1979; Emeson and Eipper 1986), POMC is initially cleaved by PC1 to generate pro-adrenocorticotrophin (pro-ACTH) and b-lipotropin. Pro-ACTH is further cleaved by PC1 to generate a 16 kDa N-terminal peptide and ACTH. ACTH is further cleaved by PC2 to generate $ACTH_{(1–17)}$ and corticotropin-like intermediate lobe peptide (CLIP) as well as γ-LPH and ß-EP1– 31. ß-LPH is processed to γ-LPH and ß-EP; N-terminal POMC is processed to γ3-MSH (Emeson and Eipper 1986; Pritchard et al. 2002). In the human, γ-LPH can be further processed to

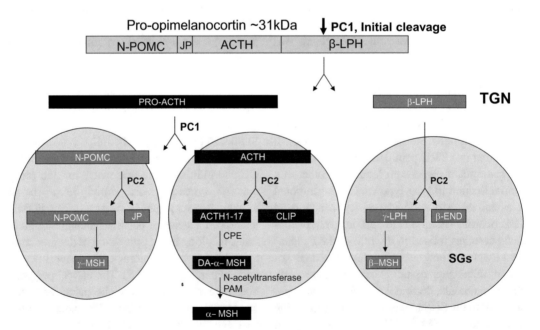

Fig. 5.8 This figure depicts a schematic representation of the POMC polypeptide and its posttranslational processing. Arrows indicate where PC1 and PC2 produce their enzymatic cleavages and the major peptide products derived from endoproteolytic cleavages. JP joining peptide, LPH lipotropin, EP endorphin, CLIP corticotropin-like-intermediate lobe peptide, da-α-MSH desacetyl α-MSH, CPE carboxypeptidase E, N-acetyltransferase, PAM peptidyl α-amidating monooxygenase

ß-MSH (Biebermann et al. 2006). Then, CPE enzyme removes C-terminal basic amino acids from ACTH $_{(1-17)}$, and the peptidyl a-amidating monooxygenase (PAM) enzyme amidates the peptide to generate desacetyl a-melanocyte-stimulating hormone (α-MSH). Acetylation of ACTH or desacetyl α-MSH to acetyl α-MSH (Wilkinson 2006; Pritchard et al. 2002; Cone et al. 1996) by the N-acetyltransferase enzyme could occur immediately after the amino-terminal side of these peptides becomes available subsequent to the PC cleavage. Fasting-induced changes in the ARC caused a significant decrease in ACTH and desacetyl α-MSH consistent with a decrease in POMC biosynthesis during fasting (Perello et al. 2007b). A similar conclusion can be drawn from the fact that POMC and ACTH levels are reduced in cerebrospinal fluid during fasting (Pritchard et al. 2002). This decrease in POMC was associated with a decrease in PC1 (Perello et al. 2007b). Leptin administration in fasted rats prevented the fasting-induced decrease in the content of all POMC-related peptides demonstrating that leptin potently regulates the biosynthesis of POMC in the ARC. While a previous report suggested an increase in acetyl α-MSH (the most active form of α-MSH) due to leptin action on the yet undefined N-acetyltransferase activity (Guo et al. 2004), studies from different laboratories concluded that leptin is not involved in the N-acetylation of hypothalamic α-MSH (Wilkinson 2006; Perello et al. 2007b; Harrold et al. 1999).

POMC neurons in the NTS innervate different areas of the brain, but not the hypothalamus (Joseph and Michael 1988; Pilcher and Joseph 1986; Ellacott and Cone 2006). They send projections primarily within the dorsal vagal complex (DVC), which includes the NTS, and to other structures in the brainstem and medulla (Palkovits and Eskay 1987; Palkovits et al. 1987). The NTS mediates some actions of leptin on energy balance. For example, administration of leptin in the fourth ventricle or directly in the DVC, area postrema, and dorsal motor nucleus (DMX) inhibits food intake (Grill et al. 2002). Leptin treatment also increases c-Fos expression in some POMC neurons in the NTS (Ellacott and Cone 2006). The administration of melanocortin agonists or antagonists directly in the DVC reduces or increases food intake, respectively (Williams et al. 2000); melanocortin agonists injected in the fourth ventricle increased uncoupling protein-1 expression in the brown fat of rats (Williams et al. 2003); and MC4-R is highly expressed in brain stem regions, such as the DVC (Kishi et al. 2003). POMC neurons present in the NTS showed to participate in energy balance. Hindbrain administration of leptin or a melanocortin receptor agonist altered energy balance in mice likely via participation of hindbrain POMC neurons (De Jonghe et al. 2012).

The biosynthesis of POMC in the NTS is unique and different from the ARC. Similar to the ARC, it produces a-MSH through the action of PC2, but the response to fasting and leptin is quite different (Perello et al. 2007b). In the NTS, a prominent peptide of about 28.1 kDa molecular mass, similar in size to POMC, has been identified, together with other POMC-derived peptides, including α-MSH. Differing from the ARC, during fasting, ACTH and desacetyl a-MSH were found to accumulate, while at the same time, *pomc* mRNA decreased. Again, differing from the ARC, these changes were not reversed by leptin (Perello et al. 2007b; Huo et al. 2006). However, it is important to point out that ObRb is present in the NTS (Shioda et al. 1998; Grill and Kaplan 2002), and the marker of leptin action phosphorylated signal transducer and activator of transcription 3 (P-STAT3) has been found in the neurons of the NTS of leptin-treated rats (Hosoi et al. 2002; Munzberg et al. 2003), and delivery of leptin into the DVC decreases food intake and body weight (Grill and Kaplan 2002). Other studies did show that peripheral administration of leptin induced P-STAT3 activation in about 30% of the POMC-EGFP neurons in the NTS (Ellacott et al. 2006). While the phosphorylation of STAT3 is valid and widely used to indicate leptin signaling, P-STAT3 could participate in other signaling pathways independent of the biosynthesis of POMC. Even though all these components of leptin signaling are present in the NTS, leptin does not regulate the *pomc* gene and POMC-related peptides (Perello et al. 2007b; Huo et al. 2006).

Similar to previous studies done in the PVN (Sanchez et al. 2004), low leptin levels resulted in a decrease in PC1 in the ARC and a decrease in PC2 in the NTS. At present, it is unknown whether this selective effect of fasting on the PCs seen in the ARC and NTS may contribute to changes in POMC processing. Still, a proper PC activity on POMC is important in maintaining a proper enzyme-substrate homeostasis. Supporting this view, as described above, PC1 and PC2 gene expression and processing of the prohormone are affected in a coordinate fashion during photoperiod changes in seasonal Siberian hamsters to control body weight (Helwig et al. 2006). In that study they compared mRNA levels and protein distribution of PC1, PC2, POMC, ACTH, α-MSH, β-endorphin, and orexin-A in selected hypothalamic areas of long day, short day, and natural day. The key finding of that study was that a major part of neuroendocrine body weight control in seasonal adaptation may be affected by posttranslational processing mediated by PC1 and PC2, in addition to regulation of gene expression of neuropeptide precursors. Therefore, the coupled regulation of POMC/processing enzymes may be a common process, by which cells generate more effective processing of the prohormone into mature peptides.

5.7 Pro-corticotropin-Releasing Hormone CRH

Corticotropin-releasing hormone (CRH) was discovered in 1981 (Vale et al. 1981), and since then, in numerous studies, it has been demonstrated to be involved in many physiological processes including energy homeostasis. CRH has a critical role in mediating the stress response (Kovacs 2013). Also CRH functions to regulate metabolic, immunologic, and homeostatic changes both basally and under various pathologic conditions (Sominsky and Spencer 2014; Chrousos 1995; Seimon et al. 2013). CRH is heterogeneously expressed in the periphery and in the brain with high expression in the PVN (Ziegler et al. 2007; Raadsheer et al. 1993; Korosi and Baram 2008). Arginine vasopressin

(AVP) is also produced in the PVN and acts to synergize CRH actions. CRH produced in the medial parvocellular division of the PVN functions as the central regulator of the HPA axis (Aguilera et al. 2008).

Similarly to pro-TRH and POMC, CRH is initially synthesized as a larger inactive precursor, prepro-CRH made of 196 amino acids (Spiess et al. 1981; Rivier et al. 1983). After removal of the signal sequence while entering the ER, pro-CRH is routed to the TGN where it undergoes enzymatic posttranslational modifications to generate several intermediate forms as well as the bioactive CRH_{1-41} peptide that is produced from the C-terminal region of the precursor polypeptide (Fig. 5.9) (Perone et al. 1998). Pro-CRH is cleaved by PC2 or PC1 at Arg 152-Arg 153, thereby releasing a 43-residue CRH peptide (Perone et al. 1998; Brar et al. 1997; Castro et al. 1991). Inactive CRH is further processed into its bioactive form via the removal of the C-terminus lysine residue by the actions of carboxypeptidase E (CPE) and is subsequently amidated at the exposed carboxyl group of a glycine residue by peptidylglycine hydroxylase (also referred to as peptidylglycine alpha-amidating monooxygenase (PAM)) (Eipper et al. 1993).

Under both basal and stimulated conditions, CRH produced in the PVN is released from nerve terminals that anteriorly juxtapose the median eminence where it is traversed into the hypothalamo-hypophyseal portal system (Rho and Swanson 1987). Upon binding to its cognate receptor (corticotropin-releasing hormone receptor 1 (CRHR1)) (Lovejoy et al. 2014), which is expressed by corticotropic cells of the adenohypophysis, CRH stimulates the synthesis and secretion of adrenocorticotropic hormone (ACTH) as well as other bioactive molecules such as β-endorphin (Solomon 1999). ACTH engages the melanocortin 2 receptor expressed by cells of the adrenal cortex and stimulates the production and secretion of steroid hormones such as CORT (Veo et al. 2011). The role of CRH and the HPA axis is fully described in Chap. 12. Both ACTH and glucocorticoids (GCs) function to regulate HPA axis activity via long and short negative feedback loops that signal at the level of the hypothalamus, the extra-hypothalamic brain

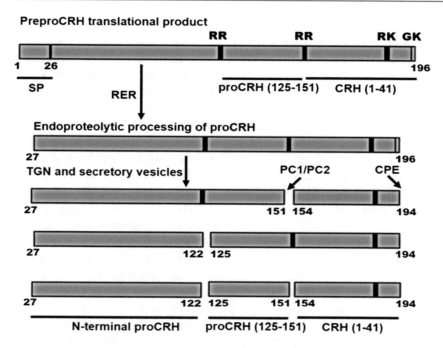

Fig. 5.9 Pro-CRH processing cascade. Pro-CRH is processed by PC1/2 and CPE to render several intermediate forms and the active CRH. Pro-CRH (125–151), pro-corticotropin-releasing hormone 125–151; CRH (1–41), corticotropin-releasing hormone 1–41; N-terminal pro-CRH, N-terminal pro-corticotropin releasing hormone; C.S. (146–159). R arginine, K lysine, and G glycine. Processing begins in the TGN and is completed within secretory vesicles

sites, and the adenohypophysis. CRH functions to promote negative energy balance by suppressing appetite and enhancing thermogenesis (Richard et al. 2000; Mastorakos and Zapanti 2004; Toriya et al. 2010), whereas GCs function to promote positive energy balance in part by affecting glucose metabolism, lipid homeostasis, and increasing appetite drive (Tataranni et al. 1996). Although a consensus on the exact role of adrenal activity in relation to energy dysfunction has yet to be reached, increased and sustained basal GC is implicated in the development of visceral obesity, insulin resistance, and metabolic disease (Laryea et al. 2013; Kong et al. 2014; Spencer and Tilbrook 2011). In a recent study, our laboratory showed the first evidence supporting the hypothesis that PVN Sirt1 activates the HPA axis and basal CORT levels by enhancing the production of CRH through an increase in the

biosynthesis of PC2, which is essential in the maturation of CRH from pro-CRH (Toorie et al. 2016). Moreover, in the DIO state, PVN Sirt1 increases basal (not stress induced) circulating CORT in a manner independent of prepro-CRH transcriptional changes. Instead, Sirt1's effects on adrenal activity are mediated via a posttranslational processing mechanism in concert with an increase in PC2 (Fig. 5.9). Increased CRH release from the PVN increased pituitary ACTH synthesis and release and in turn increased circulating basal CORT concentrations. Together these findings suggest that increasing PVN Sirt1 activity results in an increase in the amount of CRH targeted to the anterior pituitary, thereby enhancing ACTH signaling and basal CORT concentration (Toorie and Nillni 2014; Nillni 2016). Furthermore, our results demonstrated that inhibiting Sirt1 specifically in the PVN of obese rodents caused a reduction in PVN proPC2 and active forms of PC2, but not PC1. This decrease in PC2 was associated with a decrease in CRH in the ME, as well as reduced circulating GC effects that were not observed in lean individuals. More detailed information about Sirt1 is described in Chap. 7. Collectively, the findings demonstrate that Sirt1 regulates the CRH peptide by modulating the processing enzyme PC2 and FoxO1 adding a novel regulatory link between PVN Sirt1 and HPA axis activity (Toorie et al. 2016; Toorie and Nillni 2014; Nillni 2016).

5.8   Conclusions

Since the discovery of the PCs in the late 1980s and the progress made in understanding the cell biology of protein biosynthesis and secretion, a new frontier on the prohormone processing and trafficking to the RSP research had come to surface. It is particularly relevant to emphasize the concept that secretion of any particular peptide or neuropeptide hormone from a given cell involves two specific steps. The first one is the action of a specific secretagogue or ligand binding to the specific cell receptor, inducing the fusion of SGs to the plasma membrane of the cell followed by the release of the peptide. The second one, most likely triggered by the first action, is a complex set of enzymatic processes by the PCs on prohormones to generate mature bioactive peptide ready for release from the proper secretory granules. The examples described in this chapter on the action of leptin on pro-TRH and POMC and PCs demonstrate the existence of a coupled regulation pro-neuro-peptide/processing enzymes as a common mechanisms by which cells generate more efficient processing of prohormones into mature peptides ready for release. The evidence described here strongly suggests that posttrans-lational processing of neuropeptides hormones is, among many other factors, critical in the pathogenesis of obesity.

Some peptidergic systems within the central nervous system and the periphery have been investigated as potential treatments for obesity and the metabolic syndrome. Manipulation of these systems showed to reduce food intake and body weight in preclinical models. However, manipulation of peptidergic systems poses many difficulties in clinical trials. Many of them failed to meaningfully lower body weight. There are a number of challenges in using peptide hormone or analogs acting on their corresponding receptors. It is difficult to administer the desired agent at the dose required to reduce body weight without avoiding side effects. Also, several peptide hormones are widely expressed in the hypothalamus and periphery with diverse functions that makes difficult their manipulation for the treatment of obesity avoiding unwanted side effects. Another issue is the way of administration of these agents to patients. For example targeting CNS receptors that must cross the blood-brain barrier system is a strategy difficult to accomplish when dealing with long peptide forms to be administrated. However, progress has been made recently with new technological advantages. Many of these strategies and possibilities of novel drug treatments using the peptidergic system are thoroughly discussed elsewhere (Rodgers et al. 2012; Greenwood et al. 2011).

Questions

1. Define the prohormone theory and in which cases it is applicable.
2. Which are the prohormone convertases involved in the maturation of peptide hormones in the neuronal and endocrine tissue?
3. Which prohormone convertase matures in secretory granules, and through which mechanism it becomes enzymatically active?
4. Are pro-TRH-derived peptides sorted to different secretory granules, and what could be the meaning of this process?
5. What is the role of alpha-MSH and ACTH in energy balance?

References

Aguilera, G., Subburaju, S., Young, S., & Chen, J. (2008). The parvocellular vasopressinergic system and responsiveness of the hypothalamic pituitary adrenal axis during chronic stress. *Progress in Brain Research, 170,* 29–39. https://doi.org/10.1016/S0079-6123(08)00403-2.

Arvan, P., & Castle, D. (1998). Sorting and storage during secretory granule biogenesis: Looking backward and looking forward. *The Biochemical Journal, 15,* 593–610.

Ayoubi, T. A., & Van De Ven, W. J. (1996). Regulation of gene expression by alternative promoters. *The FASEB Journal, 10,* 453–460.

Bacova, Z., Najvirtova, M., Krizanova, O., Hudecova, S., Zorad, S., Strbak, V., & Benicky, J. (2005). Effect of neonatal streptozotocin and thyrotropin-releasing hormone treatments on insulin secretion in adult rats. *General Physiology and Biophysics, 24,* 181–197.

Barnea, A., Cho, G., & Porter, J. C. (1982). A reduction in the concentration of immunoreactive corticotropin, melanotropin and lipotropin in the brain of the aging rat. *Brain Research, 232,* 345–353.

Benjannet, S., Rondeau, N., Day, R., Chretien, M., & Seidah, N. G. (1991). PC1 and PC2 are proprotein convertases capable of cleaving proopiomelanocortin at distinct pairs of basic residues. *Proceedings of the National Academy of Sciences of the United States of America, 88,* 3564–3568.

Bertagna, X., Lenne, F., Comar, D., Massias, J. F., Wajcman, H., Baudin, V., Luton, J. P., & Girard, F. (1986). Human beta-melanocyte-stimulating hormone revisited. *Proceedings of the National Academy of Sciences of the United States of America, 83,* 9719–9723.

Bertagna, X., Camus, F., Lenne, F., Girard, F., & Luton, J. P. (1988). Human joining peptide: A proopiomelanocortin product secreted as a homodimer. *Molecular Endocrinology, 2,* 1108–1114.

Bianco, A. C., Sheng, X. Y., & Silva, J. E. (1988). Triiodothyronine amplifies norepinephrine stimulation of uncoupling protein gene transcription by a mechanism not requiring protein synthesis. *The Journal of Biological Chemistry, 263,* 18168–18175.

Biebermann H, Castañeda TR, van Landeghem F, von Deimling A, Escher F, Brabant G, Hebebrand J, Hinney A, Tschöp MH, Grüters A, Krude H (2006) Cell Metab 3(2):141–146. PMID:16459315.

Boler, J., Enzmann, F., Folkers, K., Bowers, Y., & Shally, V. (1969). The identity of chemical and hormonal properties of the thyrotropin releasing hormone and pyro-glutamyl-histidil-proline amide. *Biochemical and Biophysical Research Communications, 705.*

Bonnemaison, M. L., Eipper, B. A., & Mains, R. E. (2013). Role of adaptor proteins in secretory granule biogenesis and maturation. *Frontiers in Endocrinology, 4, 101.* https://doi.org/10.3389/fendo.2013.00101.

Brakch, N., Cohen, P., & Boileau, G. (1994). Processing of human prosomatostatin in AtT-20 cells: S-28 and S-14 are generated in different secretory pathways. *Biochemical and Biophysical Research Communications, 205,* 221–229.

Brakch, N., Galanopoulou, A. S., Patel, Y. C., Boileau, G., & Seidah, N. G. (1995). Comparative proteolytic processing of rat prosomatostatin by the convertases PC1, PC2, furin, PACE4, and PC5 in constitutive and regulated secretory pathways. *FEBS Letters, 362,* 143–146.

Brar, B., Sanderson, T., Wang, N., & Lowry, P. J. (1997). Post-translational processing of human procorticotrophin-releasing factor in transfected mouse neuroblastoma and Chinese hamster ovary cell lines. *The Journal of Endocrinology, 154,* 431–440

Brechler, V., Chu, W. N., Baxter, J. D., Thibault, G., & Reudelhuber, T. L. (1996). *The Journal of Biological Chemistry, 271*(34), 20636–20640..

Breslin, M. B., Lindberg, I., Benjannet, S., Mathis, J. P., Lazure, C., & Seidah, N. G. (1993). Differential processing of proenkephalin by prohormone convertases 1(3) and 2 and furin. *The Journal of Biological Chemistry, 268,* 27084–27093.

Burgess, T. L., & Kelly, R. B. (1987). Constitutive and regulated secretion of proteins. *Annual Review of Cell Biology, 3,* 243–293. https://doi.org/10.1146/annurev.cb.03.110187.001331.

Burgess, T. L., Craik, C. S., Matsuuchi, L., & Kelly, R. B. (1987). In vitro mutagenesis of trypsinogen: Role of the amino terminus in intracellular protein targeting to secretory granules. *The Journal of Cell Biology, 105,* 659–668.

Burgus, R., Dunn, T. F., Desiderio, D., Vale, W., & Guillemin, R. (1969). Derives polypeptidiques de syntheses doues d'activite hypophysiotropic. Nouvelles observations. *Comptes Rendus. Académie des Sciences,* 1870.

Castro, M., Lowenstein, P., Glynn, B., Hannah, M., Linton, E., & Lowry, P. (1991). Post-translational processing and regulated release of corticotropin-releasing hormone (CRH) in AtT20 cells expressing the human proCRH gene. *Biochemical Society Transactions, 19,* 246S.

Cawley, N. X., Li, Z., & Loh, Y. P. (2016a). 60 YEARS OF POMC: Biosynthesis, trafficking, and secretion of pro-opiomelanocortin-derived peptides. *Journal of Molecular Endocrinology,* T77–T97. https://doi.org/10.1530/JME-15-0323.

Cawley, N. X., Rathod, T., Young, S., Lou, H., Birch, N., & Loh, Y. P. (2016b). Carboxypeptidase E and Secretogranin III coordinately facilitate efficient sorting of proopiomelanocortin to the regulated secretory pathway in AtT20 cells. *Molecular Endocrinology,* 37–47. https://doi.org/10.1210/me.2015-1166.

Challis, B. G., Pritchard, L. E., Creemers, J. W., Delplanque, J., Keogh, J. M., Luan, J., Wareham, N. J., Yeo, G. S., Bhattacharyya, S., Froguel, P., White, A., Farooqi, I. S., & O'Rahilly, S. (2002). A missense mutation disrupting a dibasic prohormone processing site in pro-opiomelanocortin (POMC) increases susceptibility to early-onset obesity through a novel molecular mechanism. *Human Molecular Genetics, 11*, 1997–2004.

Chanat, E. (1993). Mechanism of sorting of secretory proteins and formation of secretory granules in neuroendocrine cells. *Comptes Rendus des Seances de la Societe de Biologie et de Ses Filiales, 187*, 697–725.

Chretien, M. (2011). The prohormone theory and the proprotein convertases: It is all about serendipity. *Methods in Molecular Biology*, 13–19. https://doi.org/10.1007/978-1-61779-204-5_2.

Chretien, M., & Li, C. H. (1967). Isolation, purification, and characterization of gamma-lipotropic hormone from sheep pituitary glands. *Canadian Journal of Biochemistry, 45*, 1163.

Chretien, M., Benjannet, S., & Gossard, F. (1979). From beta-lipotropin to beta-endorphin and 'proopiomelanocortin'. *Canadian Journal of Biochemistry, 57*, 1111–1121.

Chrousos, G. P. (1995). The hypothalamic-pituitary-adrenal axis and immune-mediated inflammation. *The New England Journal of Medicine*, 1351–1362. https://doi.org/10.1056/NEJM199505183322008.

Chun, J. Y., Korner, J., Kreiner, T., Scheller, R. H., & Axel, R. (1994). The function and differential sorting of a family of Aplysia prohormone processing enzymes. *Neuron*, 831–834.

Cone, R. D., Lu, D., Koppula, S., Vage, D. I., Klungland, H., Boston, B., Chen, W., Orth, D. N., Pouton, C., & Kesterson, R. A. (1996). The melanocortin receptors: Agonists, antagonists, and the hormonal control of pigmentation. *Recent Progress in Hormone Research*, 287–317. discussion 318.

Constam, D. B., Calfon, M., & Robertson, E. J. (1996). SPC4, SPC6 and the novel protease SPC7 are coexpressed with bone morphogenic proteins at distinct sites during embryogenesis. *The Journal of Cell Biology*, 181–191.

Cool, D. R., & Loh, Y. P. (1998). Carboxypeptidase E is a sorting receptor for prohormones: Binding and kinetic studies. *Molecular and Cellular Endocrinology*, 7–13.

Cool, D. R., Fenger, M., Snell, C. R., & Loh, P. Y. (1995). Identification of the sorting signal motif within the pro-opiomelanocortin for the regulated secretory pathway. *The Journal of Biological Chemistry*, 8723–8729.

Dannies, P. S. (1999). Protein hormone storage in secretory granules: Mechanisms for concentration and sorting. *Endocrine Reviews*, 3–21.

De Jonghe, B. C., Hayes, M. R., Zimmer, D. J., Kanoski, S. E., Grill, H. J., & Bence, K. K. (2012). Food intake reductions and increases in energetic responses by hindbrain leptin and melanotan II are enhanced in mice with POMC-specific PTP1B deficiency. *American Journal of Physiology. Endocrinology and Metabolism*, E644–E651. https://doi.org/10.1152/ajpendo.00009.2012.

Dey, A., Xhu, X., Carroll, R., Turck, C. W., Stein, J., & Steiner, D. F. (2003). Biological processing of the cocaine and amphetamine-regulated transcript precursors by prohormone convertases, PC2 and PC1/3. *The Journal of Biological Chemistry*, 15007–15014.

Dey, A., Norrbom, C., Zhu, X., Stein, J., Zhang, C., Ueda, K., & Steiner, D. F. (2004). Furin and prohormone convertase 1/3 are major convertases in the processing of mouse pro-growth hormone-releasing hormone. *Endocrinology*, 1961–1971.

Dhanvantari, S., Arnaoutova, I., Snell, C. R., Steinbach, P. J., Hammond, K., Caputo, G. A., London, E., & Loh, Y. P. (2002). Carboxypeptidase E, a prohormone sorting receptor, is anchored to secretory granules via a C-terminal transmembrane insertion. *Biochemistry*, 52–60.

Dikeakos, J. D., & Reudelhuber, T. L. (2007). Sending proteins to dense core secretory granules: Still a lot to sort out. *The Journal of Cell Biology*, 191–196.

Eipper, B. A., & Mains, R. E. (1980). Structure and biosynthesis of pro-adrenocorticotropin/endorphin and related peptides. *Endocrine Reviews*, 1–27.

Eipper, B. A., Stoffers, D. A., & Mains, R. E. (1992). The biosynthesis of neuropeptides: Peptide α-amidation. *Annual Review of Neuroscience*, 57–85.

Eipper, B. A., Milgram, S. L., Husten, E. J., Yun, H. Y., & Mains, R. E. (1993). Peptidylglycine alpha-amidating monooxygenase: A multifunctional protein with catalytic, processing, and routing domains. *Protein Science*, 489–497. https://doi.org/10.1002/pro.5560020401.

Ellacott, K. L., & Cone, R. D. (2006). The role of the central melanocortin system in the regulation of food intake and energy homeostasis: Lessons from mouse models. *Philosophical Transactions of the Royal Society of London. Series B, Biological Sciences*, 1265–1274.

Ellacott, K. L., Halatchev, I. G., & Cone, R. D. (2006). Characterization of leptin-responsive neurons in the caudal brainstem. *Endocrinology*, 3190–3195.

Emeson, R. B., & Eipper, B. A. (1986). Characterization of pro-ACTH/endorphin-derived peptides in rat hypothalamus. *The Journal of Neuroscience*, 837–849.

Ezkurdia, I., Juan, D., Rodriguez, J. M., Frankish, A., Diekhans, M., Harrow, J., Vazquez, J., Valencia, A., & Tress, M. L. (2014). Multiple evidence strands suggest that there may be as few as 19,000 human protein-coding genes. *Human Molecular Genetics*, 5866–5878. https://doi.org/10.1093/hmg/ddu309.

Farooqi, I. S., Matarese, G., Lord, G. M., Keogh, J. M., Lawrence, E., Agwu, C., Sanna, V., Jebb, S. A., Perna, F., Fontana, S., Lechler, R. I., DePaoli, A. M., & O'Rahilly, S. (2002). Beneficial effects of leptin on obesity, T cell hyporesponsiveness, and neuroendocrine/metabolic dysfunction of human congenital leptin deficiency. *The Journal of Clinical Investigation*, 1093–1103.

Fekete, C., Sarkar, S., Rand, W. M., Harney, J. W., Emerson, C. H., Bianco, A. C., & Lechan, R. M. (2002). Agouti-related protein (AGRP) has a central inhibitory action on the hypothalamic-pituitary-thyroid (HPT) axis; comparisons between the effect of AGRP and neuropeptide Y on energy homeostasis and the HPT axis. *Endocrinology*, 3846–3853.

Fink, G. (1976). The development of the releasing factor concept. *Clinical Endocrinology*, 245–260.

Fortenberry, Y., Liu, J., & Lindberg, I. (1999). The role of the 7B2 CT peptide in the inhibition of prohormone convertase 2 in endocrine cell lines. *Journal of Neurochemistry*, 994–1003.

Fricker, L. (1988). Carboxypeptidase E. *Annual Review of Physiology*, 309–321.

Fricker, L. D. (2007). Neuropeptidomics to study peptide processing in animal models of obesity. *Endocrinology*, 4185–4190.

Fricker, L. D., Berman, Y. L., Leiter, E. H., & Devi, L. A. (1996). Carboxypeptidase E activity is deficient in mice with the fat mutation. Effect on peptide processing. *The Journal of Biological Chemistry*, 30619–30624.

Friedman, T. C., Loh, Y. P., Cawley, N. X., Birch, N. P., Huang, S. S., Jackson, I. M., & Nillni, E. A. (1995). Processing of prothyrotropin-releasing hormone (pro-TRH) by bovine intermediate lobe secretory vesicle membrane PC1 and PC2 enzymes. *Endocrinology*, 4462–4472.

Furuta, M., Yano, H., Zhou, A., Rouille, Y., Holst, J. J., Carroll, R., Ravazzola, M., Orci, L., Furuta, H., & Steiner, D. F. (1997). Defective prohormone processing and altered pancreatic islet morphology in mice lacking active SPC2. *Proceedings of the National Academy of Sciences of the United States of America*, 6646–6651.

Galanopoulou, A. S., Kent, G., Rabbani, S. N., Seidah, N. G., & Patel, Y. C. (1993). Heterologous processing of Prosomatostatin in consecutive and regulated secretory pathways. *The Journal of Biological Chemistry*, 6041–6049.

Genomes Project, C., Abecasis, G. R., Auton, A., Brooks, L. D., DePristo, M. A., Durbin, R. M., Handsaker, R. E., Kang, H. M., Marth, G. T., & McVean, G. A. (2012). An integrated map of genetic variation from 1,092 human genomes. *Nature*, 56–65. https://doi.org/10.1038/nature11632.

Glombik, M. M., & Gerdes, H. H. (2000). Signal-mediated sorting of neuropeptides and prohormones: Secretory granule biogenesis revisited. *Biochimie*, 315–326.

Glombik, M. M., Kromer, A., Salm, T., Huttner, W. B., & Gerdes, H. H. (1999). The disulfide-bonded loop of chromogranin B mediates membrane binding and directs sorting from the trans-Golgi network to secretory granules. *The EMBO Journal*, 1059–1070.

Gorr, S. U., Jain, R. K., Kuehn, U., Joyce, P. B., & Cowley, D. J. (2001). Comparative sorting of neuroendocrine secretory proteins: A search for common ground in a mosaic of sorting models and mechanisms. *Molecular and Cellular Endocrinology*, 1–6.

Gramsch, C., Kleber, G., Hollt, V., Pasi, A., Mehraein, P., & Herz, A. (1980). Pro-opiocortin fragments in human and rat brain: Beta-endorphin and alpha-MSH are the predominant peptides. *Brain Research*, 109–119.

Greenwood, H. C., Bloom, S. R., & Murphy, K. G. (2011). Peptides and their potential role in the treatment of diabetes and obesity. *The Review of Diabetic Studies*, 355–368. https://doi.org/10.1900/RDS.2011.8.355.

Grill, H. J., & Kaplan, J. M. (2002). The neuroanatomical axis for control of energy balance. *Frontiers in Neuroendocrinology*, 2–40.

Grill, H. J., Schwartz, M. W., Kaplan, J. M., Foxhall, J. S., Breininger, J., & Baskin, D. G. (2002). Evidence that the caudal brainstem is a target for the inhibitory effect of leptin on food intake. *Endocrinology*, 239–246.

Guillemin, R. (1978). Peptides in the brain: The new endocrinology of the neuron. *Science*, 390–402.

Guo, L., Munzberg, H., Stuart, R. C., Nillni, E. A., & Bjorbaek, C. (2004). N-acetylation of hypothalamic alpha-melanocyte-stimulating hormone and regulation by leptin. *Proceedings of the National Academy of Sciences of the United States of America*, 11797–11802.

Hall, M. E., & Stewart, J. M. (1983). Substance P and behavior: Opposite effects of N-terminal and C-terminal fragments. *Peptides*, 763–766.

Harrold, J. A., Williams, G., & Widdowson, P. S. (1999). Changes in hypothalamic agouti-related protein (AGRP), but not alpha-MSH or pro-opiomelanocortin concentrations in dietary-obese and food-restricted rats. *Biochemical and Biophysical Research Communications*, 574–577.

Haynes, W. G., Morgan, D. A., Djalali, A., Sivitz, W. I., & Mark, A. L. (1999). Interactions between the melanocortin system and leptin in control of sympathetic nerve traffic. *Hypertension*, 542–547.

Hedstrom, L. (2002). Serine protease mechanism and specificity. *Chemical Reviews*, 4501–4524.

Helwig, M., Khorooshi, R. M., Tups, A., Barrett, P., Archer, Z. A., Exner, C., Rozman, J., Braulke, L. J., Mercer, J. G., & Klingenspor, M. (2006). PC1/3 and PC2 gene expression and post-translational endoproteolytic pro-opiomelanocortin processing is regulated by photoperiod in the seasonal Siberian hamster (Phodopus sungorus). *Journal of Neuroendocrinology*, 413–425.

Hook, V., Azaryan, A., Hwong, S., & Tezapsidis, N. (1994). Proteases and the emerging role of protease inhibitors in prohormone processing. *FASEB*, 1269–1278.

Hosoi, T., Kawagishi, T., Okuma, Y., Tanaka, J., & Nomura, Y. (2002). Brain stem is a direct target for leptin's action in the central nervous system. *Endocrinology*, 3498–3504.

Hou, J. C., Min, L., & Pessin, J. E. (2009). Insulin granule biogenesis, trafficking and exocytosis. *Vitamins and Hormones*, 473–506. https://doi.org/10.1016/S0083-6729(08)00616-X.

Howell, S. L., & Taylor, K. W. (1967). The secretion of newly synthesized insulin *in vitro*. *The Biochemical Journal*, 922–930.

Huo, L., Grill, H. J., & Bjorbaek, C. (2006). Divergent regulation of proopiomelanocortin neurons by leptin in the nucleus of the solitary tract and in the arcuate hypothalamic nucleus. *Diabetes*, 567–573.

Jackson, R., Creemers, J. W. M., Ohagi, S., Raffin-Sanson, M. L., Sanders, L., Montague, C. T., Hutton, J. C., & O'Rahilly, S. (1997). Obesity and impaired prohormone processing associated with mutations in the human convertase 1 gene. *Nature Genetics*, 303–306.

Jansen, E., Ayoubi, T. A., Meulemans, S. M., & Van de Ven, W. J. (1995). Neuroendocrine-specific expression of the human prohormone convertase 1 gene. Hormonal regulation of transcription through distinct cAMP response elements. *The Journal of Biological Chemistry*, 15391–15397.

Jensen, O. N. (2004). Modification-specific proteomics: Characterization of post-translational modifications by mass spectrometry. *Current Opinion in Chemical Biology*, 33–41. https://doi.org/10.1016/j.cbpa.2003.12.009.

Jing, E., Nillni, E. A., Sanchez, V. C., Stuart, R. C., & Good, D. J. (2004). Deletion of the Nhlh2 transcription factor decreases the levels of the anorexigenic peptides alpha melanocyte-stimulating hormone and thyrotropin-releasing hormone and implicates prohormone convertases I and II in obesity. *Endocrinology*, 1503–1513.

Joseph, S. A., & Michael, G. J. (1988). Efferent ACTH-IR opiocortin projections from nucleus tractus solitarius: A hypothalamic deafferentation study. *Peptides*, 193–201.

Kishi, T., Aschkenasi, C. J., Lee, C. E., Mountjoy, K. G., Saper, C. B., & Elmquist, J. K. (2003). Expression of melanocortin 4 receptor mRNA in the central nervous system of the rat. *The Journal of Comparative Neurology*, 213–235.

Kong, X., Yu, J., Bi, J., Qi, H., Di, W., Wu, L., Wang, L., Zha, J., Lv, S., Zhang, F., Li, Y., Hu, F., Liu, F., Zhou, H., Liu, J., & Ding, G. (2014). Glucocorticoids transcriptionally regulate miR-27b expression promoting body fat accumulation via suppressing the browning of white adipose tissue. *Diabetes*. https://doi.org/10.2337/db14-0395.

Korosi, A., & Baram, T. Z. (2008). The central corticotropin releasing factor system during development and adulthood. *European Journal of Pharmacology*, 204–214. https://doi.org/10.1016/j.ejphar.2007.11.066.

Kovacs, K. J. (2013). CRH: The link between hormonal-, metabolic- and behavioral responses to stress. *Journal of Chemical Neuroanatomy*, 25–33. https://doi.org/10.1016/j.jchemneu.2013.05.003.

Laryea, G., Schutz, G., & Muglia, L. J. (2013). Disrupting hypothalamic glucocorticoid receptors causes HPA axis hyperactivity and excess adiposity. *Molecular Endocrinology*, 1655–1665. https://doi.org/10.1210/me.2013-1187.

Laurent, V., Kimble, A., Peng, B., Zhu, P., Pintar, J. E., Steiner, D. F., & Lindber, I. (2002). Mortality in 7B2 null mice can be rescued by adrenalectomy: Involvement of dopamine in ACTH hypersecretion. *Proceedings of the National Academy of Sciences of the United States of America*, 3087–3092.

Lechan, R. M., Wu, P., Jackson, I. M. D., Wolfe, H., Cooperman, S., Mandel, G., & Goodman, R. H.

(1986a). Thyrotropin-releasing hormone precursor: Characterization in rat brain. *Science*, 159–161.

Lechan, R. M., Wu, P., & Jackson, I. M. D. (1986b). Immunolocalization of the thyrotropin-releasing hormone prohormone in the rat central nervous system. *Endocrinology*, 1210–1216.

Lee, S. L., Stewart, K., & Goodman, R. H. (1988). Structure of the gene encoding rat thyrotropin releasing hormone. *The Journal of Biological Chemistry*, 16604–16609.

Legradi, G., Rand, W. M., Hitz, S., Nillni, E. A., Jackson, I. M., & Lechan, R. M. (1996). Opiate withdrawal increases ProTRH gene expression in the ventrolateral column of the midbrain periaqueductal gray. *Brain Research*, 10–19.

Li, Q. L., Jansen, E., & Friedman, T. C. (1999). Regulation of prohormone convertase 1 (PC1) by gp130-related cytokines. *Molecular and Cellular Endocrinology*, 143–152.

Li, Q. L., Jansen, E., Brent, G. A., Naqvi, S., Wilber, J. F., & Friedman, T. C. (2000). Interactions between the prohormone convertase 2 promoter and the thyroid hormone receptor. *Endocrinology*, 3256–3266.

Li, Q. L., Jansen, E., Brent, G. A., & Friedman, T. C. (2001). Regulation of prohormone convertase 1 (PC1) by thyroid hormone. *American Journal of Physiology. Endocrinology and Metabolism*, E160–E170.

Liotta, A. S., Houghten, R., & Krieger, D. T. (1982). Identification of a beta-endorphin-like peptide in cultured human placental cells. *Nature*, 593–595.

Liston, D., Patey, G., Rossier, J., Verbanck, P., & Vanderhaeghen, J. J. (1984). Processing of proenkephalin is tissue-specific. *Science*, 734–736.

Lloyd, D. J., Bohan, S., & Gekakis, N. (2006). Obesity, hyperphagia and increased metabolic efficiency in Pc1 mutant mice. *Human Molecular Genetics*, 1884–1893.

Loh, Y. P., Maldonado, A., Zhang, C., Tam, W. H., & Cawley, N. (2002). Mechanism of sorting proopiomelanocortin and proenkephalin to the regulated secretory pathway of neuroendocrine cells. *Annals of the New York Academy of Sciences*, 416–425.

Lovejoy, D. A., Chang, B. S., Lovejoy, N. R., & del Castillo, J. (2014). Molecular evolution of GPCRs: CRH/CRH receptors. *Journal of Molecular Endocrinology*, T43–T60. https://doi.org/10.1530/JME-13-0238.

Luo, L., Luo, J. Z., & Jackson, I. M. (2008). Thyrotropin-releasing hormone (TRH) reverses hyperglycemia in rat. *Biochemical and Biophysical Research Communications*, 69–73.

Madala, P. K., Tyndall, J. D., Nall, T., & Fairlie, D. P. (2010). Update 1 of: Proteases universally recognize beta strands in their active sites. *Chemical Reviews*, PR1–P31. https://doi.org/10.1021/cr900368a.

Masanobu, Y., Radovick, S., Wondisford, F. E., Nakayama, Y., Weintraub, B. D., & Wilber, J. F. (1990). Cloning and structure of human genomic and hypothalamic cDNA and encoding human preprothyrotropin-releasing hormone. *Molecular Endocrinology*, 551–556.

Mastorakos, G., & Zapanti, E. (2004). The hypothalamic-pituitary-adrenal axis in the neuroendocrine regulation of food intake and obesity: The role of corticotropin releasing hormone. *Nutritional Neuroscience*, 271–280. https://doi.org/10.1080/10284150400020516.

Matsuuchi, L., & Kelly, R. B. (1991). Constitutive and basal secretion from the endocrine cell line, AtT-20. *The Journal of Cell Biology*, 843–852.

Morvan, J., & Tooze, S. A. (2008). Discovery and progress in our understanding of the regulated secretory pathway in neuroendocrine cells. *Histochemistry and Cell Biology*, 243–252. https://doi.org/10.1007/s00418-008-0377-z.

Mulcahy, L. R., Vaslet, C. A., & Nillni, E. A. (2005). Prohormone-convertase 1 processing enhances post-Golgi sorting of prothyrotropin-releasing hormone-derived peptides. *The Journal of Biological Chemistry*, 39818–39826.

Mulcahy, L. R., Barker, A. J., & Nillni, E. A. (2006). Disruption of disulfide bond formation alters the trafficking of prothyrotropin releasing hormone (proTRH)-derived peptides. *Regulatory Peptides*, 123–133.

Muller, L., & Lindberg, I. (1999). The cell biology of the prohormone convertases PC1 and PC2. *Progress in Nucleic Acid Research and Molecular Biology*, 69–108.

Munzberg, H., Huo, L., Nillni, E. A., Hollenberg, A. N., & Bjorbaek, C. (2003). Role of signal transducer and activator of transcription 3 in regulation of hypothalamic proopiomelanocortin gene expression by leptin. *Endocrinology*, 2121–2131.

Naggert, J. K., Fricker, L. D., Varlamov, D., Nishina, P. M., Rouillie, Y., Steiner, D. F., Carroll, R. J., Paigen, B. J., & Leiter, E. H. (1995). Hyperinsulinemia in obese fat/fat mice associated with a carboxypeptidase E mutation which reduces enzyme activity. *Nature Genetics*, 135–142.

Nie, Y., Nakashima, M., Brubaker, P. L., Li, Q. L., Perfetti, R., Jansen, E., Zambre, Y., Pipeleers, D., & Friedman, T. C. (2000). Regulation of pancreatic PC1 and PC2 associated with increased glucagon-like peptide 1 in diabetic rats. *The Journal of Clinical Investigation*, 955–965.

Nillni, E. A. (2007). Regulation of prohormone convertases in hypothalamic neurons: Implications for prothyrotropin-releasing hormone and proopiomelanocortin. *Endocrinology*, 4191–4200.

Nillni, E. A. (2010). Regulation of the hypothalamic thyrotropin releasing hormone (TRH) neuron by neuronal and peripheral inputs. *Frontiers in Neuroendocrinology, 31*(2), 134–156. S0091-3022(10)00002-6 [pii] https://doi.org/1016/j.yfrne.2010.01.001.

Nillni, E. A. (2016). The metabolic sensor Sirt1 and the hypothalamus: Interplay between peptide hormones and pro-hormone convertases. *Molecular and Cellular Endocrinology*, 77–88. https://doi.org/10.1016/j.mce.2016.09.002.

Nillni, E. A., & Sevarino, K. A. (1999). The biology of pro-thyrotropin-releasing hormone-derived peptides. *Endocrine Reviews*, 599–648.

Nillni, E. A., Sevarino, K. A., & Jackson, I. M. (1993a). Processing of proTRH to its intermediate products occurs before the packing into secretory granules of transfected AtT20 cells. *Endocrinology*, 1271–1277.

Nillni, E. A., Sevarino, K. A., & Jackson, I. M. (1993b). Identification of the thyrotropin-releasing hormone-prohormone and its posttranslational processing in a transfected AtT20 tumoral cell line. *Endocrinology*, 1260–1270.

Nillni, E. A., Friedman, T. C., Todd, R. B., Birch, N. P., Loh, Y. P., & Jackson, I. M. (1995). Pro-thyrotropin-releasing hormone processing by recombinant PC1. *Journal of Neurochemistry*, 2462–2472.

Nillni, E. A., Luo, L. G., Jackson, I. M., & McMillan, P. (1996). Identification of the thyrotropin-releasing hormone precursor, its processing products, and its coexpression with convertase 1 in primary cultures of hypothalamic neurons: Anatomic distribution of PC1 and PC2. *Endocrinology*, 5651–5661.

Nillni, E. A., Aird, F., Seidah, N. G., Todd, R. B., & Koenig, J. I. (2001). PreproTRH(178-199) and two novel peptides (pFQ7 and pSE14) derived from its processing, which are produced in the paraventricular nucleus of the rat hypothalamus, are regulated during suckling. *Endocrinology*, 896–906.

Nillni, E. A., Xie, W., Mulcahy, L., Sanchez, V. C., & Wetsel, W. C. (2002a). Deficiencies in pro-thyrotropin-releasing hormone processing and abnormalities in thermoregulation in Cpefat/fat mice. *The Journal of Biological Chemistry*, 48587–48595.

Nillni, E. A., Lee, A., Legradi, G., & Lechan, R. M. (2002b). Effect of precipitated morphine withdrawal on post-translational processing of prothyrotropin releasing hormone (proTRH) in the ventrolateral column of the midbrain periaqueductal gray. *Journal of Neurochemistry*, 874–884.

O'Rahilly, S., Farooqi, I. S., Yeo, G. S., & Challis, B. G. (2003). Minireview: Human obesity-lessons from monogenic disorders. *Endocrinology*, 3757–3764.

Orwoll, E., Kendall, J. W., Lamorena, L., & McGilvra, R. (1979). Adrenocorticotropin and melanocyte-stimulating hormone in the brain. *Endocrinology*, 1845–1852.

Palkovits, M., & Eskay, R. L. (1987). Distribution and possible origin of beta-endorphin and ACTH in discrete brainstem nuclei of rats. *Neuropeptides*, 123–137.

Palkovits, M., Mezey, E., & Eskay, R. L. (1987). Pro-opiomelanocortin-derived peptides (ACTH/beta-endorphin/alpha-MSH) in brainstem baroreceptor areas of the rat. *Brain Research*, 323–338.

Paquet, L., Massie, B., & Mains, R. E. (1996). Proneuropeptide Y processing in large dense-core vesicles: Manipulation of prohormone convertase expression in sympathetic neurons using adenoviruses. *The Journal of Neuroscience*, 964–973.

Perello, M., & Nillni, E. A. (2007). The biosynthesis and processing of neuropeptides: Lessons from prothyrotropin releasing hormone (proTRH). *Frontiers in Bioscience*, 3554–3565.

Perello, M., Friedman, T., Paez-Espinosa, V., Shen, X., Stuart, R. C., & Nillni, E. A. (2006a). Thyroid hormones selectively regulate the posttranslational processing of prothyrotropin-releasing hormone in the paraventricular nucleus of the hypothalamus. *Endocrinology*, 2705–2716.

Perello, M., Stuart, R. C., & Nillni, E. A. (2006b). The role of intracerebroventricular administration of leptin in the stimulation of prothyrotropin releasing hormone neurons in the hypothalamic paraventricular nucleus. *Endocrinology*, 3296–3306. Epub 2007 Jun 21.

Perello, M., Stuart, R. C., Varslet, C. A., & Nillni, E. A. (2007a). Cold exposure increases the biosynthesis and proteolytic processing of Prothyrotropin releasing hormone in the hypothalamic paraventricular nucleus via Beta-adrenoreceptors. *Endocrinology 148*(10): 4952–4964.

Perello, M., Stuart, R. C., & Nillni, E. A. (2007b). Differential effects of fasting and leptin on proopiomelanocortin peptides in the arcuate nucleus and in the nucleus of the solitary tract. *American Journal of Physiology. Endocrinology and Metabolism 292*(5):E1348–1357. Epub 2007 Jan 16.

Perello, M., Stuart, R., & Nillni, E. A. (2008). Prothyrotropin-releasing hormone targets its processing products to different vesicles of the secretory pathway. *The Journal of Biological Chemistry*, 19936–19947.

Perez de la Cruz, I., & Nillni, E. A. (1996). Intracellular sites of prothyrotropin-releasing hormone processing. *The Journal of Biological Chemistry*, 22736–22745.

Perone, M. J., Murray, C. A., Brown, O. A., Gibson, S., White, A., Linton, E. A., Perkins, A. V., Lowenstein, P. R., & Castro, M. G. (1998). Procorticotrophin-releasing hormone: Endoproteolytic processing and differential release of its derived peptides within AtT20 cells. *Molecular and Cellular Endocrinology*, 191–202.

Pilcher, W. H., & Joseph, S. A. (1986). Differential sensitivity of hypothalamic and medullary opiocortin and tyrosine hydroxylase neurons to the neurotoxic effects of monosodium glutamate (MSG). *Peptides*, 783–789.

Posner, S. F., Vaslet, C. A., Jurofcik, M., Lee, A., Seidah, N. G., & Nillni, E. A. (2004). Stepwise posttranslational processing of progrowth hormone-releasing hormone (proGHRH) polypeptide by furin and PC1. *Endocrine*, 199–213.

Prigge, S. T., Kolhekar, A. S., Eipper, B. A., Mains, R. E., & Amzel, L. M. (1997). Amidation of bioactive peptides: The structure of peptidylglycine alpha-hydroxylating monooxygenase. *Science*, 1300–1305.

Pritchard, L. E., Turnbull, A. V., & White, A. (2002). Pro-opiomelanocortin processing in the hypothalamus: Impact on melanocortin signalling and obesity. *The Journal of Endocrinology*, 411–421.

Pu, L. P., Ma, W., Barker, J., & Loh, Y. P. (1996). Differential expression of genes encoding proThyrotropin-releasing hormone (proTRH) and prohormone convertases (PC1 and PC2) in rat brain neurons: Implications for differential processing of proTRH. *Endocrinology*, 1233–1241.

Raadsheer, F. C., Sluiter, A. A., Ravid, R., Tilders, F. J., & Swaab, D. F. (1993). Localization of corticotropin-releasing hormone (CRH) neurons in the paraventricular nucleus of the human hypothalamus; age-dependent colocalization with vasopressin. *Brain Research*, 50–62.

Rho, J. H., & Swanson, L. W. (1987). Neuroendocrine CRF motoneurons: Intrahypothalamic axon terminals shown with a new retrograde-Lucifer-immuno method. *Brain Research*, 143–147.

Richard, D., Huang, Q., & Timofeeva, E. (2000). The corticotropin-releasing hormone system in the regulation of energy balance in obesity. *International Journal of Obesity*, S36–S39.

Rivier, J., Spiess, J., & Vale, W. (1983). Characterization of rat hypothalamic corticotropin-releasing factor. *Proceedings of the National Academy of Sciences of the United States of America*, 4851–4855.

Rodgers, R. J., Tschop, M. H., & Wilding, J. P. (2012). Anti-obesity drugs: Past, present and future. *Disease Models & Mechanisms*, 621–626. https://doi.org/10.1242/dmm.009621.

Romero, A., Cakir, I., Vaslet, C. A., Stuart, R. C., Lansari, O., Lucero, H. A., & Nillni, E. A. (2008). Role of a pro-sequence in the secretory pathway of prothyrotropin-releasing hormone. *The Journal of Biological Chemistry*, 31438–31448.

Rouille, Y., Duguay, S. J., Lund, K., Furutua, M., Gong, Q., Lipkind, G., Olive, A. A., Chan, S. J., & Steiner, D. F. (1995). Proteolytic processing mechanisms in the biosynthesis of neuroendocrine peptides: The subtilisin-like proprotein convertases. *Frontiers in Neuroendocrinology*, 332–361.

Rovere, C., Luis, J., Lissitzky, J. C., Basak, A., Marvaldi, J., Chretien, M., & Seidah, N. G. (1999). The RGD motif and the C-terminal segment of proprotein convertase 1 are critical for its cellular trafficking but not for its intracellular binding to integrin alpha5beta1. *The Journal of Biological Chemistry*, 12461–12467.

Sachs, H., & Takabatake, Y. (1964). Evidence for a precursor in vasopressin biosynthesis. *Endocrinology*, 943–952.

Sanchez, V. C., Goldstein, J., Stuart, R. C., Hovanesian, V., Huo, L., Munzberg, H., Friedman, T. C., Bjorbaek, C., & Nillni, E. A. (2004). Regulation of hypothalamic prohormone convertases 1 and 2 and effects on processing of prothyrotropin-releasing hormone. *The Journal of Clinical Investigation*, 357–369.

Sanger, F. (1959). Chemistry of insulin. *Science*, 1340–1345.

Scamuffa, N., Calvo, F., Chretien, M., Seidah, N. G., & Khatib, A. M. (2006). Proprotein convertases: Lessons from knockouts. *The FASEB Journal*, 1954–1963.

Schafer, M.-H., Day, R., Cullinan, W. E., Chretien, M., Seidah, N., & Watson, S. (1993). Gene expression of prohormone and proprotein convertases in the rat CNS: A comparative in situ hybridization analysis. *The Journal of Neuroscience*, 1258–1279.

Schally, A. (1978). Aspects of hypothalamic regulation of the pituitary gland: Its implications for the control of reproductive processes. *Science*, 18–28.

Schaner, P., Todd, R. B., Seidah, N. G., & Nillni, E. A. (1997). Processing of prothyrotropin-releasing hormone by the family of prohormone convertases. *The Journal of Biological Chemistry*, 19958–19968.

Scharrer, E., & Scharrer, B. (1940). *Secretory cells within the hypothalamus*. New York: Hafner Publishing.

Seidah, N. G. (1995). The mammalian family of subtilisin/kexin-like proprotein convertases. In U. Shinde & M. Inouye (Eds.), *Intramolecular chaperones and protein folding* (pp. 181–203). Austin: R.G. Landes Cie.

Seidah, N. G., & Chretien, M. (1992). Proprotein and prohormone convertases of the subtilisin family recent developments and future perspectives. *Trends in Endocrinology and Metabolism*, 133–140.

Seidah, N. G., & Chretien, M. (1994). Pro-protein convertases of subtilisin/kexin family. *Methods in Enzymology*, 175–188.

Seidah, N. G., & Chretien, M. (1997). Eukaryotic protein processing: Endoproteolysis of precursor proteins. *Current Opinion in Biotechnology*, 602–607.

Seidah, N. G., & Prat, A. (2012). The biology and therapeutic targeting of the proprotein convertases. *Nature Reviews, 11*, 367–383.

Seidah, N. G., Gaspar, L., Mion, P., Marcinkiewicz, M., Mbikay, M., & Chretien, M. (1990). cDNA sequence of two distinct pituitary proteins homologous to Kex2 and furin gene products: Tissue-specific mRNAs encoding candidates for pro-hormone processing proteinases. *DNA*, 415–424.

Seidah, N., Marcinkiewicz, M., Benjannet, S., Gaspar, L., Beaubien, G., Mattei, M., Lazure, C., Mbikay, M., & Chretien, M. (1991). Cloning and primary sequence of a mouse candidate prohormone convertase PC1 homologous to PC2, furin, and Kex2: Distinct chromosomal localization and messenger RNA distribution in brain and pituitary compared to PC2. *Molecular Endocrinology*, 111–122.

Seidah, N. G., Day, R., Benjannet, S., Rondeau, N., Boudreault, A., Reudelhuber, T., Schafer, M. K., Watson, S. J., & Chretien, M. (1992a). The prohormone and proprotein processing enzymes PC1 and PC2: Structure, selective cleavage of mouse POMC and human renin at pairs of basic residues, cellular expression, tissue distribution, and mRNA regulation. *NIDA Research Monograph*, 132–150.

Seidah, N. G., Day, R., Hamelin, J., Gaspar, A., Collard, M. W., & Chretien, M. (1992b). Testicular expression of PC4 in the rat: Molecular diversity of a novel germ cell-specific Kex2/subtilisin-like proprotein convertase. *Molecular Endocrinology*, 1559–1570.

Seidah, N. G., Day, R., Marcinkiewicz, M., & Chretien, M. (1993). Mammalian paired basic amino acid convertases of prohormones and proproteins. *Annals of the New York Academy of Sciences*, 135–146.

NG Seidah, M Chretien and R Day (1994) The family of subtilisin/kexin like pro-protein and pro-hormone convertases: Divergent or shared functions. Biochimie:197–209. doi: 0300-9084(94)90147-3 [pii].

Seidah, N. G., Hamelin, J., Mamarbachi, M., Dong, W., Tadros, H., Mbikay, M., Chrétien, M., & Day, R. (1996). cDNA structure, tissue distribution, and chromosomal localization of rat PC7, a novel mammalian proprotein convertase closest to yeast kexin-like proteinases. *Proceedings of the National Academy of Sciences of the United States of America*, 3388–3393.

Seimon, R. V., Hostland, N., Silveira, S. L., Gibson, A. A., & Sainsbury, A. (2013). Effects of energy restriction on activity of the hypothalamo-pituitary-adrenal axis in obese humans and rodents: Implications for diet-induced changes in body composition. *Hormone Molecular Biology and Clinical Investigation*, 71–80. https://doi.org/10.1515/hmbci-2013-0038.

Shen, X., Li, Q. L., Brent, G. A., & Friedman, T. C. (2004). Thyroid hormone regulation of prohormone convertase 1 (PC1): Regional expression in rat brain and in vitro characterization of negative thyroid hormone response elements. *Journal of Molecular Endocrinology, 33*, 21.

Shen, X., Li, Q. L., Brent, G. A., & Friedman, T. C. (2005). Regulation of regional expression in rat brain PC2 by thyroid hormone/characterization of novel negative thyroid hormone response elements in the PC2 promoter. *American Journal of Physiology. Endocrinology and Metabolism*, E236–E245.

Shioda, S., Funahashi, H., Nakajo, S., Yada, T., Maruta, O., & Nakai, Y. (1998). Immunohistochemical localization of leptin receptor in the rat brain. *Neuroscience Letters*, 41–44.

Silva, J. E. (2005). Thyroid hormone and the energetic cost of keeping body temperature. *Bioscience Reports*, 129–148.

Smeekens, S. P., & Steiner, D. F. (1990). Identification of a human insulinoma cDNA encoding a novel mammalian protein structurally related to the yeast dibasic processing protease Kex2. *The Journal of Biological Chemistry*, 2997–3000.

Smeekens, S. P., Avruch, A. S., LaMendola, J., Chan, S. J., & Steiner, D. F. (1991). Identification of a cDNA encoding a second putative prohormone convertase related to PC2 in AtT20 cells and islets of Langerhans. *Proceedings of the National Academy of Sciences of the United States of America*, 340–344.

Smeekens, S. P., Montag, A. G., Thomas, G., Albiges-Rizo, C., Carroll, R., Benig, M., Phillips, L. A., Martin, S., Ohagi, S., Gardner, P., Swift, H. H., & Steiner, D. F. (1992). Proinsulin processing by the subtilisin-related proprotein convertases furin, PC2, and PC3. *Proceedings of the National Academy of Sciences*, 8822–8826.

Solomon, S. (1999). POMC-derived peptides and their biological action. *Annals of the New York Academy of Sciences*, 22–40.

Sominsky, L., & Spencer, S. J. (2014). Eating behavior and stress: A pathway to obesity. *Frontiers in Psychology, 434*. https://doi.org/10.3389/fpsyg.2014.00434.

Spencer, S. J., & Tilbrook, A. (2011). The glucocorticoid contribution to obesity. *Stress*, 233–246. https://doi.org/10.3109/10253890.2010.534831.

Spiess, J., Rivier, J., Rivier, C., & Vale, W. (1981). Primary structure of corticotropin-releasing factor from ovine hypothalamus. *Proceedings of the National Academy of Sciences of the United States of America*, 6517–6521.

Steiner, D. F. (1998). The proprotein convertases. *Current Opinion in Chemical Biology*, 31–39.

Steiner, D. F., & Oyer, P. E. (1967). The biosynthesis of insulin and a probable precursor of insulin by a human islet cell adenoma. *Proceedings of the National Academy of Sciences of the United States of America*, 473–480.

Steiner, D. F., Cunningham, D., Spiegelman, L., & Aten, B. (1967). Insulin biosynthesis: Evidence for a precursor. *Science*, 697–700.

Steiner, D. F., Clark, J. L., Nolan, C., Rubenstein, A. H., Margoliash, E., Aten, B., & Oyer, P. E. (1969). Proinsulin and the biosynthesis of insulin. *Recent Progress in Hormone Research*, 207–215.

Steiner, D. F., Smeekens, S. P., Ohag, S., & Chan, S. J. (1992). The new enzymology of precursor processing endoproteases. *The Journal of Biological Chemistry*, 23435–23438.

Sun, M., Watanabe, T., Bochimoto, H., Sakai, Y., Torii, S., Takeuchi, T., & Hosaka, M. (2013). Multiple sorting systems for secretory granules ensure the regulated secretion of peptide hormones. *Traffic*, 205–218. https://doi.org/10.1111/tra.12029.

Tataranni, P. A., Larson, D. E., Snitker, S., Young, J. B., Flatt, J. P., & Ravussin, E. (1996). Effects of glucocorticoids on energy metabolism and food intake in humans. *The American Journal of Physiology*, E317–E325.

Thomas, L., Leduc, R., Thorne, B. A., Smeekens, S. P., Steiner, D. F., & Thomas, G. (1991). Kex2-like endoproteases PC2 and PC3 accurately cleave a model prohormone in mammalian cells: Evidence for a common core of neuroendocrine processing enzymes. *Proceedings of the National Academy of Sciences of the United States of America*, 5297–5301.

Toorie, A. M., & Nillni, E. A. (2014). Minireview: Central Sirt1 regulates energy balance via the melanocortin system and alternate pathways. *Molecular Endocrinology*, 1423–1434. https://doi.org/10.1210/me.2014-1115.

Toorie, A. M., Cyr, N. E., Steger, J. S., Beckman, R., Farah, G., & Nillni, E. A. (2016). The nutrient and energy sensor Sirt1 regulates the hypothalamic-pituitary-adrenal (HPA) axis by altering the production of the prohormone convertase 2 (PC2) essential in the maturation of corticotropin releasing hormone (CRH) from its prohormone in male rats. *The Journal of Biological Chemistry*. https://doi.org/10.1074/jbc.M115.675264.

Tooze, S. A., Chanat, E., Tooze, J., & Huttner, W. B. (1993). Secretory granule formation. In Y. P. Loh (Ed.), *Mechanisms of intracellular trafficking and processing of Proproteins* (pp. 158–177). Boca Raton: CRC Press.

Toriya, M., Maekawa, F., Maejima, Y., Onaka, T., Fujiwara, K., Nakagawa, T., Nakata, M., & Yada, T. (2010). Long-term infusion of brain-derived neurotrophic factor reduces food intake and body weight via a corticotrophin-releasing hormone pathway in the paraventricular nucleus of the hypothalamus. *Journal of Neuroendocrinology*, 987–995. https://doi.org/10.1111/j.1365-2826.2010.02039.x.

Vale, W., Spiess, J., Rivier, C., & Rivier, J. (1981). Characterization of a 41-residue ovine hypothalamic peptide that stimulates secretion of corticotropin and beta-endorphin. *Science*, 1394–1397.

van Heumen, W. R., & Roubos, E. W. (1991). Immuno-electron microscopy of sorting and release of neuropeptides in Lymnaea stagnalis. *Cell and Tissue Research*, 185–195.

Varlamov, O., Wu, F., Shields, D., & Fricker, L. D. (1999). Biosynthesis and packaging of carboxypeptidase D into nascent secretory vesicles in pituitary cell lines. *The Journal of Biological Chemistry*, 14040–14045.

Veo, K., Reinick, C., Liang, L., Moser, E., Angleson, J. K., & Dores, R. M. (2011). Observations on the ligand selectivity of the melanocortin 2 receptor. *General and Comparative Endocrinology*, 3–9. https://doi.org/10.1016/j.ygcen.2011.04.006.

Villeneuve, P., Seidah, N. G., & Beaudet, A. (2000). Immunohistochemical evidence for the implication of PC1 in the processing of proneurotensin in rat brain. *Neuroreport*, 3443-3447.

Voss-Andreae, A., Murphy, J. G., Ellacott, K. L., Stuart, R. C., Nillni, E. A., Cone, R. D., & Fan, W. (2007). Role of the central melanocortin circuitry in adaptive thermogenesis of brown adipose tissue. *Endocrinology*, 1550–1560.

Wardlaw, S. L. (2011). Hypothalamic proopiomelanocortin processing and the regulation of energy balance. *European Journal of Pharmacology*, 213–219. https://doi.org/10.1016/j.ejphar.2010.10.107.

Webb, G. C., Dey, A., Wang, J., Stein, J., Milewski, M., & Steiner, D. F. (2004). Altered proglucagon processing in an alpha-cell line derived from prohormone convertase 2 null mouse islets. *The Journal of Biological Chemistry*, 31068–31075.

Westphal, C. H., Muller, L., Zhou, A., Zhu, X., Bonner-Weir, S., Schambelan, M., Steiner, D. F., Lindberg, I., & Leder, P. (1999). The neuroendocrine protein 7B2 is required for peptide hormone processing in vivo and provides a novel mechanism for pituitary Cushing's disease. *Cell*, 689–700.

Wilkinson, C. W. (2006). Roles of acetylation and other post-translational modifications in melanocortin function and interactions with endorphins. *Peptides*, 453–471.

Williams, D. L., Kaplan, J. M., & Grill, H. J. (2000). The role of the dorsal vagal complex and the vagus nerve in feeding effects of melanocortin-3/4 receptor stimulation. *Endocrinology*, 1332–1337.

Williams, D. L., Bowers, R. R., Bartness, T. J., Kaplan, J. M., & Grill, H. J. (2003). Brainstem melanocortin 3/4 receptor stimulation increases uncoupling protein gene expression in brown fat. *Endocrinology*, 4692–4697.

Winsky-Sommerer, R., Benjannet, S., Rovere, C., Barbero, P., Seidah, N. G., Epelbaum, J., & Dournaud, P. (2000). Regional and cellular localization of the neuroendocrine prohormone convertases PC1 and PC2 in the rat central nervous system. *The Journal of Comparative Neurology*, 439–460.

Xin, X., Varlamov, O., Day, R., Dong, W., Bridgett, M. M., Leiter, E. H., & Fricker, L. D. (1997). Cloning and sequencing analysis of cDNA encoding rat carboxypeptidase D. *DNA and Cell Biology*, 897–909.

Xiong, Q. Y., Yu, C., Zhang, Y., Ling, L., Wang, L., & Gao, J. L. (2017). Key proteins involved in insulin vesicle exocytosis and secretion. *Biomedical Reports*, 134–139. https://doi.org/10.3892/br.2017.839.

Xu, H., & Shields, D. (1993). Prohormone processing in the trans-Golgi network: Endoproteolytic cleavage of prosomatostatin and formation of nascent secretory vesicles in permeabilized cells. *The Journal of Cell Biology*, 1169–1184.

Yamada, M., Saga, Y., Shibusawa, N., Hirato, J., Murakami, M., Iwasaki, T., Hashimoto, K., Satoh, T., Wakabayashi, K., Taketo, M. M., & Mori, M. (1997). Tertiary hypothyroidism and hyperglycemia in mice with targeted disruption of the thyrotropin-releasing hormone gene. *Proceedings of the National Academy of Sciences of the United States of America*, 10862–10867.

Zhang, C. F., Snell, C. R., & Loh, Y. P. (1999). Identification of a novel prohormone sorting signal-binding site on carboxypeptidase E, a regulated secretory pathway-sorting receptor. *Molecular Endocrinology*, 527–536. https://doi.org/10.1210/mend.13.4.0267.

Zhang, X., Bao, L., & Ma, G. Q. (2010). Sorting of neuropeptides and neuropeptide receptors into secretory pathways. *Progress in Neurobiology*, 276–283. https://doi.org/10.1016/j.pneurobio.2009.10.011.

Zhou, A., Bloomquist, B. T., & Mains, R. E. (1993). The prohormone convertases PC1 and PC2 mediate distinct endoproteolytic cleavages in a strict temporal order during proopiomelanocortin biosynthetic processing. *The Journal of Biological Chemistry*, 1763–1769.

Zhu, X., Zhou, A., Dey, A., Norrbom, C., Carroll, R., Zhang, C., Laurent, V., Lindberg, I., Ugleholdt, R., Holst, J. J., & Steiner, D. F. (2002a). Disruption of PC1/3 expression in mice causes dwarfism and multiple neuroendocrine peptide processing defects. *Proceedings of the National Academy of Sciences of the United States of America*, 10293–10298.

Zhu, X., Orci, L., Carroll, R., Norrbom, C., Ravazzola, M., & Steiner, D. F. (2002b). Severe block in processing of proinsulin to insulin accompanied by elevation of des-64,65 proinsulin intermediates in islets of mice lacking prohormone convertase 1/3. *Proceedings of the National Academy of Sciences of the United States of America*, 10299–10304.

Ziegler, C. G., Krug, A. W., Zouboulis, C. C., & Bornstein, S. R. (2007). Corticotropin releasing hormone and its function in the skin. *Hormone and metabolic research = Hormon- und Stoffwechselforschung = Hormones et metabolisme*, 106–109. https://doi.org/10.1055/s-2007-961809.

Nutrient Sensors Regulating Peptides

6

Isin Cakir and Eduardo A. Nillni

6.1 Introduction

The hypothalamus is an important center for coordinating mammalian physiology and maintenance of homeostasis. Accordingly, the hypothalamus regulates feeding, body temperature, energy expenditure, glucose metabolism, thirst, blood pressure, reproductive axis, and other metabolic functions associated with the overall metabolism. At the central level, the hypothalamus is the primary component of the nervous system in interpreting adiposity and nutrient-related inputs; it delivers hormonal and behavioral responses with the ultimate purpose of regulating body weight, food intake, and energy consumption (Williams et al. 2001; Wardlaw 2011; Toorie and Nillni 2014). Among the hormonal inputs that feed into the hypothalamic circuitries are adipose tissue-derived hormone leptin and adiponectin, pancreatic hormone insulin, and several hormones secreted by the gastrointestinal tract, such as ghrelin. The activity of the hypothalamic feeding centers is also responsive to basic nutrients including glucose, amino acids, and fatty acids besides other metabolites, such as ketone bodies. Much like the hypothalamus acting at the organismal level to regulate homeostasis by integrating such hormonal and nutritional signals, there are evolutionarily conserved proteins and protein complexes that act at the cellular level as "nutrient sensors" that couple cellular energetics to downstream pathways to regulate various cellular functions. The activity of these nutrient sensors in key hypothalamic feeding centers plays a major role in the regulation of energy balance and glucose metabolism.

Among the most prominent regulators within the hypothalamus, neurons in the arcuate nucleus (ARC) of the hypothalamus located in the mediobasal hypothalamus, anteriorly juxtaposing the median eminence (ME), are of critical importance for the regulation of energy balance. The ME is one of the secretory circumventricular organs (Kaur and Ling 2017) (lies outside the blood-brain barrier), which in turn confers the ARC the advantage to have access to the factors in the systemic circulation over other nuclei in the hypothalamus. The ARC receives circulating adiposity and nutritional signals and transmits responses to "second-order" neurons within and outside the hypothalamus. The ARC is sensitive to peripheral signals such as postprandial fluctuations in

I. Cakir
Life Sciences Institute and Department of Molecular and Integrative Physiology, University of Michigan, Ann Arbor, MI, USA

E. A. Nillni (✉)
Emeritus Professor of Medicine, Molecular Biology, Cell Biology & Biochemistry,
Department of Medicine, Molecular Biology, Cell Biology & Biochemistry,
The Warren Alpert Medical School of Brown University, Providence, RI, USA
e-mail: Eduardo_Nillni@Brown.edu

© Springer International Publishing AG, part of Springer Nature 2018
E. A. Nillni (ed.), *Textbook of Energy Balance, Neuropeptide Hormones, and Neuroendocrine Function*, https://doi.org/10.1007/978-3-319-89506-2_6

peripheral and central hormones as well as the nutrients such as amino acids, lipids, and glucose. The ARC neurons express the receptors for majority of the peripherally derived hormones including the adipostatic factor leptin, insulin from pancreas, ghrelin from the gut, hormones of the noradrenergic system, and thyroid hormone, which are among the major regulators of feeding and energy expenditure (Hahn et al. 1998; Morton et al. 2006). The inputs from the periphery provoke a response from these neurons by releasing neuropeptide hormones and neurotransmitters to extra-ARC hypothalamic sites such as the paraventricular nucleus (PVN) of the hypothalamus (Lu et al. 2003; Wittmann et al. 2005; Füzesi et al. 2007; Cyr et al. 2013). Within the ARC, the anorexigenic alpha-melanocyte stimulating hormone (a-MSH), derived from the posttranslational processing of its pro-opiomelanocortin (POMC) precursor, increases its levels after feeding and decreases upon fasting, whereas the orexigenic neuropeptide Y (NPY) and Agouti-related peptide (AgRP) follow an opposite pattern. AgRP expression is restricted to the ARC in the entire nervous system, and about 90% of AgRP neurons also express NPY. POMC is expressed in the ARC as well as the NTS in adult rodents, while NPY is one of the most ubiquitously expressed and abundant neuropeptides. AgRP neurons are widely distributed throughout the rostro-caudal axis of the ARC, while POMC neurons are more restricted to anterior and medial hypothalamus (Anderson et al. 2016). While the AgRP neurons are GABAergic (inhibitory), most POMC neurons are glutamatergic (excitatory). a-MSH increases energy expenditure through its action on melanocortin 3/4 receptors (MC3/4R) found in hypothalamic and extra-hypothalamic nuclei reported to bind a-MSH (Cummings and Schwartz 2000; Adan et al. 2006; Perello et al. 2007; Anderson et al. 2016). AgRP acts as an endogenous antagonist on MC3/4R receptors to block a-MSH action, while NPY acts independently through its own set of receptors (NPY1-5R) expressed throughout the nervous system. POMC and AgRP neurons and the melanocortin receptors (MC3/4R) constitute the central melanocortin signaling (Cone 2005). Anorexigenic peripheral hormones including leptin and insulin positively

regulate POMC expression and suppress AgRP and NPY expressions (Cowley et al. 2001; Benoit et al. 2002; Heuer et al. 2005).

Two of the major intracellular signaling pathways regulated by leptin are JAK2-STAT3 pathway and phosphatidylinositol-4,5-bisphosphate 3-kinase (PI3K) signaling. PI3K is required for the anorectic responses to leptin as well as insulin (11677594). Insulin exerts its effects through activation of the insulin receptor substrate (IRS) proteins. IRS phosphorylation activates a series of signaling events in the PI3K-Akt pathway resulting in the phosphorylation of Akt (also called protein kinase B), which in turn phosphorylates FoxO1. Phosphorylated FoxO1 is recruited from the nucleus to the cytosol, which leads to its degradation through the proteasome system. In the hypothalamus, FoxO1 acts as an inhibitor of POMC and activator of NPY/AgRP expressions (Kitamura et al. 2006; Kim et al. 2006; Yang et al. 2009), and STAT3-induced POMC expression appears to require release of FoxO1 repression (Ernst et al. 2009). POMC neurons are glucose responsive and express insulin receptors (Ibrahim et al. 2003), and an intact PI3K signaling in POMC neurons is required for maintenance of glucose homeostasis (Hill et al. 2009). POMC-specific ablation of PI3K through deletion of its regulatory subunits blocks acute anorectic response to central leptin infusion; however these mice exhibit normal body weight in the long term (Hill et al. 2008). Deletion of the p110β, but not p110α, catalytic subunit of PI3K has a more sustained effect in body weight with POMC-specific ablation resulting in obesity and AgRP-specific ablation conferring protection against weight gain (Al-Qassab et al. 2009). PI3K signaling also mediates leptin and insulin-induced changes neuronal activation status in other hypothalamic nuclei including the hypothalamic ventral premammillary nucleus (Williams et al. 2011) and the ventromedial hypothalamic nucleus (Sohn et al. 2016). PI3K-mediated neuronal depolarization or hyperpolarization seems to be mediated by transient receptor potential C (TRPC) channels or ATP-sensitive K$^+$ (KATP) channels, respectively (Sohn et al. 2016). Compared to the PI3K signaling, JAK2-STAT3 signaling is indispensable for leptin's effect on energy homeostasis. Neuron-specific STAT3 KO mice mimic most of the defects observed in *ob/ob* and *db/db* mice including hyperphagic obesity and infertility (Gao et al. 2004). Mutation of Tyr 1138 on LepRb, the critical residue for leptin-induced JAK2-STAT3 activation, also results in hyperphagic obesity with intact fertility (Bates et al. 2003).

Although it is not going to be discussed in detail here, the hormonal regulation of the hypothalamic feeding circuitries is inarguably more complex than summarized above and involves other components, whose regulation is spatially and temporarily distinct. The signaling cascades regulated by leptin involve additional signaling components such as STAT5, ERK/MAPK, mTOR, and AMPK pathways (Park and Ahima 2014). Furthermore, peripherally derived hormones regulate the activity of their target neurons not only by directly acting on them but also indirectly through synaptic inputs from other neurons. For example, while leptin can directly regulate AgRP neurons, these cells are subject to additional regulation by innervation through a population of leptin-responsive, GABAergic neurons in the dorsomedial nucleus of the hypothalamus (Garfield et al. 2016). This is at least one of the culprits of conditional knockout strategies heavily used in the past two decades to understand the contribution of individual hormones and their role in chemically defined neuronal populations.

Below, we start with an introduction to the biology of nutrient sensors. Next, we summarize the findings in the literature on the role of nutrient sensors in the hypothalamus and their role in the regulation of energy homeostasis. We conclude this chapter with explaining how our knowledge on metabolism is integrated at the level of nutrient sensors and prohormone processing. For a full description of the hypothalamus and hypothalamic neuropeptides controlling energy balance, we refer the readers to Chaps. 2 and 5.

6.2 Cellular Nutrient Sensors

Nutrient sensors have the ability to sense and respond to fluctuations in environmental nutrient levels, which represent a key requisite for life.

There are diverse nutrient-sensing pathways detecting intracellular and extracellular levels of sugars, amino acids, lipids, and other metabolites while also integrating hormonal and stress signals to coordinate homeostasis at the organismal level. Nutrients are simple organic compounds involved in biochemical reactions to produce energy and act as components of cellular biomass or signaling molecules. During periods of food richness, nutrient-sensing pathways engage in anabolism and energy storage. On the contrary, nutrient deprivation triggers homeostatic mechanisms including mobilization of internal stores through pathways such as autophagy to globally shut down anabolic pathways while supplying energy for vital cellular processes. Besides responding hormonal inputs, the hypothalamus is directly sensitive to nutritional changes. For example, POMC neurons are glucose responsive (Ibrahim et al. 2003), and their firing rate changes with the glucose concentration. Amino acids and lipids can also be directly sensed by hypothalamic centers; central infusion of branch chain amino acid leucine (Cota et al. 2006) or oleic acid (Obici et al. 2002) inhibits food intake and weight gain. The integration of the nutritional signals in the CNS is accomplished by a set of evolutionarily conserved proteins and protein complexes called nutrient sensors. The activity of the nutrient sensors in the hypothalamus is coupled to the regulation of energy homeostasis. While leucine-mediated suppression of food intake involves hypothalamic mechanistic target of rapamycin signaling (mTOR), fatty acids act through modulating hypothalamic malonyl-CoA levels, which is regulated by AMPK. mTOR and AMPK are nutrient/energy sensors whose activity is coupled to the cellular energy status and levels of certain nutrients. The mTOR kinase, when part of mTOR complex 1 (mTORC1), plays a role in cellular energetics by inducing numerous anabolic protein processes and lipid synthesis (Laplante and Sabatini 2012). AMPK, on the other hand, is typically activated by energy deficit and acts to suppress the anabolic pathways while activating the catabolic reactions to in general antagonize mTOR action. Another energy sensor, sirtuin 1 (SIRT1), a class III histone deacetylase, is also activated by low-energy status and acts in parallel with AMPK. Below we briefly describe the biochemistry of nutrient sensing, an overall view of the nutrient sensors in metabolic pathways, and how nutrient sensors regulate energy balance in the hypothalamus, giving a special emphasis to the connection between these enzymes and neuropeptide biosynthesis.

6.2.1 mTOR

mTOR (mechanistic target of rapamycin, or formerly called mammalian target of rapamycin) is an evolutionarily conserved serine/threonine kinase in the phosphatidylinositol 3-kinase-related kinase family. mTOR is typically activated by energy surplus and promotes anabolic processes including protein and lipid synthesis to induce cell growth and proliferation while suppressing catabolic pathways such as autophagy. Coordination of these processes involves two mTOR protein complexes: mTOR complex 1 (mTORC1) and mTOR complex 2 (mTORC2). mTOR activity is regulated by hormones and growth factors (such as insulin), as well as by nutrients including branch chain amino acids. mTORC1 kinase activity is stimulated by interaction with Ras homolog enriched in brain (Rheb), which is a ubiquitously expressed GTP-binding protein. Rheb GTPase activity is stimulated by a tumor suppressor complex formed between two proteins, tuberous sclerosis complex 1 and 2 (TSC1 and TSC2). Therefore, TSCs act as upstream inhibitors of mTORC1 signaling. Anabolic factors such as insulin and IGFs result in activation of Akt, which phosphorylates and inhibits TSC. Another negative regulator of mTORC1, PRAS40 (proline-rich Akt substrate 40 kDa), is phosphorylated by Akt, which prevents PRAS40-mTORC1 interaction (Vander Haar et al. 2007; Sancak et al. 2007; Wang et al. 2007). These PI3K-Akt-dependent phosphorylation events couple insulin and IGF signaling to mTORC1 activation. Active mTORC1, in turn, triggers the major anabolic pathways: It stimulates protein synthesis in part through phosphory-

lation of S6 K1 (S6 kinase 1), which phosphorylates ribosomal protein S6, and through phosphorylation of translational regulator eukaryotic translation initiation factor 4E binding protein 1 (4E-BP1), which results in the dissociation of 4E-BP1 from eukaryotic translation initiation factor 4E (eIF4E) and ribosomal assembly and activation of translation. mTORC1 is also activated by amino acids. It is worth to note that amino acids serve as major nitrogen source as well as the building blocks for proteins. For example, in fasting, protein breakdown mostly in the muscle provides glutamine into the circulation, where the glutamine is delivered to various tissues to be used as a major nitrogen source. Activation of mTORC1 by amino acids and amino acid sensing in general involve a set of protein complexes that are characterized only recently. Amino acid sensing involves the heterodimer of a family of four small GTPases called Rag proteins (Rag A-D). Amino acids stimulate the GTP-bound state of Rags (Sancak et al. 2008; Kim et al. 2008a). A protein complex composed of at least five distinct proteins called the Ragulator acts as the guanidine exchange factor for Rags and recruits Rags and mTORC1 to the lysosomal membrane, where Rheb is also localized (Sancak et al. 2010; Bar-Peled et al. 2012). mTORC1 senses cytosolic as well as lysosomal amino acid levels through distinct pathways that involve other protein complexes, which ultimately merge on their regulation of mTORC1 activity through Rags. Cytosolic amino acid sensing works by dissociating upstream inhibitors from an mTORC1 activating protein complex called GATOR2: Cytosolic leucine is thought to be directly sensed by Sestrin2, which acts as an upstream inhibitor of GATOR2. Leucine binding releases the Sestrin2-GATOR2 complex and results in mTORC1 activation. Likewise, cytosolic arginine binds another protein called CASTOR1, which in the absence of arginine inhibits GATOR2 (Chantranupong et al. 2016). The 11-pass lysosomal transmembrane protein, SLC38A9, interacts with the Ragulator, Rag GTPases, and vacuolar H(+)-ATPase complex in an amino acid-sensitive manner and therefore acts as the lysosomal arginine sensor (Zoncu et al. 2011; Wang et al. 2015; Rebsamen et al. 2015; Jung et al. 2015). Accordingly, upon genetic ablation of either Sestrin2, CASTOR1, or SLC38A9, amino acid-mediated activation of mTORC1 is significantly compromised.

mTORC1 acts a major metabolic hub that responds to various forms of cellular stress such as hypoxia and DNA damage, ATP level, amino acids, and growth factors to regulate glycolysis, mitochondrial and lysosome biogenesis, lipid and protein biosynthesis, proteasome formation, and autophagy (Saxton and Sabatini 2017). For example, mTORC1 activation couples cellular nutrient levels and hormonal inputs to lipid and cholesterol synthesis through activation of SREBPs (sterol regulatory element-binding proteins), which are basic helix-loop-helix (bHLH) transcription factors (Lewis et al. 2011). Accordingly, inhibition of mTORC1 activity impairs adipogenesis (Lamming and Sabatini 2013). mTORC1 also stimulates purine and pyrimidine synthesis (Ben-Sahra et al. 2013, 2016), promotes mitochondrial oxidative capacity (Cunningham et al. 2007) in skeletal muscle, and induces genes for the enzymes involved in glycolysis and pentose phosphate pathway (Düvel et al. 2010). While activating anabolism, mTORC1 activation results in suppression of major catabolic pathways. Nutrient deprivation promotes autophagy by AMPK-mediated phosphorylation and activation of ULK1, a kinase that drives autophagosome formation (Kim et al. 2011). However, in conditions of nutrient surplus, activated mTORC1 phosphorylates ULK1 on a residue different than targeted by AMPK and blocks the interaction between ULK1 and AMPK (Kim et al. 2011).

6.2.2 AMPK

AMPK is a Ser/Thr kinase composed of three subunits: catalytic α subunit, one scaffold β, and the regulatory γ subunit, which binds adenine nucleotides. Mammals have two catalytic subunits α and β and three γ isoforms. Unlike mTOR, which senses amino acid levels indirectly through its interaction partners (discussed in the section

on mTORC1), AMPK activity is directly affected by cellular AMP/ATP and ADP/ATP ratios. This regulation is accomplished through four tandem repeats of cystathionine β-synthase motifs located in the γ subunit, which form the nucleotide-binding pockets of the enzyme. AMP binding to the γ subunit triggers the phosphorylation of the activation loop at Thr172 in the kinase domain of the α subunit by the upstream activating kinase LKB1. AMP and ADP binding protects AMPK against dephosphorylation; however only AMP triggers an allosteric activation of the enzyme (Suter et al. 2006; Sanders et al. 2007; Xiao et al. 2011). These activatory effects of AMP and ADP are antagonized by ATP. Therefore, AMPK activity is coupled to fluctuations in AMP/ATP and ADP/ATP ratios, which typically rise in conditions of nutrient deprivation.

There are two well-characterized AMPK kinases: LKB1 (liver kinase B 1, also known as serine/threonine kinase 11) (Hawley et al. 2003) and CAMKKβ (the Ca^{2+}/calmodulin-dependent kinase kinase β) (Hawley et al. 2005; Woods et al. 2005). LKB1 requires heterodimerization with two other proteins, sterile20-related adaptor (STRAD) and mouse protein 25 (MO25), for full activation (Hawley et al. 2003). LKB1 phosphorylates most other kinases in the AMPK-related kinase family (Lizcano et al. 2004). CAMKK activity is coupled to Ca^{2+} levels, and its activation of AMPK is AMP-independent. Apart from LKB1 and CAMKKβ, there are other kinases proposed to phosphorylate and activate AMPK; however whether they act as endogenous AMPK kinases in vivo has yet to be established (Momcilovic et al. 2006). AMPK activation by LKB1 involves the recruitment of AMPK by Ragulator to the cytosolic surface of the lysosomal membrane under nutrient deprivation conditions. Explained in more detail in the *mTOR* section, Ragulator is a protein complex involved in the amino acid sensing of mTORC1 through a process that also requires mTORC1 recruitment to the lysosomal membrane. Activation of AMPK results in the phosphorylation of TSC1/TSC2 complex, which inhibits mTORC1 activity, thereby coupling the low-energy state-dependent AMPK activation to mTOR activation. A reciprocal regulation between AMPK and mTORC1 was also described, where activated S6 K1 phosphorylates and inhibits AMPK activity (Dagon et al. 2012).

Activation of AMPK triggers catabolic pathways aimed to restore cellular energy status. Therefore AMPK activity is negatively regulated by anabolic signals. For example, insulin-/IGF-mediated Akt activation results in the phosphorylation of a serine residue in AMPK α subunit, which in turn blocks the activatory Thr172 phosphorylation (Hawley et al. 2014). Other inhibitory phosphorylation sites on the α subunit were also reported including regulation by Fyn kinase (Yamada et al. 2016), a Src tyrosine kinase family member that also phosphorylates and inhibits LKB1 (Yamada et al. 2010), by protein kinase C (PKC) (Heathcote et al. 2016), and by Unc-51-like kinase 1 (Ulk1) (Löffler et al. 2011). AMPK plays a major role in starvation-induced autophagy induction through at least two distinct pathways: As described above, AMPK inhibits mTORC1 activity through direct phosphorylation of TSC2 and regulatory-associated protein of mTOR (Raptor) (Inoki et al. 2003; Gwinn et al. 2008), a component of the mTORC1 protein complex. In addition, AMPK directly phosphorylates and activates Unc-51-like kinase 1 (ULK1) (Lee et al. 2010; Egan et al. 2011; Kim et al. 2011; Zhao and Klionsky 2011), which triggers autophagy and mitophagy. ULK1 in turn phosphorylates and inhibits AMPK function, forming a negative feedback loop evolved to probably block excess autophagic flux (Löffler et al. 2011). Besides autophagy, AMPK induces glucose and fatty acid update, through increased membrane localization of glucose transporters, and fatty acid transporter CD36, respectively (Kurth-Kraczek et al. 1999; Fryer et al. 2002; Wu et al. 2013a). Catabolic pathways including glycolysis, mitochondrial biogenesis, and fatty acid oxidation are also activated by AMPK (Witters et al. 1991; Winder and Hardie 1996; Merrill et al. 1997; Marsin et al. 2000, 2002; Jäger et al. 2007; Cantó et al. 2009). On the other hand, synthesis of fatty acids, proteins, and glycogen is suppressed by AMPK. These are accomplished either through direct phosphorylation of AMPK

targets involved in these individual processes or indirectly through inhibition of mTORC1 and its downstream targets. Inhibition of fatty acid synthesis and promotion of fatty acid oxidation by AMPK largely depend on phosphorylation and inhibition of acetyl-CoA carboxylase (ACC) (Hardie and Pan 2002). ACC catalyzes the conversion of acetyl-CoA to malonyl-Coa, which is the precursor for fatty acid synthesis. Malonyl-CoA is also an inhibitor of carnitine palmitoyl-transferase I (CPT1), an outer mitochondrial membrane protein that transports fatty acyl CoAs from cytosol to mitochondria for their oxidation. In summary, AMPK-mediated inhibition of ACC results in a decrease in cellular malonyl-CoA pool, which in turn blocks fatty acid synthesis and triggers fatty acid oxidation. It is worth to note that manipulations altering hypothalamic malonyl-CoA levels also affect the food intake and body weight (Lane et al. 2008). For example, fatty acid synthase inhibitors C75 or cerulenin administered by intraperitoneal or intracerebroventricular route result in suppression of food intake and reduced body weight (Loftus et al. 2000). ICV C75 rapidly increases the hypothalamic malonyl-CoA levels and suppresses food intake; however, prior ICV administration of an ACC inhibitor prevents the malonyl-CoA accumulation and the decreased food intake induced by ICV C75 (Hu et al. 2003).

6.2.3 Histone Deacetylases: Sirtuins and Classical HDACs

Acetylation is a major posttranslational modification regulating protein function and contributes to metabolic homeostasis (Iyer et al. 2012; Menzies et al. 2016). Protein acetylation is regulated by acetyltransferases and deacetylases. Apart from the N-terminal acetylation (Starheim et al. 2012), which is thought to be irreversible, the coordinated action of acetyltransferases and deacetylases makes protein acetylation a reversible process. Acetyltransferase reaction involves the transfer of an acetyl group from acetyl coenzyme A (acetyl-CoA) to an amino acid residue on the substrate protein. N-terminal acetyltransferases acetylate the first amino acid in the nascent polypeptide chain following the initial methionine residue, whereas the acceptor residue is usually a lysine in other acetylation reactions catalyzed by lysine acetyltransferases. The cytosolic enzyme adenosine triphosphate (ATP)-citrate lyase converts the glucose-derived citrate into acetyl-CoA, thereby regulating a critical step that links glucose availability to histone acetylation and regulation of gene expression (Wellen et al. 2009). Histone acetyltransferases (HATs) are broadly grouped into two categories, nuclear A-type HATs and the cytoplasmic B-type HATs, which are further categorized into several subfamilies (Roth et al. 2001). The B-type HATs generally regulate the shuttling of the newly synthesized histones from the cytosol to the nucleus, while the nuclear HATs coordinate transcription-related processes.

The histone deacetylase family is grouped into four classes based on sequence homology. Class I, II, and IV are comprised of the classical HDACs (HDAC 1–11), while the sirtuin family (SIRT1–7) constitutes the class III. Classical HDACs have one deacetylase domain except HDAC6 and HDAC10, which have two catalytic domains, although one of these domains in HDAC10 is inactive (de Ruijter et al. 2003; Seto and Yoshida 2014). Classical HDACs are Zn + −dependent enzymes that deacetylate histones and other nuclear and cytosolic targets. They are subject to posttranslational modifications themselves, such as phosphorylation, which couple hormonal and nutritional inputs to their subcellular localization and therefore to their histone deacetylase activity [REF]. HDACs also have nonenzymatic functions: For example, HDAC6, which is a class IIb HDAC and the largest member of the HDAC family, has a C-terminal ubiquitin-binding domain, which renders this protein a central component of protein homeostasis. Likewise, SIRT1 has been proposed to confer, e.g., neuroprotection, independently of its deacetylase activity (Pfister et al. 2008).

Sirtuins (silent mating type information regulation 2 homolog, from here on abbreviated as SIRTs) are evolutionarily conserved enzymes that are involved in biological processes including cellular differentiation, apoptosis, metabolism, and aging. Unlike other members of the HDAC family, the activity of class III HDACs is coupled to cellular concentration of the oxidized form of nicotinamide adenine dinucleotide (NAD$^+$). Sirtuin-mediated deacetylation (and ADP-ribosylation) reaction utilizes NAD$^+$ as a coenzyme and generates nicotinamide and O-acetyl-ribose as by-products (Tong and Denu 2010). Sirtuins are insensitive to the inhibitors of classical HDACs such as trichostatin A, whereas nicotinamide inhibits the sirtuin enzymatic activity [REF].

NAD$^+$/NADH ratio typically rises in conditions of energy deficit. At the organismal level, fasting and calorie restriction result in elevated NAD+ levels in several tissues and lead to increased sirtuin activity. Mammals have seven members of the sirtuin family of deacetylases (SIRT 1–7), which differ in substrate specificity, subcellular localization, and tissue distribution (Houtkooper et al. 2012). While SIRT1 is both nuclear and cytosolic, SIRT2 is predominantly localized to the cytosol. SIRT3 is both mitochondrial and nuclear, SIRT4 and SIRT5 are mitochondrial, and SIRT6 and SIRT7 are exclusively nuclear proteins. SIRT4 and SIRT6 both have deacylase (Jiang et al. 2013; Laurent et al. 2013; Anderson et al. 2017) and ADP-ribosyltransferase activity [16,959,573, 10,381,378]. For example, SIRT6 catalyzes mono-ADP-ribosylation of PARP (poly[adenosine diphosphate (ADP)-ribose] polymerase 1) under oxidative stress to promote DNA repair (Mao et al. 2011). SIRT5 is unique in the sense that it is the only enzyme identified in mammals that can desuccinylate, demalonylate, and deglutarylate substrate proteins (Du et al. 2011; Tan et al. 2014), and the role of these additional modifications on protein function has not been completely uncovered.

SIRT1 has been the most extensively studied member of the sirtuin family. SIRT1 has the closest homology to yeast Sir2, and unlike other sirtuin knockout (KO) mice, global deletion of SIRT1 is mostly lethal at embryonic or early postnatal period [REF]. Inducible deletion of SIRT1 in adult mice does not lead to lethality, suggesting a fundamental role for SIRT1 in development (Price et al. 2012). SIRT1 is ubiquitously expressed in mammals, with the highest expression detected in the heart and nervous system in the developing embryo and in the lung, testis, and ovaries in adult rodents (Sakamoto et al. 2004).

The activity of SIRT1 is also subject to indirect regulation by a series of enzymes involved in the synthesis and degradation of the cellular NAD+. Nicotinamide phosphoribosyltransferase (NAMPT) and nicotinamide mononucleotide adenylyltransferase 1 (NMNAT1) synthesize NAD+ from nicotinamide in a two-step reaction called the NAD-salvage pathway (Revollo et al. 2004; Zhang et al. 2009). NMNAT1-mediated NAD+ biosynthesis and the resulting SIRT1 activation help prevent neurodegeneration in mice. Hydrolysis of NAD+ is predominantly regulated by a cyclic ADP-ribose hydrolase called CD38 (Aksoy et al. 2006b). CD38 knockout mice have 10–20 times elevated NAD^+ levels across various tissues. This in turn results in increased SIRT1 activity in the absence of CD38 (Aksoy et al. 2006a). The role of NAD+-dependent regulation of SIRT1 in overall physiology is further emphasized by that the CD38 knockout mice are protected from high fat diet-induced obesity, and this phenotype is dependent of SIRT1 deacetylase activity (Barbosa et al. 2007). The expression of SIRT1, NAMPT, and the cellular NAD+ levels are all subject to a circadian regulation, where SIRT1 directly controls activity of key factors involved in the regulation of circadian rhythm, such as BMAL1, CLOCK, and PER2 as well as NAMPT, thereby regulating the formation of its own coenzyme (NAD^+) and forming a transcriptional feedback loop (Asher et al. 2008; Nakahata et al. 2008, 2009; Ramsey et al. 2009). Regulation of the circadian clock by SIRT1 extends to the central pacemaker, the hypothalamic suprachiasmatic nucleus, where loss of SIRT1 activity drives the dysfunctional circadian rhythm observed in aging (Chang and Guarente 2013).

SIRT1 activity is subject to an indirect regulation by AMPK. Activation of AMPK by AICAR in the skeletal muscle leads to increased cellular NAD^+/NADH ratio and leads to SIRT1 activation and increased deacetylation of SIRT1 targets

such as PGC1α (Cantó et al. 2009). Increased SIRT1 activity then promotes mitochondrial biogenesis and oxygen consumption (Cantó et al. 2009). AMPK can directly phosphorylate and regulate SIRT1. Furthermore, increased SIRT1 activity leads to AMPK activation (Price et al. 2012) probably though deacetylation of LKB1 (Hou et al. 2008; Lan et al. 2008). Deleted in breast cancer 1 (DBC1), a tumor suppressor, acts as an endogenous inhibitor of SIRT1, while (Kim et al. 2008b) another nuclear protein called ribosomal protein S19 binding protein 1 (also called active regulator of SIRT1 – AROS) increases SIRT1 activity (Kim et al. 2007b). SIRT1 activity is also regulated by posttranscriptional modifications including phosphorylation by kinases such as AMPK, casein kinase 2 (CK2), and c-Jun N-terminal kinase 1 (JNK1) (Sasaki et al. 2008; Kang et al. 2009; Nasrin et al. 2009).

Because of its NAD^+ dependence for activity, having a large spectrum of nuclear and cytosolic targets, and its expression being regulated by changes in the nutritional status, SIRT1 functions as a major nutrient and redox sensor, which regulates metabolism at different levels (Fulco et al. 2003). High expression of SIRT1 is shown in the periphery and in neurons of the central nervous system including the hypothalamic nuclei ARC, VMH, and PVN (Ramadori et al. 2008; Zakhary et al. 2010; Toorie and Nillni 2014). SIRT1's subcellular localization is predominantly nuclear (Zakhary et al. 2010), but SIRT1 can translocate to the cytoplasm in a cell-specific and cell-autonomous manner, in response to various physiological stimuli and disease states (Tanno et al. 2007). SIRT1 regulates cellular survival and metabolism at several steps largely through deacetylation of protein targets (Brooks and Gu 2009; Peek et al. 2012; Orozco-Solis and Sassone-Corsi 2014), although there are reports on the deacetylase-independent function of SIRT1 (Pfister et al. 2008; Shah et al. 2012). During the past decade, the metabolic role of SIRT1 in mammals has been studied by tissue-specific transgenic models and pharmacological tools targeting its deacetylase activity (Blander and Guarente 2004; Chang and Guarente 2014). In the liver, SIRT1 deacetylates and activates the transcriptional coactivator PGC1-alpha and the FoxO1 to promote gluconeogenesis (Rodgers et al. 2005; Frescas et al. 2005). In adipose tissue, SIRT1 promotes fat mobilization by inhibiting peroxisome proliferator-activated receptor (PPAR-gamma) (Picard et al. 2004) and therefore functions to decrease fat storage, promote lipolysis, and protect against obesity-induced inflammation (Picard et al. 2004; Gillum et al. 2011; Chalkiadaki and Guarente 2012). SIRT1 improves glucose tolerance and positively regulates insulin secretion in pancreatic beta cells by repressing uncoupling protein 2 (UCP2) promoter (Moynihan et al. 2005; Bordone et al. 2006). In the liver SIRT1 protein levels increase upon fasting (Rodgers et al. 2005). SIRT1 deacetylates and activates PGC1α and FoxO1 and represses CREB-regulated transcription coactivator 2 (CRTC2, also known as TORC2) to regulate hepatic glucose output (Liu et al. 2008). Exposure to high-fat diet results in decreased liver SIRT1 levels (Deng et al. 2007). Compounds that lead to SIRT1 activation, such as resveratrol, have been shown to alleviate several adverse effects of high-fat diet, which otherwise results in the development of obesity and associated disorders (Baur et al. 2006; Lagouge et al. 2006). An increase in the activity and/or the expression level of SIRT1 is thought to mediate the physiology induced by calorie restriction (Cohen et al. 2004; Bordone and Guarente 2005; Bordone et al. 2007a, b). While the role of SIRT1 in the periphery is not going to be discussed further, it is important to note that by deacetylating a spectrum of metabolic regulators, SIRT1 links tissue energy status to the coordination of lipid, cholesterol, amino acid, and glucose metabolisms (Baur et al. 2006; Lagouge et al. 2006; Haigis and Sinclair 2010). Contrary to its role in the periphery, recent evidence shows that hypothalamic SIRT1 induces positive energy balance, and this activity is directly related to the regulation of hypothalamic peptide hormones by SIRT1. Therefore, this chapter will further focus on the role of central SIRT1 under different nutritional conditions on body weight, prohormone maturation, and the pro-converting enzymes involved in the processing of prohormones.

6.3 Regulation of Energy Homeostasis by the Hypothalamic Nutrient Sensors

6.3.1 mTOR

The initial studies on the role of hypothalamic mTOR signaling on energy homeostasis were conducted on rats (Cota et al. 2006; Blouet et al. 2008). mTOR is expressed in AgRP and POMC neurons (Cota et al. 2006), and fasting decreases mTOR activity in the ARC (Cota et al. 2006; Blouet et al. 2008; Cakir et al. 2009). Central administration of leucine activates central mTORC1, suppresses hypothalamic NPY expression, and inhibits fast-induced refeeding and weight gain in a rapamycin-sensitive manner (Cota et al. 2006). Rapamycin alone was also reported to activate ARC NPY expression (Shimizu et al. 2010). Furthermore, leptin activates hypothalamic mTOR signaling, likely through a PI3K-dependent pathway (Harlan et al. 2013), and this activation is required for leptin-mediated suppression of food intake (Cota et al. 2006; Ropelle et al. 2008). Accordingly, S6 K1 knockout (S6 K1 KO) mice are resistant to the anorectic action of exogenous leptin (Cota et al. 2008). Furthermore, acute and chronic exposure to high-fat diet (HFD) suppresses hypothalamic mTORC1 signaling, and leptin cannot activate hypothalamic mTORC1 in HFD-exposed animals (Cota et al. 2008). Hypothalamic mTORC1 activity is also sensitive, although indirectly, to the activity of fatty acid synthase (FASN), whose central inhibition induces potent weight loss and suppresses food intake (Loftus et al. 2000; Kumar et al. 2002). The anorectic action of FASN inhibitors is compromised upon mTORC1 inhibition (Proulx et al. 2008). These findings were supported by genetic data where activation of S6 K1 in the mediobasal hypothalamus (MBH) of rats suppressed food intake and weight gain while overexpression of a dominant negative S6 K1 increased both parameters and blunted central leptin action (Blouet et al. 2008). Leptin-induced sympathetic outflow to the periphery also depends on hypothalamic mTORC1 function

(Harlan et al. 2013). mTORC1-mediated control of food intake could also, at least in part, depend on a S6K-dependent inhibitory phosphorylation of AMPK (Dagon et al. 2012). Contrary to these findings, deletion of TSC1 from rat insulin promoter (RIP)-positive cells (by RIP1-Cre), which cover pancreatic beta cells and a subpopulation of hypothalamic neurons, results in hyperphagic obesity (Mori et al. 2009). While this study did not report a problem with neuronal viability in the absence of TSC1, neuronal TSC1 knockout mice suffer from reduced myelination and neuronal survival (Meikle et al. 2007). POMC and AgRP/NPY neurons are RIP-Cre negative; however POMC-specific TSC1 deletion mimics the obese RIP1-Cre TSC1 knockout phenotype (Mori et al. 2009; Yang et al. 2012). Importantly, rapamycin treatment could reverse the effect of TSC1 deletion in these mice (Mori et al. 2009), as well as in old WT mice with elevated mTORC1 activity in POMC neurons (22884327). MBH-specific overexpression of DEPTOR, a negative regulator of mTORC1, also protected mice from HFD-induced weight gain (Caron et al. 2016a). This effect seems to be independent of mTOR activity in POMC neurons (Caron et al. 2016b). These seemingly contradictory findings listed above are likely due to the differential regulation of hypothalamic mTORC1 signaling in response to nutrient deprivation and hormonal inputs in different subhypothalamic nuclei. For example, fasting or lack of leptin action (*ob/ob* mice) suppresses mTORC1 signaling in the VMH but activates it in the ARC of mice (19628573). Furthermore, ghrelin and insulin, two orexigenic and anorexigenic hormones, respectively, both resulted in elevated ARC mTORC1 activity (Villanueva et al. 2009; Muta et al. 2015). The positive effect of ghrelin on hypothalamic mTORC1 signaling is supported by findings from other groups, which showed that S6 K1 KO mice were not responsive to ghrelin-induced weight gain (Stevanovic et al. 2013), and the ghrelin-induced food intake was reduced upon inhibition of mTORC1 activity (Martins et al. 2012). Furthermore, rapamycin reversed the hyperphagic phenotype of hyperthyroid rats, which had had elevated hypothalamic mTOR activity,

resulting in weight loss (Varela et al. 2012), and blocked the anorectic effect of central insulin administration (Muta et al. 2015). Finally, cholecystokinin (CCK), which inhibits food intake by activating the responsive neurons in the NTS and the hypothalamus, also increases mTORC1 activity in the PVN (Lembke et al. 2011).

mTORC2 function has been less studied compared to that of mTORC1. The best-characterized role of mTORC2 is phosphorylation of Akt on Ser473, which facilitates Thr308 phosphorylation by PDK1 downstream of PI3K (Sarbassov et al. 2005). By phosphorylating PKCα (protein kinase C alpha) (Sarbassov et al. 2004; Jacinto et al. 2004), and other PKC family members (Gan et al. 2012; Li and Gao 2014), mTORC2 regulates the actin cytoskeleton and cell migration. It also activates glucose and lipid metabolism in liver and adipose tissue (Kumar et al. 2010; Hagiwara et al. 2012; Yuan et al. 2012). A recent study suggested that mTORC2 also acts as nutrient sensor: While acute glutamine deprivation activates mTORC2, prolonged lack of glutamine diminished mTORC2 activity (Moloughney et al. 2016). mTORC2 regulates the expression of GFAT1 in a glutamine-dependent manner (Moloughney et al. 2016). GFAT1 is the rate-limiting enzyme in the hexosamine biosynthetic pathway (HBP), which is involved in the hypothalamic regulation of energy balance (discussed below). The role of mTORC2 signaling in the brain and specifically in POMC and AgRP neurons was studied by targeted deletion of Rictor to inactivate mTORC2 signaling (Kocalis et al. 2014). Neuronal deletion of Rictor did not alter the food intake significantly despite increased hypothalamic NPY expression. The neuronal KO mice had higher fat mass and lower lean mass than WT counterparts (Kocalis et al. 2014). Accordingly, the mice had increased leptin levels, and resistance to the anorectic action of leptin, and glucose intolerance. While AgRP-specific Rictor KOs did not have a major phenotype, POMC-specific ablation resulted in heavier mice with increased food intake and adiposity even more than the neuronal KOs (Kocalis et al. 2014). The exact mechanism of mTORC2-mediated regulation of energy balance is not clear. While mTORC2 positively regulating Akt phosphorylation is important in the insulin action, ablation of insulin signaling in POMC neurons by targeted deletion of IRS2 (Choudhury et al. 2005) or PDK1 (Belgardt et al. 2008) does not recapitulate the effects of POMC-specific deletion of Rictor, suggesting that the effect of hypothalamic mTORC2 is mediated through an PI3K-Akt independent pathway.

6.3.2 AMPK

In the hypothalamus, AMPK acts a positive regulator of energy balance. Hypothalamic AMPK activity increases by fasting and decreases by refeeding (Minokoshi et al. 2004; Taleux et al. 2008), an effect not observed in an obesity-resistant rat strain (Taleux et al. 2008). Activation of central AMPK, either by ICV infusion of a synthetic AMP analog and AMPK activator (AICAR) (Sullivan et al. 1994) or MBH-specific overexpression of constitutively active AMPK (CA-AMPK), results in increased food intake and body weight (Andersson et al. 2004; Minokoshi et al. 2004). Conversely, overexpression of dominant negative AMPK decreases food intake and weight gain. Accordingly, hypothalamic AMPK activity is suppressed by anorectic factors including insulin, leptin, MTII (an αMSH analog), CNTF, and estradiol but induced by ghrelin, adiponectin, and orexin-A (Andersson et al. 2004; Minokoshi et al. 2004; Steinberg et al. 2006; Kubota et al. 2007; López et al. 2008; Wu et al. 2013b; Martínez de Morentin et al. 2014). Furthermore, suppression of central AMPK activity is required for leptin's anorectic effect, and hypothalamic overexpression of CA-AMPK can block central leptin action (Minokoshi et al. 2004; Gao et al. 2007). Consequently, leptin activates hypothalamic ACC activity and results in elevated central malonyl-CoA level, which is required for leptin's anorectic effect (Gao et al. 2007). Additionally, the sympathetic outflow regulated by leptin to kidney and adipose tissue (Tanida et al. 2013) but not to liver (Tanida et al. 2015) also depends on an intact hypothalamic AMPK signaling. It is worth noting that the hormonal regulation of AMPK activity in the periphery vs. the CNS is different. For example, leptin activates AMPK in the skeletal muscle and induces fatty acid oxidation (Hardie and Pan 2002), while inhibition of hypothalamic AMPK activity by central action of leptin is essential for leptin's anorectic effect (Andersson et al. 2004; Minokoshi et al. 2004).

Central infusion of glucose inhibits food intake in an AMPK-dependent manner, whereas a non-oxidizable form of glucose, 2-deoyxglucose, activates hypothalamic AMPK and induces food intake (Kim et al. 2004). This is likely due to the effect of glucose on cellular AMP/ATP ratio, which in turn alters AMPK activity. Consistent with its positive role in energy balance, hypothalamic AMPK activity positively correlates with NPY and AgRP expressions (Minokoshi et al. 2004; Lee et al. 2005). Leptin-mediated suppression of hypothalamic AMPK activity is compromised in DIO mice. Furthermore, in the medial hypothalamus, leptin activates AMPK selectively in DIO mice contrary to its effect in lean mice, which might contribute the leptin desensitization observed in obesity (Martin et al. 2006). Deletion of the α subunit of AMPK from POMC neurons results in induced weight gain and food intake and suppresses energy expenditure, whereas AgRP-specific AMPKα knockout mice were protected from weight gain (Claret et al. 2007). Central inhibition of CaMKK results in suppression of food intake and induces weight loss, and lack of CaMKKβ (CaMKK2 KO mice) confers protection from diet-induced obesity (Anderson et al. 2008). CaMKK2 knockout mice have reduced expression of AgRP and NPY and resistance to fasting-induced upregulation of NPY (Anderson et al. 2008). Furthermore, central glucagon-mediated suppression of food intake, in part, depends of suppression of hypothalamic CamKK2-AMPK signaling by glucagon (Quiñones et al. 2015).

Hypothalamic actions of AMPK extend beyond the regulation of energy balance and involve other homeostatic processes including glucose metabolism, sleep homeostasis, thyroid axis, and possibly the reproductive axis (Rougon et al. 1990; Chikahisa et al. 2009; Yang et al.

2010; Cheng et al. 2011; Claret et al. 2011; Kinote et al. 2012). BAT-mediated thermogenesis, for example, in response to agents such as GLP-1 agonists, estradiol, BMP-8B (bone morphogenetic protein 8B), and triiodothyronine (T3) also involves regulation of hypothalamic AMPK activity (Martínez de Morentin et al. 2014; Beiroa et al. 2014; Martins et al. 2016; Martínez-Sánchez et al. 2017).

As summarized above, AMPK and mTORC1 are master regulators of autophagy, which seems to play a role in the AMPK- and mTOR-mediated regulation of energy homeostasis. Inhibition of autophagy attenuates the AMPK- and rapamycin-mediated activation of NPY and suppression of POMC expressions in hypothalamic cell lines (Oh et al. 2016). Additionally, knockdown of AMPKα subunits in the ARC reduces autophagy and results in reduced food intake and weight gain (Oh et al. 2016). The role of central autophagy in the regulation of energy balance is cell type specific. Inhibition of autophagy in AgRP neurons blocks the fasting-induced AgRP upregulation and results in elevated αMSH levels (Kaushik et al. 2011). However, selective loss of autophagy in POMC neurons decreases αMSH levels and results in increased adiposity (Coupé et al. 2012; Kaushik et al. 2012; Quan et al. 2012).

6.3.3 Histone Deacetylases: Sirtuins and Classical HDACs with Special Emphasis on SIRT1

The hypothalamic role of the histone deacetylase family of proteins has recently begun to be uncovered. As noted below, the most extensively characterized HDAC member in regard to its central role has been SIRT1. We will begin this section with a brief summary of our current knowledge on the hypothalamic activity of classical HDACs and then focus most of the remainder of the chapter on SIRT1.

The expression pattern and regulation of the hypothalamic HDACs in response to nutritional challenges are not uniform. Hypothalamic histone acetylation and the expression of classical HDACs

have been reported to be nutrient-sensitive. In lean mice, fasting decreases the global histone acetylation only in the ventrolateral subdivision of the ventromedial hypothalamus (Funato et al. 2011). Among the classical HDACs, HDAC10 and HDAC11 are the most highly expressed genes in the hypothalamus (Funato et al. 2011). Upon fasting, hypothalamic expressions of HDAC3 and HDAC4 increase, while that of HDAC10 and HDAC11 decline (Funato et al. 2011), while another study suggested a fasting-induced decrease in HDAC5 expression (Kabra et al. 2016). Hypothalamic HDAC8 expression was reported to be restricted to the paraventricular nucleus, where its expression is upregulated by fasting and high-fat diet. While one study reported increased hypothalamic HDAC5 expression upon 4-week high-fat diet feeding (Funato et al. 2011), another study suggested that chronic high-fat diet decreased total hypothalamic HDAC5 mRNA levels (Kabra et al. 2016). There was no detectible change in the hypothalamic expression of other classical HDACs under high-fat diet feeding (Funato et al. 2011). HDAC5 is the only classical HDAC reported to be involved in the regulation of leptin signaling. Hypothalamic HDAC5 levels are positively regulated by leptin, and ablation of HDAC5 potentiates weight gain on high-fat diet (Kabra et al. 2016). Mechanistically, HDAC5 was suggested to deacetylate and increase the transcriptional activity of STAT3 on POMC promoter (Kabra et al. 2016). It is important to note, however, that several other studies indicated that acetylation positively regulates STAT3 phosphorylation, dimerization, and transcriptional activity (Yuan et al. 2005; Wang et al. 2005; Nie et al. 2009; Dasgupta et al. 2014; Chen et al. 2015). For example, SIRT1 inhibits hepatic STAT3 activity in the liver by deacetylation of multiple lysine residues (Nie et al. 2009), although hypothalamic SIRT1 inhibition does not alter STAT3 acetylation at least on Lys 685 (Cakir et al. 2009). The precise role of STAT3 acetylation on individual residues and their contribution to energy homeostasis remain to be further characterized.

Neuronal SIRT1 expression is predominantly nuclear in rodent and human brain (Zakhary et al. 2010). Brain nuclei associated with neurodegeneration (such as the prefrontal cortex, hippocampus, and basal ganglia) as well as the homeostatic centers including the hypothalamus and the brain stem have detectable SIRT1 expression (Ramadori et al. 2008; Zakhary et al. 2010). Activation of neuronal SIRT1 has neuroprotective effects and phenocopies some of the benefits of caloric restriction on neurodegeneration, such as the attenuation of amyloid plaques observed in Alzheimer disease (Qin et al. 2006; Kim et al. 2007a). Developmentally, neuronal SIRT1 plays a fundamental role in neural differentiation and survival. In neuronal progenitor cells (NPCs), SIRT1 couples oxidative stress to inhibition of neurogenesis and directs the progenitor cells toward an astrocyte lineage (Prozorovski et al. 2008). However, under normal conditions, SIRT1 blocks Notch signaling, which normally suppresses proneural gene expression and confers neural progenitor cell maintenance. Therefore, neuronal SIRT1 action promotes neurogenesis (Hisahara et al. 2008). Furthermore, deletion of SIRT1 in adult NPCs in rodents leads to increased progenitor oligodendrocyte population (Rafalski et al. 2013). The role of SIRT1 in neuroprotection is also supported by findings obtained from Wallerian degeneration slow (Wld^s) mice where a mutation encompassing the NAMPT gene results in increased NAD$^+$ production and elevated SIRT1 activity, which in turn delays the axonal degeneration (Araki et al. 2004).

The role of hypothalamic SIRT1 on metabolism in rodents has been studied by several groups both genetically and pharmacologically. SIRT1 expression is sensitive to cellular energy status such that increased SIRT1 expression and nuclear localization have been reported by many studies supporting its role as an energy sensor in the periphery and in the brain (Kanfi et al. 2008; Cantó et al. 2009; Cakir et al. 2009; Dietrich et al. 2010). SIRT1 protein levels in rodents increase by calorie restriction in several metabolic tissues including the brain (Cohen et al. 2004). Upon fasting, SIRT1 activity increases specifically in the hypothalamus, but not in the hindbrain (Ramadori et al. 2008; Cakir et al. 2009); this is due to the fasting-induced increase in hypothalamic SIRT1 protein level and NAD$^+$ concentration. Inhibition of central SIRT1 activ-

ity leads to cessation of feeding and results in weight loss in lean as well as obese rodents, while its acute activation by resveratrol appears to have beneficial effects in glucose homeostasis (Knight et al. 2011). Neuronal depletion of SIRT1 promotes systemic insulin sensitivity and counters diet- and age-induced weight gain, while forebrain-specific overexpression of SIRT1 results in decreased glucose tolerance, energy expenditure, and increased adiposity. In contrast, in older mice neuronal overexpression of SIRT1 seems to counteract the aging-induced gene expression profile (Oberdoerffer et al. 2008). Therefore, the specific role of SIRT1 in mammalian physiology especially in the context of its role in the regulation of neuroendocrine circuits should be evaluated based on specific tissue/cell population in consideration of the nutritional status and stress factors.

One well-known substrate of SIRT1 is the transcription factor, FoxO1, whose deacetylation augments its activity (Huang and Tindall 2007; Xia et al. 2013). FoxO1 is also a metabolic sensor in that it integrates both leptin and insulin signaling (Kitamura et al. 2006; Kim et al. 2006) pathways. It is abundantly expressed in metabolically relevant hypothalamus nuclei, including neurons of the ARC, DMH, and VMH (Toorie et al. 2016). Among its metabolic functions, FoxO1 transcriptionally regulates AgRP and NPY expression in a positive manner, while transcriptionally repressing POMC, carboxypeptidase E (CPE), and steroidogenic factor 1 SF1 expressions (Kitamura et al. 2006; Kim et al. 2006; Plum et al. 2009). In addition, FoxO1 positively regulates SIRT1 expression at the transcriptional level (Xiong et al. 2011). Physiologically, SIRT1's role as energy sensor is determined by its dependency on NAD$^+$ (Ramadori et al. 2008; Cantó et al. 2012). New studies have uncovered the involvement of hypothalamic SIRT1 in nutrient sensing (Ramadori et al. 2008; Cakir et al. 2009). As mentioned in the introduction and further developed in the following sections, SIRT1 in the ARC induces positive energy balance by affecting POMC and CPE gene expression through FoxO1. The role of hypothalamic SIRT1 and the consequences of

genetic ablation of SIRT1 from distinct hypothalamic neuronal populations are discussed extensively in the following sections. SIRT1 also
affects proCRH processing either through a
pathway of FoxO1, preproPC2/proPC2, or
directly through preproPC2/proPC2. Lastly, the
potential effect of SIRT1 action on the transcription factor nescient helix-loop-helix 2 protein
(Nhlh2) will also be considered.

6.3.4 Other Players

Apart from AMPK, mTOR, and sirtuins, our
knowledge on the role of hypothalamic nutrient
sensors on energy homeostasis has been limited.
Two recent studies addressed the role of the
hypothalamic hexosamine biosynthetic pathway
(HBP) in energy homeostasis. A small percentage of the cellular glucose is shuttled through the
HBP, where glucose is metabolized to UDP-N-
acetylglucosamine (UDP-GlcNAc) (Yang and
Qian 2017). Besides the N-linked and O-linked
glycosylation reactions, UDP-GlcNAc is the
major sugar donor in O-GlcNAcylation, which is
the posttranslational modification of proteins
with O-linked N-acetylglucosamine on serine
and threonine residues. This process involves at
least two enzymes, O-GlcNAc transferase (OGT)
and O-GlcNAcase (OGA), which collectively
regulate the cellular pool of O-GlcNAcylated
substrates in a nutrient-sensitive manner. For
example, increased glucose concentration stimulates O-GlcNAcylation in hyperglycemia (Liu
et al. 2000). Cellular O-GlcNAcylation increases
upon increased HBP flux typically observed upon
hyperglycemia, unfolded protein response, and
nutrient deprivation-induced OGT expression.

OGT is expressed in the AgRP neurons, and
its expression is induced by fasting and ghrelin
[25303527]. Deletion of OGT from AgRP neurons blocks their excitability and confers resistance to diet-induced weight gain (Ruan et al.
2014), at least in part by modulating the
O-GlcNAcylation and thus activity of the voltage-
dependent potassium channel Kcnq3. Driven by
earlier genome-wide association studies (GWAS)
indicating that glucosamine-6-phosphate deami-

nase (GNPDA2), a regulator of HBP, correlates
with obesity, another group characterized the role
of OGT in the forebrain and PVN in mice
(Lagerlöf et al. 2016). Using a tamoxifen-
inducible CaMKII-Cre model, Lagerlöf and colleagues showed that deletion of OGT from the
forebrain (including the hypothalamus) results in
hyperphagic obesity in adult mice (Lagerlöf et al.
2016). This phenotype was reproduced by deletion of OGT specifically in the PVN. O-GlcNAc
level was positively regulated by glucose in the
PVN, while fasting decreased O-GlcNAc levels.
Furthermore, these changes were restricted to
CaMKII-positive neurons (Lagerlöf et al. 2016).
These findings also point to the heterogeneous
role of the nutrient sensors in the hypothalamus
as described for SIRT1 and mTORC1 above.

Another conserved cellular nutrient sensor is a
family of proteins called Per-Arnt-Sim (PAS)
kinases (DeMille and Grose 2013). PASK is typically activated by high glucose concentrations
and regulates glucose partitioning in yeast (Smith
and Rutter 2007) and glucagon secretion in mammals (da Silva Xavier et al. 2011). PASK knockout mice are protected from HFD-induced obesity
and insulin resistance (Hao et al. 2007). The role
of central PASK in the regulation of energy balance has not been directly addressed; however,
hypothalamic PASK signaling is sensitive to glucose concentrations and GLP-1 at least in the LH
and VMH (Hurtado-Carneiro et al. 2013).
Hypothalamic PASK signaling appears to regulate AMPK and mTORC1 pathways such that the
response of AMPK and mTORC1 signaling to
fasting and refeeding is dysregulated in PASK-
deficient mice (Hurtado-Carneiro et al. 2014).

6.4 Prohormone Processing and Prohormone Convertases

The biosynthesis of mammalian neuropeptide
hormones follows the principles of the prohormone theory, which begins with a messenger
RNA (mRNA) translation into a large, inactive
precursor polypeptide, followed by a limited
posttranslational proteolysis to release different

products of processing (Steiner 1998; Nillni - Endocrinology and 2007 2007). In rodents and humans, abnormalities in prohormone processing result in deleterious health consequences, including metabolic dysfunctions (Naggert et al. 1995; Jackson et al. 1997; Challis et al. 2002; Nillni et al. 2002; Jing et al. 2004; Lloyd et al. 2006). To produce an active hormone, the prohormone is subjected to differential processing by the action of different members of the family of PCs, which results in biological and functional diversity within the central nervous system (CNS) as well as in endocrine cells in the periphery. The PCs comprise a family composed of seven subtilisin-/kexin-like endoproteases, and relevant to the neuroendocrine tissues are PC1 and PC2 (Seidah et al. 1990, 1991, 1992, 1996; Smeekens and Steiner 1990; Smeekens et al. 1991; Constam et al. 1996). These enzymes cleave at the C-terminal side of single, paired, or tetra basic amino acid residue motifs (Rouillé et al. 1995), followed by removal of remaining basic residue(s) by carboxypeptidase E and D (CPE, CPD) (Fricker et al. 1996; Xin et al. 1997; Nillni et al. 2002). The selective expression of PC1 and PC2 in endocrine and neuroendocrine cells demonstrated to be important in prohormone tissue-specific processing (Seidah et al. 1990, 1991, 1994; Schafer et al. 1993). PC1 and PC2 have been shown to process proTRH (Friedman et al. 1995; Nillni et al. 1995; Pu et al. 1996; Schaner et al. 1997; Nillni 2010), proinsulin (Smeekens et al. 1992; Steiner et al. 1992; Rouillé et al. 1995), proenkephalin (Breslin et al. 1993), prosomatostatin (Galanopoulou et al. 1993; Brakch et al. 1995), progrowth hormone-releasing hormone (proGHRH) (Posner et al. 2004; Dey et al. 2004), POMC (Thomas et al. 1991; Benjannet et al. 1991), pro-corticotropin-releasing hormone (proCRH) (Vale et al. 1981a; Rivier et al. 1983; Castro et al. 1991; Brar et al. 1997; Perone et al. 1998), proNPY (Paquet et al. 1996), proCART (Dey et al. 2003), and proneurotensin (Villeneuve et al. 2000) to various intermediates and end products.

Within the hypothalamus, PC1 and PC2 display an extensive overlapping pattern of expression (Schafer et al. 1993; Winsky-Sommerer et al. 2000). The critical role of PC1 and PC2 in prohormone processing is underscored by studies from animals lacking the genes encoding PC1 (Zhu et al. 2002) and PC2 (Furuta et al. 1997) as well as 7B2 (Westphal et al. 1999; Laurent et al. 2002) a neuropeptide essential for the maturation of PC2. Disruption of the gene-encoding mouse PC1 results in a syndrome of severe postnatal growth impairment and multiple defects in the processing of many hormone precursors, including dysfunctional processing of the hypothalamic proGHRH to mature GHRH, pituitary POMC to adrenocorticotropic hormone, islet proinsulin to insulin, and intestinal proglucagon to glucagon-like peptide-1 and peptide-2. A mouse model of PC1 deficiency generated by random mutagenesis (Lloyd et al. 2006) with a missense mutation in the PC1 catalytic domain (N222D) leads to obesity with abnormal proinsulin processing and multiple endocrine deficiencies. Although there was defective proinsulin processing leading to glucose intolerance, neither insulin resistance nor diabetes had developed despite obesity. The apparent key factor in the induction of obesity was impaired autocatalytic activation of mature PC1 causing reduced production of hypothalamic α-MSH (Lloyd et al. 2006). A patient with a compound heterozygous mutation in the PC1 gene resulting in production of nonfunctional PC1 had severe childhood obesity (Jackson et al. 1997). An analogous obese condition was found in a patient with a defect in POMC processing (Challis et al. 2002). As PC1 and PC2 are essential for the processing of a variety of proneuropeptides, alterations in the expression and protein biosynthesis of PC1 and PC2 also have profound effects on neuropeptide homeostasis.

6.4.1 SIRT1 Regulates POMC and AgRP Peptides

The interplay between POMC biosynthesis/processing to its derived peptides regulated by SIRT1 was first demonstrated in our laboratory using the Sprague-Dawley rat model (Cakir et al. 2009). This model shares with humans many characteristics of obesity physiology (Challis et al. 2002), and rats are considered to be an excellent model for this type of studies (Levin et al. 1997; Nillni et al. 2002; Cottrell and Ozanne 2008). Early studies using this rat model showed that POMC biosynthesis and processing is nutrition dependent (Perello et al. 2007). Indeed, the biosynthesis of POMC and its derived peptides including ACTH and α-MSH decrease by fasting and increase when fasted animals were treated with leptin that mimics response to refeeding. Similar correlation was observed for PC1 and PC2. These findings demonstrated that the production of POMC-derived peptides and processing enzymes in the ARC are nutrient-sensitive (Perello et al. 2007).

SIRT1 protein levels in the hypothalamus change in response to diet and appear to mediate several aspects of hypothalamic control; therefore, understanding its role in regulating the melanocortin system is timely. Indeed, we recently showed that SIRT1 expressed in the ARC participates in the regulation of AgRP and POMC neurons in an orexigenic capacity to control feeding behavior. In that study, central inhibition of SIRT1 decreased body weight and food intake as a result of a FoxO1-mediated increase in the POMC and decrease in the orexigenic AgRP expressions in the ARC (Cakir et al. 2009; Cyr et al. 2014; Toorie and Nillni 2014). This conclusion was supported by an array of different studies from our laboratory and others. For example, siRNA-mediated knockdown of ARC SIRT1 expression resulted in reduced food intake and body weight gain (Cakir et al. 2009). In addition, pharmacological inhibition of SIRT1 activity using the chemical compound EX-527 via intracerebroventricular route (ICV) suppressed fasting-induced hyperphagia and weight gain in lean rodents (Cakir et al. 2009; Cyr et al. 2014). This reduction in body weight was associated with elevated ARC POMC (Cakir et al. 2009; Cyr et al. 2014) and reduced AgRP expressions (Cakir et al. 2009). Conversely, ICV infusion of a SIRT1 activator (Sinclair and Guarente 2014) resulted in an immediate (and short-lived) increase in food intake in refed rats (Cakir et al. 2009). Co-administration of a MC3/4R antagonist, SHU9119, with the SIRT1 inhibitor (EX-527) reversed the anorectic effect of EX-527, suggesting the involvement of the central melanocortin system in mediating the effect of hypothalamic SIRT1 on energy homeostasis. Central SIRT1 inhibition resulted in reversal of fasting-induced suppression of hypothalamic mTORC1 pathway (Cakir et al. 2009), which is a negative regulator of energy balance (Cota et al. 2006), suggesting that SIRT1 acted as an upstream negative regulator of mTOR, possibly acting through a TSC2-dependent pathway (Ghosh et al. 2010). Overall, these results point to an orexigenic role of ARC SIRT1 in the lean animals (Cakir et al. 2009; Cyr et al. 2014).

Hypothalamic SIRT1 action also regulates the central effects of ghrelin, a potent orexigenic factor secreted mainly from the stomach. Dietrich and colleagues demonstrated that peripheral as well as central administration of EX-527 was sufficient to reduce food intake during the dark cycle and suppress the ghrelin-induced hyperphagia. This effect on food intake was not associated with concomitant changes in energy expenditure; rather, it was mediated through reduced melanocortin tone and was dependent on the action of uncoupling protein 2 (UCP2) (Dietrich et al. 2010). It was also demonstrated that the orexigenic action of ghrelin is mediated via a SIRT1-p53 pathway, proposed to be acting upstream of hypothalamic AMPK: Central SIRT1 inhibition blocked the ghrelin-induced hyperphagia and hypothalamic fatty acid metabolism, reducing the ghrelin's effect on FoxO1, NPY, and AgRP expressions (Velásquez et al. 2011). SIRT1 also increases the expression of the type 2 orexin receptor in the dorsomedial and lateral hypothalamic nuclei (DMH and LH, respectively) by deacetylating the homeodomain transcription factor Nkx2–1 and augmenting the response to ghrelin (Satoh et al. 2010, 2013).

Further support for the orexigenic role of neuronal SIRT1 came from a study on a neuron-specific SIRT1 knockout (SINKO) model that was generated by crossing SIRT1-floxed mice to mice expressing the Cre transgene under the control of neuron-specific synapsin I promoter (Syn1-Cre) (Lu et al. 2013). SINKO mice had a reduced neuronal SIRT1 mRNA and protein content and significantly elevated acetylation of hypothalamic p53 and STAT3. Neuronal SIRT1 deficiency resulted in increased central insulin signaling and protection against weight gain, systemic insulin resistance, and inflammation in the white adipose tissue in mice fed a high-fat diet (Lu et al. 2013). These effects were at least in part mediated by lack of an inhibitory deacetylation of IRS-1 by SIRT1 in SINKO mice (Lu et al. 2013). Additionally, AgRP-specific SIRT1 depletion phenocopies the neuronal SIRT1 KO mice in regard to their lower body weight (Dietrich et al. 2010). Another neuronal SIRT1 KO model was generated using Nestin-Cre mice: Resulting mice were smaller than WT counterparts. This was in part due to a downregulation of the growth hormone levels in the KO mice, which also had smaller pituitary mass (Cohen et al. 2009). Brain SIRT1 expression was reported to decrease by aging in the anteroventral thalamic nucleus (AV) and the ARC in mice (Lafontaine-Lacasse et al. 2010). Using mouse prion promoter, a brain-specific SIRT1 transgenic mouse was developed (Satoh et al. 2010). The mice had normal food intake and body weight compared to the WT mice at 5 and 20 months of age. One of the transgenic lines that expressed SIRT1 at higher levels in the DMH and LH compared to other hypothalamic nuclei had an extended life span. Surprisingly, another transgenic line that had uniformly higher SIRT1 expression throughout the hypothalamus did not mimic these phenotypes, suggesting a differential role of SIRT1 in different subhypothalamic nuclei at least in the context of aging. Overexpression of SIRT1 under the CaMKIIα promoter to achieve forebrain-specific expression resulted in decreased glucose tolerance and energy expenditure, decreased plasma T3 and T4 levels, and increased adiposity (Wu et al. 2011). Overall, these studies suggest that hypothalamic SIRT1 is a positive regulator of energy balance, and pharmacological or genetic inhibition of central SIRT1 function results in decreased weight gain and food intake.

In contrast to the findings summarized above, some studies suggested a negative regulation of energy balance by hypothalamic SIRT1. Hypothalamic SIRT1 expression was proposed to

be increased by short-term refeeding of fasted mice, and overexpression of SIRT1 in the mediobasal hypothalamus resulted in reduction of FoxO1-induced hyperphagia (Sasaki et al. 2010). SIRT1 overexpression rescued the obese phenotype induced by constitutively active POMC-specific FoxO1 (Susanti et al. 2014). Furthermore, POMC- or AgRP-specific SIRT1 overexpression prevented age-induced weight gain (Sasaki et al. 2014). In line with this finding, another study showed that SIRT1 and FoxO1 form a complex with necdin, a nuclear protein expressed in postmitotic neurons and located in the Prader-Willi locus (MacDonald and Wevrick 1997; Jay et al. 1997). The heterotrimeric complex promoted the SIRT1-mediated FoxO1 deacetylation (Hasegawa et al. 2012). Necdin KO mice had decreased serum levels of TSH, T3, and T4, decreased hypothalamic proTRH expression, and increased NPY and AgRP levels; however the body weight, food intake, or the energy expenditure of the mice was unaltered compared to the wild-type mice (Hasegawa et al. 2012). We would like to emphasize that the acute and chronic effects of factors that regulate energy balance could be dramatically different, and combined with the plasticity of the neuronal circuitry (Elson and Simerly 2015), these factors might account for some discrepancies in the literature. For example, PI3K is the main signaling node utilized by insulin receptor signaling; however, inhibition of PI3K pathway in rodents and primates has proven to induce weight loss and improve overall metabolism (Ortega-Molina et al. 2015), and reduced insulin/IGF-I signaling extends life span in metazoans. Orexins acutely stimulate food intake and weight gain (Sakurai et al. 1998). However, chronic orexin deficiency is associated with obesity in both rodents and humans, and transgenic overexpression of orexin confers resistance to diet-induced obesity (Funato et al. 2009). Likewise, acute inhibition of hypothalamic mTOR signaling has been associated with decreased feeding (Cota et al. 2006); however, mice with Rip2/Cre-mediated hypothalamic deletion of mTOR inhibitor Tsc1 (Rip-Tsc1cKO mice) develop hyperphagia and obesity (Mori et al. 2009).

Genetic ablation of SIRT1 from chemically distinct hypothalamic neurons resulted in phenotypes depending on the identity of the neuronal population. POMC-specific SIRT1 ablation resulted in increased high-fat diet-induced weight gain specifically in female mice but not in male mice (Ramadori et al. 2010). POMC neuronal survival, POMC mRNA expression, or hypothalamic ACTH or αMSH levels were not affected by depletion of SIRT1. These mice were reported to have decreased energy expenditure and reduced sympathetic tone to the perigonadal adipose tissue (Ramadori et al. 2010). However, central infusion of the SIRT1 inhibitor EX-527 resulted in increased POMC mitochondrial density and decreased inhibitory inputs onto POMC neurons (Dietrich et al. 2010). AgRP-specific SIRT1 KO mice showed an opposite phenotype to POMC-specific SIRT1 KOs, resulting in leaner mice with decreased food consumption (Dietrich et al. 2010). AgRP-specific ablation of PGC1α, a transcriptional coactivator positively regulated by SIRT1, results in decreased food intake, unaltered body weight with increased adiposity, but decreased lean mass (Gill et al. 2016). Deletion of SIRT1 from SF1 neurons, a subpopulation of VMH neurons, results in mice with increased body weight, normal food intake, and decreased energy expenditure (Ramadori et al. 2011). Furthermore, these mice had an altered circadian rhythm under food restriction (Orozco-Solis et al. 2015).

A full description and debate about different views on orexigenic vs. anorexigenic role of SIRT1 in the brain was previously described in a separate review (Toorie and Nillni 2014). It was concluded then that in lean animals, central SIRT1 functions as a nutrient sensor, and it is elevated during fasting state in neurons of the melanocortin system. Activated SIRT1 deacetylates and activates FoxO1, which in turn alters the biosynthesis and processing of POMC. Inhibition of central SIRT1 is sufficient to promote a negative energy balance. In the PVN, SIRT1 has also an effect on the CRH peptide that in turn also affects energy balance (see the next section). Table 6.1 depicts the impact

that SIRT1 has in ARC and PVN under different nutritional conditions.

The next question is then, how does SIRT1 regulate POMC processing as well as downstream changes in body weight and energy expenditure in diet-induced obesity (DIO) state? Indeed, central inhibition of SIRT1 in DIO decreased body weight and increased energy expenditure at higher rate than in the lean state suggesting that a different mechanism is triggered in the obese state (Cyr et al. 2014). DIO and fasted lean rodents displayed elevated SIRT1 levels in their ARC (Cyr et al. 2014), and DIO animals exhibited weight loss due to acute central SIRT1 inhibition (Table 6.1). This is consistent with a prior study in mice where SIRT1 was reported to be elevated in the hypothalamus of *db/db* mice, which are obese because of a mutation in the leptin receptor (Sasaki et al. 2010). Interestingly, whereas reduced SIRT1 activity reduced food consumption in lean animals (Cakir et al. 2009; Dietrich et al. 2010; Cyr et al. 2014), DIO animals subjected to central SIRT1 inhibition remained normophagic (Table 6.1) (Cyr et al. 2014). Results from our group and others (Dietrich et al. 2010) show that lean rodents subjected to central SIRT1 inhibition lost weight because of a decrease in food intake, without any changes in energy expenditure. However, DIO animals subjected to SIRT1 inhibition by ICV infusion of EX-527 lost significant body weight not because of decreased in food intake but instead because of an increase in oxygen consumption (elevated energy expenditure). Brain inhibition of SIRT1 in DIO increased acetylated FoxO1, which in turn increased phosphorylated FoxO1 via improved insulin/pAKT signaling (Cyr et al. 2014). Elevated acetylation and phosphorylation of FoxO1 resulted in increased POMC levels along with an increase in the expression of the a-MSH maturation enzyme CPE, which resulted in more of the bioactive POMC product α-MSH released into PVN (Fig. 6.1). Increased in a-MSH led to augmented thyrotropin-releasing hormone (TRH) levels and circulating T_3 levels (triiodothyronine, thyroid hormone). These results indicate that inhibiting hypothalamic SIRT1 in DIO enhances the activ-

Table 6.1 Central SIRT1 effect in lean and DIO state. This table depicts SIRT1's effect on hypothalamic peptide hormones, food intake, energy expenditure, weight change, melanocortin activity, and the HPT and HPA axes under different nutritional conditions, DIO state, pharmacological treatments, and SIRT1 expression inhibition by siRNA-mediated knockdown. The metabolic parameters for lean + fasted, lean + ICV EX-527, and DIO groups were compared against their appropriate lean control animals; the metabolic parameters for lean + intra-ARC SIRT1 siRNA were compared against lean + intra-ARC control siRNA; and the metabolic parameter for DIO + ICV EX-527 was compared against DIO + vehicle

Role of central SIRT1 in lean and DIO state

	Energy expenditure	Food intake	Weight change	Sirt1 expression/activity	POMC, AgRP, α-MSH expression	Processing enzymes	Endocrine axes
Lean + fasted	Decreased	Increased	Increased	Increased in ARC POMC neurons	Decreased POMC and α-MSH	Decreased PC1 and PC2 in PVN	Decreased HPT axis
Lean + ICV EX-527	Increased	Decreased	Decrease	Decrease central Sirt1 activity	Increased POMC α-MSH decreased AgRP	Increased CPE ARC	Enhanced HPT axis
Lean + intra-ARC Sirt1 siRNA	?	Decreased	Decreased	Decreased in ARC	?	?	?
DIO	Increased	Increased	Increased	Increased in ARC and PVN	Decreased α-MSH	?	Enhanced HPT and HPA axis
DIO + ICV EX-527	Increased	No change	Decrease	Decrease central Sirt1 activity	Increased POMC and α-MSH. No changes in AgRP	Increased CPE in ARC. Decrease PC2 In PVN	Enhanced HPT axis

This table is published with permission from *Molecular Endocrinology* (Toorie and Nillni 2014)

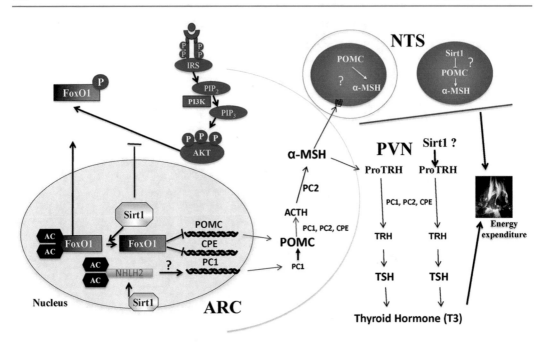

Fig. 6.1 Model depicting the regulation of hypothalamic POMC by SIRT1: In the ARC, SIRT1 deacetylates FoxO1, which decreases production of POMC and the POMC-processing enzyme CPE causing less α-MSH to reach MC3/MC4 receptors on target tissues. Lower α-MSH reduces TRH, T3, and thus energy expenditure. Inhibiting SIRT1 sensitizes Akt signaling, which increases pFox01. pFox01 promotes Fox01 nuclear exclusion, thereby increasing POMC and CPE, which increases α-MSH. SIRT1-mediated changes in POMC could affect target tissues such as the PVN and NTS that regulate energy expenditure. In the lean state, leptin and insulin signal through their respective receptors to ultimately increase Akt phosphorylation. pAKT translocates to the nucleus and phosphorylates the transcription factor FoxO1, thereby facilitating its inactivation and nuclear exclusion. Consequentially, POMC and CPE transcription is enhanced resulting in increased α-MSH, while AgRP transcription is repressed resulting in reduced AgRP. In both DIO and fasted conditions, insulin and leptin signaling are impaired via distinct mechanisms resulting in reduced pAKT levels, which promotes FoxO1 nuclear retention. In addition, SIRT1 protein con-

tent is increased within the ARC of both fasted and DIO individuals. SIRT1 via its deacetylation of FoxO1 positively regulates AgRP transcription while negatively regulating POMC and CPE transcription. Reduced POMC and CPE expression results in diminished levels of α-MSH, which signals through MC3/MC4 receptors to exert its potent anorectic effect; AgRP functions as an endogenous antagonist of the MC3/4R. SIRT1 also increases the number of GABAergic synapses onto POMC neurons by AgRP neurons, resulting in hyperpolarization of POMC neurons and reduced melanocortinergic tone. SIRT1 could deacetylate Nhlh2 in POMC neurons causing its activation of Nhlh2 that in turn has a direct role for transcriptional control of PC1. JAK, Janus-activated kinase; IRS, insulin receptor substrate; PIP$_2$, phosphatidylinositol-4,5-bisphosphate; PIP$_3$, phosphatidylinositol-3,4,5-trisphosphate; PI3K, phosphoinositide 3-kinase; STAT3, signal transducer and activator of transcription 3; Akt, protein kinase B; ac, acetyl; p, phosphate; GABA, gamma-aminobutyric acid. (This figure is adopted from our prior publication on the *Molecular and Cellular Endocrinology* (Mol Cell Endocrinol. 2016 Dec 15;438:77–88))

ity of the hypothalamic-pituitary-thyroid (HPT) axis, which stimulates energy expenditure. Therefore, pharmacological inhibition of SIRT1 in DIO state causes negative energy balance by increasing energy expenditure and HPT axis activity (Cyr et al. 2014). What remains to be determined is whether SIRT1 also regulates proTRH in the PVN and POMC in the nucleus of the solitary track (Fig. 6.1).

Collaborative studies conducted between the Good Laboratory and our group (Jing et al. 2004) revealed that targeted deletion of the neuronal basic helix-loop-helix (bHLH) transcription factor (Nhlh2) in mice results in adult-onset obesity. Nhlh2 expression can be found in rostral POMC neurons and TRH neurons of the paraventricular nucleus (PVN). We also demonstrated that in the absence of Nhlh2 in these neurons, the expression of the PC1 and PC2 mRNA and a-MSH peptides was reduced (Jing et al. 2004). In a further study, it was shown that Nhlh2 is part of the Janus kinase/STAT signaling pathway and is involved in the control of PC1 transcription by STAT3 after leptin stimulation (Fox and Good 2008). Interestingly, recent studies proposed that Nhlh2 is also involved in motivation behavior for exercise (Good et al. 2015) and it is also directly regulated by SIRT1 (Libert et al. 2011). SIRT1 deacetylates and activates Nhlh2, which in turn has a direct role for transcriptional control of PC1. Since food abundance suppresses SIRT1 and food scarcity increases SIRT1 in the ARC, it is possible that during fasting SIRT1 activation of Nhlh2 may affect PC1 in POMC neurons, an area worth exploring.

DIO results in increased hypothalamic SIRT1 expression. Central SIRT1 inhibition in DIO increases the α-MSH peptide released into PVN to stimulate TRH neurons, but these effects were not seen in lean rats (Table 6.1). In obese rats, TRH peptide levels were elevated in the PVN along with circulating levels of the active thyroid hormone T_3, a generally accepted indicator of energy expenditure (Nillni 2010), yet no changes in TRH or T_3 were detected in lean rats treated with EX-527 under the same conditions. However, EX-527 treatment resulted in a significant increase in plasma T_3 levels in lean rats upon prolonged fasting (Cakir et al. 2009). These results suggest that regulation of POMC by SIRT1 is nutrition and stress dependent. In view of these findings, it seems that physiological changes or metabolic stressors affect prohormone processing. For example, we now know that cold stress (Nillni et al. 2002) and endoplasmic reticulum stress (Cakir et al. 2013) produce alterations in the processing of proTRH and POMC, respectively. Central SIRT1 inhibition further increases ARC POMC and CPE biosynthesis in DIO rats. This change in CPE was correlative with increases in a-MSH and TRH in the PVN of DIO rats as well as circulating T_3 levels and oxygen consumption. These results collectively support the hypothesis that central SIRT1 promotes positive energy balance, as shown earlier (Cakir et al. 2009), and blocking SIRT1's hypothalamic activity can promote a negative energy balance and weight loss. Supporting the notion of SIRT1 acting on the HPT axis, a previous study showed that SIRT1 positively regulates TSH exocytosis from pituitary thyrotropes (Akieda-Asai et al. 2010). Together these results suggest that SIRT1 regulates the HPT axis at various levels and that regulation depends on the nutritional status. Since a-MSH can affect energy expenditure through several mechanisms (Xu et al. 2011), future studies should explore alternative pathways underlying the increase in energy expenditure with central SIRT1 inhibition especially in the context of metabolic syndrome and obesity.

6.4.2 SIRT1 Regulates proCRH

The major role of CRH neurons present in the PVN is to regulate the adrenal axis. Stimulation of the adrenal axis by CRH produces an increase of circulating glucocorticoids (GC) from the adrenal gland affecting energy metabolism as well as glucose metabolism. Chronic increases of basal GC are associated with increased food drive and enhanced abdominal adiposity. Early studies by Vale and others showed a role for CRH (Vale et al. 1981b) in mediating the stress response (Kovács 2013). However, CRH also regulates metabolic, immunologic, and homeostatic changes under normal and pathologic conditions (Chrousos 1995; Seimon et al. 2013; Sominsky and Spencer 2014). CRH exerts an anorexigenic effect in through its direct action within the hypothalamus (Heinrichs et al. 1993). However, glucocorticoids, whose level rises in response to CRH action in the pituitary, act as orexigenic agents. Accordingly, *ob/ob* and *db/db* mice have elevated glucocorticoid levels when compared to DIO wild-type mice, and the response of hypothalamic CRH levels to nutritional challenges in wt vs. the *ob/ob* mice is different (Jang and Romsos 1998).

Although CRH is expressed in mast cells (Cao et al. 2005), amygdala, locus, and hippocampus, the PVN-derived CRH has a major role in metabolism (Raadsheer et al. 1993; Ziegler et al. 2007; Korosi and Baram 2008). CRH is produced in the medial parvocellular division of the PVN and functions as the central regulator of the HPA axis (Aguilera et al. 2008). The levels of bioactive CRH secreted to the circulation are dependent on different peripheral and brain inputs and on the ability of the cell in performing an effective post-translational processing from its precursor pro-CRH by the PCs. Like most hypophysiotropic neurons, CRH is released from nerve terminals anteriorly juxtaposed to the median eminence where fenestrated capillaries in the hypophyseal portal system facilitate a rapid exchange between the hypothalamus and the pituitary (Rho and Swanson 1987). Upon binding to its corticotropin-releasing hormone receptor 1 (CRHR1) (Lovejoy et al. 2014) in corticotropic cells, CRH stimulates the synthesis and secretion of adrenocortico-tropic hormone (ACTH) derived from POMC as well as other bioactive molecules such as β-endorphin (Solomon 1999). ACTH engages the melanocortin 2 receptor expressed by cells of the adrenal cortex and stimulates the production and secretion of steroid hormones such as cortisol (Veo et al. 2011). Both ACTH and GC function to regulate HPA axis activity via long and short neg-ative feedback loops that signal at the level of the hypothalamus, extra-hypothalamic brain sites, and the adenohypophysis.

Among the brain circuitries involved in the control of food intake and energy expenditure capable of integrating peripheral signals is the CRH system, which has many clusters of brain neurons and the closely related peptides includ-ing urocortin (Henckens et al. 2016). The CRH system showed to have a certain degree of plas-ticity in obesity and in starvation. Based on those observations, it is possible that obesity can block or activate the expression of the CRH type 2 alpha receptor in the ventromedial hypothalamic nucleus and induce the expression of the CRH-binding protein in brain areas involved in the anorectic and thermogenic actions of CRH (Richard et al. 2000; Mastorakos and Zapanti 2004; Toriya et al. 2010). On the other hand, CRH acting in the adrenal axis stimulates the production GC that promotes positive energy bal-ance partly by affecting glucose metabolism and lipid homeostasis and increasing appetite drive

(Tataranni et al. 1996). Although a consensus on the exact role of adrenal activity in relation to energy dysfunction has yet to be reached, increased and sustained basal GC is implicated in the development of visceral obesity, insulin resis-tance, and metabolic disease (Spencer and Tilbrook 2011; Laryea et al. 2013; Kong et al. 2015).

The first evidence showing that SIRT1 could regulate CRH came from recent studies con-ducted in our laboratory (Toorie et al. 2016). We showed that PVN SIRT1 activates the HPA axis and basal corticoid levels by enhancing the bio-synthesis of CRH through an increase in the bio-chemical processing of proCRH by PC2, a key enzyme in the maturation of CRH from proCRH (Fig. 6.2), also demonstrated in an in vivo model (Dong et al. 1997). Moreover, in the DIO state, PVN SIRT1 increases basal (not stress induced) circulating corticoids in a manner independent of *preproCRH* transcriptional changes. Instead, SIRT1's effects on adrenal activity are mediated via a posttranslational processing mechanism in concert with an increase in PC2. Increased hypophysiotropic CRH release from the PVN stimulates pituitary ACTH synthesis and release and in turn increased circulating basal corticoid concentrations (Toorie et al. 2016). Cumulative findings from this study suggest that increasing PVN SIRT1 activity results in an increase in the amount of CRH targeted to the anterior pituitary, thereby enhancing ACTH signaling and basal corticoid concentration. Furthermore, our results demonstrate that inhibiting SIRT1 specifically in the PVN of obese rodents caused a reduction in PVN proPC2 and the active form of PC2, but not PC1, protein. This decrease in PC2 was associ-ated with a decrease in CRH in the ME, as well as reduced circulating corticoids, effects that were not observed in lean individuals (Toorie et al. 2016) (Fig. 6.2). It is possible that the SIRT1-CRH axis in the PVN is responsible, at least in part, for the improved glucose homeostasis observed in neuronal SIRT1 KO mice (23457303).

In the same line of studies, we identified that the transcription factor FoxO1 is responsible for the upregulation of PC2 (Toorie et al. 2016). Since FoxO1 is a positive transcriptional regulator

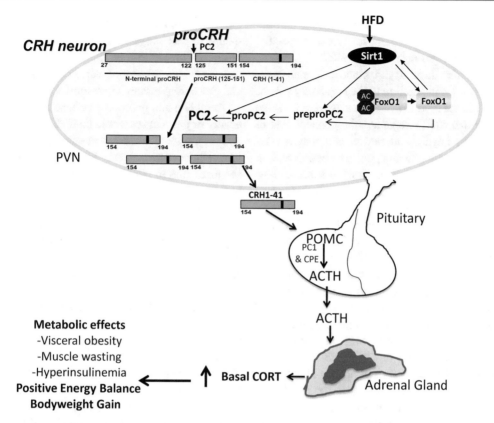

Fig. 6.2 Model depicting SIRT1's regulation on pro-CRH. During high-fat feeding, an increase of PVN SIRT1 is recorded and in turn promotes the increase proPC2 and its active form PC2. The effect of SIRT1 on preproPC2 appears to be mediated by FoxO1 or through preproPC2 directly. Since FoxO1 is a positive transcriptional regulator of SIRT1, FoxO1 could also act as upstream of SIRT1 and, in turn, affect PC2. The increased in PC2 production enhances proCRH processing, thus increasing CRH output to the hypophyseal portal system. The increase in CRH release to the circulation in turn increases POMC biosynthesis and processing in the anterior pituitary leading to more production of ACTH by corticotroph cells of the anterior pituitary. ACTH then acts on the adrenal cortex, stimulating the production and release of cortisol, thereby promoting a positive energy balance that may ultimately lead to excessive weight gain if unresolved. AC, acetyl; black bars in the proCRH molecule indicate the site of pair of basic residues; HFD, high-fat diet; PVN, paraventricular nucleus of the hypothalamus. Arrows directed to PC2 from SIRT1 and FoxO1 indicate the suggested possible actions on preproPC2 gene and proPC2 protein posttranslational processing. (This figure is adopted from our prior publication on the *Molecular and Cellular Endocrinology* (Mol Cell Endocrinol. 2016 Dec 15;438:77–88))

of SIRT1 (Xiong et al. 2011), FoxO1 could also act as upstream of SIRT1 and, in turn, affect PC2 (Fig. 6.2). In summary, SIRT1 deacetylates FoxO1 to upregulate PC2 but may also have a direct action on proPC2 to enhance the processing of proPC2 to PC2 (Toorie et al. 2016). Another observation from these studies was that inhibition of PVN SIRT1 activity did not significantly alter sated glucose levels; however, there was a trend for decreased glucose in DIO animals. Nevertheless, it resulted in a significant increase in sated serum insulin in obese rats. These results were consistent with another study showing that loss of functional SIRT1 in neurons of the central nervous system enhanced both brain and peripheral insulin sensitivity and reversed the hyperglycemia associated with obesity (Lu et al. 2013). On the other hand, an increase in the activity of SIRT1 in the medial ARC of lean rats was sufficient to improve glucose homeostasis and increase insulin sensitivity (Knight et al. 2011). This apparent contradiction shows that, as seen for many enzymes, that the role of SIRT1 in metabolism is tissue-specific. However, we can say that brain SIRT1 is involved in regulating insulin and glucose homeostasis. Interestingly, in addition to the regulatory role of SIRT1 in the thyroid and adrenal axes, it was also shown that SIRT1 regulates the gonadal axis in mediating Leydig and Sertoli cell maturity by regulating steroidogenic gene expression and, at the central level, by increasing the expression of gonadotropin-releasing hormone, thereby resulting in increased circulating luteinizing hormone and intratesticular testosterone levels (Kolthur-Seetharam et al. 2009), which accounts for the fact that the survivals of embryonic SIRT1 deletion are sterile. Finally, it is possible that other hypothalamic nutrient sensors might participate in the regulation of the CRH axis. For example, central administration of citrate suppressed the activity of hypothalamic AMPK and resulted in increased expression of CRH and POMC, but inhibited NPY expression (Stoppa et al. 2008).

In summary, SIRT1 regulates the CRH peptide by regulating the processing of PC2 and FoxO1 adding a novel regulatory link between PVN SIRT1 and HPA axis activity. Together, these results suggest that PVN SIRT1 has the capacity to activate the adrenal axis and increase basal GC levels via its positive regulation of PC2- (and possibly PC1)-mediated processing of proCRH into bioactive CRH.

6.5   Concluding Remarks

The last two decades have witnessed an enormous leap in our understanding of the central regulation of whole-body energy metabolism. With the help of recent advances in mouse genetics, electrophysiology, and optogenetic techniques, it has been possible to identify the role of peripheral hormone receptors and neuronal circuits involved in the action of these hormones on behavior and peripheral tissue functions at spatial resolution. A key component in maintaining energy balance is the hypothalamus, and the discovery of an increasing number of peptide hormones within the hypothalamus that contribute to

the process of energy homeostasis has been paramount to this progress. Furthermore, since the discovery of the PCs in the late 1980s, a new frontier on the propeptide hormone biosynthesis and processing research had come to surface.

Another major progress was done with the discovery of nutrient sensors in their ability to sense and respond to fluctuations in environmental and intracellular nutrient levels. The last nutrient sensor that came to surface as an important energy balance regulator is SIRT1. The function of SIRT1 on overall metabolism depends on the tissue of interest: While peripheral SIRT1 activity positively correlates with overall health and physiology, central SIRT1's role in energy balance showed convincing evidence that points to promoting positive energy balance. One of SIRT1's effector targets, FoxO1, plays a crucial role in determining whether SIRT1 will function to promote negative or positive energy balance. A fascinating aspect of this regulation is that SIRT1 directly regulates the biosynthesis of POMC and proCRH in the ARC and PVN, respectively, while at the same time coordinating prohormone processing, at least through the regulation of PC2.

As described above, due to the cell-specific nature of SIRT1's action in promoting either positive or negative energy balance in response to dietary excess, a more targeted therapy approach against cell-specific downstream mediators of SIRT1 signaling may prove beneficial in combating obesity and the physiological outcomes that predispose to the development of metabolic disorders. Indeed, as demonstrated by Ren and colleagues, ablation of GRP17, a G-coupled protein receptor and a downstream target of FoxO1 that is prominently expressed in AgRP neurons, results in enhanced anabolic activity and decreased food intake (Ren et al. 2012). Targeted therapeutic approaches may help to circumvent the potentially deleterious and counterproductive mechanisms of global SIRT1 activation/inactivation. Coordinating the activity of other master nutrient sensors, namely, AMPK and mTOR, SIRT1 emerges as a master regulator of energy homeostasis in the CNS. Collectively, evidence points to brain SIRT1 in mediating the adaptive responses to nutritional stress via its broad regulation of mechanisms involved in energy homeostasis. It is notable, however, that central SIRT1 inhibition or neuronal SIRT1 ablation is sufficient to promote negative energy balance. Therefore, approaches aimed at enhancing peripheral SIRT1 activity, while reducing central SIRT1 activity, may prove most beneficial in the treatment of obesity.

Questions

1. How do nutrient sensors detect the cellular energy status? What are the similarities and differences between AMPK, SIRT1, and mTOR in this regard? What are some of the inputs recognized by them?
2. What is the relationship between the anatomical location of the ARC and its sensing the circulating nutritional and hormonal signals?
3. Discuss the potential problems associated with the Cre-LoxP technology in regard to its use in the metabolic studies.
4. What are some of the possible reasons for different phenotypes reported in the literature on the role of hypothalamic nutrient sensors on energy metabolism? Discuss the advantages and disadvantages of genetic and pharmacological approaches utilized.
5. The regulation of gene expression of the hypothalamic neuropeptides is governed at various levels, including their transcription and posttranscriptional processing/modifications. At what stages do nutrient sensors interfere with the gene expression of hypothalamic neuropeptides?

References

Adan, R. A. H., Tiesjema, B., Hillebrand, J. J. G., la Fleur, S. E., Kas, M. J. H., & de Krom, M. (2006). The MC4 receptor and control of appetite. *British Journal of Pharmacology, 149*(7), 815–827.

Aguilera, G., Subburaju, S., Young, S., & Chen, J. (2008). The parvocellular vasopressinergic system and responsiveness of the hypothalamic pituitary adrenal axis during chronic stress. *Progress in Brain Research, 170*, 29–39.

Akieda-Asai, S., Zaima, N., Ikegami, K., Kahyo, T., & Yao, I. SIRT1 Regulates Thyroid-Stimulating Hormone Release by Enhancing PIP5Kγ [subscript gamma] Activity through Deacetylation of Specific Lysine Residues in …. 2010. Available at: https://dspace.mit.edu/handle/1721.1/60353.

Aksoy, P., White, T. A., Thompson, M., & Chini, E. N. (2006a). Regulation of intracellular levels of NAD: A novel role for CD38. *Biochemical and Biophysical Research Communications, 345*(4), 1386–1392.

Aksoy, P., Escande, C., White, T. A., Thompson, M., Soares, S., Benech, J. C., & Chini, E. N. (2006b). Regulation of SIRT 1 mediated NAD dependent deacetylation: A novel role for the multifunctional enzyme CD38. *Biochemical and Biophysical Research Communications, 349*(1), 353–359.

Al-Qassab, H., Smith, M. A., Irvine, E. E., Guillermet-Guibert, J., Claret, M., Choudhury, A. I., Selman, C., Piipari, K., Clements, M., Lingard, S., Chandarana, K., Bell, J. D., Barsh, G. S., Smith, A. J. H., Batterham, R. L., Ashford, M. L. J., Vanhaesebroeck, B., & Withers, D. J. (2009). Dominant role of the p110beta isoform of PI3K over p110alpha in energy homeostasis regulation by POMC and AgRP neurons. *Cell Metabolism, 10*(5), 343–354.

Anderson, K. A., Ribar, T. J., Lin, F., Noeldner, P. K., Green, M. F., Muehlbauer, M. J., Witters, L. A., Kemp, B. E., & Means, A. R. (2008). Hypothalamic CaMKK2 contributes to the regulation of energy balance. *Cell Metabolism, 7*(5), 377–388.

Anderson, E. J. P., Çakir, I., Carrington, S. J., Cone, R. D., Ghamari-Langroudi, M., Gillyard, T., Gimenez, L. E., & Litt, M. J. (2016). 60 YEARS OF POMC: Regulation of feeding and energy homeostasis by α-MSH. *Journal of Molecular Endocrinology, 56*(4), T157–T174.

Anderson, K. A., Huynh, F. K., Fisher-Wellman, K., Stuart, J. D., Peterson, B. S., Douros, J. D., Wagner, G. R., Thompson, J. W., Madsen, A. S., Green, M. F., Sivley, R. M., Ilkayeva, O. R., Stevens, R. D., Backos, D. S., Capra, J. A., Olsen, C. A., Campbell, J. E., Muoio, D. M., Grimsrud, P. A., & Hirschey, M. D. (2017). SIRT4 is a lysine Deacylase that controls leucine metabolism and insulin secretion. *Cell Metabolism, 25*(4), 838–855.e15.

Andersson, U., Filipsson, K., Abbott, C. R., Woods, A., Smith, K., Bloom, S. R., Carling, D., & Small, C. J. (2004). AMP-activated protein kinase plays a role in the control of food intake. *The Journal of Biological Chemistry, 279*(13), 12005–12008.

Araki, T., Sasaki, Y., & Milbrandt, J. (2004). Increased nuclear NAD biosynthesis and SIRT1 activation prevent axonal degeneration. *Science, 305*(5686), 1010–1013.

Asher, G., Gatfield, D., Stratmann, M., Reinke, H., Dibner, C., Kreppel, F., Mostoslavsky, R., Alt, F. W., & Schibler, U. (2008). SIRT1 regulates circadian clock gene expression through PER2 deacetylation. *Cell, 134*(2), 317–328.

Barbosa, M. T. P., Soares, S. M., Novak, C. M., Sinclair, D., Levine, J. A., Aksoy, P., & Chini, E. N. (2007). The enzyme CD38 (a NAD glycohydrolase, EC 3.2.2.5) is necessary for the development of diet-induced obesity. *The FASEB Journal, 21*(13), 3629–3639.

Bar-Peled, L., Schweitzer, L. D., Zoncu, R., & Sabatini, D. M. (2012). Ragulator is a GEF for the rag GTPases that signal amino acid levels to mTORC1. *Cell, 150*(6), 1196–1208.

Bates, S. H., Stearns, W. H., Dundon, T. A., Schubert, M., Tso, A. W. K., Wang, Y., Banks, A. S., Lavery, H. J., Haq, A. K., Maratos-Flier, E., Neel, B. G., Schwartz, M. W., & Myers, M. G., Jr. (2003). STAT3 signalling is required for leptin regulation of energy balance but not reproduction. *Nature, 421*(6925), 856–859.

Baur, J. A., Pearson, K. J., Price, N. L., Jamieson, H. A., Lerin, C., Kalra, A., Prabhu, V. V., Allard, J. S., Lopez-Lluch, G., Lewis, K., Pistell, P. J., Poosala, S., Becker, K. G., Boss, O., Gwinn, D., Wang, M., Ramaswamy, S., Fishbein, K. W., Spencer, R. G., Lakatta, E. G., Le Couteur, D., Shaw, R. J., Navas, P., Puigserver, P., Ingram, D. K., de Cabo, R., & Sinclair, D. A. (2006). Resveratrol improves health and survival of mice on a high-calorie diet. *Nature, 444*(7117), 337–342.

Beiroa, D., Imbernon, M., Gallego, R., Senra, A., Herranz, D., Villarroya, F., Serrano, M., Fernø, J., Salvador, J., Escalada, J., Dieguez, C., Lopez, M., Frühbeck, G., & Nogueiras, R. (2014). GLP-1 agonism stimulates brown adipose tissue thermogenesis and browning through hypothalamic AMPK. *Diabetes, 63*(10), 3346–3358.

Belgardt, B. F., Husch, A., Rother, E., Ernst, M. B., Wunderlich, F. T., Hampel, B., Klöckener, T., Alessi, D., Kloppenburg, P., & Brüning, J. C. (2008). PDK1 deficiency in POMC-expressing cells reveals FOXO1-dependent and -independent pathways in control of energy homeostasis and stress response. *Cell Metabolism, 7*(4), 291–301.

Benjannet, S., Rondeau, N., Day, R., Chrétien, M., & Seidah, N. G. (1991). PC1 and PC2 are proprotein convertases capable of cleaving proopiomelanocortin at distinct pairs of basic residues. *Proceedings of the National Academy of Sciences of the United States of America, 88*(9), 3564–3568.

Benoit, S. C., Air, E. L., Coolen, L. M., Strauss, R., Jackman, A., Clegg, D. J., Seeley, R. J., & Woods, S. C. (2002). The catabolic action of insulin in the brain is mediated by melanocortins. *The Journal of Neuroscience, 22*(20), 9048–9052.

Ben-Sahra, I., Howell, J. J., Asara, J. M., & Manning, B. D. (2013). Stimulation of de novo pyrimidine synthesis by growth signaling through mTOR and S6K1. *Science, 339*(6125), 1323–1328.

Ben-Sahra, I., Hoxhaj, G., Ricoult, S. J. H., Asara, J. M., & Manning, B. D. (2016). mTORC1 induces purine synthesis through control of the mitochondrial tetrahydrofolate cycle. *Science, 351*(6274), 728–733.

Blander, G., & Guarente, L. (2004). The Sir2 family of protein deacetylases. *Annual Review of Biochemistry, 73*, 417–435.

Blouet, C., Ono, H., & Schwartz, G. J. (2008). Mediobasal hypothalamic p70 S6 kinase 1 modulates the control of energy homeostasis. *Cell Metabolism, 8*(6), 459–467.

Bordone, L., & Guarente, L. (2005). Calorie restriction, SIRT1 and metabolism: Understanding longevity. *Nature Reviews. Molecular Cell Biology, 6*(4), 298–305.

Bordone, L., Motta, M. C., Picard, F., Robinson, A., Jhala, U. S., Apfeld, J., McDonagh, T., Lemieux, M., McBurney, M., Szilvasi, A., Easlon, E. J., Lin, S.-J., & Guarente, L. (2006). Sirt1 regulates insulin secretion by repressing UCP2 in pancreatic beta cells. *PLoS Biology, 4*(2), e31.

Bordone, L., Cohen, D., Robinson, A., Motta, M. C., van Veen, E., Czopik, A., Steele, A. D., Crowe, H., Marmor, S., Luo, J., Gu, W., & Guarente, L. (2007a). SIRT1 transgenic mice show phenotypes resembling calorie restriction. *Aging Cell, 6*(6), 759–767.

Bordone, L., Guarente - Diabetes L., & Metabolism O. (2007b). Sirtuins and β-cell function. Wiley Online Library 2007. Available at: http://onlinelibrary.wiley.com/doi/10.1111/j.1463-1326.2007.00769.x/full.

Brakch, N., Galanopoulou, A. S., Patel, Y. C., Boileau, G., & Seidah, N. G. (1995). Comparative proteolytic processing of rat prosomatostatin by the convertases PC1, PC2, furin, PACE4 and PC5 in constitutive and regulated secretory pathways. *FEBS Letters, 362*(2), 143–146.

Brar, B., Sanderson, T., Wang, N., & Lowry, P. J. (1997). Post-translational processing of human procorticotrophin-releasing factor in transfected mouse neuroblastoma and Chinese hamster ovary cell lines. *The Journal of Endocrinology, 154*(3), 431–440.

Breslin, M. B., Lindberg, I., Benjannet, S., & Mathis - Journal of Biological … JP. (1993). Differential processing of proenkephalin by prohormone convertases 1 (3) and 2 and furin. ASBMB 1993. Available at: http://www.jbc.org/content/268/36/27084.short.

Brooks, C. L., & Gu, W. (2009). How does SIRT1 affect metabolism, senescence and cancer? *Nature Reviews. Cancer, 9*(2), 123–128.

Cakir, I., Perello, M., Lansari, O., Messier, N. J., Vaslet, C. A., & Nillni, E. A. (2009). Hypothalamic Sirt1 regulates food intake in a rodent model system. *PLoS One, 4*(12), e8322.

Cakir, I., Cyr, N. E., Perello, M., Litvinov, B. P., Romero, A., Stuart, R. C., & Nillni, E. A. (2013). Obesity induces hypothalamic endoplasmic reticulum stress and impairs proopiomelanocortin (POMC) post-translational processing. *The Journal of Biological Chemistry, 288*(24), 17675–17688.

Cantó, C., Gerhart-Hines, Z., Feige, J. N., Lagouge, M., Noriega, L., Milne, J. C., Elliott, P. J., Puigserver, P., & Auwerx, J. (2009). AMPK regulates energy expenditure by modulating NAD+ metabolism and SIRT1 activity. *Nature, 458*(7241), 1056–1060.

Cantó, C., Houtkooper, R. H., Pirinen, E., Youn, D. Y., Oosterveer, M. H., Cen, Y., Fernandez-Marcos, P. J., Yamamoto, H., Andreux, P. A., Cettour-Rose, P., Gademann, K., Rinsch, C., Schoonjans, K., Sauve,

A. A., & Auwerx, J. (2012). The NAD(+) precursor nicotinamide riboside enhances oxidative metabolism and protects against high-fat diet-induced obesity. *Cell Metabolism, 15*(6), 838–847.

Cao, J., Papadopoulou, N., Kempuraj, D., Boucher, W. S., Sugimoto, K., Cetrulo, C. L., & Theoharides, T. C. (2005). Human mast cells express corticotropin-releasing hormone (CRH) receptors and CRH leads to selective secretion of vascular endothelial growth factor. *Journal of Immunology, 174*(12), 7665–7675.

Caron, A., Labbé, S. M., Lanfray, D., Blanchard, P.-G., Villot, R., Roy, C., Sabatini, D. M., Richard, D., & Laplante, M. (2016a). Mediobasal hypothalamic overexpression of DEPTOR protects against high-fat diet-induced obesity. *Molecular Metabolism, 5*(2), 102–112.

Caron, A., Labbé, S. M., Mouchiroud, M., Huard, R., Lanfray, D., Richard, D., & Laplante, M. (2016b). DEPTOR in POMC neurons affects liver metabolism but is dispensable for the regulation of energy balance. *American Journal of Physiology. Regulatory, Integrative and Comparative Physiology, 310*(11), R1322–R1331.

Castro, M., Lowenstein, P., Glynn, B., Hannah, M., Linton, E., & Lowry, P. (1991). Post-translational processing and regulated release of corticotropin-releasing hormone (CRH) in AtT20 cells expressing the human proCRH gene. *Biochemical Society Transactions, 19*(3), 246S.

Chalkiadaki, A., & Guarente, L. (2012). High-fat diet triggers inflammation-induced cleavage of SIRT1 in adipose tissue to promote metabolic dysfunction. *Cell Metabolism, 16*(2), 180–188.

Challis, B. G., Pritchard, L. E., & Creemers - Human molecular … J. (2002). A missense mutation disrupting a dibasic prohormone processing site in pro-opiomelanocortin (POMC) increases susceptibility to early-onset obesity through a novel …. academic.oup. com 2002. Available at: https://academic.oup.com/hmg/article-abstract/11/17/1997/589952.

Chang, H.-C., & Guarente, L. (2013). SIRT1 mediates central circadian control in the SCN by a mechanism that decays with aging. *Cell, 153*(7), 1448–1460.

Chang, H.-C., & Guarente, L. (2014). SIRT1 and other sirtuins in metabolism. *Trends in Endocrinology and Metabolism, 25*(3), 138–145.

Chantranupong, L., Scaria, S. M., Saxton, R. A., Gygi, M. P., Shen, K., Wyant, G. A., Wang, T., Harper, J. W., Gygi, S. P., & Sabatini, D. M. (2016). The CASTOR proteins are arginine sensors for the mTORC1 pathway. *Cell, 165*(1), 153–164.

Chen, Y., Wu, R., Chen, H.-Z., Xiao, Q., Wang, W.-J., He, J.-P., Li, X.-X., Yu, X.-W., Li, L., Wang, P., Wan, X.-C., Tian, X.-H., Li, S.-J., Yu, X., & Wu, Q. (2015). Enhancement of hypothalamic STAT3 acetylation by nuclear receptor Nur77 dictates leptin sensitivity. *Diabetes, 64*(6), 2069–2081.

Cheng, X.-B., Wen, J.-P., Yang, J., Yang, Y., Ning, G., & Li, X.-Y. (2011). GnRH secretion is inhibited by adiponectin through activation of AMP-activated pro-tein kinase and extracellular signal-regulated kinase. *Endocrine, 39*(1), 6–12.

Chikahisa, S., Fujiki, N., Kitaoka, K., Shimizu, N., & Séi, H. (2009). Central AMPK contributes to sleep homeostasis in mice. *Neuropharmacology, 57*(4), 369–374.

Choudhury, A. I., Heffron, H., Smith, M. A., Al-Qassab, H., Xu, A. W., Selman, C., Simmgen, M., Clements, M., Claret, M., Maccoll, G., Bedford, D. C., Hisadome, K., Diakonov, I., Moosajee, V., Bell, J. D., Speakman, J. R., Batterham, R. L., Barsh, G. S., Ashford, M. L. J., & Withers, D. J. (2005). The role of insulin receptor substrate 2 in hypothalamic and beta cell function. *The Journal of Clinical Investigation, 115*(4), 940–950.

Chrousos, G. P. (1995). The hypothalamic-pituitary-adrenal axis and immune-mediated inflammation. *The New England Journal of Medicine, 332*(20), 1351–1362.

Claret, M., Smith, M. A., Batterham, R. L., Selman, C., Choudhury, A. I., Fryer, L. G. D., Clements, M., Al-Qassab, H., Heffron, H., Xu, A. W., Speakman, J. R., Barsh, G. S., Viollet, B., Vaulont, S., Ashford, M. L. J., Carling, D., & Withers, D. J. (2007). AMPK is essential for energy homeostasis regulation and glucose sensing by POMC and AgRP neurons. *The Journal of Clinical Investigation, 117*(8), 2325–2336.

Claret, M., Smith, M. A., Knauf, C., Al-Qassab, H., Woods, A., Heslegrave, A., Piipari, K., Emmanuel, J. J., Colom, A., Valet, P., Cani, P. D., Begum, G., White, A., Mucket, P., Peters, M., Mizuno, K., Batterham, R. L., Giese, K. P., Ashworth, A., Burcelin, R., Ashford, M. L., Carling, D., & Withers, D. J. (2011). Deletion of Lkb1 in pro-opiomelanocortin neurons impairs peripheral glucose homeostasis in mice. *Diabetes, 60*(3), 735–745.

Cohen, H. Y., Miller, C., Bitterman, K. J., Wall, N. R., Hekking, B., Kessler, B., Howitz, K. T., Gorospe, M., de Cabo, R., & Sinclair, D. A. (2004). Calorie restriction promotes mammalian cell survival by inducing the SIRT1 deacetylase. *Science, 305*(5682), 390–392.

Cohen, D. E., Supinski, A. M., Bonkowski, M. S., Donmez, G., & Guarente, L. P. (2009). Neuronal SIRT1 regulates endocrine and behavioral responses to calorie restriction. *Genes & Development, 23*(24), 2812–2817.

Cone, R. D. (2005). Anatomy and regulation of the central melanocortin system. *Nature Neuroscience, 8*(5), 571–578.

Constam, D. B., Calfon, M., & Robertson, E. J. (1996). SPC4, SPC6, and the novel protease SPC7 are coexpressed with bone morphogenetic proteins at distinct sites during embryogenesis. *The Journal of Cell Biology, 134*(1), 181–191.

Cota, D., Proulx, K., Smith, K. A. B., Kozma, S. C., Thomas, G., Woods, S. C., & Seeley, R. J. (2006). Hypothalamic mTOR signaling regulates food intake. *Science, 312*(5775), 927–930.

Cota, D., Matter, E. K., Woods, S. C., & Seeley, R. J. (2008). The role of hypothalamic mammalian target of rapamycin complex 1 signaling in diet-induced obesity. *The Journal of Neuroscience, 28*(28), 7202–7208.

Cottrell, E. C., & Ozanne, S. E. (2008). Early life programming of obesity and metabolic disease. *Physiology & Behavior, 94*(1), 17–28.

Coupé, B., Ishii, Y., Dietrich, M. O., Komatsu, M., Horvath, T. L., & Bouret, S. G. (2012). Loss of autophagy in pro-opiomelanocortin neurons perturbs axon growth and causes metabolic dysregulation. *Cell Metabolism, 15*(2), 247–255.

Cowley, M. A., Smart, J. L., Rubinstein, M., Cerdán, M. G., Diano, S., Horvath, T. L., Cone, R. D., & Low, M. J. (2001). Leptin activates anorexigenic POMC neurons through a neural network in the arcuate nucleus. *Nature, 411*(6836), 480–484.

Cummings, D. E., & Schwartz, M. W. (2000). Melanocortins and body weight: A tale of two receptors. *Nature Genetics, 26*(1), 8–9.

Cunningham, J. T., Rodgers, J. T., Arlow, D. H., Vazquez, F., Mootha, V. K., & Puigserver, P. (2007). mTOR controls mitochondrial oxidative function through a YY1-PGC-1alpha transcriptional complex. *Nature, 450*(7170), 736–740.

Cyr, N. E., Toorie, A. M., Steger, J. S., Sochat, M. M., Hyner, S., Perello, M., Stuart, R., & Nillni, E. A. (2013). Mechanisms by which the orexigen NPY regulates anorexigenic α-MSH and TRH. *American Journal of Physiology. Endocrinology and Metabolism, 304*(6), E640–E650.

Cyr, N. E., Steger, J. S., Toorie, A. M., Yang, J. Z., & Stuart, R. (2014). Central Sirt1 regulates body weight and energy expenditure along with the POMC-derived peptide α-MSH and the processing enzyme CPE production in diet- …. press.endocrine.org 2014. Available at: http://press.endocrine.org/doi/pdf/10.1210/en.2013-1998.

da Silva Xavier, G., Farhan, H., Kim, H., Caxaria, S., Johnson, P., Hughes, S., Bugliani, M., Marselli, L., Marchetti, P., Birzele, F., Sun, G., Scharfmann, R., Rutter, J., Siniakowicz, K., Weir, G., Parker, H., Reimann, F., Gribble, F. M., & Rutter, G. A. (2011). Per-arnt-Sim (PAS) domain-containing protein kinase is downregulated in human islets in type 2 diabetes and regulates glucagon secretion. *Diabetologia, 54*(4), 819–827.

Dagon, Y., Hur, E., Zheng, B., Wellenstein, K., Cantley, L. C., & Kahn, B. B. (2012). p70S6 kinase phosphorylates AMPK on serine 491 to mediate leptin's effect on food intake. *Cell Metabolism, 16*(1), 104–112.

Dasgupta, M., Unal, H., Willard, B., Yang, J., Karnik, S. S., & Stark, G. R. (2014). Critical role for lysine 685 in gene expression mediated by transcription factor unphosphorylated STAT3. *The Journal of Biological Chemistry, 289*(44), 30763–30771.

de Ruijter, A. J. M., van Gennip, A. H., Caron, H. N., Kemp, S., & van Kuilenburg, A. B. P. (2003). Histone deacetylases (HDACs): Characterization of the classical HDAC family. *The Biochemical Journal, 370*(Pt 3), 737–749.

DeMille, D., & Grose, J. H. (2013). PAS kinase: A nutrient sensing regulator of glucose homeostasis. *IUBMB Life, 65*(11), 921–929.

Deng, X.-Q., Chen, L.-L., & Li, N.-X. (2007). The expression of SIRT1 in nonalcoholic fatty liver disease induced by high-fat diet in rats. *Liver International, 27*(5), 708–715.

Dey, A., Xhu, X., Carroll, R., Turck, C. W., Stein, J., & Steiner, D. F. (2003). Biological processing of the cocaine and amphetamine-regulated transcript precursors by prohormone convertases, PC2 and PC1/3. *The Journal of Biological Chemistry, 278*(17), 15007–15014.

Dey, A., Norrbom, C., Zhu, X., Stein, J., Zhang, C., Ueda, K., & Steiner, D. F. (2004). Furin and prohormone convertase 1/3 are major convertases in the processing of mouse pro-growth hormone-releasing hormone. *Endocrinology, 145*(4), 1961–1971.

Dietrich, M. O., Antunes, C., Geliang, G., Liu, Z.-W., Borok, E., Nie, Y., Xu, A. W., Souza, D. O., Gao, Q., Diano, S., Gao, X.-B., & Horvath, T. L. (2010). Agrp neurons mediate Sirt1's action on the melanocortin system and energy balance: Roles for Sirt1 in neuronal firing and synaptic plasticity. *The Journal of Neuroscience, 30*(35), 11815–11825.

Du, J., Zhou, Y., Su, X., Yu, J. J., Khan, S., Jiang, H., Kim, J., Woo, J., Kim, J. H., Choi, B. H., He, B., Chen, W., Zhang, S., Cerione, R. A., Auwerx, J., Hao, Q., & Lin, H. (2011). Sirt5 is a NAD-dependent protein lysine demalonylase and desuccinylase. *Science, 334*(6057), 806–809.

Dong W1., Seidel B., Marcinkiewicz M., Chrétien M., Seidah NG., Day R. (1997). Cellular localization of the prohormone convertases in the hypothalamic paraventricular and supraoptic nuclei: selective regulation of PC1 in corticotrophin-releasing hormone parvocellular neurons mediated by glucocorticoids. J Neurosci. 15;17(2):563–75.

Düvel, K., Yecies, J. L., Menon, S., Raman, P., Lipovsky, A. I., Souza, A. L., Triantafellow, E., Ma, Q., Gorski, R., Cleaver, S., Vander Heiden, M. G., MacKeigan, J. P., Finan, P. M., Clish, C. B., Murphy, L. O., & Manning, B. D. (2010). Activation of a metabolic gene regulatory network downstream of mTOR complex 1. *Molecular Cell, 39*(2), 171–183.

Egan, D. F., Shackelford, D. B., Mihaylova, M. M., Gelino, S., Kohnz, R. A., Mair, W., Vasquez, D. S., Joshi, A., Gwinn, D. M., Taylor, R., Asara, J. M., Fitzpatrick, J., Dillin, A., Viollet, B., Kundu, M., Hansen, M., & Shaw, R. J. (2011). Phosphorylation of ULK1 (hATG1) by AMP-activated protein kinase connects energy sensing to mitophagy. *Science, 331*(6016), 456–461.

Elson, A. E., & Simerly, R. B. (2015). Developmental specification of metabolic circuitry. *Frontiers in Neuroendocrinology, 39*, 38–51.

Ernst, M. B., Wunderlich, C. M., Hess, S., Paehler, M., Mesaros, A., Koralov, S. B., Kleinridders, A., Husch, A., Münzberg, H., Hampel, B., Alber, J., Kloppenburg, P., Brüning, J. C., & Wunderlich, F. T. (2009). Enhanced Stat3 activation in POMC neurons provokes negative feedback inhibition of leptin and insulin signaling in obesity. *The Journal of Neuroscience, 29*(37), 11582–11593.

Fox, D. L., & Good, D. J. (2008). Nescient helix-loop-helix 2 interacts with signal transducer and activator of transcription 3 to regulate transcription of prohormone convertase 1/3. *Molecular Endocrinology, 22*(6), 1438–1448.

Frescas, D., Valenti, L., & Accili, D. (2005). Nuclear trapping of the forkhead transcription factor FoxO1 via Sirt-dependent deacetylation promotes expression of glucogenetic genes. *The Journal of Biological Chemistry, 280*(21), 20589–20595.

Fricker, L. D., Berman, Y. L., Leiter, E. H., & Devi, L. A. (1996). Carboxypeptidase E activity is deficient in mice with the fat mutation: EFFECT ON PEPTIDE PROCESSING. *The Journal of Biological Chemistry, 271*(48), 30619–30624.

Friedman, T. C., Loh, Y. P., Cawley, N. X., Birch, N. P., Huang, S. S., Jackson, I. M., & Nillni, E. A. (1995). Processing of prothyrotropin-releasing hormone (pro-TRH) by bovine intermediate lobe secretory vesicle membrane PC1 and PC2 enzymes. *Endocrinology, 136*(10), 4462–4472.

Fryer, L. G. D., Foufelle, F., Barnes, K., Baldwin, S. A., Woods, A., & Carling, D. (2002). Characterization of the role of the AMP-activated protein kinase in the stimulation of glucose transport in skeletal muscle cells. *The Biochemical Journal, 363*(Pt 1), 167–174.

Fulco, M., Schiltz, R. L., Iezzi, S., King, M. T., Zhao, P., Kashiwaya, Y., Hoffman, E., Veech, R. L., & Sartorelli, V. (2003). Sir2 regulates skeletal muscle differentiation as a potential sensor of the redox state. *Molecular Cell, 12*(1), 51–62.

Funato, H., Tsai, A. L., Willie, J. T., Kisanuki, Y., Williams, S. C., Sakurai, T., & Yanagisawa, M. (2009). Enhanced orexin receptor-2 signaling prevents diet-induced obesity and improves leptin sensitivity. *Cell Metabolism, 9*(1), 64–76.

Funato, H., Oda, S., Yokofujita, J., Igarashi, H., & Kuroda, M. (2011). Fasting and high-fat diet alter histone deacetylase expression in the medial hypothalamus. *PLoS One, 6*(4), e18950.

Furuta, M., Yano, H., Zhou, A., Rouillé, Y., Holst, J. J., Carroll, R., Ravazzola, M., Orci, L., Furuta, H., & Steiner, D. F. (1997). Defective prohormone processing and altered pancreatic islet morphology in mice lacking active SPC2. *Proceedings of the National Academy of Sciences of the United States of America, 94*(13), 6646–6651.

Füzesi, T., Wittmann, G., Liposits, Z., Lechan, R. M., & Fekete, C. (2007). Contribution of noradrenergic and adrenergic cell groups of the brainstem and agouti-related protein-synthesizing neurons of the arcuate nucleus to neuropeptide-y innervation of corticotropin-releasing hormone neurons in hypothalamic paraventricular nucleus of the rat. *Endocrinology, 148*(11), 5442–5450.

Galanopoulou, A. S., Kent, G., Rabbani, S. N., Seidah, N. G., & Patel, Y. C. (1993). Heterologous processing of prosomatostatin in constitutive and regulated secretory pathways. Putative role of the endoproteases furin, PC1, and PC2. *The Journal of Biological Chemistry, 268*(8), 6041–6049.

Gan, X., Wang, J., Wang, C., Sommer, E., Kozasa, T., Srinivasula, S., Alessi, D., Offermanns, S., Simon, M. I., & Wu, D. (2012). PRR5L degradation promotes mTORC2-mediated PKC-δ phosphorylation and cell migration downstream of Gα12. *Nature Cell Biology, 14*(7), 686–696.

Gao, Q., Wolfgang, M. J., Neschen, S., Morino, K., Horvath, T. L., Shulman, G. I., & Fu, X.-Y. (2004). Disruption of neural signal transducer and activator of transcription 3 causes obesity, diabetes, infertility, and thermal dysregulation. *Proceedings of the National Academy of Sciences of the United States of America, 101*(13), 4661–4666.

Gao, S., Kinzig, K. P., Aja, S., Scott, K. A., Keung, W., Kelly, S., Strynadka, K., Chohnan, S., Smith, W. W., Tamashiro, K. L. K., Ladenheim, E. E., Ronnett, G. V., Tu, Y., Birnbaum, M. J., Lopaschuk, G. D., & Moran, T. H. (2007). Leptin activates hypothalamic acetyl-CoA carboxylase to inhibit food intake. *Proceedings of the National Academy of Sciences of the United States of America, 104*(44), 17358–17363.

Garfield, A. S., Shah, B. P., Burgess, C. R., Li, M. M., Li, C., Steger, J. S., Madara, J. C., Campbell, J. N., Kroeger, D., Scammell, T. E., Tannous, B. A., Myers, M. G., Jr., Andermann, M. L., Krashes, M. J., & Lowell, B. B. (2016). Dynamic GABAergic afferent modulation of AgRP neurons. *Nature Neuroscience, 19*(12), 1628–1635.

Ghosh, H. S., McBurney, M., & Robbins, P. D. (2010). SIRT1 negatively regulates the mammalian target of rapamycin. *PLoS One, 5*(2), e9199.

Gill, J. F., Delezie, J., Santos, G., & Handschin, C. (2016). PGC-1α expression in murine AgRP neurons regulates food intake and energy balance. *Molecular Metabolism, 5*(7), 580–588.

Gillum, M. P., Kotas, M. E., Erion, D. M., Kursawe, R., Chatterjee, P., Nead, K. T., Muise, E. S., Hsiao, J. J., Frederick, D. W., Yonemitsu, S., Banks, A. S., Qiang, L., Bhanot, S., Olefsky, J. M., Sears, D. D., Caprio, S., & Shulman, G. I. (2011). SirT1 regulates adipose tissue inflammation. *Diabetes, 60*(12), 3235–3245.

Good, D. J., Li, M., & Deater-Deckard, K. (2015). A genetic basis for motivated exercise. *Exercise and Sport Sciences Reviews, 43*(4), 231–237.

Gwinn, D. M., Shackelford, D. B., Egan, D. F., Mihaylova, M. M., Mery, A., Vasquez, D. S., Turk, B. E., & Shaw, R. J. (2008). AMPK phosphorylation of raptor mediates a metabolic checkpoint. *Molecular Cell, 30*(2), 214–226.

Hagiwara, A., Cornu, M., Cybulski, N., Polak, P., Betz, C., Trapani, F., Terracciano, L., Heim, M. H., Rüegg, M. A., & Hall, M. N. (2012). Hepatic mTORC2 activates glycolysis and lipogenesis through Akt, glucokinase, and SREBP1c. *Cell Metabolism, 15*(5), 725–738.

Hahn, T. M., Breininger, J. F., Baskin, D. G., & Schwartz, M. W. (1998). Coexpression of Agrp and NPY in fasting-activated hypothalamic neurons. *Nature Neuroscience, 1*(4), 271–272.

Haigis, M. C., & Sinclair, D. A. (2010). Mammalian sirtuins: Biological insights and disease relevance. *Annual Review of Pathology, 5*, 253–295.

Hao, H.-X., Cardon, C. M., Swiatek, W., Cooksey, R. C., Smith, T. L., Wilde, J., Boudina, S., Abel, E. D., McClain, D. A., & Rutter, J. (2007). PAS kinase is required for normal cellular energy balance. *Proceedings of the National Academy of Sciences of the United States of America, 104*(39), 15466–15471.

Hardie, D. G., & Pan, D. A. (2002). Regulation of fatty acid synthesis and oxidation by the AMP-activated protein kinase. *Biochemical Society Transactions, 30*(Pt 6), 1064–1070.

Harlan, S. M., Guo, D.-F., Morgan, D. A., Fernandes-Santos, C., & Rahmouni, K. (2013). Hypothalamic mTORC1 signaling controls sympathetic nerve activity and arterial pressure and mediates leptin effects. *Cell Metabolism, 17*(4), 599–606.

Hasegawa, K., Kawahara, T., Fujiwara, K., Shimpuku, M., Sasaki, T., Kitamura, T., & Yoshikawa, K. (2012). Necdin controls Foxo1 acetylation in hypothalamic arcuate neurons to modulate the thyroid axis. *The Journal of Neuroscience, 32*(16), 5562–5572.

Hawley, S. A., Boudeau, J., Reid, J. L., Mustard, K. J., Udd, L., Mäkelä, T. P., Alessi, D. R., & Hardie, D. G. (2003). Complexes between the LKB1 tumor suppressor, STRAD alpha/beta and MO25 alpha/beta are upstream kinases in the AMP-activated protein kinase cascade. *Journal of Biology, 2*(4), 28.

Hawley, S. A., Pan, D. A., Mustard, K. J., Ross, L., Bain, J., Edelman, A. M., Frenguelli, B. G., & Hardie, D. G. (2005). Calmodulin-dependent protein kinase kinase-beta is an alternative upstream kinase for AMP-activated protein kinase. *Cell Metabolism, 2*(1), 9–19.

Hawley, S. A., Ross, F. A., Gowans, G. J., Tibarewal, P., Leslie, N. R., & Hardie, D. G. (2014). Phosphorylation by Akt within the ST loop of AMPK-α1 down-regulates its activation in tumour cells. *The Biochemical Journal, 459*(2), 275–287.

Heathcote, H. R., Mancini, S. J., Strembitska, A., Jamal, K., Reihill, J. A., Palmer, T. M., Gould, G. W., & Salt, I. P. (2016). Protein kinase C phosphorylates AMP-activated protein kinase α1 Ser487. *The Biochemical Journal, 473*(24), 4681–4697.

Heinrichs, S. C., Menzaghi, F., Pich, E. M., Hauger, R. L., & Koob, G. F. (1993). Corticotropin-releasing factor in the paraventricular nucleus modulates feeding induced by neuropeptide Y. *Brain Research, 611*(1), 18–24.

Henckens, M. J. A. G., Deussing, J. M., & Chen, A. (2016). Region-specific roles of the corticotropin-releasing factor-urocortin system in stress. *Nature Reviews. Neuroscience, 17*(10), 636–651.

Heuer, H., Maier, M. K., Iden, S., Mittag, J., Friesema, E. C. H., Visser, T. J., & Bauer, K. (2005). The monocarboxylate transporter 8 linked to human psychomotor retardation is highly expressed in thyroid hormone-sensitive neuron populations. *Endocrinology, 146*(4), 1701–1706.

Hill, J. W., Williams, K. W., Ye, C., Luo, J., Balthasar, N., Coppari, R., Cowley, M. A., Cantley, L. C., Lowell, B. B., & Elmquist, J. K. (2008). Acute effects of leptin require PI3K signaling in hypothalamic proopiomelanocortin neurons in mice. *The Journal of Clinical Investigation, 118*(5), 1796–1805.

Hill, J. W., Xu, Y., Preitner, F., Fukuda, M., Cho, Y.-R., Luo, J., Balthasar, N., Coppari, R., Cantley, L. C., Kahn, B. B., Zhao, J. J., & Elmquist, J. K. (2009). Phosphatidyl inositol 3-kinase signaling in hypothalamic proopiomelanocortin neurons contributes to the regulation of glucose homeostasis. *Endocrinology, 150*(11), 4874–4882.

Hisahara, S., Chiba, S., Matsumoto, H., Tanno, M., Yagi, H., Shimohama, S., Sato, M., & Horio, Y. (2008). Histone deacetylase SIRT1 modulates neuronal differentiation by its nuclear translocation. *Proceedings of the National Academy of Sciences of the United States of America, 105*(40), 15599–15604.

Hou, X., Xu, S., Maitland-Toolan, K. A., Sato, K., Jiang, B., Ido, Y., Lan, F., Walsh, K., Wierzbicki, M., Verbeuren, T. J., Cohen, R. A., & Zang, M. (2008). SIRT1 regulates hepatocyte lipid metabolism through activating AMP-activated protein kinase. *The Journal of Biological Chemistry, 283*(29), 20015–20026.

Houtkooper, R. H., Pirinen, E., & Auwerx, J. (2012). Sirtuins as regulators of metabolism and healthspan. *Nature Reviews. Molecular Cell Biology, 13*(4), 225–238.

Hu, Z., Cha, S. H., Chohnan, S., & Lane, M. D. (2003). Hypothalamic malonyl-CoA as a mediator of feeding behavior. *Proceedings of the National Academy of Sciences of the United States of America, 100*(22), 12624–12629.

Huang, H., & Tindall, D. J. (2007). Dynamic FoxO transcription factors. *Journal of Cell Science, 120*(Pt 15), 2479–2487.

Hurtado-Carneiro, V., Roncero, I., Blazquez, E., Alvarez, E., & Sanz, C. (2013). PAS kinase as a nutrient sensor in neuroblastoma and hypothalamic cells required for the normal expression and activity of other cellular nutrient and energy sensors. *Molecular Neurobiology, 48*(3), 904–920.

Hurtado-Carneiro, V., Roncero, I., Egger, S. S., Wenger, R. H., Blazquez, E., Sanz, C., & Alvarez, E. (2014). PAS kinase is a nutrient and energy sensor in hypothalamic areas required for the normal function of AMPK and mTOR/S6K1. *Molecular Neurobiology, 50*(2), 314–326.

Ibrahim, N., Bosch, M. A., Smart, J. L., Qiu, J., Rubinstein, M., Rønnekleiv, O. K., Low, M. J., & Kelly, M. J. (2003). Hypothalamic proopiomelanocortin neurons are glucose responsive and express K(ATP) channels. *Endocrinology, 144*(4), 1331–1340.

Inoki, K., Zhu, T., & Guan, K.-L. (2003). TSC2 mediates cellular energy response to control cell growth and survival. *Cell, 115*(5), 577–590.

Iyer, A., Fairlie, D. P., & Brown, L. (2012). Lysine acetylation in obesity, diabetes and metabolic disease. *Immunology and Cell Biology, 90*(1), 39–46.

Jacinto, E., Loewith, R., Schmidt, A., Lin, S., Rüegg, M. A., Hall, A., & Hall, M. N. (2004). Mammalian TOR complex 2 controls the actin cytoskeleton and is rapamycin insensitive. *Nature Cell Biology, 6*(11), 1122–1128.

Jackson, R. S., Creemers, J. W., Ohagi, S., Raffin-Sanson, M. L., Sanders, L., Montague, C. T., Hutton, J. C., & O'Rahilly, S. (1997). Obesity and impaired prohormone processing associated with mutations in the human prohormone convertase 1 gene. *Nature Genetics, 16*(3), 303–306.

Jäger, S., Handschin, C., St-Pierre, J., & Spiegelman, B. M. (2007). AMP-activated protein kinase (AMPK) action in skeletal muscle via direct phosphorylation of PGC-1alpha. *Proceedings of the National Academy of Sciences of the United States of America, 104*(29), 12017–12022.

Jang, M., & Romsos, D. R. (1998). Neuropeptide Y and corticotropin-releasing hormone concentrations within specific hypothalamic regions of lean but not Ob/Ob mice respond to food-deprivation and refeeding. *The Journal of Nutrition, 128*(12), 2520–2525.

Jay, P., Rougeulle, C., Massacrier, A., Moncla, A., Mattei, M. G., Malzac, P., Roëckel, N., Taviaux, S., Lefranc, J. L., Cau, P., Berta, P., Lalande, M., & Muscatelli, F. (1997). The human necdin gene, NDN, is maternally imprinted and located in the Prader-Willi syndrome chromosomal region. *Nature Genetics, 17*(3), 357–361.

Jiang, H., Khan, S., Wang, Y., Charron, G., He, B., Sebastian, C., Du, J., Kim, R., Ge, E., Mostoslavsky, R., Hang, H. C., Hao, Q., & Lin, H. (2013). SIRT6 regulates TNF-α secretion through hydrolysis of long-chain fatty acyl lysine. *Nature, 496*(7443), 110–113.

Jing, E., Nillni, E. A., Sanchez, V. C., Stuart, R. C., & Good, D. J. (2004). Deletion of the Nhlh2 transcription factor decreases the levels of the anorexigenic peptides alpha melanocyte-stimulating hormone and thyrotropin-releasing hormone and implicates prohormone convertases I and II in obesity. *Endocrinology, 145*(4), 1503–1513.

Jung, J., Genau, H. M., & Behrends, C. (2015). Amino acid-dependent mTORC1 regulation by the lysosomal membrane protein SLC38A9. *Molecular and Cellular Biology, 35*(14), 2479–2494.

Kabra, D. G., Pfuhlmann, K., García-Cáceres, C., Schriever, S. C., Casquero García, V., Kebede, A. F., Fuente-Martin, E., Trivedi, C., Heppner, K., Uhlenhaut, N. H., Legutko, B., Kabra, U. D., Gao, Y., Yi, C.-X., Quarta, C., Clemmensen, C., Finan, B., Müller, T. D., Meyer, C. W., Paez-Pereda, M., Stemmer, K., Woods, S. C., Perez-Tilve, D., Schneider, R., Olson, E. N., Tschöp, M. H., & Pfluger, P. T. (2016). Hypothalamic leptin action is mediated by histone deacetylase 5. *Nature Communications, 7*, 10782.

Kanfi, Y., Peshti, V., Gozlan, Y. M., Rathaus, M., Gil, R., & Cohen, H. Y. (2008). Regulation of SIRT1 protein levels by nutrient availability. *FEBS Letters, 582*(16), 2417–2423.

Kang, H., Jung, J.-W., Kim, M. K., & Chung, J. H. (2009). CK2 is the regulator of SIRT1 substrate-binding affinity, deacetylase activity and cellular response to DNA-damage. *PLoS One, 4*(8), e6611.

Kaur, C., & Ling, E.-A. (2017). The circumventricular organs. *Histology and Histopathology, 32*(9), 879–892.

Kaushik, S., Rodriguez-Navarro, J. A., Arias, E., Kiffin, R., Sahu, S., Schwartz, G. J., Cuervo, A. M., & Singh, R. (2011). Autophagy in hypothalamic AgRP neurons regulates food intake and energy balance. *Cell Metabolism, 14*(2), 173–183.

Kaushik, S., Arias, E., Kwon, H., Lopez, N. M., Athonvarangkul, D., Sahu, S., Schwartz, G. J., Pessin, J. E., & Singh, R. (2012). Loss of autophagy in hypothalamic POMC neurons impairs lipolysis. *EMBO Reports, 13*(3), 258–265.

Kim, M.-S., Park, J.-Y., Namkoong, C., Jang, P.-G., Ryu, J.-W., Song, H.-S., Yun, J.-Y., Namgoong, I.-S., Ha, J., Park, I.-S., Lee, I.-K., Viollet, B., Youn, J. H., Lee, H.-K., & Lee, K.-U. (2004). Anti-obesity effects of alpha-lipoic acid mediated by suppression of hypothalamic AMP-activated protein kinase. *Nature Medicine, 10*(7), 727–733.

Kim, M.-S., Pak, Y. K., Jang, P.-G., Namkoong, C., Choi, Y.-S., Won, J.-C., Kim, K.-S., Kim, S.-W., Kim, H.-S., Park, J.-Y., Kim, Y.-B., & Lee, K.-U. (2006). Role of hypothalamic Foxo1 in the regulation of food intake and energy homeostasis. *Nature Neuroscience, 9*(7), 901–906.

Kim, E.-J., Kho, J.-H., Kang, M.-R., & Um, S.-J. (2007a). Active regulator of SIRT1 cooperates with SIRT1 and facilitates suppression of p53 activity. *Molecular Cell, 28*(2), 277–290.

Kim, D., Nguyen, M. D., Dobbin, M. M., Fischer, A., Sananbenesi, F., Rodgers, J. T., Delalle, I., Baur, J. A., Sui, G., Armour, S. M., Puigserver, P., Sinclair, D. A., & Tsai, L.-H. (2007b). SIRT1 deacetylase protects against neurodegeneration in models for Alzheimer's disease and amyotrophic lateral sclerosis. *The EMBO Journal, 26*(13), 3169–3179.

Kim, E., Goraksha-Hicks, P., Li, L., Neufeld, T. P., & Guan, K.-L. (2008a). Regulation of TORC1 by rag GTPases in nutrient response. *Nature Cell Biology, 10*(8), 935–945.

Kim, J.-E., Chen, J., & Lou, Z. (2008b). DBC1 is a negative regulator of SIRT1. *Nature, 451*(7178), 583–586.

Kim, J., Kundu, M., Viollet, B., & Guan, K.-L. (2011). AMPK and mTOR regulate autophagy through direct phosphorylation of Ulk1. *Nature Cell Biology, 13*(2), 132–141.

Kinote, A., Faria, J. A., Roman, E. A., Solon, C., Razolli, D. S., Ignacio-Souza, L. M., Sollon, C. S., Nascimento, L. F., de Araújo, T. M., Barbosa, A. P. L., Lellis-Santos, C., Velloso, L. A., Bordin, S., & Anhê, G. F. (2012). Fructose-induced hypothalamic AMPK activation stimulates hepatic PEPCK and gluconeogenesis due to increased corticosterone levels. *Endocrinology, 153*(8), 3633–3645.

Kitamura, T., Feng, Y., Kitamura, Y. I., Chua, S. C., Jr., Xu, A. W., Barsh, G. S., Rossetti, L., & Accili, D. (2006). Forkhead protein FoxO1 mediates Agrp-dependent effects of leptin on food intake. *Nature Medicine, 12*(5), 534–540.

Knight, C. M., Gutierrez-Juarez, R., Lam, T. K. T., Arrieta-Cruz, I., Huang, L., Schwartz, G., Barzilai, N., & Rossetti, L. (2011). Mediobasal hypothalamic SIRT1 is essential for resveratrol's effects on insulin action in rats. *Diabetes, 60*(11), 2691–2700.

Kocalis, H. E., Hagan, S. L., George, L., Turney, M. K., Siuta, M. A., Laryea, G. N., Morris, L. C., Muglia, L. J., Printz, R. L., Stanwood, G. D., & Niswender, K. D. (2014). Rictor/mTORC2 facilitates central regulation of energy and glucose homeostasis. *Molecular Metabolism, 3*(4), 394–407.

Kolthur-Seetharam, U., Teerds, K., de Rooij, D. G., Wendling, O., McBurney, M., Sassone-Corsi, P., & Davidson, I. (2009). The histone deacetylase SIRT1 controls male fertility in mice through regulation of hypothalamic-pituitary gonadotropin signaling. *Biology of Reproduction, 80*(2), 384–391.

Kong, X., Yu, J., Bi, J., Qi, H., Di, W., Wu, L., Wang, L., Zha, J., Lv, S., Zhang, F., Li, Y., Hu, F., Liu, F., Zhou, H., Liu, J., & Ding, G. (2015). Glucocorticoids transcriptionally regulate miR-27b expression promoting body fat accumulation via suppressing the browning of white adipose tissue. *Diabetes, 64*(2), 393–404.

Korosi, A., & Baram, T. Z. (2008). The central corticotropin releasing factor system during development and adulthood. *European Journal of Pharmacology, 583*(2–3), 204–214.

Kovács, K. J. (2013). CRH: The link between hormonal-, metabolic- and behavioral responses to stress. *Journal of Chemical Neuroanatomy, 54*, 25–33.

Kubota, N., Yano, W., Kubota, T., Yamauchi, T., Itoh, S., Kumagai, H., Kozono, H., Takamoto, I., Okamoto, S., Shiuchi, T., Suzuki, R., Satoh, H., Tsuchida, A., Moroi, M., Sugi, K., Noda, T., Ebinuma, H., Ueta, Y., Kondo, T., Araki, E., Ezaki, O., Nagai, R., Tobe, K., Terauchi, Y., Ueki, K., Minokoshi, Y., & Kadowaki, T. (2007). Adiponectin stimulates AMP-activated protein kinase in the hypothalamus and increases food intake. *Cell Metabolism, 6*(1), 55–68.

Kumar, M. V., Shimokawa, T., Nagy, T. R., & Lane, M. D. (2002). Differential effects of a centrally acting fatty acid synthase inhibitor in lean and obese mice. *Proceedings of the National Academy of Sciences of the United States of America, 99*(4), 1921–1925.

Kumar, A., Lawrence, J. C., Jr., Jung, D. Y., Ko, H. J., Keller, S. R., Kim, J. K., Magnuson, M. A., & Harris, T. E. (2010). Fat cell-specific ablation of rictor in mice impairs insulin-regulated fat cell and whole-body glucose and lipid metabolism. *Diabetes, 59*(6), 1397–1406.

Kurth-Kraczek, E. J., Hirshman, M. F., Goodyear, L. J., & Winder, W. W. (1999). 5' AMP-activated protein kinase activation causes GLUT4 translocation in skeletal muscle. *Diabetes, 48*(8), 1667–1671.

Lafontaine-Lacasse, M., Richard, D., & Picard, F. (2010). Effects of age and gender on Sirt 1 mRNA expressions in the hypothalamus of the mouse. *Neuroscience Letters, 480*(1), 1–3.

Lagerlöf, O., Slocomb, J. E., Hong, I., Aponte, Y., Blackshaw, S., Hart, G. W., & Huganir, R. L. (2016). The nutrient sensor OGT in PVN neurons regulates feeding. *Science, 351*(6279), 1293–1296.

Lagouge, M., Argmann, C., Gerhart-Hines, Z., Meziane, H., Lerin, C., Daussin, F., Messadeq, N., Milne, J., Lambert, P., Elliott, P., Geny, B., Laakso, M., Puigserver, P., & Auwerx, J. (2006). Resveratrol improves mitochondrial function and protects against metabolic disease by activating SIRT1 and PGC-1alpha. *Cell, 127*(6), 1109–1122.

Lamming, D. W., & Sabatini, D. M. (2013). A central role for mTOR in lipid homeostasis. *Cell Metabolism, 18*(4), 465–469.

Lan, F., Cacicedo, J. M., Ruderman, N., & Ido, Y. (2008). SIRT1 modulation of the acetylation status, cytosolic localization, and activity of LKB1. Possible role in AMP-activated protein kinase activation. *The Journal of Biological Chemistry, 283*(41), 27628–27635.

Lane, M. D., Wolfgang, M., Cha, S.-H., & Dai, Y. (2008). Regulation of food intake and energy expenditure by hypothalamic malonyl-CoA. *International Journal of Obesity, 32*(Suppl 4), S49–S54.

Laplante, M., & Sabatini, D. M. (2012). mTOR signaling in growth control and disease. *Cell, 149*(2), 274–293.

Laryea, G., Schütz, G., & Muglia, L. J. (2013). Disrupting hypothalamic glucocorticoid receptors causes HPA axis hyperactivity and excess adiposity. *Molecular Endocrinology, 27*(10), 1655–1665.

Laurent, V., Kimble, A., Peng, B., Zhu, P., Pintar, J. E., Steiner, D. F., & Lindberg, I. (2002). Mortality in 7B2 null mice can be rescued by adrenalectomy: Involvement of dopamine in ACTH hypersecretion. *Proceedings of the National Academy of Sciences of the United States of America, 99*(5), 3087–3092.

Laurent, G., German, N. J., Saha, A. K., de Boer, V. C. J., Davies, M., Koves, T. R., Dephoure, N., Fischer, F., Boanca, G., Vaitheesvaran, B., Lovitch, S. B., Sharpe, A. H., Kurland, I. J., Steegborn, C., Gygi, S. P., Muoio, D. M., Ruderman, N. B., & Haigis, M. C. (2013). SIRT4 coordinates the balance between lipid synthesis and catabolism by repressing malonyl CoA decarboxylase. *Molecular Cell, 50*(5), 686–698.

Lee, K., Li, B., Xi, X., Suh, Y., & Martin, R. J. (2005). Role of neuronal energy status in the regulation of adenosine 5'-monophosphate-activated protein kinase, orexigenic neuropeptides expression, and feeding behavior. *Endocrinology, 146*(1), 3–10.

Lee, J. W., Park, S., Takahashi, Y., & Wang, H.-G. (2010). The association of AMPK with ULK1 regulates autophagy. *PLoS One, 5*(11), e15394.

Lembke, V., Goebel, M., Frommelt, L., Inhoff, T., Lommel, R., Stengel, A., Taché, Y., Grötzinger,

C., Bannert, N., Wiedenmann, B., Klapp, B. F., & Kobelt, P. (2011). Sulfated cholecystokinin-8 activates phospho-mTOR immunoreactive neurons of the paraventricular nucleus in rats. *Peptides, 32*(1), 65–70.

Levin, B. E., Dunn-Meynell, A. A., Balkan, B., & Keesey, R. E. (1997). Selective breeding for diet-induced obesity and resistance in Sprague-Dawley rats. *The American Journal of Physiology, 273*(2 Pt 2), R725–R730.

Lewis, C. A., Griffiths, B., Santos, C. R., Pende, M., & Schulze, A. (2011). Regulation of the SREBP transcription factors by mTORC1. *Biochemical Society Transactions, 39*(2), 495–499.

Li, X., & Gao, T. (2014). mTORC2 phosphorylates protein kinase Cζ to regulate its stability and activity. *EMBO Reports, 15*(2), 191–198.

Libert, S., Pointer, K., Bell, E. L., Das, A., Cohen, D. E., Asara, J. M., Kapur, K., Bergmann, S., Preisig, M., Otowa, T., Kendler, K. S., Chen, X., Hettema, J. M., van den Oord, E. J., Rubio, J. P., & Guarente, L. (2011). SIRT1 activates MAO-A in the brain to mediate anxiety and exploratory drive. *Cell, 147*(7), 1459–1472.

Liu, K., Paterson, A. J., Chin, E., & Kudlow, J. E. (2000). Glucose stimulates protein modification by O-linked GlcNAc in pancreatic beta cells: Linkage of O-linked GlcNAc to beta cell death. *Proceedings of the National Academy of Sciences of the United States of America, 97*(6), 2820–2825.

Liu, Y., Dentin, R., Chen, D., Hedrick, S., Ravnskjaer, K., Schenk, S., Milne, J., Meyers, D. J., Cole, P., Yates, J., 3rd, Olefsky, J., Guarente, L., & Montminy, M. (2008). A fasting inducible switch modulates gluconeogenesis via activator/coactivator exchange. *Nature, 456*(7219), 269–273.

Lizcano, J. M., Göransson, O., Toth, R., Deak, M., Morrice, N. A., Boudeau, J., Hawley, S. A., Udd, L., Mäkelä, T. P., Hardie, D. G., & Alessi, D. R. (2004). LKB1 is a master kinase that activates 13 kinases of the AMPK subfamily, including MARK/PAR-1. *The EMBO Journal, 23*(4), 833–843.

Lloyd, D. J., Bohan, S., & Gekakis, N. (2006). Obesity, hyperphagia and increased metabolic efficiency in Pc1 mutant mice. *Human Molecular Genetics, 15*(11), 1884–1893.

Löffler, A. S., Alers, S., Dieterle, A. M., Keppeler, H., Franz-Wachtel, M., Kundu, M., Campbell, D. G., Wesselborg, S., Alessi, D. R., & Stork, B. (2011). Ulk1-mediated phosphorylation of AMPK constitutes a negative regulatory feedback loop. *Autophagy, 7*(7), 696–706.

Loftus, T. M., Jaworsky, D. E., Frehywot, G. L., Townsend, C. A., Ronnett, G. V., Lane, M. D., & Kuhajda, F. P. (2000). Reduced food intake and body weight in mice treated with fatty acid synthase inhibitors. *Science, 288*(5475), 2379–2381.

López, M., Lage, R., Saha, A. K., Pérez-Tilve, D., Vázquez, M. J., Varela, L., Sangiao-Alvarellos, S., Tovar, S., Raghay, K., Rodríguez-Cuenca, S., Deoliveira, R. M., Castañeda, T., Datta, R., Dong, J. Z., Culler, M., Sleeman, M. W., Alvarez, C. V.,

Gallego, R., Lelliott, C. J., Carling, D., Tschöp, M. H., Diéguez, C., & Vidal-Puig, A. (2008). Hypothalamic fatty acid metabolism mediates the orexigenic action of ghrelin. *Cell Metabolism, 7*(5), 389–399.

Lovejoy, D. A., Chang, B. S. W., Lovejoy, N. R., & del Castillo, J. (2014). Molecular evolution of GPCRs: CRH/CRH receptors. *Journal of Molecular Endocrinology, 52*(3), T43–T60.

Lu, X.-Y., Barsh, G. S., Akil, H., & Watson, S. J. (2003). Interaction between alpha-melanocyte-stimulating hormone and corticotropin-releasing hormone in the regulation of feeding and hypothalamo-pituitary-adrenal responses. *The Journal of Neuroscience, 23*(21), 7863–7872.

Lu, M., Sarruf, D. A., Li, P., Osborn, O., Sanchez-Alavez, M., Talukdar, S., Chen, A., Bandyopadhyay, G., Xu, J., Morinaga, H., Dines, K., Watkins, S., Kaiyala, K., Schwartz, M. W., & Olefsky, J. M. (2013). Neuronal Sirt1 deficiency increases insulin sensitivity in both brain and peripheral tissues. *The Journal of Biological Chemistry, 288*(15), 10722–10735.

MacDonald, H. R., & Wevrick, R. (1997). The necdin gene is deleted in Prader-Willi syndrome and is imprinted in human and mouse. *Human Molecular Genetics, 6*(11), 1873–1878.

Mao, Z., Hine, C., Tian, X., Van Meter, M., Au, M., Vaidya, A., Seluanov, A., & Gorbunova, V. (2011). SIRT6 promotes DNA repair under stress by activating PARP1. *Science, 332*(6036), 1443–1446.

Marsin, A. S., Bertrand, L., Rider, M. H., Deprez, J., Beauloye, C., Vincent, M. F., Van den Berghe, G., Carling, D., & Hue, L. (2000). Phosphorylation and activation of heart PFK-2 by AMPK has a role in the stimulation of glycolysis during ischaemia. *Current Biology, 10*(20), 1247–1255.

Marsin, A.-S., Bouzin, C., Bertrand, L., & Hue, L. (2002). The stimulation of glycolysis by hypoxia in activated monocytes is mediated by AMP-activated protein kinase and inducible 6-phosphofructo-2-kinase. *The Journal of Biological Chemistry, 277*(34), 30778–30783.

Martin, T. L., Alquier, T., Asakura, K., Furukawa, N., Preitner, F., & Kahn, B. B. (2006). Diet-induced obesity alters AMP kinase activity in hypothalamus and skeletal muscle. *The Journal of Biological Chemistry, 281*(28), 18933–18941.

Martínez de Morentin, P. B., González-García, I., Martins, L., Lage, R., Fernández-Mallo, D., Martínez-Sánchez, N., Ruíz-Pino, F., Liu, J., Morgan, D. A., Pinilla, L., Gallego, R., Saha, A. K., Kalsbeek, A., Fliers, E., Bisschop, P. H., Diéguez, C., Nogueiras, R., Rahmouni, K., Tena-Sempere, M., & López, M. (2014). Estradiol regulates brown adipose tissue thermogenesis via hypothalamic AMPK. *Cell Metabolism, 20*(1), 41–53.

Martínez-Sánchez, N., Seoane-Collazo, P., Contreras, C., Varela, L., Villarroya, J., Rial-Pensado, E., Buqué, X., Aurrekoetxea, I., Delgado, T. C., Vázquez-Martínez, R., González-García, I., Roa, J., Whittle, A. J., Gomez-Santos, B., Velagapudi, V., YCL, T., Morgan, D. A., Voshol, P. J., Martínez de Morentin, P. B.,

López-González, T., Liñares-Pose, L., Gonzalez, F., Chatterjee, K., Sobrino, T., Medina-Gómez, G., Davis, R. J., Casals, N., Orešič, M., Coll, A. P., Vidal-Puig, A., Mittag, J., Tena-Sempere, M., Malagón, M. M., Diéguez, C., Martínez-Chantar, M. L., Aspichueta, P., Rahmouni, K., Nogueiras, R., Sabio, G., Villarroya, F., & López, M. (2017). Hypothalamic AMPK-ER stress-JNK1 Axis mediates the central actions of thyroid hormones on energy balance. *Cell Metabolism, 26*(1), 212–229.e12.

Martins, L., Fernández-Mallo, D., Novelle, M. G., Vázquez, M. J., Tena-Sempere, M., Nogueiras, R., López, M., & Diéguez, C. (2012). Hypothalamic mTOR signaling mediates the orexigenic action of ghrelin. *PLoS One, 7*(10), e46923.

Martins, L., Seoane-Collazo, P., Contreras, C., González-García, I., Martínez-Sánchez, N., González, F., Zalvide, J., Gallego, R., Diéguez, C., Nogueiras, R., Tena-Sempere, M., & López, M. (2016). A functional link between AMPK and orexin mediates the effect of BMP8B on energy balance. *Cell Reports, 16*(8), 2231–2242.

Mastorakos, G., & Zapanti, E. (2004). The hypothalamic-pituitary-adrenal axis in the neuroendocrine regulation of food intake and obesity: The role of corticotropin releasing hormone. *Nutritional Neuroscience, 7*(5–6), 271–280.

Meikle, L., Talos, D. M., Onda, H., Pollizzi, K., Rotenberg, A., Sahin, M., Jensen, F. E., & Kwiatkowski, D. J. (2007). A mouse model of tuberous sclerosis: Neuronal loss of Tsc1 causes dysplastic and ectopic neurons, reduced myelination, seizure activity, and limited survival. *The Journal of Neuroscience, 27*(21), 5546–5558.

Menzies, K. J., Zhang, H., Katsyuba, E., & Auwerx, J. (2016). Protein acetylation in metabolism - metabolites and cofactors. *Nature Reviews. Endocrinology, 12*(1), 43–60.

Merrill, G. F., Kurth, E. J., Hardie, D. G., & Winder, W. W. (1997). AICA riboside increases AMP-activated protein kinase, fatty acid oxidation, and glucose uptake in rat muscle. *The American Journal of Physiology, 273*(6 Pt 1), E1107–E1112.

Minokoshi, Y., Alquier, T., Furukawa, N., Kim, Y.-B., Lee, A., Xue, B., Mu, J., Foufelle, F., Ferré, P., Birnbaum, M. J., Stuck, B. J., & Kahn, B. B. (2004). AMP-kinase regulates food intake by responding to hormonal and nutrient signals in the hypothalamus. *Nature, 428*(6982), 569–574.

Moloughney, J. G., Kim, P. K., Vega-Cotto, N. M., Wu, C.-C., Zhang, S., Adlam, M., Lynch, T., Chou, P.-C., Rabinowitz, J. D., Werlen, G., & Jacinto, E. (2016). mTORC2 responds to glutamine catabolite levels to modulate the Hexosamine biosynthesis enzyme GFAT1. *Molecular Cell, 63*(5), 811–826.

Momcilovic, M., Hong, S.-P., & Carlson, M. (2006). Mammalian TAK1 activates Snf1 protein kinase in yeast and phosphorylates AMP-activated protein kinase in vitro. *The Journal of Biological Chemistry, 281*(35), 25336–25343.

Mori, H., Inoki, K., Münzberg, H., Opland, D., Faouzi, M., Villanueva, E. C., Ikenoue, T., Kwiatkowski, D., MacDougald, O. A., Myers, M. G., Jr., & Guan, K.-L. (2009). Critical role for hypothalamic mTOR activity in energy balance. *Cell Metabolism, 9*(4), 362–374.

Morton, G. J., Cummings, D. E., Baskin, D. G., Barsh, G. S., & Schwartz, M. W. (2006). Central nervous system control of food intake and body weight. *Nature, 443*(7109), 289–295.

Moynihan, K. A., Grimm, A. A., Plueger, M. M., Bernal-Mizrachi, E., Ford, E., Cras-Méneur, C., Permutt, M. A., & Imai, S.-I. (2005). Increased dosage of mammalian Sir2 in pancreatic beta cells enhances glucose-stimulated insulin secretion in mice. *Cell Metabolism, 2*(2), 105–117.

Muta, K., Morgan, D. A., & Rahmouni, K. (2015). The role of hypothalamic mTORC1 signaling in insulin regulation of food intake, body weight, and sympathetic nerve activity in male mice. *Endocrinology, 156*(4), 1398–1407.

Naggert, J. K., Fricker, L. D., Varlamov, O., Nishina, P. M., Rouille, Y., Steiner, D. F., Carroll, R. J., Paigen, B. J., & Leiter, E. H. (1995). Hyperproinsulinaemia in obese fat/fat mice associated with a carboxypeptidase E mutation which reduces enzyme activity. *Nature Genetics, 10*(2), 135–142.

Nakahata, Y., Kaluzova, M., Grimaldi, B., Sahar, S., Hirayama, J., Chen, D., Guarente, L. P., & Sassone-Corsi, P. (2008). The NAD+−dependent deacetylase SIRT1 modulates CLOCK-mediated chromatin remodeling and circadian control. *Cell, 134*(2), 329–340.

Nakahata, Y., Sahar, S., Astarita, G., Kaluzova, M., & Sassone-Corsi, P. (2009). Circadian control of the NAD+ salvage pathway by CLOCK-SIRT1. *Science, 324*(5927), 654–657.

Nasrin, N., Kaushik, V. K., Fortier, E., Wall, D., Pearson, K. J., de Cabo, R., & Bordone, L. (2009). JNK1 phosphorylates SIRT1 and promotes its enzymatic activity. *PLoS One, 4*(12), e8414.

Nie, Y., Erion, D. M., Yuan, Z., Dietrich, M., Shulman, G. I., Horvath, T. L., & Gao, Q. (2009). STAT3 inhibition of gluconeogenesis is downregulated by SirT1. *Nature Cell Biology, 11*(4), 492–500.

Nillni, E. A. (2010). Regulation of the hypothalamic thyrotropin releasing hormone (TRH) neuron by neuronal and peripheral inputs. *Frontiers in Neuroendocrinology, 31*(2), 134–156.

Nillni - Endocrinology EA. (2007). Regulation of prohormone convertases in hypothalamic neurons: implications for prothyrotropin-releasing hormone and proopiomelanocortin. press.endocrine.org 2007. Available at: http://press.endocrine.org/doi/abs/10.1210/en.2007-0173.

Nillni, E. A., Friedman, T. C., Todd, R. B., Birch, N. P., Loh, Y. P., & Jackson, I. M. (1995). Pro-thyrotropin-releasing hormone processing by recombinant PC1. *Journal of Neurochemistry, 65*(6), 2462–2472.

Nillni, E. A., Xie, W., Mulcahy, L., Sanchez, V. C., & Wetsel, W. C. (2002). Deficiencies in pro-thyrotropin-releasing hormone processing and abnormalities in thermoregulation in Cpefat/fat mice. *The Journal of Biological Chemistry, 277*(50), 48587–48595.

Oberdoerffer, P., Michan, S., McVay, M., Mostoslavsky, R., Vann, J., Park, S.-K., Hartlerode, A., Stegmuller, J., Hafner, A., Loerch, P., Wright, S. M., Mills, K. D., Bonni, A., Yankner, B. A., Scully, R., Prolla, T. A., Alt, F. W., & Sinclair, D. A. (2008). SIRT1 redistribution on chromatin promotes genomic stability but alters gene expression during aging. *Cell, 135*(5), 907–918.

Obici, S., Feng, Z., Morgan, K., Stein, D., Karkanias, G., & Rossetti, L. (2002). Central administration of oleic acid inhibits glucose production and food intake. *Diabetes, 51*(2), 271–275.

Oh, T. S., Cho, H., Cho, J. H., Yu, S.-W., & Kim, E.-K. (2016). Hypothalamic AMPK-induced autophagy increases food intake by regulating NPY and POMC expression. *Autophagy, 12*(11), 2009–2025.

Orozco-Solis, R., & Sassone-Corsi, P. (2014). Epigenetic control and the circadian clock: Linking metabolism to neuronal responses. *Neuroscience, 264*, 76–87.

Orozco-Solis, R., Ramadori, G., Coppari, R., & Sassone-Corsi, P. (2015). SIRT1 relays nutritional inputs to the circadian clock through the Sf1 neurons of the ventromedial hypothalamus. *Endocrinology, 156*(6), 2174–2184.

Ortega-Molina, A., Lopez-Guadamillas, E., Mattison, J. A., Mitchell, S. J., Muñoz-Martin, M., Iglesias, G., Gutierrez, V. M., Vaughan, K. L., Szarowicz, M. D., González-García, I., López, M., Cebrián, D., Martinez, S., Pastor, J., de Cabo, R., & Serrano, M. (2015). Pharmacological inhibition of PI3K reduces adiposity and metabolic syndrome in obese mice and rhesus monkeys. *Cell Metabolism, 21*(4), 558–570.

Paquet, L., Massie, B., & Mains, R. E. (1996). Proneuropeptide Y processing in large dense-core vesicles: Manipulation of prohormone convertase expression in sympathetic neurons using adenoviruses. *The Journal of Neuroscience, 16*(3), 964–973.

Park, H.-K., & Ahima, R. S. (2014). Leptin signaling. *F1000prime Reports, 6*, 73.

Peek, C. B., Ramsey, K. M., Marcheva, B., & Bass, J. (2012). Nutrient sensing and the circadian clock. *Trends in Endocrinology and Metabolism, 23*(7), 312–318.

Perello, M., Stuart, R. C., & Nillni, E. A. (2007). Differential effects of fasting and leptin on proopiomelanocortin peptides in the arcuate nucleus and in the nucleus of the solitary tract. *American Journal of Physiology. Endocrinology and Metabolism, 292*(5), E1348–E1357.

Perone, M. J., Murray, C. A., Brown, O. A., Gibson, S., White, A., Linton, E. A., Perkins, A. V., Lowenstein, P. R., & Castro, M. G. (1998). Procorticotrophin-releasing hormone: Endoproteolytic processing and differential release of its derived peptides within AtT20 cells. *Molecular and Cellular Endocrinology, 142*(1–2), 191–202.

Pfister, J. A., Ma, C., Morrison, B. E., & D'Mello, S. R. (2008). Opposing effects of sirtuins on neuronal survival: SIRT1-mediated neuroprotection is independent of its deacetylase activity. *PLoS One, 3*(12), e4090.

Picard, F., Kurtev, M., Chung, N., Topark-Ngarm, A., Senawong, T., Machado De Oliveira, R., Leid, M., McBurney, M. W., & Guarente, L. (2004). Sirt1 promotes fat mobilization in white adipocytes by repressing PPAR-gamma. *Nature, 429*(6993), 771–776.

Plum, L., Lin, H. V., Dutia, R., Tanaka, J., Aizawa, K. S., Matsumoto, M., Kim, A. J., Cawley, N. X., Paik, J.-H., Loh, Y. P., DePinho, R. A., Wardlaw, S. L., & Accili, D. (2009). The obesity susceptibility gene Cpe links FoxO1 signaling in hypothalamic pro-opiomelanocortin neurons with regulation of food intake. *Nature Medicine, 15*(10), 1195–1201.

Posner, S. F., Vaslet, C. A., Jurofcik, M., Lee, A., Seidah, N. G., & Nillni, E. A. (2004). Stepwise posttranslational processing of progrowth hormone-releasing hormone (proGHRH) polypeptide by furin and PC1. *Endocrine, 23*(2–3), 199–213.

Price, N. L., Gomes, A. P., Ling, A. J. Y., Duarte, F. V., Martin-Montalvo, A., North, B. J., Agarwal, B., Ye, L., Ramadori, G., Teodoro, J. S., Hubbard, B. P., Varela, A. T., Davis, J. G., Varamini, B., Hafner, A., Moaddel, R., Rolo, A. P., Coppari, R., Palmeira, C. M., de Cabo, R., Baur, J. A., & Sinclair, D. A. (2012). SIRT1 is required for AMPK activation and the beneficial effects of resveratrol on mitochondrial function. *Cell Metabolism, 15*(5), 675–690.

Proulx, K., Cota, D., Woods, S. C., & Seeley, R. J. (2008). Fatty acid synthase inhibitors modulate energy balance via mammalian target of rapamycin complex 1 signaling in the central nervous system. *Diabetes, 57*(12), 3231–3238.

Prozorovski, T., Schulze-Topphoff, U., Glumm, R., Baumgart, J., Schröter, F., Ninnemann, O., Siegert, E., Bendix, I., Brüstle, O., Nitsch, R., Zipp, F., & Aktas, O. (2008). Sirt1 contributes critically to the redox-dependent fate of neural progenitors. *Nature Cell Biology, 10*(4), 385–394.

Pu, L. P., Ma, W., Barker, J. L., & Loh - Endocrinology YP. (1996). Differential coexpression of genes encoding prothyrotropin-releasing hormone (pro-TRH) and prohormone convertases (PC1 and PC2) in rat brain neurons: …. academic.oup.com 1996. Available at: https://academic.oup.com/endo/article-abstract/137/4/1233/3037181.

Qin, W., Yang, T., Ho, L., Zhao, Z., Wang, J., Chen, L., Zhao, W., Thiyagarajan, M., MacGrogan, D., Rodgers, J. T., Puigserver, P., Sadoshima, J., Deng, H., Pedrini, S., Gandy, S., Sauve, A. A., & Pasinetti, G. M. (2006). Neuronal SIRT1 activation as a novel mechanism underlying the prevention of Alzheimer disease amyloid neuropathology by calorie restriction. *The Journal of Biological Chemistry, 281*(31), 21745–21754.

Quan, W., Kim, H.-K., Moon, E.-Y., Kim, S. S., Choi, C. S., Komatsu, M., Jeong, Y. T., Lee, M.-K., Kim, K.-W., Kim, M.-S., & Lee, M.-S. (2012). Role of hypothalamic proopiomelanocortin neuron autophagy in the control of appetite and leptin response. *Endocrinology, 153*(4), 1817–1826.

Quiñones, M., Al-Massadi, O., Gallego, R., Fernø, J., Diéguez, C., López, M., & Nogueiras, R. (2015).

Hypothalamic CaMKKβ mediates glucagon anorectic effect and its diet-induced resistance. *Molecular Metabolism, 4*(12), 961–970.

Raadsheer, F. C., Sluiter, A. A., Ravid, R., Tilders, F. J. H., & Swaab, D. F. (1993). Localization of corticotropin-releasing hormone (CRH) neurons in the paraventricular nucleus of the human hypothalamus; age-dependent colocalization with vasopressin. *Brain Research, 615*(1), 50–62.

Rafalski, V. A., Ho, P. P., Brett, J. O., Ucar, D., Dugas, J. C., Pollina, E. A., Chow, L. M. L., Ibrahim, A., Baker, S. J., Barres, B. A., Steinman, L., & Brunet, A. (2013). Expansion of oligodendrocyte progenitor cells following SIRT1 inactivation in the adult brain. *Nature Cell Biology, 15*(6), 614–624.

Ramadori, G., Lee, C. E., Bookout, A. L., Lee, S., Williams, K. W., Anderson, J., Elmquist, J. K., & Coppari, R. (2008). Brain SIRT1: Anatomical distribution and regulation by energy availability. *The Journal of Neuroscience, 28*(40), 9989–9996.

Ramadori, G., Fujikawa, T., Fukuda, M., Anderson, J., Morgan, D. A., Mostoslavsky, R., Stuart, R. C., Perello, M., Vianna, C. R., Nillni, E. A., Rahmouni, K., & Coppari, R. (2010). SIRT1 deacetylase in POMC neurons is required for homeostatic defenses against diet-induced obesity. *Cell Metabolism, 12*(1), 78–87.

Ramadori, G., Fujikawa, T., Anderson, J., Berglund, E. D., Frazao, R., Michán, S., Vianna, C. R., Sinclair, D. A., Elias, C. F., & Coppari, R. (2011). SIRT1 deacetylase in SF1 neurons protects against metabolic imbalance. *Cell Metabolism, 14*(3), 301–312.

Ramsey, K. M., Yoshino, J., Brace, C. S., Abrassart, D., Kobayashi, Y., Marcheva, B., Hong, H.-K., Chong, J. L., Buhr, E. D., Lee, C., Takahashi, J. S., Imai, S.-I., & Bass, J. (2009). Circadian clock feedback cycle through NAMPT-mediated NAD+ biosynthesis. *Science, 324*(5927), 651–654.

Rebsamen, M., Pochini, L., Stasyk, T., de Araújo, M. E. G., Galluccio, M., Kandasamy, R. K., Snijder, B., Fauster, A., Rudashevskaya, E. L., Bruckner, M., Scorzoni, S., Filipek, P. A., Huber, K. V. M., Bigenzahn, J. W., Heinz, L. X., Kraft, C., Bennett, K. L., Indiveri, C., Huber, L. A., & Superti-Furga, G. (2015). SLC38A9 is a component of the lysosomal amino acid sensing machinery that controls mTORC1. *Nature, 519*(7544), 477–481.

Ren, H., Orozco, I. J., Su, Y., Suyama, S., Gutiérrez-Juárez, R., Horvath, T. L., Wardlaw, S. L., Plum, L., Arancio, O., & Accili, D. (2012). FoxO1 target Gpr17 activates AgRP neurons to regulate food intake. *Cell, 149*(6), 1314–1326.

Revollo, J. R., Grimm, A. A., & Imai, S.-I. (2004). The NAD biosynthesis pathway mediated by nicotinamide phosphoribosyltransferase regulates Sir2 activity in mammalian cells. *The Journal of Biological Chemistry, 279*(49), 50754–50763.

Rho, J. H., & Swanson, L. W. (1987). Neuroendocrine CRF motoneurons: Intrahypothalamic axon terminals shown with a new retrograde-Lucifer-immuno method. *Brain Research, 436*(1), 143–147.

Richard, D., Huang, Q., & Timofeeva, E. (2000). The corticotropin-releasing hormone system in the regulation of energy balance in obesity. *International Journal of Obesity, 24*, S36–S39.

Rivier, J., Rivier, C., Spiess, J., & Vale, W. (1983). High-performance liquid chromatographic purification of peptide hormones: Ovine hypothalamic Amunine (Corticotropin releasing factor)1. In M. T. W. Hearn, F. E. Regnier, & C. T. Wehr (Eds.), *High-performance liquid chromatography of proteins and peptides* (pp. 233–241). Academic Press. Cambridge, Massachusetts

Rodgers, J. T., Lerin, C., Haas, W., Gygi, S. P., Spiegelman, B. M., & Puigserver, P. (2005). Nutrient control of glucose homeostasis through a complex of PGC-1alpha and SIRT1. *Nature, 434*(7029), 113–118.

Ropelle, E. R., Fernandes, M. F. A., Flores, M. B. S., Ueno, M., Rocco, S., Marin, R., Cintra, D. E., Velloso, L. A., Franchini, K. G., Saad, M. J. A., & Carvalheira, J. B. C. (2008). Central exercise action increases the AMPK and mTOR response to leptin. *PLoS One, 3*(12), e3856.

Roth, S. Y., Denu, J. M., & Allis, C. D. (2001). Histone acetyltransferases. *Annual Review of Biochemistry, 70*, 81–120.

Rougon, G., Nédélec, J., Malapert, P., Goridis, C., & Chesselet, M. F. (1990). Post-translation modifications of neural cell surface molecules. *Acta Histochemica. Supplementband, 38*, 51–57.

Rouillé, Y., Duguay, S. J., Lund, K., Furuta, M., Gong, Q., Lipkind, G., Oliva, A. A., Jr., Chan, S. J., & Steiner, D. F. (1995). Proteolytic processing mechanisms in the biosynthesis of neuroendocrine peptides: The subtilisin-like proprotein convertases. *Frontiers in Neuroendocrinology, 16*(4), 322–361.

Ruan, H.-B., Dietrich, M. O., Liu, Z.-W., Zimmer, M. R., Li, M.-D., Singh, J. P., Zhang, K., Yin, R., Wu, J., Horvath, T. L., & Yang, X. (2014). O-GlcNAc transferase enables AgRP neurons to suppress browning of white fat. *Cell, 159*(2), 306–317.

Sakamoto, J., Miura, T., Shimamoto, K., & Horio, Y. (2004). Predominant expression of Sir2alpha, an NAD-dependent histone deacetylase, in the embryonic mouse heart and brain. *FEBS Letters, 556*(1–3), 281–286.

Sakurai, T., Amemiya, A., Ishii, M., Matsuzaki, I., Chemelli, R. M., Tanaka, H., Williams, S. C., Richardson, J. A., Kozlowski, G. P., Wilson, S., Arch, J. R., Buckingham, R. E., Haynes, A. C., Carr, S. A., Annan, R. S., McNulty, D. E., Liu, W. S., Terrett, J. A., Elshourbagy, N. A., Bergsma, D. J., & Yanagisawa, M. (1998). Orexins and orexin receptors: A family of hypothalamic neuropeptides and G protein-coupled receptors that regulate feeding behavior. *Cell, 92*(4), 573–585.

Sancak, Y., Thoreen, C. C., Peterson, T. R., Lindquist, R. A., Kang, S. A., Spooner, E., Carr, S. A., & Sabatini, D. M. (2007). PRAS40 is an insulin-regulated inhibitor of the mTORC1 protein kinase. *Molecular Cell, 25*(6), 903–915.

Sancak, Y., Peterson, T. R., Shaul, Y. D., Lindquist, R. A., Thoreen, C. C., Bar-Peled, L., & Sabatini, D. M. (2008). The rag GTPases bind raptor and mediate amino acid signaling to mTORC1. *Science, 320*(5882), 1496–1501.

Sancak, Y., Bar-Peled, L., Zoncu, R., Markhard, A. L., Nada, S., & Sabatini, D. M. (2010). Ragulator-rag complex targets mTORC1 to the lysosomal surface and is necessary for its activation by amino acids. *Cell, 141*(2), 290–303.

Sanders, M. J., Grondin, P. O., Hegarty, B. D., Snowden, M. A., & Carling, D. (2007). Investigating the mechanism for AMP activation of the AMP-activated protein kinase cascade. *The Biochemical Journal, 403*(1), 139–148.

Sarbassov, D. D., Ali, S. M., Kim, D.-H., Guertin, D. A., Latek, R. R., Erdjument-Bromage, H., Tempst, P., & Sabatini, D. M. (2004). Rictor, a novel binding partner of mTOR, defines a rapamycin-insensitive and raptor-independent pathway that regulates the cytoskeleton. *Current Biology, 14*(14), 1296–1302.

Sarbassov, D. D., Guertin, D. A., Ali, S. M., & Sabatini, D. M. (2005). Phosphorylation and regulation of Akt/PKB by the rictor-mTOR complex. *Science, 307*(5712), 1098–1101.

Sasaki, T., Maier, B., Koclega, K. D., Chruszcz, M., Gluba, W., Stukenberg, P. T., Minor, W., & Scrable, H. (2008). Phosphorylation regulates SIRT1 function. *PLoS One, 3*(12), e4020.

Sasaki, T., Kim, H.-J., Kobayashi, M., Kitamura, Y.-I., Yokota-Hashimoto, H., Shiuchi, T., Minokoshi, Y., & Kitamura, T. (2010). Induction of hypothalamic Sirt1 leads to cessation of feeding via agouti-related peptide. *Endocrinology, 151*(6), 2556–2566.

Sasaki, T., Kikuchi, O., Shimpuku, M., Susanti, V. Y., Yokota-Hashimoto, H., Taguchi, R., Shibusawa, N., Sato, T., Tang, L., Amano, K., Kitazumi, T., Kuroko, M., Fujita, Y., Maruyama, J., Lee, Y.-S., Kobayashi, M., Nakagawa, T., Minokoshi, Y., Harada, A., Yamada, M., & Kitamura, T. (2014). Hypothalamic SIRT1 prevents age-associated weight gain by improving leptin sensitivity in mice. *Diabetologia, 57*(4), 819–831.

Satoh, A., Brace, C. S., Ben-Josef, G., West, T., Wozniak, D. F., Holtzman, D. M., Herzog, E. D., & Imai, S.-I. (2010). SIRT1 promotes the central adaptive response to diet restriction through activation of the dorsomedial and lateral nuclei of the hypothalamus. *The Journal of Neuroscience, 30*(30), 10220–10232.

Satoh, A., Brace, C. S., Rensing, N., Cliften, P., Wozniak, D. F., Herzog, E. D., Yamada, K. A., & Imai, S.-I. (2013). Sirt1 extends life span and delays aging in mice through the regulation of Nk2 homeobox 1 in the DMH and LH. *Cell Metabolism, 18*(3), 416–430.

Saxton, R. A., & Sabatini, D. M. (2017). mTOR signaling in growth, metabolism, and disease. *Cell, 168*(6), 960–976.

Schafer, M. K., Day, R., Cullinan, W. E., Chretien, M., Seidah, N. G., & Watson, S. J. (1993). Gene expression of prohormone and proprotein convertases in the rat CNS: A comparative in situ hybridization analysis. *The Journal of Neuroscience, 13*(3), 1258–1279.

Schaner, P., Todd, R. B., Seidah, N. G., & Nillni, E. A. (1997). Processing of prothyrotropin-releasing hormone by the family of prohormone convertases. *The Journal of Biological Chemistry, 272*(32), 19958–19968.

Seidah, N. G., Gaspar, L., & Mion - DNA and cell … P. (1990). cDNA sequence of two distinct pituitary proteins homologous to Kex2 and furin gene products: tissue-specific mRNAs encoding candidates for prohormone …. online.liebertpub.com 1990. Available at: http://online.liebertpub.com/doi/abs/10.1089/dna.1990.9.415.

Seidah, N. G., Marcinkiewicz, M., & Benjannet - Molecular … S. (1991). Cloning and primary sequence of a mouse candidate prohormone convertase PC1 homologous to PC2, Furin, and Kex2: distinct chromosomal localization and …. academic.oup.com 1991. Available at: https://academic.oup.com/mend/article-abstract/5/1/111/2714242.

Seidah, N. G., Day, R., Hamelin, J., Gaspar, A., Collard, M. W., & Chrétien, M. (1992). Testicular expression of PC4 in the rat: Molecular diversity of a novel germ cell-specific Kex2/subtilisin-like proprotein convertase. *Molecular Endocrinology, 6*(10), 1559–1570.

Seidah, N. G., Chrétien, M., & Day, R. (1994). The family of subtilisin/kexin like pro-protein and pro-hormone convertases: Divergent or shared functions. *Biochimie, 76*(3–4), 197–209.

Seidah, N. G., Hamelin, J., Mamarbachi, M., Dong, W., Tardos, H., Mbikay, M., Chretien, M., & Day, R. (1996). cDNA structure, tissue distribution, and chromosomal localization of rat PC7, a novel mammalian proprotein convertase closest to yeast kexin-like proteinases. *Proceedings of the National Academy of Sciences of the United States of America, 93*(8), 3388–3393.

Seimon, R. V., Hostland, N., Silveira, S. L., Gibson, A. A., & Sainsbury, A. (2013). Effects of energy restriction on activity of the hypothalamo-pituitary-adrenal axis in obese humans and rodents: Implications for diet-induced changes in body composition. |*Hormone Molecular Biology and Clinical Investigation, 15*(2), 71–80.

Seto, E., & Yoshida, M. (2014). Erasers of histone acetylation: The histone deacetylase enzymes. *Cold Spring Harbor Perspectives in Biology, 6*(4), a018713.

Shah, Z. H., Ahmed, S. U., Ford, J. R., Allison, S. J., Knight, J. R. P., & Milner, J. (2012). A deacetylase-deficient SIRT1 variant opposes full-length SIRT1 in regulating tumor suppressor p53 and governs expression of cancer-related genes. *Molecular and Cellular Biology, 32*(3), 704–716.

Shimizu, H., Arima, H., Ozawa, Y., Watanabe, M., Banno, R., Sugimura, Y., Ozaki, N., Nagasaki, H., & Oiso, Y. (2010). Glucocorticoids increase NPY gene expression in the arcuate nucleus by inhibiting mTOR signaling in rat hypothalamic organotypic cultures. *Peptides, 31*(1), 145–149.

Sinclair, D. A., & Guarente, L. (2014). Small-molecule allosteric activators of sirtuins. *Annual Review of Pharmacology and Toxicology, 54*, 363–380.

Smeekens, S. P., & Steiner, D. F. (1990). Identification of a human insulinoma cDNA encoding a novel mammalian protein structurally related to the yeast dibasic processing protease Kex2. *The Journal of Biological Chemistry, 265*(6), 2997–3000.

Smeekens, S. P., Avruch, A. S., LaMendola, J., Chan, S. J., & Steiner, D. F. (1991). Identification of a cDNA encoding a second putative prohormone convertase related to PC2 in AtT20 cells and islets of Langerhans. *Proceedings of the National Academy of Sciences of the United States of America, 88*(2), 340–344.

Smeekens, S. P., Montag, A. G., Thomas, G., Albiges-Rizo, C., Carroll, R., Benig, M., Phillips, L. A., Martin, S., Ohagi, S., & Gardner, P. (1992). Proinsulin processing by the subtilisin-related proprotein convertases furin, PC2, and PC3. *Proceedings of the National Academy of Sciences of the United States of America, 89*(18), 8822–8826.

Smith, T. L., & Rutter, J. (2007). Regulation of glucose partitioning by PAS kinase and Ugp1 phosphorylation. *Molecular Cell, 26*(4), 491–499.

Sohn, J.-W., Oh, Y., Kim, K. W., Lee, S., Williams, K. W., & Elmquist, J. K. (2016). Leptin and insulin engage specific PI3K subunits in hypothalamic SF1 neurons. *Molecular Metabolism, 5*(8), 669–679.

Solomon, S. (1999). POMC-derived peptides and their biological action. *Annals of the New York Academy of Sciences, 885*, 22–40.

Sominsky, L., & Spencer, S. J. (2014). Eating behavior and stress: A pathway to obesity. *Frontiers in Psychology, 5*, 434.

Spencer, S. J., & Tilbrook, A. (2011). The glucocorticoid contribution to obesity. *Stress, 14*(3), 233–246.

Starheim, K. K., Gevaert, K., & Arnesen, T. (2012). Protein N-terminal acetyltransferases: When the start matters. *Trends in Biochemical Sciences, 37*(4), 152–161.

Steinberg, G. R., Watt, M. J., Fam, B. C., Proietto, J., Andrikopoulos, S., Allen, A. M., Febbraio, M. A., & Kemp, B. E. (2006). Ciliary neurotrophic factor suppresses hypothalamic AMP-kinase signaling in leptin-resistant obese mice. *Endocrinology, 147*(8), 3906–3914.

Steiner, D. F. (1998). The proprotein convertases. *Current Opinion in Chemical Biology, 2*(1), 31–39.

Steiner, D. F., Smeekens, S. P., Ohagi, S., & Chan, S. J. (1992). The new enzymology of precursor processing endoproteases. *The Journal of Biological Chemistry, 267*(33), 23435–23438.

Stevanovic, D., Trajkovic, V., Müller-Lühlhoff, S., Brandt, E., Abplanalp, W., Bumke-Vogt, C., Liehl, B., Wiedmer, P., Janjetovic, K., Starcevic, V., Pfeiffer, A. F. H., Al-Hasani, H., Tschöp, M. H., & Castañeda, T. R. (2013). Ghrelin-induced food intake and adiposity depend on central mTORC1/S6K1 signaling. *Molecular and Cellular Endocrinology, 381*(1–2), 280–290.

Stoppa, G. R., Cesquini, M., Roman, E. A., Prada, P. O., Torsoni, A. S., Romanatto, T., Saad, M. J., Velloso, L. A., & Torsoni, M. A. (2008). Intracerebroventricular injection of citrate inhibits hypothalamic AMPK and modulates feeding behavior and peripheral insulin signaling. *The Journal of Endocrinology, 198*(1), 157–168.

Sullivan, J. E., Brocklehurst, K. J., Marley, A. E., Carey, F., Carling, D., & Beri, R. K. (1994). Inhibition of lipolysis and lipogenesis in isolated rat adipocytes with AICAR, a cell-permeable activator of AMP-activated protein kinase. *FEBS Letters, 353*(1), 33–36.

Susanti, V. Y., Sasaki, T., Yokota-Hashimoto, H., Matsui, S., Lee, Y.-S., Kikuchi, O., Shimpuku, M., Kim, H.-J., Kobayashi, M., & Kitamura, T. (2014). Sirt1 rescues the obesity induced by insulin-resistant constitutively-nuclear FoxO1 in POMC neurons of male mice. *Obesity, 22*(10), 2115–2119.

Suter, M., Riek, U., Tuerk, R., Schlattner, U., Wallimann, T., & Neumann, D. (2006). Dissecting the role of 5'-AMP for allosteric stimulation, activation, and deactivation of AMP-activated protein kinase. *The Journal of Biological Chemistry, 281*(43), 32207–32216.

Taleux, N., De Potter, I., Deransart, C., Lacraz, G., Favier, R., Leverve, X. M., Hue, L., & Guigas, B. (2008). Lack of starvation-induced activation of AMP-activated protein kinase in the hypothalamus of the Lou/C rats resistant to obesity. *International Journal of Obesity, 32*(4), 639–647.

Tan, M., Peng, C., Anderson, K. A., Chhoy, P., Xie, Z., Dai, L., Park, J., Chen, Y., Huang, H., Zhang, Y., Ro, J., Wagner, G. R., Green, M. F., Madsen, A. S., Schmiesing, J., Peterson, B. S., Xu, G., Ilkayeva, O. R., Muehlbauer, M. J., Braulke, T., Mühlhausen, C., Backos, D. S., Olsen, C. A., McGuire, P. J., Pletcher, S. D., Lombard, D. B., Hirschey, M. D., & Zhao, Y. (2014). Lysine glutarylation is a protein posttranslational modification regulated by SIRT5. *Cell Metabolism, 19*(4), 605–617.

Tanida, M., Yamamoto, N., Shibamoto, T., & Rahmouni, K. (2013). Involvement of hypothalamic AMP-activated protein kinase in leptin-induced sympathetic nerve activation. *PLoS One, 8*(2), e56660.

Tanida, M., Yamamoto, N., Morgan, D. A., Kurata, Y., Shibamoto, T., & Rahmouni, K. (2015). Leptin receptor signaling in the hypothalamus regulates hepatic autonomic nerve activity via phosphatidylinositol 3-kinase and AMP-activated protein kinase. *The Journal of Neuroscience, 35*(2), 474–484.

Tanno, M., Sakamoto, J., Miura, T., Shimamoto, K., & Horio, Y. (2007). Nucleocytoplasmic shuttling of the NAD+−dependent histone deacetylase SIRT1. *The Journal of Biological Chemistry, 282*(9), 6823–6832.

Tataranni, P. A., Larson, D. E., Snitker, S., Young, J. B., Flatt, J. P., & Ravussin, E. (1996). Effects of glucocorticoids on energy metabolism and food intake in humans. *The American Journal of Physiology, 271*(2 Pt 1), E317–E325.

Thomas, L., Leduc, R., & Thorne - Proceedings of the … BA. (1991). Kex2-like endoproteases PC2 and PC3 accurately cleave a model prohormone in mammalian cells: evidence for a common core of neuroendocrine processing …. National Acad Sciences 1991. Available at: http://www.pnas.org/content/88/12/5297.short.

Tong, L., & Denu, J. M. (2010). Function and metabolism of sirtuin metabolite O-acetyl-ADP-ribose. *Biochimica et Biophysica Acta, 1804*(8), 1617–1625.

Toorie, A. M., & Nillni, E. A. (2014). Minireview: Central Sirt1 regulates energy balance via the melanocortin system and alternate pathways. *Molecular Endocrinology, 28*(9), 1423–1434.

Toorie, A. M., Cyr, N. E., Steger, J. S., Beckman, R., Farah, G., & Nillni, E. A. (2016). The nutrient and energy sensor Sirt1 regulates the hypothalamic-pituitary-adrenal (HPA) Axis by altering the production of the prohormone convertase 2 (PC2) essential in the maturation of Corticotropin-releasing hormone (CRH) from its prohormone in male rats. *The Journal of Biological Chemistry, 291*(11), 5844–5859.

Toriya, M., Maekawa, F., Maejima, Y., Onaka, T., Fujiwara, K., Nakagawa, T., Nakata, M., & Yada, T. (2010). Long-term infusion of brain-derived neurotrophic factor reduces food intake and body weight via a corticotrophin-releasing hormone pathway in the paraventricular nucleus of the hypothalamus. *Journal of Neuroendocrinology, 22*(9), 987–995.

Vale, W., Spiess, J., Rivier, C., & Rivier, J. (1981a). Characterization of a 41-residue ovine hypothalamic peptide that stimulates secretion of corticotropin and beta-endorphin. *Science, 213*(4514), 1394–1397.

Vale, W., Spiess, J., Rivier, C., & Rivier, J. (1981b). Characterization of a 41-residue ovine hypothalamic peptide that stimulates secretion of Corticotropin and β-endorphin. *Science, 213*(4514), 1394–1397.

Vander Haar, E., Lee, S.-I., Bandhakavi, S., Griffin, T. J., & Kim, D.-H. (2007). Insulin signalling to mTOR mediated by the Akt/PKB substrate PRAS40. *Nature Cell Biology, 9*(3), 316–323.

Varela, L., Martínez-Sánchez, N., Gallego, R., Vázquez, M. J., Roa, J., Gándara, M., Schoenmakers, E., Nogueiras, R., Chatterjee, K., Tena-Sempere, M., Diéguez, C., & López, M. (2012). Hypothalamic mTOR pathway mediates thyroid hormone-induced hyperphagia in hyperthyroidism. *The Journal of Pathology, 227*(2), 209–222.

Velásquez, D. A., Martínez, G., Romero, A., Vázquez, M. J., Boit, K. D., Dopeso-Reyes, I. G., López, M., Vidal, A., Nogueiras, R., & Diéguez, C. (2011). The central Sirtuin 1/p53 pathway is essential for the orexigenic action of ghrelin. *Diabetes, 60*(4), 1177–1185.

Veo, K., Reinick, C., Liang, L., Moser, E., Angleson, J. K., & Dores, R. M. (2011). Observations on the ligand selectivity of the melanocortin 2 receptor. *General and Comparative Endocrinology, 172*(1), 3–9.

Villanueva, E. C., Münzberg, H., Cota, D., Leshan, R. L., Kopp, K., Ishida-Takahashi, R., Jones, J. C., Fingar, D. C., Seeley, R. J., & Myers, M. G., Jr. (2009). Complex regulation of mammalian target of rapamycin complex 1 in the basomedial hypothalamus by leptin and nutritional status. *Endocrinology, 150*(10), 4541–4551.

Villeneuve, P., Seidah, N. G., & Beaudet, A. (2000). Immunohistochemical evidence for the implication of PCI in the processing of proneurotensin in rat brain. *Neuroreport, 11*(16), 3443.

Wang, R., Cherukuri, P., & Luo, J. (2005). Activation of Stat3 sequence-specific DNA binding and transcription by p300/CREB-binding protein-mediated acetylation. *The Journal of Biological Chemistry, 280*(12), 11528–11534.

Wang, L., Harris, T. E., Roth, R. A., & Lawrence, J. C., Jr. (2007). PRAS40 regulates mTORC1 kinase activity by functioning as a direct inhibitor of substrate binding. *The Journal of Biological Chemistry, 282*(27), 20036–20044.

Wang, S., Tsun, Z.-Y., Wolfson, R. L., Shen, K., Wyant, G. A., Plovanich, M. E., Yuan, E. D., Jones, T. D., Chantranupong, L., Comb, W., Wang, T., Bar-Peled, L., Zoncu, R., Straub, C., Kim, C., Park, J., Sabatini, B. L., & Sabatini, D. M. (2015). Metabolism. Lysosomal amino acid transporter SLC38A9 signals arginine sufficiency to mTORC1. *Science, 347*(6218), 188–194.

Wardlaw, S. L. (2011). Hypothalamic proopiomelanocortin processing and the regulation of energy balance. *European Journal of Pharmacology, 660*(1), 213–219.

Wellen, K. E., Hatzivassiliou, G., Sachdeva, U. M., Bui, T. V., Cross, J. R., & Thompson, C. B. (2009). ATP-citrate lyase links cellular metabolism to histone acetylation. *Science, 324*(5930), 1076–1080.

Westphal, C. H., Muller, L., Zhou, A., Zhu, X., Bonner-Weir, S., Schambelan, M., Steiner, D. F., Lindberg, I., & Leder, P. (1999). The neuroendocrine protein 7B2 is required for peptide hormone processing in vivo and provides a novel mechanism for pituitary Cushing's disease. *Cell, 96*(5), 689–700.

Williams, G., Bing, C., Cai, X. J., Harrold, J. A., King, P. J., & Liu, X. H. (2001). The hypothalamus and the control of energy homeostasis: Different circuits, different purposes. *Physiology & Behavior, 74*(4–5), 683–701.

Williams, K. W., Sohn, J.-W., Donato, J., Jr., Lee, C. E., Zhao, J. J., Elmquist, J. K., & Elias, C. F. (2011). The acute effects of leptin require PI3K signaling in the hypothalamic ventral premammillary nucleus. *The Journal of Neuroscience, 31*(37), 13147–13156.

Winder, W. W., & Hardie, D. G. (1996). Inactivation of acetyl-CoA carboxylase and activation of AMP-activated protein kinase in muscle during exercise. *The American Journal of Physiology, 270*(2 Pt 1), E299–E304.

Winsky-Sommerer, R., Benjannet, S., Rovère, C., Barbero, P., Seidah, N. G., Epelbaum, J., & Dournaud, P. (2000). Regional and cellular localization of the neuroendocrine prohormone convertases PC1 and PC2 in the rat central nervous system. *The Journal of Comparative Neurology, 424*(3), 439–460.

Witters, L. A., Nordlund, A. C., & Marshall, L. (1991). Regulation of intracellular acetyl-CoA carboxylase by ATP depletors mimics the action of the 5'-AMP-activated protein kinase. *Biochemical and Biophysical Research Communications, 181*(3), 1486–1492.

Wittmann, G., Lechan, R. M., Liposits, Z., & Fekete, C. (2005). Glutamatergic innervation of corticotropin-releasing hormone- and thyrotropin-releasing hormone-synthesizing neurons in the hypothalamic paraventricular nucleus of the rat. *Brain Research, 1039*(1–2), 53–62.

Woods, A., Dickerson, K., Heath, R., Hong, S.-P., Momcilovic, M., Johnstone, S. R., Carlson, M., & Carling, D. (2005). Ca2+/calmodulin-dependent protein kinase kinase-beta acts upstream of AMP-activated protein kinase in mammalian cells. *Cell Metabolism, 2*(1), 21–33.

Wu, D., Qiu, Y., Gao, X., Yuan, X.-B., & Zhai, Q. (2011). Overexpression of SIRT1 in mouse forebrain impairs lipid/glucose metabolism and motor function. *PLoS One, 6*(6), e21759.

Wu, N., Zheng, B., Shaywitz, A., Dagon, Y., Tower, C., Bellinger, G., Shen, C.-H., Wen, J., Asara, J., McGraw, T. E., Kahn, B. B., & Cantley, L. C. (2013a). AMPK-dependent degradation of TXNIP upon energy stress leads to enhanced glucose uptake via GLUT1. *Molecular Cell, 49*(6), 1167–1175.

Wu, W.-N., Wu, P.-F., Zhou, J., Guan, X.-L., Zhang, Z., Yang, Y.-J., Long, L.-H., Xie, N., Chen, J.-G., & Wang, F. (2013b). Orexin-a activates hypothalamic AMP-activated protein kinase signaling through a Ca2+–dependent mechanism involving voltage-gated L-type calcium channel. *Molecular Pharmacology, 84*(6), 876–887.

Xia, N., Strand, S., Schlufter, F., Siuda, D., Reifenberg, G., Kleinert, H., Förstermann, U., & Li, H. (2013). Role of SIRT1 and FOXO factors in eNOS transcriptional activation by resveratrol. *Nitric Oxide, 32*, 29–35.

Xiao, B., Sanders, M. J., Underwood, E., Heath, R., Mayer, F. V., Carmena, D., Jing, C., Walker, P. A., Eccleston, J. F., Haire, L. F., Saiu, P., Howell, S. A., Aasland, R., Martin, S. R., Carling, D., & Gamblin, S. J. (2011). Structure of mammalian AMPK and its regulation by ADP. *Nature, 472*(7342), 230–233.

Xin, X., Varlamov, O., Day, R., Dong, W., Bridgett, M. M., Leiter, E. H., & Fricker, L. D. (1997). Cloning and sequence analysis of cDNA encoding rat carboxypeptidase D. *DNA and Cell Biology, 16*(7), 897–909.

Xiong, S., Salazar, G., Patrushev, N., & Alexander, R. W. (2011). FoxO1 mediates an autofeedback loop regulating SIRT1 expression. *The Journal of Biological Chemistry, 286*(7), 5289–5299.

Xu, Y., Elmquist, J. K., & Fukuda, M. (2011). Central nervous control of energy and glucose balance: Focus on the central melanocortin system. *Annals of the New York Academy of Sciences, 1243*, 1–14.

Yamada, E., Pessin, J. E., Kurland, I. J., Schwartz, G. J., & Bastie, C. C. (2010). Fyn-dependent regulation of energy expenditure and body weight is mediated by tyrosine phosphorylation of LKB1. *Cell Metabolism, 11*(2), 113–124.

Yamada, E., Okada, S., Bastie, C. C., Vatish, M., Nakajima, Y., Shibusawa, R., Ozawa, A., Pessin, J. E., & Yamada, M. (2016). Fyn phosphorylates AMPK to inhibit AMPK activity and AMP-dependent activation of autophagy. *Oncotarget, 7*(46), 74612–74629.

Yang, X., & Qian, K. (2017). Protein O-GlcNAcylation: Emerging mechanisms and functions. *Nature Reviews. Molecular Cell Biology, 18*(7), 452–465.

Yang, G., Lim, C.-Y., Li, C., Xiao, X., Radda, G. K., Li, C., Cao, X., & Han, W. (2009). FoxO1 inhibits leptin regulation of pro-opiomelanocortin promoter activity by blocking STAT3 interaction with specificity protein 1. *The Journal of Biological Chemistry, 284*(6), 3719–3727.

Yang, C. S., Lam, C. K. L., Chari, M., Cheung, G. W. C., Kokorovic, A., Gao, S., Leclerc, I., Rutter, G. A., & Lam, T. K. T. (2010). Hypothalamic AMP-activated protein kinase regulates glucose production. *Diabetes, 59*(10), 2435–2443.

Yang, S.-B., Tien, A.-C., Boddupalli, G., Xu, A. W., Jan, Y. N., & Jan, L. Y. (2012). Rapamycin ameliorates age-dependent obesity associated with increased mTOR signaling in hypothalamic POMC neurons. *Neuron, 75*(3), 425–436.

Yuan, Z.-L., Guan, Y.-J., Chatterjee, D., & Chin, Y. E. (2005). Stat3 dimerization regulated by reversible acetylation of a single lysine residue. *Science, 307*(5707), 269–273.

Yuan, M., Pino, E., Wu, L., Kacergis, M., & Soukas, A. A. (2012). Identification of Akt-independent regulation of hepatic lipogenesis by mammalian target of rapamycin (mTOR) complex 2. *The Journal of Biological Chemistry, 287*(35), 29579–29588.

Zakhary, S. M., Ayubcha, D., Dileo, J. N., Jose, R., Leheste, J. R., Horowitz, J. M., & Torres, G. (2010). Distribution analysis of deacetylase SIRT1 in rodent and human nervous systems. *The Anatomical Record, 293*(6), 1024–1032.

Zhang, T., Berrocal, J. G., Frizzell, K. M., Gamble, M. J., DuMond, M. E., Krishnakumar, R., Yang, T., Sauve, A. A., & Kraus, W. L. (2009). Enzymes in the NAD+ salvage pathway regulate SIRT1 activity at target gene promoters. *The Journal of Biological Chemistry, 284*(30), 20408–20417.

Zhao, M., & Klionsky, D. J. (2011). AMPK-dependent phosphorylation of ULK1 induces autophagy. *Cell Metabolism, 13*(2), 119–120.

Zhu, X., Zhou, A., Dey, A., Norrbom, C., Carroll, R., Zhang, C., Laurent, V., Lindberg, I., Ugleholdt, R., Holst, J. J., & Steiner, D. F. (2002). Disruption of PC1/3 expression in mice causes dwarfism and multiple neuroendocrine peptide processing defects. *Proceedings of the National Academy of Sciences of the United States of America, 99*(16), 10293–10298.

Ziegler, C. G., Krug, A. W., Zouboulis, C. C., & Bornstein, S. R. (2007). Corticotropin releasing hormone and its function in the skin. *Hormone and Metabolic Research, 39*(2), 106–109.

Zoncu, R., Bar-Peled, L., Efeyan, A., Wang, S., Sancak, Y., & Sabatini, D. M. (2011). mTORC1 senses lysosomal amino acids through an inside-out mechanism that requires the vacuolar H(+)-ATPase. *Science, 334*(6056), 678–683.

Part III

Peripheral Contributors Participating in Energy Homeostasis and Obesity

Gastrointestinal Hormones Controlling Energy Homeostasis and Their Potential Role in Obesity

7

María F. Andreoli, Pablo N. De Francesco, and Mario Perello

Abbreviations

ARC   Arcuate nucleus
cAMP   Cyclic adenosine monophosphate
CCK   Cholecystokinin
CCK-1R   Cholecystokinin receptor 1
CCK-2R   Cholecystokinin receptor 2
CNS   Central nervous system
DAG   Diacylglycerol

DPP-4   Dipeptidyl peptidase 4
GCGR   Glucagon receptors
GHSR   Growth hormone secretagogue receptor type 1a
GI   Gastrointestinal
GIP   Gastric inhibitory polypeptide
GLP-1   Glucagon-like peptide-1
GLP1R   Glucagon-like peptide-1 receptor
GLP-2   Glucagon-like peptide-2
GOAT   Ghrelin O-acyltransferase
GPCRs   G protein-coupled receptors
GRPP   Glicentin-related pancreatic peptide
GTP   Guanosine triphosphate
IP3   Inositol (1,4,5)-trisphosphate
NPY   Neuropeptide Y
NTS   Nucleus of the solitary tract
OXM   Oxyntomodulin
PAM   Peptidylglycine α-amidating mono-oxygenase enzyme
PC   Prohormone convertase
PP   Pancreatic polypeptide
PYY   Peptide tyrosine-tyrosine
RYGB   Roux-en-Y gastric bypass
SST   Somatostatin
SSTR   Somatostatin receptor
YR   Neuropeptide Y receptor

M. F. Andreoli
School of Biochemistry and Biological Sciences, National University of Litoral (UNL) and Institute of Environmental Health [ISAL, Argentine Research Council (CONICET)- (UNL)], Santa Fe, Argentina

P. N. De Francesco
Laboratory of Neurophysiology, Multidisciplinary Institute of Cell Biology [IMBICE, Argentine Research Council (CONICET), National University of La Plata and Scientific Research Commission, Province of Buenos Aires (CIC-PBA)], La Plata, Buenos Aires, Argentina

M. Perello (✉)
Laboratory of Neurophysiology, Multidisciplinary Institute of Cell Biology [IMBICE, Argentine Research Council (CONICET), National University of La Plata and Scientific Research Commission, Province of Buenos Aires (CIC-PBA)], La Plata, Buenos Aires, Argentina
e-mail: mperello@imbice.gov.ar

© Springer International Publishing AG, part of Springer Nature 2018
E. A. Nillni (ed.), *Textbook of Energy Balance, Neuropeptide Hormones, and Neuroendocrine Function*, https://doi.org/10.1007/978-3-319-89506-2_7

7.1 Introduction: General Aspects of Gastrointestinal Hormones

Gastrointestinal (GI) hormones are produced by specialized enteroendocrine cells that are located in the epithelial layer throughout the whole GI tract. All together, these cells form the biggest endocrine organ of the body and produce the largest number of hormones (Ahlman and Nilsson 2001; Valle 2014). Enteroendocrine cells are characterized as being either open- or closed-type. Open-type cells display direct contact with the GI lumen and are able to sense molecules present in the luminal content, such as nutrients, which play a major regulatory role on hormone secretion. In contrast, closed-type cells lack connection with the lumen because their apical side is enclosed by epithelial cells. Closed-type cells are mainly regulated by molecules coming from the GI capillaries or by autonomic activity. Interestingly, some enteroendocrine cells display long cytoplasmic basal processes that are filled with secretory granules and extend beneath the absorptive epithelium in order to facilitate the communication with adjacent cells.

The biosynthesis of GI hormones is a complex cellular process, which involves the initial biosynthesis of polypeptide precursors that are further cleaved, sorted, and post-translationally modified within the regulated secretory pathway in order to generate the bioactive peptides (Perello and Nillni 2007; Wren and Bloom 2007). Initially, pre-prohormones are synthesized on membrane-bound ribosomes, by which they are translocated into the lumen of the rough endoplasmic reticulum. In the endoplasmic reticulum, the signal peptide of the pre-prohormone is removed by a signal peptidase. Then, the newly synthesized prohormone is transported to the Golgi complex, which is the branch point where trafficking pathways emanate in order to generate the mature secretory granules that store the mature GI hormones. During this vectorial transport, the prohormones are subjected to post-translational modifications, which vary for the different GI hormones, in order to generate the bioactive peptides. Some frequent post-translational modifications affecting GI hormones include (1) the cleavage of the prohormones at the C-terminus of paired basic residues by proteases named prohormone convertases (PC), such as PC1/3 or PC2; (2) trimming of the C-terminal basic residues of the prohormone-derived peptides by exopeptidases, such as carboxypeptidases; and (3) carboxyterminal amidation by the peptidylglycine α-amidating monooxygenase enzyme (PAM). In addition, the biosynthesis of some GI hormones involves specific post-translational modifications, such as sulfation or octanoylation, which are described below for each case. The post-translational modifications of the GI hormones are essential not only for peptide stability but also to ensure specific binding to their corresponding receptors.

GI hormones stored in secretory granules are released upon cell stimulation. GI hormone secretion mainly occurs in the basolateral side of enteroendocrine cells and seems to be regulated by nutrients, which activate specific nutrient receptors or transporters, as well as by different hormones and neurotransmitters (Steinert et al. 2017; Valle 2014). The secretion of all GI hormones is a reciprocally coordinated process, and their local action strongly controls the secretion of each other, as described below. In addition, the GI motility affects the secretion of some GI hormones. Secreted GI hormones can act locally on either vagal afferents and/or nearby cells or diffuse into GI capillaries to further reach the systemic circulation and act on distant tissues. Notably, the GI hormone levels acting locally are much higher than the levels in other organs, such as the brain. In addition, it is important to note that the half-life of most GI hormones is very short (ranging between 1 and 20 min), suggesting that their inactivation is another level that regulates their concentration.

The action of the GI hormones described in this chapter is mediated through G protein-coupled receptors (GPCRs), which contain an extracellular amino terminus, seven lipophilic transmembrane domains, and an intracellular carboxyl terminus (Mace et al. 2015; Valle 2014). Upon binding of the specific GI hormone, the

GPCR increases its affinity for a downstream G protein that transduces the signal to an intracellular event. G proteins consist of a trimer of alpha, beta, and gamma subunits and are classified according to their alpha subunit type in $G_{\alpha s}$, $G_{\alpha i/o}$, $G_{\alpha q/11}$, and $G_{\alpha 12/13}$ G proteins. Upon GPCR activation, the beta and gamma subunits dissociate from the alpha subunit, which exchanges a molecule of guanosine diphosphate for guanosine triphosphate (GTP). The GTP-bound alpha subunit regulates specific downstream signaling cascades. The $G_{\alpha s}$ or $G_{\alpha i/o}$ pathways display a stimulatory or inhibitory effect, respectively, on the activity of adenylate cyclases that catalyze the conversion of cytosolic adenosine triphosphate to cyclic adenosine monophosphate (cAMP), a second messenger that activates the protein kinase A. The target of the $G_{\alpha q/11}$ pathway is phospholipase C, which cleavages phosphatidylinositol 4,5-bisphosphate into the second messengers inositol (1,4,5)-trisphosphate (IP3) and diacylglycerol (DAG). IP3 acts at the endoplasmic reticulum to elicit Ca^{2+} release, while DAG diffuses along the plasma membrane where it activates protein kinase C. Elevated intracellular Ca^{2+} also activates calmodulins, which in turn activate Ca^{2+}/calmodulin-dependent kinases. The $G_{\alpha 12/13}$ pathway regulates cell processes through guanine nucleotide exchange factors that activate the cytosolic small Rho GTPase. Usually each GI hormone mainly activates one type of signaling cascade; however, some GI hormones activate multiple signaling cascades, which greatly increase the complexity of their function. In addition to the classical signaling pathways, some GI hormones have been shown to activate G protein-independent pathways, including the mitogen-activated protein kinase and extracellular signal-regulated kinase signaling pathways.

The GPCRs for GI hormones mediate their actions in the peripheral tissues as well as in the central nervous system (CNS). In addition to integrating GI functions, some GI hormones play an important role in the regulation of body energy homeostasis. The mechanisms by which some GI hormones regulate energy homeostasis are diverse and involve control of meal size, meal timing, hedonic aspects of eating, and adiposity

as well as the regulation of the meal-related glycemia. Here, we focus on four GI hormones: ghrelin, cholecystokinin (CCK), glucagon-like peptide-1 (GLP-1), and peptide tyrosine-tyrosine (PYY), which likely play a relevant role in the control of the energy homeostasis. In addition, we briefly review the role of oxyntomodulin (OXM), pancreatic polypeptide (PP), and somatostatin (SST), which may also play a role on energy balance.

7.2 Ghrelin

Ghrelin is a 28-residue octanoylated peptide hormone discovered by Kojima in 1999 (Kojima et al. 1999). Ghrelin is the only known secreted peptide modified by an O-octanoylation; in addition, ghrelin is the only peptide hormone known to increase food intake (Kojima and Kangawa 2010). In contrast to other GI hormones that mainly impact on meal size and frequency, ghrelin seems also relevant for the long-term control of energy intake and body weight regulation as its plasma levels are inversely related to body adiposity in healthy-weight, obese, and weight-reduced humans (Cummings 2006; Perello and Zigman 2012). Ghrelin regulates both homeostatic and hedonic aspects of eating (Perello and Zigman 2012). Ghrelin may also contribute to glycemic control and, to a lesser extent, to the regulation of the GI motility (Steinert et al. 2017). Figure 7.1 depicts the GI regions producing ghrelin as well as the key targets and the main actions of this GI hormone.

Ghrelin is predominantly synthesized by closed-type enteroendocrine ghrelin cells located within the gastric oxyntic fundic mucosa, previously referred to as P/D1-type cells in humans and A like-type cells in rodents (Kojima and Kangawa 2010). The synthesis of ghrelin is atypical as it involves the enzyme ghrelin O-acyltransferase (GOAT), which catalyzes ghrelin octanoylation. GOAT is present in the endoplasmic reticulum and acylates proghrelin before being translocated to the Golgi (Gutierrez et al. 2008; Yang et al. 2008). In the Golgi complex, the octanoylated proghrelin is cleaved by

↑ food intake
↑ gastric emptying
▪ induces hyperglycemic
   mechanisms

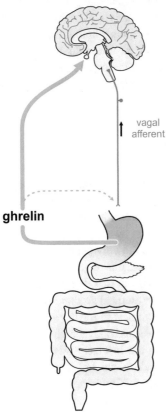

vagal
afferent

ghrelin

Fig. 7.1 Ghrelin production and effects on energy homeostasis. Ghrelin is produced in the gastric mucosa (extent indicated by intensity of light blue shading). The pathways by which ghrelin exerts its effects on energy balance are mainly central but might also involve signaling through GI vagal afferents (solid and dashed light blue arrows, respectively)

PC1/3. Since the 28 residues of ghrelin sequence immediately follow the signal peptide, the N-terminal fragment generated by cleavage of proghrelin produces bioactive ghrelin, whereas the C-terminal fragment further yields a peptide named obestatin (Takahashi et al. 2009; Zhu et al. 2006). Initial reports indicated that obestatin had anorexigenic effects; however, further studies could not confirm such observations, and its role on energy balance is still under debate (Gourcerol et al. 2007; Nogueiras et al. 2007; Zhang et al. 2005). Ghrelin is also secreted as a des-

octanoylated version, named des-acyl ghrelin, which circulates even at higher levels than ghrelin because it is also produced in plasma by ghrelin deacylation (De Vriese et al. 2004). Some studies suggest that des-acyl ghrelin displays some actions in a GHSR-independent manner; however, no specific receptor for des-acyl ghrelin has yet been reported, and, as a consequence, the role of this peptide remains uncertain (Callaghan and Furness 2014).

Ghrelin levels display a surge before meals, decline after meals, and then increase gradually until the next preprandial peak (Cummings 2006). In addition, ghrelin levels are elevated in energy deficit conditions, such as fasting, malnutrition, anorexia nervosa, or cachexia (Cummings 2006). Since the half-life of plasma ghrelin is very short (Akamizu et al. 2004), the dynamics of ghrelin levels highly depends on its secretion. Ghrelin levels after meals are reduced by the consumption of all three macronutrients, although carbohydrates and proteins are more potent than isoenergetic lipid loads (Foster-Schubert et al. 2008; Steinert et al. 2017). The post-meal inhibition of ghrelin secretion is not due to signals arriving from the gastric lumen, since ghrelin cells are closed-type. Still, ghrelin cells express several nutrient-sensing receptors suggesting that ghrelin secretion is regulated by circulating nutrients (Steinert et al. 2017). In support of this notion, it has been shown that ghrelin secretion in humans is unaffected by intraduodenal glucose infusion but is inhibited when glucose reaches distal to the duodenum and proximal jejunum (Little et al. 2006). Post-meal ghrelin secretion is also inhibited by GI hormones that are recruited by nutrient ingestion. In particular, intravenous infusions of CCK or PYY reduce ghrelin levels in humans, whereas GLP-1 infusions fail to affect it (Batterham et al. 2003; Brennan et al. 2007; Degen et al. 2005). Other hormones that contribute to the post-meal inhibition of ghrelin secretion include insulin that decreases ghrelin levels and is necessary for the post-meal ghrelin reduction (Flanagan et al. 2003; Mohlig et al. 2002). In contrast, leptin fails to affect ghrelin levels (Chan et al. 2004). The autonomic nervous system does not seem to mediate the post-meal ghrelin inhibition but is critical for stimulating ghrelin secretion during fasting in both humans and animals (Broglio et al. 2004; Hosoda and Kangawa 2008; Zhao et al. 2010).

There is a single ghrelin receptor, named the growth hormone secretagogue receptor type 1a (GHSR) (Cruz and Smith 2008; Howard et al. 1996). GHSR displays some uncommon features that greatly increase the complexity of its function. GHSR is the GPCR with the highest known constitutive activity, signaling about 50% of the maximum activity in the absence of ghrelin (Holst et al. 2003). Interestingly, ghrelin-evoked activation of GHSR mainly recruits $G_{\alpha q/11}$ signaling pathway while the ghrelin-independent activity of GHSR not only activates $G_{\alpha q/11}$ signaling but also involves $G_{\alpha i/o}$ signaling (Holst et al. 2003; Lopez Soto et al. 2015). Additionally, GHSR heterodimerizes with other GPCRs, and such interaction mutually impacts on the signaling pathways of each receptor (Schellekens et al. 2015). GHSR is mainly expressed in the pituitary as well as in some regions of the CNS, which mediates ghrelin actions on food intake, gastric function, and glucose homeostasis (Cruz and Smith 2008).

Ghrelin administration increases gastric emptying; however, it is uncertain if endogenous patterns of ghrelin levels are sufficient to affect gastric emptying in humans (Jones et al. 2012; Levin et al. 2006; Steinert et al. 2017). Pre-meal ghrelin levels correlate with hunger sensations and meal size, and administration of ghrelin to humans strongly increases food intake; still, the causal role of ghrelin in hunger remains unclear (Cummings 2006). Studies in rodents have shown that ghrelin plays a key role in the regulation of food intake (Perello and Zigman 2012). The main target of the orexigenic actions of ghrelin are the neuropeptide Y (NPY)-producing neurons of the arcuate nucleus (ARC) that express high levels of GHSR (Cabral et al. 2016; Perello and Raingo 2014). In contrast to rodents, intact vagal afferents seem to be required for ghrelin-induced food intake in humans (Arnold et al. 2006; le Roux et al. 2005). Additionally, ghrelin affects rewarding aspects of eating by modulating the activity of the mesolimbic pathway (Perello and

Zigman 2012). By acting at the central level, ghrelin also promotes adiposity (Al Massadi et al. 2017). On the other hand, ghrelin affects the glycemic control by activating a variety of mechanisms that tend to increase glucose levels. Intravenous infusion of ghrelin reduces insulin levels in response to glucose infusion and increases growth hormone and cortisol levels, without any effect on glucagon or epinephrine levels (Tong et al. 2013). Studies in mice indicate that the insulin-inhibitory and glucagon-stimulatory effects of ghrelin involve a direct pancreatic action (Chuang et al. 2011; Kurashina et al. 2015).

7.3 Cholecystokinin

CCK was discovered at the beginning of the twentieth century by Bayliss and Starling (Bayliss and Starling 1902). CCK is structurally related to gastrin: they share a 5-residue sequence at the C-terminal that includes an amidated phenylalanine that is key to their bioactivity (Eysselein et al. 1986). Although gastrin is believed to have evolved from CCK by gene duplication, it does not regulate food intake in mammals but rather gastric acid secretion. CCK is secreted after meals in response to the products of carbohydrate, lipid, and protein digestion, and it is recognized as the best-established GI endocrine satiation signal in humans. In addition, CCK contributes to the control of meal-related glycemia indirectly, via its effect on gastric emptying, as well as directly via control of hepatic glucose production (Steinert et al. 2017). Figure 7.2 depicts the GI regions producing CCK as well as the key targets and the main actions on energy balance of this GI hormone.

CCK is produced by open-type cells traditionally denominated I cells, which are found in the proximal part of the small intestine (duodenum and jejunum). For the CCK synthesis, pre-proCCK is initially cleaved and yields the signal peptide and the proCCK polypeptide. Then, three out of four tyrosine residues in proCCK are sulfated by protein tyrosine sulfotransferase in the Golgi complex, and sulfated proCCK is sorted

↓↓ food intake
↓ gastric emptying

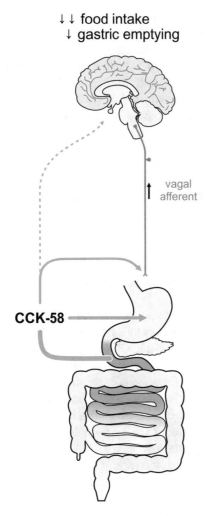

vagal
afferent

CCK-58

Fig. 7.2 CCK production and effects on energy homeostasis. GI-derived CCK is secreted in the proximal small intestine (production level indicated by intensity of green shading) and exerts its effects on energy balance primarily by direct action on the stomach and also by signaling through vagal afferents (solid green arrows). Potential direct central actions might also be present (dashed green arrow). The thickness of the arrows indicates the relative importance of each pathway

into immature secretory granules (Beinfeld 2003). The structure of proCCK allows cleavage by PC1/3, PC2, and PC5 at six monobasic motifs, leading to the formation of CCK-58, CCK-39, CCK-33, CCK-22, and CCK-8 (Beinfeld 2003). These CCK forms share the sulfated 8 residues at their C-terminal side, which is trimmed by carboxypeptidases and amidated by PAM. The most abundant CCK form in the intestine is CCK-58,

followed by CCK-39, CCK-33, and CCK-8, whereas in the brain CCK-8 is the most abundant form (Eysselein et al. 1986). The fact that CCK-8 is the major form in the brain may allow a fast degradation after release, while the larger forms might be more difficult to degrade by the kidney and liver leading to a longer biological half-life in circulation (Beinfeld 2003).

CCK secretion is directly and indirectly stimulated by intraluminal nutrients, which are sensed by a variety of nutrient receptors expressed on the apical surface of I cells (Steinert et al. 2017). In humans, CCK secretion is highly elicited by lipids and proteins and to a lesser extent by carbohydrates (Dockray 2009). Additionally, CCK secretion is stimulated by the CCK-releasing factors "pancreatic monitor peptide" and "intestinal luminal CCK-releasing factor" (Steinert et al. 2017; Wang et al. 2002).

Two CCK receptors have been identified and named CCK-1R and CCK-2R (Dufresne et al. 2006). Interestingly, the ligand binding domain of CCK receptors is different. CCK-1R is highly specific for sulfated CCK forms, whereas the CCK-2R shows only a 10- to 20-fold higher preference for the sulfated forms and binds to gastrin with similar affinity (Huang et al. 1989; Miller and Desai 2016). CCK receptors predominantly couple to $G_{\alpha q/11}$ signaling pathway (Miller and Desai 2016; Wank 1995). In rats, CCK-1R is mainly expressed in the GI tract and participates in the digestive process, while CCK-2R is mostly present in the CNS, where it is involved in satiety (Wank 1995). Both CCK receptors are present in vagal afferent fibers projecting to the nucleus of the solitary tract (NTS) in the brainstem, which controls food intake and autonomic functions (Corp et al. 1993).

CCK is a potent regulator of the GI function. In particular, CCK reduces gastric acid secretion and gastric emptying (Whited et al. 2006), while it stimulates gallbladder contraction (Liddle et al. 1985) and exocrine pancreas secretion (Adler et al. 1991). In addition, CCK is the best-established GI endocrine satiation signal in humans playing a major role in the regulation of meal size (Lieverse et al. 1995). Studies in rodents suggest that CCK inhibits eating via both local and endocrine mechanisms; however, it seems that CCK mainly stimulates satiation in humans via activation of CCK-1R on afferent vagal sensory neurons (Lieverse et al. 1995; Steinert et al. 2017; Schwartz et al. 1993). The role of CCK in meal termination does not seem to be translated into regulation of long-term energy intake since the reduction in meal size induced by CCK before every meal is compensated for by an increase in meal frequency (Woods 2004). Notably, many actions of CCK have been studied using synthetic CCK-8, which is relatively easy to synthesize and strongly binds to CCK-1R. However, the dominant circulating form of CCK is CCK-58 (Eysselein et al. 1986), and several studies have revealed that CCK-8 and CCK-58 elicit a different pattern of biological actions on pancreatic secretion, gastric motility (Reeve et al. 1996; Yamamoto et al. 2005), meal size reduction (Owyang and Heldsinger 2011), and extension of the interval between meals (Overduin et al. 2014). Animal studies indicate that CCK affects glucose metabolism by reducing hepatic glucose production via a vagal-vagal reflex (Cheung et al. 2009). Nonetheless, CCK does not seem to play a major role in the glycemic control in humans (Steinert et al. 2017).

7.4 Glucagon-Like Peptide-1

GLP-1 is a GI hormone derived from the pre-proglucagon gene that increases in response to meals. GLP-1 inhibits food intake and gastric emptying; however, it is best known for its potent inhibitory effect on the meal-related increases in glycemia (Holst 2007). Remarkably, the pre-proglucagon gene is also expressed in the pancreas and the brain, where proglucagon is subject to a different post-translational processing and generates different products (Orskov et al. 1986). Specifically, proglucagon is cleaved by PC2 in pancreatic α cells, whereas it is cleaved by PC1/3 in the GI and the brain (Holst 2007). In the pancreas, the main products of proglucagon are proglucagon$_{1-30}$ (or glicentin-related pancreatic peptide, GRPP, which is inactive), proglucagon$_{33-61}$ (glucagon itself), and proglucagon$_{72-158}$

(or major proglucagon fragment, which is also inactive). In the brain, proglucagon is produced by a subset of neurons of the NTS and gives rise to GLP-1 in the same fashion as described below for the GI tract (Trapp and Hisadome 2011). Figure 7.3 depicts the GI regions producing GLP-1 as well as the key targets, the extracellular cleavage, and the main actions of this GI hormone.

In the GI tract, GLP-1 is produced by open-type L cells located in the distal jejunum and ileum (Mojsov et al. 1986). In L cells, proglucagon is processed to generate GRPP, proglucagon$_{33-69}$ (also known as OXM), proglucagon$_{78-107}$ (GLP-1), and proglucagon$_{126-158}$ (GLP-2) (Holst 2007). In humans, the GLP-1 peptide corresponds to the sequence of proglucagon$_{78-107}$ amide, and it is designated GLP-1 (7–36amide), which is the main circulating form (Holst 2007). In rodents, however, about half of the GLP-1 appears as the Gly-extended form or GLP-1 (7–37). The Gly corresponding to proglucagon$_{108}$ serves as substrate for amidation, but the biological consequences of this reaction are unclear, as GLP-1 (7–36amide) and GLP-1 (7–37) display similar bioactivities and overall metabolism. After release, both GLP-1 forms are cleaved to the inactive forms GLP-1(9–36amide) and GLP-1(9–37) by the enzyme dipeptidyl peptidase 4 (DPP-4), which is a membrane glycoprotein found locally in the GI tract, in circulation, and in the liver. Due to its inactivation, only about 10–15% of the newly secreted GLP-1 reaches the systemic circulation in its intact form, suggesting that GLP-1 mainly acts locally rather than in an endocrine manner (Holst 2007).

The L cells have microvilli protruding into the intestinal lumen that sense nutrients, which regulate GLP-1 secretion (Holst 2007). GLP-1 secretion is stimulated by dietary carbohydrates and fats (Parker et al. 2010; Hansen et al. 2011). Human studies revealed that the lowest GLP-1 levels can be seen after overnight fasting; they increase rapidly during meals and usually do not return to the morning level between meals. While oral loads of carbohydrates usually result in monophasic increases in GLP-1 levels, mixed-nutrient meals induce biphasic patterns, with

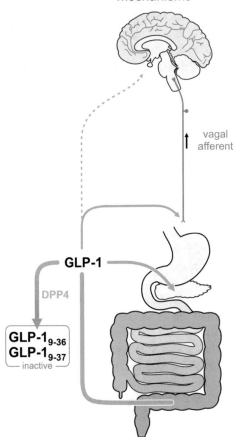

↓ food intake
• induces hypoglycemic
  mechanisms

↑ vagal
  afferent

GLP-1

DPP4

GLP-1$_{9-36}$
GLP-1$_{9-37}$
— inactive —

Fig. 7.3 GLP-1 production and effects on energy homeostasis. GLP-1 is secreted in the distal small intestine (level of production indicated by intensity of orange shading). GLP-1 effects on energy homeostasis are exerted by direct action on the endocrine pancreas but also by signaling through vagal afferents (solid orange arrows). Potential direct central actions have also been suggested (dashed orange arrows). Note that upon secretion GLP-1 is heavily degraded to inactive forms by DPP-4 (fading orange arrow). The thickness of the arrows indicates the relative importance of each pathway

secondary peaks after 60–120 min (Carr et al. 2010; Steinert et al. 2017).

The GLP-1 receptor (GLP1R) belongs to the glucagon receptor family, which is a group of GPCRs that also includes the glucagon receptor (GCGR), the GLP-2 receptor, and the gastric inhibitory polypeptide (GIP) receptor (Mayo et al. 2003). GLP1R binds both GLP-1 and gluca-

gon. GLP1R predominantly couples to the $G_{\alpha s}$ pathways (Drucker et al. 1987; Fehmann et al. 1995) but may also activate other G_{α} pathways (Wheeler et al. 1993). GLP1R is expressed in pancreatic islets, throughout the whole GI tract (Holst 2007) and in the brain, especially in vagal afferents and other areas implicated in food intake and energy balance regulation (Alvarez et al. 2005).

The most acknowledged effect of intestinal GLP-1 is the reduction of meal-related increases in glycemia, an effect that mainly takes place by stimulating insulin secretion, inhibiting glucagon secretion, and slowing gastric emptying (Holst 2007; Sandoval and D'Alessio 2015; Drucker et al. 1987). GLP-1, together with GIP, plays a major role in the incretin effect, which refers to the direct effect of some GI hormones on the β-cells to enhance glucose-stimulated insulin secretion and accounts for around 50–70% of the insulin release after carbohydrate intake (Mayo et al. 2003). In healthy subjects, GLP-1 may also contribute to glycemic control in the fasting state mainly via the regulation of gastric emptying and glucagon secretion (Steinert et al. 2017; Vilsboll et al. 2003; Nauck et al. 1997). Regarding food intake control, peripheral administration of GLP-1 inhibits eating in many species, including humans (Barrera et al. 2011; Williams 2009). The mechanism by which intestinal GLP-1 reduces food intake in humans remains unsettled, but likely involves its action at both vagal sensory afferents and CNS targets (Orskov et al. 1996). In rodents, GLP1R is expressed in both the ARC and the NTS, and GLP-1 seems to directly recruit NTS neurons and indirectly act on ARC neurons via ascending pathways (Wren and Bloom 2007). As described above, the CNS also produces GLP-1 that likely contributes to the regulation of energy homeostasis (Barrera et al. 2011; Trapp and Hisadome 2011).

7.5 PYY

PYY was first described in 1980 (Tatemoto and Mutt 1980) and belongs to the PP-fold family that also includes PP and NPY, all of which present a similar U-shaped tertiary structure and sequence homology (Wynne and Bloom 2006). PYY is released in response to meals and takes part in the local regulation of GI activity by slowing GI motility. PYY may also play a role in reducing food intake and controlling meal-related glycemia (Batterham et al. 2002; Chandarana et al. 2013). Figure 7.4 depicts the GI regions producing PYY as well as the key targets, the extracellular cleavage and the main actions of this GI hormone.

PYY is produced by L cells, which can also co-secrete CCK, GLP-1, glicentin, OXM, and GIP to a varying extent (Spreckley and Murphy 2015; Steinert et al. 2017). L cells are located in the distal intestine, mainly in the rectum, followed by the colon and ileum (Wynne and Bloom 2006). Additionally, PYY is expressed in the endocrine pancreas and in some neurons of the CNS (Spreckley and Murphy 2015). As described for the other GI hormone precursors, pre-proPYY is initially cleaved by the signal peptidase and generates proPYY, which contains the PYY sequence at its N-terminal end. Thus, proPYY is further cleaved by PCs and the two basic residues trimmed by carboxypeptidases in order to generate PYY_{1-36}-Gly. Finally, the C-terminal end of PYY is amidated, a modification required for its bioactivity, in order to generate mature PYY_{1-36} (Stanley et al. 2004). Two major circulating forms of PYY exist: the full-length PYY_{1-36}, which is the secreted form, and a truncated PYY_{3-36} form that results from the DPP-4-mediated N-terminal cleavage, which mainly occurs in the intestinal lamina propria, liver, capillary endothelium, and blood (Spreckley and Murphy 2015; Wynne and Bloom 2006). PYY_{3-36} is the predominant form of PYY in circulation and is usually regarded as the active form (Manning and Batterham 2014; Wynne and Bloom 2006).

PYY exhibits a two-phase release profile following a meal. The first-phase response is mediated via neuroendocrine reflexes and likely involves CCK and GLP-1, while the second is likely driven by sensing of nutrients in the distal GI tract (Spreckley and Murphy 2015; Svendsen et al. 2015). PYY levels start rising few minutes

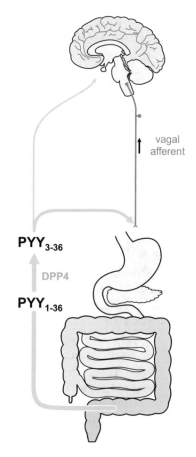

↓ food intake
↓ gastric emptying

Fig. 7.4 PYY production and effects on energy homeostasis. PYY is secreted from the distal intestine (extent indicated by intensity of yellow shading) and promptly undergoes a cleavage to its main active form by DPP-4 (intermediate yellow arrow). PYY effects on energy balance are primarily exerted by signaling through vagal afferents but also involve direct central signaling (solid outgoing yellow arrows). The thickness of the arrows indicates the relative importance of each pathway

after meals and reach the peak 60–90 min after meals, with levels proportional to caloric intake and dependent on ingested macronutrient composition, as larger and more sustained PYY elevations are induced by lipid ingestion as compared to carbohydrate ingestion (Manning and Batterham 2014; Spreckley and Murphy 2015; Steinert et al. 2017). The half-life of PYY in humans is short, but the sustained postprandial release increases its levels for several hours

(Spreckley and Murphy 2015). Thus, it takes several hours after evening meals to reach typical morning fasting levels (Steinert et al. 2017). Interestingly, postprandial profiles of active PYY and GLP-1 are often dissimilar because DPP-4 activates PYY, but inactivates GLP-1 (Mortensen et al. 2003).

PYY-related peptides bind to several members of the Y receptor (YR) family, all of which are GPCRs that couple to $G_{\alpha i/o}$ pathway (Stadlbauer et al. 2015). Specifically, PYY_{1-36} activates Y1R, Y2R, Y4R, and Y5R, while PYY_{3-36} selectively targets Y2R (Dumont et al. 1995). Importantly, Y2R is expressed in a number of regions along the GI tract and in vagal afferents, as well as in several brain areas including the ARC (Steinert et al. 2017). In neurons, Y2R is located at presynaptic terminals, and its activation inhibits neurotransmitter release (Stadlbauer et al. 2015).

PYY reduces gastric emptying and acid secretion, increases ileal absorption, and delays gallbladder and pancreatic secretion (Kirchner et al. 2010; Spreckley and Murphy 2015; Wynne and Bloom 2006). Thus, PYY is part of the feedback mechanism known as the "ileal break," which is elicited by the presence of unabsorbed dietary components in the distal GI tract and is aimed to slow proximal GI motility in order to facilitate efficient digestion and uptake of nutrients (Imamura 2002; Spreckley and Murphy 2015). PYY may also act as a satiety signal (Manning and Batterham 2014; Stanley et al. 2004; Wynne and Bloom 2006) since PYY_{3-36} administration reduces food intake in both rodents and humans (Batterham et al. 2002; Manning and Batterham 2014). This anorexigenic effect seems to occur via Y2R and involves the inhibition of both orexigenic ARC NPY neurons and vagal afferents fibers (Stadlbauer et al. 2015; Stanley et al. 2004; Wynne and Bloom 2006; Chaudhri et al. 2006). PYY administration affects glycemia, but the physiological relevance of endogenous PYY on glucose homeostasis in humans is still a matter of debate (Manning and Batterham 2014; Persaud and Bewick 2014; Steinert et al. 2017).

7.6 Other Gastrointestinal Hormones

Oxyntomodulin First reported in 1981 (Bataille et al. 1981), OXM is a 37-residue peptide derived from the post-translational processing of the proglucagon precursor. OXM is produced by L cells of the small intestine, and its distribution along the GI tract mirrors that of GLP-1 (Holst 2007; Wynne and Bloom 2006; Chaudhri et al. 2006). Similarly to GLP-1, OXM is also produced in the pancreas and the brain (Stanley et al. 2004). OXM binds to both GLP1R and GCGR, although its affinity is almost two orders of magnitude lower than their native ligands (Spreckley and Murphy 2015). Furthermore, OXM elicits a biased GLP1R response as compared with GLP-1, as it has less preference toward the $G_{\alpha s}$ pathway as compared to the ERK pathway (Pocai 2014). Similarly to PYY and GLP-1, OXM is released after meals in an extent that depends on the caloric intake and macronutrient composition, being mainly stimulated by fat content (Huda et al. 2006). The half-life of OXM is short, being rapidly degraded in circulation by DPP-4, in a similar fashion as GLP-1 (Yi et al. 2015). OXM inhibits gastric secretion, pancreatic exocrine secretion, and gastric emptying (Field et al. 2008). OXM also inhibits food intake, in part, due to the suppression of ghrelin levels (Wren and Bloom 2007). OXM might be involved in long-term energy balance in humans since its repeated administration reduces body weight as a result of both reduction in food intake and increase in energy expenditure (Field et al. 2008). Despite OXM binds to both GLP1R and GCGR, it mainly inhibits appetite via the GLP1R since co-administration of OXM and a GLP1R antagonist blocks the anorectic actions of OXM (Dakin et al. 2001; Pocai 2014). In contrast to GLP-1, OXM is thought to inhibit food intake by acting directly at the ARC level (Wren and Bloom 2007). This possibility together with the additional effect of OXM on GCGR may explain the reason why this GI hormone is as potent as GLP-1 to reduce food intake despite its lower affinity for the GLP1R (Wynne and Bloom 2006).

OXM also exhibits incretin activity albeit modest compared to that of GLP-1 (Du et al. 2012).

Pancreatic Polypeptide PP is an amidated 36-residue peptide that belongs to the PP-fold family and was discovered in 1975 from pancreatic extracts (Kimmel et al. 1975; Chaudhri et al. 2006). PP is mainly produced in specialized pancreatic islets cells, called F cells, and to a lesser extent in the exocrine pancreas, colon, and rectum (Huda et al. 2006; Khandekar et al. 2015; Wren and Bloom 2007). As all members of the PP-fold family, PP binds to YR showing the highest affinity for Y4R but also binding to Y5R and Y1R, all of which are expressed in the CNS (Huda et al. 2006). The PP production is under vagal control (Lean and Malkova 2016), and its effects are also mainly mediated by the vagus nerve (Choudhury et al. 2016). PP is released after meals, in an amount proportional to the calories ingested, and in response to hypoglycemia, exercise, gastric distension, and elevations in gastrin, secretin, and motilin (Field et al. 2008; Huda et al. 2006). Similarly to PYY, PP has a short half-life but its levels remain elevated for several hours after meals (Chaudhri et al. 2006; Choudhury et al. 2016). In humans, peripheral administration of PP inhibits gastric emptying, exocrine pancreatic secretion, and gallbladder motility and acutely decreases food intake (Lean and Malkova 2016; Wren and Bloom 2007). In rodents, peripheral administration of PP also reduces food intake and gastric emptying, and these effects seem to occur via hypothalamic actions but also require the integrity of the vagal system (Asakawa et al. 2003; Khandekar et al. 2015). PP inhibits glucagon release through the activation of Y1R in pancreatic α cells, and delays the postprandial rise in insulin (Aragon et al. 2015; Chaudhri et al. 2006).

Somatostatin SST is a cyclic peptide that was first identified in 1973 in hypothalamic extracts (Brazeau et al. 1973). SST displays a widespread distribution in the body, being produced in the GI tract, as well as in the brain, peripheral nerves, the pancreas, and the retina (Kumar and Grant 2010). Most of the circulating SST derives from the GI tract, where it is released from D cells as well as from intrinsic neurons located in the stomach, intestines, and pancreas (Low 2004). Interestingly, proSST can yield two bioactive products: a short 14-residue form (SST-14) and a longer 28-residue form (SST-28) that contain an extension at the N-terminus (Kumar and Grant 2010). Both SST forms present a disulfide bond between cysteine residues at positions 1 and 12. SST-28 predominates in the intestinal mucosal cells, while SST-14 predominates in the pancreas, the stomach, and neural tissues. SST binds to five GPCRs, named SSTR1–5. SSTRs 1 to 4 bind SST-14 and SST-28 with similar affinity, while SSTR5 binds SST-28 with five- to tenfold higher affinity (Rai et al. 2015). All SSTRs are coupled to the $G_{\alpha i}$ pathway and are widely distributed in the body, although SSTR2 and SSTR5 are predominantly expressed in the peripheral tissues (Low 2004). SST secretion is increased by a combination of nutritional, hormonal, and neural signals after meals (Low 2004), and SST levels can remain elevated postprandially for few hours, despite SST in plasma displays a very short half-life (Rai et al. 2015). Well known for its inhibitory physiological actions in multiple targets, SST can function via endocrine, paracrine, or neurocrine pathways. SST is involved in a variety of effects, and it is unclear whether its effects on the energy balance are direct or indirect, as SST inhibits the release of numerous hormones (Rai et al. 2015). SST inhibits the secretion of some GI hormones (CCK, ghrelin, GLP-1, GIP, secretin), gastric emptying, and GI motility while also inhibiting insulin and glucagon secretion from the pancreas (Rai et al. 2015). The effect of SST on food intake is unclear as studies in rodents have found inconsistent results (Fenske et al. 2012; Rai et al. 2015). Still, the relevance of SST in the regulation of energy balance remains uncertain (Rai et al. 2015; Steinert et al. 2017).

7.7 Gastrointestinal Hormones and Obesity

The understanding of the role of the GI hormones on the regulation of the mechanisms controlling energy balance has notably improved over the

Table 7.1 Summary of some key features of the GI hormones described in this chapter

GI hormone	Ghrelin	CCK	GLP-1	PYY
Secreting enteroendocrine cells	Ghrelin cells	I cells	L cells	L cells
Location of enteroendocrine cells	Stomach	Duodenum and proximal part of jejunum	Jejunum, ileum, and colon	Distal part of the jejunum, ileum, rectum, and colon
Secretion pattern	Before meals	After meals	After meals	After meals
Peptides derived from the precursor (activity)	Ghrelin (active) Des-acyl ghrelin Obestatin	CCK-8 CCK-58 Others	GRPP Oxyntomodulin GLP-1 (7–36amide) GLP-1 (7–37) GLP-2	PYY_{3-36} (active) PYY_{1-36} (active)
Effect on food intake	Increases	Potently decreases	Decreases	Decreases
Effect on gastric emptying	Increases	Decreases	Decreases	Decreases
Effect on glucose homeostasis	Promotes hyperglycemic mechanism. (decreases insulin secretion. Increases glucagon, glucocorticoids, growth hormone)	No major effects	Increases insulin secretion, decreases glucagon	No major effects
Known Receptors (ligand)	GHSR-1a (ghrelin)	CCK-1R (CCK) CCK-2R (gastrin, CCK)	GLP1R (GLP-1, OXM)	Y2R (PYY_{3-36}) Y1R, Y2R, Y4R, Y5R (PYY_{1-36})

last decades. As summarized in Table 7.1, it is now clear that CCK, GLP-1, and PYY, which are secreted after eating, reduce food intake mainly by affecting the timing and size of individual meals, while ghrelin, which is secreted before eating, may also play a tonic role controlling the long-term body weight. Additionally, some GI hormones also regulate glucose homeostasis by affecting insulin secretion and/or sensitivity as well as by affecting gastric emptying, which contributes to the regulation of meal-related glycemia. Thus, GLP-1 strongly controls meal-related glycemia via a variety of mechanisms, while CCK and PYY decrease gastric emptying but do not seem to impact on other aspects of glycemic control. In contrast, ghrelin promotes mechanisms that increase glycemia and only slightly increases gastric emptying. Given the key role of these GI hormones in the regulation of the energy homeostasis, it is likely that they are also involved in the pathophysiology of obesity, which is the most significant growing health concern worldwide. Currently, the only treatment that produces long-term weight loss and improves obesity-

related comorbid conditions in severely obese patients is bariatric surgery, particularly Roux-en-Y gastric bypass (RYGB) (Sjostrom et al. 2007). Although the mechanisms involved in the bariatric surgery-mediated remission of obesity are still uncertain, it is well-established that the levels of various GI hormones are altered and that such changes correlate with the beneficial effects of the procedure (Steinert et al. 2017). These observations have highlighted the necessity to improve our understanding of the role of the GI hormones in the pathophysiology of obesity in order to further explore their therapeutic manipulation. In this section, we briefly review possible links between the above described GI hormones and obesity as well as their potential as pharmaceutical targets.

Little association has been found between ghrelin or GHSR gene mutations and obesity in humans (Gueorguiev et al. 2009; Liu et al. 2011); however, ghrelin gene polymorphisms were associated with body mass index variation in some human populations (Li et al. 2014). Most obese patients display low ghrelin levels and a blunted increase of nocturnal ghrelin levels, as compared to normal subjects (Hillman et al. 2011; Tschop et al. 2001). In addition, some studies report that obese patients show a blunted postprandial decrease of ghrelin levels, which likely increases the time they feel hungry and contributes to the pathophysiology of obesity (le Roux et al. 2005; Morpurgo et al. 2003; Yang et al. 2009). On the other hand, ghrelin levels rise in obese patients after diet-induced weight loss, and such increase seems to be involved in the rebound weight gain usually observed in dieters (Cummings et al. 2002). Patients that undergo RYGB display reduced fasting and post-meal ghrelin levels in the first weeks after surgery, but the longer-term effects are controversial (Cummings and Shannon 2003; Steinert et al. 2017; Beckman et al. 2010). Patients that go through sleeve gastrectomy show a dramatic decrease of body weight and display almost undetectable ghrelin levels (Peterli et al. 2012). These evidences suggested that pharmacological manipulations of ghrelin signaling may be a potential anti-obesity strategy. However, no therapy targeting this system has been shown to be successful up to date. Among other reasons, the complexity of the GHSR biology, including the fact that ligands of the receptor can function as agonist, antagonist, or inverse agonists in a biased fashion (M'Kadmi et al. 2015), has become an intrinsic limitation hard to overcome.

Human CCK-1R polymorphisms have been associated with meal size, total food intake, and body weight alterations, suggesting that CCK system is involved in the pathophysiology of obesity (Steinert et al. 2017). However, it is currently controversial if CCK levels are altered in obesity, as one study found reduced fasting CCK levels in obese patients, which could contribute to overeating, while other studies could not confirm these findings (Baranowska et al. 2000; Brennan et al. 2012; Stewart et al. 2011). Interestingly, postprandial CCK levels have been reported to be normal or increased following RYGB in different studies indicating that this system may contribute to the early satiation seen in these patients (Steinert et al. 2017). The reason for this observation is unclear, as the ingested nutrients fail to contact the majority of the I cells after the surgi-

cal procedure, although this suggests that CCK system may represent a target for obesity treatment (Steinert et al. 2017). Many attempts have been performed in order to reduce appetite by pharmacological manipulations of the CCK system. In rats, continuous CCK administration rapidly becomes ineffective, while intermittent CCK administration before each meal effectively reduces meal size; nevertheless, animals compensate daily food intake by increasing meal frequency (Crawley and Beinfeld 1983; Woods 2004). Similarly, intravenous CCK reduces food intake in humans, but repeated administration of an orally available CCK-1R agonist to obese patients failed to reduce body weight (Jordan et al. 2008; Kissileff et al. 1981). Although several CCK-1R agonists have been developed, research is still ongoing and no active agents have yet reached clinical practice (Miller and Desai 2016).

A polymorphism in GLP1R has been associated with elevated body mass index in some populations, suggesting that defects in GLP-1 signaling could contribute to obesity risk (Li et al. 2014; Steinert et al. 2017). Most, but not all, studies have shown that post-meal GLP-1 levels are reduced in obese patients, suggesting that defects in GLP-1 system may contribute to overeating in obesity (Steinert et al. 2017). Interestingly, the incretin effect of GLP-1 is diminished, but still present, in obese patients, and GLP-1 is a crucial contributor for glycemic control in individuals with insulin resistance or diabetes (Aulinger et al. 2015; Bagger et al. 2011). Human and animal studies have shown that RYGB fails to affect fasting GLP-1 levels but substantially increases postprandial GLP-1 levels, which seem to play a major beneficial effect of RYGB (Rhee et al. 2012). It is still a matter of debate if GLP-1 is involved in the reduction of eating and weight loss seen after RYGB; however, compelling evidence indicates that GLP-1 contributes to the beneficial effects of RYGB on glycemic regulation in humans (Steinert et al. 2017). GLP-1 is currently the most successful GI hormone exploited as a therapeutic target, mainly in relation with glycemic control. The pharmacological strategies include the DPP-4-resistant

GLP1R agonist exendin-4, which is a natural peptide component of Gila monster saliva, and its synthetic versions exenatide and liraglutide, which are GLP-1 analogs with an acylated side chain (Troke et al. 2014). Alternatively, DPP-4 inhibitors, such as sitagliptin and vildagliptin, are used to increase the endogenous GLP-1 levels. These DPP-4 inhibitors are currently licensed for the treatment of type 2 diabetes mellitus but fail to reduce body weight, likely because DPP-4 inhibition also impacts on the generation of bio-active PYY(3–36) (Field et al. 2008). Thus, the utility of the GLP-1 system as a target for weight loss therapies remains limited, and it seems unlikely that GLP-1-related monotherapy will be used for the treatment of obesity (Troke et al. 2014).

The relationship between PYY levels and obesity is uncertain. Some studies found that obese patients display lower basal PYY levels and a blunted meal-induced increase; however, this findings could not be confirmed by others (Stanley et al. 2004; Stadlbauer et al. 2015). As seen for GLP-1, obese patients that underwent RYGB show markedly increased meal-stimulated PYY levels (Manning and Batterham 2014) that, together with GLP-1 elevations, may contribute to the reduction in food intake following surgery (Karra et al. 2009; Manning and Batterham 2014). This observations as well as many rodent studies have supported the notion that the PYY system may be pharmacologically targeted as an anti-obesity therapy (Karra et al. 2009). Interestingly, PYY equally reduced food intake in lean or obese subjects in clinical trials; however, the rapid rise of PYY levels induced by either intranasal or oral preparations of PYY(3–36) was shown to induce nausea and vomiting (Field et al. 2008). Thus, it remains unclear if PYY-based therapies will be available to treat obesity.

The potential link between OXM, PP, or SST and obesity in humans is currently uncertain. Few data are available regarding the OXM levels or effects in obese patients (Huda et al. 2006). Similarly to PYY and GLP-1, OXM is exaggeratedly increased after meals in patients that underwent RYGB surgery, and such increase has been linked to the loss of appetite found after surgery (Pocai 2014; Huda et al. 2006). The beneficial actions of OXM on food intake, energy expenditure, and body weight make it a promising target for the treatment of obesity. Given the short half-life of OXM, future modified versions of this peptide, resistant to degradation, may be potential therapeutic candidates to improve glycemic control and suppress appetite in obese patients (Pocai 2014; Spreckley and Murphy 2015). Regarding PP, some studies found that obese patients display lower fasting PP levels, as compared to lean subjects, but this observation could not be confirmed by other studies (Lean and Malkova 2016). The effect of weight loss on PP levels in obese individuals has also been inconsistent (Lean and Malkova 2016). Currently, there is no anti-obesity treatment targeting the PP system. However, the ability of PP to strongly suppress appetite and promote weight loss in humans has made this GI hormone an attractive therapeutic target, and several PP analogs have been generated and successfully tested in animal models (Choudhury et al. 2016; Troke et al. 2014). Thus, some novel compounds targeting the PP system may be soon tested for human use. Notably, the potent anorectic effect of PP seen in rodents suggests that Y4R agonists may be useful to treat obesity (Troke et al. 2014). Regarding the SST system, it seems unlikely to develop an anti-obesity therapy based on this system given its pleiotropic functions (Rai et al. 2015).

In summary, no anti-obesity drug based on the manipulation of GI hormones currently exists on the market mainly because of lack of evidence of sustained body weight loss. Despite the potent effect of some GI hormones on appetite, their failure as therapeutic drugs is most likely due to the existence of many associated and redundant compensatory mechanisms that control energy homeostasis. In addition, the short-term effect of most GI hormones in physiological conditions adds some pharmacological challenges that need to be overcome in long-acting compounds. More recently, the use of combined therapies that take advantage of two or even more compounds targeting different GI hormonal systems has emerged as an exciting possibility that is currently under study (Troke et al. 2014). Hopefully, future innovative investigations will be able to develop GI hormone-based therapies to provide an effective treatment for obesity.

Didactic Elements

Q1. Which are the main GI hormones that regulate energy balance?

Q2. In which part of the GI are ghrelin, PYY, GLP-1, and CCK mainly produced?

Q3. Which GI hormones are more important in the regulation of the appetite?

Q4. Which GI hormones are more important in the regulation of the postprandial glycaemia?

Q5. Which GI hormones increase their bioactivity by proteolysis once they are secreted?

Disclosure Statement The authors have nothing to disclose.

References

Adler, G., Beglinger, C., Braun, U., Reinshagen, M., Koop, I., Schafmayer, A., Rovati, L., & Arnold, R. (1991). Interaction of the cholinergic system and cholecystokinin in the regulation of endogenous and exogenous stimulation of pancreatic secretion in humans. *Gastroenterology, 100*(2), 537–543.

Ahlman, H. N. (2001). The gut as the largest endocrine organ in the body. *Annals of Oncology: Official Journal of the European Society for Medical Oncology, 12*(Suppl 2), S63–S68.

Akamizu, T., Takaya, K., Irako, T., Hosoda, H., Teramukai, S., Matsuyama, A., Tada, H., Miura, K., Shimizu, A., Fukushima, M., Yokode, M., Tanaka, K., & Kangawa, K. (2004). Pharmacokinetics, safety, and endocrine and appetite effects of ghrelin administration in young healthy subjects. *European Journal of Endocrinology, 150*(4), 447–455.

Al Massadi, O., Lopez, M., Tschop, M., Dieguez, C., & Nogueiras, R. (2017). Current understanding of the hypothalamic ghrelin pathways inducing appetite and adiposity. *Trends in Neurosciences.* https://doi.org/10.1016/j.tins.2016.12.003.

Alvarez, E., Martinez, M. D., Roncero, I., Chowen, J. A., Garcia-Cuartero, B., Gispert, J. D., Sanz, C., Vazquez, P., Maldonado, A., de Caceres, J., Desco, M., Pozo, M. A., & Blazquez, E. (2005). The expression of GLP-1 receptor mRNA and protein

allows the effect of GLP-1 on glucose metabolism in the human hypothalamus and brainstem. *Journal of Neurochemistry, 92*(4), 798–806. https://doi.org/10.1111/j.1471-4159.2004.02914.x.

Aragon, F., Karaca, M., Novials, A., Maldonado, R., Maechler, P., & Rubi, B. (2015). Pancreatic polypeptide regulates glucagon release through PPYR1 receptors expressed in mouse and human alpha-cells. *Biochimica et Biophysica Acta, 1850*(2), 343–351. https://doi.org/10.1016/j.bbagen.2014.11.005.

Arnold, M., Mura, A., Langhans, W., & Geary, N. (2006). Gut vagal afferents are not necessary for the eating-stimulatory effect of intraperitoneally injected ghrelin in the rat. *The Journal of Neuroscience: The Official Journal of the Society for Neuroscience, 26*(43), 11052–11060. https://doi.org/10.1523/JNEUROSCI.2606-06.2006.

Asakawa, A., Inui, A., Yuzuriha, H., Ueno, N., Katsuura, G., Fujimiya, M., Fujino, M. A., Niijima, A., Meguid, M. M., & Kasuga, M. (2003). Characterization of the effects of pancreatic polypeptide in the regulation of energy balance. *Gastroenterology, 124*(5), 1325–1336.

Aulinger, B. A., Vahl, T. P., Wilson-Perez, H. E., Prigeon, R. L., & D'Alessio, D. A. (2015). Beta-cell sensitivity to GLP-1 in healthy humans is variable and proportional to insulin sensitivity. *The Journal of Clinical Endocrinology and Metabolism, 100*(6), 2489–2496. https://doi.org/10.1210/jc.2014-4009.

Bagger, J. I., Knop, F. K., Lund, A., Vestergaard, H., Holst, J. J., & Vilsboll, T. (2011). Impaired regulation of the incretin effect in patients with type 2 diabetes. *The Journal of Clinical Endocrinology and Metabolism, 96*(3), 737–745. https://doi.org/10.1210/jc.2010-2435.

Baranowska, B., Radzikowska, M., Wasilewska-Dziubinska, E., Roguski, K., & Borowiec, M. (2000). Disturbed release of gastrointestinal peptides in anorexia nervosa and in obesity. *Diabetes, Obesity & Metabolism, 2*(2), 99–103.

Barrera, J. G., Sandoval, D. A., D'Alessio, D. A., & Seeley, R. J. (2011). GLP-1 and energy balance: An integrated model of short-term and long-term control. *Nature Reviews Endocrinology, 7*(9), 507–516. https://doi.org/10.1038/nrendo.2011.77.

Bataille, D., Gespach, C., Tatemoto, K., Marie, J. C., Coudray, A. M., Rosselin, G., & Mutt, V. (1981). Bioactive enteroglucagon (oxyntomodulin): Present knowledge on its chemical structure and its biological activities. *Peptides, 2*(Suppl 2), 41–44.

Batterham, R. L., Cohen, M. A., Ellis, S. M., Le Roux, C. W., Withers, D. J., Frost, G. S., Ghatei, M. A., & Bloom, S. R. (2003). Inhibition of food intake in obese subjects by peptide YY3-36. *The New England Journal of Medicine, 349*(10), 941–948. https://doi.org/10.1056/NEJMoa030204.

Batterham, R. L., Cowley, M. A., Small, C. J., Herzog, H., Cohen, M. A., Dakin, C. L., Wren, A. M., Brynes, A. E., Low, M. J., Ghatei, M. A., Cone, R. D., & Bloom, S. R. (2002). Gut hormone PYY(3-36) physiologically inhibits food intake. *Nature, 418*(6898), 650–654. https://doi.org/10.1038/nature02666.

Bayliss, W. M., & Starling, E. H. (1902). The mechanism of pancreatic secretion. *The Journal of Physiology, 28*(5), 325–353.

Beckman, L. M., Beckman, T. R., & Earthman, C. P. (2010). Changes in gastrointestinal hormones and leptin after roux-en-Y gastric bypass procedure: A review. *Journal of the American Dietetic Association, 110*(4), 571–584. https://doi.org/10.1016/j.jada.2009.12.023.

Beinfeld, M. C. (2003). Biosynthesis and processing of pro CCK: Recent progress and future challenges. *Life Sciences, 72*(7), 747–757.

Brazeau, P., Vale, W., Burgus, R., Ling, N., Butcher, M., Rivier, J., & Guillemin, R. (1973). Hypothalamic polypeptide that inhibits the secretion of immunoreactive pituitary growth hormone. *Science, 179*(4068), 77–79.

Brennan, I. M., Luscombe-Marsh, N. D., Seimon, R. V., Otto, B., Horowitz, M., Wishart, J. M., & Feinle-Bisset, C. (2012). Effects of fat, protein, and carbohydrate and protein load on appetite, plasma cholecystokinin, peptide YY, and ghrelin, and energy intake in lean and obese men. *American Journal of Physiology Gastrointestinal and Liver Physiology, 303*(1), G129–G140. https://doi.org/10.1152/ajpgi.00478.2011.

Brennan, I. M., Otto, B., Feltrin, K. L., Meyer, J. H., Horowitz, M., & Feinle-Bisset, C. (2007). Intravenous CCK-8, but not GLP-1, suppresses ghrelin and stimulates PYY release in healthy men. *Peptides, 28*(3), 607–611. https://doi.org/10.1016/j.peptides.2006.10.014.

Broglio, F., Gottero, C., Van Koetsveld, P., Prodam, F., Destefanis, S., Benso, A., Gauna, C., Hofland, L., Arvat, E., van der Lely, A. J., & Ghigo, E. (2004). Acetylcholine regulates ghrelin secretion in humans. *The Journal of Clinical Endocrinology and Metabolism, 89*(5), 2429–2433. https://doi.org/10.1210/jc.2003-031517.

Cabral, A., Portiansky, E., Sanchez-Jaramillo, E., Zigman, J. M., & Perello, M. (2016). Ghrelin activates hypophysiotropic corticotropin-releasing factor neurons independently of the arcuate nucleus. *Psychoneuroendocrinology, 67*, 27–39. https://doi.org/10.1016/j.psyneuen.2016.01.027.

Callaghan, B., & Furness, J. B. (2014). Novel and conventional receptors for ghrelin, desacyl-ghrelin, and pharmacologically related compounds. *Pharmacological Reviews, 66*(4), 984–1001. https://doi.org/10.1124/pr.113.008433.

Carr, R. D., Larsen, M. O., Jelic, K., Lindgren, O., Vikman, J., Holst, J. J., Deacon, C. F., & Ahren, B. (2010). Secretion and dipeptidyl peptidase-4-mediated metabolism of incretin hormones after a mixed meal or glucose ingestion in obese compared to lean, nondiabetic men. *The Journal of Clinical Endocrinology and Metabolism, 95*(2), 872–878. https://doi.org/10.1210/jc.2009-2054.

Corp, E. S., McQuade, J., Moran, T. H., & Smith, G. P. (1993). Characterization of type a and type B CCK receptor binding sites in rat vagus nerve. *Brain Research, 623*(1), 161–166.

Crawley, J. N., & Beinfeld, M. C. (1983). Rapid development of tolerance to the behavioural actions of cholecystokinin. *Nature, 302*(5910), 703–706.

Cruz, C. R., & Smith, R. G. (2008). The growth hormone secretagogue receptor. *Vitamins and Hormones, 77*, 47–88. https://doi.org/10.1016/S0083-6729(06)77004-2.

Cummings, D. E. (2006). Ghrelin and the short- and long-term regulation of appetite and body weight. *Physiology & Behavior, 89*(1), 71–84. https://doi.org/10.1016/j.physbeh.2006.05.022.

Cummings, D. E., & Shannon, M. H. (2003). Ghrelin and gastric bypass: Is there a hormonal contribution to surgical weight loss? *The Journal of Clinical Endocrinology and Metabolism, 88*(7), 2999–3002. https://doi.org/10.1210/jc.2003-030705.

Cummings, D. E., Weigle, D. S., Frayo, R. S., Breen, P. A., Ma, M. K., Dellinger, E. P., & Purnell, J. Q. (2002). Plasma ghrelin levels after diet-induced weight loss or gastric bypass surgery. *The New England Journal of Medicine, 346*(21), 1623–1630. https://doi.org/10.1056/NEJMoa012908.

Chan, J. L., Bullen, J., Lee, J. H., Yiannakouris, N., & Mantzoros, C. S. (2004). Ghrelin levels are not regulated by recombinant leptin administration and/or three days of fasting in healthy subjects. *The Journal of Clinical Endocrinology and Metabolism, 89*(1), 335–343. https://doi.org/10.1210/jc.2003-031412.

Chandarana, K., Gelegen, C., Irvine, E. E., Choudhury, A. I., Amouyal, C., Andreelli, F., Withers, D. J., & Batterham, R. L. (2013). Peripheral activation of the Y2-receptor promotes secretion of GLP-1 and improves glucose tolerance. *Molecular Metabolism, 2*(3), 142–152. https://doi.org/10.1016/j.molmet.2013.03.001.

Chaudhri, O., Small, C., & Bloom, S. (2006). Gastrointestinal hormones regulating appetite. *Philosophical Transactions of the Royal Society of London. Series B, Biological Sciences, 361*(1471), 1187–1209. https://doi.org/10.1098/rstb.2006.1856.

Cheung, G. W., Kokorovic, A., Lam, C. K., Chari, M., & Lam, T. K. (2009). Intestinal cholecystokinin controls glucose production through a neuronal network. *Cell Metabolism, 10*(2), 99–109. https://doi.org/10.1016/j.cmet.2009.07.005.

Choudhury, S. M., Tan, T. M., & Bloom, S. R. (2016). Gastrointestinal hormones and their role in obesity. *Current Opinion in Endocrinology, Diabetes, and Obesity, 23*(1), 18–22. https://doi.org/10.1097/MED.0000000000000216.

Chuang, J. C., Sakata, I., Kohno, D., Perello, M., Osborne-Lawrence, S., Repa, J. J., & Zigman, J. M. (2011). Ghrelin directly stimulates glucagon secretion from pancreatic alpha-cells. *Molecular Endocrinology, 25*(9), 1600–1611. https://doi.org/10.1210/me.2011-1001.

Dakin, C. L., Gunn, I., Small, C. J., Edwards, C. M., Hay, D. L., Smith, D. M., Ghatei, M. A., & Bloom, S. R. (2001). Oxyntomodulin inhibits food intake in the rat. *Endocrinology, 142*(10), 4244–4250. https://doi.org/10.1210/endo.142.10.8430.

De Vriese, C., Gregoire, F., Lema-Kisoka, R., Waelbroeck, M., Robberecht, P., & Delporte, C. (2004). Ghrelin degradation by serum and tissue homogenates: Identification of the cleavage sites. *Endocrinology, 145*(11), 4997–5005. https://doi.org/10.1210/en.2004-0569.

Degen, L., Oesch, S., Casanova, M., Graf, S., Ketterer, S., Drewe, J., & Beglinger, C. (2005). Effect of peptide YY3-36 on food intake in humans. *Gastroenterology, 129*(5), 1430–1436. https://doi.org/10.1053/j.gastro.2005.09.001.

Dockray, G. J. (2009). Cholecystokinin and gut-brain signalling. *Regulatory Peptides, 155*(1–3), 6–10. https://doi.org/10.1016/j.regpep.2009.03.015.

Drucker, D. J., Philippe, J., Mojsov, S., Chick, W. L., & Habener, J. F. (1987). Glucagon-like peptide I stimulates insulin gene expression and increases cyclic AMP levels in a rat islet cell line. *Proceedings of the National Academy of Sciences of the United States of America, 84*(10), 3434–3438.

Du, X., Kosinski, J. R., Lao, J., Shen, X., Petrov, A., Chicchi, G. G., Eiermann, G. J., & Pocai, A. (2012). Differential effects of oxyntomodulin and GLP-1 on glucose metabolism. *American Journal of Physiology. Endocrinology and Metabolism, 303*(2), E265–E271. https://doi.org/10.1152/ajpendo.00142.2012.

Dufresne, M., Seva, C., & Fourmy, D. (2006). Cholecystokinin and gastrin receptors. *Physiological Reviews, 86*(3), 805–847. https://doi.org/10.1152/physrev.00014.2005.

Dumont, Y., Fournier, A., St-Pierre, S., & Quirion, R. (1995). Characterization of neuropeptide Y binding sites in rat brain membrane preparations using [125I][Leu31,Pro34]peptide YY and [125I]peptide YY3-36 as selective Y1 and Y2 radioligands. *The Journal of Pharmacology and Experimental Therapeutics, 272*(2), 673–680.

Eysselein, V. E., Reeve, J. R., Jr., & Eberlein, G. (1986). Cholecystokinin--gene structure, and molecular forms in tissue and blood. *Zeitschrift fur Gastroenterologie, 24*(10), 645–659.

Fehmann, H. C., Goke, R., & Goke, B. (1995). Cell and molecular biology of the incretin hormones glucagon-like peptide-I and glucose-dependent insulin releasing polypeptide. *Endocrine Reviews, 16*(3), 390–410. https://doi.org/10.1210/edrv-16-3-390.

Fenske, W. K., Bueter, M., Miras, A. D., Ghatei, M. A., Bloom, S. R., & le Roux, C. W. (2012). Exogenous peptide YY3-36 and Exendin-4 further decrease food intake, whereas octreotide increases food intake in rats after roux-en-Y gastric bypass. *International Journal of Obesity, 36*(3), 379–384. https://doi.org/10.1038/ijo.2011.126.

Field, B. C. T., Wren, A. M., Cooke, D., & Bloom, S. R. (2008). Gut hormones as potential new targets for appetite regulation and the treatment of obesity. *Drugs, 68*(2), 147–163.

Flanagan, D. E., Evans, M. L., Monsod, T. P., Rife, F., Heptulla, R. A., Tamborlane, W. V., & Sherwin, R. S. (2003). The influence of insulin on circulating ghrelin. *American Journal of Physiology. Endocrinology and Metabolism, 284*(2), E313–E316. https://doi.org/10.1152/ajpendo.00569.2001.

Foster-Schubert, K. E., Overduin, J., Prudom, C. E., Liu, J., Callahan, H. S., Gaylinn, B. D., Thorner, M. O., & Cummings, D. E. (2008). Acyl and total ghrelin are suppressed strongly by ingested proteins, weakly by lipids, and biphasically by carbohydrates. *The Journal of Clinical Endocrinology and Metabolism, 93*(5), 1971–1979. https://doi.org/10.1210/jc.2007-2289.

Gourcerol, G., Coskun, T., Craft, L. S., Mayer, J. P., Heiman, M. L., Wang, L., Million, M., St-Pierre, D. H., & Tache, Y. (2007). Preproghrelin-derived peptide, obestatin, fails to influence food intake in lean or obese rodents. *Obesity (Silver Spring), 15*(11), 2643–2652. https://doi.org/10.1038/oby.2007.316.

Gueorguiev, M., Lecoeur, C., Meyre, D., Benzinou, M., Mein, C. A., Hinney, A., Vatin, V., Weill, J., Heude, B., Hebebrand, J., Grossman, A. B., Korbonits, M., & Froguel, P. (2009). Association studies on ghrelin and ghrelin receptor gene polymorphisms with obesity. *Obesity (Silver Spring), 17*(4), 745–754. https://doi.org/10.1038/oby.2008.589.

Gutierrez, J. A., Solenberg, P. J., Perkins, D. R., Willency, J. A., Knierman, M. D., Jin, Z., Witcher, D. R., Luo, S., Onyia, J. E., & Hale, J. E. (2008). Ghrelin octanoylation mediated by an orphan lipid transferase. *Proceedings of the National Academy of Sciences of the United States of America, 105*(17), 6320–6325. https://doi.org/10.1073/pnas.0800708105.

Hansen, K. B., Rosenkilde, M. M., Knop, F. K., Wellner, N., Diep, T. A., Rehfeld, J. F., Andersen, U. B., Holst, J. J., & Hansen, H. S. (2011). 2-Oleoyl glycerol is a GPR119 agonist and signals GLP-1 release in humans. *The Journal of Clinical Endocrinology and Metabolism, 96*(9), E1409–E1417. https://doi.org/10.1210/jc.2011-0647.

Hillman, J. B., Tong, J., & Tschop, M. (2011). Ghrelin biology and its role in weight-related disorders. *Discovery Medicine, 11*(61), 521–528.

Holst, B., Cygankiewicz, A., Jensen, T. H., Ankersen, M., & Schwartz, T. W. (2003). High constitutive signaling of the ghrelin receptor--identification of a potent inverse agonist. *Molecular Endocrinology, 17*(11), 2201–2210. https://doi.org/10.1210/me.2003-0069.

Holst, J. J. (2007). The physiology of glucagon-like peptide 1. *Physiological Reviews, 87*(4), 1409–1439. https://doi.org/10.1152/physrev.00034.2006.

Hosoda, H., & Kangawa, K. (2008). The autonomic nervous system regulates gastric ghrelin secretion in rats. *Regulatory Peptides, 146*(1–3), 12–18. https://doi.org/10.1016/j.regpep.2007.07.005.

Howard, A. D., Feighner, S. D., Cully, D. F., Arena, J. P., Liberator, P. A., Rosenblum, C. I., Hamelin, M., Hreniuk, D. L., Palyha, O. C., Anderson, J., Paress, P. S., Diaz, C., Chou, M., Liu, K. K., McKee, K. K., Pong, S. S., Chaung, L. Y., Elbrecht, A., Dashkevicz, M., Heavens, R., Rigby, M., Sirinathsinghji, D. J., Dean, D. C., Melillo, D. G., Patchett, A. A., Nargund, R., Griffin, P. R., DeMartino, J. A., Gupta, S. K.,

Schaeffer, J. M., Smith, R. G., & Van der Ploeg, L. H. (1996). A receptor in pituitary and hypothalamus that functions in growth hormone release. *Science, 273*(5277), 974–977.

Huang, S. C., Yu, D. H., Wank, S. A., Mantey, S., Gardner, J. D., & Jensen, R. T. (1989). Importance of sulfation of gastrin or cholecystokinin (CCK) on affinity for gastrin and CCK receptors. *Peptides, 10*(4), 785–789.

Huda, M. S. B., Wilding, J. P. H., & Pinkney, J. H. (2006). Gut peptides and the regulation of appetite. *Obesity Reviews, 7*(2), 163–182. https://doi.org/10.1111/j.1467-789X.2006.00245.x.

Imamura, M. (2002). Effects of surgical manipulation of the intestine on peptide YY and its physiology. *Peptides, 23*(2), 403–407.

Jones, R. B., McKie, S., Astbury, N., Little, T. J., Tivey, S., Lassman, D. J., McLaughlin, J., Luckman, S., Williams, S. R., Dockray, G. J., & Thompson, D. G. (2012). Functional neuroimaging demonstrates that ghrelin inhibits the central nervous system response to ingested lipid. *Gut, 61*(11), 1543–1551. https://doi.org/10.1136/gutjnl-2011-301323.

Jordan, J., Greenway, F. L., Leiter, L. A., Li, Z., Jacobson, P., Murphy, K., Hill, J., Kler, L., & Aftring, R. P. (2008). Stimulation of cholecystokinin-a receptors with GI181771X does not cause weight loss in overweight or obese patients. *Clinical Pharmacology and Therapeutics, 83*(2), 281–287. https://doi.org/10.1038/sj.clpt.6100272.

Karra, E., Chandarana, K., & Batterham, R. L. (2009). The role of peptide YY in appetite regulation and obesity. *The Journal of Physiology, 587*(1), 19–25. https://doi.org/10.1113/jphysiol.2008.164269.

Khandekar, N., Berning, B. A., Sainsbury, A., & Lin, S. (2015). The role of pancreatic polypeptide in the regulation of energy homeostasis. *Molecular and Cellular Endocrinology, 418*(Pt 1), 33–41. https://doi.org/10.1016/j.mce.2015.06.028.

Kimmel, J. R., Hayden, L. J., & Pollock, H. G. (1975). Isolation and characterization of a new pancreatic polypeptide hormone. *The Journal of Biological Chemistry, 250*(24), 9369–9376.

Kirchner, H., Tong, J., Tschöp, M. H., & Pfluger, P. T. (2010). Ghrelin and PYY in the regulation of energy balance and metabolism: Lessons from mouse mutants. *American Journal of Physiology. Endocrinology and Metabolism, 298*(5), E909–E919. https://doi.org/10.1152/ajpendo.00191.2009.

Kissileff, H. R., Pi-Sunyer, F. X., Thornton, J., & Smith, G. P. (1981). C-terminal octapeptide of cholecystokinin decreases food intake in man. *The American Journal of Clinical Nutrition, 34*(2), 154–160.

Kojima, M., Hosoda, H., Date, Y., Nakazato, M., Matsuo, H., & Kangawa, K. (1999). Ghrelin is a growth-hormone-releasing acylated peptide from stomach. *Nature, 402*(6762), 656–660. https://doi.org/10.1038/45230.

Kojima, M., & Kangawa, K. (2010). Ghrelin: More than endogenous growth hormone secretagogue. *Annals of the New York Academy of Sciences, 1200*, 140–148. https://doi.org/10.1111/j.1749-6632.2010.05516.x.

Kumar, U., & Grant, M. (2010). Somatostatin and somatostatin receptors. *Results and Problems in Cell Differentiation, 50*, 137–184. https://doi.org/10.1007/400_2009_29.

Kurashina, T., Dezaki, K., Yoshida, M., Sukma Rita, R., Ito, K., Taguchi, M., Miura, R., Tominaga, M., Ishibashi, S., Kakei, M., & Yada, T. (2015). The beta-cell GHSR and downstream cAMP/TRPM2 signaling account for insulinostatic and glycemic effects of ghrelin. *Scientific Reports, 5*, 14041. https://doi.org/10.1038/srep14041.

le Roux, C. W., Patterson, M., Vincent, R. P., Hunt, C., Ghatei, M. A., & Bloom, S. R. (2005). Postprandial plasma ghrelin is suppressed proportional to meal calorie content in normal-weight but not obese subjects. *The Journal of Clinical Endocrinology and Metabolism, 90*(2), 1068–1071. https://doi.org/10.1210/jc.2004-1216.

Lean, M. E., & Malkova, D. (2016). Altered gut and adipose tissue hormones in overweight and obese individuals: Cause or consequence? *International Journal of Obesity, 40*(4), 622–632. https://doi.org/10.1038/ijo.2015.220.

Levin, F., Edholm, T., Schmidt, P. T., Gryback, P., Jacobsson, H., Degerblad, M., Hoybye, C., Holst, J. J., Rehfeld, J. F., Hellstrom, P. M., & Naslund, E. (2006). Ghrelin stimulates gastric emptying and hunger in normal-weight humans. *The Journal of Clinical Endocrinology and Metabolism, 91*(9), 3296–3302. https://doi.org/10.1210/jc.2005-2638.

Li, P., Tiwari, H. K., Lin, W. Y., Allison, D. B., Chung, W. K., Leibel, R. L., Yi, N., & Liu, N. (2014). Genetic association analysis of 30 genes related to obesity in a European American population. *International Journal of Obesity, 38*(5), 724–729. https://doi.org/10.1038/ijo.2013.140.

Liddle, R. A., Goldfine, I. D., Rosen, M. S., Taplitz, R. A., & Williams, J. A. (1985). Cholecystokinin bioactivity in human plasma. Molecular forms, responses to feeding, and relationship to gallbladder contraction. *The Journal of Clinical Investigation, 75*(4), 1144–1152. https://doi.org/10.1172/JCI111809.

Lieverse, R. J., Jansen, J. B., Masclee, A. A., & Lamers, C. B. (1995). Satiety effects of a physiological dose of cholecystokinin in humans. *Gut, 36*(2), 176–179.

Little, T. J., Doran, S., Meyer, J. H., Smout, A. J., O'Donovan, D. G., Wu, K. L., Jones, K. L., Wishart, J., Rayner, C. K., Horowitz, M., & Feinle-Bisset, C. (2006). The release of GLP-1 and ghrelin, but not GIP and CCK, by glucose is dependent upon the length of small intestine exposed. *American Journal of Physiology. Endocrinology and Metabolism, 291*(3), E647–E655. https://doi.org/10.1152/ajpendo.00099.2006.

Liu, B., Garcia, E. A., & Korbonits, M. (2011). Genetic studies on the ghrelin, growth hormone secretagogue receptor (GHSR) and ghrelin O-acyl transferase

(GOAT) genes. *Peptides, 32*(11), 2191–2207. https://doi.org/10.1016/j.peptides.2011.09.006.

Lopez Soto, E. J., Agosti, F., Cabral, A., Mustafa, E. R., Damonte, V. M., Gandini, M. A., Rodriguez, S., Castrogiovanni, D., Felix, R., Perello, M., & Raingo, J. (2015). Constitutive and ghrelin-dependent GHSR1a activation impairs CaV2.1 and CaV2.2 currents in hypothalamic neurons. *The Journal of General Physiology, 146*(3), 205–219. https://doi.org/10.1085/jgp.201511383.

Low, M. J. (2004). Clinical endocrinology and metabolism. The somatostatin neuroendocrine system: Physiology and clinical relevance in gastrointestinal and pancreatic disorders. *Best Practice & Research Clinical Endocrinology & Metabolism, 18*(4), 607–622. https://doi.org/10.1016/j.beem.2004.08.005.

M'Kadmi, C., Leyris, J. P., Onfroy, L., Gales, C., Sauliere, A., Gagne, D., Damian, M., Mary, S., Maingot, M., Denoyelle, S., Verdie, P., Fehrentz, J. A., Martinez, J., Baneres, J. L., & Marie, J. (2015). Agonism, antagonism, and inverse Agonism Bias at the ghrelin receptor signaling. *The Journal of Biological Chemistry, 290*(45), 27021–27039. https://doi.org/10.1074/jbc.M115.659250.

Mace, O. J., Tehan, B., & Marshall, F. (2015). Pharmacology and physiology of gastrointestinal enteroendocrine cells. *Pharmacology Research & Perspectives, 3*(4), e00155. https://doi.org/10.1002/prp2.155.

Manning, S., & Batterham, R. L. (2014). The role of gut hormone peptide YY in energy and glucose homeostasis: Twelve years on. *Annual Review of Physiology, 76*, 585–608. https://doi.org/10.1146/annurev-physiol-021113-170404.

Mayo, K. E., Miller, L. J., Bataille, D., Dalle, S., Goke, B., Thorens, B., & Drucker, D. J. (2003). International Union of Pharmacology. XXXV. The glucagon receptor family. *Pharmacological Reviews, 55*(1), 167–194. https://doi.org/10.1124/pr.55.1.6.

Miller, L. J., & Desai, A. J. (2016). Metabolic actions of the type 1 cholecystokinin receptor: Its potential as a therapeutic target. *Trends in Endocrinology and Metabolism: TEM, 27*(9), 609–619. https://doi.org/10.1016/j.tem.2016.04.002.

Mohlig, M., Spranger, J., Otto, B., Ristow, M., Tschop, M., & Pfeiffer, A. F. (2002). Euglycemic hyperinsulinemia, but not lipid infusion, decreases circulating ghrelin levels in humans. *Journal of Endocrinological Investigation, 25*(11), RC36–RC38. https://doi.org/10.1007/BF03344062.

Mojsov, S., Heinrich, G., Wilson, I. B., Ravazzola, M., Orci, L., & Habener, J. F. (1986). Preproglucagon gene expression in pancreas and intestine diversifies at the level of post-translational processing. *The Journal of Biological Chemistry, 261*(25), 11880–11889.

Morpurgo, P. S., Resnik, M., Agosti, F., Cappiello, V., Sartorio, A., & Spada, A. (2003). Ghrelin secretion in severely obese subjects before and after a 3-week integrated body mass reduction program. *Journal*

of Endocrinological Investigation, 26(8), 723–727. https://doi.org/10.1007/BF03347353.

Mortensen, K., Christensen, L. L., Holst, J. J., & Orskov, C. (2003). GLP-1 and GIP are colocalized in a subset of endocrine cells in the small intestine. *Regulatory Peptides, 114*(2–3), 189–196.

Nauck, M. A., Niedereichholz, U., Ettler, R., Holst, J. J., Orskov, C., Ritzel, R., & Schmiegel, W. H. (1997). Glucagon-like peptide 1 inhibition of gastric emptying outweighs its insulinotropic effects in healthy humans. *The American Journal of Physiology, 273*(5 Pt 1), E981–E988.

Nogueiras, R., Pfluger, P., Tovar, S., Arnold, M., Mitchell, S., Morris, A., Perez-Tilve, D., Vazquez, M. J., Wiedmer, P., Castaneda, T. R., DiMarchi, R., Tschop, M., Schurmann, A., Joost, H. G., Williams, L. M., Langhans, W., & Dieguez, C. (2007). Effects of obestatin on energy balance and growth hormone secretion in rodents. *Endocrinology, 148*(1), 21–26. https://doi.org/10.1210/en.2006-0915.

Orskov, C., Holst, J. J., Knuhtsen, S., Baldissera, F. G., Poulsen, S. S., & Nielsen, O. V. (1986). Glucagon-like peptides GLP-1 and GLP-2, predicted products of the glucagon gene, are secreted separately from pig small intestine but not pancreas. *Endocrinology, 119*(4), 1467–1475. https://doi.org/10.1210/endo-119-4-1467.

Orskov, C., Poulsen, S. S., Moller, M., & Holst, J. J. (1996). Glucagon-like peptide I receptors in the subfornical organ and the area postrema are accessible to circulating glucagon-like peptide I. *Diabetes, 45*(6), 832–835.

Overduin, J., Gibbs, J., Cummings, D. E., & Reeve, J. R., Jr. (2014). CCK-58 elicits both satiety and satiation in rats while CCK-8 elicits only satiation. *Peptides, 54*, 71–80. https://doi.org/10.1016/j.peptides.2014.01.008.

Owyang, C., & Heldsinger, A. (2011). Vagal control of satiety and hormonal regulation of appetite. *Journal of Neurogastroenterology and Motility, 17*(4), 338–348. https://doi.org/10.5056/jnm.2011.17.4.338.

Parker, H. E., Reimann, F., & Gribble, F. M. (2010). Molecular mechanisms underlying nutrient-stimulated incretin secretion. *Expert Reviews in Molecular Medicine, 12*, e1. https://doi.org/10.1017/S146239940900132X.

Perello, M., & Nillni, E. A. (2007). The biosynthesis and processing of neuropeptides: Lessons from prothyrotropin releasing hormone (proTRH). *Frontiers in Bioscience: A Journal and Virtual Library, 12*, 3554–3565.

Perello, M., & Raingo, J. (2014). Central ghrelin receptors and food intake. In J. Portelli & I. Smolders (Eds.), *Central functions of the ghrelin receptor* (pp. 65–88). New York: Springer. https://doi.org/10.1007/978-1-4939-0823-3_5.

Perello, M., & Zigman, J. M. (2012). The role of ghrelin in reward-based eating. *Biological Psychiatry, 72*(5), 347–353. https://doi.org/10.1016/j.biopsych.2012.02.016.

Persaud, S. J., & Bewick, G. A. (2014). Peptide YY: More than just an appetite regulator. *Diabetologia, 57*(9), 1762–1769. https://doi.org/10.1007/s00125-014-3292-y.

Peterli, R., Steinert, R. E., Woelnerhanssen, B., Peters, T., Christoffel-Courtin, C., Gass, M., Kern, B., von Fluee, M., & Beglinger, C. (2012). Metabolic and hormonal changes after laparoscopic roux-en-Y gastric bypass and sleeve gastrectomy: A randomized, prospective trial. *Obesity Surgery, 22*(5), 740–748. https://doi.org/10.1007/s11695-012-0622-3.

Pocai, A. (2014). Action and therapeutic potential of oxyntomodulin. *Molecular Metabolism, 3*(3), 241–251. https://doi.org/10.1016/j.molmet.2013.12.001.

Rai, U., Thrimawithana, T. R., Valery, C., & Young, S. A. (2015). Therapeutic uses of somatostatin and its analogues: Current view and potential applications. *Pharmacology & Therapeutics, 152*, 98–110. https://doi.org/10.1016/j.pharmthera.2015.05.007.

Reeve, J. R., Jr., Eysselein, V. E., Rosenquist, G., Zeeh, J., Regner, U., Ho, F. J., Chew, P., Davis, M. T., Lee, T. D., Shively, J. E., Brazer, S. R., & Liddle, R. A. (1996). Evidence that CCK-58 has structure that influences its biological activity. *The American Journal of Physiology, 270*(5 Pt 1), G860–G868.

Rhee, N. A., Vilsboll, T., & Knop, F. K. (2012). Current evidence for a role of GLP-1 in roux-en-Y gastric bypass-induced remission of type 2 diabetes. *Diabetes, Obesity & Metabolism, 14*(4), 291–298. https://doi.org/10.1111/j.1463-1326.2011.01505.x.

Sandoval, D. A., & D'Alessio, D. A. (2015). Physiology of proglucagon peptides: Role of glucagon and GLP-1 in health and disease. *Physiological Reviews, 95*(2), 513–548. https://doi.org/10.1152/physrev.00013.2014.

Schellekens, H., De Francesco, P. N., Kandil, D., Theeuwes, W. F., McCarthy, T., van Oeffelen, W. E., Perello, M., Giblin, L., Dinan, T. G., & Cryan, J. F. (2015). Ghrelin's Orexigenic effect is modulated via a serotonin 2C receptor interaction. *ACS Chemical Neuroscience, 6*(7), 1186–1197. https://doi.org/10.1021/cn500318q.

Schwartz, G. J., McHugh, P. R., & Moran, T. H. (1993). Gastric loads and cholecystokinin synergistically stimulate rat gastric vagal afferents. *The American Journal of Physiology, 265*(4 Pt 2), R872–R876.

Sjostrom, L., Narbro, K., Sjostrom, C. D., Karason, K., Larsson, B., Wedel, H., Lystig, T., Sullivan, M., Bouchard, C., Carlsson, B., Bengtsson, C., Dahlgren, S., Gummesson, A., Jacobson, P., Karlsson, J., Lindroos, A. K., Lonroth, H., Naslund, I., Olbers, T., Stenlof, K., Torgerson, J., Agren, G., & Carlsson, L. M. (2007). Effects of bariatric surgery on mortality in Swedish obese subjects. *The New England Journal of Medicine, 357*(8), 741–752. https://doi.org/10.1056/NEJMoa066254.

Spreckley, E., & Murphy, K. G. (2015). The L-cell in nutritional sensing and the regulation of appetite. *Frontiers in Nutrition, 2*, 23. https://doi.org/10.3389/fnut.2015.00023.

Stadlbauer, U., Woods, S. C., Langhans, W., & Meyer, U. (2015). PYY3-36: Beyond food intake. *Frontiers in Neuroendocrinology, 38*, 1–11. https://doi.org/10.1016/j.yfrne.2014.12.003.

Stanley, S., Wynne, K., & Bloom, S. (2004). Gastrointestinal satiety signals III. Glucagon-like peptide 1, oxyntomodulin, peptide YY, and pancreatic polypeptide. *American Journal of Physiology Gastrointestinal and Liver Physiology, 286*(5), G693–G697. https://doi.org/10.1152/ajpgi.00536.2003.

Steinert, R. E., Feinle-Bisset, C., Asarian, L., Horowitz, M., Beglinger, C., & Geary, N. (2017). Ghrelin, CCK, GLP-1, and PYY(3-36): Secretory controls and physiological roles in eating and Glycemia in health, obesity, and after RYGB. *Physiological Reviews, 97*(1), 411–463. https://doi.org/10.1152/physrev.00031.2014.

Stewart, J. E., Seimon, R. V., Otto, B., Keast, R. S., Clifton, P. M., & Feinle-Bisset, C. (2011). Marked differences in gustatory and gastrointestinal sensitivity to oleic acid between lean and obese men. *The American Journal of Clinical Nutrition, 93*(4), 703–711. https://doi.org/10.3945/ajcn.110.007583.

Svendsen, B., Pedersen, J., Albrechtsen, N. J., Hartmann, B., Torang, S., Rehfeld, J. F., Poulsen, S. S., & Holst, J. J. (2015). An analysis of cosecretion and coexpression of gut hormones from male rat proximal and distal small intestine. *Endocrinology, 156*(3), 847–857. https://doi.org/10.1210/en.2014-1710.

Takahashi, T., Ida, T., Sato, T., Nakashima, Y., Nakamura, Y., Tsuji, A., & Kojima, M. (2009). Production of n-octanoyl-modified ghrelin in cultured cells requires prohormone processing protease and ghrelin O-acyltransferase, as well as n-octanoic acid. *Journal of Biochemistry, 146*(5), 675–682. https://doi.org/10.1093/jb/mvp112.

Tatemoto, K., & Mutt, V. (1980). Isolation of two novel candidate hormones using a chemical method for finding naturally occurring polypeptides. *Nature, 285*(5764), 417–418.

Tong, J., Prigeon, R. L., Davis, H. W., Bidlingmaier, M., Tschop, M. H., & D'Alessio, D. (2013). Physiologic concentrations of exogenously infused ghrelin reduces insulin secretion without affecting insulin sensitivity in healthy humans. *The Journal of Clinical Endocrinology and Metabolism, 98*(6), 2536–2543. https://doi.org/10.1210/jc.2012-4162.

Trapp, S., & Hisadome, K. (2011). Glucagon-like peptide 1 and the brain: Central actions-central sources? *Autonomic Neuroscience: Basic & Clinical, 161*(1–2), 14–19. https://doi.org/10.1016/j.autneu.2010.09.008.

Troke, R. C., Tan, T. M., & Bloom, S. R. (2014). The future role of gut hormones in the treatment of obesity. *Therapeutic Advances in Chronic Disease, 5*(1), 4–14. https://doi.org/10.1177/2040622313506730.

Tschop, M., Weyer, C., Tataranni, P. A., Devanarayan, V., Ravussin, E., & Heiman, M. L. (2001). Circulating ghrelin levels are decreased in human obesity. *Diabetes, 50*(4), 707–709.

Valle, J. D. (2014). Gastrointestinal hormones in the regulation of gut function in health and disease. In *Gastrointestinal anatomy and physiology* (pp. 15–32). John Wiley & Sons, Ltd. https://doi.org/10.1002/9781118833001.ch2.

Vilsboll, T., Krarup, T., Madsbad, S., & Holst, J. J. (2003). Both GLP-1 and GIP are insulinotropic at basal and postprandial glucose levels and contribute nearly equally to the incretin effect of a meal in healthy subjects. *Regulatory Peptides, 114*(2–3), 115–121.

Wang, Y., Prpic, V., Green, G. M., Reeve, J. R., Jr., & Liddle, R. A. (2002). Luminal CCK-releasing factor stimulates CCK release from human intestinal endocrine and STC-1 cells. *American Journal of Physiology Gastrointestinal and Liver Physiology, 282*(1), G16–G22.

Wank, S. A. (1995). Cholecystokinin receptors. *The American Journal of Physiology, 269*(5 Pt 1), G628–G646.

Wheeler, M. B., Lu, M., Dillon, J. S., Leng, X. H., Chen, C., & Boyd, A. E., 3rd. (1993). Functional expression of the rat glucagon-like peptide-I receptor, evidence for coupling to both adenylyl cyclase and phospholipase-C. *Endocrinology, 133*(1), 57–62. https://doi.org/10.1210/endo.133.1.8391428.

Whited, K. L., Thao, D., Lloyd, K. C., Kopin, A. S., & Raybould, H. E. (2006). Targeted disruption of the murine CCK1 receptor gene reduces intestinal lipid-induced feedback inhibition of gastric function. *American Journal of Physiology Gastrointestinal and Liver Physiology, 291*(1), G156–G162. https://doi.org/10.1152/ajpgi.00569.2005.

Williams, D. L. (2009). Minireview: Finding the sweet spot: Peripheral versus central glucagon-like peptide 1 action in feeding and glucose homeostasis. *Endocrinology, 150*(7), 2997–3001. https://doi.org/10.1210/en.2009-0220.

Woods, S. C. (2004). Gastrointestinal satiety signals I. An overview of gastrointestinal signals that influence food intake. *American Journal of Physiology Gastrointestinal and Liver Physiology, 286*(1), G7–G13. https://doi.org/10.1152/ajpgi.00448.2003.

Wren, A. M., & Bloom, S. R. (2007). Gut hormones and appetite control. *Gastroenterology, 132*(6), 2116–2130. https://doi.org/10.1053/j.gastro.2007.03.048.

Wynne, K., & Bloom, S. R. (2006). The role of oxyntomodulin and peptide tyrosine-tyrosine (PYY) in appetite control. *Nature Clinical Practice. Endocrinology & Metabolism, 2*(11), 612–620. https://doi.org/10.1038/ncpendmet0318.

Yamamoto, M., Reeve, J. R., Jr., Keire, D. A., & Green, G. M. (2005). Water and enzyme secretion are tightly coupled in pancreatic secretion stimulated by food or CCK-58 but not by CCK-8. *American Journal of Physiology Gastrointestinal and Liver Physiology, 288*(5), G866–G879. https://doi.org/10.1152/ajpgi.00389.2003.

Yang, J., Brown, M. S., Liang, G., Grishin, N. V., & Goldstein, J. L. (2008). Identification of the acyltransferase that octanoylates ghrelin, an appetite-stimulating peptide hormone. *Cell, 132*(3), 387–396. https://doi.org/10.1016/j.cell.2008.01.017.

Yang, N., Liu, X., Ding, E. L., Xu, M., Wu, S., Liu, L., Sun, X., & Hu, F. B. (2009). Impaired ghrelin response after high-fat meals is associated with decreased satiety in obese and lean Chinese young adults. *The Journal of Nutrition, 139*(7), 1286–1291. https://doi.org/10.3945/jn.109.104406.

Yi, J., Warunek, D., & Craft, D. (2015). Degradation and stabilization of peptide hormones in human blood specimens. *PLoS One, 10*(7), e0134427. https://doi.org/10.1371/journal.pone.0134427.

Zhang, J. V., Ren, P. G., Avsian-Kretchmer, O., Luo, C. W., Rauch, R., Klein, C., & Hsueh, A. J. (2005). Obestatin, a peptide encoded by the ghrelin gene, opposes ghrelin's effects on food intake. *Science, 310*(5750), 996–999. https://doi.org/10.1126/science.1117255.

Zhao, T. J., Sakata, I., Li, R. L., Liang, G., Richardson, J. A., Brown, M. S., Goldstein, J. L., & Zigman, J. M. (2010). Ghrelin secretion stimulated by {beta}1-adrenergic receptors in cultured ghrelinoma cells and in fasted mice. *Proceedings of the National Academy of Sciences of the United States of America, 107*(36), 15868–15873. https://doi.org/10.1073/pnas.1011116107.

Zhu, X., Cao, Y., Voogd, K., & Steiner, D. F. (2006). On the processing of proghrelin to ghrelin. *The Journal of Biological Chemistry, 281*(50), 38867–38870. https://doi.org/10.1074/jbc.M607955200.

The Complexity of Adipose Tissue

8

Katie M. Troike, Kevin Y. Lee, Edward O. List, and Darlene E. Berryman

8.1 Types of Adipose Tissue and Adipocytes

WAT vs *BAT* – AT is generally classified as one of two discrete types: white adipose tissue (WAT) or brown adipose tissue (BAT). Despite being subsumed under a single category, these tissues have little in common, differing drastically in lineage, composition, and function. WAT, primarily composed of white adipocytes, is present in far greater amounts than BAT and acts as the primary reserve for surplus energy in the body, storing excess nutrients as triacylglycerol (TAG) and releasing energy as free fatty acids (FFAs) and glycerol. Both white and brown adipocytes originate from multipotent mesenchymal stem cells but have different precursors. White adipocytes are derived mainly from myogenic factor 5 (Myf5)-negative cells, while brown adipocytes share a Myf5-positive precursor with myocytes (Peirce et al. 2014) (Fig. 8.1).

BAT is predominately composed of brown adipocytes, which employ a thermogenic process to dissipate excess energy through the production of heat. Brown adipocytes contain several, small lipid droplets distributed throughout the cell. They also house large numbers of iron-containing mitochondria, which, along with a dense vascular network, are responsible for the brown color of BAT (Kwok et al. 2016). The unique thermogenic property of brown fat is due to the presence of uncoupling protein-1 (UCP1), a protein found in the inner membrane of BAT mitochondria. UCP1 permits the reentry of protons into the mitochondrial matrix independent ATP synthesis, so that the potential energy generated by the proton gradient is lost as heat. The importance of UCP1 is evident in studies of mice with targeted UCP1 ablation, which results in cold intolerance and increased WAT accumulation to help combat heat loss (Enerback et al. 1997; Feldmann et al. 2009). The thermogenic program is initiated in response to cold exposure or other stimuli, causing the release of catecholamines from sympathetic nerves. The initial role of the nervous system in thermogenesis

K. M. Troike
The Diabetes Institute, Konneker Research Labs, Ohio University, Athens, OH, USA
e-mail: kt305408@ohio.edu

K. Y. Lee
Department of Biomedical Sciences, The Diabetes Institute, Heritage College of Osteopathic Medicine, Athens, OH, USA
e-mail: leek2@ohio.edu

E. O. List
Edison Biotechnology Institute, Konneker Research Labs, Ohio University, Athens, OH, USA
e-mail: list@ohio.edu

D. E. Berryman (✉)
The Diabetes Institute, Department of Biomedical Sciences, Heritage College of Osteopathic Medicine, Ohio University, Athens, OH, USA
e-mail: berrymad@ohio.edu

© Springer International Publishing AG, part of Springer Nature 2018
E. A. Nillni (ed.), *Textbook of Energy Balance, Neuropeptide Hormones, and Neuroendocrine Function*, https://doi.org/10.1007/978-3-319-89506-2_8

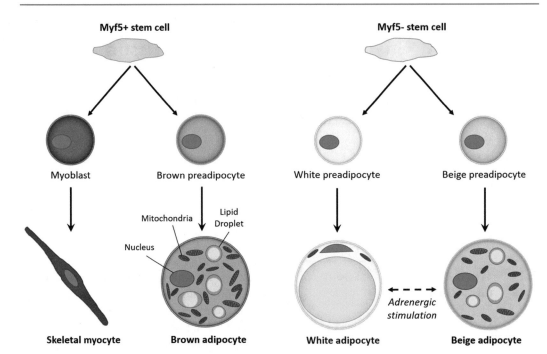

Fig. 8.1 Adipocyte lineages. Adipocytes of BAT and WAT both originate from mesenchymal stem cells but are derived from separate precursors. Brown adipocytes share a common Myf5-positive progenitor with skeletal muscle cells. Conversely, Myf5-negative precursors give rise to the majority of white and beige adipocytes. Beige adipocytes can also arise through transdifferentiation of existing white adipocytes. (Adapted with permission (Rosen and Spiegelman 2014))

likely explains why BAT boasts such extensive sympathetic innervation (Bamshad et al. 1999). Catecholamines then bind β-adrenergic receptors, inducing lipolysis, and the resultant FFAs lead to a marked mitochondrial fission that is critical for full BAT activation (Wikstrom et al. 2014).

Beige adipocytes – The identification of a third adipocyte type once again challenged traditional understanding of this tissue and undermined the binary perspective of white and brown. Beige AT, also termed "brown-in-white," or "brite," is an intermediary of WAT and BAT, as it retains many functional characteristics of BAT while being dispersed throughout WAT depots. Like its brown counterpart, beige AT has the capacity for thermogenesis, expresses UCP1, and can be activated in response to cold exposure or treatment with certain chemicals (Giralt and Villarroya 2013). However, beige adipocytes do not share the same lineage as brown adipocytes. Unlike brown adipocytes, beige adipocytes have no history of Myf5 expression. Evidence exists that beige adipocytes are formed from both trans-

differentiation of unilocular white adipocytes and from a unique precursor population within at least subcutaneous depots (Vitali et al. 2012; Wang et al. 2013). However, more recent evidence suggests the presence of functionally distinct populations of beige adipocytes (Wang et al. 2016) that are molecularly distinct from brown and white adipocytes in both mice and humans (Wu et al. 2012; Shinoda et al. 2015). Regardless, increasing beige AT has been the target of many studies aimed at increasing energy expenditure in an effort to more effectively combat obesity. The morphological and functional spectrum of AT reveals an intricate organ with diverse roles, proving that not all fat is created equal.

8.2 Depots and Depot Differences in AT

Unlike other organs that have a single anatomical location, AT is distributed in multiple depots throughout the body, a feature that further

A   B

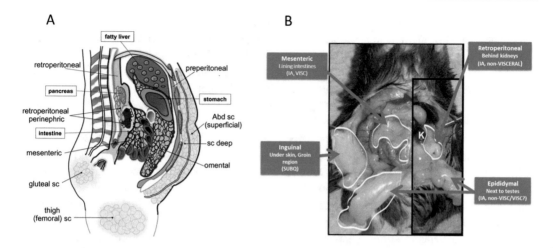

Fig. 8.2 WAT depots in humans and mice. (**a**) Several prominent human WAT depots are depicted. SubQ depots commonly studied in humans include two distinct upper body depots, superficial and deep, as well as gluteal and femoral fat pads. Several depots are defined as intra-abdominal, including perinephric around the kidneys, retroperitoneal in the retroperitoneal space behind the kidneys, and mesenteric and omental WAT lining organs of the digestive tract. (**b**) Four commonly studied WAT depots in a male mouse. Inguinal fat is a subQ depot that lies just beneath the skin and is similar to the gluteal and femoral subQ in humans. Intra-abdominal depots include mesenteric fat associated with the intestine, retroperitoneal behind the kidneys (K), and epididymal WAT next to the testes (paraovarian surrounds the ovaries of female mice). According to the stricter rules of nomenclature, only mesenteric fat is a true visceral depot in mice, as it is the only one that drains into the portal vein. Figure **a** was reused with permission (Lee et al. 2013b). Figure **b** was adapted and reused with permission (Sackmann-Sala et al. 2012; Berryman et al. 2011))

contributes to its complex and heterogeneous nature (Fig. 8.2). Defining WAT depots is of particular importance, as the properties of this tissue can vary considerably depending on its location. The major WAT depots in humans and mice are broadly defined as either subcutaneous (subQ) or intra-abdominal. SubQ WAT, the largest depot in humans, occupies the superficial space beneath the skin, while intra-abdominal WAT is associated with internal organs. In humans, most commonly defined and studied subQ depots are the abdominal, gluteal, and femoral (Lee et al. 2013b) (Fig. 8.2a). The abdominal portion can be further divided into superficial or deep, depending on its position relative to the fascia. Similarly, mice have two distinct sites of subQ deposition: an anterior, or subscapular, depot and a posterior, or inguinal, depot (Fig. 8.2b). The inguinal depot is comparable to lower body subQ found in humans and is therefore more commonly studied than the anterior depot (Chusyd et al. 2016). Intra-abdominal WAT is positioned within the cavities of the abdomen and thorax. In humans, this includes fat located around the greater omen-

tum (omental), along the intestines (mesenteric), and behind the kidneys (retroperitoneal). Additionally, several smaller WAT depots are associated with organs like the heart (pericardial) and stomach (epigastric). Intra-abdominal depots are often collectively referred to as "visceral" although some authors are more conservative with their use of the visceral designation, reserving it only for those depots that drain into the portal vein. Using the latter definition, humans possess two "true" visceral depots (mesenteric and omental), whereas rodents, which lack a significant omental depot, have only the mesenteric depot. A depot commonly studied in mice is the perigonadal depot, which surrounds the testes in males (epididymal) and ovaries in females (paraovarian). The use of this fat pad stems from its large size and ease of dissection, yet caution is advised in interpretation of findings from this depot as there is no analogous depot in humans.

Perhaps the most striking example of the differential effects exhibited by WAT depots is their impact on metabolic health. Accumulation of WAT in visceral depots, defined here as omental and mesenteric, is associated with inflammation and greater risk of cardiometabolic disease, while subQ WAT deposition does not confer disease risk and may instead be protective (Ibrahim 2010). In 1990, the "portal theory" was proposed to account for these differences, suggesting that drainage of visceral depots by the portal vein may cause accumulation of FFAs in the liver, stimulating increased glucose production and culminating in hepatic insulin resistance (Bjorntorp 1990). By comparison, FFAs from subQ WAT, drained by the superior vena cava, are diluted in the circulation and therefore do not cause metabolic abnormalities (Kwok et al. 2016). When measured, however, FFAs are not consistently higher in portal blood relative to peripheral circulation, and upper body subQ depots in humans appear to be the primary source of FFAs in the blood (Lee et al. 2013b; Kwok et al. 2016). As such, the portal theory is likely inadequate to explain this observation, and a combination of factors, including depot-dependent production and secretion of inflammatory molecules, has been proposed as a potential mechanism (Lee et al. 2013b). Evidence from WAT transplantation studies support this idea, as transplantation of visceral WAT into subQ depots has little effect, but improved glucose homeostasis and decreased adiposity are observed when subQ WAT is transplanted into the visceral depot (Tran and Kahn 2010). Further supporting the idea of intrinsic depot heterogeneity between adipocytes is the observation that preadipocytes derived from different fat depots have unique gene expression signatures, suggesting these cells arise from different developmental origins (Gesta et al. 2006; Tchkonia et al. 2007; Lee et al. 2013a). These differences in gene expression persist after numerous passages in culture indicating that they are cell autonomous. Supporting this concept, recent lineage tracing studies have determined that subQ WAT, but not visceral WAT, is derived from a paired related homeobox 1 (Prx1)-positive and zinc finger protein 423 (Zfp423)-positive cell lineages (Sanchez-Gurmaches et al. 2015; Shao et al. 2017).

The lack of uniformity among WAT depots is also readily apparent when considering factors that differentially impact regional fat distribution, such as sex, age, and race (Lee et al. 2013b). WAT distribution exhibits sexual dimorphism, with men tending to amass more fat centrally in visceral depots and women accumulating greater amounts of subQ fat in the gluteofemoral region, as well as having more fat mass overall (Palmer and Clegg 2015; Geer and Shen 2009). Depot-specific effects of sex hormones have been identified as contributors to these patterns, and disease states that disrupt the normal balance of androgens and estrogens provide further evidence for their role. For example, women with increased androgen levels in polycystic ovarian syndrome (PCOS) have greater visceral WAT accumulation, while men with reduced testosterone, resulting from hypogonadism or androgen resistance, experience similar increases in visceral depots (Baptiste et al. 2010; Mammi et al. 2012). Another proposed explanation for variations in WAT distribution is sex-dependent control of insulin sensitivity and lipolysis. Adipocytes derived from female subjects have increased insulin sensitivity (Macotela et al. 2009) and lipolysis (Pujol et al. 2003) compared to those derived from males. These differences are, at least in part, due to the differential actions of sex hormones on adipocyte biology (Mittendorfer et al. 2001; Shen et al. 2014; D'Eon et al. 2005; Fan et al. 2005; Heine et al. 2000) and varying enzyme and receptor levels (Stubbins et al. 2012; Ramis et al. 2006).

Age also acts as a determinant of fat distribution, evidenced by the shift in adiposity from subQ to visceral depots in mid-life and general loss of fat mass accompanied by WAT dysfunction and ectopic deposition in elderly populations (Tchkonia et al. 2010; Palmer and Kirkland 2016). Additionally, differences in race, by virtue of both genetic and environmental variables, also influence regional accumulation of WAT. Data from studies comparing WAT distribution in different races report that African American individuals have reduced visceral WAT deposition compared to Hispanic and Caucasian populations, while South Asians have relatively greater visceral adiposity compared to Caucasians (Carroll et al. 2008; Katzmarzyk et al. 2010).

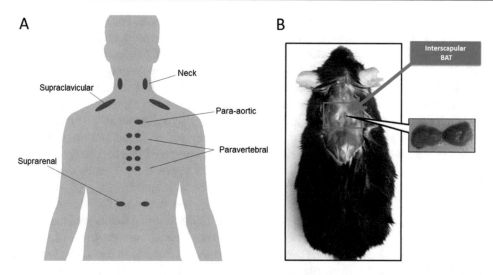

Fig. 8.3 BAT depots in humans and mice. (**a**) Several human BAT depots are illustrated. BAT in adult humans is present in several different locations, including the neck, heart, spinal cord, and kidneys. However, the majority is contained within the supraclavicular depot. (**b**) The location of the interscapular BAT depot in mice. Figure **a** was adapted with permission (Nedergaard et al. 2010). Figure **b** was reused with permission (Troike et al. 2017).

Although it is present in far less quantities than WAT, BAT is similarly a multi-depot organ. Small amounts of BAT are associated with organs such as the heart and kidneys in humans and mice, but the majority is concentrated in a single fat pad (Townsend and Tseng 2015) (Fig. 8.3). In mice and human infants, BAT develops in utero and presents as a large interscapular depot, which is highly active after birth (Lidell et al. 2013). It is believed that the thermogenic activity of newborns, in coordination with their large subQ depots, evolved as an ideal protective mechanism against cold stress, providing a heat source as well as insulation (Merkestein et al. 2014). The interscapular BAT depot is retained throughout life in rodents but diminishes with age in humans and becomes undetectable by adulthood (Townsend and Tseng 2015). This led to the belief that adult humans lacked a significant BAT depot, until recent utilization of fluorodeoxyglucose positron emission tomography with computed tomography scans allowed for the identification of active BAT depots in adult subjects (Nedergaard et al. 2007; Saito et al. 2009). This "adult BAT" does not share the same interscapular location seen in babies, however, and is mainly clustered near the spinal cord, neck, and clavicle, with the most prominent depot located in the supraclavicular region (Townsend and Tseng 2015). Like WAT, BAT deposition may be influenced by sex, as it is more commonly detected in females although data are inconclusive (Nedergaard et al. 2010). Additionally, the likelihood of BAT detection decreases with advancing age (Cypess et al. 2009). It is important to note that the authenticity of adult BAT has recently been called into question, and although many scientists rush to defend its characteristics as resembling those of true "classical" interscapular BAT, evidence has revealed that it is more similar to beige AT (Cypess et al. 2013). Although a seemingly trivial detail, it is important to consider due to the morphological and functional differences exhibited by brown and beige AT.

8.3 Cellular and Noncellular Composition of WAT/BAT

Adipocytes The cellular makeup of AT is highly variable across depots and is inextricably linked to tissue function. Because of this, a thorough understanding of tissue cellular composition is essential for the study of AT. Adipocytes, or fat cells, are the building blocks of AT and occupy the greatest amount of space in white fat depots, approximately 90% of the tissue volume (Lee et al. 2013b). White adipocytes are distinguishable by their unilocular lipid droplet, which can expand in response to overfeeding and is the primary determinant of cell size. Not unexpectedly, adipocyte size is depot-dependent, with data from rodent studies suggesting that adipocytes in subQ and mesenteric fat pads are smaller compared to those found in perigonadal WAT (Sackmann-Sala et al. 2012). Additionally, studies of human WAT reveal smaller visceral adipocytes compared to subQ in women, but similar adipocyte size between depots in men (Lee et al. 2013b). Although adipocytes dominate WAT in terms of size, they fall short in number compared to other cell types.

Cells within the stromal vascular fraction (SVF) The SVF refers to the collective assortment of non-adipocyte cells in WAT, which include immune cells, endothelial cells, neural cells, preadipocytes, and fibroblasts. Depot differences are also observable at the level of SVF composition. A study characterizing gene expression profiles of preadipocytes from human WAT found that those isolated from subQ, mesenteric, and omental depots all had distinct expression patterns, with mesenteric resembling subQ more closely than omental (Tchkonia et al. 2007). Differences in immune cell populations are reported, including greater numbers of macrophages in visceral WAT than subQ in the lean state and greater macrophage infiltration in visceral depots during obesity (Harman-Boehm et al. 2007). It is important to acknowledge that the immune cell population can be altered with environmental stimuli. However, lean adipose tissue harbors various anti-inflammatory immune cells, such as eosinophils, M2 macrophages, Th2 cells, iNKT cells, and Treg cells. These immune cells help in maintaining insulin sensitivity and store extra energy in the form of TAGs.

Lymph and vasculature Lymph nodes and vessels, which are embedded throughout some WAT depots, contribute to heterogeneity both within

and between depots. For example, visceral WAT is more vascularized and innervated than subQ WAT (Ibrahim 2010). Interestingly, adipocytes in close proximity to lymph nodes are unaffected by fasting and feeding signals but respond after stimulation of the immune system, suggesting that they may aid immune function by serving as an energy source (Pond and Mattacks 2002). It has been proposed that these adipocytes are sensitive to lipolysis by lymphoid-secreted cytokines, such as interleukin 6 (IL-6) and tumor necrosis factor alpha (TNF-α) (Mattacks and Pond 1999). It would follow then that adipocytes near lymph nodes have more variable patterns of lipolysis and that depots containing lymph nodes, including subQ and mesenteric, may be more heterogeneous than those without (Pond and Mattacks 1998; Mattacks and Pond 1999).

Senescent cells Senescent cells are present in adipose tissue, and tissue dysfunction is linked to their accumulation (Tchkonia et al. 2010). Senescent cells are those cells that undergo irreversible cell cycle arrest. Their accumulation is common during natural aging and in response to other factors such as oncogene activation, oxidative damage, or exposure to other metabolic stressors, such as high glucose. While senescent cells can originate from most cell types, the majority of research in AT has focused on the preadipocyte precursor pool although there is also evidence for endothelial cells and mature adipocytes to senesce. These cells, along with their inherent senescence-associated secretary phenotype (SASP), promote a potent pro-inflammatory state, which can alter the tissue inflammation, angiogenesis, and fibrosis. Importantly, the accumulation of these cells can have a major impact on AT homeostasis irrespective of AT mass (Crewe et al. 2017; Stout et al. 2017).

Extracellular matrix (ECM) An important noncellular component of WAT is the ECM, which surrounds individual adipocytes and provides structural support for the tissue. Adipocytes and cells within the SVF secrete ECM proteins, as well as the enzymes necessary for their degradation, thus allowing for extensive remodeling of the ECM to accommodate changes in adipocyte size (Maquoi et al. 2002). Matrix proteins include collagens, laminin complexes, proteoglycans, and fibronectin (Mariman and Wang 2010). Collagens are the most abundant ECM component, and collagens V and VI, in particular, appear to have significant impact on the health of WAT during tissue expansion and obesity. In a study of obese and lean human subjects, obese individuals are shown to have increased collagen V, which co-localizes with blood vessels and inhibits endothelial budding, suggesting a role in inhibition of angiogenesis (Spencer et al. 2011). Notably, collagen VI is preferentially expressed in WAT. The ECM has been the target of investigation in many obesity studies due to its role in the development of fibrosis (Pasarica et al. 2009). Obese mice with targeted deletion of the collagen VI gene (*col6KOob/ob*) have larger adipocytes than controls, due to a more elastic ECM and exhibit marked improvements in insulin sensitivity (Khan et al. 2009). Conversely, control mice (*ob/ob*) become insulin resistant and have higher collagen content in their ECM, reducing plasticity and restricting adipocyte expansion, ultimately resulting in smaller adipocytes (Khan et al. 2009). Furthermore, the C5 domain of collagen VIα3 (also known as endotrophin) is a secreted factor upregulated during obesity and serves as a powerful adipokine that drives the formation of fibrosis and inflammation locally in adipose tissue (Sun et al. 2014). Other physiologic changes in WAT accompany this ECM remodeling during obesity, which is more thoroughly addressed in a subsequent section. Differences in ECM composition between fat pads may provide further explanation for functional depot differences observed in lean and obese states. For example, Wistar rats have greater quantities of fibronectin and collagen IV in visceral depots and higher collagen I expression in subQ WAT (Mori et al. 2014). Thus, the ECM is an important component of AT and plays a crucial role in the overall health of the tissue.

Beige adipocytes As an intermediary tissue, adipocytes of beige fat are given the designation "paucilocular," a term which describes UCP1-positive adipocytes with lipid droplet size and distribution between that of white and brown. Just as in BAT, the presence of UCP1-containing mitochondria imparts this tissue with calorie-burning potential. Unlike BAT, however, beige adipocytes do not express UCP1 and other thermogenic genes under basal conditions. Instead, clusters of beige adipocytes are identifiable in WAT depots only upon stimulation. Although some studies have suggested a positive correlation between sympathetic innervation density and beige adipocyte development in white fat pads of mice (Vitali et al. 2012), WAT is traditionally known to be sparsely innervated. Within WAT, beige adipocytes display an increased cell-to-cell coupling via connexin 43 (Cx43) gap junction channels that are critical for the propagation of limited neuronal inputs in WAT and the induction of beige adipocytes (Zhu et al. 2016). Additionally, expression of PR domain containing 16 (Prdm16), a transcriptional regulator of brown and beige adipocyte development, results in greater density of nerve fibers in the WAT of mice independent of cold exposure (Seale et al. 2011). This finding indicates that nerve remodeling in WAT may occur to support thermogenesis of these newly activated beige adipocytes. Vasculature in WAT also appears to aid in the regulation of beige AT function (Wang and Seale 2016). Immune cells, such as alternatively activated macrophages, have also been shown to activate thermogenesis within BAT and beige cells (Nguyen et al. 2011; Brestoff and Artis 2015). The relative abundance of beige adipocytes varies according to depot, with greater numbers found in subQ and mesenteric depots compared to perigonadal (Rosen and Spiegelman 2014).

Intra-depot adipocyte heterogeneity While much attention has been paid to identifying differences among white, brown, and beige adipocytes, recent studies have begun to suggest potential heterogeneity among white adipocytes themselves. Heterogeneity among adipocyte subpopulations is observed in selected genetic manipulations, such as fat-specific knockout of the insulin receptor (Bluher et al. 2004) and ablation of hormone-sensitive lipase (Wang et al. 2001), which unmask intrinsic heterogeneity of WAT leading to a bimodal distribution of adipocyte cell size. Lineage tracing techniques have shown that white adipocytes can come from different developmental origins. A subpopulation of white adipocytes in the visceral cavity is derived from the mesothelium (Chau et al. 2014), while the anterior subQ and retroperitoneal WAT are derived from a Myf5-positive lineage. Furthermore, precursors that express an upstream transcription factor of Myf5, paired box 3 (Pax3), give rise to tissue regions that overlap with Myf-5 in the subQ fat (Sanchez-Gurmaches and Guertin 2014). Interestingly, a subpopulation of adipocytes in the visceral adipose tissue may also be derived from bone marrow progenitors (Majka et al. 2010). The physiological contribution of these adipocyte subpopulations and how they relate to adipocyte physiology and function is not yet known. However, several studies have suggested that white adipocytes also have functional heterogeneity in physiological processes including insulin-stimulated glucose uptake (Salans and Dougherty 1971), maximal lipogenic rate (Gliemann and Vinten 1974), response to catecholamines (Seydoux et al. 1996), and uptake of FFA (Varlamov et al. 2015). Interestingly, adipogenesis in response to high-fat diet (HFD) is mediated by the WAT microenvironment and not by cell-intrinsic mechanisms (Jeffery et al. 2016). Similar to WAT, clonal analysis of beige adipocytes indicates intrinsic heterogeneity of these cells, even within a single AT depot (Xue et al. 2015). Furthermore, differential gene expression between beige adipocyte subpopulations suggests the presence of metabolic heterogeneity with distinct anabolic and catabolic subpopulations (Lee et al. 2017).

8.4   Endocrine Function of AT

While another chapter in this book will provide a broader overview of the endocrine functions of AT, we would be remiss if we did not mention the depot-dependent adipokine profile of AT. As an endocrine organ, WAT secretes and responds to a variety of hormones and cytokines, also known as "adipokines." Two such adipokines, leptin and adiponectin, were discovered almost simultane-

ously (leptin in 1994 and adiponectin in 1995), capturing the interest of scientists around the world and bringing about a new age of AT research. Leptin is a satiety hormone primarily secreted by adipocytes that acts on the hypothalamus to decrease food intake and increase energy expenditure among other functions. As such, mice and humans with mutations of leptin or its receptor exhibit marked obesity (Montague et al. 1997a; Clement et al. 1998; Chen et al. 1996). Leptin secretion is positively correlated with fat mass, and obese individuals may develop leptin resistance due to increased adiposity (Frederich et al. 1995). Secretion of leptin appears to be depot-dependent, with subQ WAT producing greater amounts than visceral (Masuzaki et al. 1995; Montague et al. 1997b). In contrast, adiponectin is negatively correlated with fat mass, although it too is secreted in greater quantities by subQ WAT compared to visceral (Arita et al. 1999; Matsubara et al. 2002). Adiponectin is an anti-inflammatory and insulin-sensitizing hormone and can be used to predict the development of insulin resistance and metabolic syndrome in overweight and obese patients (Hara et al. 2006; Hirose et al. 2010). Many other adipokines have been identified that appear to be preferentially secreted by certain WAT depots. As its name suggests, the adipokine omentin is mainly produced by SVF cells of visceral depots. Reduced omentin levels have been implicated in obesity and insulin resistance in humans, suggesting that it may play a role in promoting insulin sensitivity (de Souza Batista et al. 2007; Hana et al. 2004). Similarly, visfatin is predominantly secreted by the macrophages of visceral depots (Curat et al. 2006) in humans and mice and is present at high levels in patients with obesity, type 2 diabetes, metabolic syndrome, and cardiovascular disease (CVD) (Chang et al. 2011).

The endocrine properties of brown and beige fat are not as thoroughly understood, but it is clear that they too are capable of adipokine secretion. BAT has been shown to secrete a number of factors, including IL-6, insulin-like growth factor 1 (IGF-1), angiotensin, and nitric oxide, among others (Villarroya et al. 2013). BAT transplantation experiments result in improved glucose uptake and insulin sensitivity, effects which are believed to be mediated through responses of the heart and liver to BAT secretion of IGF-1, IL-6, and fibroblast growth factor 21 (FGF21) (Villarroya et al. 2013). BAT also produces the "batokine" neuregulin 4 (NRG4), which targets the liver to inhibit de novo lipogenesis (Wang et al. 2014).

Noncoding RNAs are 19–22 nucleotides in length and can be found within cells as well as in circulation, where they regulate many cellular processes and disease states. Although not classically considered endocrine products, AT (both WAT and BAT) has recently been shown to be a major source of circulating microRNAs or miR-NAs (Thomou et al. 2017). These circulating AT-derived miRNAs appear to regulate gene expression and metabolism in distant tissues and thereby function as a previously undescribed type of adipokine (Thomou et al. 2017).

8.5 Control of Lipolysis and Lipogenesis

Adipocytes respond to nutritional and hormonal cues to accommodate changes in nutrient availability, either by storing surplus FFAs in the form of TAGs (lipogenesis) or through the breakdown of TAGs into FFAs (lipolysis) for fuel. As might be expected, both lipogenesis and lipolysis are highly responsive to changes in nutritional status, including diet composition. High-carbohydrate diets promote lipogenesis, while intake of polyunsaturated fatty acids has the opposite effect (Kersten 2001). Additionally, certain dietary compounds, such as calcium and caffeine, have been shown to increase lipolysis, while others, like ethanol, are inhibitory (Melanson et al. 2003; Duncan et al. 2007; Abramson and Arky 1968). Notably, fasting has an antilipogenic effect and simultaneously induces lipolysis. Upon refeeding, lipolysis is reduced. These effects are mediated through the potent action of insulin, which stimulates glucose uptake in adipocytes by recruiting glucose transporters to the cell surface and activating enzymes related to glycolysis and lipogenesis via covalent modification (Kersten

2001). Insulin can also exert its effects by acting on transcriptional regulators of lipogenic genes (Wong and Sul 2010). A number of other factors can alter lipolysis and lipogenesis. For example, growth hormone (GH) is a key regulator of these processes, acting to stimulate lipolysis and impede lipogenesis, although the exact mechanisms of this are still unknown. The effects of GH are probably due, in part, to its inhibition of insulin's action, which results in downregulation of FA synthase, a key enzyme in de novo lipogenesis (Yin et al. 1998). Adipokines such as leptin also promote the release of glycerol from adipocytes through downregulation of genes involved in FFA and TAG synthesis and through the induction of FFA oxidation (Kersten 2001; Soukas et al. 2000).

Rates of lipolysis and lipogenesis are impacted by depot and sex. In mice, basal FFA and glycerol release are higher in mesenteric WAT than perigonadal, a difference that disappears when the mice are fed a HFD (Wueest et al. 2012b). A similar study reveals that lipolysis is higher in subQ adipocytes compared to mesenteric under both basal and HFD conditions, and perigonadal lipolysis is significantly increased with HFD feeding (Wueest et al. 2012a). Moreover, subQ adipocytes retain their sensitivity to insulin on a HFD, while perigonadal do not (Wueest et al. 2012a). In humans, adipocytes from the omental depot have similar or lower rates of lipolysis under basal conditions, but higher rates in response to lipolytic stimuli compared to subQ adipocytes (Tchernof et al. 2006). When examining sex differences, female mice appear to have increased insulin sensitivity and lipogenic response than males in both perigonadal and subQ depots (Macotela et al. 2009). In studies of obese men and women, lipolytic capacity per adipocyte is higher in obese women and decreases with weight loss, while lipolytic capacity is similar between obese and lean men and does not change with weight loss (Lofgren et al. 2002). A study of lean and obese men and women also reports greater lipolysis in women than men relative to resting energy expenditure (Nielsen et al. 2003). Data from normal weight patients reveal that women may be more sensitive to the lipolytic effects of catecholamines compared to men, which would explain how women maintain similar levels of lipid mobilization during exercise, despite having lower catecholamine levels than men (Horton et al. 2009). The complex regulation of lipolysis and lipogenesis by various factors, as well as sex- and depot-dependent effects, further highlights the heterogeneity of AT and its importance in the maintenance of metabolic homeostasis.

8.6 Ectopic Fat Deposition/ Lipodystrophy

Lipodystrophies are groups of heterogeneous disorders characterized by loss of AT, which affect the whole body (generalized) or specific regions (partial), and are either inherited or acquired. While many forms of lipodystrophy have been identified, the most common is an acquired form that occurs in an estimated 40–70% of patients treated with highly active antiretroviral therapy (HAART) for human immunodeficiency virus (HIV) (Fiorenza et al. 2011). These individuals usually present with subQ fat loss of the face, arms, and legs, which can be accompanied by accumulation of WAT in the visceral or cervical regions (Bindlish et al. 2015; Fiorenza et al. 2011). Increased fat deposition near the back of the neck results in a characteristic "buffalo hump" and serves as an indicator of lipodystrophy in HAART treated patients. Inherited forms are far less common and are defined by their specific gene mutations that determine patterns of fat loss in these individuals. For example, mutations of two genes in congenital generalized lipodystrophy (CGL), 1-acylglycerol-3-phosphate O-acyltransferase 2 (*AGPAT2*) and Berardinelli-Seip congenital lipodystrophy 2 (*BSCL2*), result in almost complete absence of WAT in virtually all depots, accompanied by muscle hypertrophy (Fiorenza et al. 2011). Mutations in other genes, such as lamin A/C (*LMNA*) and peroxisome proliferator-activated receptor gamma (*PPARG*), which hallmark familial partial lipodystrophy (FPL), cause wasting of trunk and limb WAT, but do not affect the face and neck (Garg 2011).

Several mechanisms to explain these extreme phenotypes have been identified and include increased rates of adipocyte death, impaired adipocyte differentiation, and abnormal preadipocyte development (Bindlish et al. 2015). Thus, both the clinical presentations and underlying disease mechanisms of different lipodystrophies can vary widely.

Highlighting the critical role of AT in homeostasis, the metabolic consequences that often accompany excess fat accumulation (e.g., insulin resistance, type 2 diabetes, hepatic steatosis) are also observed in patients with lipodystrophy. Several theories have emerged to help explain this paradox. One such theory suggests that restricted storage capacity of WAT in lipodystrophy leads to ectopic fat deposition in other tissues, such as liver and muscle, causing both hepatic and peripheral insulin resistance (Garg 2006). It has also been postulated that increased FA flux in patients with partial lipodystrophies may cause increased production of glucose by the liver, resulting in metabolic syndrome (Garg 2006). Regardless, alterations in adipokine production may play a role in these metabolic perturbations. Individuals with generalized lipodystrophy often have voracious appetites resulting from decreased leptin levels, as well as severe glucose intolerance, hypertriglyceridemia, and hepatic steatosis (Haque et al. 2002). In fact, metreleptin, a recombinant methionyl human leptin, has been used successfully to improve the metabolic disturbances associated with lipodystrophy (Vatier et al. 2016). Additionally, many patients with CGL have reduced adiponectin, which exacerbates insulin resistance (Haque et al. 2002; Antuna-Puente et al. 2010). Elucidating the mechanisms that result in HAART-related lipodystrophy is far more difficult, as a number of metabolic disturbances can occur as side effects of the drugs used in treatment, including increased production of reactive oxygen species and inflammatory adipokines and impaired insulin signaling, leptin, and adiponectin secretion (Fiorenza et al. 2011). Thus, treatment of lipodystrophies is dependent on the type and may include lifestyle modification or the use of various medications, such as metformin to improve insulin sensitivity, statins to aid in dyslipidemia and inflammation, or GH to help reduce truncal adiposity (Fiorenza et al. 2011). The complex presentation and drastic impact of lipodystrophies underscore the importance of WAT, and the metabolic similarities of these disorders to those seen in obesity reveal a very basic truth: when it comes to fat, balance is everything.

8.7 AT Remodeling

Sex Differences During periods of nutrient excess, storage of TAG in AT increases, resulting in tissue expansion. As a result, the ECM undergoes remodeling to ameliorate mechanical stress caused by this growth. This tissue growth can occur in two ways: (1) through hypertrophy, which is an increase in the size of existing adipocytes, and (2) through hyperplasia, which involves recruitment of new adipocytes from a pool of progenitors. Depot and sex are major determinants of which expansion method predominates in WAT. In humans, excessive caloric intake appears to induce hypertrophy in abdominal subQ depots but results in hyperplasia in femoral subQ (Tchoukalova et al. 2010). Male Wistar rats fed a HFD exhibit similar hyperplastic growth of the subQ depot and hypertrophic growth of the mesenteric and epididymal depots (DiGirolamo et al. 1998). Conversely, hyperplasia in visceral WAT of male mice and subQ WAT of female mice after HFD feeding has been reported, highlighting sex-specific differences in depot growth (Jeffery et al. 2016). This observation is further supported by data showing expansion by hypertrophy in the subQ depots of male mice (Wang et al. 2013). Thus, the preferred method of tissue growth varies in a depot-dependent manner, an effect that may be further impacted by sex differences.

Obesity The plasticity and remodeling capacity of WAT is perhaps best exemplified by the changes that occur with obesity (Fig. 8.4). WAT expansion can be characterized as "healthy" or pathological. Healthy WAT expansion is marked by hyperplasia, marginal ECM deposition, minor inflammation, and adequate vascularization to the tissue

(Kusminski et al. 2016; Sun et al. 2011). In contrast, pathological tissue growth occurs via hypertrophy, and, as the adipocytes enlarge, they increase their secretion of pro-inflammatory cytokines TNF-α, IL-6, and MCP-1 (Sun et al. 2011; Kwok et al. 2016). Pathological expansion also results in a dramatic immune population shift, with neutrophils, M1 macrophages, mast cells, Th1 cells, and CD8 T cells being greatly elevated. The switch from anti-inflammatory M2 to pro-inflammatory M1 macrophages with obesity is probably the best characterized immune cell shift (Lumeng et al. 2007; Kwok et al. 2016). Other immune cellular changes, like increased natural killer cells and decreased T regulatory cells and other anti-inflammatory factors, like adiponectin, exacerbate the inflammatory state of the tissue (Sun et al. 2011; Lumeng et al. 2007). The ever-increasing size of WAT also results in dysregulation of the ECM, mainly through upregulation of collagens, notably collagen VI (Choe et al. 2016). The increasingly fibrotic ECM acts to inhibit uncontrolled adipocyte expansion. Rapid expansion and remodeling may limit the ability of WAT to recruit new blood vessels, resulting in poor oxygenation and hypoxia. Activation of a transcriptional regulator of hypoxic response, hypoxia-inducible factors 1 α (HIF-1α), leads to increased fibrosis and local inflammation (Choe et al. 2016). Although changes in immune cell populations, ECM deposition, and vascularization are all well-documented, cause and effect relationships between these shifts have not been established, and the order in which they occur is still debated. Adipocyte necrosis and death occur subsequently to hypoxia, promoting additional macrophage infiltration and the formation of crown-like structures surrounding dead adipocytes (Cinti et al. 2005). Rates of adipocyte death in HFD-fed mice are only 3% in subQ depots, compared to 80% in epididymal, suggesting a depot-dependent effect on the susceptibility of adipocytes to die (Strissel et al. 2007). Furthermore, female mice appear to be protected from HFD-induced adipocyte death (Grove et al. 2010). BAT also appears to be negatively impacted by obesity, and studies in human subjects have shown less BAT and decreased thermogenic activity in

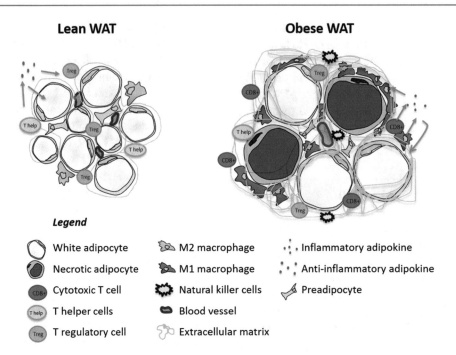

Fig. 8.4 WAT composition in lean and obese states. In a lean state (left), WAT exhibits a predominantly anti-inflammatory profile, characterized by greater amounts of M2 macrophages, T regulatory cells, and T helper cells. The ECM is loose and elastic, and sufficient vascularization provides oxygen to support adipocytes. The pathological state of WAT in obesity (right) exhibits a characteristic pro-inflammatory profile which is hall-marked by increased M1 macrophages, natural killer cells, and cytotoxic T cells, and decreased T regulatory cells and T helper cells. Increased expression of collagens, notably collagen VI, causes the ECM to become rigid, and further ECM secretion results in fibrotic WAT. The growing demand for vascularization by hypertrophic adipocytes leads to hypoxia and death. (Adapted and reused with permission (Berryman et al. 2015; Troike et al. 2017))

obese individuals (Vijgen et al. 2011; Cypess et al. 2009; Cuthbertson et al. 2017).

Adipose tissue growth in obesity does not always result in these pathological states, however, and the identification of "metabolically healthy" individuals within obese populations introduces an intriguing paradigm shift about the importance of WAT quantity versus quality. While there are currently no standardized criteria for defining and identifying metabolically healthy obese (MHO) patients, the condition is generally classified as obesity that is not accompanied by metabolic or cardiovascular diseases such as type 2 diabetes, atherosclerosis, and hypertension (Munoz-Garach et al. 2016; Denis and Obin 2013). The favorable metabolic profile described in MHO populations may be attributed to the changes at the level of WAT, including decreased adiposity in visceral depots, smaller adipocytes, reduced inflammation and fibrosis within WAT, and increased adiponectin compared to other obese individuals (Rosen and Spiegelman 2014). Increases in hyperplasia, rather than hypertrophy, may account for the smaller adipocytes observed in this state. The MHO designation is not intended to imply, however, that these individuals are the picture of health, as they still present with troubling conditions such as gallbladder disease, arthritis, impaired endothelial function, and other comorbidities (Denis and Obin 2013). In contrast, individuals referred to as "metabolically obese but normal weight" (MONW) have normal BMI and adipocyte volume, yet exhibit metabolically unhealthy phenotypes, such as hyperglycemia, insulin resistance, and hypercholesterolemia (Denis and Obin 2013). The mechanisms underlying metabolic abnormalities seen in MONW individuals are not well understood, although genetic factors have been suggested to play a role, and gene variants associated with insulin resistance are also found to be associated with reduced adiponectin and an increased visceral-to-subcutaneous AT ratio (Yaghootkar et al. 2014).

8.8 Conclusion

As highlighted in this review, AT is a complex organ whose action is governed by a dynamic interaction of numerous cell types, endocrine signals, locally produced factors, and nutritional status. Over the past two decades, the field of adipocyte biology has made great progress in understanding the metabolic contribution of white, beige, and brown adipose tissue to whole-body physiology. The ability to control the numbers and activity of each of the various adipocyte cell types could provide new treatments for obesity and the metabolic syndrome. Likewise, the physiological significance of both inter- and intra-depot heterogeneity of both adipocytes and cellular populations of the SVF has yet to be fully ascertained. A better understanding of AT and its cellular composition is an important area for research that holds promise for developing targeted interventions to combat the comorbidities associated with obesity.

Chapter 8 Questions

1. What are the functional, developmental, and distribution differences among white, beige, and brown adipocytes?
2. Which adipose depots are directly comparable and which are distinct between mouse and man?
3. Adipose tissue depots are distinct in cellular composition and function. Can you describe three differences in subcutaneous versus visceral adipose depots?
4. Are the same adipokines produced by brown and white adipose tissue?
5. What metabolic features do lipodystrophy and obesity have in common? Explain how this might occur.

6. Obesity results in a number of changes in adipose tissue. What are the cellular, adipokine, and structural changes in adipose tissue when going from a lean to an obese state?

References

Abramson, E. A., & Arky, R. A. (1968). Acute antilipolytic effects of ethyl alcohol and acetate in man. *The Journal of Laboratory and Clinical Medicine, 72*(1), 105–117.

Antuna-Puente, B., Boutet, E., Vigouroux, C., Lascols, O., Slama, L., Caron-Debarle, M., Khallouf, E., Levy-Marchal, C., Capeau, J., Bastard, J. P., & Magre, J. (2010). Higher adiponectin levels in patients with Berardinelli-Seip congenital lipodystrophy due to seipin as compared with 1-acylglycerol-3-phosphate-o-acyltransferase-2 deficiency. *The Journal of Clinical Endocrinology and Metabolism, 95*(3), 1463–1468. https://doi.org/10.1210/jc.2009-1824.

Arita, Y., Kihara, S., Ouchi, N., Takahashi, M., Maeda, K., Miyagawa, J., Hotta, K., Shimomura, I., Nakamura, T., Miyaoka, K., Kuriyama, H., Nishida, M., Yamashita, S., Okubo, K., Matsubara, K., Muraguchi, M., Ohmoto, Y., Funahashi, T., & Matsuzawa, Y. (1999). Paradoxical decrease of an adipose-specific protein, adiponectin, in obesity. *Biochemical and Biophysical Research Communications, 257*(1), 79–83.

Bamshad, M., Song, C. K., & Bartness, T. J. (1999). CNS origins of the sympathetic nervous system outflow to brown adipose tissue. *The American Journal of Physiology, 276*(6 Pt 2), R1569–R1578.

Baptiste, C. G., Battista, M. C., Trottier, A., & Baillargeon, J. P. (2010). Insulin and hyperandrogenism in women with polycystic ovary syndrome. *The Journal of Steroid Biochemistry and Molecular Biology, 122*(1–3), 42–52. https://doi.org/10.1016/j.jsbmb.2009.12.010.

Berryman, D., Householder, L., Lesende, V., List, E., & Kopchick, J. J. (2015). Living large: What mouse models reveal about growth hormone. In N. A. Berger (Ed.), *Murine models, energy, balance, and cancer* (pp. 65–95). New York: Springer.

Berryman, D. E., List, E. O., Sackmann-Sala, L., Lubbers, E., Munn, R., & Kopchick, J. J. (2011). Growth hormone and adipose tissue: Beyond the adipocyte. *Growth Hormone & IGF Research, 21*(3), 113–123. https://doi.org/10.1016/j.ghir.2011.03.002.

Bindlish, S., Presswala, L. S., & Schwartz, F. (2015). Lipodystrophy: Syndrome of severe insulin resistance. *Postgraduate Medicine, 127*(5), 511–516. https://doi.org/10.1080/00325481.2015.1015927.

Bjorntorp, P. (1990). "Portal" adipose tissue as a generator of risk factors for cardiovascular disease and diabetes. *Arteriosclerosis, 10*(4), 493–496.

Bluher, M., Patti, M. E., Gesta, S., Kahn, B. B., & Kahn, C. R. (2004). Intrinsic heterogeneity in adipose tissue of fat-specific insulin receptor knock-out mice is associated with differences in patterns of gene expression. *The Journal of Biological Chemistry, 279*(30), 31891–31901. https://doi.org/10.1074/jbc.M404569200.

Brestoff, J. R., & Artis, D. (2015). Immune regulation of metabolic homeostasis in health and disease. *Cell, 161*(1), 146–160. https://doi.org/10.1016/j.cell.2015.02.022.

Carroll, J. F., Chiapa, A. L., Rodriquez, M., Phelps, D. R., Cardarelli, K. M., Vishwanatha, J. K., Bae, S., & Cardarelli, R. (2008). Visceral fat, waist circumference, and BMI: Impact of race/ethnicity. *Obesity, 16*(3), 600–607. https://doi.org/10.1038/oby.2007.92.

Chang, Y. H., Chang, D. M., Lin, K. C., Shin, S. J., & Lee, Y. J. (2011). Visfatin in overweight/obesity, type 2 diabetes mellitus, insulin resistance, metabolic syndrome and cardiovascular diseases: A meta-analysis and systemic review. *Diabetes/Metabolism Research and Reviews, 27*(6), 515–527. https://doi.org/10.1002/dmrr.1201.

Chau, Y. Y., Bandiera, R., Serrels, A., Martinez-Estrada, O. M., Qing, W., Lee, M., Slight, J., Thornburn, A., Berry, R., McHaffie, S., Stimson, R. H., Walker, B. R., Chapuli, R. M., Schedl, A., & Hastie, N. (2014). Visceral and subcutaneous fat have different origins and evidence supports a mesothelial source. *Nature Cell Biology, 16*(4), 367–375. https://doi.org/10.1038/ncb2922.

Chen, H., Charlat, O., Tartaglia, L. A., Woolf, E. A., Weng, X., Ellis, S. J., Lakey, N. D., Culpepper, J., Moore, K. J., Breitbart, R. E., Duyk, G. M., Tepper, R. I., & Morgenstern, J. P. (1996). Evidence that the diabetes gene encodes the leptin receptor: Identification of a mutation in the leptin receptor gene in db/db mice. *Cell, 84*(3), 491–495.

Choe, S. S., Huh, J. Y., Hwang, I. J., Kim, J. I., & Kim, J. B. (2016). Adipose tissue remodeling: Its role in energy metabolism and metabolic disorders. *Frontiers in Endocrinology (Lausanne), 7*, 30. https://doi.org/10.3389/fendo.2016.00030.

Chusyd, D. E., Wang, D., Huffman, D. M., & Nagy, T. R. (2016). Relationships between rodent white adipose fat pads and human white adipose fat depots. *Frontiers in Nutrition, 3*, 10. https://doi.org/10.3389/fnut.2016.00010.

Cinti, S., Mitchell, G., Barbatelli, G., Murano, I., Ceresi, E., Faloia, E., Wang, S., Fortier, M., Greenberg, A. S., & Obin, M. S. (2005). Adipocyte death defines macrophage localization and function in adipose tissue of obese mice and humans. *Journal of Lipid Research, 46*(11), 2347–2355.

Clement, K., Vaisse, C., Lahlou, N., Cabrol, S., Pelloux, V., Cassuto, D., Gourmelen, M., Dina, C., Chambaz, J., Lacorte, J. M., Basdevant, A., Bougneres, P., Lebouc, Y., Froguel, P., & Guy-Grand, B. (1998). A mutation in the human leptin receptor gene causes obesity and pituitary dysfunction. *Nature, 392*(6674), 398–401. https://doi.org/10.1038/32911.

Crewe, C., An, Y. A., & Scherer, P. E. (2017). The ominous triad of adipose tissue dysfunction: Inflammation, fibrosis, and impaired angiogenesis. *The Journal of Clinical Investigation, 127*(1), 74–82. https://doi.org/10.1172/JCI88883.

Curat, C. A., Wegner, V., Sengenes, C., Miranville, A., Tonus, C., Busse, R., & Bouloumie, A. (2006). Macrophages in human visceral adipose tissue: Increased accumulation in obesity and a source of resistin and visfatin. *Diabetologia, 49*(4), 744–747. https://doi.org/10.1007/s00125-006-0173-z.

Cuthbertson, D. J., Steele, T., Wilding, J. P., Halford, J. C., Harrold, J. A., Hamer, M., & Karpe, F. (2017). What have human experimental overfeeding studies taught us about adipose tissue expansion and susceptibility to obesity and metabolic complications? *International Journal of Obesity*. https://doi.org/10.1038/ijo.2017.4.

Cypess, A. M., Lehman, S., Williams, G., Tal, I., Rodman, D., Goldfine, A. B., Kuo, F. C., Palmer, E. L., Tseng, Y. H., Doria, A., Kolodny, G. M., & Kahn, C. R. (2009). Identification and importance of brown adipose tissue in adult humans. *The New England Journal of Medicine, 360*(15), 1509–1517.

Cypess, A. M., White, A. P., Vernochet, C., Schulz, T. J., Xue, R., Sass, C. A., Huang, T. L., Roberts-Toler, C., Weiner, L. S., Sze, C., Chacko, A. T., Deschamps, L. N., Herder, L. M., Truchan, N., Glasgow, A. L., Holman, A. R., Gavrila, A., Hasselgren, P. O., Mori, M. A., Molla, M., & Tseng, Y. H. (2013). Anatomical localization, gene expression profiling and functional characterization of adult human neck brown fat. *Nature Medicine, 19*(5), 635–639. https://doi.org/10.1038/nm.3112.

D'Eon, T. M., Souza, S. C., Aronovitz, M., Obin, M. S., Fried, S. K., & Greenberg, A. S. (2005). Estrogen regulation of adiposity and fuel partitioning. Evidence of genomic and non-genomic regulation of lipogenic and oxidative pathways. *The Journal of Biological Chemistry, 280*(43), 35983–35991. https://doi.org/10.1074/jbc.M507339200.

de Souza Batista, C. M., Yang, R. Z., Lee, M. J., Glynn, N. M., Yu, D. Z., Pray, J., Ndubuizu, K., Patil, S., Schwartz, A., Kligman, M., Fried, S. K., Gong, D. W., Shuldiner, A. R., Pollin, T. I., & McLenithan, J. C. (2007). Omentin plasma levels and gene expression are decreased in obesity. *Diabetes, 56*(6), 1655–1661. https://doi.org/10.2337/db06-1506.

Denis, G. V., & Obin, M. S. (2013). Metabolically healthy obesity: Origins and implications. *Molecular Aspects of Medicine, 34*(1), 59–70. https://doi.org/10.1016/j.mam.2012.10.004.

DiGirolamo, M., Fine, J. B., Tagra, K., & Rossmanith, R. (1998). Qualitative regional differences in adipose tissue growth and cellularity in male Wistar rats fed ad libitum. *The American Journal of Physiology, 274*(5 Pt 2), R1460–R1467.

Duncan, R. E., Ahmadian, M., Jaworski, K., Sarkadi-Nagy, E., & Sul, H. S. (2007). Regulation of lipolysis in adipocytes. *Annual Review of Nutrition,* *27,* 79–101. https://doi.org/10.1146/annurev. nutr.27.061406.093734.

Enerback, S., Jacobsson, A., Simpson, E. M., Guerra, C., Yamashita, H., Harper, M. E., & Kozak, L. P. (1997). Mice lacking mitochondrial uncoupling protein are cold-sensitive but not obese. *Nature, 387*(6628), 90–94. https://doi.org/10.1038/387090a0.

Fan, W., Yanase, T., Nomura, M., Okabe, T., Goto, K., Sato, T., Kawano, H., Kato, S., & Nawata, H. (2005). Androgen receptor null male mice develop late-onset obesity caused by decreased energy expenditure and lipolytic activity but show normal insulin sensitivity with high adiponectin secretion. *Diabetes, 54*(4), 1000–1008.

Feldmann, H. M., Golozoubova, V., Cannon, B., & Nedergaard, J. (2009). UCP1 ablation induces obesity and abolishes diet-induced thermogenesis in mice exempt from thermal stress by living at thermoneutrality. *Cell Metabolism, 9*(2), 203–209. https://doi. org/10.1016/j.cmet.2008.12.014.

Fiorenza, C. G., Chou, S. H., & Mantzoros, C. S. (2011). Lipodystrophy: Pathophysiology and advances in treatment. *Nature Reviews Endocrinology, 7*(3), 137–150. https://doi.org/10.1038/nrendo.2010.199.

Frederich, R. C., Hamann, A., Anderson, S., Lollmann, B., Lowell, B. B., & Flier, J. S. (1995). Leptin levels reflect body lipid content in mice: Evidence for diet-induced resistance to leptin action. *Nature Medicine, 1*(12), 1311–1314.

Garg, A. (2006). Adipose tissue dysfunction in obesity and lipodystrophy. *Clinical Cornerstone, 8*(Suppl 4), S7–S13.

Garg, A. (2011). Clinical review#: Lipodystrophies: Genetic and acquired body fat disorders. *The Journal of Clinical Endocrinology and Metabolism, 96*(11), 3313–3325. https://doi.org/10.1210/jc.2011-1159.

Geer, E. B., & Shen, W. (2009). Gender differences in insulin resistance, body composition, and energy balance. *Gender Medicine, 6*(Suppl 1), 60–75. https://doi. org/10.1016/j.genm.2009.02.002.

Gesta, S., Bluher, M., Yamamoto, Y., Norris, A. W., Berndt, J., Kralisch, S., Boucher, J., Lewis, C., & Kahn, C. R. (2006). Evidence for a role of developmental genes in the origin of obesity and body fat distribution. *Proceedings of the National Academy of Sciences of the United States of America, 103*(17), 6676–6681.

Giralt, M., & Villarroya, F. (2013). White, brown, beige/brite: Different adipose cells for different functions? *Endocrinology, 154*(9), 2992–3000. https://doi. org/10.1210/en.2013-1403.

Gliemann, J., & Vinten, J. (1974). Lipogenesis and insulin sensitivity of single fat cells. *The Journal of Physiology, 236*(3), 499–516.

Grove, K. L., Fried, S. K., Greenberg, A. S., Xiao, X. Q., & Clegg, D. J. (2010). A microarray analysis of sexual dimorphism of adipose tissues in high-fat-diet-induced obese mice. *International Journal of Obesity, 34*(6), 989–1000. https://doi.org/10.1038/ijo.2010.12.

Hana, V., Silha, J. V., Justova, V., Lacinova, Z., Stepan, J. J., & Murphy, L. J. (2004). The effects of GH replacement in adult GH-deficient patients: Changes in body composition without concomitant changes in the adipokines and insulin resistance. *Clinical Endocrinology, 60*(4), 442–450.

Haque, W. A., Shimomura, I., Matsuzawa, Y., & Garg, A. (2002). Serum adiponectin and leptin levels in patients with lipodystrophies. *The Journal of Clinical Endocrinology and Metabolism, 87*(5), 2395. https:// doi.org/10.1210/jcem.87.5.8624.

Hara, K., Horikoshi, M., Yamauchi, T., Yago, H., Miyazaki, O., Ebinuma, H., Imai, Y., Nagai, R., & Kadowaki, T. (2006). Measurement of the high-molecular weight form of adiponectin in plasma is useful for the prediction of insulin resistance and metabolic syndrome. *Diabetes Care, 29*(6), 1357–1362.

Harman-Boehm, I., Bluher, M., Redel, H., Sion-Vardy, N., Ovadia, S., Avinoach, E., Shai, I., Kloting, N., Stumvoll, M., Bashan, N., & Rudich, A. (2007). Macrophage infiltration into omental versus subcutaneous fat across different populations: Effect of regional adiposity and the comorbidities of obesity. *The Journal of Clinical Endocrinology and Metabolism, 92*(6), 2240–2247. https://doi.org/10.1210/jc.2006-1811.

Heine, P. A., Taylor, J. A., Iwamoto, G. A., Lubahn, D. B., & Cooke, P. S. (2000). Increased adipose tissue in male and female estrogen receptor-alpha knockout mice. *Proceedings of the National Academy of Sciences of the United States of America, 97*(23), 12729–12734. https://doi.org/10.1073/pnas.97.23.12729.

Hirose, H., Yamamoto, Y., Seino-Yoshihara, Y., Kawabe, H., & Saito, I. (2010). Serum high-molecular-weight adiponectin as a marker for the evaluation and care of subjects with metabolic syndrome and related disorders. *Journal of Atherosclerosis and Thrombosis, 17*(12), 1201–1211.

Horton, T. J., Dow, S., Armstrong, M., & Donahoo, W. T. (2009). Greater systemic lipolysis in women compared with men during moderate-dose infusion of epinephrine and/or norepinephrine. *Journal of Applied Physiology, 107*(1), 200–210. https://doi.org/10.1152/ japplphysiol.90812.2008.

Ibrahim, M. M. (2010). Subcutaneous and visceral adipose tissue: Structural and functional differences. *Obesity Reviews: An Official Journal of the International Association for the Study of Obesity, 11*(1), 11–18. https://doi.org/10.1111/j.1467-789X.2009.00623.x.

Jeffery, E., Wing, A., Holtrup, B., Sebo, Z., Kaplan, J. L., Saavedra-Pena, R., Church, C. D., Colman, L., Berry, R., & Rodeheffer, M. S. (2016). The adipose tissue microenvironment regulates depot-specific adipogenesis in obesity. *Cell Metabolism, 24*(1), 142–150. https://doi.org/10.1016/j.cmet.2016.05.012.

Kajimura, S., Spiegelman, B. M., & Seale, P. (2015). Brown and beige fat: Physiological roles beyond heat generation. *Cell Metabolism, 22*(4), 546–559. https:// doi.org/10.1016/j.cmet.2015.09.007.

Katzmarzyk, P. T., Bray, G. A., Greenway, F. L., Johnson, W. D., Newton, R. L., Jr., Ravussin, E., Ryan, D. H.,

Smith, S. R., & Bouchard, C. (2010). Racial differences in abdominal depot-specific adiposity in white and African American adults. *The American Journal of Clinical Nutrition, 91*(1), 7–15. https://doi.org/10.3945/ajcn.2009.28136.

Kersten, S. (2001). Mechanisms of nutritional and hormonal regulation of lipogenesis. *EMBO Reports, 2*(4), 282–286. https://doi.org/10.1093/embo-reports/kve071.

Khan, T., Muise, E. S., Iyengar, P., Wang, Z. V., Chandalia, M., Abate, N., Zhang, B. B., Bonaldo, P., Chua, S., & Scherer, P. E. (2009). Metabolic dysregulation and adipose tissue fibrosis: Role of collagen VI. *Molecular and Cellular Biology, 29*(6), 1575–1591.

Kusminski, C. M., Bickel, P. E., & Scherer, P. E. (2016). Targeting adipose tissue in the treatment of obesity-associated diabetes. *Nature Reviews Drug Discovery.* https://doi.org/10.1038/nrd.2016.75.

Kwok, K. H., Lam, K. S., & Xu, A. (2016). Heterogeneity of white adipose tissue: Molecular basis and clinical implications. *Experimental & Molecular Medicine, 48*, e215. https://doi.org/10.1038/emm.2016.5.

Lee, K. Y., Yamamoto, Y., Boucher, J., Winnay, J. N., Gesta, S., Cobb, J., Bluher, M., & Kahn, C. R. (2013a). Shox2 is a molecular determinant of depot-specific adipocyte function. *Proceedings of the National Academy of Sciences of the United States of America, 110*(28), 11409–11414. https://doi.org/10.1073/pnas.1310331110.

Lee, M. J., Wu, Y., & Fried, S. K. (2013b). Adipose tissue heterogeneity: Implication of depot differences in adipose tissue for obesity complications. *Molecular Aspects of Medicine, 34*(1), 1–11. https://doi.org/10.1016/j.mam.2012.10.001.

Lee, Y. H., Kim, S. N., Kwon, H. J., & Granneman, J. G. (2017). Metabolic heterogeneity of activated beige/brite adipocytes in inguinal adipose tissue. *Scientific Reports, 7*, 39794. https://doi.org/10.1038/srep39794.

Lidell, M. E., Betz, M. J., Dahlqvist Leinhard, O., Heglind, M., Elander, L., Slawik, M., Mussack, T., Nilsson, D., Romu, T., Nuutila, P., Virtanen, K. A., Beuschlein, F., Persson, A., Borga, M., & Enerback, S. (2013). Evidence for two types of brown adipose tissue in humans. *Nature Medicine, 19*(5), 631–634. https://doi.org/10.1038/nm.3017.

Lofgren, P., Hoffstedt, J., Ryden, M., Thorne, A., Holm, C., Wahrenberg, H., & Arner, P. (2002). Major gender differences in the lipolytic capacity of abdominal subcutaneous fat cells in obesity observed before and after long-term weight reduction. *The Journal of Clinical Endocrinology and Metabolism, 87*(2), 764–771. https://doi.org/10.1210/jcem.87.2.8254.

Lumeng, C. N., Bodzin, J. L., & Saltiel, A. R. (2007). Obesity induces a phenotypic switch in adipose tissue macrophage polarization. *The Journal of Clinical Investigation, 117*(1), 175–184. https://doi.org/10.1172/JCI29881.

Macotela, Y., Boucher, J., Tran, T. T., & Kahn, C. R. (2009). Sex and depot differences in adipocyte insulin sensitivity and glucose metabolism. *Diabetes, 58*(4), 803–812. https://doi.org/10.2337/db08-1054.

Majka, S. M., Fox, K. E., Psilas, J. C., Helm, K. M., Childs, C. R., Acosta, A. S., Janssen, R. C., Friedman, J. E., Woessner, B. T., Shade, T. R., Varella-Garcia, M., & Klemm, D. J. (2010). De novo generation of white adipocytes from the myeloid lineage via mesenchymal intermediates is age, adipose depot, and gender specific. *Proceedings of the National Academy of Sciences of the United States of America, 107*(33), 14781–14786. https://doi.org/10.1073/pnas.1003512107.

Mammi, C., Calanchini, M., Antelmi, A., Cinti, F., Rosano, G. M., Lenzi, A., Caprio, M., & Fabbri, A. (2012). Androgens and adipose tissue in males: A complex and reciprocal interplay. *International Journal of Endocrinology, 2012*, 789653. https://doi.org/10.1155/2012/789653.

Maquoi, E., Munaut, C., Colige, A., Collen, D., & Lijnen, H. R. (2002). Modulation of adipose tissue expression of murine matrix metalloproteinases and their tissue inhibitors with obesity. *Diabetes, 51*(4), 1093–1101.

Mariman, E. C., & Wang, P. (2010). Adipocyte extracellular matrix composition, dynamics and role in obesity. *Cellular and Molecular Life Sciences, 67*(8), 1277–1292.

Masuzaki, H., Ogawa, Y., Isse, N., Satoh, N., Okazaki, T., Shigemoto, M., Mori, K., Tamura, N., Hosoda, K., Yoshimasa, Y., et al. (1995). Human obese gene expression. Adipocyte-specific expression and regional differences in the adipose tissue. *Diabetes, 44*(7), 855–858.

Matsubara, M., Maruoka, S., & Katayose, S. (2002). Inverse relationship between plasma adiponectin and leptin concentrations in normal-weight and obese women. *European Journal of Endocrinology, 147*(2), 173–180.

Mattacks, C. A., & Pond, C. M. (1999). Interactions of noradrenalin and tumour necrosis factor alpha, interleukin 4 and interleukin 6 in the control of lipolysis from adipocytes around lymph nodes. *Cytokine, 11*(5), 334–346. https://doi.org/10.1006/cyto.1998.0442.

Melanson, E. L., Sharp, T. A., Schneider, J., Donahoo, W. T., Grunwald, G. K., & Hill, J. O. (2003). Relation between calcium intake and fat oxidation in adult humans. *International Journal of Obesity and Related Metabolic Disorders: Journal of the International Association for the Study of Obesity, 27*(2), 196–203. https://doi.org/10.1038/sj.ijo.802202.

Merkestein, M., Cagampang, F. R., & Sellayah, D. (2014). Fetal programming of adipose tissue function: An evolutionary perspective. *Mammalian Genome: Official Journal of the International Mammalian Genome Society, 25*(9–10), 413–423. https://doi.org/10.1007/s00335-014-9528-9.

Mittendorfer, B., Horowitz, J. F., & Klein, S. (2001). Gender differences in lipid and glucose kinetics during short-term fasting. *American Journal of Physiology Endocrinology and Metabolism, 281*(6), E1333–E1339.

Montague, C. T., Farooqi, I. S., Whitehead, J. P., Soos, M. A., Rau, H., Wareham, N. J., Sewter, C. P., Digby, J. E., Mohammed, S. N., Hurst, J. A., Cheetham,

C. H., Earley, A. R., Barnett, A. H., Prins, J. B., & O'Rahilly, S. (1997a). Congenital leptin deficiency is associated with severe early-onset obesity in humans. *Nature, 387*(6636), 903–908.

Montague, C. T., Prins, J. B., Sanders, L., Digby, J. E., & O'Rahilly, S. (1997b). Depot- and sex-specific differences in human leptin mRNA expression: Implications for the control of regional fat distribution. *Diabetes, 46*(3), 342–347.

Mori, S., Kiuchi, S., Ouchi, A., Hase, T., & Murase, T. (2014). Characteristic expression of extracellular matrix in subcutaneous adipose tissue development and adipogenesis; comparison with visceral adipose tissue. *International Journal of Biological Sciences, 10*(8), 825–833. https://doi.org/10.7150/ijbs.8672.

Munoz-Garach, A., Cornejo-Pareja, I., & Tinahones, F. J. (2016). Does metabolically healthy obesity exist? *Nutrients, 8*(6). https://doi.org/10.3390/nu8060320.

Nedergaard, J., Bengtsson, T., & Cannon, B. (2007). Unexpected evidence for active brown adipose tissue in adult humans. *American Journal of Physiology Endocrinology and Metabolism, 293*(2), E444–E452. https://doi.org/10.1152/ajpendo.00691.2006.

Nedergaard, J., Bengtsson, T., & Cannon, B. (2010). Three years with adult human brown adipose tissue. *Annals of the New York Academy of Sciences, 1212*, E20–E36. https://doi.org/10.1111/j.1749-6632.2010.05905.x.

Nguyen, K. D., Qiu, Y., Cui, X., Goh, Y. P., Mwangi, J., David, T., Mukundan, L., Brombacher, F., Locksley, R. M., & Chawla, A. (2011). Alternatively activated macrophages produce catecholamines to sustain adaptive thermogenesis. *Nature, 480*(7375), 104–108. https://doi.org/10.1038/nature10653.

Nielsen, S., Guo, Z., Albu, J. B., Klein, S., O'Brien, P. C., & Jensen, M. D. (2003). Energy expenditure, sex, and endogenous fuel availability in humans. *The Journal of Clinical Investigation, 111*(7), 981–988. https://doi.org/10.1172/JCI16253.

Palmer, A. K., & Kirkland, J. L. (2016). Aging and adipose tissue: Potential interventions for diabetes and regenerative medicine. *Experimental Gerontology*. https://doi.org/10.1016/j.exger.2016.02.013.

Palmer, B. F., & Clegg, D. J. (2015). The sexual dimorphism of obesity. *Molecular and Cellular Endocrinology, 402*, 113–119. https://doi.org/10.1016/j.mce.2014.11.029.

Pasarica, M., Gowronska-Kozak, B., Burk, D., Remedios, I., Hymel, D., Gimble, J., Ravussin, E., Bray, G. A., & Smith, S. R. (2009). Adipose tissue collagen VI in obesity. *The Journal of Clinical Endocrinology and Metabolism, 94*(12), 5155–5162. https://doi.org/10.1210/jc.2009-0947.

Peirce, V., Carobbio, S., & Vidal-Puig, A. (2014). The different shades of fat. *Nature, 510*(7503), 76–83. https://doi.org/10.1038/nature13477.

Pond, C. M., & Mattacks, C. A. (1998). In vivo evidence for the involvement of the adipose tissue surrounding lymph nodes in immune responses. *Immunology Letters, 63*(3), 159–167.

Pond, C. M., & Mattacks, C. A. (2002). The activation of the adipose tissue associated with lymph nodes during the early stages of an immune response. *Cytokine, 17*(3), 131–139. https://doi.org/10.1006/cyto.2001.0999.

Pujol, E., Rodriguez-Cuenca, S., Frontera, M., Justo, R., Llado, I., Kraemer, F. B., Gianotti, M., & Roca, P. (2003). Gender- and site-related effects on lipolytic capacity of rat white adipose tissue. *Cellular and Molecular Life Sciences: CMLS, 60*(9), 1982–1989. https://doi.org/10.1007/s00018-003-3125-5.

Ramis, J. M., Salinas, R., Garcia-Sanz, J. M., Moreiro, J., Proenza, A. M., & Llado, I. (2006). Depot- and gender-related differences in the lipolytic pathway of adipose tissue from severely obese patients. *Cellular Physiology and Biochemistry: International Journal of Experimental Cellular Physiology, Biochemistry, and Pharmacology, 17*(3–4), 173–180. https://doi.org/10.1159/000092079.

Rosen, E. D., & Spiegelman, B. M. (2014). What we talk about when we talk about fat. *Cell, 156*(1–2), 20–44. https://doi.org/10.1016/j.cell.2013.12.012.

Sackmann-Sala, L., Berryman, D. E., Munn, R. D., Lubbers, E. R., & Kopchick, J. J. (2012). Heterogeneity among white adipose tissue depots in male C57BL/6J mice. *Obesity (Silver Spring), 20*(1), 101–111. https://doi.org/10.1038/oby.2011.235.

Saito, M., Okamatsu-Ogura, Y., Matsushita, M., Watanabe, K., Yoneshiro, T., Nio-Kobayashi, J., Iwanaga, T., Miyagawa, M., Kameya, T., Nakada, K., Kawai, Y., & Tsujisaki, M. (2009). High incidence of metabolically active brown adipose tissue in healthy adult humans: Effects of cold exposure and adiposity. *Diabetes, 58*(7), 1526–1531.

Salans, L. B., & Dougherty, J. W. (1971). The effect of insulin upon glucose metabolism by adipose cells of different size. Influence of cell lipid and protein content, age, and nutritional state. *The Journal of Clinical Investigation, 50*(7), 1399–1410. https://doi.org/10.1172/JCI106623.

Sanchez-Gurmaches, J., & Guertin, D. A. (2014). Adipocytes arise from multiple lineages that are heterogeneously and dynamically distributed. *Nature Communications, 5*, 4099. https://doi.org/10.1038/ncomms5099.

Sanchez-Gurmaches, J., Hsiao, W. Y., & Guertin, D. A. (2015). Highly selective in vivo labeling of subcutaneous white adipocyte precursors with Prx1-Cre. *Stem Cell Reports, 4*(4), 541–550. https://doi.org/10.1016/j.stemcr.2015.02.008.

Seale, P., Conroe, H. M., Estall, J., Kajimura, S., Frontini, A., Ishibashi, J., Cohen, P., Cinti, S., & Spiegelman, B. M. (2011). Prdm16 determines the thermogenic program of subcutaneous white adipose tissue in mice. *The Journal of Clinical Investigation, 121*(1), 96–105. https://doi.org/10.1172/JCI44271.

Seydoux, J., Muzzin, P., Moinat, M., Pralong, W., Girardier, L., & Giacobino, J. P. (1996). Adrenoceptor heterogeneity in human white adipocytes differentiated in culture as assessed by cytosolic free calcium measurements. *Cellular Signalling, 8*(2), 117–122.

Shao, M., Hepler, C., Vishvanath, L., MacPherson, K. A., Busbuso, N. C., & Gupta, R. K. (2017). Fetal development of subcutaneous white adipose tissue is dependent on Zfp423. *Molecular Metabolism, 6*(1), 111–124. https://doi.org/10.1016/j.molmet.2016.11.009.

Shen, M., Kumar, S. P., & Shi, H. (2014). Estradiol regulates insulin signaling and inflammation in adipose tissue. *Hormone Molecular Biology and Clinical Investigation, 17*(2), 99–107. https://doi.org/10.1515/hmbci-2014-0007.

Shinoda, K., Luijten, I. H., Hasegawa, Y., Hong, H., Sonne, S. B., Kim, M., Xue, R., Chondronikola, M., Cypess, A. M., Tseng, Y. H., Nedergaard, J., Sidossis, L. S., & Kajimura, S. (2015). Genetic and functional characterization of clonally derived adult human brown adipocytes. *Nature Medicine, 21*(4), 389–394. https://doi.org/10.1038/nm.3819.

Soukas, A., Cohen, P., Socci, N. D., & Friedman, J. M. (2000). Leptin-specific patterns of gene expression in white adipose tissue. *Genes & Development, 14*(8), 963–980.

Spencer, M., Unal, R., Zhu, B., Rasouli, N., McGehee, R. E., Jr., Peterson, C. A., & Kern, P. A. (2011). Adipose tissue extracellular matrix and vascular abnormalities in obesity and insulin resistance. *The Journal of Clinical Endocrinology and Metabolism, 96*(12), E1990–E1998. https://doi.org/10.1210/jc.2011-1567.

Stout, M. B., Justice, J. N., Nicklas, B. J., & Kirkland, J. L. (2017). Physiological aging: Links among adipose tissue dysfunction, diabetes, and frailty. *Physiology (Bethesda), 32*(1), 9–19. https://doi.org/10.1152/physiol.00012.2016.

Strissel, K. J., Stancheva, Z., Miyoshi, H., Perfield, J. W., 2nd, DeFuria, J., Jick, Z., Greenberg, A. S., & Obin, M. S. (2007). Adipocyte death, adipose tissue remodeling, and obesity complications. *Diabetes, 56*(12), 2910–2918. https://doi.org/10.2337/db07-0767.

Stubbins, R. E., Holcomb, V. B., Hong, J., & Nunez, N. P. (2012). Estrogen modulates abdominal adiposity and protects female mice from obesity and impaired glucose tolerance. *European Journal of Nutrition, 51*(7), 861–870. https://doi.org/10.1007/s00394-011-0266-4.

Sun, K., Kusminski, C. M., & Scherer, P. E. (2011). Adipose tissue remodeling and obesity. *The Journal of Clinical Investigation, 121*(6), 2094–2101. https://doi.org/10.1172/JCI45887.

Sun, K., Park, J., Gupta, O. T., Holland, W. L., Auerbach, P., Zhang, N., Goncalves Marangoni, R., Nicoloro, S. M., Czech, M. P., Varga, J., Ploug, T., An, Z., & Scherer, P. E. (2014). Endotrophin triggers adipose tissue fibrosis and metabolic dysfunction. *Nature Communications, 5*, 3485. https://doi.org/10.1038/ncomms4485.

Tchernof, A., Belanger, C., Morisset, A. S., Richard, C., Mailloux, J., Laberge, P., & Dupont, P. (2006). Regional differences in adipose tissue metabolism in women: Minor effect of obesity and body fat distribution. *Diabetes, 55*(5), 1353–1360.

Tchkonia, T., Lenburg, M., Thomou, T., Giorgadze, N., Frampton, G., Pirtskhalava, T., Cartwright, A., Cartwright, M., Flanagan, J., Karagiannides, I., Gerry, N., Forse, R. A., Tchoukalova, Y., Jensen, M. D., Pothoulakis, C., & Kirkland, J. L. (2007). Identification of depot-specific human fat cell progenitors through distinct expression profiles and developmental gene patterns. *American Journal of Physiology. Endocrinology and Metabolism, 292*(1), E298–E307.

Tchkonia, T., Morbeck, D. E., Von Zglinicki, T., Van Deursen, J., Lustgarten, J., Scrable, H., Khosla, S., Jensen, M. D., & Kirkland, J. L. (2010). Fat tissue, aging, and cellular senescence. *Aging Cell, 9*(5), 667–684. https://doi.org/10.1111/j.1474-9726.2010.00608.x.

Tchoukalova, Y. D., Votruba, S. B., Tchkonia, T., Giorgadze, N., Kirkland, J. L., & Jensen, M. D. (2010). Regional differences in cellular mechanisms of adipose tissue gain with overfeeding. *Proceedings of the National Academy of Sciences of the United States of America, 107*(42), 18226–18231. https://doi.org/10.1073/pnas.1005259107.

Thomou, T., Mori, M. A., Dreyfuss, J. M., Konishi, M., Sakaguchi, M., Wolfrum, C., Rao, T. N., Winnay, J. N., Garcia-Martin, R., Grinspoon, S. K., Gorden, P., & Kahn, C. R. (2017). Adipose-derived circulating miRNAs regulate gene expression in other tissues. *Nature.* https://doi.org/10.1038/nature21365.

Townsend, K. L., & Tseng, Y. H. (2015). Of mice and men: Novel insights regarding constitutive and recruitable brown adipocytes. *International Journal of Obesity Supplements, 5*(Suppl 1), S15–S20. https://doi.org/10.1038/ijosup.2015.5.

Tran, T. T., & Kahn, C. R. (2010). Transplantation of adipose tissue and stem cells: Role in metabolism and disease. *Nature Reviews Endocrinology, 6*(4), 195–213. https://doi.org/10.1038/nrendo.2010.20.

Troike, K. M., Henry, B. E., Jensen, E. A., Young, J. A., List, E. O., Kopchick, J. J., & Berryman, D. E. (2017). Impact of growth hormone on regulation of adipose tissue. *Comprehensive Physiology.* Jun 18;7(3): 819–840. https://doi.org/10.1002/cphy.c160027.

Varlamov, O., Chu, M., Cornea, A., Sampath, H., & Roberts, C. T., Jr. (2015). Cell-autonomous heterogeneity of nutrient uptake in white adipose tissue of rhesus macaques. *Endocrinology, 156*(1), 80–89. https://doi.org/10.1210/en.2014-1699.

Vatier, C., Fetita, S., Boudou, P., Tchankou, C., Deville, L., Riveline, J. P., Young, J., Mathivon, L., Travert, F., Morin, D., Cahen, J., Lascols, O., Andreelli, F., Reznik, Y., Mongeois, E., Madelaine, I., Vantyghem, M. C., Gautier, J. F., & Vigouroux, C. (2016). One-year metreleptin improves insulin secretion in patients with diabetes linked to genetic lipodystrophic syndromes. *Diabetes, Obesity and Metabolism, 18*(7), 693–697. https://doi.org/10.1111/dom.12606.

Vijgen, G. H., Bouvy, N. D., Teule, G. J., Brans, B., Schrauwen, P., & van Marken Lichtenbelt, W. D. (2011). Brown adipose tissue in morbidly obese subjects. *PLoS One, 6*(2), e17247. https://doi.org/10.1371/journal.pone.0017247.

Villarroya, J., Cereijo, R., & Villarroya, F. (2013). An endocrine role for brown adipose tissue? *American Journal of Physiology Endocrinology and Metabolism, 305*(5), E567–E572. https://doi.org/10.1152/ajpendo.00250.2013.

Vitali, A., Murano, I., Zingaretti, M. C., Frontini, A., Ricquier, D., & Cinti, S. (2012). The adipose organ of obesity-prone C57BL/6J mice is composed of mixed white and brown adipocytes. *Journal of Lipid Research, 53*(4), 619–629. https://doi.org/10.1194/jlr.M018846.

Wang, G. X., Zhao, X. Y., Meng, Z. X., Kern, M., Dietrich, A., Chen, Z., Cozacov, Z., Zhou, D., Okunade, A. L., Su, X., Li, S., Bluher, M., & Lin, J. D. (2014). The brown fat-enriched secreted factor Nrg4 preserves metabolic homeostasis through attenuation of hepatic lipogenesis. *Nature Medicine, 20*(12), 1436–1443. https://doi.org/10.1038/nm.3713.

Wang, H., Liu, L., Lin, J. Z., Aprahamian, T. R., & Farmer, S. R. (2016). Browning of white adipose tissue with roscovitine induces a distinct population of UCP1+ adipocytes. *Cell Metabolism, 24*(6), 835–847. https://doi.org/10.1016/j.cmet.2016.10.005.

Wang, Q. A., Tao, C., Gupta, R. K., & Scherer, P. E. (2013). Tracking adipogenesis during white adipose tissue development, expansion and regeneration. *Nature Medicine, 19*(10), 1338–1344. https://doi.org/10.1038/nm.3324.

Wang, S. P., Laurin, N., Himms-Hagen, J., Rudnicki, M. A., Levy, E., Robert, M. F., Pan, L., Oligny, L., & Mitchell, G. A. (2001). The adipose tissue phenotype of hormone-sensitive lipase deficiency in mice. *Obesity Research, 9*(2), 119–128.

Wang, W., & Seale, P. (2016). Control of brown and beige fat development. *Nature Reviews. Molecular Cell Biology, 17*(11), 691–702. https://doi.org/10.1038/nrm.2016.96.

Wikstrom, J. D., Mahdaviani, K., Liesa, M., Sereda, S. B., Si, Y., Las, G., Twig, G., Petrovic, N., Zingaretti, C., Graham, A., Cinti, S., Corkey, B. E., Cannon, B., Nedergaard, J., & Shirihai, O. S. (2014). Hormone-induced mitochondrial fission is utilized by brown adipocytes as an amplification pathway for energy expenditure. *The EMBO Journal, 33*(5), 418–436. https://doi.org/10.1002/embj.201385014.

Wong, R. H., & Sul, H. S. (2010). Insulin signaling in fatty acid and fat synthesis: A transcriptional perspective. *Current Opinion in Pharmacology, 10*(6), 684–691. https://doi.org/10.1016/j.coph.2010.08.004.

Wu, J., Bostrom, P., Sparks, L. M., Ye, L., Choi, J. H., Giang, A. H., Khandekar, M., Virtanen, K. A., Nuutila, P., Schaart, G., Huang, K., Tu, H., van Marken Lichtenbelt, W. D., Hoeks, J., Enerback, S., Schrauwen, P., & Spiegelman, B. M. (2012). Beige adipocytes are a distinct type of thermogenic fat cell in mouse and human. *Cell, 150*(2), 366–376. https://doi.org/10.1016/j.cell.2012.05.016.

Wueest, S., Schoenle, E. J., & Konrad, D. (2012a). Depot-specific differences in adipocyte insulin sensitivity in mice are diet- and function-dependent. *Adipocytes, 1*(3), 153–156. https://doi.org/10.4161/adip.19910.

Wueest, S., Yang, X., Liu, J., Schoenle, E. J., & Konrad, D. (2012b). Inverse regulation of basal lipolysis in perigonadal and mesenteric fat depots in mice. *American Journal of Physiology Endocrinology and Metabolism, 302*(1), E153–E160. https://doi.org/10.1152/ajpendo.00338.2011.

Xue, R., Lynes, M. D., Dreyfuss, J. M., Shamsi, F., Schulz, T. J., Zhang, H., Huang, T. L., Townsend, K. L., Li, Y., Takahashi, H., Weiner, L. S., White, A. P., Lynes, M. S., Rubin, L. L., Goodyear, L. J., Cypess, A. M., & Tseng, Y. H. (2015). Clonal analyses and gene profiling identify genetic biomarkers of the thermogenic potential of human brown and white preadipocytes. *Nature Medicine, 21*(7), 760–768. https://doi.org/10.1038/nm.3881.

Yaghootkar, H., Scott, R. A., White, C. C., Zhang, W., Speliotes, E., Munroe, P. B., Ehret, G. B., Bis, J. C., Fox, C. S., Walker, M., Borecki, I. B., Knowles, J. W., Yerges-Armstrong, L., Ohlsson, C., Perry, J. R., Chambers, J. C., Kooner, J. S., Franceschini, N., Langenberg, C., Hivert, M. F., Dastani, Z., Richards, J. B., Semple, R. K., & Frayling, T. M. (2014). Genetic evidence for a normal-weight "metabolically obese" phenotype linking insulin resistance, hypertension, coronary artery disease, and type 2 diabetes. *Diabetes, 63*(12), 4369–4377. https://doi.org/10.2337/db14-0318.

Yin, D., Clarke, S. D., Peters, J. L., & Etherton, T. D. (1998). Somatotropin-dependent decrease in fatty acid synthase mRNA abundance in 3T3-F442A adipocytes is the result of a decrease in both gene transcription and mRNA stability. *The Biochemical Journal, 331*(Pt. 3), 815–820.

Zhu, Y., Gao, Y., Tao, C., Shao, M., Zhao, S., Huang, W., Yao, T., Johnson, J. A., Liu, T., Cypess, A. M., Gupta, O., Holland, W. L., Gupta, R. K., Spray, D. C., Tanowitz, H. B., Cao, L., Lynes, M. D., Tseng, Y. H., Elmquist, J. K., Williams, K. W., Lin, H. V., & Scherer, P. E. (2016). Connexin 43 mediates white adipose tissue beiging by facilitating the propagation of sympathetic neuronal signals. *Cell Metabolism, 24*(3), 420–433. https://doi.org/10.1016/j.cmet.2016.08.005.

Adipokines, Inflammation, and Insulin Resistance in Obesity

Hyokjoon Kwon and Jeffrey E. Pessin

9.1 General

As calorie availability is unpredictable in nature, humans have developed multiple mechanisms for efficient energy storage and utilization. In modern society, however, excess nutrition and sedentary lifestyle are major factors responsible for excessive lipid accumulation in adipose and peripheral tissues, resulting in obesity. Obesity has become a pandemic health problem in which more than 60% of American adults are overweight or obese and is closely associated with diverse metabolic diseases such as type 2 diabetes (T2D), cardiovascular disease (CV), nonalcoholic fatty liver disease (NAFLD), and polycystic ovarian diseases (Finkelstein et al. 2012). Thus, the financial cost to manage obesity-related metabolic diseases is problematic in public healthcare system. T2D is a quickly growing global metabolic disease characterized by impaired insulin secretion from pancreatic β cells and insulin resistance in the liver, muscle, and adipose tissue. In T2D, pancreatic β cells are continuously activated to synthesize and secret insulin due to uncontrolled hyperglycemia, and

this cellular stress gradually induces deterioration and cell death of pancreatic β cells (Ashcroft and Rorsman 2012; Butler et al. 2003). Thus, both impaired pancreatic β cell function and insulin resistance further exacerbate physiological consequences of T2D. The muscle and adipose tissue show impaired insulin-stimulated glucose uptake with enhanced glucose production (gluconeogenesis) in the liver. These dysregulated tissue-specific pathophysiologies result in increased fasting glucose levels and the inability to uptake glucose from the blood circulation in the postprandial state.

The molecular mechanisms of obesity-induced pathogenesis of T2D are not clear. However, recent studies have suggested that low-grade chronic inflammation is one of the important factors in the pathogenesis of T2D in humans and rodent animal models (Hotamisligil 2006; Ouchi et al. 2011; Schenk et al. 2008; Shoelson et al. 2006). Although the liver and muscle show very mild obesity-induced inflammatory responses without significant changes in the number and composition of immune cells, adipose tissues especially visceral white adipose tissues are the most vulnerable target to mediate significant immune cell enrichment and inflammatory responses to mediate systemic inflammation and insulin resistance in obese rodents and humans (Odegaard and Chawla 2013). Adipose tissue is a major tissue to store excess nutrient as triglycerides and also produces various secreted

H. Kwon · J. E. Pessin (✉)
Department of Medicine and Molecular
Pharmacology, Albert Einstein College of Medicine,
Bronx, NY, USA
e-mail: jeffrey.pessin@einstein.yu.edu

© Springer International Publishing AG, part of Springer Nature 2018
E. A. Nillni (ed.), *Textbook of Energy Balance, Neuropeptide Hormones,*
and Neuroendocrine Function, https://doi.org/10.1007/978-3-319-89506-2_9

proteins called adipokines as an endocrine organ (Waki and Tontonoz 2007). Adipose tissues have been shown to produce adipokines such as leptin and adiponectin and also generate diverse pro- and anti-inflammatory adipokines to modulate inflammatory responses in peripheral organs including adipose tissue itself. Adipocytes, the most abundant cell population of adipose tissue, are differentiated from pluripotent mesenchymal stem cells through various transcriptional regulations such as PPARγ and C/EBPα (Rosen and MacDougald 2006; Tang and Lane 2012). Mature white adipocytes contain a single large lipid droplet (uniocular) surrounded by a thin cytoplasm and lipid droplet-associated proteins such as perilipin. These lipid droplet-associated proteins function as part of the control mechanism to regulate lipolysis (Miyoshi et al. 2006). As adipocytes provide reversible energy storage depot in adipose tissue, excess nutrition overload initiates adipocyte hypertrophy and hyperplasia, resulting in cellular stress that in turn initiates oxidative stress and inflammatory responses in adipose tissue. Inflammatory responses in adipose tissues become self-generating that eventually leads to increased local and systemic levels of various pro-inflammatory cytokines including tumor necrosis factor-α (TNF-α), interleukin-6 (IL-6), IL-1β, and CC chemokine ligand 2 (CCL2) that have all been shown to drive insulin resistance. Along with inflammatory adipokine production in adipose tissues, obesity-related hyperlipidemia, hyperglycemia, hypoxia, hyperendotoxemia, oxidative stress, and endoplasmic reticulum (ER) stress can also induce insulin resistance in peripheral tissues and can amplify the activation of inflammatory signaling cascades in adipose tissues.

The current main objective in this field is to understand the initiating factors responsible for induction of obesity-induced adipose tissue inflammation and the complex signaling cascade that continues to amplify and maintain the pro-inflammatory state. Understanding of the cellular and molecular cross talk between adipocytes and the immune system in obesity will be able to develop specific therapies to prevent obesity-induced inflammation and to restore insulin sensitivity in an effective manner without inducing secondary complications such as ectopic lipid accumulation and further exacerbating obesity. In this chapter, we will focus on the recent progress regarding the physiological and molecular functions of adipokines in the obesity-induced inflammation and insulin resistance.

9.2 Insulin Signaling and Resistance

Glucose homeostasis is maintained by delicate regulation of pancreatic endocrine systems. The pancreas has exocrine and endocrine functions and is essential for nutrient metabolism. The pancreatic exocrine cells are composed of acinar and ductal cells that secrete digestive enzymes into the duodenum through the ductal system. In addition, pancreatic acinar cells are considered as major targets for pancreatic β cell replacement therapy because it retains pluripotent ability for β cell reprograming (Zhou et al. 2008). The pancreatic endocrine cells localized to the islets of Langerhans secrete hormones into the blood circulation to regulate nutrient metabolism. Pancreatic islets have several endocrine cells including glucagon-producing α cells, insulin-producing β cells, somatostatin-producing δ cells, pancreatic polypeptide-producing PP cells, and ghrelin-producing ε cells. In particular, insulin secretion is enhanced in response to increased circulating glucose and amino acids. In peripheral tissues, insulin stimulates glucose uptake (skeletal muscle and adipose tissue) and glycogen synthesis (skeletal muscle and liver) and inhibits gluconeogenesis and glycogenolysis (liver). Insulin also increases lipogenesis in hepatocytes and adipocytes and diminishes lipolysis in adipocytes to reduce circulating free fatty acid and glycerol (Pessin and Saltiel 2000). Thus, the definition of insulin resistance is the perturbation of insulin-mediated signaling pathway, resulting in systemic hyperglycemia. As insulin has pleiotropic functions, insulin resistance is closely linked with other metabolic symptoms such as hypertension and hyperlipidemia (Cornier et al. 2008).

9.2.1 Insulin Signaling

At the molecular level, insulin binds to the cell surface insulin receptor to initiate intracellular signaling cascades that ultimately results in specific cellular biological responses (Taniguchi et al. 2006). The insulin receptor is composed of $\alpha_2\beta_2$ heterotetramers, and the extracellular α subunits directly bind insulin that allows for a transmembrane conformational change that activates the intracellular tyrosine kinase domain of β subunits (Ebina et al. 1985; Ullrich et al. 1985). The activated tyrosine kinase domain of β subunits phosphorylates itself in a transphosphorylation reaction that activates its intrinsic kinase activity to proximal substrates such as insulin receptor substrate (IRS) family (IRS1–IRS4), Src homology 2-containing (Shc) adaptor proteins, signal regulatory protein (SIRP) family, and Grb2-associated binder-1 (Gab1) (Fig. 9.1). IRS1/2 phosphorylated on specific tyrosine residues activates two major signaling pathways: (i) the phosphatidylinositol 3-kinase (PI3K)-AKT/protein kinase B (PKB) pathway and (ii) Ras-mitogen-activated protein kinase (MAPK) pathway. In addition, there are inhibitory molecules for insulin signaling such as protein tyrosine phosphatase 1B (PTP1B), suppressor of cytokine signaling

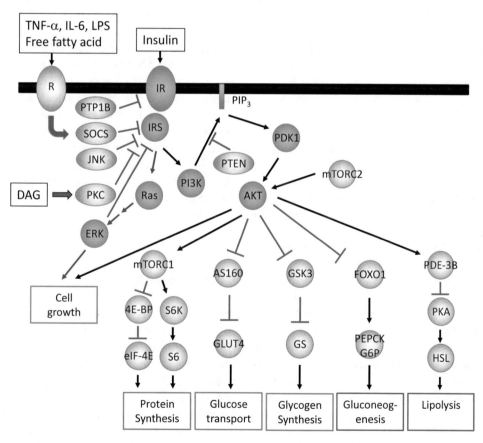

Fig. 9.1 Inflammatory adipokines suppress insulin signaling to mediate insulin resistance. IRS1/2 phosphorylated on specific tyrosine residues activates the phosphatidylinositol 3-kinase (PI3K)-AKT/protein kinase B (PKB) pathway and Ras-mitogen-activated protein kinase (MAPK) pathway. PI3K-AKT signaling pathway regulates metabolic processes such as glycogen synthesis (muscle and liver), glucose uptake (muscle and adipocytes), protein synthesis (muscle and liver), and gluconeogenesis (liver). Inflammatory signals such as TNF-α, saturated free fatty acid, IL-6, LPS, and DAG activate inhibitory molecules such as SOCS and JNK to suppress insulin signaling, resulting in insulin resistance

(SOCS), and growth factor receptor-bound protein 10 (Grb10) that suppress insulin signaling by inducing insulin receptor dephosphorylation, physical blocking of substrate phosphorylation, and degradation of the insulin receptor and/or IRS substrates. Metabolites such as diacylglycerol (DAG) also mediate the inhibition of insulin signaling.

PI3K-AKT/PKB signaling pathway modulates most metabolic functions of insulin. IRS binds insulin receptor through phosphotyrosine-binding domain (PTB) and has several tyrosine residues, which are phosphorylated by activated insulin receptor, for the interaction with SH2 domain containing regulatory subunit of PI3K (Myers et al. 1992). PI3K, regulatory and catalytic heterodimer (p85α/p110β), generates phosphatidylinositol (3,4,5)-triphosphate (PIP$_3$) on the plasma membrane followed by the activation of Ser/Thr protein kinase including 3-phosphoinositide-dependent protein kinase 1 (PDK1). PI3K-dependent PDK1 activation is negatively regulated by phospholipid phosphatases such as phosphatase and tensin homologue (PTEN) and SH2-containing inositol 5′-phosphatase-2 (SHIP2) that degrade PIP$_3$ (Sleeman et al. 2005; Wijesekara et al. 2005).

AKT/PKB has three isoforms (AKT1/PKBα, AKT2/PKBβ, AKT3/PKBγ) and is the major substrate target of PDK1. The PH domain of AKT/PKB interacts with PIP$_3$ to bring it in close proximity to the plasma membrane in which PDK1 is localized, resulting in AKT/PKB phosphorylation on Thr-308. AKT/PKB is also phosphorylated on Ser-473 by the mammalian target of rapamycin (mTOR) complex 2, and this dual phosphorylation results in full activation of AKT/PKB kinase activity (Sarbassov et al. 2005). Interestingly, the AKT isoforms show differential functions according to the distinctive expression profiles in tissues. AKT1 and AKT2 are broadly expressed with AKT2 that is more closely linked to metabolic processes. AKT1-deficient mice have growth retardation without defects in metabolism. In contrast, AKT2-deficient mice show insulin resistance due to interrupted insulin signaling (Cho et al. 2001). Thus, activated AKT/PKB regulates diverse insulin-mediated meta-bolic pathways such as glucose transport, glycogen synthesis, gluconeogenesis, protein synthesis, and cell growth. AKT phosphorylates AKT substrate of 160 kDa (AS160) to activate Rab small GTPase that initiates the translocation of the glucose transporter 4 (GLUT4), resulting in the glucose uptake in the muscle and adipocytes (Fig. 9.1). AKT also suppresses glycogen synthase kinase-3 (GSK3) via phosphorylation on Ser-21 or Ser-9 to activate glycogen synthase, resulting in the glycogen synthesis in the muscle and liver (Cross et al. 1995). AKT also phosphorylates forkhead box O1 (FOXO1) that induces FOXO1 association with 14-3-3 proteins to exclude FOXO1 from the nucleus. In the liver, this suppresses gluconeogenic gene expression and thereby inhibits hepatic glucose output. AKT phosphorylates tuberous sclerosis complex 1 and 2 (TSC1/2), which release the inhibition of Ras homologue enriched in the brain (Rheb) for the activation of mTORC1 complex, which in turn enhances protein synthesis through the activation of eukaryotic translation initiation factor 4E-binding protein-1 (4E-BP) and p70 ribosomal protein S6 kinase 1 (p70S6K1). AKT-mediated phosphorylation also activates phosphodieterase-3B (PDE-3B) to decrease the cAMP, resulting in the suppression of PKA activity and thereby suppressing hormone-sensitive lipase activity in adipocytes.

9.2.2   Insulin Resistance

Although there is substantial progress in our molecular understanding of insulin signaling, the mechanisms accounting for insulin resistance are still unclear. Insulin resistance is the integral result of alterations in insulin secretion in pancreatic β cells, insulin receptor expression, ligand binding, phosphorylation, and kinase activity and affects the downstream of insulin signaling, resulting in diverse clinical syndromes such as T2D, cardiovascular disease, type A syndrome, leprechaunism, and Rabson-Mendenhall syndrome. Cellular and molecular stress such as obesity decreases insulin sensitivity in peripheral tissues and stimulates pancreatic β cells to produce more insulin to maintain glucose homeostasis. However, when the pancreatic β cells are not able to produce sufficient amounts of insulin due to cellular stress such as inflammation, reactive oxygen, and ER stress, T2D develops. Thus, significant reduction in both pancreatic β cells mass and function is observed in T2D patients. Insulin receptor alteration is also one of the mechanisms to induce insulin resistance. Insulin receptor gene (*INSR*) mutations are rare, but at least more than 30 *INSR* mutations have been shown to mediate insulin receptor dysfunction, and these mutations may induce insulin resistance with polygenic defects in insulin signaling (Hegele 2003). In addition, mutations of DM1 kinase gene cause defective alternative splicing of *INSR* (Savkur et al. 2001), and mutations of high mobility group A1 (HMGA1) gene suppress the expression of *INSR*, resulting in the insulin resistance (Chiefari et al. 2011).

Impaired insulin signaling mediates insulin resistance. Decreased IRS protein levels contribute insulin resistance in rodents and humans (Shimomura et al. 2000). Although complete molecular understanding and mechanisms of reduced IRS levels are still under investigation, excess insulin suppresses the expression of IRS2. Furthermore, SOCS1/3 induced by inflammatory adipokines such as TNF-α, IL-6, and IL-1β enhances the degradation of IRS1/2 through E3 ubiquitin ligase activation (Rui et al. 2002) (Fig. 9.1). IRS phosphorylation on serine residues is another mechanism to induce insulin resistance. IRS contains several serine residues that are phosphorylated by kinases such as extracellular signal-regulated kinase (ERK), c-Jun N-terminal kinase (JNK), protein kinase Cζ (PKCζ), and p70S6K (Boura-Halfon and Zick 2009). The phosphorylation of IRS on Ser-307 is a typical inhibitory signal to suppress insulin signaling as Ser-307 locates in phosphotyrosine-binding (PTB) domain of IRS (Hirosumi et al. 2002). Thus, increased TNF-α and saturated free fatty acids in obese individuals activate JNK and inhibitor of nuclear factor κB kinase β (IKKβ) to phosphorylate Ser-307 of IRS. In addition, ERK activated by insulin also phosphorylates IRS1 on Ser-612 to attenuate AKT activation (Bard-Chapeau et al. 2005). Serine phosphorylation of IRS1 mediated by mTOR-S6K1 pathway induces IRS1 subcellular redistribution, potentially reducing insulin signaling through an additional mechanism (Shah and Hunter 2006).

Ectopic lipid accumulation in the liver and muscle is also associated with obesity-induced insulin resistance. High levels of diacylglycerol (DAG) generated by incomplete synthesis to triacylglycerol or breakdown of triacylglycerol to DAG has been proposed to inhibit insulin signaling through protein kinase C activation in the muscle (Badin et al. 2013; Chin et al. 1994; Griffin et al. 1999). Similarly, DAG accumulation in the liver is also associated with hepatic insulin resistance (Jornayvaz and Shulman 2012). In this regard, ATGL-deficient mice that have reduced ability to convert triacylglycerol to DAG show enhanced glucose tolerance and insulin sensitivity (Haemmerle et al. 2006). More recently, an alternative model of increased ceramide levels has also been shown to associate with insulin resistance (Chavez and Summers 2012). However, whether DAGs or ceramides mediate a cell autonomous insulin resistance or are part of the complex pathways responsible for obesity-induced inflammation has not been resolved.

The liver is one of the major tissues to regulate glucose homeostasis in response to pancreatic hormones such as insulin and glucagon. Insulin-mediated AKT activation regulates hepatic glucose and lipid metabolism. In insulin-sensitive individuals, regular meals increase the blood glucose, and then released pancreatic insulin stimulates insulin signaling and glucose uptake through insulin receptor and GLUT2, respectively. AKT activation enhances glycogen synthesis and de novo lipogenesis and also suppresses gluconeogenesis through FOXO inactivation (Fig. 9.2a). Thus, hepatic insulin signaling may suppress gluconeogenesis through transcriptional suppression of gluconeogenic genes for long term. However, the ability of insulin to acutely suppress gluconeogenesis is mostly mediated by an inhibition of adipose tissue lipolysis (Rebrin et al. 1996). First, insulin reduces

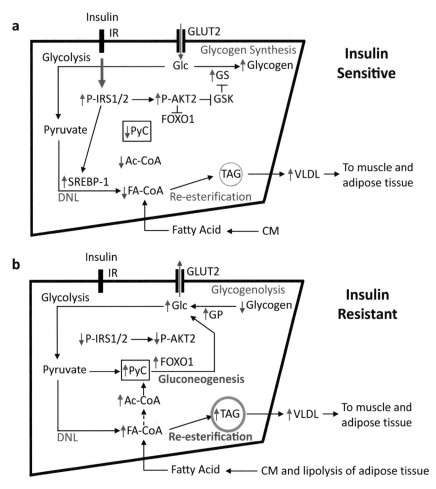

Fig. 9.2 Insulin resistance in the liver. (**a**) In insulin-sensitive individuals, insulin derived from the pancreas activates AKT to enhance glycogen synthesis and de novo lipogenesis and to suppress gluconeogenesis through FOXO inactivation. (**b**) In insulin resistance, impaired insulin signaling increases glycogenolysis and gluconeogenesis, and flux of fatty acid and glycerol from the adi-pose tissue and intestine results in glucose production and fatty liver. DNL de novo lipogenesis, GS glycogen synthase, GP glycogen phosphorylase, PyC pyruvate carboxylase, Ac-CoA acetyl-CoA, FA-CoA fatty acyl-CoA, TAG triacylglycerol, CM chylomicron, VLDL very low density lipoprotein

hepatic glucose production within 30 min but does not reduce gluconeogenic proteins levels (Ramnanan et al. 2010). Second, insulin suppresses adipose tissue lipolysis to regulate hepatic gluconeogenesis. Suppressed adipose tissue lipolysis lowers hepatic acetyl-CoA and glycerol contents to reduce pyruvate carboxylase (PC) activity and glucose production (Perry et al. 2015, 2014; Rebrin et al. 1996). In contrast, in insulin resistance status, glycogenolysis and gluconeogenesis are enhanced due to dysfunctional insulin signaling and flux of fatty acid and

glycerol from adipose tissue and intestine, resulting in the glucose production and fat accumulation in liver (Fig. 9.2b).

9.3 Inflammation in Adipose Tissue

Since Hotamisligil et al. suggest the role of TNF-α in insulin resistance (Hotamisligil et al. 1996, 1993), many groups show that obesity-induced inflammatory immune response in adipose tissue is one of the critical molecular mechanisms in obesity-induced insulin resistance (Gregor and Hotamisligil 2011; Schenk et al. 2008; Shoelson et al. 2006). In contrast to the muscle and liver, adipose tissues are highly vulnerable to obesity-induced inflammation. Accordingly, inflammatory responses in adipose tissues are mediated by the amplification of cellular stress-induced inflammatory signaling pathways in rodents and humans. Hyperlipidemia and hyperglycemia caused by excess nutrients, lipolysis, and unsuppressed gluconeogenesis induce mitochondrial dysfunction, ER stress, and oxidative stress, resulting in the activation of stress-responsive signaling pathways such as JNK, protein kinase C (PKC), protein kinase R (PKR), and inhibitor of NF-κB kinase β (IKKβ) (Nakamura et al. 2010; Solinas and Karin 2010). In addition to IRS Ser-307 phosphorylation to suppress insulin signaling, JNK, PKC, and IKKβ activation enhances inflammatory gene expression in target tissues, resulting in systemic inflammation and insulin resistance (Han et al. 2013; Samuel and Shulman 2012). Saturated free fatty acid derived from excess nutrients and lipolysis and gut-derived bacterial lipopolysaccharide (LPS) in obesity also stimulate downstream signaling of Toll-like receptor 4 (TLR4) to stimulate NF-κB and JNK (Ghoshal et al. 2009; Shi et al. 2006). Furthermore, inflammatory immune responses in adipose tissue are mediated by inflammatory adipokines, cytokines, and chemokines produced by adipocytes and infiltrated pro-inflammatory immune cells. In the following sections, we will focus on the cellular and molecular immune responses in adipose tissues of obese rodents and humans to highlight the role of diverse adipokines in obesity-induced inflammation and insulin resistance.

9.3.1 Adipose Tissues

There are two functionally and developmentally defined types of adipose tissue, white and brown. Brown adipose tissue (BAT) is the main site of non-shivering thermogenesis in mammals, whereas white adipose tissue (WAT) is the main depot to store metabolic energy in the form of triglycerides. Brown adipose tissue is found in newborn humans and hibernating mammals. Although it was previously believed that adult humans do not express brown adipocytes, more recently studies have documented the presence of brown adipose tissue that distributes in cervical-supraclavicular regions in humans and shows polygonal shape with multi-ocular lipid droplets with higher number of mitochondria and capillaries (Ravussin and Galgani 2011). Brown adipose tissue expresses unique uncoupling protein 1 (UCP1) in inner membrane of mitochondria to mediate uncoupling the respiratory chain from oxidative phosphorylation, resulting in the heat generation. Although brown adipose tissue has been shown to mediate non-shivering thermogenesis, it also produces various adipokines such as vascular endothelial growth factor A (VEGFA), bone morphogenetic proteins (BMPs), tumor necrosis factor (TNF), IL-6, and fibroblast growth factor 21 (FGF21) to mediate diverse physiologies in obesity and cold stimulation (Villarroya et al. 2017). More recently, it has been shown that a third form of adipose tissue also present in rodent models termed beige or brite adipocytes (Wu et al. 2012a). Similar to brown adipocytes, this recently identified adipocyte subtype is derived from a distinct progenitor cell population that resides within classical white adipose tissue. Classical brown adipocytes in mice are derived from a $Myf5^+$ myogenic progenitor cells (Seale et al. 2008; Timmons et al. 2007). However, beige adipocytes are differentiated from $Myf5^-$ endothelial and perivascular progenitor cells in white adipose tissue depot upon cold stimulation and exercise (Gupta et al. 2012; Tran et al. 2012). Thus, brown adipocytes, beige adipocytes, and white adipocytes show distinct origin for differentiation.

In contrast to brown and beige adipocytes, white adipocytes contain a single large lipid droplet. White adipocytes have been shown as a primary storage depot to store excess fuel and as an endocrine organ to release adipokines such as leptin and adiponectin that regulate systemic energy homeostasis. White adipose tissue is broadly located throughout the body, and anatomically distinct subcutaneous and visceral adipose tissues are major depots for white adipose tissues. Subcutaneous and visceral adipose tissues have differences in gene expression, hypertrophy, and hyperplasia in obesity and differentially contribute to obesity-induced inflammation and insulin resistance (Hardy et al. 2012). Subcutaneous adipose tissue has high capacity for adipocytes differentiation and cell size expansion to store large amounts of triglycerides. This storage capacity serves to reduce visceral adipose tissue mass and lipotoxicity mediated by lipid deposition in the liver and muscle. The impaired conversion of excess carbohydrate to lipid in subcutaneous adipose tissue due to decreased gene expression of SREBP-1 and ChREBP is associated with diabetes in obese humans (Kursawe et al. 2013). In contrast, visceral adipose tissue is positively associated with risk of obesity-induced inflammation and insulin resistance and shows higher monocytes infiltration and IL-6 production than subcutaneous adipose tissue to induce inflammation in obese subjects (Cancello et al. 2006; Fontana et al. 2007).

9.3.2 Immune Responses in Adipose Tissues

Obesity-induced inflammatory immune responses in adipose tissues are one of the major mechanisms to mediate insulin resistance in rodents and humans, and dynamic changes of immune cell numbers and composition in adipose tissues regulate inflammatory responses in obesity (Fig. 9.3). White adipose tissue consists of a variety of cell types including adipocytes, macrophages, granulocytes, lymphocytes, fibroblasts, and endothelial cells. Macrophages have been shown as a major cell population to mediate inflammatory innate immune responses in adipose tissue, resulting in insulin resistance. Macrophages constitute about 5% of immune cells in adipose tissue of lean mice, whereas it exceeds to 50% in obese mice. Humans also show similarly increased macrophage enrichment in adipose tissues of obese individuals to produce pro-inflammatory mediators such as TNF-α (Harman-Boehm et al. 2007). Exact

mechanism of macrophage enrichment is not clear, but it has been suggested that obesity-induced adipocytes hyperplasia and necrosis regulate the infiltration of macrophages. Macrophages are generically classified into two functionally distinct populations, M1 and M2 macrophages. In rodents, Th1 cytokines, IFN-γ, enhance nitric oxide synthase (NOS2) expression in classically activated macrophages (M1), whereas the Th2 cytokines such as IL-4 and IL-13 induce arginase-1 (ARG1) in alternatively activated macrophages (M2) (Lumeng et al. 2007a, b; Mantovani et al. 2004; Martinez et al. 2009; Mosser and Edwards 2008). $F4/80^+CD206^-CD11c^+$ inflammatory M1 macrophages are increased in adipose tissue and secrete pro-inflammatory cytokines such as TNF-α, IL-6, and IL-1β. Thus, macrophage composition diverts from M2 to M1 in obesity. Accordingly, TNF-α levels are increased in obese diabetic humans and rodents, and neutralization of TNF-α improves insulin sensitivity in obese rodents (Hotamisligil et al. 1993). TNF-α further enhances the expression of inflammatory cytokines (TNF-α and IL-6) and chemokines (CCL2 and RANTES) in adipocytes through the cross talk between enriched macrophages and adipocytes. TNF-α also induces Ser-307 phosphorylation of IRS1 to modulate the downstream effectors of the insulin receptor, resulting in insulin resistance (Hotamisligil et al. 1996). IL-1β is elevated in circulation (Spranger et al. 2003) and pancreatic islets of obese type 2 diabetic humans and rodents and induces the loss of pancreatic β cell mass, resulting in hyperglycemia (Donath et al. 1999; Ehses et al. 2009; Sauter et al. 2008). IL-1β mainly produced by monocytes and macrophages is synthesized as a IL-1β precursor in the cytosol, and activation-induced NALP3 (cryopyrin) inflammasome activates caspase-1 to mediate active IL-1β secretion (Dinarello 2009). Thus, inflammasome is critical for obesity-induced insulin resistance (Stienstra et al. 2012).

Granulocytes are also closely linked with obesity-induced inflammation in adipose tissues. Neutrophils, mast cells (Liu et al. 2009), eosinophils (Wu et al. 2011), and dendritic cells (Bertola et al. 2012) are involved in obesity-induced

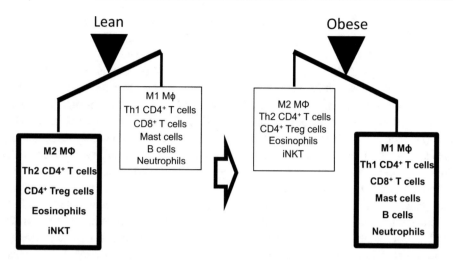

Fig. 9.3 Dynamic alternation of immune cell composition in adipose tissue. Alternatively activated M2 macrophages, Th2 CD4+ T cells, regulatory CD4+ T cells (Treg), eosinophils, and iNKT cells are dominant immune cells in adipose tissue of lean mice. In obese mice, the composition of immune cells is dynamically shifted to enhance pro-inflammatory responses in adipose tissue. Classically activated M1 macrophages, Th1 CD4+ T cells, effector CD8+ T cells, mast cells, B cells, and neutrophils are increased

inflammation and insulin resistance through the production of pro- and anti-inflammatory cytokines in adipose tissues. Eosinophils are motile phagocytic immune cells to protect multicellular parasites such as worms. Although Th2 CD4+ T cells are typical immune cells to produce anti-inflammatory IL-4, eosinophils are major cells to produce IL-4 in epididymal adipose tissue of lean mice to maintain Th2 immune responses in adipose tissue. Obesity decreases the number and proportion of eosinophils in epididymal adipose tissue to enhance Th1 immune responses, exacerbating insulin resistance (Wu et al. 2011). Obesity-induced insulin resistance is mediated by chronic low-grade inflammation in adipose tissue and shows different enrichment kinetics of diverse immune cells. Neutrophils are the first immune cells to respond to high-fat diet (HFD)-induced inflammation and involved in the trafficking of other immune cells into inflammatory sites. Neutrophils quickly infiltrate into adipose tissue within 3 days after HFD in mice model and produce neutrophil elastase to accelerate inflammatory responses. Thus, deletion of neutrophil elastase shows less inflammation and improved insulin sensitivity in obese rodents with reduced neutrophils and macrophages in adipose tissues (Talukdar et al. 2012).

Adaptive immune responses have been shown to be a critical factor for obesity-induced inflammation and insulin resistance in humans and rodents. Lymphocytes including T cells, B cells, natural killer (NK) cells, and natural killer T (NKT) cells are dynamically changed in adipose tissues to modulate obesity-induced insulin resistance. T cells are originated from the bone marrow and maturated in thymus. T cells have diverse immunological functions and classified by CD4 and CD8 surface makers. CD4+ T cells in adipose tissues of obese rodents and humans mediate HFD-induced insulin resistance. IFN-γ-producing Th1 CD4+ T cells are increased in adipose tissues of obese mice, overwhelming the anti-inflammatory Th2 CD4+ T cells and Foxp3+ regulatory CD4+ T cells. IFN-γ produced by enriched Th1 CD4+ T cells induces pro-inflammatory M1 macrophage differentiation and further Th1 CD4+ T cell amplification in adipose tissue. Interestingly, adoptive transfer of Th2 CD4+ T cells, which produce IL-4 and IL-13, rescues HFD-induced obesity and insulin resistance in *Rag1*-deficient mice, suggesting that Th2 cytokines such as IL-4 and

IL-13 suppress HFD-induced inflammation to improve insulin sensitivity (Winer et al. 2009a). Foxp3+ regulatory CD4+ T cells (T$_{reg}$), anti-inflammatory IL-10 producer, are unique cell population that suppresses inflammation. T$_{reg}$ cells are decreased in HFD-induced and genetically modified obese mice, resulting in insulin resistance (Feuerer et al. 2009). T$_{reg}$ cells suppress pro-inflammatory M1 macrophage differentiation and promote anti-inflammatory M2 macrophage differentiation by producing IL-4 and IL-10 (Feuerer et al. 2009; Tiemessen et al. 2007). Thiazolidinedione (TZD) treatment also enhances T$_{reg}$ development through the cooperation between peroxisome proliferator-activated receptor (PPAR)-γ and FOXP3, resulting in the improved inflammation and insulin sensitivity in HFD-fed mice (Cipolletta et al. 2012). Th17 CD4+ T cells, IL-17 and IL-23 producers, mediate diverse autoimmune diseases, and HFD predisposes autoimmune diseases such as trinitrobenzene sulfonic acid (TNBS)-induced colitis and experimental autoimmune encephalomyelitis (EAE) (Winer et al. 2009b). The role of Th17 CD4+ T cells in obesity-induced inflammation in adipose tissues is still controversial. Winer et al. show that Th17 CD4+ T cells are not involved in the obesity-induced inflammation in adipose tissue (Winer et al. 2009a). However, Th17 CD4+ T cells are increased in the blood of T2D patients (Jagannathan-Bogdan et al. 2011), and obesity-induced human CD11c+CD1c+ and mouse CD11c^{high}F4/80^{low} dendritic cells increase Th17 CD4+ T cells in adipose tissue to mediate inflammation (Bertola et al. 2012). CD8+ T cells are activated by antigens presented by MHC class I and produce IFN-γ to enhance Th1 CD4+ T cells and M1 macrophages differentiation. Thus, cytotoxic CD8+ T cells are also significantly increased in adipose tissues of obese mice, and depletion of CD8+ T cells reverses inflammation and insulin resistance, suggesting that obesity-induced infiltration of CD8+ T cells deteriorates systemic insulin sensitivity (Nishimura et al. 2009).

B cells are exclusive immune cells that produce antigen-specific antibodies. B cells are accumulated in adipose tissues of obese rodents and induce inflammation and insulin resistance

along with macrophages and T cells. B cells produce IgG2c autoantibodies to induce systemic inflammation, and B cell-deficient mice and depletion of B cells using anti-CD20 antibody administration suppress systemic inflammation and enhance insulin sensitivity (Winer et al. 2011). B cells are also involved in M1 macrophage differentiation in adipose tissue, and IgG2c is detected in crown-like structure in adipose tissues of HFD-fed mice, suggesting that B cells are involved in the clearance of necrotic adipocytes and inflammation (DeFuria et al. 2013). The role of innate invariant natural killer T (iNKT) cells in obesity-induced insulin resistance has also been shown. As iNKT cells rapidly response to its ligands, iNKT cells generate Th1 IFN-γ and Th2 IL-4 cytokines to regulate immune responses promptly (Bendelac et al. 2007). iNKT cells also produce IL-17 after TGF-β and IL-1β stimulation, resulting in neutrophilic airway inflammation (Monteiro et al. 2013). iNKT cells are highly enriched in adipose tissue of lean rodents and humans. However, iNKT cells are decreased in adipose tissues of obese rodents and humans, and the number of iNKT cells is recovered after weight loss. iNKT cell-deficient mice show that iNKT cells protect inflammation and insulin resistance in both lean and obese rodents and humans as adipose tissue-derived iNKT cells produce anti-inflammatory cytokines (Lynch et al. 2012; Schipper et al. 2012). However, the role of iNKT cells in HFD-induced inflammation and insulin resistance is still controversial as several previous reports show that iNKT cells are not necessary to suppress the HFD-induced inflammation and insulin resistance (Mantell et al. 2011; Wu et al. 2012b). Further investigation is necessary to understand the precise role of iNKT cells in obesity.

9.4 Adipokines

Adipose tissue is a storage organ for excess energy in the form of triglycerides and also functions as an active endocrine organ to modulate physiological metabolic processes. As adipose tissue contains various cell types such as adipocytes, immune cells, endothelial cells, and fibroblasts, it releases diverse secretory proteins called adipokines into the systemic circulation. Visceral and subcutaneous white adipose tissues produce unique profiles of adipokines to mediate inflammation and insulin resistance in obese rodents and humans. In addition, brown adipose tissue produces diverse adipokines to regulate different metabolic processes including thermogenesis and inflammation. In the past two decades, adipsin (complement factor C) (Cook et al. 1987), TNF-α, and leptin are identified as bona fide adipokines, and those studies have facilitated the identification of other adipokines. Although adipokines have multiple metabolic functions, this chapter will mainly discuss the inflammatory functions of adipokines that play important roles in mediating obesity-induced insulin resistance. Thus, adipokines are classified as pro- and anti-inflammatory adipokines according to their effects on inflammatory immune responses in adipose tissues. In general, the pro-inflammatory adipokines are increased, whereas the anti-inflammatory adipokines are decreased in obese rodents and humans that are associated with insulin resistance.

9.4.1 Pro-inflammatory Adipokines

9.4.1.1 Leptin

Leptin is a non-glycosylated 167 amino acid protein in humans with high homology in mammalian species. Obese mutation *ob*, an autosomal recessive mutation, increases food intake five times compared to wild-type mice, resulting in T2D. Parabiosis experiments show that Lep^{ob}/Lep^{ob} mice have defects in circulating factor(s), and obese gene identified by positional cloning has been shown to regulate feeding behavior through the hypothalamic regulation in the central nervous system (Zhang et al. 1994). Thus, leptin-deficient Lep^{ob}/Lep^{ob} and leptin receptor-deficient $Lepr^{db}/Lepr^{db}$ mice display marked hyperphagia, obesity, and insulin resistance along with neuroendocrine dysfunctions. Importantly, exogenous administration of leptin to Lep^{ob}/Lep^{ob} mice reduces obesity and restores insulin sensi-

tivity. However, leptin levels in circulation are increased in obese rodents and humans, suggesting that obese subjects display leptin resistance (Friedman and Halaas 1998). Leptin resistance has been reported as mediated by impaired leptin transport in blood-brain barrier, hyperleptinemia-induced SOCS3 expression (Kievit et al. 2006), defective autophagy (Quan et al. 2012), and ER stress (Ozcan et al. 2009). Obesity-induced chronic inflammation also induces leptin resistance through the activation of TLR4, JNK, and IKKβ (Kleinridders et al. 2009; Zhang et al. 2008). However, leptin resistance is complex and can develop at many points in the neural circuit that regulates feeding behavior. In obese humans, leptin administration has a minor effect on body weight reduction, probably due to the development of central leptin resistance in obese patients with very high circulating leptin (Heymsfield et al. 1999). In contrast, leptin analogues such as metreleptin have been approved to treat dysfunctional metabolism in lipodystrophy.

Leptin level is regulated by food intake and the endocrine system. However, innate immune responses also regulate the leptin production in adipose tissue. Pro-inflammatory signals such as TNF-α, IL-1β, LPS, and turpentine acutely stimulate the expression of leptin and leptin receptor (Gan et al. 2012; Grunfeld et al. 1996). Although leptin modulates feeding behavior through neuroendocrine axis, it also regulates inflammatory responses in various target cells. The structure of leptin is similar to pro-inflammatory long-chain helical cytokines including IL-6 and IL-12, and leptin indeed induces inflammatory responses through the long isoform of the leptin receptor b ($LepR_b$) and its proximal signaling molecules such as Janus kinase 2 (JAK2) and signal transducer and activator of transcription 3 (STAT3). Leptin activates monocytes and macrophages to produce pro-inflammatory IL-6, TNF-α, and IL-12 (Gainsford et al. 1996) and stimulates the production of CCL2 and vascular endothelial growth factor in human hepatic stellate cells (Aleffi et al. 2005). Leptin also enhances the production of pro-inflammatory Th1 cytokines, whereas suppresses the production of anti-inflammatory Th2 cytokines such as IL-4 in

CD4[+] T cells (Lord et al. 1998). Thus, *Lep^{ob}/Lep^{ob}* and *Lepr^{db}/Lepr^{db}* mice are resistant to concanavalin-A (Con-A)-induced hepatitis and experimental autoimmune encephalomyelitis (EAE) as these mice are skewed to anti-inflammatory Th2 immune responses (Faggioni et al. 2000; Matarese et al. 2001), suggesting that leptin may regulate Th1/Th2 CD4[+] T cell differentiation. Accordingly, leptin-deficient and starved individuals show abnormal shift from TH1 to TH2 immune responses and an increased susceptibility to infectious disease (Ahima et al. 1996). Furthermore, leptin induces collagen-induced arthritis through the differentiation of Th17 CD4[+] T cells to enhance joint inflammation (Deng et al. 2012).

9.4.1.2 Tumor Necrosis Factor-α (TNF-α)

TNF-α mainly expressed in monocytes and macrophages is a 26 kDa (233 amino acids) transmembrane protein and then is converted to active trimer by TNF-α converting enzyme. TNF-α is a typical pro-inflammatory cytokine that is increased in obese humans and rodents, suggesting that TNF-α contributes to insulin resistance. TNF-α treatment in cell lines and rodents induces insulin resistance, and neutralization of TNF-α in obese *fa/fa* rats enhances insulin sensitivity (Hotamisligil et al. 1993). Accordingly, TNF-α or its receptor-deficient mice show improved insulin sensitivity in white adipose tissues and skeletal muscles of HFD-fed and *Lep^{ob}/Lep^{ob}* mice (Uysal et al. 1997). As TNF-α stimulates the phosphorylation of IRS1 on PTB domain serine-307 residues (serine-312 in human IRS1), TNF-α suppresses insulin-induced IRS1 tyrosine phosphorylation and proximal signaling molecules, resulting in impaired glucose uptake (Hotamisligil et al. 1996). Although TNF-α levels in the circulation is positively correlated with insulin resistance, and neutralization of TNF-α improved the insulin sensitivity in rodents, clinical effects of TNF-α neutralization in humans are still controversial. Short-term administration of TNF-α blocking reagents to obese T2D patients suppresses inflammation but does not show improved insulin sensitivity (Ofei et al. 1996). In contrast, long-term treatment of TNF-α blocking reagents

in obese patients with severe inflammatory diseases such as rheumatoid arthritis improves insulin sensitivity (Gonzalez-Gay et al. 2006; Stanley et al. 2011). TNF-α also suppresses the expression of phosphodiesterase 3B (PDE3B) and perilipin. As PDE3B reduces cAMP after insulin stimulation and perilipin regulates the access of hormone-sensitive lipase in adipocytes, these events account for TNF-α-induced lipolysis and the release free fatty acid (Souza et al. 1998; Zhang et al. 2002). Free fatty acid in turn binds to TLR4 to induce the expression of pro-inflammatory mediators through NF-κB activation (Lee et al. 2001). Consistent with this model, TLR4-deficient mice show improved HFD-induced insulin resistance (Kim et al. 2007a).

9.4.1.3 Interleukin-6

IL-6 is a 212 amino acid peptide produced by the immune cells, adipose tissues, and skeletal muscle and displays molecular ranges between 21 and 28 kDa due to posttranslational modification. Adipose tissue in fact contributes 15–35% of the basal circulating IL-6 (Mohamed-Ali et al. 1997). Thus, IL-6 is highly expressed in adipose tissue and positively correlated with obesity in humans. Peripheral administration of IL-6 interrupts insulin signaling due to enhanced expression of SOCS3 in hepatocytes, suggesting that obesity-induced IL-6 expression mediates liver insulin resistance (Senn et al. 2003). Gastric surgery of morbidly obese patients demonstrates a significantly improved glucose tolerance along with a significant decrease in serum IL-6 levels (Kopp et al. 2003). In contrast, IL-6-deficient mice show mature-onset obesity and hepatic inflammation, and IL-6 administration reverses insulin resistance (Matthews et al. 2010; Wallenius et al. 2002). As central administration of IL-6 enhances energy expenditure and decreases obesity, IL-6 can also influence obesity and insulin sensitivity through a central nervous system mechanism. Although resting the muscle expresses very little IL-6, exercise induces a 100-fold increase of plasma IL-6 levels to enhance glucose uptake in the skeletal muscle (Pedersen et al. 2004). IL-6 treatment of L6 myotube cultured cells also shows enhanced GLUT4 translocation in an AMP-activated protein kinase (AMPK)-dependent manner (Carey et al. 2006). Human genetic studies also do not identify reliable correlation between the common polymorphisms in the *IL6* gene/promoter and the risk of T2D (Qi et al. 2006a). Furthermore, IL-6 is required for the induction of "beige" cells in subcutaneous white adipose tissue in response to a cold environment as IL-6-deficient mice show blunted expression of UCP1 in subcutaneous white adipose tissue after cold challenge (Knudsen et al. 2014). Thus, the role IL-6 in obesity and insulin resistance is controversial and likely depends upon the specific sites of expression that is integrated with other factors in a systems integrated manner.

9.4.1.4 Interleukin-1β

IL-1 family such as IL-1α, IL-1β, and IL-1 receptor antagonist (IL-1Ra) is increased in obese humans and rodents (Meier et al. 2002), and weight loss significantly diminishes the expression of IL-1β in adipose tissue. Large human cohort studies also show that increased plasma IL-1β and IL-6 levels augment the risk of T2D (Spranger et al. 2003). Both IL-1α and IL-1β are pro-inflammatory cytokines and antagonized by IL-1Ra. Although IL-1α and IL-1β are of the same family, their expression and activation are different. First, IL-1β precursor is produced by innate immune cells such as macrophages, and IL-1α precursor is constitutively expressed in resting cells. Second, intracellular processing of the IL-1β precursor is tightly controlled by inflammasomes, whereas IL-1α precursor is present as a biologically active form. IL-1α precursor can also be processed by calpain, a membrane-bound calcium-activated cysteine protease, to generate more potent IL-1α. Interestingly, IL-1α has another mechanism to induce inflammatory cytokines through the translocation into nucleus as an IL-1α/HAX-1 transcription factor complex (Dinarello 2009). It has been shown that IL-1β induces pancreatic β cell inflammation and apoptosis for type 1 diabetes, and many groups are working to uncover the role of IL-1β in the pathogenesis of T2D. In fact, IL-1β treatment diminishes the expression of IRS1 in 3T3-L1 adipocytes and results in the impaired insulin-mediated GLUT-4 translocation. Thus, inflammasome-

defective mice models show improved HFD-induced inflammation and insulin resistance, emphasizing the role of IL-1β in the development of insulin resistance (Wen et al. 2011). Treatment of obese mice and T2D patients with recombinant IL-1R antagonist (anakinra) or inhibitors of caspase-1 also improves insulin sensitivity and pancreatic β cell function (Larsen et al. 2007; Stienstra et al. 2010). Although there is strong evidence for the role of IL-1β to mediate obesity-induced insulin resistance, the effects of IL-1α in glucose metabolism remain to be determined.

9.4.1.5 CC Chemokine Ligand Type 2 (CCL2) and CC Chemokine Receptor Type 5 (CCR5)

Chemokines and their receptors play essential roles in mediating infiltration of immune cells into adipose tissue. CCL2 (MCP1) and CCR5 are typical chemokine and chemokine receptor that mediate inflammatory responses and are significantly increased in obese rodents and humans. Human adipocytes secrete CCL2 and CCL3 to recruit macrophages to adipose tissues, resulting in the inflammation and insulin resistance. Accordingly, CCR2 (the receptor of CCL2)-deficient mice show attenuated macrophage infiltration, inflammation, and insulin resistance (Weisberg et al. 2006). The treatment with CCR2 inhibitors, INCD3344 and propagermanium, also attenuates HFD-induced insulin resistance and nonalcoholic steatohepatitis (NASH) (Mulder et al. 2017; Weisberg et al. 2006). Intestinal epithelial cell-specific CCR2-deficient mice show diminished HFD-induced insulin resistance due to decreased macrophage infiltration and inflammation in gut, suggesting the possible cross talk between the gut and systemic insulin resistance (Kawano et al. 2016). CCR5 (the receptor of CCL5 (RANTES) and CCL3 (MIP-1α))-deficient mice show improved inflammation, insulin sensitivity, and NASH with reduced macrophage infiltration and preferred anti-inflammatory M2 macrophage differentiation (Kitade et al. 2012). However, the role of CCL2 in obesity-induced inflammation and insulin resistance is still controversial. CCL2-deficient mice show decreased macrophage infiltration and inflammation in adipose tissues (Kanda et al. 2006), whereas in another study, CCL2-deficient mice show no differences in macrophage accumulation and inflammation in adipose tissue of obese mice (Kirk et al. 2008). Although the basis for this difference is not clear, it is possible that CCL2 deficiency might be compensated by other related chemokines in certain genetic background.

9.4.1.6 Retinol-Binding Protein 4 (RBP4)

RBP4 expressed in the liver, white/brown adipocytes, and macrophages functions as the vitamin A transporter, and many clinical and large epidemiologic studies demonstrate that elevated circulating RBP4 is linked to insulin resistance and cardiovascular disease (Sun et al. 2013; Yang et al. 2005). For example, RBP4 inhibits insulin-induced IRS1 phosphorylation and is inversely correlated with GLUT4 expression in adipocytes, and administration of recombinant RBP4 to normal mice induces insulin resistance (Yang et al. 2005). Clinical studies also show that increased RBP4 levels are closely associated with high blood pressure, high levels of triglyceride, high body mass index (BMI) (Graham et al. 2006), subclinical inflammation, and nephropathy (Akbay et al. 2010). In fact, RBP4 stimulates human primary endothelial cells to produce proinflammatory molecules such as vascular cell adhesion molecule 1 (VCAM1), CCL2, and IL-6, resulting in the progression of endothelial inflammation in cardiovascular disease and microvascular complication in diabetes (Farjo et al. 2012). RBP4 induces antigen presentation by macrophages through MyD88 and NF-κB activation, which activates T cells toward a pro-inflammatory Th1 profile to induce the insulin resistance (Moraes-Vieira et al. 2014).

9.4.1.7 Chemerin

Chemerin is a ligand of the G protein-coupled receptor ChemR23 (Wittamer et al. 2003) and expressed in most tissues except leukocytes. Chemerin is a chemoattractant that induces the infiltration of macrophages, immature dendritic cells, and NK cells in inflammatory diseases such as ulcerative colitis and skin lupus (Albanesi et al. 2009). In addition, chemerin has been shown as an adipokine to regulate adipogenesis and adipocytes metabolism (Goralski et al. 2007) although molecular mechanisms are still controversial (Bondue et al. 2011). Chemerin level is positively correlated with BMI, fasting glucose, triglycerides, and inflammatory cytokines in obese subjects, and administration of chemerin exacerbates glucose intolerance in obese mice (Ernst et al. 2010). Obesity-induced free fatty acid also enhances the production of chemerin in adipocytes (Bauer et al. 2011). Recently, it has been shown that adipocyte-derived chemerin recruits plasma dendritic cells (pDC) to produce type 1 IFN and then type 1 IFN induces M1 macrophage differentiation in adipose tissues, resulting in the systemic insulin resistance (Ghosh et al. 2016). However, chemerin suppresses the zymosan-induced peritonitis, suggesting that chemerin can also have anti-inflammatory activity (Cash et al. 2008).

9.4.1.8 Dipeptidyl Peptidase-4 (DPP-4)

DPP-4 (776 amino acids), originally known as a T cell marker CD26, is wildly expressed glycoproteins in the immune cells, adipose tissue, salivary gland, prostate, kidney, and intestine and degrades various chemokines and peptide hormones to modulate T cell activation. As DPP-4 also degrades incretin proteins such as glucagon-like peptide 1 (GLP1) and glucose-dependent insulinotropic polypeptide (GIP) to suppress incretin-induced insulin secretion in pancreatic β cells, DPP4 inhibitors including alogliptin, sitagliptin, and linagliptin are effective to treat T2D (Mulvihill and Drucker 2014). T2D patients show increased DDP-4 levels in visceral adipose tissues and systemic circulation (Sell et al. 2013). DPP-4 in diabetes is not restricted to its incretin degradation properties. Soluble DPP-4 directly affects primary human adipocytes and smooth muscle cells to suppress insulin-induced AKT phosphorylation, resulting in the insulin resistance. Furthermore, DPP-4 inhibitor linagliptin suppresses HFD-induced macrophage infiltration and pro-inflammatory M1 macrophage polarization, suggesting that DPP-4 directly enhances M1 macrophage differentiation to induce HFD-inflammation and insulin resistance (Zhuge et al. 2016).

9.4.1.9 Resistin

Resistin (ADSF/FIZZ3/XCP1) is a 12 kDa polypeptide with 114 amino acids in rodents and was identified as an inducer of pulmonary inflammation (Holcomb et al. 2000) and insulin resistance (Steppan et al. 2001). The expression pattern and function of resistin are different between rodents and humans. Adipocytes stimulated by high glucose are the major source of resistin in rodents, whereas monocytes and macrophages activated by inflammatory stimulation such as LPS, TNF-α, and IL-1β produce resistin in humans (Schwartz and Lazar 2011). Resistin belongs to the cysteine-rich family and circulates as a disulfide-linked hexamer and trimer. High molecular weight hexamer is more abundant, but trimer shows more potent effect to induce insulin resistance in mice. To mediate insulin resistance in rodents, resistin is involved in the modulation of SOCS3 and GLUT4 expression, resulting in the suppression of insulin-mediated signaling in adipocytes (Steppan et al. 2005). Resistin-deficient Lep^{ob}/Lep^{ob} mice show improved glucose tolerance and insulin sensitivity (Qi et al. 2006b). Rosiglitazone, a PPARγ agonist, suppresses the resistin expression in adipose tissues, resulting in the attenuation of inflammatory responses (Bokarewa et al. 2005). In contrast, the function of resistin in humans T2D is not clear, as resistin levels in blood circulation are not correlated with obesity and insulin resistance. Resistin stimulates human peripheral mononuclear cells, macrophages, and hepatic stellate cells to produce IL-6 and TNF-α through the NF-κB signaling pathway to drive inflammation. Resistin also regulates brain inflammation through the activation of JNK and p38 MAPK in the hypothalamus that in turn affects hepatic insulin resistance (Benomar et al. 2012).

9.4.2 Anti-inflammatory Adipokines

9.4.2.1 Adiponectin

Adiponectin (244 amino acids) was found shortly after the identification of leptin and is highly expressed by adipocytes with potent anti-inflammatory, insulin-sensitizing, and anti-apoptotic properties (Scherer et al. 1995). Adiponectin has four domains including a signaling peptide region, an N-terminal collagen-like domain, a cysteine-rich domain for oligomerization, and a C-terminal complement factor C1q-like globular domain for receptor binding. Adiponectin circulates as trimers (90 kDa), hexamers (180 kDa), and a high molecular weight (HMW, 18-36mers) form, and hexamer and HMW are the major forms in circulation. As pro-inflammatory factors such as TNF-α, IL-6, ROS, and hypoxia suppress the expression of adiponectin in adipocytes, adiponectin levels are diminished in obese rodents and humans (Li et al. 2009). In contrast, PPARγ antagonists stimulate the expression of adiponectin in adipocytes (Maeda et al. 2001). Adiponectin activates its receptor and adaptor complex, ADIPOR1/2 and phosphotyrosine interacting with PH domain and leucine zipper 1 (APPL1), to activate AMP-dependent protein kinase (AMPK), resulting in the enhancement of fatty acid oxidation and glucose uptake in the muscle and suppression of gluconeogenesis in the liver (Yamauchi et al. 2002). Accordingly, adiponectin administration or overexpression in transgenic mice results in improved insulin sensitivity, whereas adiponectin-deficient mice develop HFD-induced inflammation and insulin resistance (Kim et al. 2007b; Maeda et al. 2002). In fact, orally active small synthetic adiponectin agonist, AdipoRon, treatment improves insulin sensitivity and glucose tolerance through AMPK and PPAR-α activation (Okada-Iwabu et al. 2013). In addition, adiponectin induces autophagy to reduce HFD-induced oxidative stress through the induction of antioxidant enzymes (*Gpxs*, *Prdx*, and *Sod*) in the skeletal muscle (Liu et al. 2015). Recently, it has been shown that adiponectin mediates the effect of FGF21 on metabolism and insulin sensitivity in the skeletal muscle and liver as beneficial effects of FGF21 are abrogated in adiponectin knockout mice, suggesting that adiponectin induced by FGF21 treatment mediates FGF21 effects to decrease obesity and to improve insulin sensitivity (Holland et al. 2013; Lin et al. 2013).

Adiponectin also directly modulates immune responses in immune cells. Adiponectin inhibits

LPS-induced TNF-α production in macrophages through inhibition of NF-κB activation and stimulates the production of anti-inflammatory IL-10, suggesting that adiponectin promotes the differentiation of anti-inflammatory M2 macrophages (Kumada et al. 2004; Mandal et al. 2011; Yokota et al. 2000). Adiponectin also modulates the activation of T cells and NK cells. Adiponectin receptors are upregulated on the surface of human T cells after antigen/MHC stimulation to mediate apoptosis of antigen-specific T cells, resulting in the suppression of antigen-specific T cell expansion (Wilk et al. 2011). Furthermore, adiponectin suppresses TLR-mediated IFN-γ production in NK cells without affecting cytotoxicity of NK cells (Wilk et al. 2013). Adiponectin can also suppress the development of atherosclerosis and fatty liver diseases (Okamoto et al. 2002; Xu et al. 2003).

9.4.2.2 Fibroblast Growth Factor 21 (FGF 21)

FGF21 is a 209 amino acid peptide and an atypical member of the FGF superfamily. FGF21 is produced by the liver, adipose tissue, and skeletal muscle and displays profound effects on glucose and lipid homeostasis (Itoh 2014). FGF21 expression is strongly induced in the liver by fasting through a PPAR-α-dependent mechanism (Badman et al. 2007), resulting in the enhanced hepatic fatty acid oxidation and gluconeogenesis. FGF21 overexpression suppresses HFD-induced obesity (Kharitonenkov et al. 2005), and FGF21-deficient mice show exacerbated HFD-induced insulin resistance with weight gain (Badman et al. 2009). Recombinant FGF21 and LY2405319 treatment of obese humans also results in improved metabolic status (Gaich et al. 2013). FGF receptor and βKlotho co-receptor mediate FGF21 signaling and are highly expressed in the adipose tissue and liver, thus responding to FGF21 to improve glucose and lipid metabolism (Ding et al. 2012). Although the liver is the major organ to produce FGF21, white adipose tissue and brown adipose tissue also express and release high amount of FGF21 during fasting and cold challenge, respectively. In addition, subcutaneous white adipose tissue expresses FGF21

after cold challenge and β₃ adrenergic receptor activation with increased UCP-1 expression and PGC-1α protein stabilization, resulting in the browning of white adipose tissue (Fisher et al. 2012). FGF21 treatment suppresses the HFD-induced inflammation and insulin resistance in adipose tissue during diabetes-induced renal inflammation (Zhang et al. 2013) and septic shock in obesity (Feingold et al. 2012). However, molecular mechanisms of FGF21 as an anti-inflammatory mediator in immune cells are unknown.

9.4.2.3 Secreted Frizzled-Related Protein 5 (SFRP5)

SFRPs have an N-terminal cysteine-rich domain that is homologous to frizzled proteins, the cell surface receptors for wingless-type MMTV integration site (WNT) binding, and comprise five members in humans, SFRP1–SFRP5. As SFRPs have homology with the WNT-binding domain of frizzled proteins, SFRPs antagonize WNT signaling (Leyns et al. 1997; Wang et al. 1997). WNT signaling pathways are mediated by canonical β-catenin activation, noncanonical planar cell polarity, and noncanonical Ca⁺⁺ activation, and in particular Wnt5a expression is closely linked to inflammatory responses. SFRP5 that is highly expressed in adipocytes of mouse white adipose tissues may prevent the binding of WNT proteins to its receptors. In fact, the expression of SFRP5 is decreased, but the expression of Wnt5a is increased in white adipose tissues of obese rodents and humans, suggesting that SFRP5 have potential to attenuate inflammatory effect of Wnt5a in adipose tissues. Accordingly, HFD-fed SFRP5-deficient *Sfrp5⁻/⁻* mice have insulin resistance and fatty liver along with enhanced inflammatory macrophage accumulation to produce IL-6, TNF-α, and CCL2, suggesting that SFRP5 is an anti-inflammatory adipokine (Ouchi et al. 2010). Wnt5a protein induces the noncanonical activation of JNK1, and SFRP5-deficient mice show highly activated JNK1 in HFD-fed mice, indicating that SFRP5 inhibits Wnt5a-mediated noncanonical JNK1 activation in adipose tissues to suppress obesity-induced inflammation and insulin resistance. SFRP5 also suppresses ischemia/

reperfusion-induced heart inflammation and failure through the suppression of Wnt5a positive macrophage infiltration and inflammatory cytokine production (Nakamura et al. 2016). In contrast to mice models, human adipose tissues hardly produce SFRP5, and the anti-inflammatory function of SFRP5 is unclear in several human epidemiological studies (Schulte et al. 2012).

9.4.2.4 Bone Morphogenetic Protein (BMP)

BMP is a member of the transforming growth factor (TGF)-β superfamily that regulates embryonic development. BMP has diverse biological functions, but it has recently been shown that BMP modulates inflammatory and metabolic disease such as fibrosis, inflammatory bowel disease, rheumatoid arthritis, and T2D (Grgurevic et al. 2016). BMP expressed by diverse tissues such as the skin, thyroid, adipose tissue, and intestine is secreted after protease digestion to activate receptors, BMPR1 and BMPR2, and activated receptors in turn recruit SMAD1/5/8 to induce the translocation to nucleus. BMP7 (49 kDa) shows diverse anti-inflammatory effects. In liver fibrosis, BMP7 suppresses TGF-β-dependent epithelial to mesenchymal transition (EMT) and inflammation (Zeisberg et al. 2003). BMP7 treatment also suppresses TNBS-induced colitis as BMP7 treatment suppresses IL-6 production, suggesting that BMP7 has anti-inflammatory property to suppress inflammatory disease. Systemic BMP7 treatment also reduces body weight and improves metabolic parameters due to increased energy expenditure and decreased food intake, resulting in the improved insulin sensitivity (Townsend et al. 2012). BMP4 (46 kDa) expressed by differentiated pre-adipocytes increases PPAR-γ expression in pre-adipocytes to drive browning of white adipose tissue with mitochondria biogenesis and PGC-1α expression, suggesting that BMP4 is a regulator for the transition from white to beige adipocytes differentiation (Gustafson et al. 2015).

9.4.2.5 Omentin-1

Omentin-1 (313 amino acids) identified as a soluble galactofructose-binding lectin is expressed by human omental adipose tissues in which omentin-1 is preferentially expressed by the cells associated with the omental stromal vascular fraction (Schaffler et al. 2005). Omentin-1 levels in blood circulation are inversely correlated with obesity and suppressed by glucose and insulin (de Souza Batista et al. 2007). Furthermore, omentin-1 level is increased after weight loss and exercise with improved insulin sensitivity. Omentin-1 enhances the insulin-induced glucose uptake in human visceral and subcutaneous adipocytes through increased phosphorylation of AKT/PKB (Yang et al. 2006), indicating that omentin-1 may improve insulin sensitivity. Omentin-1 also attenuates C-reactive protein (CRP) and TNF-α-induced NF-κB activation in human endothelial cells and vascular smooth muscle cells to suppress inflammatory signaling (Kazama et al. 2012; Tan et al. 2010). Thus, omentin-1 suppresses LPS-induced acute respiratory distress syndrome through the AKT-eNOS-dependent suppression of pulmonary inflammation (Qi et al. 2016). In addition, omentin-1 overexpression under aP2 promoter in apoE knockout mice attenuates atherosclerosis lesion formation and suppresses macrophage accumulation and the expression of pro-inflammatory cytokines in aorta. When human macrophages are treated with omentin-1, LPS-induced pro-inflammatory gene expression is diminished through AKT phosphorylation. This study suggests that omentin-1 directly suppresses pro-inflammatory macrophage activation to suppress atherosclerosis (Hiramatsu-Ito et al. 2016). It has been speculated that visceral adipose tissue-derived omentin-1 may be a potential candidate to treat inflammatory disease such as T2D, atherosclerosis, and lung inflammation.

9.4.2.6 Apelin

Apelin (36 amino acids) expressed in many tissues such as lung, adipose tissue, hypothalamus, and skeletal muscle was identified as the endogenous ligand of orphan G protein-coupled receptor termed APJ (Tatemoto et al. 1998). Although apelin has diverse physiological functions to regulate fluid homeostasis, heart rate, and angiogenesis (Carpene et al. 2007), increasing evidence suggests that apelin improves insulin sensitivity, glucose utilization, and brown adipogenesis in different tissues associated with diabetes. Plasma level of apelin is increased in obese and insulin-resistant humans and rodents (Boucher et al. 2005). As apelin enhances glucose uptake through AMPK-dependent manner and suppresses lipolysis, HFD-fed apelin-deficient mice show insulin resistance (Yue et al. 2010, 2011). Acute and chronic apelin treatments also stimulate glucose uptake and fatty acid oxidation in adipose tissue and skeletal muscle through AMPK activation, resulting in the improved insulin sensitivity (Attane et al. 2012). Brown adipose tissue participates non-shivering thermogenesis, and chronic apelin-13 treatment increases the expression of UCP-1 and PGC-1α to induce mitochondria biogenesis, thermogenesis, and reduced lipolysis (Than et al. 2015). Apelin is also involved in inflammatory responses in obese diabetic subjects. Apelin administration reduces inflammation in the kidney to ameliorate diabetic nephropathy through the suppression of CCL2 expression, monocyte infiltration, and NF-κB activation (Day et al. 2013). However, apelin activates JNK and NF-κB to induce inflammatory adhesion molecules such as ICAM in human umbilical vein endothelial cells (Lu et al. 2012). Thus, the precise role for apelin in regulating inflammatory responses remains undefined.

9.4.2.7 Visceral Adipose Tissue-Derived Serine Protease Inhibitor (Vaspin)

Vaspin is a 415 amino acid protein that was identified from visceral white adipose tissues of Otsuka Long-Evans Tokushima Fatty (OLETF) rat as an insulin-sensitizing adipokine because vaspin suppresses the expression of pro-inflammatory adipokines such as resistin, leptin, and TNF-α (Hida et al. 2005). Although vaspin expression was originally identified from adipose tissue, several tissues including skin, stomach, and hypothalamus also express vaspin. Increased vaspin levels in circulation are associated with obesity and impaired insulin sensitivity, whereas type 2 diabetes abrogates the correlation, and exercise increases vaspin levels (Youn et al. 2008). Recombinant vaspin administration also improves glucose tolerance and insulin sensitiv-

ity. Inflammatory stimulators including TNF-α play an important role in the development of atherosclerosis. Vaspin suppresses TNF-α-induced ROS production and monocyte adhesion to smooth muscle cells by inhibiting the activation of NF-κB and PKCθ (Phalitakul et al. 2011).

9.5 Adipokines and the Neuroendocrine System

As obesity and related metabolic disease are pandemic health problem in modern society, the control of appetite and energy balance are key biological issue in humans and higher animals, and uncovering the molecular mechanisms of systemic metabolism is essential to resolve problems. To control appetite and energy balance, several tissues such as brain, liver, pancreas, muscle, and adipose tissue are closely linked to communicate with diverse signaling mediators in the nerve system and endocrine/exocrine systems. Central neuroendocrine pathway regulates energy intake and expenditure. Appetite stimulation (orexigenic) pathway is regulated by neuropeptide Y (NPY), melanin-concentrating hormone (MCH), and agouti-related peptide (AgRP), and losing appetite (anorexia) is modulated by pro-opiomelanocortin (POMC) and corticotropin-releasing hormone (CRH) (Cornejo et al. 2016). Peripheral tissues generate signals to control appetite and energy balance, and gastrointestinal track-derived molecules such as ghrelin, peptide YY, and cholecystokinin are involved in the modulation of appetite and energy balance. In addition, adipose tissue also produces adipokines to mediate cross talk between the adipose tissue and central nervous system especially the hypothalamus to control appetite and energy balance. Leptin is a typical adipokine to regulate appetite and energy balance and described by other chapters.

In addition to leptin, it has recently been shown that other adipokines such as BMP7, FGF21, and vaspin also modulate food intake and energy balance. Systemic treatment of HFD-fed mice with BMP7 shows decreased food intake and increased energy expenditure, resulting in the decreased body weight and improved metabolic parameters (Townsend et al. 2012). BMP7-induced body weight loss in HFD-fed mice is primarily due to loss of fat mass along with decreased serum leptin levels. Suppressed food intake with BMP7 treatment is leptin-independent pathway as BMP7-treated Lep^{ob}/Lep^{ob} mice show reduced food intake. Intracerebroventricular (ICV) injection of BMP7 also shows that BMP7 reduces acute food intake through the rapamycin-sensitive mTORC1-p70S6 kinase pathway. Recently, it has been shown that FGF21 regulates sugar ingestion (von Holstein-Rathlou et al. 2016). Sugar intake increases FGF21 expression in the liver through ChREBP-dependent pathway and then results in increased serum FGF21 levels in rodents and humans. Thus, FGF21-deficient mice show increased sucrose consumption, whereas acute administration or overexpression of FGF21 suppresses the intake of both sugar and noncaloric sweeteners. FGF21 does not affect chorda tympani nerve responses to sweet-tasting nutrients as the expression of FGF21 receptors, FGFR1c and βKlotho, is undetectable in taste epithelium. Instead, FGF21 reduces sweet-seeking behavior and meal size through the modulation of neurons in the hypothalamus. Thus, FGF21 mediates negative feedback circuit to modulate sugar ingestion in rodents and humans. Molecular mechanisms of vaspin in the regulation of food intake are not clear. However, peripheral and central vaspin administration decreases food intake in obese Lep^{db}/Lep^{db} and lean C57BL/6 mice, suggesting that vaspin suppresses food intake through leptin-independent pathway. In addition, vaspin treatment maintains glucose-lowering effects for at least 6 days after injection. Leptin has been studied to understand detailed molecular mechanisms of the interaction between adipose tissue and central nerve systems to regulate appetite and energy balances. However, further studies are necessary for other adipokines to fully understand the mechanisms such as target neurons, signaling pathways, and positive/negative feedback pathways in the modulation of appetite and energy balances.

Questions

- What are the cellular and molecular mechanisms in the initiation of obesity-induced inflammation at adipose tissue?
- What is the molecular mechanism of adipokine resistance?
- Does benign obesity, obesity without insulin resistance, show different adipokine profiles in humans?
- Can adipokine profile be established as early diagnosis markers of insulin resistance?
- What is the possible adipokine-based pharmacological treatment for metabolic disease?

References

Ahima, R. S., Prabakaran, D., Mantzoros, C., Qu, D., Lowell, B., Maratos-Flier, E., & Flier, J. S. (1996). Role of leptin in the neuroendocrine response to fasting. *Nature, 382*, 250–252.

Akbay, E., Muslu, N., Nayir, E., Ozhan, O., & Kiykim, A. (2010). Serum retinol binding protein 4 level is related with renal functions in type 2 diabetes. *Journal of Endocrinological Investigation, 33*, 725–729.

Albanesi, C., Scarponi, C., Pallotta, S., Daniele, R., Bosisio, D., Madonna, S., Fortugno, P., Gonzalvo-Feo, S., Franssen, J. D., Parmentier, M., et al. (2009). Chemerin expression marks early psoriatic skin lesions and correlates with plasmacytoid dendritic cell recruitment. *The Journal of Experimental Medicine, 206*, 249–258.

Aleffi, S., Petrai, I., Bertolani, C., Parola, M., Colombatto, S., Novo, E., Vizzutti, F., Anania, F. A., Milani, S., Rombouts, K., et al. (2005). Upregulation of proinflammatory and proangiogenic cytokines by leptin in human hepatic stellate cells. *Hepatology, 42*, 1339–1348.

Ashcroft, F. M., & Rorsman, P. (2012). Diabetes mellitus and the beta cell: The last ten years. *Cell, 148*, 1160–1171.

Attane, C., Foussal, C., Le Gonidec, S., Benani, A., Daviaud, D., Wanecq, E., Guzman-Ruiz, R., Dray, C., Bezaire, V., Rancoule, C., et al. (2012). Apelin treatment increases complete fatty acid oxidation, mitochondrial oxidative capacity, and biogenesis in muscle of insulin-resistant mice. *Diabetes, 61*, 310–320.

Badin, P. M., Vila, I. K., Louche, K., Mairal, A., Marques, M. A., Bourlier, V., Tavernier, G., Langin, D., & Moro, C. (2013). High-fat diet-mediated lipotoxicity and insulin resistance is related to impaired lipase expression in mouse skeletal muscle. *Endocrinology, 154*, 1444–1453.

Badman, M. K., Pissios, P., Kennedy, A. R., Koukos, G., Flier, J. S., & Maratos-Flier, E. (2007). Hepatic fibroblast growth factor 21 is regulated by PPARalpha and is a key mediator of hepatic lipid metabolism in ketotic states. *Cell Metabolism, 5*, 426–437.

Badman, M. K., Koester, A., Flier, J. S., Kharitonenkov, A., & Maratos-Flier, E. (2009). Fibroblast growth factor 21-deficient mice demonstrate impaired adaptation to ketosis. *Endocrinology, 150*, 4931–4940.

Bard-Chapeau, E. A., Hevener, A. L., Long, S., Zhang, E. E., Olefsky, J. M., & Feng, G. S. (2005). Deletion of Gab1 in the liver leads to enhanced glucose tolerance and improved hepatic insulin action. *Nature Medicine, 11*, 567–571.

Bauer, S., Wanninger, J., Schmidhofer, S., Weigert, J., Neumeier, M., Dorn, C., Hellerbrand, C., Zimara, N., Schaffler, A., Aslanidis, C., et al. (2011). Sterol regulatory element-binding protein 2 (SREBP2) activation after excess triglyceride storage induces chemerin in hypertrophic adipocytes. *Endocrinology, 152*, 26–35.

Bendelac, A., Savage, P. B., & Teyton, L. (2007). The biology of NKT cells. *Annual Review of Immunology, 25*, 297–336.

Benomar, Y., Gertler, A., De Lacy, P., Crepin, D., Hamouda, H. O., Riffault, L., & Taouis, M. (2012). Central resistin overexposure induces insulin resistance through toll-like receptor 4. *Diabetes*.

Bertola, A., Ciucci, T., Rousseau, D., Bourlier, V., Duffaut, C., Bonnafous, S., Blin-Wakkach, C., Anty, R., Iannelli, A., Gugenheim, J., et al. (2012). Identification of adipose tissue dendritic cells correlated with obesity-associated insulin-resistance and inducing Th17 responses in mice and patients. *Diabetes, 61*, 2238–2247.

Bokarewa, M., Nagaev, I., Dahlberg, L., Smith, U., & Tarkowski, A. (2005). Resistin, an adipokine with potent proinflammatory properties. *Journal of Immunology, 174*, 5789–5795.

Bondue, B., Wittamer, V., & Parmentier, M. (2011). Chemerin and its receptors in leukocyte trafficking, inflammation and metabolism. *Cytokine & Growth Factor Reviews, 22*, 331–338.

Boucher, J., Masri, B., Daviaud, D., Gesta, S., Guigne, C., Mazzucotelli, A., Castan-Laurell, I., Tack, I., Knibiehler, B., Carpene, C., et al. (2005). Apelin, a newly identified adipokine up-regulated by insulin and obesity. *Endocrinology, 146*, 1764–1771.

Boura-Halfon, S., & Zick, Y. (2009). Phosphorylation of IRS proteins, insulin action, and insulin resistance. *American Journal of Physiology. Endocrinology and Metabolism, 296*, E581–E591.

Butler, A. E., Janson, J., Bonner-Weir, S., Ritzel, R., Rizza, R. A., & Butler, P. C. (2003). Beta-cell deficit and increased beta-cell apoptosis in humans with type 2 diabetes. *Diabetes, 52*, 102–110.

Cancello, R., Tordjman, J., Poitou, C., Guilhem, G., Bouillot, J. L., Hugol, D., Coussieu, C., Basdevant, A., Bar Hen, A., Bedossa, P., et al. (2006). Increased infiltration of macrophages in omental adipose tissue

is associated with marked hepatic lesions in morbid human obesity. *Diabetes, 55*, 1554–1561.

Carey, A. L., Steinberg, G. R., Macaulay, S. L., Thomas, W. G., Holmes, A. G., Ramm, G., Prelovsek, O., Hohnen-Behrens, C., Watt, M. J., James, D. E., et al. (2006). Interleukin-6 increases insulin-stimulated glucose disposal in humans and glucose uptake and fatty acid oxidation in vitro via AMP-activated protein kinase. *Diabetes, 55*, 2688–2697.

Carpene, C., Dray, C., Attane, C., Valet, P., Portillo, M. P., Churruca, I., Milagro, F. I., & Castan-Laurell, I. (2007). Expanding role for the apelin/APJ system in physiopathology. *Journal of Physiology and Biochemistry, 63*, 359–373.

Cash, J. L., Hart, R., Russ, A., Dixon, J. P., Colledge, W. H., Doran, J., Hendrick, A. G., Carlton, M. B., & Greaves, D. R. (2008). Synthetic chemerin-derived peptides suppress inflammation through ChemR23. *The Journal of Experimental Medicine, 205*, 767–775.

Chavez, J. A., & Summers, S. A. (2012). A ceramide-centric view of insulin resistance. *Cell Metabolism, 15*, 585–594.

Chiefari, E., Tanyolac, S., Paonessa, F., Pullinger, C. R., Capula, C., Iiritano, S., Mazza, T., Forlin, M., Fusco, A., Durlach, V., et al. (2011). Functional variants of the HMGA1 gene and type 2 diabetes mellitus. *JAMA, 305*, 903–912.

Chin, J. E., Liu, F., & Roth, R. A. (1994). Activation of protein kinase C alpha inhibits insulin-stimulated tyrosine phosphorylation of insulin receptor substrate-1. *Molecular Endocrinology, 8*, 51–58.

Cho, H., Mu, J., Kim, J. K., Thorvaldsen, J. L., Chu, Q., Crenshaw, E. B., 3rd, Kaestner, K. H., Bartolomei, M. S., Shulman, G. I., & Birnbaum, M. J. (2001). Insulin resistance and a diabetes mellitus-like syndrome in mice lacking the protein kinase Akt2 (PKB beta). *Science, 292*, 1728–1731.

Cipolletta, D., Feuerer, M., Li, A., Kamei, N., Lee, J., Shoelson, S. E., Benoist, C., & Mathis, D. (2012). PPAR-gamma is a major driver of the accumulation and phenotype of adipose tissue Treg cells. *Nature, 486*, 549–553.

Cook, K. S., Min, H. Y., Johnson, D., Chaplinsky, R. J., Flier, J. S., Hunt, C. R., & Spiegelman, B. M. (1987). Adipsin: A circulating serine protease homolog secreted by adipose tissue and sciatic nerve. *Science, 237*, 402–405.

Cornejo, M. P., Hentges, S. T., Maliqueo, M., Coirini, H., Becu-Villalobos, D., & Elias, C. F. (2016). Neuroendocrine regulation of metabolism. *Journal of Neuroendocrinology, 28*.

Cornier, M. A., Dabelea, D., Hernandez, T. L., Lindstrom, R. C., Steig, A. J., Stob, N. R., Van Pelt, R. E., Wang, H., & Eckel, R. H. (2008). The metabolic syndrome. *Endocrine Reviews, 29*, 777–822.

Cross, D. A., Alessi, D. R., Cohen, P., Andjelkovich, M., & Hemmings, B. A. (1995). Inhibition of glycogen synthase kinase-3 by insulin mediated by protein kinase B. *Nature, 378*, 785–789.

Day, R. T., Cavaglieri, R. C., & Feliers, D. (2013). Apelin retards the progression of diabetic nephropathy. *American Journal of Physiology. Renal Physiology, 304*, F788–F800.

de Souza Batista, C. M., Yang, R. Z., Lee, M. J., Glynn, N. M., Yu, D. Z., Pray, J., Ndubuizu, K., Patil, S., Schwartz, A., Kligman, M., et al. (2007). Omentin plasma levels and gene expression are decreased in obesity. *Diabetes, 56*, 1655–1661.

DeFuria, J., Belkina, A. C., Jagannathan-Bogdan, M., Snyder-Cappione, J., Carr, J. D., Nersesova, Y. R., Markham, D., Strissel, K. J., Watkins, A. A., Zhu, M., et al. (2013). B cells promote inflammation in obesity and type 2 diabetes through regulation of T-cell function and an inflammatory cytokine profile. *Proceedings of the National Academy of Sciences of the United States of America, 110*, 5133–5138.

Deng, J., Liu, Y., Yang, M., Wang, S., Zhang, M., Wang, X., Ko, K. H., Hua, Z., Sun, L., Cao, X., et al. (2012). Leptin exacerbates collagen-induced arthritis via enhancement of Th17 cell response. *Arthritis and Rheumatism, 64*, 3564–3573.

Dinarello, C. A. (2009). Immunological and inflammatory functions of the interleukin-1 family. *Annual Review of Immunology, 27*, 519–550.

Ding, X., Boney-Montoya, J., Owen, B. M., Bookout, A. L., Coate, K. C., Mangelsdorf, D. J., & Kliewer, S. A. (2012). betaKlotho is required for fibroblast growth factor 21 effects on growth and metabolism. *Cell Metabolism, 16*, 387–393.

Donath, M. Y., Gross, D. J., Cerasi, E., & Kaiser, N. (1999). Hyperglycemia-induced beta-cell apoptosis in pancreatic islets of Psammomys obesus during development of diabetes. *Diabetes, 48*, 738–744.

Ebina, Y., Ellis, L., Jarnagin, K., Edery, M., Graf, L., Clauser, E., Ou, J. H., Masiarz, F., Kan, Y. W., Goldfine, I. D., et al. (1985). The human insulin receptor cDNA: The structural basis for hormone-activated transmembrane signalling. *Cell, 40*, 747–758.

Ehses, J. A., Lacraz, G., Giroix, M. H., Schmidlin, F., Coulaud, J., Kassis, N., Irminger, J. C., Kergoat, M., Portha, B., Homo-Delarche, F., et al. (2009). IL-1 antagonism reduces hyperglycemia and tissue inflammation in the type 2 diabetic GK rat. *Proceedings of the National Academy of Sciences of the United States of America, 106*, 13998–14003.

Ernst, M. C., Issa, M., Goralski, K. B., & Sinal, C. J. (2010). Chemerin exacerbates glucose intolerance in mouse models of obesity and diabetes. *Endocrinology, 151*, 1998–2007.

Faggioni, R., Jones-Carson, J., Reed, D. A., Dinarello, C. A., Feingold, K. R., Grunfeld, C., & Fantuzzi, G. (2000). Leptin-deficient (ob/ob) mice are protected from T cell-mediated hepatotoxicity: Role of tumor necrosis factor alpha and IL-18. *Proceedings of the National Academy of Sciences of the United States of America, 97*, 2367–2372.

Farjo, K. M., Farjo, R. A., Halsey, S., Moiseyev, G., & Ma, J. X. (2012). Retinol-binding protein 4 induces inflammation in human endothelial cells by a NADPH oxidase- and nuclear factor kappa B-dependent and retinol-independent mechanism. *Molecular and Cellular Biology*.

Feingold, K. R., Grunfeld, C., Heuer, J. G., Gupta, A., Cramer, M., Zhang, T., Shigenaga, J. K., Patzek, S. M., Chan, Z. W., Moser, A., et al. (2012). FGF21 is increased by inflammatory stimuli and protects leptin-deficient ob/ob mice from the toxicity of sepsis. *Endocrinology, 153*, 2689–2700.

Feuerer, M., Herrero, L., Cipolletta, D., Naaz, A., Wong, J., Nayer, A., Lee, J., Goldfine, A. B., Benoist, C., Shoelson, S., et al. (2009). Lean, but not obese, fat is enriched for a unique population of regulatory T cells that affect metabolic parameters. *Nature Medicine, 15*, 930–939.

Finkelstein, E. A., Khavjou, O. A., Thompson, H., Trogdon, J. G., Pan, L., Sherry, B., & Dietz, W. (2012). Obesity and severe obesity forecasts through 2030. *American Journal of Preventive Medicine, 42*, 563–570.

Fisher, F. M., Kleiner, S., Douris, N., Fox, E. C., Mepani, R. J., Verdeguer, F., Wu, J., Kharitonenkov, A., Flier, J. S., Maratos-Flier, E., et al. (2012). FGF21 regulates PGC-1alpha and browning of white adipose tissues in adaptive thermogenesis. *Genes & Development, 26*, 271–281.

Fontana, L., Eagon, J. C., Trujillo, M. E., Scherer, P. E., & Klein, S. (2007). Visceral fat adipokine secretion is associated with systemic inflammation in obese humans. *Diabetes, 56*, 1010–1013.

Friedman, J. M., & Halaas, J. L. (1998). Leptin and the regulation of body weight in mammals. *Nature, 395*, 763–770.

Gaich, G., Chien, J. Y., Fu, H., Glass, L. C., Deeg, M. A., Holland, W. L., Kharitonenkov, A., Bumol, T., Schilske, H. K., & Moller, D. E. (2013). The effects of LY2405319, an FGF21 analog, in obese human subjects with type 2 diabetes. *Cell Metabolism, 18*, 333–340.

Gainsford, T., Willson, T. A., Metcalf, D., Handman, E., McFarlane, C., Ng, A., Nicola, N. A., Alexander, W. S., & Hilton, D. J. (1996). Leptin can induce proliferation, differentiation, and functional activation of hemopoietic cells. *Proceedings of the National Academy of Sciences of the United States of America, 93*, 14564–14568.

Gan, L., Guo, K., Cremona, M. L., McGraw, T. E., Leibel, R. L., & Zhang, Y. (2012). TNF-alpha up-regulates protein level and cell surface expression of the leptin receptor by stimulating its export via a PKC-dependent mechanism. *Endocrinology*.

Ghosh, A. R., Bhattacharya, R., Bhattacharya, S., Nargis, T., Rahaman, O., Duttagupta, P., Raychaudhuri, D., Liu, C. S., Roy, S., Ghosh, P., et al. (2016). Adipose recruitment and activation of plasmacytoid dendritic cells fuel metaflammation. *Diabetes, 65*, 3440–3452.

Ghoshal, S., Witta, J., Zhong, J., de Villiers, W., & Eckhardt, E. (2009). Chylomicrons promote intestinal absorption of lipopolysaccharides. *Journal of Lipid Research, 50*, 90–97.

Gonzalez-Gay, M. A., De Matias, J. M., Gonzalez-Juanatey, C., Garcia-Porrua, C., Sanchez-Andrade, A., Martin, J., & Llorca, J. (2006). Anti-tumor necrosis factor-alpha blockade improves insulin resistance in patients with rheumatoid arthritis. *Clinical and Experimental Rheumatology, 24*, 83–86.

Goralski, K. B., McCarthy, T. C., Hanniman, E. A., Zabel, B. A., Butcher, E. C., Parlee, S. D., Muruganandan, S., & Sinal, C. J. (2007). Chemerin, a novel adipokine that regulates adipogenesis and adipocyte metabolism. *The Journal of Biological Chemistry, 282*, 28175–28188.

Graham, T. E., Yang, Q., Bluher, M., Hammarstedt, A., Ciaraldi, T. P., Henry, R. R., Wason, C. J., Oberbach, A., Jansson, P. A., Smith, U., et al. (2006). Retinol-binding protein 4 and insulin resistance in lean, obese, and diabetic subjects. *The New England Journal of Medicine, 354*, 2552–2563.

Gregor, M. F., & Hotamisligil, G. S. (2011). Inflammatory mechanisms in obesity. *Annual Review of Immunology, 29*, 415–445.

Grgurevic, L., Christensen, G. L., Schulz, T. J., & Vukicevic, S. (2016). Bone morphogenetic proteins in inflammation, glucose homeostasis and adipose tissue energy metabolism. *Cytokine & Growth Factor Reviews, 27*, 105–118.

Griffin, M. E., Marcucci, M. J., Cline, G. W., Bell, K., Barucci, N., Lee, D., Goodyear, L. J., Kraegen, E. W., White, M. F., & Shulman, G. I. (1999). Free fatty acid-induced insulin resistance is associated with activation of protein kinase C theta and alterations in the insulin signaling cascade. *Diabetes, 48*, 1270–1274.

Grunfeld, C., Zhao, C., Fuller, J., Pollack, A., Moser, A., Friedman, J., & Feingold, K. R. (1996). Endotoxin and cytokines induce expression of leptin, the ob gene product, in hamsters. *The Journal of Clinical Investigation, 97*, 2152–2157.

Gupta, R. K., Mepani, R. J., Kleiner, S., Lo, J. C., Khandekar, M. J., Cohen, P., Frontini, A., Bhowmick, D. C., Ye, L., Cinti, S., et al. (2012). Zfp423 expression identifies committed preadipocytes and localizes to adipose endothelial and perivascular cells. *Cell Metabolism, 15*, 230–239.

Gustafson, B., Hammarstedt, A., Hedjazifar, S., Hoffmann, J. M., Svensson, P. A., Grimsby, J., Rondinone, C., & Smith, U. (2015). BMP4 and BMP antagonists regulate human white and beige adipogenesis. *Diabetes, 64*, 1670–1681.

Haemmerle, G., Lass, A., Zimmermann, R., Gorkiewicz, G., Meyer, C., Rozman, J., Heldmaier, G., Maier, R., Theussl, C., Eder, S., et al. (2006). Defective lipolysis and altered energy metabolism in mice lacking adipose triglyceride lipase. *Science, 312*, 734–737.

Han, M. S., Jung, D. Y., Morel, C., Lakhani, S. A., Kim, J. K., Flavell, R. A., & Davis, R. J. (2013). JNK expression by macrophages promotes obesity-induced insulin resistance and inflammation. *Science, 339*, 218–222.

Hardy, O. T., Czech, M. P., & Corvera, S. (2012). What causes the insulin resistance underlying obesity? *Current Opinion in Endocrinology, Diabetes, and Obesity, 19*, 81–87.

Harman-Boehm, I., Bluher, M., Redel, H., Sion-Vardy, N., Ovadia, S., Avinoach, E., Shai, I., Kloting, N., Stumvoll, M., Bashan, N., et al. (2007). Macrophage infiltration into omental versus subcutaneous fat across different populations: Effect of regional adiposity and the comorbidities of obesity. *The Journal of Clinical Endocrinology and Metabolism, 92*, 2240–2247.

Hegele, R. A. (2003). Monogenic forms of insulin resistance: Apertures that expose the common metabolic syndrome. *Trends in Endocrinology and Metabolism, 14*, 371–377.

Heymsfield, S. B., Greenberg, A. S., Fujioka, K., Dixon, R. M., Kushner, R., Hunt, T., Lubina, J. A., Patane, J., Self, B., Hunt, P., et al. (1999). Recombinant leptin for weight loss in obese and lean adults: A randomized, controlled, dose-escalation trial. *JAMA, 282*, 1568–1575.

Hida, K., Wada, J., Eguchi, J., Zhang, H., Baba, M., Seida, A., Hashimoto, I., Okada, T., Yasuhara, A., Nakatsuka, A., et al. (2005). Visceral adipose tissue-derived serine protease inhibitor: A unique insulin-sensitizing adipocytokine in obesity. *Proceedings of the National Academy of Sciences of the United States of America, 102*, 10610–10615.

Hiramatsu-Ito, M., Shibata, R., Ohashi, K., Uemura, Y., Kanemura, N., Kambara, T., Enomoto, T., Yuasa, D., Matsuo, K., Ito, M., et al. (2016). Omentin attenuates atherosclerotic lesion formation in apolipoprotein E-deficient mice. *Cardiovascular Research, 110*, 107–117.

Hirosumi, J., Tuncman, G., Chang, L., Gorgun, C. Z., Uysal, K. T., Maeda, K., Karin, M., & Hotamisligil, G. S. (2002). A central role for JNK in obesity and insulin resistance. *Nature, 420*, 333–336.

Holcomb, I. N., Kabakoff, R. C., Chan, B., Baker, T. W., Gurney, A., Henzel, W., Nelson, C., Lowman, H. B., Wright, B. D., Skelton, N. J., et al. (2000). FIZZ1, a novel cysteine-rich secreted protein associated with pulmonary inflammation, defines a new gene family. *The EMBO Journal, 19*, 4046–4055.

Holland, W. L., Adams, A. C., Brozinick, J. T., Bui, H. H., Miyauchi, Y., Kusminski, C. M., Bauer, S. M., Wade, M., Singhal, E., Cheng, C. C., et al. (2013). An FGF21-adiponectin-ceramide axis controls energy expenditure and insulin action in mice. *Cell Metabolism, 17*, 790–797.

Hotamisligil, G. S. (2006). Inflammation and metabolic disorders. *Nature, 444*, 860–867.

Hotamisligil, G. S., Shargill, N. S., & Spiegelman, B. M. (1993). Adipose expression of tumor necrosis factor-alpha: Direct role in obesity-linked insulin resistance. *Science, 259*, 87–91.

Hotamisligil, G. S., Peraldi, P., Budavari, A., Ellis, R., White, M. F., & Spiegelman, B. M. (1996). IRS-1-mediated inhibition of insulin receptor tyrosine kinase activity in TNF-alpha- and obesity-induced insulin resistance. *Science, 271*, 665–668.

Itoh, N. (2014). FGF21 as a hepatokine, adipokine, and myokine in metabolism and diseases. *Frontiers in Endocrinology, 5*, 107.

Jagannathan-Bogdan, M., McDonnell, M. E., Shin, H., Rehman, Q., Hasturk, H., Apovian, C. M., & Nikolajczyk, B. S. (2011). Elevated proinflammatory cytokine production by a skewed T cell compartment requires monocytes and promotes inflammation in type 2 diabetes. *Journal of Immunology, 186*, 1162–1172.

Jornayvaz, F. R., & Shulman, G. I. (2012). Diacylglycerol activation of protein kinase Cepsilon and hepatic insulin resistance. *Cell Metabolism, 15*, 574–584.

Kanda, H., Tateya, S., Tamori, Y., Kotani, K., Hiasa, K., Kitazawa, R., Kitazawa, S., Miyachi, H., Maeda, S., Egashira, K., et al. (2006). MCP-1 contributes to macrophage infiltration into adipose tissue, insulin resistance, and hepatic steatosis in obesity. *The Journal of Clinical Investigation, 116*, 1494–1505.

Kawano, Y., Nakae, J., Watanabe, N., Kikuchi, T., Tateya, S., Tamori, Y., Kaneko, M., Abe, T., Onodera, M., & Itoh, H. (2016). Colonic pro-inflammatory macrophages cause insulin resistance in an intestinal Ccl2/Ccr2-dependent manner. *Cell Metabolism, 24*, 295–310.

Kazama, K., Usui, T., Okada, M., Hara, Y., & Yamawaki, H. (2012). Omentin plays an anti-inflammatory role through inhibition of TNF-alpha-induced superoxide production in vascular smooth muscle cells. *European Journal of Pharmacology, 686*, 116–123.

Kharitonenkov, A., Shiyanova, T. L., Koester, A., Ford, A. M., Micanovic, R., Galbreath, E. J., Sandusky, G. E., Hammond, L. J., Moyers, J. S., Owens, R. A., et al. (2005). FGF-21 as a novel metabolic regulator. *The Journal of Clinical Investigation, 115*, 1627–1635.

Kievit, P., Howard, J. K., Badman, M. K., Balthasar, N., Coppari, R., Mori, H., Lee, C. E., Elmquist, J. K., Yoshimura, A., & Flier, J. S. (2006). Enhanced leptin sensitivity and improved glucose homeostasis in mice lacking suppressor of cytokine signaling-3 in POMC-expressing cells. *Cell Metabolism, 4*, 123–132.

Kim, J. Y., van de Wall, E., Laplante, M., Azzara, A., Trujillo, M. E., Hofmann, S. M., Schraw, T., Durand, J. L., Li, H., Li, G., et al. (2007a). Obesity-associated improvements in metabolic profile through expansion of adipose tissue. *The Journal of Clinical Investigation, 117*, 2621–2637.

Kim, F., Pham, M., Luttrell, I., Bannerman, D. D., Tupper, J., Thaler, J., Hawn, T. R., Raines, E. W., & Schwartz, M. W. (2007b). Toll-like receptor-4 mediates vascular inflammation and insulin resistance in diet-induced obesity. *Circulation Research, 100*, 1589–1596.

Kirk, E. A., Sagawa, Z. K., McDonald, T. O., O'Brien, K. D., & Heinecke, J. W. (2008). Monocyte chemoattractant protein deficiency fails to restrain macrophage infiltration into adipose tissue [corrected]. *Diabetes, 57*, 1254–1261.

Kitade, H., Sawamoto, K., Nagashimada, M., Inoue, H., Yamamoto, Y., Sai, Y., Takamura, T., Yamamoto, H., Miyamoto, K., Ginsberg, H. N., et al. (2012). CCR5 plays a critical role in obesity-induced adipose tissue inflammation and insulin resistance by regulating both macrophage recruitment and M1/M2 status. *Diabetes, 61*, 1680–1690.

Kleinridders, A., Schenten, D., Konner, A. C., Belgardt, B. F., Mauer, J., Okamura, T., Wunderlich, F. T., Medzhitov, R., & Bruning, J. C. (2009). MyD88 signaling in the CNS is required for development of fatty acid-induced leptin resistance and diet-induced obesity. *Cell Metabolism, 10*, 249–259.

Knudsen, J. G., Murholm, M., Carey, A. L., Bienso, R. S., Basse, A. L., Allen, T. L., Hidalgo, J., Kingwell, B. A., Febbraio, M. A., Hansen, J. B., et al. (2014). Role of IL-6 in exercise training- and cold-induced UCP1 expression in subcutaneous white adipose tissue. *PLoS One, e84910*, 9.

Kopp, H. P., Kopp, C. W., Festa, A., Krzyzanowska, K., Kriwanek, S., Minar, E., Roka, R., & Schernthaner, G. (2003). Impact of weight loss on inflammatory proteins and their association with the insulin resistance syndrome in morbidly obese patients. *Arteriosclerosis, Thrombosis, and Vascular Biology, 23*, 1042–1047.

Kumada, M., Kihara, S., Ouchi, N., Kobayashi, H., Okamoto, Y., Ohashi, K., Maeda, K., Nagaretani, H., Kishida, K., Maeda, N., et al. (2004). Adiponectin specifically increased tissue inhibitor of metalloproteinase-1 through interleukin-10 expression in human macrophages. *Circulation, 109*, 2046–2049.

Kursawe, R., Caprio, S., Giannini, C., Narayan, D., Lin, A., D'Adamo, E., Shaw, M., Pierpont, B., Cushman, S. W., & Shulman, G. I. (2013). Decreased transcription of ChREBP-alpha/beta isoforms in abdominal subcutaneous adipose tissue of obese adolescents with prediabetes or early type 2 diabetes: Associations with insulin resistance and hyperglycemia. *Diabetes, 62*, 837–844.

Larsen, C. M., Faulenbach, M., Vaag, A., Volund, A., Ehses, J. A., Seifert, B., Mandrup-Poulsen, T., & Donath, M. Y. (2007). Interleukin-1-receptor antagonist in type 2 diabetes mellitus. *The New England Journal of Medicine, 356*, 1517–1526.

Lee, J. Y., Sohn, K. H., Rhee, S. H., & Hwang, D. (2001). Saturated fatty acids, but not unsaturated fatty acids, induce the expression of cyclooxygenase-2 mediated through toll-like receptor 4. *The Journal of Biological Chemistry, 276*, 16683–16689.

Leyns, L., Bouwmeester, T., Kim, S. H., Piccolo, S., & De Robertis, E. M. (1997). Frzb-1 is a secreted antagonist of Wnt signaling expressed in the Spemann organizer. *Cell, 88*, 747–756.

Li, S., Shin, H. J., Ding, E. L., & van Dam, R. M. (2009). Adiponectin levels and risk of type 2 diabetes: A systematic review and meta-analysis. *JAMA, 302*, 179–188.

Lin, Z., Tian, H., Lam, K. S., Lin, S., Hoo, R. C., Konishi, M., Itoh, N., Wang, Y., Bornstein, S. R., Xu, A., et al. (2013). Adiponectin mediates the metabolic effects of FGF21 on glucose homeostasis and insulin sensitivity in mice. *Cell Metabolism, 17*, 779–789.

Liu, J., Divoux, A., Sun, J., Zhang, J., Clement, K., Glickman, J. N., Sukhova, G. K., Wolters, P. J., Du,

J., Gorgun, C. Z., et al. (2009). Genetic deficiency and pharmacological stabilization of mast cells reduce diet-induced obesity and diabetes in mice. *Nature Medicine, 15*, 940–945.

Liu, Y., Palanivel, R., Rai, E., Park, M., Gabor, T. V., Scheid, M. P., Xu, A., & Sweeney, G. (2015). Adiponectin stimulates autophagy and reduces oxidative stress to enhance insulin sensitivity during high-fat diet feeding in mice. *Diabetes, 64*, 36–48.

Lord, G. M., Matarese, G., Howard, J. K., Baker, R. J., Bloom, S. R., & Lechler, R. I. (1998). Leptin modulates the T-cell immune response and reverses starvation-induced immunosuppression. *Nature, 394*, 897–901.

Lu, Y., Zhu, X., Liang, G. X., Cui, R. R., Liu, Y., Wu, S. S., Liang, Q. H., Liu, G. Y., Jiang, Y., Liao, X. B., et al. (2012). Apelin-APJ induces ICAM-1, VCAM-1 and MCP-1 expression via NF-kappaB/JNK signal pathway in human umbilical vein endothelial cells. *Amino Acids, 43*, 2125–2136.

Lumeng, C. N., Bodzin, J. L., & Saltiel, A. R. (2007a). Obesity induces a phenotypic switch in adipose tissue macrophage polarization. *The Journal of Clinical Investigation, 117*, 175–184.

Lumeng, C. N., Deyoung, S. M., Bodzin, J. L., & Saltiel, A. R. (2007b). Increased inflammatory properties of adipose tissue macrophages recruited during diet-induced obesity. *Diabetes, 56*, 16–23.

Lynch, L., Nowak, M., Varghese, B., Clark, J., Hogan, A. E., Toxavidis, V., Balk, S. P., O'Shea, D., O'Farrelly, C., & Exley, M. A. (2012). Adipose tissue invariant NKT cells protect against diet-induced obesity and metabolic disorder through regulatory cytokine production. *Immunity, 37*, 574–587.

Maeda, N., Takahashi, M., Funahashi, T., Kihara, S., Nishizawa, H., Kishida, K., Nagaretani, H., Matsuda, M., Komuro, R., Ouchi, N., et al. (2001). PPARgamma ligands increase expression and plasma concentrations of adiponectin, an adipose-derived protein. *Diabetes, 50*, 2094–2099.

Maeda, N., Shimomura, I., Kishida, K., Nishizawa, H., Matsuda, M., Nagaretani, H., Furuyama, N., Kondo, H., Takahashi, M., Arita, Y., et al. (2002). Diet-induced insulin resistance in mice lacking adiponectin/ACRP30. *Nature Medicine, 8*, 731–737.

Mandal, P., Pratt, B. T., Barnes, M., McMullen, M. R., & Nagy, L. E. (2011). Molecular mechanism for adiponectin-dependent M2 macrophage polarization: Link between the metabolic and innate immune activity of full-length adiponectin. *The Journal of Biological Chemistry, 286*, 13460–13469.

Mantell, B. S., Stefanovic-Racic, M., Yang, X., Dedousis, N., Sipula, I. J., & O'Doherty, R. M. (2011). Mice lacking NKT cells but with a complete complement of CD8+ T-cells are not protected against the metabolic abnormalities of diet-induced obesity. *PLoS One, 6*, e19831.

Mantovani, A., Sica, A., Sozzani, S., Allavena, P., Vecchi, A., & Locati, M. (2004). The chemokine system in diverse forms of macrophage activation and polarization. *Trends in Immunology, 25*, 677–686.

Martinez, F. O., Helming, L., & Gordon, S. (2009). Alternative activation of macrophages: An immunologic functional perspective. *Annual Review of Immunology, 27*, 451–483.

Matarese, G., Di Giacomo, A., Sanna, V., Lord, G. M., Howard, J. K., Di Tuoro, A., Bloom, S. R., Lechler, R. I., Zappacosta, S., & Fontana, S. (2001). Requirement for leptin in the induction and progression of autoimmune encephalomyelitis. *Journal of Immunology, 166*, 5909–5916.

Matthews, V. B., Allen, T. L., Risis, S., Chan, M. H., Henstridge, D. C., Watson, N., Zaffino, L. A., Babb, J. R., Boon, J., Meikle, P. J., et al. (2010). Interleukin-6-deficient mice develop hepatic inflammation and systemic insulin resistance. *Diabetologia, 53*, 2431–2441.

Meier, C. A., Bobbioni, E., Gabay, C., Assimacopoulos-Jeannet, F., Golay, A., & Dayer, J. M. (2002). IL-1 receptor antagonist serum levels are increased in human obesity: A possible link to the resistance to leptin? *The Journal of Clinical Endocrinology and Metabolism, 87*, 1184–1188.

Miyoshi, H., Souza, S. C., Zhang, H. H., Strissel, K. J., Christoffolete, M. A., Kovsan, J., Rudich, A., Kraemer, F. B., Bianco, A. C., Obin, M. S., et al. (2006). Perilipin promotes hormone-sensitive lipase-mediated adipocyte lipolysis via phosphorylation-dependent and -independent mechanisms. *The Journal of Biological Chemistry, 281*, 15837–15844.

Mohamed-Ali, V., Goodrick, S., Rawesh, A., Katz, D. R., Miles, J. M., Yudkin, J. S., Klein, S., & Coppack, S. W. (1997). Subcutaneous adipose tissue releases interleukin-6, but not tumor necrosis factor-alpha, in vivo. *The Journal of Clinical Endocrinology and Metabolism, 82*, 4196–4200.

Monteiro, M., Almeida, C. F., Agua-Doce, A., & Graca, L. (2013). Induced IL-17-producing invariant NKT cells require activation in presence of TGF-beta and IL-1beta. *Journal of Immunology, 190*, 805–811.

Moraes-Vieira, P. M., Yore, M. M., Dwyer, P. M., Syed, I., Aryal, P., & Kahn, B. B. (2014). RBP4 activates antigen-presenting cells, leading to adipose tissue inflammation and systemic insulin resistance. *Cell Metabolism, 19*, 512–526.

Mosser, D. M., & Edwards, J. P. (2008). Exploring the full spectrum of macrophage activation. *Nature Reviews. Immunology, 8*, 958–969.

Mulder, P., van den Hoek, A. M., & Kleemann, R. (2017). The CCR2 inhibitor propagermanium attenuates diet-induced insulin resistance, adipose tissue inflammation and non-alcoholic steatohepatitis. *PLoS One, 12*, e0169740.

Mulvihill, E. E., & Drucker, D. J. (2014). Pharmacology, physiology, and mechanisms of action of dipeptidyl peptidase-4 inhibitors. *Endocrine Reviews, 35*, 992–1019.

Myers, M. G., Jr., Backer, J. M., Sun, X. J., Shoelson, S., Hu, P., Schlessinger, J., Yoakim, M., Schaffhausen,

B., & White, M. F. (1992). IRS-1 activates phosphatidylinositol 3′-kinase by associating with src homology 2 domains of p85. *Proceedings of the National Academy of Sciences of the United States of America, 89*, 10350–10354.

Nakamura, T., Furuhashi, M., Li, P., Cao, H., Tuncman, G., Sonenberg, N., Gorgun, C. Z., & Hotamisligil, G. S. (2010). Double-stranded RNA-dependent protein kinase links pathogen sensing with stress and metabolic homeostasis. *Cell, 140*, 338–348.

Nakamura, K., Sano, S., Fuster, J. J., Kikuchi, R., Shimizu, I., Ohshima, K., Katanasaka, Y., Ouchi, N., & Walsh, K. (2016). Secreted frizzled-related protein 5 diminishes cardiac inflammation and protects the heart from ischemia/reperfusion injury. *The Journal of Biological Chemistry, 291*, 2566–2575.

Nishimura, S., Manabe, I., Nagasaki, M., Eto, K., Yamashita, H., Ohsugi, M., Otsu, M., Hara, K., Ueki, K., Sugiura, S., et al. (2009). CD8+ effector T cells contribute to macrophage recruitment and adipose tissue inflammation in obesity. *Nature Medicine, 15*, 914–920.

Odegaard, J. I., & Chawla, A. (2013). Pleiotropic actions of insulin resistance and inflammation in metabolic homeostasis. *Science, 339*, 172–177.

Ofei, F., Hurel, S., Newkirk, J., Sopwith, M., & Taylor, R. (1996). Effects of an engineered human anti-TNF-alpha antibody (CDP571) on insulin sensitivity and glycemic control in patients with NIDDM. *Diabetes, 45*, 881–885.

Okada-Iwabu, M., Yamauchi, T., Iwabu, M., Honma, T., Hamagami, K., Matsuda, K., Yamaguchi, M., Tanabe, H., Kimura-Someya, T., Shirouzu, M., et al. (2013). A small-molecule AdipoR agonist for type 2 diabetes and short life in obesity. *Nature, 503*, 493–499.

Okamoto, Y., Kihara, S., Ouchi, N., Nishida, M., Arita, Y., Kumada, M., Ohashi, K., Sakai, N., Shimomura, I., Kobayashi, H., et al. (2002). Adiponectin reduces atherosclerosis in apolipoprotein E-deficient mice. *Circulation, 106*, 2767–2770.

Ouchi, N., Higuchi, A., Ohashi, K., Oshima, Y., Gokce, N., Shibata, R., Akasaki, Y., Shimono, A., & Walsh, K. (2010). Sfrp5 is an anti-inflammatory adipokine that modulates metabolic dysfunction in obesity. *Science, 329*, 454–457.

Ouchi, N., Parker, J. L., Lugus, J. J., & Walsh, K. (2011). Adipokines in inflammation and metabolic disease. *Nature Reviews. Immunology, 11*, 85–97.

Ozcan, L., Ergin, A. S., Lu, A., Chung, J., Sarkar, S., Nie, D., Myers, M. G., Jr., & Ozcan, U. (2009). Endoplasmic reticulum stress plays a central role in development of leptin resistance. *Cell Metabolism, 9*, 35–51.

Pedersen, B. K., Steensberg, A., Fischer, C., Keller, C., Keller, P., Plomgaard, P., Wolsk-Petersen, E., & Febbraio, M. (2004). The metabolic role of IL-6 produced during exercise: Is IL-6 an exercise factor? *The Proceedings of the Nutrition Society, 63*, 263–267.

Perry, R. J., Zhang, X. M., Zhang, D., Kumashiro, N., Camporez, J. P., Cline, G. W., Rothman, D. L., &

Shulman, G. I. (2014). Leptin reverses diabetes by suppression of the hypothalamic-pituitary-adrenal axis. *Nature Medicine, 20*, 759–763.

Perry, R. J., Camporez, J. P., Kursawe, R., Titchenell, P. M., Zhang, D., Perry, C. J., Jurczak, M. J., Abudukadier, A., Han, M. S., Zhang, X. M., et al. (2015). Hepatic acetyl CoA links adipose tissue inflammation to hepatic insulin resistance and type 2 diabetes. *Cell, 160*, 745–758.

Pessin, J. E., & Saltiel, A. R. (2000). Signaling pathways in insulin action: Molecular targets of insulin resistance. *The Journal of Clinical Investigation, 106*, 165–169.

Phalitakul, S., Okada, M., Hara, Y., & Yamawaki, H. (2011). Vaspin prevents TNF-alpha-induced intracellular adhesion molecule-1 via inhibiting reactive oxygen species-dependent NF-kappaB and PKCtheta activation in cultured rat vascular smooth muscle cells. *Pharmacological Research, 64*, 493–500.

Qi, L., van Dam, R. M., Meigs, J. B., Manson, J. E., Hunter, D., & Hu, F. B. (2006a). Genetic variation in IL6 gene and type 2 diabetes: Tagging-SNP haplotype analysis in large-scale case-control study and meta-analysis. *Human Molecular Genetics, 15*, 1914–1920.

Qi, Y., Nie, Z., Lee, Y. S., Singhal, N. S., Scherer, P. E., Lazar, M. A., & Ahima, R. S. (2006b). Loss of resistin improves glucose homeostasis in leptin deficiency. *Diabetes, 55*, 3083–3090.

Qi, D., Tang, X., He, J., Wang, D., Zhao, Y., Deng, W., Deng, X., Zhou, G., Xia, J., Zhong, X., et al. (2016). Omentin protects against LPS-induced ARDS through suppressing pulmonary inflammation and promoting endothelial barrier via an Akt/eNOS-dependent mechanism. *Cell Death & Disease, e2360, 7*.

Quan, W., Kim, H. K., Moon, E. Y., Kim, S. S., Choi, C. S., Komatsu, M., Jeong, Y. T., Lee, M. K., Kim, K. W., Kim, M. S., et al. (2012). Role of hypothalamic proopiomelanocortin neuron autophagy in the control of appetite and leptin response. *Endocrinology, 153*, 1817–1826.

Ramnanan, C. J., Edgerton, D. S., Rivera, N., Irimia-Dominguez, J., Farmer, B., Neal, D. W., Lautz, M., Donahue, E. P., Meyer, C. M., Roach, P. J., et al. (2010). Molecular characterization of insulin-mediated suppression of hepatic glucose production in vivo. *Diabetes, 59*, 1302–1311.

Ravussin, E., & Galgani, J. E. (2011). The implication of brown adipose tissue for humans. *Annual Review of Nutrition, 31*, 33–47.

Rebrin, K., Steil, G. M., Mittelman, S. D., & Bergman, R. N. (1996). Causal linkage between insulin suppression of lipolysis and suppression of liver glucose output in dogs. *The Journal of Clinical Investigation, 98*, 741–749.

Rosen, E. D., & MacDougald, O. A. (2006). Adipocyte differentiation from the inside out. *Nature Reviews. Molecular Cell Biology, 7*, 885–896.

Rui, L., Yuan, M., Frantz, D., Shoelson, S., & White, M. F. (2002). SOCS-1 and SOCS-3 block insulin signaling by ubiquitin-mediated degradation of IRS1 and IRS2. *The Journal of Biological Chemistry, 277*, 42394–42398.

Samuel, V. T., & Shulman, G. I. (2012). Mechanisms for insulin resistance: Common threads and missing links. *Cell, 148*, 852–871.

Sarbassov, D. D., Guertin, D. A., Ali, S. M., & Sabatini, D. M. (2005). Phosphorylation and regulation of Akt/PKB by the rictor-mTOR complex. *Science, 307*, 1098–1101.

Sauter, N. S., Schulthess, F. T., Galasso, R., Castellani, L. W., & Maedler, K. (2008). The antiinflammatory cytokine interleukin-1 receptor antagonist protects from high-fat diet-induced hyperglycemia. *Endocrinology, 149*, 2208–2218.

Savkur, R. S., Philips, A. V., & Cooper, T. A. (2001). Aberrant regulation of insulin receptor alternative splicing is associated with insulin resistance in myotonic dystrophy. *Nature Genetics, 29*, 40–47.

Schaffler, A., Neumeier, M., Herfarth, H., Furst, A., Scholmerich, J., & Buchler, C. (2005). Genomic structure of human omentin, a new adipocytokine expressed in omental adipose tissue. *Biochimica et Biophysica Acta, 1732*, 96–102.

Schenk, S., Saberi, M., & Olefsky, J. M. (2008). Insulin sensitivity: Modulation by nutrients and inflammation. *The Journal of Clinical Investigation, 118*, 2992–3002.

Scherer, P. E., Williams, S., Fogliano, M., Baldini, G., & Lodish, H. F. (1995). A novel serum protein similar to C1q, produced exclusively in adipocytes. *The Journal of Biological Chemistry, 270*, 26746–26749.

Schipper, H. S., Rakhshandehroo, M., van de Graaf, S. F., Venken, K., Koppen, A., Stienstra, R., Prop, S., Meerding, J., Hamers, N., Besra, G., et al. (2012). Natural killer T cells in adipose tissue prevent insulin resistance. *The Journal of Clinical Investigation, 122*, 3343–3354.

Schulte, D. M., Muller, N., Neumann, K., Oberhauser, F., Faust, M., Gudelhofer, H., Brandt, B., Krone, W., & Laudes, M. (2012). Pro-inflammatory wnt5a and anti-inflammatory sFRP5 are differentially regulated by nutritional factors in obese human subjects. *PLoS One, 7*, e32437.

Schwartz, D. R., & Lazar, M. A. (2011). Human resistin: Found in translation from mouse to man. *Trends in Endocrinology and Metabolism, 22*, 259–265.

Seale, P., Bjork, B., Yang, W., Kajimura, S., Chin, S., Kuang, S., Scime, A., Devarakonda, S., Conroe, H. M., Erdjument-Bromage, H., et al. (2008). PRDM16 controls a brown fat/skeletal muscle switch. *Nature, 454*, 961–967.

Sell, H., Bluher, M., Kloting, N., Schlich, R., Willems, M., Ruppe, F., Knoefel, W. T., Dietrich, A., Fielding, B. A., Arner, P., et al. (2013). Adipose dipeptidyl peptidase-4 and obesity: Correlation with insulin resistance and depot-specific release from adipose tissue in vivo and in vitro. *Diabetes Care, 36*, 4083–4090.

Senn, J. J., Klover, P. J., Nowak, I. A., Zimmers, T. A., Koniaris, L. G., Furlanetto, R. W., & Mooney, R. A. (2003). Suppressor of cytokine signaling-3 (SOCS-3), a potential mediator of interleukin-6-dependent insu-

lin resistance in hepatocytes. *The Journal of Biological Chemistry, 278*, 13740–13746.

Shah, O. J., & Hunter, T. (2006). Turnover of the active fraction of IRS1 involves raptor-mTOR- and S6K1-dependent serine phosphorylation in cell culture models of tuberous sclerosis. *Molecular and Cellular Biology, 26*, 6425–6434.

Shi, H., Kokoeva, M. V., Inouye, K., Tzameli, I., Yin, H., & Flier, J. S. (2006). TLR4 links innate immunity and fatty acid-induced insulin resistance. *The Journal of Clinical Investigation, 116*, 3015–3025.

Shimomura, I., Matsuda, M., Hammer, R. E., Bashmakov, Y., Brown, M. S., & Goldstein, J. L. (2000). Decreased IRS-2 and increased SREBP-1c lead to mixed insulin resistance and sensitivity in livers of lipodystrophic and ob/ob mice. *Molecular Cell, 6*, 77–86.

Shoelson, S. E., Lee, J., & Goldfine, A. B. (2006). Inflammation and insulin resistance. *The Journal of Clinical Investigation, 116*, 1793–1801.

Sleeman, M. W., Wortley, K. E., Lai, K. M., Gowen, L. C., Kintner, J., Kline, W. O., Garcia, K., Stitt, T. N., Yancopoulos, G. D., Wiegand, S. J., et al. (2005). Absence of the lipid phosphatase SHIP2 confers resistance to dietary obesity. *Nature Medicine, 11*, 199–205.

Solinas, G., & Karin, M. (2010). JNK1 and IKKbeta: Molecular links between obesity and metabolic dysfunction. *The FASEB Journal, 24*, 2596–2611.

Souza, S. C., de Vargas, L. M., Yamamoto, M. T., Lien, P., Franciosa, M. D., Moss, L. G., & Greenberg, A. S. (1998). Overexpression of perilipin A and B blocks the ability of tumor necrosis factor alpha to increase lipolysis in 3T3-L1 adipocytes. *The Journal of Biological Chemistry, 273*, 24665–24669.

Spranger, J., Kroke, A., Mohlig, M., Hoffmann, K., Bergmann, M. M., Ristow, M., Boeing, H., & Pfeiffer, A. F. (2003). Inflammatory cytokines and the risk to develop type 2 diabetes: Results of the prospective population-based European Prospective Investigation into Cancer and Nutrition (EPIC)-Potsdam Study. *Diabetes, 52*, 812–817.

Stanley, T. L., Zanni, M. V., Johnsen, S., Rasheed, S., Makimura, H., Lee, H., Khor, V. K., Ahima, R. S., & Grinspoon, S. K. (2011). TNF-alpha antagonism with etanercept decreases glucose and increases the proportion of high molecular weight adiponectin in obese subjects with features of the metabolic syndrome. *The Journal of Clinical Endocrinology and Metabolism, 96*, E146–E150.

Steppan, C. M., Bailey, S. T., Bhat, S., Brown, E. J., Banerjee, R. R., Wright, C. M., Patel, H. R., Ahima, R. S., & Lazar, M. A. (2001). The hormone resistin links obesity to diabetes. *Nature, 409*, 307–312.

Steppan, C. M., Wang, J., Whiteman, E. L., Birnbaum, M. J., & Lazar, M. A. (2005). Activation of SOCS-3 by resistin. *Molecular and Cellular Biology, 25*, 1569–1575.

Stienstra, R., Joosten, L. A., Koenen, T., van Tits, B., van Diepen, J. A., van den Berg, S. A., Rensen, P. C., Voshol, P. J., Fantuzzi, G., Hijmans, A., et al. (2010).

The inflammasome-mediated caspase-1 activation controls adipocyte differentiation and insulin sensitivity. *Cell Metabolism, 12*, 593–605.

Stienstra, R., Tack, C. J., Kanneganti, T. D., Joosten, L. A., & Netea, M. G. (2012). The inflammasome puts obesity in the danger zone. *Cell Metabolism, 15*, 10–18.

Sun, Q., Kiernan, U. A., Shi, L., Phillips, D. A., Kahn, B. B., Hu, F. B., Manson, J. E., Albert, C. M., & Rexrode, K. M. (2013). Plasma retinol-binding protein 4 (RBP4) levels and risk of coronary heart disease: A prospective analysis among women in the nurses' health study. *Circulation, 127*, 1938–1947.

Talukdar, S., Oh, D. Y., Bandyopadhyay, G., Li, D., Xu, J., McNelis, J., Lu, M., Li, P., Yan, Q., Zhu, Y., et al. (2012). Neutrophils mediate insulin resistance in mice fed a high-fat diet through secreted elastase. In *Nat Med.*

Tan, B. K., Adya, R., Farhatullah, S., Chen, J., Lehnert, H., & Randeva, H. S. (2010). Metformin treatment may increase omentin-1 levels in women with polycystic ovary syndrome. *Diabetes, 59*, 3023–3031.

Tang, Q. Q., & Lane, M. D. (2012). Adipogenesis: From stem cell to adipocyte. *Annual Review of Biochemistry, 81*, 715–736.

Taniguchi, C. M., Emanuelli, B., & Kahn, C. R. (2006). Critical nodes in signalling pathways: Insights into insulin action. *Nature Reviews. Molecular Cell Biology, 7*, 85–96.

Tatemoto, K., Hosoya, M., Habata, Y., Fujii, R., Kakegawa, T., Zou, M. X., Kawamata, Y., Fukusumi, S., Hinuma, S., Kitada, C., et al. (1998). Isolation and characterization of a novel endogenous peptide ligand for the human APJ receptor. *Biochemical and Biophysical Research Communications, 251*, 471–476.

Than, A., He, H. L., Chua, S. H., Xu, D., Sun, L., Leow, M. K., & Chen, P. (2015). Apelin enhances brown adipogenesis and browning of white adipocytes. *The Journal of Biological Chemistry, 290*, 14679–14691.

Tiemessen, M. M., Jagger, A. L., Evans, H. G., van Herwijnen, M. J., John, S., & Taams, L. S. (2007). CD4+CD25+Foxp3+ regulatory T cells induce alternative activation of human monocytes/macrophages. *Proceedings of the National Academy of Sciences of the United States of America, 104*, 19446–19451.

Timmons, J. A., Wennmalm, K., Larsson, O., Walden, T. B., Lassmann, T., Petrovic, N., Hamilton, D. L., Gimeno, R. E., Wahlestedt, C., Baar, K., et al. (2007). Myogenic gene expression signature establishes that brown and white adipocytes originate from distinct cell lineages. *Proceedings of the National Academy of Sciences of the United States of America, 104*, 4401–4406.

Townsend, K. L., Suzuki, R., Huang, T. L., Jing, E., Schulz, T. J., Lee, K., Taniguchi, C. M., Espinoza, D. O., McDougall, L. E., Zhang, H., et al. (2012). Bone morphogenetic protein 7 (BMP7) reverses obesity and regulates appetite through a central mTOR pathway. *The FASEB Journal, 26*, 2187–2196.

Tran, K. V., Gealekman, O., Frontini, A., Zingaretti, M. C., Morroni, M., Giordano, A., Smorlesi, A., Perugini, J.,

De Matteis, R., Sbarbati, A., et al. (2012). The vascular endothelium of the adipose tissue gives rise to both white and brown fat cells. *Cell Metabolism, 15*, 222–229.

Ullrich, A., Bell, J. R., Chen, E. Y., Herrera, R., Petruzzelli, L. M., Dull, T. J., Gray, A., Coussens, L., Liao, Y. C., Tsubokawa, M., et al. (1985). Human insulin receptor and its relationship to the tyrosine kinase family of oncogenes. *Nature, 313*, 756–761.

Uysal, K. T., Wiesbrock, S. M., Marino, M. W., & Hotamisligil, G. S. (1997). Protection from obesity-induced insulin resistance in mice lacking TNF-alpha function. *Nature, 389*, 610–614.

Villarroya, F., Cereijo, R., Villarroya, J., & Giralt, M. (2017). Brown adipose tissue as a secretory organ. *Nature Reviews. Endocrinology, 13*, 26–35.

von Holstein-Rathlou, S., BonDurant, L. D., Peltekian, L., Naber, M. C., Yin, T. C., Claflin, K. E., Urizar, A. I., Madsen, A. N., Ratner, C., Holst, B., et al. (2016). FGF21 mediates endocrine control of simple sugar intake and sweet taste preference by the liver. *Cell Metabolism, 23*, 335–343.

Waki, H., & Tontonoz, P. (2007). Endocrine functions of adipose tissue. *Annual Review of Pathology, 2*, 31–56.

Wallenius, V., Wallenius, K., Ahren, B., Rudling, M., Carlsten, H., Dickson, S. L., Ohlsson, C., & Jansson, J. O. (2002). Interleukin-6-deficient mice develop mature-onset obesity. *Nature Medicine, 8*, 75–79.

Wang, S., Krinks, M., Lin, K., Luyten, F. P., & Moos, M., Jr. (1997). Frzb, a secreted protein expressed in the Spemann organizer, binds and inhibits Wnt-8. *Cell, 88*, 757–766.

Weisberg, S. P., Hunter, D., Huber, R., Lemieux, J., Slaymaker, S., Vaddi, K., Charo, I., Leibel, R. L., & Ferrante, A. W., Jr. (2006). CCR2 modulates inflammatory and metabolic effects of high-fat feeding. *The Journal of Clinical Investigation, 116*, 115–124.

Wen, H., Gris, D., Lei, Y., Jha, S., Zhang, L., Huang, M. T., Brickey, W. J., & Ting, J. P. (2011). Fatty acid-induced NLRP3-ASC inflammasome activation interferes with insulin signaling. *Nature Immunology, 12*, 408–415.

Wijesekara, N., Konrad, D., Eweida, M., Jefferies, C., Liadis, N., Giacca, A., Crackower, M., Suzuki, A., Mak, T. W., Kahn, C. R., et al. (2005). Muscle-specific Pten deletion protects against insulin resistance and diabetes. *Molecular and Cellular Biology, 25*, 1135–1145.

Wilk, S., Scheibenbogen, C., Bauer, S., Jenke, A., Rother, M., Guerreiro, M., Kudernatsch, R., Goerner, N., Poller, W., Elligsen-Merkel, D., et al. (2011). Adiponectin is a negative regulator of antigen-activated T cells. *European Journal of Immunology, 41*, 2323–2332.

Wilk, S., Jenke, A., Stehr, J., Yang, C. A., Bauer, S., Goldner, K., Kotsch, K., Volk, H. D., Poller, W., Schultheiss, H. P., et al. (2013). Adiponectin modulates NK-cell function. *European Journal of Immunology, 43*, 1024–1033.

Winer, S., Paltser, G., Chan, Y., Tsui, H., Engleman, E., Winer, D., & Dosch, H. M. (2009a). Obesity predisposes to Th17 bias. *European Journal of Immunology, 39*, 2629–2635.

Winer, S., Chan, Y., Paltser, G., Truong, D., Tsui, H., Bahrami, J., Dorfman, R., Wang, Y., Zielenski, J., Mastronardi, F., et al. (2009b). Normalization of obesity-associated insulin resistance through immunotherapy. *Nature Medicine, 15*, 921–929.

Winer, D. A., Winer, S., Shen, L., Wadia, P. P., Yantha, J., Paltser, G., Tsui, H., Wu, P., Davidson, M. G., Alonso, M. N., et al. (2011). B cells promote insulin resistance through modulation of T cells and production of pathogenic IgG antibodies. *Nature Medicine, 17*, 610–617.

Wittamer, V., Franssen, J. D., Vulcano, M., Mirjolet, J. F., Le Poul, E., Migeotte, I., Brezillon, S., Tyldesley, R., Blanpain, C., Detheux, M., et al. (2003). Specific recruitment of antigen-presenting cells by chemerin, a novel processed ligand from human inflammatory fluids. *The Journal of Experimental Medicine, 198*, 977–985.

Wu, D., Molofsky, A. B., Liang, H. E., Ricardo-Gonzalez, R. R., Jouihan, H. A., Bando, J. K., Chawla, A., & Locksley, R. M. (2011). Eosinophils sustain adipose alternatively activated macrophages associated with glucose homeostasis. *Science, 332*, 243–247.

Wu, J., Bostrom, P., Sparks, L. M., Ye, L., Choi, J. H., Giang, A. H., Khandekar, M., Virtanen, K. A., Nuutila, P., Schaart, G., et al. (2012a). Beige adipocytes are a distinct type of thermogenic fat cell in mouse and human. *Cell, 150*, 366–376.

Wu, L., Parekh, V. V., Gabriel, C. L., Bracy, D. P., Marks-Shulman, P. A., Tamboli, R. A., Kim, S., Mendez-Fernandez, Y. V., Besra, G. S., Lomenick, J. P., et al. (2012b). Activation of invariant natural killer T cells by lipid excess promotes tissue inflammation, insulin resistance, and hepatic steatosis in obese mice. *Proceedings of the National Academy of Sciences of the United States of America, 109*, E1143–E1152.

Xu, A., Wang, Y., Keshaw, H., Xu, L. Y., Lam, K. S., & Cooper, G. J. (2003). The fat-derived hormone adiponectin alleviates alcoholic and nonalcoholic fatty liver diseases in mice. *The Journal of Clinical Investigation, 112*, 91–100.

Yamauchi, T., Kamon, J., Minokoshi, Y., Ito, Y., Waki, H., Uchida, S., Yamashita, S., Noda, M., Kita, S., Ueki, K., et al. (2002). Adiponectin stimulates glucose utilization and fatty-acid oxidation by activating AMP-activated protein kinase. *Nature Medicine, 8*, 1288–1295.

Yang, Q., Graham, T. E., Mody, N., Preitner, F., Peroni, O. D., Zabolotny, J. M., Kotani, K., Quadro, L., & Kahn, B. B. (2005). Serum retinol binding protein 4 contributes to insulin resistance in obesity and type 2 diabetes. *Nature, 436*, 356–362.

Yang, R. Z., Lee, M. J., Hu, H., Pray, J., Wu, H. B., Hansen, B. C., Shuldiner, A. R., Fried, S. K., McLenithan, J. C., & Gong, D. W. (2006). Identification of omentin as a novel depot-specific adipokine in human adipose tissue: Possible role in modulating insulin action.

American Journal of Physiology. Endocrinology and Metabolism, 290, E1253–E1261.

Yokota, T., Oritani, K., Takahashi, I., Ishikawa, J., Matsuyama, A., Ouchi, N., Kihara, S., Funahashi, T., Tenner, A. J., Tomiyama, Y., et al. (2000). Adiponectin, a new member of the family of soluble defense collagens, negatively regulates the growth of myelomonocytic progenitors and the functions of macrophages. *Blood, 96*, 1723–1732.

Youn, B. S., Kloting, N., Kratzsch, J., Lee, N., Park, J. W., Song, E. S., Ruschke, K., Oberbach, A., Fasshauer, M., Stumvoll, M., et al. (2008). Serum vaspin concentrations in human obesity and type 2 diabetes. *Diabetes, 57*, 372–377.

Yue, P., Jin, H., Aillaud, M., Deng, A. C., Azuma, J., Asagami, T., Kundu, R. K., Reaven, G. M., Quertermous, T., & Tsao, P. S. (2010). Apelin is necessary for the maintenance of insulin sensitivity. *American Journal of Physiology. Endocrinology and Metabolism, 298*, E59–E67.

Yue, P., Jin, H., Xu, S., Aillaud, M., Deng, A. C., Azuma, J., Kundu, R. K., Reaven, G. M., Quertermous, T., & Tsao, P. S. (2011). Apelin decreases lipolysis via G(q), G(i), and AMPK-dependent mechanisms. *Endocrinology, 152*, 59–68.

Zeisberg, M., Hanai, J., Sugimoto, H., Mammoto, T., Charytan, D., Strutz, F., & Kalluri, R. (2003). BMP-7 counteracts TGF-beta1-induced epithelial-to-mesenchymal transition and reverses chronic renal injury. *Nature Medicine, 9*, 964–968.

Zhang, Y., Proenca, R., Maffei, M., Barone, M., Leopold, L., & Friedman, J. M. (1994). Positional cloning of the mouse obese gene and its human homologue. *Nature, 372*, 425–432.

Zhang, H. H., Halbleib, M., Ahmad, F., Manganiello, V. C., & Greenberg, A. S. (2002). Tumor necrosis factor-alpha stimulates lipolysis in differentiated human adipocytes through activation of extracellular signal-related kinase and elevation of intracellular cAMP. *Diabetes, 51*, 2929–2935.

Zhang, X., Zhang, G., Zhang, H., Karin, M., Bai, H., & Cai, D. (2008). Hypothalamic IKKbeta/NF-kappaB and ER stress link overnutrition to energy imbalance and obesity. *Cell, 135*, 61–73.

Zhang, C., Shao, M., Yang, H., Chen, L., Yu, L., Cong, W., Tian, H., Zhang, F., Cheng, P., Jin, L., et al. (2013). Attenuation of hyperlipidemia- and diabetes-induced early-stage apoptosis and late-stage renal dysfunction via administration of fibroblast growth factor-21 is associated with suppression of renal inflammation. *PLoS One, e82275, 8*.

Zhou, Q., Brown, J., Kanarek, A., Rajagopal, J., & Melton, D. A. (2008). In vivo reprogramming of adult pancreatic exocrine cells to beta-cells. *Nature, 455*, 627–632.

Zhuge, F., Ni, Y., Nagashimada, M., Nagata, N., Xu, L., Mukaida, N., Kaneko, S., & Ota, T. (2016). DPP-4 inhibition by linagliptin attenuates obesity-related inflammation and insulin resistance by regulating M1/M2 macrophage polarization. *Diabetes, 65*, 2966–2979.

Part IV

Neuroendocrine Axes and Obesity

The Thyroid Hormone Axis: Its Roles in Body Weight Regulation, Obesity, and Weight Loss

10

Kristen Rachel Vella

10.1 Introduction: The Hypothalamic-Pituitary-Thyroid Axis

The hypothalamus-pituitary-thyroid (HPT) axis is the central mechanism by which the body maintains thyroid hormone homeostasis (Costa-e-Sousa and Hollenberg 2012). Thyrotropin-releasing hormone (TRH) is released from neurons in the paraventricular nucleus (PVN) of the hypothalamus (Fig. 10.1) (Chiamolera and Wondisford 2009). TRH binds to the TRH receptor at the levels of the pituitary to stimulate the secretion of thyroid-stimulating hormone (TSH) (Bowers et al. 1965; Geras and Gershengorn 1982; Greer 1951; Guillemin et al. 1965, 1963; Schally et al. 1969; Shupnik et al. 1996; Yamada et al. 1995). TSH triggers the release of thyroid hormone, both the predominant prohormone thyroxine (T4) and the active form triiodothyronine (T3), from the thyroid gland into the bloodstream. The HPT axis operates in a negative feedback loop as T3 suppresses TRH and TSH at several levels including gene transcription and prohormone processing (Lechan and Hollenberg 2003; Lechan et al. 1986; Vella and Hollenberg 2009).

K. R. Vella (✉)
Weill Cornell Medical College, New York, NY, USA
e-mail: krv2005@med.cornell.edu

10.1.1 The Regulation of Thyroid Hormone Levels

As discussed in previous chapters, TRH neurons in the PVN of the hypothalamus (PVN) are believed to represent the regulatory core of the HPT axis. Early evidence suggested impairment of TRH neurons severely affects the regulation of TSH secretion (Martin et al. 1970). Additionally, TRH is key to determining TSH bioactivity (Beck-Peccoz et al. 1985; Menezes-Ferreira et al. 1986; Nikrodhanond et al. 2006; Taylor et al. 1986). However, research has demonstrated that the pituitary can increase TSH synthesis during severe hypothyroidism, even in the absence of TRH (Nikrodhanond et al. 2006; Schaner et al. 1997; Yamada et al. 1997). Mature TRH is a tripeptide that is derived from pro-TRH by the action of prohormone convertases (Perello and Nillni 2007; Schaner et al. 1997). Remarkably, both transcription of TRH and its posttranslational processing are suppressed by T3 (Perello et al. 2006; Segerson et al. 1987; Sugrue et al. 2010).

Circulating TSH is a universally accepted biomarker of thyroid hormone action in humans. Given the tight negative regulation of TSH by thyroid hormone, a high TSH level is indicative of low T4 and T3 levels and hypothyroidism. Conversely, a low TSH measurement signals high T4 and T3 levels and hyperthyroidism. Clinicians measure TSH levels in their patients to assess their thyroid hormone status.

© Springer International Publishing AG, part of Springer Nature 2018
E. A. Nillni (ed.), *Textbook of Energy Balance, Neuropeptide Hormones, and Neuroendocrine Function*, https://doi.org/10.1007/978-3-319-89506-2_10

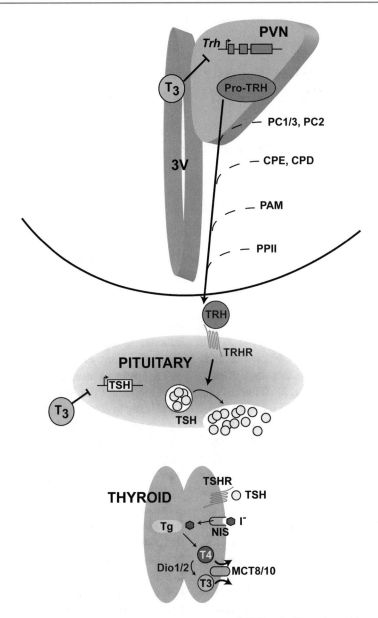

Fig. 10.1 The HPT axis is tightly regulated by a negative feedback system. Thyrotropin-releasing hormone (TRH) is transcribed and processed by several enzymes as it travels down the axon including prohormone convertases 1/3 and 2 (PC1/3 and PC2), carboxypeptidases E and D (CPE and CPD), and peptidyl-amidating monooxygenase (PAM) and pyroglutamyl peptidase II (PPII). TRH is released from the median eminence at the bottom of the hypothalamus where it binds to the TRH receptor (TRHR) and stimulates the release of thyroid-stimulating hormone (TSH). TSH circulates through the blood and binds to the TSH receptor (TSH) to facilitate thyroid hormone release from the thyroid. Iodine (I⁻) is taken up by the thyroid through the sodium-iodide symporter (NIS) and pendrin (not shown) transporters. Thyroglobulin (Tg) is processed in the colloid (not shown) and coupled to I⁻ to create T4 and T3. Then, T4 and T3 are shuttled back into the cell, where T4 can be metabolized to T3 by deiodinases. Finally, T4 and T3 are released into the bloodstream via the monocarboxylate transporters (MCT8 or MCT10). T3 represses the transcriptional and posttranslational processing of TSH in the pituitary and TRH in the PVN of the hypothalamus

However, there are several cases where the levels of TSH in the blood do not accurately interpret T4 and T3 levels (Refetoff et al. 1993).

TSH acts on the thyroid gland to synthesize and secrete thyroid hormone (Fig. 10.1). The thyroid gland produces thyroid hormone through thyroid follicular cells (Kopp 2005). Iodine is essential to thyroid hormone. The follicular cells concentrate iodine within the colloid by using sodium-iodide symporter (NIS) and pendrin transporters. These transporters capture circulating iodine (I^-) and translocate it to the cytoplasm, where I^- is then secreted into the colloid. In the colloid, thyroglobulin incorporates iodine after the iodine is oxidized by peroxide. Iodinated thyroglobulin is then endocytosed back into the cytoplasm of the follicular cell through the actions of TSH on the TSH receptor. In the vesicles containing iodinated thyroglobulin, enzymes cleave T4 off the thyroglobulin.

The discovery of thyroid hormone transporters has challenged the previous theory that T4 and T3 passively cross cell membranes (Friesema et al. 2005). The monocarboxylate transporters 8 and 10 (MCT8 and MCT10) transport T4 and T3 into and out of the cell. MCT8 does have a preference for transporting T3. Additionally, the organic anion transporter (OATP1) and the L-type amino acid transporter (LAT) also transport thyroid hormones across membranes. OATP1 preferentially transports T4 and rT3, whereas LAT transports both T4 and T3 at a lower affinity. Mutations in MCT8 have been found in patients with the Allan-Herndon-Dudley syndrome (AHDS) (Friesema et al. 2010). AHDS is an X-linked disorder. In addition to neurological complications, hypotonia and muscle hypoplasia, patients with AHDS exhibit abnormal thyroid function tests including normal TSH, low free T4, and high levels of circulating T3. The primary structure of MCT8 forms 12 transmembrane subunits that facilitate the bidirectional transport of T3 in favor of its gradient of concentration (Friesema et al. 2010). Interestingly, dimerization of MCT8 also appears to be necessary for its function.

In the bloodstream, T4 and T3 circulate attached to serum proteins including thyroxine-binding globulin (TBG), transthyretin (TTR), thyroglobulin (TBG), and albumin (Benvenga et al. 1994). A small fraction of circulating T_4 is free (FT_4) to be transported into the cytoplasm. Once in the cells of the target tissue, T4 is converted into the bioactive hormone T3 by either the type 1 or type 2 iodothyronine deiodinase (Dio1 or Dio2) (St Germain et al. 2009). Indeed, the deiodinases ultimately regulate intracellular availability of T3. These enzymes are expressed in a tissue-specific pattern where Dio1 is the predominant form in the liver and kidney and Dio2 is expressed in the central nervous system, pituitary, brown adipose tissue, and muscle. Dio1 and Dio2 metabolize T4 to its active form T3 by outer-ring deiodination. T3 binds to thyroid hormone receptors in the nucleus to activate or suppress gene transcription.

T3 and T4 can be inactivated and metabolized by several mechanisms locally in the cell including inner ring deiodination by the type 3 deiodinase (Dio3), glucuronidation, and sulfation. In adults, Dio3 is expressed at low levels in all tissues that rely on thyroid hormone, but Dio3 is particularly active during fetal development (Gereben et al. 2015). It degrades T4 into reverse T3 (rT3) and T3 to 3,3'-T2. Dio3 plays a role in the developing cochlea by preventing the premature response to thyroid hormone (Ng et al. 2004, 2009). Silencing D3 in zebrafish during development results in delayed hatching, significantly smaller size, and decreased inflation of the swim bladder (Heijlen et al. 2014).

Sulfation and glucuronidation are phase II detoxification reactions, which increase the water solubility of thyroid hormone to facilitate its clearance through the urine or bile, respectively. Sulfotransferases tag T4 and T3 with a sulfate (T4S and T3S) (Visser 1996). These conjugates are rapidly degraded by Dio1 and then excreted through the urine (Mol and Visser 1985). UDP-glucuronosyltransferases transfer the glucuronic acid component of uridine diphosphate glucuronic acid to T3 and T4 such that thyroid hormone can be excreted through the bile and then feces (Vansell and Klaassen 2001, 2002). Sulfatransferases and glucuronidases are regulated by several factors including thyroid hormone, fasting, and xenobiotics (Maglich et al. 2004; Qatanani et al. 2005; Visser 1996).

10.1.2 The Role of Thyroid Hormone in Gene Regulation

At the level of transcription, rodents and humans possess two different thyroid hormone receptor-encoding genes termed THRA and THRB. The THRA locus is located on human chromosome 17 and expresses two major isoforms, TRα1 and TRα2. These two isoforms differ at their C-terminal region due to the presence of an alternative exon and only TRα1 binds T3 (Lazar 1993). The THRB locus on chromosome 3 also leads to the expression of two major isoforms TRβ1 and TRβ2 who differ at their amino-termini based upon alternative exon use. Both of the TRβ isoforms bind T_3.

TRα1 and TRβ differ in their tissue expression, but they are homologous in function and their molecular structure is conserved in across species (Brent 2012). All thyroid hormone-binding TR isoforms contain three domains that include highly conserved DNA and ligand-binding domains. The most diverse region of the TR isoforms is the amino-terminal or A/B domains, whose functions have not been well clarified. The thyroid hormone receptors are ligand-activated transcription factor that exist in the nucleus in the presence and absence of thyroid hormone.

To study the functions of the different TR isoforms, researchers have relied on studying knockout mouse models and resistance to thyroid hormone syndrome (RTH) in humans. Mouse knockout studies have demonstrated a unique role for the TRβ isoforms in the regulation of TSH production by the pituitary (Abel et al. 2001; Forrest et al. 1996; Forrest and Vennstrom 2000; Ng et al. 2015). Additionally, the TRβ2 isoform plays a specialized role in the retina allowing for the expression of the opsin photopigments in the retina of mice and thus allowing color

vision development (Ng et al. 2001). Interestingly, both TRβ isoforms are important in cochlear development and accordingly hearing development, while TRβ1 is required for adult hearing (Forrest et al. 1996; Ng et al. 2015). In the liver, the TRβ1 isoform is the principal mediator of thyroid hormone action, particularly in mediating cholesterol metabolism (Gullberg et al. 2000, 2002). Similarly, both isoforms target thyroid hormone action in the brain, but TRα1 has clear actions in hypothalamic neurons that regulate sympathetic function (Mittag et al. 2013). TRα1 has the majority of actions in the skeleton, heart, and intestine, but TRβ1 may play a role in certain cell types in these tissues. Taken together, mouse genetic studies have well outlined the actions of the TR isoforms. However, studies using global knockouts of the TR isoforms have their limitations and conditional alleles will allow for the tissue and cell-specific functions of the TR isoforms.

In humans, there are two distinct RTH syndromes due to mutations in the respective TR isoforms. RTHβ was first described in the late 1960s and identified as being secondary to mutations in the TRβ isoforms in the 1980s. Patients with RTHβ present with inappropriately high TSH secretion in the face of elevated thyroid hormone levels proving that this isoform regulates the hypothalamic-pituitary-thyroid (HPT) axis (Refetoff et al. 1993). The clinical signs and symptoms of the disorder align with the TR isoform tissue distribution including goiter. Although TRβ-expressing tissues such as the liver and pituitary are resistant to thyroid hormone, TRα-expressing tissues such as the heart and skeleton sense elevated circulating thyroid hormone levels and are hyperthyroid. As such, RTHβ patients have tachycardia and short stature. Some of the clinical findings in RTHβ may be the result of a combination of effects of resistant TRβ signaling and activated TRα signaling such as attention deficit hyperactivity syndrome (Refetoff et al. 1993).

With the help of mice expressing TRα mutations to model RTHα, TRα isoform mutations and RTHα were not identified in humans until 2012 (Bochukova et al. 2012; Kaneshige et al. 2001).

The first TRα patient had features consistent with relative hypothyroidism in TRα-expressing tissues including a skeletal phenotype, short stature, constipation, bradycardia, and neurodevelopmental issues. All of the TRα mutations found to date impair T3 binding and lead to the recruitment of a repressive complex that cannot be released. Certain features like macrocephaly and constipation tend to be uniform across all TRα mutations. Strikingly, mutations in regions of TRα that are common to both the TRα1 and TRα2 isoforms have not revealed any unique biochemical or syndrome-specific features, which suggests that TRα2 may not play an important role in thyroid hormone action.

The TR transcriptional complex recruits coregulatory factors including the corepressors nuclear receptor corepressor 1 (NCoR1) and silencing mediator for retinoid or thyroid hormone receptors (SMRT, also known as NCoR2) and the coactivators SRC-1, SRC-2, and SRC-3 (Alland et al. 1997; Halachmi et al. 1994; Heinzel et al. 1997; Lonard and O'Malley B 2007; Nagy et al. 1997; Onate et al. 1995). Both NCoR1 and SMRT interact with the TR isoforms via C-terminal domains, the nuclear receptor interacting domains (RIDs) (Hu and Lazar 1999; Nagy et al. 1999; Perissi et al. 1999). However, NCoR1 prefers to interact with the TR via its more N-terminal RIDs. The steroid receptor coactivators 1, 2, and 3 (SRC-1, SRC-2, and SRC-3) share structural homology but appear to have a variety of different functions (Lonard and O'Malley B 2007). The SRCs interact with liganded nuclear receptors including the TR isoforms via a central interacting domain that contains a number of LxxLL motifs. A previous study demonstrated that the SRC isoforms can interact with the TRβ2 amino-terminus and that this interaction could be important in thyroid hormone action (Yang and Privalsky 2001). Like the corepressors, members of the SRC family can be differentially expressed in a variety of cell types. Additionally, they play nonredundant roles in physiology with SRC-1 having the most significant role in thyroid hormone action (Vella et al. 2014; Weiss et al. 1999, 2002). Numerous other proteins with coactivator-like activity have been identified and can interact with the TR isoforms,

but their roles in thyroid hormone action remain to be determined.

The roles of co-regulators in thyroid hormone action first came to light in the SRC-1 KO mice. Among other steroid receptor signaling deficits, SRC-1 KO mice have RTH elevated TSH levels in the presence of elevated circulating thyroid hormone levels (Weiss et al. 1999, 2002). Until conditional alleles were generated, the roles of NCoR1 and SMRT in vivo were impossible to determine because deleting either paralog led to embryonic lethality (Jepsen et al. 2000, 2008). To address the role of NCoR1 in thyroid hormone action, Astapova et al. developed a mouse model that expressed a Cre-driven hypomorphic NCoR1 allele (NCoRΔID), which lacked the two principal RIDs that interacted with the TR (Astapova et al. 2008). These liver-specific L-NCoRΔID mice had a number of derepressed hepatic TRβ1 targets in the hypothyroid setting consistent with the classic role predicted for NCoR1. A global NCoRΔID mouse had low levels of circulating T4 and T3 with normal TSH levels and normal levels of TRH mRNA in the hypothalamus (Astapova et al. 2011). NCoRΔID mice were not small and had evidence of increased energy expenditure. Furthermore, T3 targets in the liver had normal expression. Thus, global removal of a functional NCoR1 molecule in vivo appears to increase sensitivity to thyroid hormone at the level of the HPT axis and the liver.

Combining the contrasting roles of SRC-1 and NCoR in managing thyroid hormone levels, Vella et al. developed a mouse model that combined both of these genetic alterations (Vella et al. 2014). As expected, deletion of SRC-1 led to RTH at the level of the HPT axis. When NCoRΔID was introduced on this background, normal thyroid hormone sensitivity was reestablished. In the liver, positively regulated T3 target genes in SRC-1 KO/NCoRΔID mice had normal sensitivity and response to T3.

While the in vivo models have clarified the role of co-regulators in thyroid hormone action, many questions remain. Key insight into co-regulator function has only been established in the HPT axis and in the liver. The role of co-regulators in other thyroid hormone-responsive tissues remains unknown.

10.2   Thyroid Hormone's Role in Body Weight Regulation

The interplay between thyroid disease, body weight, and metabolism has been studied for a long time in both humans and other vertebrates (Barker 1951; Du Bois 1936). Indeed, the first studies linking thyroid hormone and energy expenditure were conducted over 100 years ago by Magnus-Levy starting in 1895 (Mangus-Levy 1895). Metabolism or energy expenditure (EE) can be defined by the amount of oxygen used by the body over a specific amount of time. Resting energy expenditure (REE) is the energy required to maintain basic cell and organ function while the body is at rest. Prior to thyroid function tests (measurement of T4, T3, and TSH), REE was one of the earliest indications to test a patient's thyroid status. Poor thyroid gland function was associated with low REE, whereas overactive thyroid gland function was associated with high REE (Fig. 10.2). Upon further study when thyroid hormone measurements were available, it was found that low thyroid hormone levels are linked to low REE and conversely high thyroid hormones linked to high REE. Due to the complexity of the test and the number of factors that can affect REE, testing REE to determine thyroid function is no longer a favorable test.

Hypothyroidism is a disease marked by low T4 and T3 and higher TSH. There are several diseases that can lead to hypothyroidism including an autoimmune disease such as Hashimoto's thyroiditis, where the immune system attacks the thyroid gland. Other causes of hypothyroidism include too little or too much iodine in the diet, illness, medicines, congenital hypothyroidism, and treatments for thyroid disease and thyroid cancer where the thyroid is surgically removed or treated with radioactive iodine (I-131). Hypothyroidism has long been associated with a small weight gain, usually about 5–10 pounds. This is due to the lower REE in patients with hypothyroidism. In cases of severe hypothyroidism, the weight gain is often greater. The cause of the weight gain in hypothyroid individuals is also complex and can be attributed to several factors including excess fat accumulation or excess accumulation of salt and water. Interestingly, thyroid hormone levels have no root cause in the

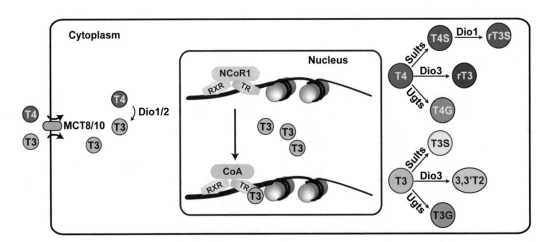

Fig. 10.2 At the cellular level, T4 and T3 enter through thyroid hormone transporters such as the monocarboxylate transporters (MCT8, MCT10). T4 is converted in the bioactive thyroid hormone, T3. Once in the nucleus, T3 binds to the thyroid hormone receptor to recruit transcriptional machinery on positive T3 target genes including coactivators like SRC-1. In the absence of T3, transcriptional repression machinery is recruited as TR binds NCoR1 or other corepressors. The TR forms dimers, sometimes with RXR. T4 and T3 are metabolized by several factors including the type 3 deiodinase (Dio3). Dio3 converts T4 into reverse T3 (rT3) and T3 into 3,3'T2. Sulfotransferases tag T4 and T3 with sulfates. T4S is converted to rT3S by Dio1. Sulfation of thyroid hormone increases its solubility and excretion via urine (T4S, rT3S, and T3S). UDP-glucuronosyltransferases (Ugts) increase thyroid hormone solubility and excretion through the bile and feces (T4G and T3G)

obesity and weight gain seen in modern times. However, a study comparing treatment of hypothyroid patients with T3 demonstrated significant weight loss and reduction in total cholesterol without adverse cardiovascular outcomes (Celi et al. 2011).

In hyperthyroidism, T4 and T3 are elevated thus elevating REE. Weight loss can be a symptom as patients can experience elevated body temperature and increased REE despite increased food intake in some cases. Severe cases of hyperthyroidism can cause extreme weight loss. Graves' disease is a cause of hyperthyroidism, where the immune system causes overstimulation and growth of the thyroid. Other possibilities include goiter and a temporary condition called thyroiditis, which is normally linked to a viral infection.

Given the small influence of thyroid disease on body weight, the focus of this next chapter will be on the effects of thyroid hormone on whole body metabolism, how it influences EE, and how researchers hope to harness its effects as a treatment for obesity.

10.2.1 Thyroid Hormone Has Direct and Indirect Effects on Metabolism

At the cellular level, thyroid hormone has effects on several processes including glycolysis and glucogenesis, fatty acid oxidation and lipogenesis, and protein turnover (Mullur et al. 2014; Vaitkus et al. 2015). With a small effect on EE, thyroid hormone has been shown to reduce reactive oxygen species while increasing energy expenditure (Grant 2007). Additionally, thyroid hormone has been shown in increase ion leakage over the cellular membrane through the Na+/K+ ATPase ion leak and the sarco/endoplasmic reticulum Ca^{2+} ATPase (Haber et al. 1988; Silva 2006).

Thyroid hormone has a large effect on stimulating mitochondrial biogenesis through several mechanisms including effects on mitochondrial genes like cytochrome c, promoting mitochon-

dria gene transcription through p43, p28, and a truncated TRα1 and establishing a positive feedback loop in which thyroid hormone increases nuclear expression of intermediate factors that in turn increase mitochondrial transcription such as PGC-1α (Psarra et al. 2006; Rodgers et al. 2008; Thijssen-Timmer et al. 2006; Wrutniak et al. 1995; Wulf et al. 2008). Thyroid hormone can also work within the mitochondria in processes like non-shivering thermogenesis. Here chemical energy is converted directly into heat. Thyroid hormone plays an important role in the regulation of uncoupling protein 1 (UCP1), which renders the inner membrane of the mitochondria permeable to electrons and allows for the generation of heat. UCP1 is expressed primarily in brown adipose tissue (BAT). Until recently, human BAT was thought to only exist in infants and not significant in adults. Recent PET and CT imaging studies have shown a significant amount of BAT exists in adults, especially in the subscapular and chest region (Cypess et al. 2009; van Marken Lichtenbelt et al. 2009).

Fatty acid oxidation, the catabolic breakdown of fatty acids in the cell, is regulated in part by thyroid hormone. In the heart, thyroid hormone has been shown to affect carnitine/acylcarnitine transporter (CACT), the mitochondrial carrier protein involved in fatty acid metabolism (Paradies et al. 1996). In hyperthyroid rats, researchers found an increased rate of palmitoylcarnitine/carnitine exchange and increased fatty acid oxidation in heart mitochondria. In hypothyroid rats, fatty acid oxidation is reduced in rat heart mitochondria due to a decreased CACT activity. Hypothyroid rats supplemented with T3 restored normal CACT activity (Paradies et al. 1997). In the liver, T3 stimulates several genes involved with fatty acid oxidation including in the transcription of carnitine palmitoyltransferase 1 (CPT1) gene (Flores-Morales et al. 2002; Jackson-Hayes et al. 2003; Santillo et al. 2013). CPT1 transforms fatty acids to carnitine esters when it is localized to the outer mitochondrial membrane. Also in the liver, thyroid hormone stimulates the citrate carrier (CiC) gene expression and activity, another inner mitochondrial membrane carrier protein (Giudetti et al. 2006; Paradies and Ruggiero 1990). More work has to be done to determine how T3 influences the fatty acid oxidation process, especially in terms of carrier gene expression and activity.

Thyroid hormone has been shown to stimulate gluconeogenesis, the metabolic process that generates glucose from substrates such as lactate, glycerol, and glucogenic amino acids. Thyroid hormone treatment increases genes involved in gluconeogenesis including phosphoenolpyruvate carboxykinase (PEPCK), the rate-limiting step in gluconeogenesis (Park et al. 1999). In the liver, PEPCK mRNA was stimulated 3.5-fold in hyperthyroid rats (Klieverik et al. 2008). The affect of thyroid hormone on liver gluconeogenesis might have a central component as administration of T3 to the PVN of the hypothalamus in rats increased glucose production (Klieverik et al. 2009).

Cholesterol synthesis can be regulated by thyroid hormone through multiple mechanisms. Thyroid hormone stimulates the low-density lipoprotein receptor (LDLR) gene, which increases uptake of cholesterol and enhanced cholesterol synthesis (Lopez et al. 2007). Patients with hypothyroidism experience mild to severe hypercholesterolemia (Klein and Danzi 2007; Thompson et al. 1981). Thyroid hormone replacement reverses the increased serum levels of cholesterol (Klein and Danzi 2007). Thyroid hormone also regulates cholesterol through the sterol response element-binding protein (SREBP)-2, which regulates LDLR (Goldstein et al. 2006). SREBP-2 is a member of the transcription factor family that regulates glucose metabolism, fatty acid synthesis, and cholesterol metabolism. In hypothyroid rats, SREBP-2 mRNA is suppressed, but this is reversed when T3 levels are restored (Shin and Osborne 2003).

The physiological benefit of thyroid hormone is multicellular and across most tissues. Indeed, hypothyroidism is linked to poor gluconeogenesis, fatty acid oxidation, and cholesterol synthesis. Treatment with thyroid hormone reverses these conditions. As obesity and metabolic syndrome have similar deficits as above, there is great potential in thyroid hormone or thyroid hormone analogs as therapeutics.

10.2.2 Thyroid Hormone as a Treatment for Obesity

Due to thyroid hormone's effects on metabolism, researchers asked if it would be a potential therapeutic for obesity and dyslipidemia. While T3 has many beneficial effects, supraphysiologic thyroid hormone levels induce tachycardia, bone loss, muscle wasting, and neuropsychiatric disturbances (Burch and Wartofsky 1993). Several have pursued thyroid hormone derivatives that are tissue or TR isoform specific with hopes that this would affect metabolism and have no response in the heart and bone. The synthetic thyroid hormone analog GC-1 (sobetirome) has been shown to prevent or reduce hepatosteatosis and reduce serum triglyceride levels and cholesterol levels without significant side effects on heart rate (Perra et al. 2008; Trost et al. 2000). In a separate study, GC-1 has been shown to increase EE and prevent fat accumulation in female rats (Villicev et al. 2007). Another thyroid hormone analog MB07811 exhibits increased TR activation in the liver and reduces hepatic triglyceride levels and increases hepatic fatty acid oxidation in both chow-fed and high-fat diet rodent models (Cable et al. 2009; Erion et al. 2007). In human clinical trials, MB07811 reduced LDL cholesterol and triglyceride levels without severe adverse effects (Baxter and Webb 2009).

Thyroid hormone derivatives including 3,3′,5-triiodothyronine (rT3), thyronamines (TAMs), and 3,5-diiodothyronine (T2) have been found to have some effects on body weight in rodents. The effects of TAMs, primarily 3-iodothyronamine (3-T1 AM) and thyronamine (T0 AM), have been studied in rodent and include hypothermia, reduction in energy expenditure, hyperglycemia, and reduction of fat mass (Piehl et al. 2011). Studies using rT3 in rodents have found that it can enhance actin cytoskeleton repair in neurons and astrocytes in the hypothalamus following a hypothyroid-induced decline (Farwell et al. 1990; Siegrist-Kaiser et al. 1990). Also, studies in chickens revealed that rT3 inhibits the surge in free fatty acids seen after treatment with dexamethasone or adrenaline (Bobek et al. 2002). A larger number of studies have been done using T2. T2 will stimulate mitochondrial activity and elevate resting EE in rats (Lombardi et al. 1998; Moreno et al. 1997). In high-fat diet rat models, T2 administration prevented hepatosteatosis, insulin resistance, and obesity while stimulating mitochondrial uncoupling (Grasselli et al. 2008; Lanni et al. 2005; Mollica et al. 2009; Moreno et al. 2011). When T2 was administered in human studies, euthyroid subjects had a significant elevation in REE, reduced body weight, and normal thyroid and cardiac function (Antonelli et al. 2011).

Thyroid hormone analogs and alternatives show some promise as obesity therapeutics. GC-1 and T2 both challenge EE and have beneficial effects in the liver. However, more studies must be done to determine efficacy and safety of TAMs, rT3, and MB07811.

10.3   The Effects of Weight Loss on Thyroid Hormone

As global obesity rates rise, understanding the complex neurocircuitry regulating energy expenditure is increasingly pivotal to finding preventative measures and therapies. Thyroid hormone levels affect the molecular mechanisms governing basal metabolic rate, energy expenditure, and temperature regulation. Individuals trying to lose weight experience a decrease in thyroid hormone, both the prohormone, thyroxine (T4), and the bioactive form, triiodothyronine (T3) (Katzeff et al. 1990). This is accompanied by decreases in leptin levels, energy expenditure, and sympathetic nervous system tone (Rosenbaum et al. 2002, 2005). These effects are counterproductive to weight loss maintenance. Thyroid hormone replacement alleviates these effects but has ramifications including muscle wasting, atrial fibrillation, and osteoporosis. Understanding the molecular mechanisms that regulate thyroid hormone will uncover therapeutic targets that alleviate metabolic disorders and assist in weight loss maintenance.

In rodents, a similar physiology exists where T4 and T3 are suppressed following 24–48 h of fasting along with reduced energy expenditure (Ahima et al. 1996; Connors et al. 1985; Legradi et al. 1997). Indeed, the entire hypothalamic-pituitary-thyroid axis is suppressed in fasted rodents including thyrotropin-releasing hormone (Trh) mRNA in the paraventricular nucleus of the hypothalamus and thyroid-stimulating hormone (TSH) in the pituitary (Blake et al. 1991, 1992; Spencer et al. 1983). In previous studies, melanocortin and neuropeptide Y (NPY) signals are important for communicating fasting signals to the hypothalamic-pituitary-thyroid (HPT) axis (Bjorbaek and Hollenberg 2002; Fekete et al. 2000, 2001, 2002; Legradi and Lechan 1998). Using NPY/melanocortin 4 receptor ($Npy^{-/-}Mc4r^{-/-}$) whole-body double-knockout mice, NPY was found necessary to suppress the hypothalamic-pituitary-thyroid (HPT) axis during fasting (Vella et al. 2011).

In an association study between TSH and cardiometabolic risk factors in $n = 1167$ euthyroid adolescents, TSH was found to strongly correlate with BMI, systolic blood pressure, total cholesterol, high fasting glucose, and insulin resistance (Le et al. 2016). Additionally, the free T3 to free T4 ratio was found to correlate with BMI, systolic blood pressure, triglycerides, fasting glucose, and insulin resistance suggesting that this ratio could be a useful marker of higher cardiometabolic risk. This study highlights again that although T4 and T3 are not altered with obesity in adults and adolescents, other factors that are regulated by T4 and T3 are altered including TSH. This suggests that while homeostatic T4 and T3 are normal, their physiological functions may be impaired in obesity (Fig. 10.3).

Several studies have explored the relationship between weight loss and decreased thyroid hormone levels. Indeed, in adults following a 10% body weight loss and reduced leptin levels, thyroid hormone levels, REE, sympathetic and parasympathetic nervous system tone, and TSH are all suppressed 20–30% (Katzeff et al. 1990; Rosenbaum et al. 2002, 2005). This is counterproductive to the maintenance of weight loss. Interestingly, leptin replacement restores sympathetic nervous system

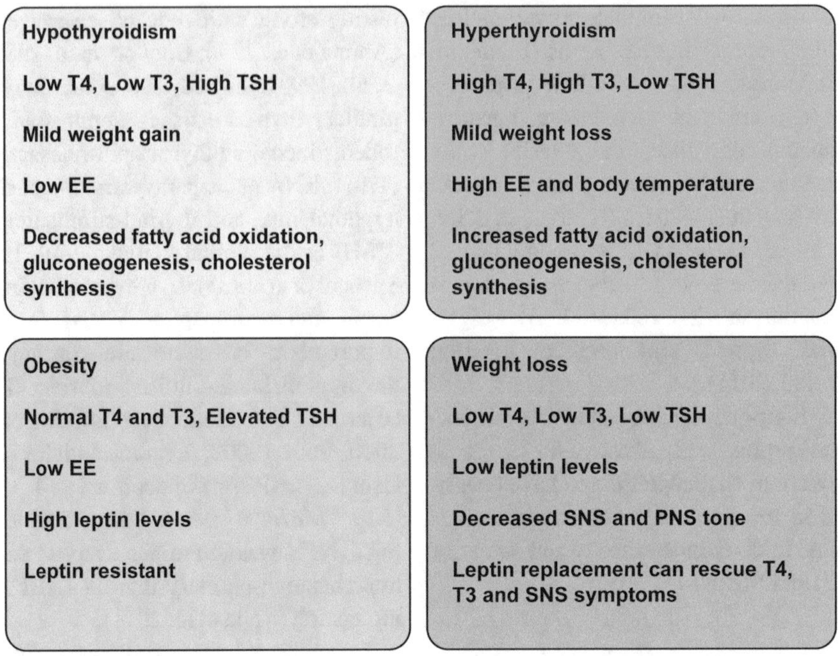

Hypothyroidism	Hyperthyroidism
Low T4, Low T3, High TSH	High T4, High T3, Low TSH
Mild weight gain	Mild weight loss
Low EE	High EE and body temperature
Decreased fatty acid oxidation, gluconeogenesis, cholesterol synthesis	Increased fatty acid oxidation, gluconeogenesis, cholesterol synthesis

Obesity	Weight loss
Normal T4 and T3, Elevated TSH	Low T4, Low T3, Low TSH
Low EE	Low leptin levels
High leptin levels	Decreased SNS and PNS tone
Leptin resistant	Leptin replacement can rescue T4, T3 and SNS symptoms

Fig. 10.3 A summary of the differences in thyroid hormone and TSH levels, body weight levels, energy expenditure, leptin levels, and thyroid hormone-responsive cellular functions like fatty acid oxidation, gluconeogenesis, and cholesterol synthesis between hypothyroidism, hyperthyroidism, obesity, and weight loss

tone and T3 and T4 levels, but not parasympathetic nervous system tone or TSH levels to pre-weight loss levels. The incomplete reversal of the weight- reduced phenotype following leptin repletion suggests that there are non-leptin-dependent mechanisms affected by weight loss. Further study is required to understand what causes the decrease in thyroid hormone levels? Is it just centrally? Or are thyroid hormone metabolic mechanisms activated in peripheral tissues? If peripheral mechanisms are activated, are they also responsive to leptin? Furthermore, how does leptin rescues thyroid hormone levels and sympathetic nervous system tone? Is this centrally through the hypothalamus or through some peripheral tissue? Studies have shown that thyroid hormone levels can still be suppressed 2–5 years following weight loss. Is leptin a long-term therapeutic for weight loss? Or will leptin resistance develop? Are there other methods for elevating thyroid hormone after weight loss?

10.4 Conclusions

Research continues to understand the relationships between thyroid hormone, body weight, and energy expenditure. Indeed there is a link between the central axis and whole body weight maintenance. Thyroid hormone has direct and indirect effects on mitochondrial function, gluconeogenesis, and fatty acid oxidation. While T3 is a poor therapeutic due to complications in the heart, muscle, and bone, thyroid hormone analogs and metabolites have promise as treatments in obesity. GC-1 and T2 have been shown to prevent hepatosteatosis and promote weight loss. While there is little evidence to support hypothyroidism as a cause of obesity, thyroid hormone does play a very important role in the maintenance of weight loss and this is directly linked to leptin. Future studies should explore this further as this would help the 75–80% of individuals who regain weight after weight loss.

Review Questions

1. Describe the mechanisms by which HPT axis is a negative feedback loop. Are there conditions when this negative feedback is disrupted (i.e., low thyroid hormone levels, low TRH and TSH)?
2. Are hypothyroidism and hyperthyroid strongly associated with large changes in body weight?
3. Thyroid hormone has diverse effects in many tissues in the body. Describe how the thyroid hormone works in the mitochondria and how this affects energy expenditure.
4. Weight loss in humans results in decreases in thyroid hormone levels among others. Is there a rodent model in which this can be studied?
5. What are the detrimental effects of supplementing with T3? What are the beneficial effects?

References

Abel, E. D., Ahima, R. S., Boers, M. E., Elmquist, J. K., & Wondisford, F. E. (2001). Critical role for thyroid hormone receptor beta2 in the regulation of paraventricular thyrotropin-releasing hormone neurons. *The Journal of Clinical Investigation, 107*, 1017–1023.

Ahima, R. S., Prabakaran, D., Mantzoros, C., Qu, D., Lowell, B., Maratos-Flier, E., & Flier, J. S. (1996). Role of leptin in the neuroendocrine response to fasting. *Nature, 382*, 250–252.

Alland, L., Muhle, R., Hou, H., Jr., Potes, J., Chin, L., Schreiber-Agus, N., & DePinho, R. A. (1997). Role for N-CoR and histone deacetylase in Sin3-mediated transcriptional repression. *Nature, 387*, 49–55.

Antonelli, A., Fallahi, P., Ferrari, S. M., Di Domenicantonio, A., Moreno, M., Lanni, A., & Goglia, F. (2011). 3,5-diiodo-L-thyronine increases resting metabolic rate and reduces body weight without undesirable side effects. *Journal of Biological Regulators and Homeostatic Agents, 25*, 655–660.

Astapova, I., Lee, L. J., Morales, C., Tauber, S., Bilban, M., & Hollenberg, A. N. (2008). The nuclear corepressor, NCoR, regulates thyroid hormone action in vivo. *Proceedings of the National Academy of Sciences of the United States of America, 105*, 19544–19549.

Astapova, I., et al. (2011). The nuclear receptor corepressor (NCoR) controls thyroid hormone sensitivity and the set point of the hypothalamic-pituitary-thyroid axis. *Molecular Endocrinology (Baltimore, MD), 25*, 212–224.

Barker, S. B. (1951). Mechanism of action of the thyroid hormone. *Physiological Reviews, 31*, 205–243.

Baxter, J. D., & Webb, P. (2009). Thyroid hormone mimetics: Potential applications in atherosclerosis, obesity and type 2 diabetes. *Nature Reviews. Drug Discovery, 8*, 308–320. https://doi.org/10.1038/nrd2830.

Beck-Peccoz, P., Amr, S., Menezes-Ferreira, M. M., Faglia, G., & Weintraub, B. D. (1985). Decreased receptor binding of biologically inactive thyrotropin in central hypothyroidism. Effect of treatment with thyrotropin-releasing hormone. *The New England Journal of Medicine, 312*, 1085–1090. https://doi.org/10.1056/NEJM198504253121703.

Benvenga, S., Cahnmann, H. J., Rader, D., Kindt, M., Facchiano, A., & Robbins, J. (1994). Thyroid hormone binding to isolated human apolipoproteins A-II, C-I, C-II, and C-III: Homology in thyroxine binding sites. *Thyroid, 4*, 261–267. https://doi.org/10.1089/thy.1994.4.261.

Bjorbaek, C., & Hollenberg, A. N. (2002). Leptin and melanocortin signaling in the hypothalamus. *Vitamins and Hormones, 65*, 281–311.

Blake, N. G., Eckland, D. J., Foster, O. J., & Lightman, S. L. (1991). Inhibition of hypothalamic thyrotropin-releasing hormone messenger ribonucleic acid during food deprivation. *Endocrinology, 129*, 2714–2718.

Blake, N. G., Johnson, M. R., Eckland, D. J., Foster, O. J., & Lightman, S. L. (1992). Effect of food deprivation and altered thyroid status on the hypothalamic-pituitary-thyroid axis in the rat. *The Journal of Endocrinology, 133*, 183–188.

Bobek, S., Sechman, A., Niezgoda, J., & Jacek, T. (2002). Reverse 3,3′,5′-triiodothyronine suppresses increase in free fatty acids in chickens elicited by dexamethasone or adrenaline. *Journal of Veterinary Medicine. A, Physiology, Pathology, Clinical Medicine, 49*, 121–124.

Bochukova, E., et al. (2012). A mutation in the thyroid hormone receptor alpha gene. *The New England Journal of Medicine, 366*, 243–249. https://doi.org/10.1056/NEJMoa1110296.

Bowers, C. R., Redding, T. W., & Schally, A. V. (1965). Effect of thyrotropin releasing factor (TRF) of ovine, bovine, porcine and human origin on thyrotropin release in vitro and in vivo. *Endocrinology, 77*, 609–616. https://doi.org/10.1210/endo-77-4-609.

Brent, G. A. (2012). Mechanisms of thyroid hormone action. *The Journal of Clinical Investigation, 122*, 3035–3043. https://doi.org/10.1172/JCI60047.

Burch, H. B., & Wartofsky, L. (1993). Life-threatening thyrotoxicosis. *Thyroid storm Endocrinol Metab Clin North Am, 22*, 263–277.

Cable, E. E., et al. (2009). Reduction of hepatic steatosis in rats and mice after treatment with a liver-targeted thyroid hormone receptor agonist. *Hepatology, 49*, 407–417. https://doi.org/10.1002/hep.22572.

Celi, F. S., et al. (2011). Metabolic effects of liothyronine therapy in hypothyroidism: A randomized, double-blind, crossover trial of liothyronine versus levothyroxine. *The Journal of Clinical Endocrinology and Metabolism, 96*, 3466–3474. https://doi.org/10.1210/jc.2011-1329.

Chiamolera, M. I., & Wondisford, F. E. (2009). Minireview: Thyrotropin-releasing hormone and the thyroid hormone feedback mechanism. *Endocrinology, 150*, 1091–1096.

Connors, J. M., DeVito, W. J., & Hedge, G. A. (1985). Effects of food deprivation on the feedback regulation of the hypothalamic-pituitary-thyroid axis of the rat. *Endocrinology, 117*, 900–906.

Costa-e-Sousa, R. H., & Hollenberg, A. N. (2012). Minireview: The neural regulation of the hypothalamic-pituitary-thyroid axis. *Endocrinology, 153*, 4128–4135.

Cypess, A. M., et al. (2009). Identification and importance of brown adipose tissue in adult humans. *The New England Journal of Medicine, 360*, 1509–1517. https://doi.org/10.1056/NEJMoa0810780.

Du Bois, E. F. (1936). *Basal metabolism in health and disease* (3rd ed.). Philadelphia: Lea & Febiger.

Erion, M. D., et al. (2007). Targeting thyroid hormone receptor-beta agonists to the liver reduces cholesterol and triglycerides and improves the therapeutic index. *Proceedings of the National Academy of Sciences of the United States of America, 104*, 15490–15495. https://doi.org/10.1073/pnas.0702759104.

Farwell, A. P., Lynch, R. M., Okulicz, W. C., Comi, A. M., & Leonard, J. L. (1990). The actin cytoskeleton mediates the hormonally regulated translocation of type II iodothyronine 5′-deiodinase in astrocytes. *The Journal of Biological Chemistry, 265*, 18546–18553.

Fekete, C., et al. (2001). Neuropeptide Y has a central inhibitory action on the hypothalamic-pituitary-thyroid axis. *Endocrinology, 142*, 2606–2613.

Fekete, C., et al. (2000). alpha-Melanocyte-stimulating hormone is contained in nerve terminals innervating thyrotropin-releasing hormone-synthesizing neurons in the hypothalamic paraventricular nucleus and prevents fasting-induced suppression of prothyrotropin-releasing hormone gene expression. *The Journal of Neuroscience, 20*, 1550–1558.

Fekete, C., Sarkar, S., Rand, W. M., Harney, J. W., Emerson, C. H., Bianco, A. C., & Lechan, R. M. (2002). Agouti-related protein (AGRP) has a central inhibitory action on the hypothalamic-pituitary-thyroid (HPT) axis; comparisons between the effect of AGRP and neuropeptide Y on energy homeostasis and the HPT axis. *Endocrinology, 143*, 3846–3853.

Flores-Morales, A., Gullberg, H., Fernandez, L., Stahlberg, N., Lee, N. H., Vennstrom, B., & Norstedt, G. (2002). Patterns of liver gene expression governed by TRbeta. *Molecular Endocrinology (Baltimore, MD), 16*, 1257–1268. https://doi.org/10.1210/mend.16.6.0846.

Forrest, D., Erway, L. C., Ng, L., Altschuler, R., & Curran, T. (1996). Thyroid hormone receptor beta is essential for development of auditory function. *Nature Genetics, 13*, 354–357. https://doi.org/10.1038/ng0796-354.

Forrest, D., & Vennstrom, B. (2000). Functions of thyroid hormone receptors in mice. *Thyroid, 10*, 41–52. https://doi.org/10.1089/thy.2000.10.41.

Friesema, E. C., Jansen, J., & Visser, T. J. (2005). Thyroid hormone transporters. *Biochemical Society Transactions, 33*, 228–232. https://doi.org/10.1042/BST0330228.

Friesema, E. C., Visser, W. E., & Visser, T. J. (2010). Genetics and phenomics of thyroid hormone transport by MCT8. *Molecular and Cellular Endocrinology, 322*, 107–113. https://doi.org/10.1016/j.mce.2010.01.016.

Geras, E. J., & Gershengorn, M. C. (1982). Evidence that TRH stimulates secretion of TSH by two calcium-mediated mechanisms. *The American Journal of Physiology, 242*, E109–E114.

Gereben, B., McAninch, E. A., Ribeiro, M. O., & Bianco, A. C. (2015). Scope and limitations of iodothyronine deiodinases in hypothyroidism. *Nature Reviews. Endocrinology, 11*, 642–652. https://doi.org/10.1038/nrendo.2015.155.

Giudetti, A. M., Leo, M., Siculella, L., & Gnoni, G. V. (2006). Hypothyroidism down-regulates mitochondrial citrate carrier activity and expression in rat liver. *Biochimica et Biophysica Acta, 1761*, 484–491. https://doi.org/10.1016/j.bbalip.2006.03.021.

Goldstein, J. L., DeBose-Boyd, R. A., & Brown, M. S. (2006). Protein sensors for membrane sterols. *Cell, 124*, 35–46. https://doi.org/10.1016/j.cell.2005.12.022.

Grant, N. (2007). The role of triiodothyronine-induced substrate cycles in the hepatic response to overnutrition: Thyroid hormone as an antioxidant. *Medical Hypotheses, 68*, 641–649. https://doi.org/10.1016/j.mehy.2006.07.045.

Grasselli, E., Canesi, L., Voci, A., De Matteis, R., Demori, I., Fugassa, E., & Vergani, L. (2008). Effects of 3,5-diiodo-L-thyronine administration on the liver of high fat diet-fed rats. *Experimental Biology and Medicine (Maywood, N.J.), 233*, 549–557. https://doi.org/10.3181/0710-RM-266.

Greer, M. A. (1951). Evidence of hypothalamic control of the pituitary release of thyrotropin. *Proceedings of the Society for Experimental Biology and Medicine, 77*, 603–608.

Guillemin, R., Sakiz, E., & Ward, D. N. (1965). Further purification of Tsh-releasing factor (Trf) from sheep hypothalamic tissues, with observations on the amino acid composition. *Proceedings of the Society for Experimental Biology and Medicine, 118*, 1132–1137.

Guillemin, R., Yamazaki, E., Gard, D. A., Jutisz, M., & Sakiz, E. (1963). In vitro secretion of thyrotropin (Tsh): Stimulation by a hypothalamic peptide (Trf). *Endocrinology, 73*, 564–572. https://doi.org/10.1210/endo-73-5-564.

Gullberg, H., Rudling, M., Forrest, D., Angelin, B., & Vennstrom, B. (2000). Thyroid hormone receptor beta-deficient mice show complete loss of the normal cholesterol 7alpha-hydroxylase (CYP7A) response to thyroid hormone but display enhanced resistance to dietary cholesterol. *Molecular Endocrinology (Baltimore, MD), 14*, 1739–1749. https://doi.org/10.1210/mend.14.11.0548.

Gullberg, H., Rudling, M., Salto, C., Forrest, D., Angelin, B., & Vennstrom, B. (2002). Requirement for thyroid hormone receptor beta in T3 regulation of cholesterol metabolism in mice. *Molecular Endocrinology (Baltimore, MD), 16*, 1767–1777. https://doi.org/10.1210/me.2002-0009.

Haber, R. S., Ismail-Beigi, F., & Loeb, J. N. (1988). Time course of na,k transport and other metabolic responses to thyroid hormone in clone 9 cells. *Endocrinology, 123*, 238–247. https://doi.org/10.1210/endo-123-1-238.

Halachmi, S., Marden, E., Martin, G., MacKay, H., Abbondanza, C., & Brown, M. (1994). Estrogen receptor-associated proteins: Possible mediators of hormone-induced transcription. *Science (New York, NY), 264*, 1455–1458.

Heijlen, M., et al. (2014). Knockdown of type 3 iodothyronine deiodinase severely perturbs both embryonic and early larval development in zebrafish. *Endocrinology, 155*, 1547–1559. https://doi.org/10.1210/en.2013-1660.

Heinzel, T., et al. (1997). A complex containing N-CoR, mSin3 and histone deacetylase mediates transcriptional repression. *Nature, 387*, 43–48.

Hu, X., & Lazar, M. A. (1999). The CoRNR motif controls the recruitment of corepressors by nuclear hormone receptors. *Nature, 402*, 93–96. https://doi.org/10.1038/47069.

Jackson-Hayes, L., et al. (2003). A thyroid hormone response unit formed between the promoter and first intron of the carnitine palmitoyltransferase-Ialpha gene mediates the liver-specific induction by thyroid hormone. *The Journal of Biological Chemistry, 278*, 7964–7972. https://doi.org/10.1074/jbc.M211062200.

Jepsen, K., Gleiberman, A. S., Shi, C., Simon, D. I., & Rosenfeld, M. G. (2008). Cooperative regulation in development by SMRT and FOXP1. *Genes & Development, 22*, 740–745. https://doi.org/10.1101/gad.1637108.

Jepsen, K., et al. (2000). Combinatorial roles of the nuclear receptor corepressor in transcription and development. *Cell, 102*, 753–763.

Kaneshige, M., et al. (2001). A targeted dominant negative mutation of the thyroid hormone alpha 1 receptor causes increased mortality, infertility, and dwarfism in mice. *Proceedings of the National Academy of Sciences of the United States of America, 98*, 15095–15100. https://doi.org/10.1073/pnas.261565798.

Katzeff, H. L., Yang, M. U., Presta, E., Leibel, R. L., Hirsch, J., & Van Itallie, T. B. (1990). Calorie restriction and iopanoic acid effects on thyroid hormone metabolism. *The American Journal of Clinical Nutrition, 52*, 263–266.

Klein, I., & Danzi, S. (2007). Thyroid disease and the heart. *Circulation, 116*, 1725–1735. https://doi.org/10.1161/CIRCULATIONAHA.106.678326.

Klieverik, L. P., et al. (2009). Thyroid hormone modulates glucose production via a sympathetic pathway from the hypothalamic paraventricular nucleus to the liver. *Proceedings of the National Academy of Sciences of the United States of America, 106*, 5966–5971. https://doi.org/10.1073/pnas.0805355106.

Klieverik, L. P., Sauerwein, H. P., Ackermans, M. T., Boelen, A., Kalsbeek, A., & Fliers, E. (2008). Effects of thyrotoxicosis and selective hepatic autonomic denervation on hepatic glucose metabolism in rats. *American Journal of Physiology, 294*, E513–E520. https://doi.org/10.1152/ajpendo.00659.2007.

Kopp, P. (2005). Thyroid hormone synthesis: Thyroid iodine metabolism. In L. U. R. Braverman (Ed.), *Wegner and Ingbar's the thyroid: A fundamental and clinical text* (pp. 52–76). USA: Lippincott Williams & Wilkins.

Lanni, A., et al. (2005). 3,5-diiodo-L-thyronine powerfully reduces adiposity in rats by increasing the burning of fats. *The FASEB Journal, 19*, 1552–1554. https://doi.org/10.1096/fj.05-3977fje.

Lazar, M. A. (1993). Thyroid hormone receptors: Multiple forms, multiple possibilities. *Endocrine Reviews, 14*, 184–193.

Le, T. N., Celi, F. S., & Wickham, E. P., 3rd. (2016). Thyrotropin levels are associated with cardiometabolic risk factors in euthyroid adolescents. *Thyroid, 26*, 1441–1449. https://doi.org/10.1089/thy.2016.0055.

Lechan, R. M., & Hollenberg, A. N. (2003). Thyrotropin-releasing hormone (TRH). In H. L. Henry & A. W. Norman (Eds.), *Encyclopedia of hormones* (pp. 510–524). New York: Elsevier Science.

Lechan, R. M., Wu, P., Jackson, I. M., Wolf, H., Cooperman, S., Mandel, G., & Goodman, R. H. (1986). Thyrotropin-releasing hormone precursor: Characterization in rat brain. *Science (New York, NY), 231*, 159–161.

Legradi, G., Emerson, C. H., Ahima, R. S., Flier, J. S., & Lechan, R. M. (1997). Leptin prevents fasting-induced suppression of prothyrotropin-releasing hormone messenger ribonucleic acid in neurons of the hypothalamic paraventricular nucleus. *Endocrinology, 138*, 2569–2576.

Legradi, G., & Lechan, R. M. (1998). The arcuate nucleus is the major source for neuropeptide Y-innervation of thyrotropin-releasing hormone neurons in the hypothalamic paraventricular nucleus. *Endocrinology, 139*, 3262–3270.

Lombardi, A., Lanni, A., Moreno, M., Brand, M. D., & Goglia, F. (1998). Effect of 3,5-di-iodo-L-thyronine on the mitochondrial energy-transduction apparatus. *The Biochemical Journal, 330*(Pt 1), 521–526.

Lonard, D. M., & O'Malley, B. W. (2007). Nuclear receptor coregulators: Judges, juries, and executioners of cellular regulation. *Molecular Cell, 27*, 691–700. https://doi.org/10.1016/j.molcel.2007.08.012.

Lopez, D., Abisambra Socarras, J. F., Bedi, M., & Ness, G. C. (2007). Activation of the hepatic LDL receptor promoter by thyroid hormone. *Biochimica et*

Biophysica Acta, 1771, 1216–1225. https://doi.org/10.1016/j.bbalip.2007.05.001.

Maglich, J. M., Watson, J., McMillen, P. J., Goodwin, B., Willson, T. M., & Moore, J. T. (2004). The nuclear receptor CAR is a regulator of thyroid hormone metabolism during caloric restriction. *The Journal of Biological Chemistry, 279,* 19832–19838.

Mangus-Levy, A. (1895). Uber den respiratorischen Gaswechsel unter dem Einfluss der Thyroiden sowie unter verschiedenen physiologischen. *Zustanden Berlin klinische Wochenschrift, 32,* 650–652.

Martin, J. B., Boshans, R., & Reichlin, S. (1970). Feedback regulation of TSH secretion in rats with hypothalamic lesions. *Endocrinology, 87,* 1032–1040. https://doi.org/10.1210/endo-87-5-1032.

Menezes-Ferreira, M. M., Petrick, P. A., & Weintraub, B. D. (1986). Regulation of thyrotropin (TSH) bioactivity by TSH-releasing hormone and thyroid hormone. *Endocrinology, 118,* 2125–2130. https://doi.org/10.1210/endo-118-5-2125.

Mittag, J., et al. (2013). Thyroid hormone is required for hypothalamic neurons regulating cardiovascular functions. *The Journal of Clinical Investigation, 123,* 509–516. https://doi.org/10.1172/JCI65252.

Mol, J. A., & Visser, T. J. (1985). Rapid and selective inner ring deiodination of thyroxine sulfate by rat liver deiodinase. *Endocrinology, 117,* 8–12.

Mollica, M. P., et al. (2009). 3,5-diiodo-l-thyronine, by modulating mitochondrial functions, reverses hepatic fat accumulation in rats fed a high-fat diet. *Journal of Hepatology, 51,* 363–370. https://doi.org/10.1016/j.jhep.2009.03.023.

Moreno, M., Lanni, A., Lombardi, A., & Goglia, F. (1997). How the thyroid controls metabolism in the rat: Different roles for triiodothyronine and diiodothyronines. *The Journal of Physiology, 505*(Pt 2), 529–538.

Moreno, M., et al. (2011). 3,5-Diiodo-L-thyronine prevents high-fat-diet-induced insulin resistance in rat skeletal muscle through metabolic and structural adaptations. *The FASEB Journal, 25,* 3312–3324. https://doi.org/10.1096/fj.11-181982.

Mullur, R., Liu, Y. Y., & Brent, G. A. (2014). Thyroid hormone regulation of metabolism. *Physiological Reviews, 94,* 355–382. https://doi.org/10.1152/physrev.00030.2013.

Nagy, L., et al. (1997). Nuclear receptor repression mediated by a complex containing SMRT, mSin3A, and histone deacetylase. *Cell, 89,* 373–380.

Nagy, L., et al. (1999). Mechanism of corepressor binding and release from nuclear hormone receptors. *Genes & Development, 13,* 3209–3216.

Ng, L., Cordas, E., Wu, X., Vella, K. R., Hollenberg, A. N., & Forrest, D. (2015). Age-related hearing loss and degeneration of Cochlear hair cells in mice lacking thyroid hormone receptor beta1. *Endocrinology, 156,* 3853–3865. https://doi.org/10.1210/en.2015-1468.

Ng, L., et al. (2004). Hearing loss and retarded cochlear development in mice lacking type 2 iodothyronine deiodinase. *Proceedings of the National Academy of Sciences of the United States of America, 101,* 3474–3479. https://doi.org/10.1073/pnas.0307402101.

Ng, L., et al. (2009). A protective role for type 3 deiodinase, a thyroid hormone-inactivating enzyme, in cochlear development and auditory function. *Endocrinology, 150,* 1952–1960. https://doi.org/10.1210/en.2008-1419.

Ng, L., et al. (2001). A thyroid hormone receptor that is required for the development of green cone photoreceptors. *Nature Genetics, 27,* 94–98. https://doi.org/10.1038/83829.

Nikrodhanond, A. A., et al. (2006). Dominant role of thyrotropin-releasing hormone in the hypothalamic-pituitary-thyroid axis. *The Journal of Biological Chemistry, 281,* 5000–5007.

Onate, S. A., Tsai, S. Y., Tsai, M. J., & O'Malley, B. W. (1995). Sequence and characterization of a coactivator for the steroid hormone receptor superfamily. *Science (New York, NY), 270,* 1354–1357.

Paradies, G., & Ruggiero, F. M. (1990). Enhanced activity of the tricarboxylate carrier and modification of lipids in hepatic mitochondria from hyperthyroid rats. *Archives of Biochemistry and Biophysics, 278,* 425–430.

Paradies, G., Ruggiero, F. M., Petrosillo, G., & Quagliariello, E. (1996). Stimulation of carnitine acyl-carnitine translocase activity in heart mitochondria from hyperthyroid rats. *FEBS Letters, 397,* 260–262.

Paradies, G., Ruggiero, F. M., Petrosillo, G., & Quagliariello, E. (1997). Alterations in carnitine-acylcarnitine translocase activity and in phospholipid composition in heart mitochondria from hypothyroid rats. *Biochimica et Biophysica Acta, 1362,* 193–200.

Park, E. A., Song, S., Vinson, C., & Roesler, W. J. (1999). Role of CCAAT enhancer-binding protein beta in the thyroid hormone and cAMP induction of phosphoenolpyruvate carboxykinase gene transcription. *The Journal of Biological Chemistry, 274,* 211–217.

Perello, M., Friedman, T., Paez-Espinosa, V., Shen, X., Stuart, R. C., & Nillni, E. A. (2006). Thyroid hormones selectively regulate the posttranslational processing of prothyrotropin-releasing hormone in the paraventricular nucleus of the hypothalamus. *Endocrinology, 147,* 2705–2716.

Perello, M., & Nillni, E. A. (2007). The biosynthesis and processing of neuropeptides: Lessons from prothyrotropin releasing hormone (proTRH). *Frontiers in Bioscience, 12,* 3554–3565.

Perissi, V., et al. (1999). Molecular determinants of nuclear receptor-corepressor interaction. *Genes & Development, 13,* 3198–3208.

Perra, A., et al. (2008). Thyroid hormone (T3) and TRbeta agonist GC-1 inhibit/reverse nonalcoholic fatty liver in rats. *The FASEB Journal, 22,* 2981–2989. https://doi.org/10.1096/fj.08-108464.

Piehl, S., Hoefig, C. S., Scanlan, T. S., & Kohrle, J. (2011). Thyronamines--past, present, and future. *Endocrine Reviews, 32,* 64–80. https://doi.org/10.1210/er.2009-0040.

Psarra, A. M., Solakidi, S., & Sekeris, C. E. (2006). The mitochondrion as a primary site of action of steroid and thyroid hormones: Presence and action of steroid and thyroid hormone receptors in mitochondria of animal cells. *Molecular and Cellular Endocrinology, 246*, 21–33. https://doi.org/10.1016/j.mce.2005.11.025.

Qatanani, M., Zhang, J., & Moore, D. D. (2005). Role of the constitutive androstane receptor in xenobiotic-induced thyroid hormone metabolism. *Endocrinology, 146*, 995–1002.

Refetoff, S., Weiss, R. E., & Usala, S. J. (1993). The syndromes of resistance to thyroid hormone. *Endocrine Reviews, 14*, 348–399.

Rodgers, J. T., Lerin, C., Gerhart-Hines, Z., & Puigserver, P. (2008). Metabolic adaptations through the PGC-1 alpha and SIRT1 pathways. *FEBS Letters, 582*, 46–53. https://doi.org/10.1016/j.febslet.2007.11.034.

Rosenbaum, M., et al. (2005). Low-dose leptin reverses skeletal muscle, autonomic, and neuroendocrine adaptations to maintenance of reduced weight. *The Journal of Clinical Investigation, 115*, 3579–3586.

Rosenbaum, M., Murphy, E. M., Heymsfield, S. B., Matthews, D. E., & Leibel, R. L. (2002). Low dose leptin administration reverses effects of sustained weight-reduction on energy expenditure and circulating concentrations of thyroid hormones. *The Journal of Clinical Endocrinology and Metabolism, 87*, 2391–2394.

Santillo, A., Burrone, L., Falvo, S., Senese, R., Lanni, A., & Chieffi Baccari, G. (2013). Triiodothyronine induces lipid oxidation and mitochondrial biogenesis in rat Harderian gland. *The Journal of Endocrinology, 219*, 69–78. https://doi.org/10.1530/JOE-13-0127.

Schally, A. V., Redding, T. W., Bowers, C. Y., & Barrett, J. F. (1969). Isolation and properties of porcine thyrotropin-releasing hormone. *The Journal of Biological Chemistry, 244*, 4077–4088.

Schaner, P., Todd, R. B., Seidah, N. G., & Nillni, E. A. (1997). Processing of prothyrotropin-releasing hormone by the family of prohormone convertases. *The Journal of Biological Chemistry, 272*, 19958–19968.

Segerson, T. P., Kauer, J., Wolfe, H. C., Mobtaker, H., Wu, P., Jackson, I. M., & Lechan, R. M. (1987). Thyroid hormone regulates TRH biosynthesis in the paraventricular nucleus of the rat hypothalamus. *Science (New York, NY), 238*, 78–80.

Shin, D. J., & Osborne, T. F. (2003). Thyroid hormone regulation and cholesterol metabolism are connected through sterol regulatory element-binding Protein-2 (SREBP-2). *The Journal of Biological Chemistry, 278*, 34114–34118. https://doi.org/10.1074/jbc.M305417200.

Shupnik, M. A., Weck, J., & Hinkle, P. M. (1996). Thyrotropin (TSH)-releasing hormone stimulates TSH beta promoter activity by two distinct mechanisms involving calcium influx through L type Ca2+ channels and protein kinase C. *Molecular Endocrinology (Baltimore, MD), 10*, 90–99. https://doi.org/10.1210/mend.10.1.8838148.

Siegrist-Kaiser, C. A., Juge-Aubry, C., Tranter, M. P., Ekenbarger, D. M., & Leonard, J. L. (1990). Thyroxine-dependent modulation of actin polymerization in cultured astrocytes. A novel, extranuclear action of thyroid hormone. *The Journal of Biological Chemistry, 265*, 5296–5302.

Silva, J. E. (2006). Thermogenic mechanisms and their hormonal regulation. *Physiological Reviews, 86*, 435–464.

Spencer, C. A., Lum, S. M., Wilber, J. F., Kaptein, E. M., & Nicoloff, J. T. (1983). Dynamics of serum thyrotropin and thyroid hormone changes in fasting. *The Journal of Clinical Endocrinology and Metabolism, 56*, 883–888.

St Germain, D. L., Galton, V. A., & Hernandez, A. (2009). Minireview: Defining the roles of the iodothyronine deiodinases: Current concepts and challenges. *Endocrinology, 150*, 1097–1107. https://doi.org/10.1210/en.2008-1588.

Sugrue, M. L., Vella, K. R., Morales, C., Lopez, M. E., & Hollenberg, A. N. (2010). The thyrotropin-releasing hormone gene is regulated by thyroid hormone at the level of transcription in vivo. *Endocrinology, 151*, 793–801.

Taylor, T., Gesundheit, N., & Weintraub, B. D. (1986). Effects of in vivo bolus versus continuous TRH administration on TSH secretion, biosynthesis, and glycosylation in normal and hypothyroid rats. *Molecular and Cellular Endocrinology, 46*, 253–261.

Thijssen-Timmer, D. C., Schiphorst, M. P., Kwakkel, J., Emter, R., Kralli, A., Wiersinga, W. M., & Bakker, O. (2006). PGC-1alpha regulates the isoform mRNA ratio of the alternatively spliced thyroid hormone receptor alpha transcript. *Journal of Molecular Endocrinology, 37*, 251–257. https://doi.org/10.1677/jme.1.01914.

Thompson, G. R., Soutar, A. K., Spengel, F. A., Jadhav, A., Gavigan, S. J., & Myant, N. B. (1981). Defects of receptor-mediated low density lipoprotein catabolism in homozygous familial hypercholesterolemia and hypothyroidism in vivo. *Proceedings of the National Academy of Sciences of the United States of America, 78*, 2591–2595.

Trost, S. U., et al. (2000). The thyroid hormone receptor-beta-selective agonist GC-1 differentially affects plasma lipids and cardiac activity. *Endocrinology, 141*, 3057–3064. https://doi.org/10.1210/endo.141.9.7681.

Vaitkus, J. A., Farrar, J. S., & Celi, F. S. (2015). Thyroid hormone mediated modulation of energy expenditure. *International Journal of Molecular Sciences, 16*, 16158–16175. https://doi.org/10.3390/ijms160716158.

van Marken Lichtenbelt, W. D., et al. (2009). Cold-activated brown adipose tissue in healthy men. *The New England Journal of Medicine, 360*, 1500–1508. https://doi.org/10.1056/NEJMoa0808718.

Vansell, N. R., & Klaassen, C. D. (2001). Increased biliary excretion of thyroxine by microsomal enzyme inducers. *Toxicology and Applied Pharmacology, 176*, 187–194.

Vansell, N. R., & Klaassen, C. D. (2002). Increase in rat liver UDP-glucuronosyltransferase mRNA by microsomal enzyme inducers that enhance thyroid hormone glucuronidation. *Drug Metabolism and Disposition, 30*, 240–246.

Vella, K. R., & Hollenberg, A. N. (2009). The ups and downs of thyrotropin-releasing hormone. *Endocrinology, 150*, 2021–2023.

Vella, K. R., et al. (2014). Thyroid hormone signaling in vivo requires a balance between coactivators and corepressors. *Molecular and Cellular Biology, 34*, 1564–1575. https://doi.org/10.1128/MCB.00129-14.

Vella, K. R. et al. (2011) NPY and MC4R signaling regulate thyroid hormone levels during fasting through both central and peripheral pathways Cell Metabolism 14:780-790 doi:S1550–4131(11)00403–7 [pii] https://doi.org/10.1016/j.cmet.2011.126009.

Villicev, C. M., et al. (2007). Thyroid hormone receptor beta-specific agonist GC-1 increases energy expenditure and prevents fat-mass accumulation in rats. *The Journal of Endocrinology, 193*, 21–29. https://doi.org/10.1677/joe.1.07066.

Visser, T. J. (1996). Pathways of thyroid hormone metabolism. *Acta Medica Austriaca, 23*, 10–16.

Weiss, R. E., Gehin, M., Xu, J., Sadow, P. M., O'Malley, B. W., Chambon, P., & Refetoff, S. (2002). Thyroid function in mice with compound heterozygous and homozygous disruptions of SRC-1 and TIF-2 coactivators: Evidence for haploinsufficiency. *Endocrinology, 143*, 1554–1557.

Weiss, R. E., Xu, J., Ning, G., Pohlenz, J., O'Malley, B. W., & Refetoff, S. (1999). Mice deficient in the steroid receptor co-activator 1 (SRC-1) are resistant to thyroid hormone. *The EMBO Journal, 18*, 1900–1904.

Wrutniak, C., et al. (1995). A 43-kDa protein related to c-Erb A alpha 1 is located in the mitochondrial matrix of rat liver. *The Journal of Biological Chemistry, 270*, 16347–16354.

Wulf, A., Harneit, A., Kroger, M., Kebenko, M., Wetzel, M. G., & Weitzel, J. M. (2008). T3-mediated expression of PGC-1alpha via a far upstream located thyroid hormone response element. *Molecular and Cellular Endocrinology, 287*, 90–95. https://doi.org/10.1016/j.mce.2008.01.017.

Yamada, M., et al. (1995). Activation of the thyrotropin-releasing hormone (TRH) receptor by a direct precursor of TRH, TRH-Gly. *Neuroscience Letters, 196*, 109–112.

Yamada, M., et al. (1997). Tertiary hypothyroidism and hyperglycemia in mice with targeted disruption of the thyrotropin-releasing hormone gene. *Proceedings of the National Academy of Sciences of the United States of America, 94*, 10862–10867.

Yang, Z., & Privalsky, M. L. (2001). Isoform-specific transcriptional regulation by thyroid hormone receptors: Hormone-independent activation operates through a steroid receptor mode of co-activator interaction. *Molecular Endocrinology (Baltimore, MD), 15*, 1170–1185. https://doi.org/10.1210/mend.15.7.0656.

Obesity and Stress: The Melanocortin Connection

11

Sara Singhal and Jennifer W. Hill

The role of the melanocortin system in energy homeostasis, feeding behavior, and metabolism has been a focus of intense study since its discovery in 1979 (Crine et al. 1979). The ability of melanocortins to suppress feeding and increase energy expenditure has made melanocortin receptors (MCRs) a major target of anti-obesity drugs in development (Fani et al. 2014). In addition, the melanocortin system's influence on circulating glucose levels suggests it could also be targeted to treat obesity-related type 2 diabetes (Morgan et al. 2015; Parton et al. 2007). While very promising in theory, problematic side effects have plagued pharmaceutical trials for such medications, preventing FDA approval (Ericson et al. 2017). These adverse effects are due to other systemic and central functions of the melanocortin system. To understand and overcome these challenges, a more comprehensive understanding is needed of the role melanocortin peptides play and how they perform their diverse functions.

The melanocortin system can coordinate a wide variety of behavioral and physiological responses to internal and environmental cues. The number of known roles that melanocortins play continues to proliferate, ranging from the control of adrenal function, pain, and inflammation to surprising behavioral outputs such as grooming. As we will see, an organism's need to respond to stressors may be the most useful context for understanding the actions of this system. The ability to rank-order threats is critical to survival. Melanocortins play a critical role in enabling "fight or flight" responses to immediate danger. Later, endogenous opioids, AgRP/NPY circuitry, and other systems permit animals to focus on recovery, obtaining food to restore energy reserves, and activities of lesser importance. The interplay between these systems allows the animal to deal successfully with most stressors and return to physiological equilibrium.

This chapter will review the variety of roles played by melanocortins in the response to stress using insights from evolutionary development to understand their integration. Finally, we will discuss the lessons for obesity prevention and treatment arising from a holistic view of the actions of melanocortins.

11.1 The Melanocortin System: Proopiomelanocortin

Melanocortin peptides are generated from the polypeptide proopiomelanocortin (POMC) via successive posttranslational cleavage events. POMC is abundantly expressed in the pituitary and hypothalamus but also in other sites; its processing varies between tissues (Chen et al. 1986;

S. Singhal · J. W. Hill (✉)
University of Toledo, Toledo, OH, USA
e-mail: JenniferW.Hill@utoledo.edu

© Springer International Publishing AG, part of Springer Nature 2018
E. A. Nillni (ed.), *Textbook of Energy Balance, Neuropeptide Hormones,
and Neuroendocrine Function*, https://doi.org/10.1007/978-3-319-89506-2_11

Smith and Funder 1988; Mechanick et al. 1992; Forman and Bagasra 1992; Hummel and Zuhlke 1994; Ottaviani et al. 1997; Tsatmali et al. 2000; Iqbal et al. 2010; Alam et al. 2012). Generally, POMC is cleaved to form β-LPH and pro-ACTH, which prohormone convertase (PC)1/3 cleaves to form ACTH1–39. ACTH1–39 is the main product of corticotrophs in the pituitary, but in the hypothalamus, PC2 cleaves it to form ACTH1–17. Carboxypeptidase E (CPE) then removes the C-terminal basic residues to produce ACTH1–13. The C-terminus of ACTH1–13 is then amidated by peptidyl α-amidating monooxygenase (PAM) to create ACTH(1–3)NH2, also known as desacetyl α-MSH. In humans, an additional N-terminal cleavage site results in production of β-MSH, γ –MSH, and α-MSH (Pritchard et al. 2002). PC2 also cleaves β-LPH to form the endogenous opioid β-endorphin1–31. This posttranslational processing of the POMC preprohormone has been remarkably well conserved (Vallarino et al. 2012).

The genetic sequence for POMC appears across many species, from the earliest vertebrates such as lampreys to mammals. All the sequences have shown the same structural organization, suggesting that POMC was present in common ancestors 5–700 million years ago (Heinig et al. 1995). In the sea lamprey, separate genes named proopiocortin (POC) and proopiomelanotropin (POM) produce ACTH and MSH, respectively. Even invertebrates such as the leech have POMC-related sequences possessing over 80% homology in its melanocortin domain (Duvaux-Miret and Capron 1992; Salzet et al. 1997; Stefano et al. 1999). Indeed, tetrapods, mussel, and leech have the same sequentially arranged hormonal segments of this gene (Kawauchi and Sower 2006). It appears that α-, β-, and γ-MSH arose during the early evolution of invertebrates from intramolecular duplication of an ancestral MSH.

In the rodent brain, there are two recognized neuronal populations expressing POMC, although low levels of POMC mRNA have been reported in other CNS regions (Zhou et al. 2013). The largest population resides in the arcuate nucleus of the hypothalamus (ARC) and co-expresses the cocaine amphetamine-related transcript (CART) peptide (Elias et al. 1998). A second smaller population is located in the brain stem, in the nucleus of the solitary tract (NTS) (Khachaturian et al. 1986).

The melanocortin receptor family is unique in having its activity regulated by both agonists and antagonists. Two naturally occurring antagonists to MCRs exist, agouti and AgRP. Mice express agouti protein primarily in the skin where it influences pigmentation (Bultman et al. 1992). The human homolog, agouti signaling protein (ASIP), may also regulate pigmentation (Voisey et al. 2003). ASIP expression has been found in the skin and other tissues including the heart, ovary, testis, foreskin, adipose tissue, liver, and kidney (Wilson et al. 1995).

AgRP was discovered based on its sequence homology to agouti (Ollmann et al. 1997; Shutter et al. 1997). In contrast to that protein, AgRP is mainly expressed in the adrenal gland and arcuate nucleus of the hypothalamus (ARC). AgRP-producing neurons co-express two generally inhibitory neurotransmitters: neuropeptide Y (NPY) (Broberger et al. 1998; Hahn et al. 1998) and γ-aminobutyric acid (GABA) (Wu and Palmiter 2011). POMC neurons receive direct input from these NPY/AgRP neurons (Cowley et al. 2001; Atasoy et al. 2012; Tong et al. 2008; Smith et al. 2007) and other neurons inhibited by AgRP (Corander et al. 2011; Tolle and Low 2008). The mammalian central melanocortin system is defined as the neurons expressing POMC, AgRP neurons, which antagonize the effects of POMC neurons, and the downstream CNS circuits they collectively influence via MCRs.

As detailed below, POMC and AgRP neurons send and receive projections from many CNS regions (Broberger et al. 1998; Tsou et al. 1986; Palkovits et al. 1987; Zheng et al. 2005a; Haskell-Luevano et al. 1999; Bagnol et al. 1999; Schwartz 2000; Odonohue and Dorsa 1982; Cone 2005; Rinaman 2010; Magoul et al. 1993). ARC POMC neurons project most heavily throughout the hypothalamus, including to the anterior hypothalamus, medial preoptic area, medial preoptic nucleus (MPON), lateral hypothalamus, dorsomedial hypothalamus (DMH), ventromedial hypothalamic nucleus (VMH), paraventricular nucleus of the hypothalamus (PVH), parasubthalamic nucleus (PSThN), and the posterior hypothalamus (PH). Projections to the forebrain target the bed nucleus of the stria terminalis (BNST), lateral septum (LS), nucleus of the diagonal band, medial amygdala (MeA), and the nucleus accumbens (NAc). In the brain stem, the periaqueductal gray (PAG), superior colliculus, deep mesencephalic nucleus, NTS, medial lemniscus, substantia nigra, dorsal raphe, and locus coeruleus receive projections. Finally, the spinal cord also receives input from ARC POMC neurons.

Overall, ARC POMC and AgRP neurons have similar connectivity. However, AgRP neuron projections appear to be sparser, with fewer synapses. This population projects heavily to areas from which it receives the most incoming connections. Robust projections exist to the PVH, DMH, LH, MPON, septal areas of the anterior commissure, paraventricular nucleus of the thalamus, and the LS. These neurons also send axons to the BNST, the organum vasculosum of the lamina terminalis, and the perifornical nucleus. One might expect that AgRP release directly opposes melanocortins released at these sites, but the microcircuitry involves a complex summation of both inputs. In the PVH, AgRP synapses contact cell bodies, while POMC synapses contact distal dendrites (Atasoy et al. 2012; Bouyer and Simerly 2013). Areas of volume release of AgRP and POMC, however, are likely to overlap. Finally, AgRP projections are notably absent to the brain stem, hippocampus, amygdala, corpus striatum, and olfactory cortical tract, all of which have dense POMC innervation (Palkovits et al. 1987; Bagnol et al. 1999; Watson et al. 1978; Jacobowitz and Odonohue 1978; Nilaver et al. 1979; Joseph et al. 1983).

POMC fibers of the NTS have less widespread connections, projecting sparsely to the PVH and PSThN, but more strongly within the brain stem. Targeted regions include the subcoeruleus nucleus, parvicellular reticular nucleus, medullary reticular nucleus (both dorsal and ventral), magnocellular reticular nucleus, pontine reticular nucleus, intermediate reticular nucleus, supratrigeminal nucleus, and the lateral parabrachial nucleus. Interestingly, ARC and NTS POMC neurons have reciprocal projections to each other.

11.2 Melanocortin Receptors

Five MCRs have been identified in humans, named in the order they were cloned: MC1R, MC2R, MC3R, MC4R, and MC5R (Girardet and Butler 2014). The MCRs coevolved with the POMC gene early in chordate evolution. During the multiplication of the chordates, genome duplications occurred that resulted in the ancestral MCR differentiating into an MC1/2 receptor precursor and an MC3/4 receptor precursor. (The origin of the MC5R is still debated (Cortes et al. 2014)). Evidence suggests that all MCRs responded to ACTH (and MSH) and caused release of glucocorticoids. For example, a primitive CRH-ACTH-corticosterone axis exists in the jawless hagfish (Amano et al. 2016).

Comparisons of the elephant shark, Japanese sting ray, and bony fish suggest that the MC2R gradually lost its ability to act without the melanocortin receptor 2 accessory protein (MRAP1) (Takahashi et al. 2016; Reinick et al. 2012). However, once paired with MRAP1, MC2R became the most efficient ACTH receptor, allowing the other MCRs to develop differing affinities to melanocortins and differing expression levels in tissues (Schiöth et al. 2005; Dores et al. 2014). These processes allowed unique roles for MCRs to develop without disturbing glucocorticoid production (Cone 2006; Kobayashi et al. 2012). In the lamprey, MCRs appear in the skin, liver, heart, and skeletal muscle, but not in the brain (Young 1935; Eddy and Strahan 1968). Some cartilaginous fish, which arose 450 million years ago, express α-MSH and β-endorphin in the brain and melanocortin (MC) receptors in the hypothalamus, brain stem, and telencephalon (Vallarino et al. 1988, 1989; Chiba 2001; Klovins et al. 2004). These findings suggest the period when POMC products ceased to act in the periphery alone and became neurotransmitters or neuromodulators. In mammals, melanocortins acting through MC3R and MC4Rs in the hypothalamus, telencephalon, brain stem, and olfactory bulb came to reinforce the endocrine control of stress hormones (Liang et al. 2013; Haitina et al. 2007).

While MRAP1 is strongly expressed in the adrenal, gonadal, and adipose tissue, a related protein named MRAP2 is highly expressed in the hypothalamus, including the PVN (Chan et al. 2009). Interestingly, MRAP2 interacts with other MCRs in mammals. MRAP2 knockout mice and humans with MRAP2 mutations display severe obesity, changes in cholesterol metabolism, and Sim1 deficiency without hyperphagia or reduced energy expenditure (Novoselova et al. 2016; Asai et al. 2013). These effects suggest MC4Rs and other hypothalamic receptors interact with the MRAP2 protein (Clark and Chan 2017).

In humans, the MC1R melanocyte receptor regulates melanogenesis and pigmentation of the skin and hair. Upon activation, this receptor functions by promoting eumelanin and downregulating pheomelanin (Cone 2006). Sun sensitivity and risk of skin cancer increase with mutations in the MC1R gene (Rees 2000). Many immune cells also express MC1R, suggesting that the MC1R also has an anti-inflammatory role (Catania et al. 2010).

The ACTH receptor, MC2R, is primarily expressed in the adrenal cortex (Mountjoy et al. 1992). The major role of MC2R is to regulate steroidogenesis in the adrenal gland. Gene mutations of MC2R contribute to 25% of familial glucocorticoid deficiency cases, a rare autosomal recessive disorder. MC2R knockout mice (Chida et al. 2007) share characteristics of these patients, including severe glucocorticoid deficiency and failure of the adrenal gland to respond to ACTH (Thistlethwaite et al. 1975; Chung et al. 2008; Clark and Weber 1998). Human skin cells (Slominski et al. 1996) and mouse adipocytes (Norman et al. 2003; Boston and Cone 1996; Cho et al. 2005; Moller et al. 2011) express the MC2R receptor, suggesting it may have a role in lipolysis regulation (Boston 1999). In adipocytes, ACTH and α-MSH are strong inhibitors of expression of the adipokine leptin (Norman et al. 2003).

The MC5R receptor, the newest member of this receptor family, has the most diverse expression pattern of all the MCRs (Chen et al. 1997). It is involved with exocrine gland secretion, immunomodulation in B and T cells, and adipocyte cytokine release (Chen et al. 1997; Zhang et al. 2011; Lee and Taylor 2011; Taylor and Lee 2010; Taylor and Namba 2001; Buggy 1998; Jun et al. 2010). It also alters fatty acid oxidation control in skeletal muscle, enhances lipolysis, and suppresses fatty acid reesterification (Moller et al. 2011; An et al. 2007; Rodrigues et al. 2013). Recently, the MC5R has been implicated in regulating glucose uptake by skeletal muscle and thermogenesis (Enriori et al. 2016). These results suggest that MC5R agonists may offer a new target for obesity treatment in the periphery.

The primary regulators of energy homeostasis, MC3R and MC4R, are called neural MCRs due to their high expression in the CNS (Mountjoy 2010). Both receptors interact with melanocortins and are antagonized by AgRP. MC3Rs have an expression pattern limited primarily to hypothalamic and limbic structures, with highest expression in the ARC, VMH, ventral tegmental area (VTA), and the medial habenula (MHb) (Rosellirehfuss et al. 1993). MC3Rs promote body weight regulation and sensitize NPY/AgRP neurons to the metabolic state of the animal (Butler et al. 2017). Indeed, fasted $Mc3r-/-$ mice fail to increase lipolysis or activate the HPA axis (Renquist et al. 2012). MC3Rs also have roles in the periphery. MC3Rs expressed on macrophages have anti-inflammatory immune functions (Getting et al. 1999a; Getting et al. 1999b). Renal MC3Rs promote urinary excretion of sodium, reducing blood pressure on high sodium diets (Mayan et al. 1996; Ni et al. 2003, 2006; Chandramohan et al. 2009).

MC4Rs play critical roles in metabolic regulation, pain, and reproduction, including erectile function and sexual behavior in both sexes (Starowicz and Przewlocka 2003; Starowicz et al. 2009; Pfaus et al. 2004; Martin and MacIntyre 2004; Wikberg and Mutulis 2008). MC4Rs have a wide distribution in the CNS, existing in over one hundred brain nuclei. MC4Rs are most concentrated in the brain stem and the hypothalamus. Importantly, MC4Rs are found in neurons of the PVH that produce corticotrophin-releasing hormone (CRH), oxytocin, and thyrotropin-releasing hormone (TRH) (Liu et al. 2003; Lu et al. 2003). Preganglionic neurons in the intermediolateral cell column (IML) of the spinal cord also show MC4R expression and receive direct inputs from POMC fibers (Elias et al. 1998).

The MCR family is a member of the G-protein-coupled receptor (GPCR) superfamily that maintains a high level of constitutive activity (Srinivasan et al. 2004). These receptors generally couple to Gαs proteins, which activate adenylate cyclase, increase intracellular cyclic 3′,5′-adenosine monophosphate (cAMP), and activate protein kinase A (PKA). These signaling molecules can increase neuronal excitability, facilitate neurotransmitter release, regulate how neurons integrate synaptic input, and alter synaptic strength and connectivity (Grueter et al. 2012; Kreitzer and Malenka 2008; Russo et al. 2010). That melanocortins can alter synaptic strength has implications for their downstream functions, including their influence on body weight and reward pathways (Caruso et al. 2014). However, under continuous stimulation, the MC4R undergoes desensitization and internalization (Shinyama et al. 2003).

AgRP inhibits the basal activity of MC3Rs (Tao et al. 2010) and MC4Rs (Haskell-Luevano and Monck 2001; Nijenhuis et al. 2001) and acts as a competitive antagonist that prevents the binding of melanocortins. In contrast to α-MSH, AgRP stimulates the coupling of the MC4R receptor to the Gαi/o subunit, which inhibits adenylate cyclase and decreases intracellular cAMP levels (Büch et al. 2009; Fu and van den Pol 2008). Recently, it has also been shown that AgRP can hyperpolarize neurons by binding to MC4R and opening Kir7.1, an inwardly rectifying potassium channel, independently of its inhibition of α-MSH binding (Ghamari-Langroudi et al. 2015).

The MC3R appears to be the only melanocortin receptor expressed by ARC POMC neurons (Bagnol et al. 1999; Jegou et al. 2000; Mounien et al. 2005). In contrast, AgRP/NPY neurons express both MC3Rs and MC4Rs

(Bagnol et al. 1999; Mounien et al. 2005). The activation of MCRs on AgRP neurons may allow these neurons to sense the level of POMC activity and regulate AgRP release in a short feedback loop. In addition, MC3R activation of AgRP neurons increases their release of inhibitory neurotransmitters onto POMC neurons and POMC projection sites (Cowley et al. 2001). Indeed, electrophysiological, immunohistochemical, and behavioral evidence shows activation of MC3Rs diminishes POMC neuronal activity and suppresses POMC mRNA expression (Cowley et al. 2001; Lee et al. 2008; Marks et al. 2006). More research is needed to understand the role of these regulatory mechanisms.

11.3 Beta-Endorphin

The production of an opioid peptide, β-endorphin, from the POMC gene adds complexity to this neuronal system. In all chordates, POMC encodes a core melanocortin sequence and a core opioid sequence for β-endorphin. β-endorphin$_{1-31}$ is the sole opioid sequence encoded in the POMC gene in humans and rodents, although some ancient species process it into smaller opioids (Takahashi et al. 1995, 2001, 2006; Shoureshi et al. 2007). Further cleavage of β-endorphin$_{1-31}$ by PC2 and CPE to form β-endorphin$_{1-27}$ and β-endorphin$_{1-26}$ abolishes its ability to bind to opioid receptors. This effect shifts the balance in favor of MSH-related actions (Wardlaw 2011).

In many, but not all cases, melanocortins and endorphins produce opposing physiological and behavioral effects that ensure a coordinated and balanced response to changing environmental demands and stressors (Table 11.1, modified from (Bertolini and Ferrari 1982)). The result is a form of functional reciprocity. For instance, melanocortins upregulate attention and pain sensitivity, promoting arousal and adaptation to external challenges, while simultaneous release of opioids favor de-arousal and shifting to self-directed behavior (Bertolini and Ferrari 1982; De Wied and Jolles 1982; Sandman and Kastin 1981).

The primordial role of opioids is the control of protective reactions. Even in protozoa, opiate

Table 11.1 A comparison of the effects of opioids and melanocortins

	Function	Melanocortins	Opioids
HPA and stress response	Physiological stress response	↑ ACTH, cortisone/corticosterone	↓ CRH release ↓ Development of stress adaptation ↑ Suppression of HPA axis Anti-stress via kappa receptor pathway
	Shock Hypotensive, hypovolemia	↓	↑
	Stress-induced anxious/depressive behavior	↑ Via MC4R MC4R KO and antagonists = reduced anxiety/depression behaviors	↓ Attenuates anxious behaviors Reduced hypercorticosterone response

CNS actions	CNS activity
	  Neuronal firing
	  Adenylate cyclase/cAMP
	  Ca2+ uptake at synapse
		↑	↓

	Neurotransmitter release
	  Norepinephrine (stress)
	  Dopamine (behavior)
	  Acetylcholine (immunity)
		↑	↓

	Neurotransmitter turnover Serotonin (behavior-depression anxiety)	↓	↑
	POMC neurons	↓ Autoinhibition via MC3R Inhibition via MC3R/AgRP/NPY pathway	↓/↑ ↓ At high concentration Presynaptic via low sensitivity receptor on POMC neurons ↑ At low concentration Via disinhibition Postsynaptic via high sensitivity receptor near GABA synapses
	Glial expression	Express MCRs	N/A
Pain	Pain threshold	↓ Increase hypersensitivity Antagonize opioid-induced analgesia Reduce opioid tolerance *MC4R antagonism synergizes with opioid pain reduction	↑

(continued)

Table 11.1 (continued)

	Function	Melanocortins	Opioids
Immunity and inflammation	Inflammation (general)	↓ Anti-inflammatory Immune suppressive Central immune modulation (via vagus nerve-cholinergic AND glucocorticoid release) Neuroprotective	↓ Anti-inflammatory Immune suppressive
	Temperature regulation	↓ Antipyretic	↑
	Neuroinflammation	↓ Neuroinflammation ↓ Excitotoxicity	N/A
	Neuroprotection	↑ Via Oligodendrocyte development Activate astrocyte-/microglia-mediated protection	N/A
	Immune cell expression	MCR expressed on: Macrophages, B and T lymphocytes	N/A
Behavior	Arousal	↑	↓
	Attention	↑	↓
	Motivation	↑	↓
	Learning/memory	↑	↓
	Yawning/stretching	↑	↓
	Grooming	↑ Induces all components	↑ Increase/instigate duration Prolong sensitivity of grooming

ligands suppress a protective contractile response and reduce growth and motility (Dyakonova 2001; Zagon and Mclaughlin 1992), although the receptors responsible for such actions are unclear (Lesouhaitier et al. 2009; Stefano and Kream 2008). In invertebrates like mollusks and arthropods, many functions of opioids (e.g., stress-induced analgesia, deactivating immune responses, and regulation of feeding, mating, and social behavior) resemble those in vertebrate species (Dyakonova 2001). A single ancestral opioid receptor duplicated itself twice early in vertebrate evolution to create the four known opioid receptor types (Sundstrom et al. 2010; Larhammar et al. 2009). The addition of an opioid sequence to the POMC gene likely occurred around this time (Duvaux-Miret and Capron 1992; Salzet et al. 1997; Stefano et al. 1999).

Opioid receptors are found throughout the central and peripheral nervous system and the immune system (Stein and Machelska 2011; Zollner and Stein 2007). These receptors use the Gi/o signaling cascade; so, like AgRP and in opposition to melanocortins, opioid receptors inhibit adenylate cyclase activity and lower cAMP levels (Collier 1980; Tao 2010; Rene et al. 1998). β-endorphin binds to mu (μ), delta (δ), and kappa (κ) opioid receptors, with highest affinity for mu and δ types (Katritch et al. 2013; Cox 2013). Opioids may interact with ion channels as well (Luscher and Slesinger 2010; Tedford and Zamponi 2006).

β-endorphin regulates POMC neuronal activity and gene transcription through a complex feedback mechanism. Hyperpolarization of POMC neurons occurs when hypothalamic explants are treated with opioid agonists (Kelly et al. 1990), while antagonists increase secretion of both β-endorphin and γ-MSH (Jaffe et al. 1994; Nikolarakis et al. 1987). By binding to the μ-opioid receptors they express, β-endorphin inhibits POMC activity and gene expression (Kelly et al. 1990; Zheng et al. 2005b; Markowitz et al. 1992; Pennock and Hentges 2011). This mechanism serves as a form of autoinhibition of POMC neurons (Bouret et al. 1999). In addition, opioid receptors exist on the numerous GABAergic terminals that synapse onto POMC neurons. The sensitivity to opioids is much greater at the presynaptic μ-opioid receptors than the postsynaptic μ-opioid receptors (Pennock and Hentges 2011). So, at low concentrations, β-endorphin may inhibit the presynaptic release of GABA, disinhibiting POMC neurons (Pennock and Hentges 2011, 2016). The interplay between these mechanisms likely provides fine-tuning of melanocortin and endorphin release by POMC neurons.

Optogenetic experiments suggest that the differential release of β-endorphin and α-MSH may be key to POMC neuronal actions (Yang et al. 2011; Aponte et al. 2011). Under default conditions, AgRP inhibits POMC neuron activity to promote feeding in mice. However, if leptin release by adipocytes rises in response to a long-term energy surplus, POMC neurons release β-endorphin, shutting off this inhibitory circuit. At the same time, activation of POMC neurons reduces food intake in an MCR-dependent manner. So, by simultaneously releasing melanopeptides and opioid peptides in variable ratios, the POMC system may respond to physiological or external changes with a variety of tailored responses.

Several mechanisms may underlie this differential release. Differential enzymatic inactivation of α-MSH or β-endorphin can modify the action of POMC products (Dutia et al. 2012). The enzymes which process POMC products can alter the ratio of various forms of β-endorphin and melanocortins in response to different neuronal, hormonal, environmental, and pharmacological stimuli (Wilkinson and Dorsa 1986; Cangemi et al. 1995; Young et al. 1993). Also, because POMC is posttranslationally processed to ACTH and MSH peptides in secretory vesicles, packaging of POMC in secretory granules controls the extent of POMC cleavage. Alternative methods of sorting POMC products may produce heterogeneity in secretory granule content (Pritchard and White 2007). As hinted at by earlier work (Perello et al. 2007, 2008; Petervari et al. 2011; Mercer et al. 2014), Koch and colleagues recently used electron microscopy to show that β-endorphin and α-MSH exist in separate vesicles within individual neurons of the PVH (Koch et al. 2015). In a third of POMC synaptic boutons in the PVH, β-endorphin and α-MSH did not overlap. The authors also identified hypothalamic UCP2 as being crucial for the switch from α-MSH to β-endorphin release triggered by endocannabinoids. As we shall see, the flexibility inherent in the ability of POMC neurons to release either a melanocortin or an opioid has large repercussions for the CNS response to stress.

11.4 The Stress Response: Overview

Stress is the experience of coping with a physical or emotional threat. Common physical stressors include visceral or somatic pain, hemorrhage, respiratory distress, and inflammation from illness or injury. Psychological or emotional stress may result from circumstances the individual perceives as negative or threatening, such as interpersonal conflict or financial problems. Afferent sensory information from peripheral receptors alert the CNS to physical stressors. Forebrain limbic structures like the prefrontal cortex (PFC), hippocampus, and amygdala receive input about psychogenic and emotional stressors (Ulrich-Lai and Herman 2009; Ulrich-Lai and Ryan 2014). The physical and psychological signals of the brain stem and limbic system converge at the paraventricular nucleus of the hypothalamus (PVH). Here, they integrate to engage effector mechanisms to regulate the body's physiological response (Ulrich-Lai and Herman 2009).

Two systems regulate the stress response. First, the sympathetic nervous system (SNS) releases the catecholamines epinephrine and norepinephrine. Activation of the SNS during acute stress elicits a rapid physiological response by two mechanisms: direct innervation of peripheral organs and the systemic release of catecholamines by the adrenal medulla (Ulrich-Lai and Herman 2009; Ulrich-Lai and Engeland 2002). This system mobilizes energetic stores of both glucose and free fatty acids, increases blood pressure and heart rate, and downregulates physiological processes unnecessary in the short term, such as digestion and reproduction (Bartness and Song 2007; Yamaguchi 1992). Therefore, the sympathetic "fight or flight" system (countered by the parasympathetic "rest and digest" system) allows fast modulation of energy allocation to respond to an immediate threat to survival.

The second system regulating the stress response, the hypothalamic-pituitary-adrenal (HPA) axis, releases glucocorticoids such as cortisol. Activation of the HPA axis yields a slower, sustained, and amplified physiological response to acute stress. The HPA axis becomes activated when stress-related internal and external sensory input converges on corticotropin-releasing hormone neurons in the paraventricular nucleus of the hypothalamus (PVH) (Dent et al. 2000; Ma et al. 1997). CRH stimulates synthesis and release of melanocortin peptides by the anterior pituitary, specifically adrenocorticotropic hormone (ACTH) from corticotrophs and MSH from melanotrophs (Gagner and Drouin 1985; Gagner and Drouin 1987; Eberwine et al. 1987). ACTH, produced from POMC, then stimulates the adrenal cortex to produce glucocorticoids (cortisol in humans or corticosterone in rodents). Once released, glucocorticoids mediate many systemic and neurological effects. Transcription of the POMC gene in corticotrophs undergoes feedback repression by glucocorticoids via the glucocorticoid receptors (GR) (Gagner and Drouin 1985). This negative feedback loop is essential for the HPA axis to maintain homeostasis. During long-term stress, however, the HPA axis can become chronically activated (Ulrich-Lai and Herman 2009; Ulrich-Lai and Ryan 2014).

CRH also has other mechanisms to trigger a stress response. It influences the sympathetic stress response by acting on the locus ceruleus, adrenal medulla, and the peripheral SNS (Valentino et al. 1993; Brown et al. 1982). In addition, CRH directly affects behavioral states of anxiety (Bale and Vale 2004; Reul and Holsboer 2002). Many brain areas involved in stress perception and response express CRH, including the amygdala (Roozendaal et al. 2002; Gallagher et al. 2008; Regev et al. 2012), hippocampus (Lee et al. 1993; Chen et al. 2001, 2013; Refojo et al. 2011), inferior olive (Chang et al. 1996), locus ceruleus (Valentino and Van Bockstaele 2008), bed nucleus of the stria terminalis (Dabrowska et al. 2011), and the cortex (De Souza et al. 1986; Behan et al. 1995; Gallopin et al. 2006). Central overexpression of CRH induces an anxious behavioral phenotype in rodents (Dunn and Berridge 1990; Dautzenberg et al. 2004; van Gaalen et al. 2002), while suppressing CRH expression has anxiolytic effects under both stressed and unstressed conditions (Skutella et al. 1994a, b). CRH can also suppress GnRH release (Sirinathsinghji 1987; Traslavina and Franci 2012), sleep (Romanowski et al. 2010), and appetite (Glowa et al. 1992).

The evolution of CRH in chordate ancestors played an essential role in the success of vertebrates and ultimately mammals and humans (Endsin et al. 2017) by permitting a robust stress response. In fish, hypothalamic CRH released during stress triggers pituitary ACTH release, which stimulates the interrenal tissue to secrete glucocorticoids. The pituitary-interrenal axis is also responsible for other biological processes such as metabolism of carbohydrates, amino acids, and free fatty acids; mineral balance; immune function; and growth (Wendelaar Bonga 1997). The multiple roles for glucocorticoids in this distant relative highlight the relationship between energy use and the stress response. The stress axis seems to have diverged from the reproductive axis around the time of the evolution of jawless fish, the most ancient of vertebrates; in the lamprey, GnRH and CRH cause release of 11-deoxycortisol, a putative stress steroid (Roberts et al. 2014). Likewise, GRs diverged from estrogen receptors (Thornton 2001). So, the trade-off between stressor survival and fertility predates and shaped the development of mammalian systems.

Once the HPA axis is activated, glucocorticoids levels rise. Glucocorticoids raise blood glucose levels to meet elevated energy needs under stressful conditions. To do this, glucocorticoids promote gluconeogenesis by increasing protein catabolism and lipolysis while decreasing glucose use and insulin sensitivity of adipose tissue. At the same time, they suppress the immune system and aid metabolism of macronutrients. They increase blood pressure by promoting the sensitivity of the vasculature to epinephrine and norepinephrine (Smart et al. 2007; Pavlov and Tracey 2006). Glucocorticoids also impair liver sensitivity to growth hormone, leading to low levels of circulating insulin-like growth factor 1 (IGF-1), which may cause a decrease in neural plasticity (Mechanick et al. 1992), hippocampal learning (Lutter and Nestler 2009), neuroprotection, and neuronal replacement (Lutter and Nestler 2009; Fletcher and Kim 2017).

Glucocorticoids also have a direct impact on the brain and spinal cord. Glucocorticoids cross the blood-brain barrier and bind to glucocorticoid (GR) and mineralocorticoid receptors (MR) on both neurons and glial cells. In the brain, MRs show high affinity for glucocorticoids (de Kloet and Sarabdjitsingh 2008; De Kloet et al. 1998). Therefore, they may sense basal glucocorticoid levels and mediate physiological responses to low glucocorticoid levels (de Kloet et al. 2005). In contrast, GR remains unbound at low glucocorticoid levels due to its lower binding affinity; it therefore mediates responses to elevated levels of glucocorticoids (de Kloet et al. 2005; Reul and Dekloet 1985). The expression patterns of these receptors reflect their roles. GR is abundantly expressed throughout the brain, including in key sites for stress regulation like the medial prefrontal cortex (mPFC), hippocampus, amygdala, BNST, hypothalamus, and the hindbrain (Reul and Dekloet 1986; Meaney et al. 1985; Fuxe et al. 1987)). MR expression, in contrast, is more limited. Interestingly, MR and GR co-localize in several areas important to the

behavioral response to stress, including the hippocampus, amygdala, and the mPFC (Reul and Dekloet 1985; Dekloet and Reul 1987).

Alterations of the HPA axis are associated with anxiety (Pego et al. 2010). The chronic actions of glucocorticoids in the mPFC increase the use of habitual strategies over goal-oriented decision-making (Dias-Ferreira et al. 2009). This reversion to habitual strategies during chronic stress may improve efficiency in predictable tasks by creating an instinctual response instead of wasting resources on the appraisal process (Dias-Ferreira et al. 2009; Schwabe et al. 2013). The amygdala underlies anxiety and fear responses (Davis et al. 2010; LeDoux 2012). When administered to the central amygdala, glucocorticoids enhance anxious behavior in rodents (Shepard et al. 2000). Application of GR and MR antagonists to the central amygdala abolishes this behavior (Myers and Greenwood-Van Meerveld 2007).

By acting on the hippocampus and amygdala, glucocorticoids enhance memory consolidation but impair working memory (Barsegyan et al. 2010; Roozendaal et al. 2004; Mizoguchi et al. 2000). In the hippocampus, glucocorticoids help form emotionally powerful short-term memories (de Kloet et al. 1999; Smeets et al. 2009) but cause memory retrieval disruption for already assimilated items (Roozendaal 2002). Chronic stress conditions weaken performance on hippocampal-dependent tasks (Conrad et al. 1996; Kleen et al. 2006) and cause spatial reference memory deficits (Oliveira et al. 2013). Glucocorticoids acting in the basolateral amygdalar complex (BLA) affect the learning and memory of aversive stimuli (Roozendaal et al. 1996). They also alter the reconsolidation of auditory fear-based conditioning and memory consolidation (Jin et al. 2007). These mechanisms allow consolidation of emotionally prominent events while diminishing input from competing information.

Glucocorticoids also have direct actions on neuroendocrine functions of the hypothalamus, including mimicking the negative feedback effects of sex hormones. They inhibit GnRH release and may, as a result, impair fertility and delay the onset of puberty during chronic stress (Calogero et al. 1999; Gore et al. 2006). In addition, high glucocorticoid levels directly impede male sexual behavior (Rivier and Vale 1984; Pednekar et al. 1993; Retana-Marquez et al. 2009). Glucocorticoids also suppress the secretion of other hypothalamic hormones, such as thyrotropin-releasing hormone (TRH) (Brabant et al. 1987), leading to decreased metabolic demands during stress and contributing to weight gain.

While excess glucocorticoids are harmful, levels that are too low also lead to physiological dysfunction (de Kloet et al. 1999; Herman 2013; Myers et al. 2014). The psychological and physiological adaptations induced by glucocorticoids allow an organism to survive and regain normal equilibrium. Specifically, when energy availability is low, restriction of growth, reproduction, immune processes, and other vegetative functions become the "biological cost" of adapting to the stressor (Moberg 2000). Whether these trade-offs prove to be helpful or damaging depends on the organism's environmental context and on how long they last (Sinha 2008; McEwen 2007).

11.5 Melanocortins and the Physiologic Stress Response

A unifying function underlying the actions of the melanocortin system and its doppelgänger, the opioid system, is to produce a suitable response to acute stress. Melanocortins promote pain sensing, the energy releasing powers of glucocorticoids, and anxious behavior while inhibiting inflammation so that the individual can combat a threat to survival. These functions complement those of the ACTH-CORT endocrine axis from which the melanocortin system evolved. The role that POMC products play in this aspect of the stress response is under-recognized.

Three main stages to the stress response have been identified: alarm, resistance, and compensation (Schreck 2000). To handle a stressor effectively, the threat must first be perceived. This alarm stage involves the induction of the secretion of glucocorticoids and catecholamines to make energy available for resistance strategies, such as fighting or flight. During the resistance stage, the organism rations energy and attention in such a way that disease resistance, reproduction, growth, learning, and other functions are impaired. Compensation or recovery consists of adaptive changes to restore homeostasis and physiological equilibrium.

Under conditions of stress, ARC POMC neurons show rapid activation. For example, acute restraint and forced-swim treatments in rodents cause increased activity of ARC POMC neurons projecting to the PVH (Liu et al. 2007). In addition, rats subjected to foot shock showed increased POMC mRNA in the hypothalamus, increased CRH mRNA in the amygdala, and increased MC4R mRNA in both locations (Yamano et al. 2004). Another study found that psychological stress increases expression of MC4R mRNA in the ARC of rats in a glucocorticoid-independent manner (Ryan et al. 2014). Intracerebroventricular (ICV) injection of an MC4R agonist produces a dose-dependent increase in renal and lumbar sympathetic nerve activity, which is reversed by an MC3R/MC4R antagonist (Haynes et al. 1999). This SNS activation exacerbates the response to stress; sympathetic innervation of the adrenal cortex enhances sensitivity to ACTH, which promotes glucocorticoid secretion (Engeland and Arnhold 2005; Edwards and Jones 1993).

Tonic suppression by neuronal POMC peptides keeps CRH levels within physiological limits. In the hypothalamus, the ARC POMC neuron population has abundant synaptic projections to PVH CRH neurons (Lu et al. 2003), many of which express MC4R (Lu et al. 2003; Dhillo et al. 2002; Sarkar et al. 2002). A chronic reduction or absence of hypothalamic POMC leads to elevated CRH specifically in the PVH, elevated basal but attenuated stress-induced ACTH secretion, and elevated basal plasma corticosterone (Smart et al. 2007). Whether suppression of CRH levels by POMC peptides occurs via direct actions in the PVH remains to be determined. Some evidence suggests β-endorphin suppresses CRH levels in vivo as well (Plotsky 1986), but selective loss of β-endorphin beginning during development does not alter CRH mRNA or glucocorticoid levels (Smart et al. 2007; Rubinstein et al. 1996).

α-MSH release by POMC neurons may also act via the MC4R to augment CRH release during a psychological or physiological challenge. Short-term ICV infusion of α-MSH can increase CRH expression, ACTH, and corticosterone levels after a stressful event (Lu et al. 2003; Dhillo et al. 2002; Kas et al. 2005). Loss of MC4R signaling prevents stress-induced activation of neurons in the PVH and MeA and lowers ACTH and corticosterone release (Ryan et al. 2014; Karami Kheirabad et al. 2015). Pharmacological stimulation of MC4R in the PVH also increases CRH mRNA expression, circulating ACTH, and corticosterone (Lu et al. 2003; Dhillo et al. 2002). In contrast, the MC3R influences the response of the HPA axis to fasting but no other stressors (Renquist et al. 2012).

Glucocorticoids, the final products of HPA axis activation, regulate the HPA axis through negative feedback. They suppress ACTH release at the level of the pituitary and CRH synthesis and release in the PVH (Lim et al. 2000). ACTH may also exert a regional form of feedback control of the HPA axis. ACTH suppresses CRH expression in the medial amygdala and hippocampus, but these effects are not seen in the hypothalamus (Brunson et al. 2001; Wang et al. 2012).

Glucocorticoids appear to exert feedback control of the melanocortin system in a similarly region-specific manner (Beaulieu et al. 1988; Wardlaw et al. 1998). Glucocorticoids downregulate POMC gene expression in the pituitary (Pritchard et al. 2002; Krude and Gruters 2000). While most POMC neurons in the ARC also contain GR, glucocorticoids seem to upregulate POMC gene expression in the hypothalamus of rats. ARC POMC gene expression falls after adrenalectomy and returns to normal after replacing glucocorticoids at physiological concentrations (Wardlaw et al. 1998; Pelletier 1993). The number of inhibitory synaptic inputs onto POMC neuron cell bodies also fell in adrenalectomized animals; replacement of corticosterone reversed this effect (Gyengesi et al. 2010). These mechanisms allow regulation of the melanocortin response to stress.

These studies show that melanocortins have powerful and integrative effects on both systems that mediate classic stress responses, the SAM system and the HPA axis. However, stressors often involve pain, immune reactions, and psychological challenges; melanocortins have a foundational role reestablishing homeostatic balance in each of these situations as well.

11.6 Melanocortins and Immune Function

Illness is a classic condition of physiological stress. As described below, the melanocortin system acts to inhibit inflammatory processes both centrally and peripherally. In addition, it combats fever and the cardiovascular processes related to shock. These functions play an essential role in appropriately calibrating the physiological response to illness. Indeed, melanocortin agonists are an underexploited anti-inflammatory therapy that holds great promise for treatment of immune disorders (Montero-Melendez 2015).

The systemic effects of melanocortins include direct action by circulating or local melanocortins on their receptors. The MC1R has an anti-inflammatory role in a wide variety of immune cells (Catania et al. 2004, 2010). In addition, the MC5R plays a role in lymphocyte modulation, specifically the activation of regulatory T-cell lymphocytes in ocular immunity (Taylor and Lee 2010; Taylor et al. 2006) and B-lymphocyte immunomodulation (Buggy 1998). In adipocytes, the MC5R stimulates cytokine secretion (Jun et al. 2010).

The central melanocortin system also has suppressive effects on the systemic immune response. ICV α-MSH inhibits transcription factor nuclear factor kappa B (NF-κB) activation at peripheral inflammatory sites (Ceriani et al. 1994; Ichiyama et al. 1999a; Lipton et al. 1991). By directly sensing cytokines at the circumventricular organs, highly vascularized areas of the brain with a leaky blood-brain barrier, the CNS can detect peripheral inflammation. The ARC, NTS, and dorsal motor nucleus of the vagus (DMX) are near circumventricular organs. Acute inflammation such as that caused by bacterial lipopolysaccharide (LPS) activates TNF-α, IL-1β, and IL-6 and induces Pomc expression (Kariagina et al. 2004). Under these conditions, NF-κB activation directly promotes Pomc transcription independent of STAT3 activation (Shi et al. 2013). (Chronic inflammatory conditions, however, can impair the direct activation of Pomc promoter (Shi et al. 2013).) The CNS also receives information on the systemic inflammatory status via ascending sensory pathways from affected tissues through the vagus nerve and pain afferent fibers (Goehler et al. 1997, 2000; Watkins et al. 1995). By promoting glucocorticoid release in response to sensory input, the melanocortin system suppresses immune responses in the periphery (Besedovsky et al. 1986; Hu et al. 1991).

If augmented pharmacologically, the anti-inflammatory effects of α-MSH can counter the life-threatening vasodilatory actions of endogenous opioids and histamine that are massively released during shock (Bernton et al. 1985; Carmignani et al. 2005; Chernow et al. 1986; Elam et al. 1984; Schadt 1989; Guarini et al. 1997, 2004; Bertolini et al. 1986a, 1986b; Giuliani et al. 2007; Bitto et al. 2011). Melanocortins can reverse this immune response by acting on parasympathetic preganglionic neurons of the DMV that produce acetylcholine and express MC4Rs (Catania et al. 2004; Sohn et al. 2013). These neurons form part of the efferent arm of the cholinergic anti-inflammatory pathway (Guarini et al. 1989; Bertolini et al. 2009). The vagus nerve can induce rapid release of acetylcholine to inhibit pro-inflammatory cytokine release from macrophages in target organs (especially in the liver, spleen, gastrointestinal tract, and heart) (Guarini et al. 2004; Pavlov and Tracey 2006; Tracey 2002, 2007). The release of anti-inflammatory cytokines is unaffected (Borovikova et al. 2000). Bilateral injury or dissection of the vagus nerve or

inhibition of primary afferent nociceptive nerve fibers compromises the ability of melanocortins to inhibit harmful and unnecessary inflammatory responses to hypoxic conditions, such as during hemorrhagic shock (Bertolini et al. 1989, 2009).

Melanocortins also combat fever and have a broad, suppressive effect on body temperature. ICV α-MSH inhibits systemic inflammatory reactions, including fever (Delgado Hernandez et al. 1999; Murphy et al. 1983). Melanocortins reduce body temperature acutely (reaching a nadir at 40 minutes) through several mechanisms, including reducing brown adipose tissue thermogenesis, lessening vasodilation, promoting active seeking of a cool environment, reducing physical activity, and suppressing compensatory shivering (Lute et al. 2014).

Melanocortins also induce stretching and yawning (Wessells et al. 2000, 2003; Vergoni et al. 1998), which we suggest are part of this program of body temperature reduction. Strong evidence from humans and other warm-blooded animals shows that both stretching and yawning are part of a coordinated physiological program to alter brain and body temperature (Gallup and Eldakar 2013; Eguibar et al. 2017). When body temperature is both excessive and higher than air temperature, yawning and stretching increase to allow increased airflow within the mouth and around the limbs to cool circulating blood. Co-injection of an MC4R antagonist inhibits yawning produced by microinjection of ACTH into the PVH (Argiolas et al. 2000), although other target areas may also be involved (Argiolas et al. 1987). In contrast, opioids suppress yawning and stretching; for example, β-endorphin inhibits ACTH-induced yawning (Fratta et al. 1981; Vergoni et al. 1989; Himmelsbach 1939; Seevers 1936; Zharkovsky et al. 1993). ACTH- and α-MSH-induced yawning correspond to an increase in the turnover rate of acetylcholine in the hippocampus; central cholinergic antagonists impede yawns (Ferrari et al. 1963; Fujikawa et al. 1995; Wood et al. 1978). The circuit underlying these behaviors may include α-MSH-activated neurons in the PVH that project to medial septum cholinergic neurons that, in turn, project to the hippocampus (Collins and Eguibar 2010).

Melanocortins may have direct anti-inflammatory and neuroprotective actions within the brain. For example, in a traumatic brain injury mouse model, a single application of α-MSH (Hummel and Zuhlke 1994; Ottaviani et al. 1997; Tsatmali et al. 2000) reduced inflammation, apoptosis, and brain damage (Schaible et al. 2013). Similarly, MC4R activation had anti-apoptotic effects during cerebral ischemia (Giuliani et al. 2006) and in the hippocampus where excitotoxicity induced neuronal cell death (Forslin Aronsson et al. 2007).

Evidence suggests that these neuroprotective effects may result from melanocortin actions in glial cells. Multiple types of glial cells show MCR expression. Human microglia express MCRs (MC1R, MC3R, MC4R, and MC5R) (Lindberg et al. 2005). Melanocortins can directly suppress the activation of nuclear factor kappa B (NF-κB), tumor necrosis factor-α (TNF-α), and inducible nitric oxide synthase (iNOS) expression in activated microglia (Catania et al. 2004; Delgado et al. 1998; Galimberti et al. 1999). Oligodendrocytes, the CNS glial cells responsible for myelination, also express MC4R (Arnason et al. 2013; Selkirk et al. 2007; Lisak et al. 2016; Benjamins et al. 2014). $ACTH_{1-39}$ increases proliferation, differentiation, and maturation of oligodendrocyte progenitor cells. It also reduces apoptosis in both progenitor and mature oligodendrocytes. At the same time, $ACTH_{1-39}$ protects against excitotoxicity and inflammation (Lisak et al. 2016; Benjamins et al. 2014). In multiple sclerosis, $ACTH_{1-39}$ has been used clinically to treat immune system-induced myelin damage. Finally, astrocytes express MC4Rs (but not MC3Rs) in rats (Selkirk et al. 2007; Caruso et al. 2007). In reaction to hypoxia or other stressors, reactive astrocytes produce nitric oxide (NO) and pro-inflammatory cytokines and chemokines (Dong and Benveniste 2001). MC4R activation blocks apoptosis of astrocytes (Giuliani et al. 2006), reduces their secretion of NO and prostaglandin G2 (PEG2), and inhibits their expression of iNOS and COX-2 (Giuliani et al. 2006; Caruso et al. 2007). Although not involved in anti-inflammatory responses of α-MSH, the MC1R is expressed in astrocytes as well

(Ichiyama et al. 1999b). The activation of astrocytes plays a role in the pathology of many neurodegenerative conditions, so controlling astrocyte activation through melanocortin receptor activation may be an effective avenue for decreasing the severity of such diseases.

11.7 Melanocortins and Pain Pathways

Pain is a potent stressor. Pain is a noxious sensory or emotional experience caused by actual or potential tissue damage (Leeson et al. 2014). The ability of an organism to sense noxious stimuli is essential for preventing physical injury. Afferent fibers in the spinal cord carry nociceptive signals to higher brain centers through spinothalamic, spinobulbar, spinopontine, and spinomesencephalic tracts (Al-Chaer 2013; Boadas-Vaello et al. 2016). Sensory-discriminative signals mediating pain localization propagate through the dorsal root of the spinal cord to thalamic nuclei and the PAG in the midbrain. The brain stem reticular formation, thalamus, and hypothalamus contribute the affective-motivational components of pain (Ab Aziz and Ahmad 2006). At the level of the thalamus, third-order neurons that receive both sensory and affective information ascend to terminate in the somatosensory cortex.

A diffuse, multisynaptic descending pathway produces analgesia. It originates from higher brain centers, such as the cerebral cortex, hypothalamus, and amygdala, and projects to the PAG. From here, projections synapse at the rostral ventromedial medulla (RVM) and locus ceruleus. They then project down the spinal cord and terminate on the initial pain sensing spinal dorsal root to inhibit incoming signals (Fardin et al. 1984; Pagano et al. 2012; Kerman et al. 2006). This PAG-RVM system plays a key role in pain sensation and modulation. These pathways use predominantly catecholaminergic, serotonergic, and opioid systems. The PAG was the first brain area where activation of an endogenous pain inhibition system was described; electrical stimulation and opioid injections into the PAG produce analgesia, which is reversible by application of naloxone (Reynolds 1969; Hosobuchi et al. 1977; Lewis and Gebhart 1977). The RVM is the final common relay point in the modulation of the descending pain pathway. It can both enhance and lessen pain (Heinricher et al. 2009).

The melanocortin system can modulate pain sensitivity. For example, MC4R signaling amplifies neuropathic pain in rats (Vrinten et al. 2001; Nijenhuis et al. 2003). MC4R expression is extensive throughout both ascending and descending nociceptive circuits. As well as being found in primary afferent neurons, MC4Rs are located in the reticular formation, somatosensory and motor cortex, PAG, RVM, and dorsal horn of the spinal cord (Kishi et al. 2003; Gautron et al. 2012; Ye et al. 2014). The MC4R-positive neurons of the RVM are 10% catecholaminergic and 50–75% serotonergic, suggesting that MC4R signaling modulates nociceptive serotonergic sympathetic outflow (Pan et al. 2013).

MCRs share a neuroanatomical distribution pattern with μ-opioid receptors (Arvidsson et al. 1995; Kalyuzhny et al. 1996; Matthes et al. 1996). Anatomically, α-MSH and β-endorphin are both released in response to painful stimuli at the same site (Adan and Gispen 2000). As mentioned before, melanocortins and endorphins generally produce opposing responses in their common targets (Bertolini et al. 1979; Amir and Amit 1979). This effect applies in the modulation of pain through interaction at the level of the brain and spinal cord. For example, ICV injection of α-MSH in rodents induces hypersensitivity to pain, reversing the analgesia produced by endogenous opioids and morphine (Sandman and Kastin 1981; Contreras and Takemori 1984; Kalange et al. 2007). MC4R agonists enhance hypersensitivity in a neuropathic pain model (Starowicz et al. 2002; Vrinten et al. 2000).

Melanocortin antagonists work synergistically with opioids, enhancing the analgesic effect produced by opioid agonists (Kalange et al. 2007; Vrinten et al. 2000; Ercil et al. 2005). For example, the opioid antagonist naloxone lessened the analgesic effects of a melanocortin antagonist administered to the spinal theca in a model of pain hypersensitivity (Vrinten et al. 2000). Also, targeted delivery of an MC4R antagonist to the PAG diminished neuropathic hyperalgesia (Chu et al. 2012), to an even greater extent than morphine (Starowicz et al. 2002; Chu et al. 2012). MC4R and POMC mRNA expression rises in the PAG along with heightened sensitivity in a rat model of neuropathic pain (Chu et al. 2012). Consistent with this, morphine downregulated MC4R mRNA expression in various brain areas, including the PAG, NuA, and striatum, in a time-dependent manner (Alvaro et al. 1996). This response may be, in part, an adaptive mechanism of opioids to cause tolerance and dependency. MC4R antagonists also prevent opioid tolerance when administered to the brain or spinal cord (Kalange et al. 2007; Niu et al. 2012). In morphine-tolerant rats, a single administration of melanocortin antagonists restored the analgesic potency of morphine (Starowicz et al. 2005). If MC4Rs participate in opioid tolerance, they make a logical target for its prevention. Developing pharmacological treatments targeting MCRs may improve pain management by preventing tolerance and dependency.

11.8   Melanocortins and the Behavioral Responses to Stress

Along with increasing sensitivity to pain, the melanocortin system, and particularly MC4R signaling, promotes stress, anxiety, and depression-related behaviors. In mice, chronic social defeat results in social avoidance associated with reduced expression of POMC in the hypothalamus (Chuang et al. 2010). Administration of an MC3R/MC4R agonist increased this avoidance. Conversely, MC4R-null mice showed less anxiety and depression and more social behaviors (Chuang et al. 2010). Similarly, the ICV administration of a selective MC4R antagonist to rats before stressful restraint reduces depressive behavior (Goyal et al. 2006; Chaki and Okubo 2007) and stress-elicited anorexia (Chaki et al. 2003; Vergoni et al. 1999a). An intranasal MC4R antagonist,

HS014, also prevented anxious and depressive behavior in rats (Serova et al. 2013). Further, the intranasal MC4R agonist, HS014, led to improved resilience in rats after traumatic stress (Serova et al. 2013). These behavioral responses involve the medial amygdala (MeA), which receives ARC POMC projections and expresses high levels of MC4R. Acute restraint activates MC4R-expressing neurons in the MeA in rats as shown by c-fos induction (Liu et al. 2013). Pharmacological stimulation of MC4Rs in the MeA before restraint stress test induced anxiety-associated behaviors, increased plasma corticosterone, and reduced food intake. Conversely, MC4R loss or inhibition abolished these stress-induced responses (Ryan et al. 2014; Liu et al. 2013). Taken together, these data show that MC4R signaling in a POMC-MeA circuit strongly regulates behavioral responses to stress in rodents. Strategies targeting this melanocortin pathway could therefore lead to treatments in humans suffering from post-traumatic stress disorders.

Another behavior altered by stress is grooming. Self-grooming is an essential behavior present in arthropods, birds, and mammals to care for the body surface. In rodents, grooming comprises a highly stereotyped sequence of behaviors. It begins with licking the paws, then head, body, legs, genitals, and tail, interrupted occasionally by scratching and whole body shaking (Fentress 1988; Berridge et al. 2005). Humans also exhibit self-grooming behavior (Prokop et al. 2014; Cohen-Mansfield and Jensen 2007). While the brain stem initiates self-grooming movements and regulates the assembly of the sequential patterning in rodents, control of its sequencing requires the basal ganglia, particularly dopaminergic inputs to the striatum (Kalueff et al. 2016). However, in times of stress, grooming can occur at inappropriate times or with inappropriate intensity. The amygdala and other limbic regions modulate this context-specific behavior (Kalueff et al. 2016; Hong et al. 2014; Roeling et al. 1993). In rats, stressful conditions result in an excessive, aberrant form of grooming that damages the fur (Adan et al. 1999; Mul et al. 2012; Willemse et al. 1994).

Under these conditions, grooming loses its precise temporal patterning (Kalueff and Tuohimaa 2004, 2005). Aberrant rodent self-grooming resembles human disorders with abnormal self-grooming and other compulsive or stereotyped behaviors that do not require sensory feedback (Kalueff et al. 2007).

In dogs, rabbits, cats, rats, mice, and monkeys, central or cerebrospinal administration of melanocortins (or CRH) potently induces behavior similar to spontaneous grooming (Vergoni et al. 1998; Argiolas et al. 2000; Ferrari et al. 1955, 1963; Ferrari 1958; Gessa et al. 1967; Aloyo et al. 1983; Spruijt et al. 1985; Gispen et al. 1975; Dunn 1988; Dunn et al. 1987). The MC4R mediates this induced and spontaneous grooming behavior (Nijenhuis et al. 2003; Adan et al. 1994, 1999; for example, melanocortin peptides did not elicit any grooming response in rats deficient for MC4R (Mul et al. 2012). Additionally, administration of an MC4R antagonist reverses excessive grooming behavior (Adan et al. 1999).

Connections between the hypothalamus, amygdala, and mesolimbic reward system may allow melanocortins to alter stress-related grooming and its patterning (Hong et al. 2014; Roeling et al. 1993; Kruk et al. 1998; Homberg et al. 2002). It is known that dopaminergic cell bodies in the VTA receive input from GABAergic neurons, and ACh input can modulate their activity. A key study showed that α-MSH administration stimulates cholinergic neurons in the VTA to cause excessive grooming. When a GABA antagonist was injected before α-MSH, excessive grooming behavior increased (De Barioglio et al. 1991), suggesting the presynaptic actions of GABA can promote these melanocortin effects (Sanchez et al. 2001; Debarioglio et al. 1991). Thus, melanocortins can promote anxiety and stress-related behaviors including disordered self-care.

Interestingly, opioids also promote grooming and lead to excessive and obsessive grooming (Willemse et al. 1994; Ayhan and Randrup 1973). In addition, opioids like β-endorphin extend grooming bouts and prolong sensitivity to the grooming-inducing effects of melanocortins

(Jolles et al. 1978). α-MSH has no affinity for opiate receptors (Terenius et al. 1975); however, naloxone, a high-affinity μ-opioid antagonist, can still block α-MSH-induced grooming (Aloyo et al. 1983; Walker et al. 1982; van Wimersma Greidanus et al. 1986), suggesting α-MSH and β-endorphin target similar neural circuits. Additional research is needed to understand the interaction of melanocortins and opioids in grooming circuits. At a conceptual level, since opioids ease the ability to cope with stress, grooming may also act as a coping mechanism through which the organism lessens arousal.

11.9 Stress and Body Weight Regulation

Melanocortins are intimately involved in the effects of stress on body weight. Three mechanisms underlie this influence. First, stressors induce a sympathetic and HPA axis stress response, in which, as we have seen, melanocortins play an integral role. Second, food stress interacts with hypothalamic circuits that include POMC neurons regulating caloric intake and expenditure. Finally, physical and psychological stressors act on mesolimbic dopaminergic pathways that express melanocortin receptors to influence hedonic feeding (Lutter and Nestler 2009). During acute stress, these mechanisms allow melanocortins to suppress feeding and promote energy expenditure. However, chronic stress can oppose and undermine these actions. In addition to these topics, two situations deserve special attention: the stress of food scarcity and the stress of obesity itself. Finally, we will discuss recent progress in investigating how sensitivity to melanocortins and their downstream effectors can be restored when adaptive responses have failed.

11.9.1 Food Insecurity and Obesity

Since most threats are difficult to anticipate, complex organisms must have a set of responses that will be appropriate for any attack, injury, or illness an animal is facing. As we have seen, melanocortins play a key role in coordinating the body's perception of and response to acute stress by releasing stored energy, increasing pain sensitivity, increasing anxious behavior, and preventing the diversion of resources for inflammation and related recuperative processes. This array of physiological processes does not require the precise nature of the threat to be identified. These reactions are generally useful regardless of the threat, although they may fail to deal adequately with chronic stressors.

Food scarcity, in contrast, is a specific and predictable threat. It might well have been the first stressor encountered by organisms. Coping mechanisms for famine predate the development of the melanocortin system, the HPA axis, and the SNS. These later systems were later incorporated into the overall response for preventing starvation.

An organism accustomed to food insecurity will often take advantage of temporary abundance by maturing and reproducing quickly. When food is scarce, two strategies are available: spending additional energy to find and digest food or suppressing the metabolic rate as much as possible to extend life span. These choices are dramatically demonstrated by the nematode *C. elegans*, where an insulin-like signaling pathway regulates reproduction, life span, and entry into a dormant state (Fletcher and Kim 2017; Ren et al. 1996; Schackwitz et al. 1996). Specifically, insulin, cGMP, and TGF-β pathways signal a favorable environment and encourage continued growth and reproduction (Riddle et al. 1981; Kenyon et al. 1993; Vowels and Thomas 1992; Gottlieb and Ruvkun 1994). When food is limited, young worms assume a nonreproductive form specialized for long-term survival instead of developing to adulthood (Riddle et al. 1981). Thus, insulin acquired an important role in energy allocation and food intake before the development of the neuronal melanocortin system.

In mammals, arcuate POMC neurons and associated circuits play an essential role in the control of food intake (Hill and Faulkner 2017). While stimulation of POMC neurons inhibits feeding behavior, stimulating AgRP/NPY neurons provokes feeding (Aponte et al. 2011; Zhan et al. 2013; Krashes et al. 2011). The inhibition of POMC or AgRP/NPY neurons can lead to obesity or anorexia, respectively (Yaswen et al. 1999; Gropp et al. 2005; Luquet et al. 2005). Fasting activates NPY/AgRP neurons and suppresses the activity of POMC neurons. MC3Rs may reinforce this pattern of neuronal activity. MC3R knockout mice show no adjustment of circadian corticosterone secretion or orexigenic neuropeptide expression to food restriction (Girardet et al. 2017). When access to food is restricted to a brief window each day, MC3Rs are required for binge feeding, anticipatory activity, and entrainment to nutrient availability (Butler et al. 2017; Begriche et al. 2012; Girardet et al. 2017; Mavrikaki et al. 2016). Thus, while MC3Rs in the CNS have a minor impact on feeding behavior in mice when food is plentiful, they regulate the motivation to feed and possibly the discomfort of hunger during food restriction (Girardet et al. 2017).

Both POMC and AgRP/NPY neurons can sense circulating metabolic factors such as leptin and insulin, thought to allow them to regulate food intake and energy expenditure appropriately (Varela and Horvath 2012). However, the ability of insulin to regulate food intake, energy balance, and glucose homeostasis may depend primarily on its actions in a more ancient set of NPY neurons (only some of which express AgRP) in both rodents and fruit flies (Loh et al. 2017; Konner et al. 2007). If true, insulin sensing by POMC neurons may primarily regulate adipose tissue lipolysis and prevent hepatic fat storage during exposure to high-caloric diets in adult animals (Shin et al. 2017). In addition, insulin signaling in POMC neurons may reinforce the actions of leptin in this neuronal population; both contribute to systemic insulin sensitivity and the browning of white fat (Hill et al. 2010; Dodd et al. 2015). This adjustment can occur prenatally or in early infancy; hyperinsulinemia influences the formation of POMC circuits postnatally, resulting in hyperphagia and an obese phenotype in adulthood (Vogt et al. 2014).

Leptin, in contrast, took on its role in energy balance more recently. *C. elegans* has no appar-

ent leptin ortholog; instead, it may use products of the fat metabolism pathway to regulate feeding behavior (Hyun et al. 2016). Although found in numerous vertebrate species, leptin appears to have evolved its role as an adiposity signal in tetrapods (Cui et al. 2014; Prokop et al. 2012). Mammalian leptin shows particularly high sequence conservation (Doyon et al. 2001), which we suggest is due to the critical nature of fat depot regulation in warm-blooded animals. The development of the arcuate melanocortin system and its ability to sense leptin and insulin no doubt added robustness and precision to the control of body weight and metabolic homeostasis in mammals.

While missing a meal is not an acute threat to a healthy individual, food scarcity or insecurity acts as a psychological stressor. Fasting increases cortisol levels (Nakamura et al. 2016). Placing mice accustomed to high-fat chow on a "diet" of low-fat food induced stress, anxiety, depression, and high motivation to consume both sucrose and fatty food (Sharma et al. 2013; Avena et al. 2008; Cottone et al. 2009). Low food security combined with plentiful high-calorie, energy-dense foods causes weight gain in humans to increase (Wilde and Peterman 2006). Remarkably, just the perception of scarcity in resources can increase the desire for calories and anxious behavior (Briers and Laporte 2013). As a result, many social solutions for addressing food insecurity do not reduce and can even increase obesity (Leroy et al. 2013; Jones and Frongillo 2006; Townsend et al. 2001). These findings suggest increased energy intake among those of low socioeconomic status may be a fundamental response to threats to food security, which persists regardless of the actual food supply (Dhurandhar 2016).

11.9.2 Acute Stress and Body Weight

The stress endocrine axis arose to divert energy from nonessential functions to life-sustaining energy conservation. Coping with an acute stress requires potentially high levels of energy expenditure. Under acute stress brought about by an imminent threat, HPA axis and sympathetic activation serve to liberate energy into the bloodstream for use by the muscles and cardiovascular system. Thus, glucocorticoid hormones promote gluconeogenesis, lipolysis, and insulin resistance to raise circulating levels of glucose and fatty acids acutely (Ottosson et al. 2000; Bjorntorp 1996). When combined with energy use, these actions result in weight loss.

Behavioral changes accompany these hormonal effects. Attention and effort cannot be expended on restoring energy reserves until the immediate danger has passed and the compensatory stage begins. So, it is not surprising that acute and intense stressors, including illness, inhibit feeding (Krahn et al. 1990; Rybkin et al. 1997). For example, intense emotional stress suppresses appetite in humans and laboratory rodents (Valles et al. 2000; Laurent et al. 2013; De Souza et al. 2000). Anticipatory fear promotes hypophagia and anorexia in otherwise hungry rats. These effects depend on activity in the central amygdala, likely working with the ventromedial prefrontal cortex and lateral hypothalamic area (Land et al. 2014; Mena et al. 2013; Petrovich et al. 2009).

Melanocortins are very important for appetite suppression under stressful conditions. Using double immunolabeling techniques, it has been shown that most POMC neurons in the arcuate nucleus are glucocorticoid receptor positive (Cintra and Bortolotti 1992). Thirty minutes of restraint stress activates ARC POMC neurons and MC4R expressing neurons in the MeA, stimulates the HPA axis, induces anxious behavior, and reduces food intake (Ryan et al. 2014; Liu et al. 2013; Baubet et al. 1994). Anorexia and weight loss induced by stress were reversed by MC4 receptor blockade (Vergoni et al. 1999b). Similar effects were seen with infusion of an MC4R agonist to the MeA, while blockade of MC4R in this brain region attenuated restraint stress-induced anorectic effects and endocrine responses (Liu et al. 2013). Therefore, enhanced arcuate melanocortinergic input to the MeA during stress may contribute to anorexia and HPA axis activation.

Mice subjected to chronic stress, such as restraint stress, are also anhedonic as demonstrated by a loss of preference for a sucrose solution over water (Lim et al. 2012; Nestler and Hyman 2010). Reduced activation of D1 medium spiny neurons in the nucleus accumbens may underlie this effect. The loss of sucrose preference requires MC4R activation in NAc D1-MSNs, since knockdown of MC4R in the NAc or specifically in D1-MSNs prevented this loss (Lim et al. 2012). Therefore, release of α-MSH by POMC neurons can suppress activity in dopaminergic neurons in the nucleus accumbens and lead to the loss of appetite that is associated with stress and depression.

POMC neurons may also induce stress-related anorexia by acting directly or indirectly on CRH neurons. CRH neurons affect food intake (Bale et al. 2002; Menzaghi et al. 1993); chronic administration of CRH into the hypothalamus or activation of CRH-2 receptors decreases food intake and body weight gain in rats (Tempel and Leibowitz 1994; Fekete and Zorrilla 2007). Injecting a CRH-2 receptor blocker into the BNST attenuated restraint-induced anorexia (Ohata and Shibasaki 2011). The effects of CRH may be due to it suppressing NPY synthesis and release, thus reducing food intake (Tempel and Leibowitz 1994; White 1993). Melanocortin-sensitive MeA neurons project to the vicinity of the PVH where projections to CRH neurons can influence HPA output (Herman and Morrison 1996; Cullinan 2000; Miklos and Kovacs 2002). In addition, the MeA projects to BNST CRH neurons that directly innervate PVN CRH neurons (Ohata and Shibasaki 2011; Coolen and Wood 1998; Ciccocioppo et al. 2003). By increasing CRH release, melanocortins may suppress feeding.

In addition, the effect of glucocorticoids on circulating leptin levels may play a role. Glucocorticoids can directly increase leptin levels (Mostyn et al. 2001; Zakrzewska et al. 1999; Dagogo-Jack et al. 1997). In normal humans, administration of dexamethasone can increase plasma leptin almost threefold compared to controls (Miell et al. 1996). Similarly, repeated injection of dexamethasone for 4 days in rats dramatically increased plasma leptin levels, reduced body weight, and suppressed food intake (Jahng et al. 2008). In response to increased leptin levels, POMC neuronal activity increases to promote satiety; this mechanism may contribute to the suppressive effect of acute stress on the appetite.

Acute illness is another stressor with suppressive effects on appetite. LPS stimulates insulin and leptin secretion in peripheral tissues and secretion of other pro-inflammatory cytokines in microglial cells and periphery (Grunfeld et al. 1996). Altered leptin and cytokine levels during an inflammatory challenge suppress food intake (Borges et al. 2016a). A recent study found that this effect requires activation of the PI3K/Akt pathway in hypothalamic neurons (Borges et al. 2016b). Acute inflammation, like that induced by LPS and IL-1β, leads to activation of POMC neurons (Ellacott and Cone 2006), increases expression of MC4R (Borges et al. 2011), and increases POMC expression (Jang et al. 2010; Endo et al. 2007). MC4R antagonism can prevent LPS-induced anorexia (Jang et al. 2010; Sartin et al. 2008; Huang et al. 1999). Interestingly, data from pharmacogenetically activated AgRP neurons in LPS-treated mice show that AgRP-DREADD neuronal activation does not prevent LPS hypophagia in mice (Liu et al. 2016). Thus, leptin activation of PI3K and Jak-STAT signaling after LPS administration may stimulate transcription of POMC, which inhibits food intake and promotes weight loss (Borges et al. 2016a). Even so, studies have found no detectable changes in LPS-induced c-Fos expression in POMC neurons (Liu et al. 2016; Gautron et al. 2005). The precise role of POMC neurons in LPS-induced hypophagia remains inconclusive and further studies are warranted.

The mechanism inducing cachexia in longer-term illnesses, such as cancer (Michalaki et al. 2004; Okada et al. 1998; Andersson et al. 2014), HIV (Roberts et al. 2010), heart failure (Rauchhaus et al. 2000; Pan et al. 2004), and COPD (Humbert et al. 1995), is unclear and likely to be multifactorial (Ezeoke and Morley 2015). In such cases leptin levels drop (Lopez-Soriano et al. 1999) and POMC expression decreases (Suzuki et al. 2011; Hashimoto et al. 2007; Dwarkasing et al. 2014; Wisse et al. 2003). IL-1β activates and depolarizes POMC neurons in the ARC, suggesting that this cytokine takes part in the hypophagia during these diseases (Scarlett et al. 2007). However, blocking cytokines in the presence of cancer (Arruda et al. 2010; Strassmann et al. 1992; Fujimoto-Ouchi et al. 1995; Gelin et al. 1991) or HIV (Ting and Koo 2006) only results in a partial, though significant, reduction of anorexia-cachexia. AgRP inhibits anorexia in mice carrying sarcomas (Marks et al. 2001). In addition, melanocortin antagonists increase food intake in several cancer models (Tran et al. 2007; Chen et al. 2008; Jiang et al. 2007; Markison et al. 2005; Dallmann et al. 2011; Weyermann et al. 2009; Wisse et al. 2001). Furthermore, MC4R knockout mice show no decrease in food intake when they carry lung adenocarcinoma (Wisse et al. 2001). In contrast, an MC4R antagonist did not restore feeding in rats with a methylcholanthrene-induced sarcoma (Chance et al. 2003). Therefore, the melanocortin system mediates the cachexia produced by some, but not all, cancers (Ezeoke and Morley 2015).

11.9.3 Chronic Stress and Body Weight

The stress that humans encounter on a daily basis is generally prolonged and mild, unlike the intense stressors that laboratory animals undergo to induce appetite suppression. In humans, the effect of chronic stress on feeding is highly variable. This type of stressor can induce either weight gain or anorexia (Oliver et al. 2000; Zellner et al. 2006; Pollard et al. 1995; Adam and Epel 2007; Serlachius et al. 2007). Evidence suggests that lean individuals may be more prone to weight loss, while overweight individuals tend to increase body weight in response to chronic stress (Kivimaki et al. 2006). On average, chronic psychological life stress induces weight gain

(Torres and Nowson 2007); in a meta-analysis of 13 studies, job strain positively correlated with BMI (Nyberg et al. 2012).

Chronic stress in humans (Bjorntorp and Rosmond 2000; Peeke and Chrousos 1995; Wallerius et al. 2003) and rodents (Rebuffescrive et al. 1992) increases glucocorticoid levels, adipocyte size, and abdominal fat. Greater responsiveness of the HPA axis generally correlates with abdominal obesity (Rodriguez et al. 2015). In one study, women were subjected to three sessions of stressful activities such as public speaking, math tests, and visuospatial puzzles over the course of 3 days. Unlike lean women, the women with the most central fat secreted high levels of cortisol on the first day. In addition, they failed to show cortisol habituation or a drop in cortisol secretion on subsequent days once the tests were familiar (Epel et al. 2000). Poverty is also associated higher basal cortisol levels and a lack of cortisol habituation (Adler et al. 2000; Gruenewald et al. 2006; Hellhammer et al. 1997; Kirschbaum et al. 1995). These findings likely indicate greater exposure to repeated challenges in these individuals that results in dysregulation of the stress response (Adler and Snibbe 2003).

Chronic cortisol exposure promotes the conversion of preadipocytes to mature adipocytes, expanding the adipose tissue and promoting the secretion of pro-inflammatory cytokines and adipokines (Peckett et al. 2011; Andrews and Walker 1999). These actions contrast with the lipolysis induced by acute glucocorticoid release. It is possible that GRα mediates the lipolytic effects of glucocorticoids, while MR and GRβ mediate adipogenesis during chronic glucocorticoid exposure (John et al. 2016). Excess glucocorticoid secretion may be amplified locally within adipose tissue by the activating enzyme 11β HSD1. 11β HSD1 is elevated in the adipose tissue of people with morbid obesity and metabolic syndrome (Baudrand et al. 2010; Luisella et al. 2007; Valsamakis et al. 2004; Constantinopoulos et al. 2015) and normalized by bariatric surgery (Methlie et al. 2013; Woods et al. 2015). Despite highly promising animal

studies (Morton et al. 2004; Tiwari 2010; Morgan et al. 2014), inhibitors of 11β-HSD1 in humans have shown inconsistent results for treating metabolic syndrome in clinical trials (Walker et al. 1995; Andrews et al. 2003; Shah et al. 2011; Feig et al. 2011; Rosenstock et al. 2010). New inhibitors with higher specificity for the enzyme and a preference for adipose tissue may be required before this treatment strategy is viable.

In addition to directly promoting adiposity, stressors can lead to alterations in energy intake. Chronic life stress leads to increased appetite, binge eating, and craving energy-dense foods, snacks, and fast foods (Epel et al. 2001; Steptoe et al. 1998; Oliver and Wardle 1999; Gluck et al. 2004). In female rhesus monkeys, social subordinates under social stress eat more when offered unlimited access to rich foods than social dominants (Arce et al. 2010; Michopoulos et al. 2012). In rodents, anorexia from restraint stress later leads to increased intake of food high in fat and sugar (Foster et al. 2009; la Fleur et al. 2005). Likewise, animals stressed by repeated mild pinch exhibited hyperphagia of sweet food and a large gain in body weight (Pecoraro et al. 2004). These behaviors are under the control of the dopaminergic mesolimbic reward pathways and the HPA axis (Dallman et al. 2006). They are used to calm and sooth emotions to recover from the recurring stressors (Dallman et al. 2006). This strategy is, in fact, effective in reducing HPA activation (Foster et al. 2009; la Fleur et al. 2005; Pecoraro et al. 2004; Ortolani et al. 2011). Eating often improves mood, reduces irritability, and increases calmness (Gibson 2006). The opioid system, which interacts with the mesolimbic dopamine pathway, is a key mediator of this hedonic feeding; mu-opioid receptors mediate the rewarding properties of food and some drugs of abuse (Blasio et al. 2014; Zhang and Kelley 2000; Nathan and Bullmore 2009). So, β-endorphin production by POMC neurons may promote the hyperevaluation of palatable foods, leading to the loss of control during overeating.

Interestingly, a modest amount of sucrose intake can reduce behavioral and physiological stress responses without leading to obesity. The basolateral amygdala, a key reward- and stress-regulatory brain region, is necessary for sucrose-induced stress relief and undergoes synaptic remodeling following sucrose intake (Ulrich-Lai et al. 2010, 2015). Overall, stress reduction occurs in rats with voluntary intake of limited amounts of sugar or carbohydrates with no increase in body weight (Ulrich-Lai et al. 2015). These results suggest that using small amounts of sweet treats to reduce stress can align with healthy body weight goals (Ulrich-Lai et al. 2010).

Altered CRH levels could mediate some of these effects. Stress induces CRH release by cells in the PVN as well as the medial amygdala. CRH-1 receptor activation increases palatable food consumption and binge eating (Koob 2010; Parylak et al. 2011). Indeed, antagonism of CRH-1 receptors in socially subordinate female rhesus macaques blocks increased palatable food consumption (Moore et al. 2015). Therefore, increased CRH-1 signaling induced by stress could promote excess food intake. In addition, ghrelin, a peptide produced by gastrointestinal endocrine cells that induces feeding and anxious behavior (Currie et al. 2005; Seoane et al. 2004; Kojima et al. 1999), rises in response to stress. In animal models, circulating ghrelin levels increase in response to social defeat (Lutter et al. 2008), restraint stress (Zheng et al. 2009), and chronic stress (Ochi et al. 2008). Mice subjected to social defeat had increased ghrelin levels and consumed more of a high-fat diet (Chuang et al. 2011). Ghrelin appears to increase preference for sweet food independent of calorie content since ghrelin administration increased consumption of a saccharin solution (Disse et al. 2010). Blocking or ablating the ghrelin receptor decreases intake of palatable food compared to standard chow (Egecioglu et al. 2010). Ghrelin directly activates AgRP/NPY neurons to stimulate feeding and increase inhibitory GABAergic input on POMC cells to suppress release of melanocortins (Briggs et al. 2010; Andrews et al. 2008; Andrews 2011). Peripheral and central ghrelin administration also activates CRH neurons (Cabral et al. 2012;

Asakawa et al. 2001), which may promote binge eating. Thus, increased ghrelin levels may partly mediate stress-induced feeding.

Glucocorticoids may also have direct actions on food intake. The glucocorticoid receptor is widely expressed in the CNS. It is found in reward areas as well as in key appetite regulatory regions like the arcuate nucleus, lateral hypothalamus, and paraventricular nucleus of the hypothalamus (Morimoto et al. 1996; Reul and de Kloet 1986; Aronsson et al. 1988; Cintra et al. 1987; McEwen et al. 1986). In contrast, the mineralocorticoid receptor in the CNS is mainly restricted to the septum, hippocampus, and amygdala (Sanchez et al. 2000). Glucocorticoids therefore have direct access to brain sites that regulate energy metabolism and reward. In humans, individuals with a strong cortisol response consumed the most food during an experimental stress session (Epel et al. 2001). In addition, glucocorticoid administration caused higher food intake in subjects allowed ad libitum food selection (Tataranni et al. 1996).

Several mechanisms have been suggested for how glucocorticoids induce feeding. Glucocorticoids could stimulate feeding responses by inhibiting CRH release in the hypothalamus (Cavagnini et al. 2000). CRH and related stress peptides like urocortin can act through the CRH-2 receptor to suppress feeding (Stengel and Tache 2014). However, as noted above, a reduction in CRH would also suppress CRH-1 signaling that promotes food intake. Alternatively, glucocorticoids may directly stimulate feeding responses by increasing the release of NPY and/or AgRP in the hypothalamus. Glucocorticoids increase AgRP and NPY expression (Goto et al. 2006; Sato et al. 2005). Likewise, adrenalectomy decreases NPY levels and corticosterone replacement restores them (White et al. 1990). An important recent study found that deletion of GR on AgRP neurons resulted in leanness on chow diet in females and resistance to diet induced obesity in both sexes (Shibata et al. 2016). Interestingly, food intake was unchanged, but metabolic rate was increased due to brown adipose tissue activity. These results suggest that glucocorticoids can promote obesity by acting in on AgRP neurons to suppress energy expenditure (Shibata et al. 2016). Additional research will be needed to fully understand how glucocorticoids interact with homeostatic feeding circuits.

11.9.4 Obesity-Induced "Stress"

Obesity is increasingly being described as a state of "energetic stress." Overconsumption of a high-fat, high-sugar diet can in essence serve as a physiological challenge (Gibson 2006; Anderson et al. 1987; Barr et al. 1999; Decastro 1987; Deuster et al. 1992; Dube et al. 2005; Fernandez et al. 2003; Lieberman et al. 1986; Utter et al. 1999). Multiple mechanisms allow energetic stress to interact with neuroendocrine stress response systems, including by impacting the sympathetic nervous system and altering the gut microbiota (Harrell et al. 2016). Although melanocortins normally suppress energy use for inflammation and food seeking to permit a fast response to danger, a chronic rise in inflammation can undermine the ability of POMC neurons to modulate energy use and intake. The result is failure of allostasis or homeostatic adaptation.

High-fat diet feeding rapidly activates multiple inflammatory and stress response pathways in the hypothalamus (De Souza et al. 2005). High-fat diet exposure induces hypothalamic inflammation before body weight gain (Thaler et al. 2012) and before peripheral tissues like the liver develop inflammation (Tolle and Low 2008). For example, saturated fats, but not monounsaturated fats, induce the TLR4 and MyD88 inflammatory signaling cascades within days, compromising hypothalamic function (Lee et al. 2001; Kleinridders et al. 2009; Valdearcos et al. 2014). A high-calorie diet rapidly stimulates microglial reactivity in the mediobasal hypothalamus (Thaler et al. 2012; Gao et al. 2014), leading the microglia to increase TNF-α production.

These findings suggest that the loss of sensitivity of POMC neurons to signals of adiposity caused by inflammation can perpetuate overeating. These pathways cause neuronal insulin and leptin resistance, which leads to the failure of anorexigenic melanocortin circuits to suppress more feeding. In parallel to the early occurrence of inflammation, 3 days of HFD feeding is enough to reduce hypothalamic insulin sensitivity in rodents substantially (Corander et al. 2011). Specifically, brain-specific activation of IKKβ interrupts central insulin and leptin signaling and results in increased food intake and body weight gain (Bouyer and Simerly 2013). Activation of NF-kB induces expression of suppressor of cytokine signaling 3 (SOCS3), which then inhibits neuronal insulin signaling (Bouyer and Simerly 2013). Pharmacologic inhibition of neuronal TLR4 signaling inhibits fatty acid-induced insulin (Schwartz 2000) and leptin resistance (Magoul et al. 1993). In the same way, mice with CNS-specific ablation of MyD88 resist HFD-induced weight gain and deterioration of glucose metabolism (Rinaman 2010).

The ER system further amplifies these HFD-induced perturbations by activating unfolded protein response (UPR) signaling pathways (Jacobowitz and Odonohue 1978; Young 1935; Eddy and Strahan 1968). ER stress and IKK/NF-kB promote each other during HFD feeding and worsen the energy imbalance underlying obesity (Bouyer and Simerly 2013). Central induction of ER stress inhibits the ability of leptin and insulin to reduce food intake and body weight (Vallarino et al. 1988). Conversely, mice with neuron-specific deletion of ER stress activator Xbpl show increased leptin resistance and adiposity (Young 1935). Constitutive expression of Xbpls selectively in POMC neurons represses Socs3 and protein tyrosine phosphatase IB (PtpIB) expression and protects against HFD-induced obesity (Vallarino et al. 1989). Therefore, ER stress and the UPR are potent regulators of POMC neurons.

Central inflammatory processes and weight gain lead to low-grade activation of the immune system throughout the body. Obesity tightly correlates with elevations in inflammatory factors, such as tumor necrosis factor (TNF)-α and interleukin-6 (IL-6) (Hotamisligil et al. 1993; Xu et al. 2003). Prolonged low-grade systemic inflammation results in tissue damage and exacerbates disease processes, such as insulin resistance. TNF-α and IL-6 inhibit serine phosphorylation in the insulin receptor substrate-1 (IRS-1), disrupting insulin signaling transduction and causing insulin resistance (Wellen and Hotamisligil 2005). Low-grade inflammation is an independent risk factor for heart disease (Hansson 2005) stroke (Corrado et al. 2006), diabetes (Pradhan et al. 2001; Spranger et al. 2003), and all-cause mortality (Ford 2005). For example, chronic inflammation that develops within atherosclerotic plaques can cause stroke or myocardial infarction by leading to plaque rupture (Libby 2002). Age-related macular degeneration (Telander 2011) and Alzheimer's disease (Wyss-Coray 2006) and osteoarthritis (Sokolove and Lepus 2013) associate with innate immune activation and low-grade inflammation.

Extended overnutrition perpetuates hypothalamic inflammatory interactions between neurons and non-neuronal cell populations. These effects ultimately lead to overeating and further weight gain (Jais and Bruning 2017). Persistent microglial reactivity and TNF-α production have a specific harmful effect on POMC neurons (Thaler et al. 2012). Recently it was reported that TNF-α released by microglia induces mitochondrial stress in POMC neurons; TNF-α acts on POMC neurons to promote mitochondrial ATP production, cause mitochondrial fusion in neurites, and elevate neuronal excitability and firing rates (Yi et al. 2017). In the long run, these actions may disrupt the ability of POMC neurons to suppress feeding and increase energy use, leading to obesity.

11.10 Conclusions

The current obesity crisis is being driven by increased consumption of widely available, palatable, high-calorie food coupled with decreased activity in daily life. The neural pathways underlying the motivation for and

enjoyment of foods high in fat and sugar have been well studied (Castro et al. 2015). These include dopaminergic pathways projecting from the nucleus accumbens to the ventral tegmental area and areas of the NA and ventral pallidum sensitive to endogenous opioids. Arcuate POMC neurons can influence this system at several levels. POMC neurons innervate key neural nodes of the mesocorticolimbic system, including the VTA and NAc (Lim et al. 2012; King and Hentges 2011). While β-endorphin has only a minor impact on the enjoyment of foods (Mendez et al. 2015), melanocortins like α-MSH can influence the motivation to obtain food. Intra-VTA α-MSH acts through the MC4R to increase NAc dopamine levels (Lindblom et al. 2001).

As we have seen, chronic stress increases the consumption of certain palatable foods ("comfort foods") in both animals and humans (Pecoraro et al. 2004; Dallman et al. 2003; Fairburn 1997). It can also precipitate eating disorders like binge eating (Cifani et al. 2009; Hagan et al. 2003). In fact, binge eating can be induced in rodents with a combination of stress and caloric restriction (Hagan et al. 2002; Boggiano and Chandler 2006). No pharmaceuticals have been approved for reducing common forms of emotional eating in response to chronic life stress. However, binge eating disorder shows improvement when treated with amphetamines, which regulate dopamine release, as well as off-label antidepressants and anti-seizure medications. Developing technologies may permit pharmaceuticals that specifically target emotional eating to be designed in the future (Caruso et al. 2014; Hill and Faulkner 2017). Until such drugs are available, obesity treatment should be individualized using tailored strategies to address the type of hedonic eating in each patient. For instance, some patients may benefit from becoming more selective in the food they eat, demanding higher quality and eating slowly to enable them to maintain the same satisfaction while eating less food (Scarinci 2004). Learning alternative methods for coping with stress (such as exercise, focused breathing, progressive muscle relaxation, mediation) may assist patients in avoiding stressed-induced overeating.

A wise health professional will also address the underlying causes of stress to promote the overall well-being of the patient. As previously mentioned, this approach is more effective in reducing obesity than efforts to improve diets directly in at-risk populations. For example, low socioeconomic status populations may not use exercise facilities made available to them (Giles-Corti and Donovan 2002). Likewise, giving money or food to a low socioeconomic population in rural Mexico causes weight gain rather than loss (Leroy et al. 2013). Another study found that increasing government food vouchers to $2000 per year had no effect on BMI disparities between social strata (Jones and Frongillo 2006). Hoarding calories appears to be a psychological mechanism to buffer against the stress of low socioeconomic status (Dhurandhar 2016). Instead, interventions focused on improving socioeconomic opportunity, with no focus on nutrition or physical activity, may improve rates of obesity and diabetes. For example, randomizing families to move to a more well-off neighborhood reduced average BMI without additional assistance (Ludwig et al. 2011). These data demonstrate that, unlike nutrition programs, social interventions can reduce obesity. Therefore, obese patients with the most stress-filled lives, including those in poverty or recovering from trauma, require referral to assistance programs that focus on the underlying causes of insecurity.

Equilibrium in body weight is described as a "set point" of adiposity that the body defends against intentional or unintentional weight loss or gain. By definition, homeostatic processes cannot initiate obesity. However, the homeostatic processes suppressing body weight gain seem weaker than those preventing drops in body weight. Whether this fact is due to beneficial effects of storing additional energy in case of famine or because modern humans face few negative short-term consequences of obesity is unclear (Speakman 2008; Sellayah et al. 2014). In many individuals, the hedonic drive to overconsume in a food environment of easily obtainable, palatable, and energy-dense foods succeeds in increasing body weight, which the homeostatic system then defends against weight loss. Increased body weight leads to cellular leptin resistance in arcuate circuits regulating feeding that diminishes the ability of hyperleptinemia to act on the melanocortin system to suppress food intake and increase energy expenditure (Myers et al. 2010).

Intentional weight loss causes leptin and insulin levels to decrease (Rosenbaum and Leibel 2014). Interestingly, leptin falls more than expected from the magnitude of fat loss (Myers et al. 2010) and remains low if weight loss is maintained (Kissileff et al. 2012; Naslund et al. 2000). In response, arcuate melanocortin and NPY circuitry increase the drive for food and to reduce energy expenditure. In addition, circulating levels of the orexigenic hormone ghrelin increase while the anorexigenic hormones CCK, PYY, and GLP-1 fall (Melby et al. 2017). These changes result in increased hunger (Chaput et al. 2007), food cravings (Gilhooly et al. 2007), and less satiation after eating (Cornier et al. 2004). Weight loss also chronically suppresses energy expenditure, including resting metabolic rate, the thermic effect of food, exercise energy expenditure, and non-exercise activity thermogenesis (Kissileff et al. 2012; Melby et al. 2017; Fothergill et al. 2016; Martin et al. 2007; Byrne et al. 2012; Knuth et al. 2014). Because of these effects, current approaches to substantial weight loss maintenance require constant vigilance and motivation on the part of the patient (Melby et al. 2017). The frequent failure of individuals to maintain weight loss discourages patients from attempting to lose weight. A method of altering the body weight set point would be transformative for patient care.

Recently, an important study has made advances in understanding the biological basis of the set point. Exogenous leptin normally suppresses food intake and induces weight loss. In obese humans (Zelissen et al. 2005) and animals (Enriori et al. 2007; Frederich et al. 1995), leptin administration fails to have this effect, likely as a result of leptin resistance induced by chronic exposure to hyperleptinemia (Knight et al. 2010; Gamber et al. 2012). Weight loss reverses this resistance, allowing leptin to assist in the maintenance of weight loss (Rosenbaum and Leibel 2014; Chhabra et al. 2016).

Similarly, weight-reduced MC4R-null mice respond to leptin treatment (Marsh et al. 1999). Previous work had shown that mice lacking POMC expression develop obesity, hyperleptinemia, and leptin resistance (Bumaschny et al. 2012). Interestingly, reducing the weight of these mice through food restriction did not restore the ability of leptin to inhibit feeding. In other words, simply restoring intracellular leptin signaling was insufficient to restore the effects of leptin; rather, a second defect downstream of the leptin receptor exists in these mice. Given that both MC4RKO (responsive to leptin when lean) and the arc POMCKO mice (not responsive to leptin when lean) have no activation of MC4R pathways, another receptor responsive to POMC products is responsible for conveying leptin responsiveness. Chhabra and coworkers next examined how to restore leanness to arc POMCKO mice. They found that reactivating POMC expression after to the establishment of obesity did not normalize body weight. However, if the mice were first calorie restricted to reduce their body weight, POMC reexpression permitted them to maintain that weight at a new, lower set point (Chhabra et al. 2016). Critically, this normalization could be prevented by inducing hyperleptinemia with exogenous leptin. Therefore, both hypothalamic leptin sensitivity and *Pomc* gene expression regulate the body weight set point. If true, weight loss in the obese patient restores the effects of leptin (Rosenbaum and Leibel 2014; Chhabra et al. 2016; Quarta et al. 2016) both due to improved intracellular signaling by leptin and also increased activation of a receptor for POMC products other than MC4R, such as the MC3R or mu-opioid receptor.

Work described in the previous section has led to the concept of MC3Rs sensitizing AgRP neurons to the metabolic state of the animal and regulating hunger (Girardet et al. 2017). If MC3Rs modulate the metabolic "set point" in conjunction with leptin, effective leptin signaling induced by relatively low levels of leptin needs to be synchronized with a normal level of hunger and energy expenditure through modulation of MC3R action or expression. In theory, this combination can restore a set point in the normal body weight range. Therapies targeting melanocortin signaling may restore normal body weight only when plasma leptin levels are below a critical threshold. Regular exercise may also heighten the brain's sensitivity to leptin (MacLean et al. 2009), suggesting it could also be useful in combination therapy.

Pharmaceuticals targeting the melanocortin system hold promise for numerous disorders that range from opioid addiction to shock to PTSD. In the case of ischemic or neurodegenerative disorders, they are already showing exciting clinical potential (Arnason et al. 2013; Leone et al. 2013; Spaccapelo et al. 2013). As described above, targeting this system to alter the body weight set point could also be enormously useful for combating rising rates of obesity. This potential has led to many preclinical and clinical studies investigating how melanocortins can be harnessed to stimulate weight loss. Targeting melanocortin receptors for the treatment of obesity, however, has proven challenging. Clinical trial has revealed problematic side effects of MCR agonists, including cardiovascular actions like tachycardia and elevated arterial pressure (Royalty et al. 2014; Greenfield et al. 2009; Girardet and Butler 2014; Skibicka and Grill 2009; Kuo et al. 2003). Indeed, melanocortins promote hypertension (Harrell et al. 2016); POMC neuron stimulation by leptin leads to SNS hyperactivity (da Silva et al. 2013), likely via activation of MC4Rs in the VMH (Lim et al. 2016). The extensive role of melanocortins in the stress response makes these findings unsurprising.

A recent MCR agonist that just entered phase 3 clinical trials for patients with POMC deficiency has thus far avoided such side effects. Setmelanotide is an eight-amino acid cyclic peptide that acts as a full agonist of human MC4R. It binds with ~10-fold selectivity over human MC3R (Fani et al. 2014). Preclinical studies in obese rhesus macaques indicated subcutaneous setmelanotide reduced overall food intake, decreased body weight, improved glucose tolerance, and did not induce negative cardiac effects (Kievit et al. 2013). Phase 1 and 2 studies have successfully evaluated the safety, efficacy, tolerability, pharmacokinetics, and pharmacodynamics of the octapeptide in obese volunteers (Ericson et al. 2017; Chen et al. 2015; Kuhnen et al. 2016). The reason for a lack of cardiovascular side effects has not been established, but several possible explanations exist (Kievit et al. 2013). These include (1) differing receptor pharmacology or mechanism for activating the MC4R; (2) higher affinity for the MC3R than previous drugs, since MC3R activity may counteract sympathetic stimulation mediated by MC4R signaling (Wikberg and Mutulis 2008); or (3) lack of penetration by setmelanotide to brain regions controlling heart rate and blood pressure. Until the cause of the lack of side effects in this drug is clear, it will be hard to replicate its success. Future techniques that allow targeting of MC3R or MC4R receptors in specific brain regions such as the VTA may also have clinical potential (Vogel et al. 2016).

The evolution of melanocortins from serving solely as stress hormones to also serving as critical anorexigenic neuropeptides demonstrates the opportunistic nature of biology. Yet, this system remains profoundly integrated with the physiological stress response. This knowledge should guide clinical care and pharmaceutical development. Overall, a critical need exists for studies that focus more broadly on how the CNS coordinates behavioral, endocrine, and autonomic responses to stressors. Investigating the melanocortin system in this light may hold the key to future medical advances.

Summative Questions

1. *What peripheral actions of melanocortins can affect adiposity? Which MC receptors are involved?*
2. *How do glucocorticoids affect behavior?*
3. *How does α-MSH treat shock?*
4. *Contrast the effects of melanocortins and beta-endorphin on pain.*
5. *Why are MCs a promising target in the treatment of PTSD?*
6. *How does the consumption of sweet treats interact with stress?*
7. *What are the most effective treatment options for obese patients facing food insecurity?*

References

Ab Aziz, C. B., & Ahmad, A. H. (2006). The role of the thalamus in modulating pain. *The Malaysian Journal of Medical Sciences : MJMS, 13*(2), 11–18.

Adam, T. C., & Epel, E. S. (2007). Stress, eating and the reward system. *Physiology & Behavior, 91*(4), 449–458.

Adan, R. A., & Gispen, W. H. (2000). Melanocortins and the brain: From effects via receptors to drug targets. *European Journal of Pharmacology, 405*(1–3), 13–24.

Adan, R. A. H., et al. (1994). Differential-effects of melanocortin peptides on neural melanocortin receptors. *Molecular Pharmacology, 46*(6), 1182–1190.

Adan, R. A. H., et al. (1999). Characterization of melanocortin receptor ligands on cloned brain melanocortin receptors and on grooming behavior in the rat. *European Journal of Pharmacology, 378*(3), 249–258.

Adler, N. E., & Snibbe, A. C. (2003). The role of psychosocial processes in explaining the gradient between socioeconomic status and health. *Current Directions in Psychological Science, 12*(4), 119–123.

Adler, N. E., et al. (2000). Relationship of subjective and objective social status with psychological and physiological functioning: Preliminary data in healthy white women. *Health Psychology, 19*(6), 586–592.

Alam, T., et al. (2012). Expression of genes involved in energy homeostasis in the duodenum and liver of Holstein-Friesian and Jersey cows and their F-1 hybrid. *Physiological Genomics, 44*(2), 198–209.

Al-Chaer, E. D. (2013). Neuroanatomy of pain and pain pathways. In *Handbook of pain and palliative care* (pp. 273–294). New York: Springer.

Aloyo, V. J., et al. (1983). Peptide-induced excessive grooming in the rat – The role of opiate receptors. *Peptides, 4*(6), 833–836.

Alvaro, J. D., et al. (1996). Morphine down-regulates melanocortin-4 receptor expression in brain regions that mediate opiate addiction. *Molecular Pharmacology, 50*(3), 583–591.

Amano, M., et al. (2016). Immunohistochemical detection of corticotropin-releasing hormone (CRH) in the brain and pituitary of the hagfish, Eptatretus burgeri. *General and Comparative Endocrinology, 236*, 174–180.

Amir, S., & Amit, Z. (1979). The pituitary gland mediates acute and chronic pain responsiveness in stressed and non-stressed rats. *Life Sciences, 24*(5), 439–448.

An, J. J., et al. (2007). Peripheral effect of alpha-melanocyte-stimulating hormone on fatty acid oxidation in skeletal muscle. *Journal of Biological Chemistry, 282*(5), 2862–2870.

Anderson, K. E., et al. (1987). Diet-hormone interactions – Protein carbohydrate ratio alters reciprocally the plasma-levels of testosterone and cortisol and their respective binding globulins in man. *Life Sciences, 40*(18), 1761–1768.

Andersson, B. A., et al. (2014). Plasma tumor necrosis factor-alpha and C-reactive protein as biomarker for survival in head and neck squamous cell carcinoma. *Journal of Cancer Research and Clinical Oncology, 140*(3), 515–519.

Andrews, Z. B. (2011). Central mechanisms involved in the orexigenic actions of ghrelin. *Peptides, 32*(11), 2248–2255.

Andrews, R. C., & Walker, B. R. (1999). Glucocorticoids and insulin resistance: Old hormones, new targets. *Clinical Science, 96*(5), 513–523.

Andrews, R. C., Rooyackers, O., & Walker, B. R. (2003). Effects of the 11 beta-hydroxysteroid dehydrogenase inhibitor carbenoxolone on insulin sensitivity in men with type 2 diabetes. *The Journal of Clinical Endocrinology and Metabolism, 88*(1), 285–291.

Andrews, Z. B., et al. (2008). UCP2 mediates ghrelin's action on NPY/AgRP neurons by lowering free radicals. *Nature, 454*(7206), 846–851.

Aponte, Y., Atasoy, D., & Sternson, S. M. (2011). AGRP neurons are sufficient to orchestrate feeding behavior rapidly and without training. *Nature Neuroscience, 14*(3), 351–355.

Arce, M., et al. (2010). Diet choice, cortisol reactivity, and emotional feeding in socially housed rhesus monkeys. *Physiology & Behavior, 101*(4), 446–455.

Argiolas, A., et al. (1987). Paraventricular nucleus lesion prevents yawning and penile erection induced by apomorphine and oxytocin but not by ACTH in rats. *Brain Research, 421*(1–2), 349–352.

Argiolas, A., et al. (2000). ACTH- and alpha-MSH-induced grooming, stretching, yawning and penile erection in male rats: Site of action in the brain and role of melanocortin receptors. *Brain Research Bulletin, 51*(5), 425–431.

Arnason, B. G., et al. (2013). Mechanisms of action of adrenocorticotropic hormone and other melanocortins relevant to the clinical management of patients with multiple sclerosis. *Multiple Sclerosis Journal, 19*(2), 130–136.

Aronsson, M., et al. (1988). Localization of glucocorticoid receptor mRNA in the male rat brain by in situ hybridization. *Proceedings of the National Academy of Sciences of the United States of America, 85*(23), 9331–9335.

Arruda, A. P., et al. (2010). Hypothalamic actions of tumor necrosis factor alpha provide the thermogenic core for the wastage syndrome in cachexia. *Endocrinology, 151*(2), 683–694.

Arvidsson, U., et al. (1995). Distribution and targeting of a mu-opioid receptor (MOR1) in brain and spinalcord. *Journal of Neuroscience, 15*(5), 3328–3341.

Asai, M., et al. (2013). Loss of function of the melanocortin 2 receptor accessory protein 2 is associated with mammalian obesity. *Science, 341*(6143), 275–278.

Asakawa, A., et al. (2001). A role of ghrelin in neuroendocrine and behavioral responses to stress in mice. *Neuroendocrinology, 74*(3), 143–147.

Atasoy, D., et al. (2012). Deconstruction of a neural circuit for hunger. *Nature, 488*(7410), 172-+.

Avena, N. M., Rada, P., & Hoebel, B. G. (2008). Evidence for sugar addiction: Behavioral and neurochemical effects of intermittent, excessive sugar intake. *Neuroscience and Biobehavioral Reviews, 32*(1), 20–39.

Ayhan, I., & Randrup, A. (1973). Behavioural and pharmacological studies on morphine-induced excitation of rats. Possible relation to brain catecholamines. *Psychopharmacologia, 29*(4), 317–328.

Bagnol, D., et al. (1999). Anatomy of an endogenous antagonist: Relationship between Agouti-related protein and proopiomelanocortin in brain. *Journal of Neuroscience, 19*(18), RC26.

Bale, T. L., & Vale, W. W. (2004). CRF and CRF receptors: Role in stress responsivity and other behaviors. *Annual Review of Pharmacology and Toxicology, 44*, 525–557.

Bale, T. L., Lee, K. F., & Vale, W. W. (2002). The role of corticotropin-releasing factor receptors in stress and anxiety. *Integrative and Comparative Biology, 42*(3), 552–555.

Barr, R. G., et al. (1999). The response of crying newborns to sucrose: Is it a "sweetness" effect? *Physiology & Behavior, 66*(3), 409–417.

Barsegyan, A., et al. (2010). Glucocorticoids in the prefrontal cortex enhance memory consolidation and impair working memory by a common neural mechanism. *Proceedings of the National Academy of Sciences of the United States of America, 107*(38), 16655–16660.

Bartness, T. J., & Song, C. K. (2007). Thematic review series: Adipocyte biology. Sympathetic and sensory innervation of white adipose tissue. *Journal of Lipid Research, 48*(8), 1655–1672.

Baubet, V., et al. (1994). Effects of an acute immobilization stress upon Proopiomelanocortin (Pomc) messenger-Rna levels in the mediobasal hypothalamus – A quantitative in-situ hybridization study. *Molecular Brain Research, 26*(1–2), 163–168.

Baudrand, R., et al. (2010). Overexpression of 11 beta-Hydroxysteroid dehydrogenase type 1 in hepatic and visceral adipose tissue is associated with metabolic disorders in morbidly obese patients. *Obesity Surgery, 20*(1), 77–83.

Beaulieu, S., Gagne, B., & Barden, N. (1988). Glucocorticoid regulation of proopiomelanocortin messenger ribonucleic acid content of rat hypothalamus. *Molecular Endocrinology, 2*(8), 727–731.

Begriche, K., et al. (2012). Melanocortin-3 receptors are involved in adaptation to restricted feeding. *Genes Brain and Behavior, 11*(3), 291–302.

Behan, D. P., et al. (1995). Displacement of Corticotropin-releasing factor from its binding-protein as a possible treatment for Alzheimers-disease. *Nature, 378*(6554), 284–287.

Benjamins, J. A., Nedelkoska, L., & Lisak, R. P. (2014). Adrenocorticotropin hormone 1-39 promotes proliferation and differentiation of oligodendroglial progenitor cells and protects from excitotoxic and inflammation-related damage. *Journal of Neuroscience Research, 92*(10), 1243–1251.

Bernton, E. W., Long, J. B., & Holaday, J. W. (1985). Opioids and neuropeptides – Mechanisms in circulatory shock. *Federation Proceedings, 44*(2), 290–299.

Berridge, K. C., et al. (2005). Sequential super-stereotypy of an instinctive fixed action pattern in hyper-dopaminergic mutant mice: A model of obsessive compulsive disorder and Tourette's. *BMC Biology, 3*, 4.

Bertolini, A., & Ferrari, W. (1982). Evidence and implications of a melanocortins-endorphins homeostatic system. In *Neuropeptides and psychosomatic process* (pp. 245–261). Budapest: Akadémiai Kiadó.

Bertolini, A., Poggioli, R., & Ferrari, W. (1979). ACTH-induced Hyperalgesia in rats. *Experientia, 35*(9), 1216–1217.

Bertolini, A., et al. (1986a). Alpha-msh and other acth fragments improve cardiovascular function and survival in experimental hemorrhagic-shock. *European Journal of Pharmacology, 130*(1–2), 19–26.

Bertolini, A., et al. (1986b). Adrenocorticotropin reversal of experimental hemorrhagic-shock is antagonized by morphine. *Life Sciences, 39*(14), 1271–1280.

Bertolini, A., Ferrari, W., & Guarini, S. (1989). The adrenocorticotropic hormone (ACTH)-induced reversal of hemorrhagic-shock. *Resuscitation, 18*(2–3), 253–267.

Bertolini, A., Tacchi, R., & Vergoni, A. V. (2009). Brain effects of melanocortins. *Pharmacological Research, 59*(1), 13–47.

Besedovsky, H., et al. (1986). Immunoregulatory feedback between interleukin-1 and glucocorticoid hormones. *Science, 233*(4764), 652–654.

Bitto, A., et al. (2011). Melanocortins protect against multiple organ dysfunction syndrome in mice. *British Journal of Pharmacology, 162*(4), 917–928.

Bjorntorp, P. (1996). The regulation of adipose tissue distribution in humans. *International Journal of Obesity and Related Metabolic Disorders, 20*(4), 291–302.

Bjorntorp, P., & Rosmond, R. (2000). Obesity and cortisol. *Nutrition, 16*(10), 924–936.

Blasio, A., et al. (2014). Opioid system in the medial prefrontal cortex mediates binge-like eating. *Addiction Biology, 19*(4), 652–662.

Boadas-Vaello, P., et al. (2016). Neuroplasticity of ascending and descending pathways after somatosensory system injury: Reviewing knowledge to identify neuropathic pain therapeutic targets. *Spinal Cord, 54*(5), 330–340.

Boggiano, M. M., & Chandler, P. C. (2006). Binge eating in rats produced by combining dieting with stress. *Current Protocols in Neuroscience, Chapter 9*, Unit9 23A.

Borges, B. C., et al. (2011). Leptin resistance and desensitization of hypophagia during prolonged inflammatory challenge. *American Journal of Physiology. Endocrinology and Metabolism, 300*(5), E858–E869.

Borges, B. C., Elias, C. F., & Elias, L. L. (2016a). PI3K signaling: A molecular pathway associated with acute hypophagic response during inflammatory challenges. *Molecular and Cellular Endocrinology, 438*, 36–41.

Borges, B. C., et al. (2016b). PI3K p110beta subunit in leptin receptor expressing cells is required for the acute hypophagia induced by endotoxemia. *Molecular Metabolism, 5*(6), 379–391.

Borovikova, L. V., et al. (2000). Vagus nerve stimulation attenuates the systemic inflammatory response to endotoxin. *Nature, 405*(6785), 458–462.

Boston, B. A. (1999). The role of melanocortins in adipocyte function. In T. A. Luger et al. (Eds.), *Cutaneous Neuroimmunomodulation: The proopiomelanocortin system* (pp. 75–84).

Boston, B. A., & Cone, R. D. (1996). Characterization of melanocortin receptor subtype expression in murine adipose tissues and in the 3T3-L1 cell line. *Endocrinology, 137*(5), 2043–2050.

Bouret, S., et al. (1999). mu-Opioid receptor mRNA expression in proopiomelanocortin neurons of the rat arcuate nucleus. *Molecular Brain Research, 70*(1), 155–158.

Bouyer, K., & Simerly, R. B. (2013). Neonatal leptin exposure specifies innervation of presympathetic hypothalamic neurons and improves the metabolic status of leptin-deficient mice. *The Journal of Neuroscience, 33*(2), 840–851.

Brabant, G., et al. (1987). Circadian and pulsatile thyrotropin secretion in euthyroid man under the influence of thyroid-hormone and glucocorticoid administration. *Journal of Clinical Endocrinology & Metabolism, 65*(1), 83–88.

Briers, B., & Laporte, S. (2013). A wallet full of calories: The effect of financial dissatisfaction on the desire for food energy. *Journal of Marketing Research, 50*(6), 767–781.

Briggs, D. I., et al. (2010). Diet-induced obesity causes ghrelin resistance in Arcuate NPY/AgRP neurons. *Endocrinology, 151*(10), 4745–4755.

Broberger, C., et al. (1998). The neuropeptide Y agouti gene-related protein (AGRP) brain circuitry in normal, anorectic, and monosodium glutamate-treated mice. *Proceedings of the National Academy of Sciences of the United States of America, 95*(25), 15043–15048.

Brown, M. R., et al. (1982). Corticotropin-releasing factor – Actions on the sympathetic nervous-system and metabolism. *Endocrinology, 111*(3), 928–931.

Brunson, K. L., et al. (2001). Corticotropin (ACTH) acts directly on amygdala neurons to down-regulate corticotropin-releasing hormone gene expression. *Annals of Neurology, 49*(3), 304–312.

Büch, T. R., et al. (2009). Pertussis toxin-sensitive signaling of melanocortin-4 receptors in hypothalamic GT1-7 cells defines agouti-related protein as a biased agonist. *Journal of Biological Chemistry, 284*(39), 26411–26420.

Buggy, J. J. (1998). Binding of alpha-melanocyte-stimulating hormone to its G-protein-coupled receptor on B-lymphocytes activates the Jak/STAT pathway. *Biochemical Journal, 331*, 211–216.

Bultman, S. J., Michaud, E. J., & Woychik, R. P. (1992). Molecular characterization of the mouse agouti locus. *Cell, 71*(7), 1195–1204.

Bumaschny, V. F., et al. (2012). Obesity-programmed mice are rescued by early genetic intervention. *Journal of Clinical Investigation, 122*(11), 4203–4212.

Butler, A. A., et al. (2017). A life without hunger: The ups (and downs) to modulating melanocortin-3 receptor signaling. *Frontiers in Neuroscience, 11*, 128.

Byrne, N. M., et al. (2012). Does metabolic compensation explain the majority of less-than-expected weight loss in obese adults during a short-term severe diet and exercise intervention? *International Journal of Obesity, 36*(11), 1472–1478.

Cabral, A., et al. (2012). Ghrelin indirectly activates hypophysiotropic CRF neurons in rodents. *PLoS One, 7*(2), e31462.

Calogero, A. E., et al. (1999). Glucocorticoids inhibit gonadotropin-releasing hormone by acting directly at the hypothalamic level. *Journal of Endocrinological Investigation, 22*(9), 666–670.

Cangemi, L., et al. (1995). N-Acetyltransferase mechanism for alpha-melanocyte stimulating hormone regulation in rat ageing. *Neuroscience Letters, 201*(1), 65–68.

Carmignani, M., et al. (2005). Shock induction by arterial hypoperfusion of the gut involves synergistic interactions between the peripheral enkephalin and nitric oxide systems. *International Journal of Immunopathology and Pharmacology, 18*(1), 33–48.

Caruso, C., et al. (2007). Activation of melanocortin 4 receptors reduces the inflammatory response and prevents apoptosis induced by lipopolysaccharide and interferon-gamma in astrocytes. *Endocrinology, 148*(10), 4918–4926.

Caruso, V., et al. (2014). Synaptic changes induced by melanocortin signalling. *Nature Reviews. Neuroscience, 15*(2), 98–110.

Castro, D. C., Cole, S. L., & Berridge, K. C. (2015). Lateral hypothalamus, nucleus accumbens, and ventral pallidum roles in eating and hunger:Interactions between homeostatic and reward circuitry. *Frontiers in Systems Neuroscience, 9*, 90.

Catania, A., et al. (2004). Targeting melanocortin receptors as a novel strategy to control inflammation. *Pharmacological Reviews, 56*(1), 1–29.

Catania, A., et al. (2010). The melanocortin system in control of inflammation. *The Scientific World Journal, 10*, 1840–1853.

Cavagnini, F., et al. (2000). Glucocorticoids and neuroendocrine function. *International Journal of Obesity and Related Metabolic Disorders, 24*(Suppl 2), S77–S79.

Ceriani, G., et al. (1994). Central neurogenic antiinflammatory action of alpha-msh - modulation of peripheral inflammation-induced by cytokines and other mediators of inflammation. *Neuroendocrinology, 59*(2), 138–143.

Chaki, S., & Okubo, T. (2007). Melanocortin-4 receptor antagonists for the treatment of depression and anxiety disorders. *Current Topics in Medicinal Chemistry, 7*(11), 1145–1151.

Chaki, S., et al. (2003). Involvement of the melanocortin MC4 receptor in stress-related behavior in rodents. *European Journal of Pharmacology, 474*(1), 95–101.

Chan, L. F., et al. (2009). MRAP and MRAP2 are bidirectional regulators of the melanocortin receptor family.

Proceedings of the National Academy of Sciences of the United States of America, 106(15), 6146–6151.

Chance, W. T., et al. (2003). Refractory hypothalamic alpha-MSH satiety and AGRP feeding systems in rats bearing MCA sarcomas. *Peptides, 24*(12), 1909–1919.

Chandramohan, G., et al. (2009). Role of γ melanocyte-stimulating hormone–renal melanocortin 3 receptor system in blood pressure regulation in salt-resistant and salt-sensitive rats. *Metabolism, 58*(10), 1424–1429.

Chang, D., Yi, S. J., & Baram, T. Z. (1996). Developmental profile of corticotropin releasing hormone messenger RNA in the rat inferior olive. *International Journal of Developmental Neuroscience, 14*(1), 69–76.

Chaput, J. P., et al. (2007). Psychobiological effects observed in obese men experiencing body weight loss plateau. *Depression and Anxiety, 24*(7), 518–521.

Chen, C. L. C., et al. (1986). Expression and regulation of proopiomelanocortin-like gene in the ovary and placenta – Comparison with the testis. *Endocrinology, 118*(6), 2382–2389.

Chen, W. B., et al. (1997). Exocrine gland dysfunction in MC5-R-deficient mice: Evidence for coordinated regulation of exocrine gland function by melanocortin peptides. *Cell, 91*(6), 789–798.

Chen, Y. C., et al. (2001). Novel and transient populations of corticotropin-releasing hormone-expressing neurons in developing hippocampus suggest unique functional roles: A quantitative spatiotemporal analysis. *Journal of Neuroscience, 21*(18), 7171–7181.

Chen, C., et al. (2008). Pharmacological and pharmacokinetic characterization of 2-piperazine-alpha-isopropyl benzylamine derivatives as melanocortin-4 receptor antagonists. *Bioorganic & Medicinal Chemistry, 16*(10), 5606–5618.

Chen, Y., et al. (2013). Impairment of synaptic plasticity by the stress mediator CRH involves selective destruction of thin dendritic spines via RhoA signaling. *Molecular Psychiatry, 18*(4), 485–496.

Chen, K. Y., et al. (2015). RM-493, a melanocortin-4 receptor (MC4R) agonist, increases resting energy expenditure in obese individuals. *The Journal of Clinical Endocrinology and Metabolism, 100*(4), 1639–1645.

Chernow, B., et al. (1986). Hemorrhagic hypotension increases plasma beta-endorphin concentrations in the nonhuman primate. *Critical Care Medicine, 14*(5), 505–507.

Chhabra, K. H., et al. (2016). Reprogramming the body weight set point by a reciprocal interaction of hypothalamic leptin sensitivity and Pomc gene expression reverts extreme obesity. *Molecular Metabolism, 5*(10), 869–881.

Chiba, A. (2001). Marked distributional difference of alpha-melanocyte-stimulating hormone (alpha-MSH)-like immunoreactivity in the brain between two elasmobranchs (Scyliorhinus torazame and Etmopterus brachyurus): An immunohistochemical study. *General and Comparative Endocrinology, 122*(3), 287–295.

Chida, D., et al. (2007). Melanocortin 2 receptor is required for adrenal gland development, steroidogenesis, and neonatal gluconeogenesis. *Proceedings of the National Academy of Sciences of the United States of America, 104*(46), 18205–18210.

Cho, K. J., et al. (2005). Signaling pathways implicated in alpha-melanocyte stimulating hormone-induced lipolysis in 3T3-L1 adipocytes. *Journal of Cellular Biochemistry, 96*(4), 869–878.

Chu, H. C., et al. (2012). Effect of periaqueductal gray melanocortin 4 receptor in pain facilitation and glial activation in rat model of chronic constriction injury. *Neurological Research, 34*(9), 871–888.

Chuang, J.-C., et al. (2010). A β3-adrenergic-Leptin-Melanocortin circuit regulates behavioral and metabolic changes induced by chronic stress. *Biological Psychiatry, 67*(11), 1075–1082.

Chuang, J. C., et al. (2011). Ghrelin mediates stress-induced food-reward behavior in mice. *The Journal of Clinical Investigation, 121*(7), 2684–2692.

Chung, T. T., et al. (2008). The majority of adrenocorticotropin receptor (melanocortin 2 receptor) mutations found in familial glucocorticoid deficiency type 1 lead to defective trafficking of the receptor to the cell surface. *Journal of Clinical Endocrinology & Metabolism, 93*(12), 4948–4954.

Ciccocioppo, R., et al. (2003). The bed nucleus is a neuroanatomical substrate for the anorectic effect of corticotropin-releasing factor and for its reversal by nociceptin/orphanin FQ. *The Journal of Neuroscience, 23*(28), 9445–9451.

Cifani, C., et al. (2009). A preclinical model of binge eating elicited by yo-yo dieting and stressful exposure to food: Effect of sibutramine, fluoxetine, topiramate, and midazolam. *Psychopharmacology, 204*(1), 113–125.

Cintra, A., & Bortolotti, F. (1992). Presence of strong glucocorticoid receptor immunoreactivity within hypothalamic and hypophyseal cells containing pro-opiomelanocortic peptides. *Brain Research, 577*(1), 127–133.

Cintra, A., et al. (1987). Presence of glucocorticoid receptor immunoreactivity in corticotrophin releasing factor and in growth hormone releasing factor immunoreactive neurons of the rat di- and telencephalon. *Neuroscience Letters, 77*(1), 25–30.

Clark, A. J. L., & Chan, L. F. (2017). Promiscuity among the MRAPs. *Journal of Molecular Endocrinology, 58*(3), F1–F4.

Clark, A. J. L., & Weber, A. (1998). Adrenocorticotropin insensitivity syndromes. *Endocrine Reviews, 19*(6), 828–843.

Cohen-Mansfield, J., & Jensen, B. (2007). Dressing and grooming – Preferences of community-dwelling older adults. *Journal of Gerontological Nursing, 33*(2), 31–39.

Collier, H. O. (1980). Cellular site of opiate dependence. *Nature, 283*(5748), 625–629.

Collins, G. T., & Eguibar, J. R. (2010). Neurophamacology of yawning. *Frontiers of Neurology and Neuroscience, 28*, 90–106.

Cone, R. D. (2005). Anatomy and regulation of the central melanocortin system. *Nature Neuroscience, 8*(5), 571–578.

Cone, R. D. (2006). Studies on the physiological functions of the melanocortin system. *Endocrine Reviews, 27*(7), 736–749.

Conrad, C. D., et al. (1996). Chronic stress impairs rat spatial memory on the Y maze, and this effect is blocked by tianeptine pretreatment. *Behavioral Neuroscience, 110*(6), 1321–1334.

Constantinopoulos, P., et al. (2015). Cortisol in tissue and systemic level as a contributing factor to the development of metabolic syndrome in severely obese patients. *European Journal of Endocrinology, 172*(1), 69–78.

Contreras, P. C., & Takemori, A. E. (1984). Antagonism of morphine-induced analgesia, tolerance and dependence by alpha-melanocyte-stimulating hormone. *Journal of Pharmacology and Experimental Therapeutics, 229*(1), 21–26.

Coolen, L. M., & Wood, R. I. (1998). Bidirectional connections of the medial amygdaloid nucleus in the Syrian hamster brain: Simultaneous anterograde and retrograde tract tracing. *The Journal of Comparative Neurology, 399*(2), 189–209.

Corander, M. P., et al. (2011). Loss of Agouti-related peptide does not significantly impact the phenotype of murine POMC deficiency. *Endocrinology, 152*(5), 1819–1828.

Cornier, M. A., et al. (2004). Effects of short-term overfeeding on hunger, satiety, and energy intake in thin and reduced-obese individuals. *Appetite, 43*(3), 253–259.

Corrado, E., et al. (2006). Markers of inflammation and infection influence the outcome of patients with baseline asymptomatic carotid lesions: A 5-year follow-up study. *Stroke, 37*(2), 482–486.

Cortes, R., et al. (2014). Evolution of the melanocortin system. *General and Comparative Endocrinology, 209*, 3–10.

Cottone, P., et al. (2009). Consummatory, anxiety-related and metabolic adaptations in female rats with alternating access to preferred food. *Psychoneuroendocrinology, 34*(1), 38–49.

Cowley, M. A., et al. (2001). Leptin activates anorexigenic POMC neurons through a neural network in the arcuate nucleus. *Nature, 411*(6836), 480–484.

Cox, B. M. (2013). Recent developments in the study of opioid receptors. *Molecular Pharmacology, 83*(4), 723–728.

Crine, P., et al. (1979). Concomitant synthesis of beta-endorphin and alpha-melanotropin from two forms of pro-opiomelanocortin in the rat pars intermedia. *Proceedings of the National Academy of Sciences of the United States of America, 76*(10), 5085–5089.

Cui, M. Y., et al. (2014). Ancient origins and evolutionary conservation of intracellular and neural signaling pathways engaged by the Leptin receptor. *Endocrinology, 155*(11), 4202–4214.

Cullinan, W. E. (2000). GABA(A) receptor subunit expression within hypophysiotropic CRH neurons: A dual hybridization histochemical study. *The Journal of Comparative Neurology, 419*(3), 344–351.

Currie, P. J., et al. (2005). Ghrelin is an orexigenic and metabolic signaling peptide in the arcuate and paraventricular nuclei. *American Journal of Physiology. Regulatory, Integrative and Comparative Physiology, 289*(2), R353–R358.

da Silva, A. A., do Carmo, J. M., & Hall, J. E. (2013). Role of leptin and central nervous system melanocortins in obesity hypertension. *Current Opinion in Nephrology and Hypertension, 22*(2), 135–140.

Dabrowska, J., et al. (2011). Neuroanatomical evidence for reciprocal regulation of the corticotrophin-releasing factor and oxytocin systems in the hypothalamus and the bed nucleus of the stria terminalis of the rat: Implications for balancing stress and affect. *Psychoneuroendocrinology, 36*(9), 1312–1326.

Dagogo-Jack, S., et al. (1997). Robust leptin secretory responses to dexamethasone in obese subjects. *The Journal of Clinical Endocrinology and Metabolism, 82*(10), 3230–3233.

Dallman, M. F., et al. (2003). Chronic stress and obesity: A new view of "comfort food". *Proceedings of the National Academy of Sciences of the United States of America, 100*(20), 11696–11701.

Dallman, M. F., et al. (2006). Glucocorticoids, chronic stress, and obesity. In A. Kalsbeek et al. (Eds.), *Hypothalamic integration of energy metabolism* (pp. 75–105).

Dallmann, R., et al. (2011). The orally active melanocortin-4 receptor antagonist BL-6020/979: A promising candidate for the treatment of cancer cachexia. *Journal of Cachexia, Sarcopenia and Muscle, 2*(3), 163–174.

Dautzenberg, F. M., et al. (2004). Cell-type specific calcium signaling by corticotropin-releasing factor type 1 (CRF1) and 2a (CRF2(a)) receptors: Phospholipase C-mediated responses in human embryonic kidney 293 but not SK-N-MC neuroblastoma cells. *Biochemical Pharmacology, 68*(9), 1833–1844.

Davis, M., et al. (2010). Phasic vs sustained fear in rats and humans: Role of the extended amygdala in fear vs anxiety. *Neuropsychopharmacology, 35*(1), 105–135.

De Barioglio, S. R., Lezcano, N., & Celis, M. E. (1991). Alpha MSH-induced excessive grooming behavior involves a GABAergic mechanism. *Peptides, 12*(1), 203–205.

de Kloet, E. R., & Sarabdjitsingh, R. A. (2008). Everything has rhythm: Focus on glucocorticoid pulsatility. *Endocrinology, 149*(7), 3241–3243.

De Kloet, E. R., et al. (1998). Brain corticosteroid receptor balance in health and disease. *Endocrine Reviews, 19*(3), 269–301.

de Kloet, E. R., Oitzl, M. S., & Joels, M. (1999). Stress and cognition: Are corticosteroids good or bad guys? *Trends in Neurosciences, 22*(10), 422–426.

de Kloet, E. R., Joels, M., & Holsboer, F. (2005). Stress and the brain: From adaptation to disease. *Nature Reviews Neuroscience, 6*(6), 463–475.

De Souza, E. B., et al. (1986). Reciprocal changes in corticotropin-releasing factor (CRF)-like immunoreactivity and CRF receptors in cerebral cortex of Alzheimer's disease. *Nature, 319*(6054), 593–595.

De Souza, J., Butler, A. A., & Cone, R. D. (2000). Disproportionate inhibition of feeding in A(y) mice by certain stressors: A cautionary note. *Neuroendocrinology, 72*(2), 126–132.

De Souza, C. T., et al. (2005). Consumption of a fat-rich diet activates a proinflammatory response and induces insulin resistance in the hypothalamus. *Endocrinology, 146*(10), 4192–4199.

De Wied, D., & Jolles, J. (1982). Neuropeptides derived from pro-opiocortin: Behavioral, physiological, and neurochemical effects. *Physiological Reviews, 62*(3), 976–1059.

Decastro, J. M. (1987). Macronutrient relationships with meal patterns and mood in the spontaneous feeding-behavior of humans. *Physiology & Behavior, 39*(5), 561–569.

Dekloet, E. R., & Reul, J. (1987). Feedback action and tonic influence of corticosteroids on brain-function – A concept arising from the heterogeneity of brain receptor systems. *Psychoneuroendocrinology, 12*(2), 83–105.

Delgado Hernandez, R., et al. (1999). Inhibition of systemic inflammation by central action of the neuropeptide alpha-melanocyte-stimulating hormone. *Neuroimmunomodulation, 6*(3), 187–192.

Delgado, R., et al. (1998). Melanocortin peptides inhibit production of proinflammatory cytokines and nitric oxide by activated microglia. *Journal of Leukocyte Biology, 63*(6), 740–745.

Dent, G. W., Smith, M. A., & Levine, S. (2000). Rapid induction of corticotropin-releasing hormone gene transcription in the paraventricular nucleus of the developing rat. *Endocrinology, 141*(5), 1593–1598.

Deuster, P. A., et al. (1992). Hormonal responses to ingesting water or a carbohydrate beverage during a 2-H run. *Medicine and Science in Sports and Exercise, 24*(1), 72–79.

Dhillo, W. S., et al. (2002). The hypothalamic melanocortin system stimulates the hypothalamo-pituitary-adrenal axis in vitro and in vivo in male rats. *Neuroendocrinology, 75*(4), 209–216.

Dhurandhar, E. J. (2016). The food-insecurity obesity paradox: A resource scarcity hypothesis. *Physiology & Behavior, 162*, 88–92.

Dias-Ferreira, E., et al. (2009). Chronic stress causes frontostriatal reorganization and affects decision-making. *Science, 325*(5940), 621–625.

Disse, E., et al. (2010). Peripheral ghrelin enhances sweet taste food consumption and preference, regardless of its caloric content. *Physiology & Behavior, 101*(2), 277–281.

Dodd, G. T., et al. (2015). Leptin and insulin act on POMC neurons to promote the Browning of white fat. *Cell, 160*(1–2), 88–104.

Dong, Y., & Benveniste, E. N. (2001). Immune function of astrocytes. *Glia, 36*(2), 180–190.

Dores, R. M., et al. (2014). Molecular evolution of GPCRs: Melanocortin/melanocortin receptors. *Journal of Molecular Endocrinology, 52*(3), T29–T42.

Doyon, C., et al. (2001). Molecular evolution of leptin. *General and Comparative Endocrinology, 124*(2), 188–198.

Dube, L., LeBel, J. L., & Lu, J. (2005). Affect asymmetry and comfort food consumption. *Physiology & Behavior, 86*(4), 559–567.

Dunn, A. J. (1988). Studies on the neurochemical mechanisms and significance of ACTH-induced grooming. *Annals of the New York Academy of Sciences, 525,* 150–168.

Dunn, A. J., & Berridge, C. W. (1990). Is Corticotropin-releasing factor a mediator of stress responses. *Annals of the New York Academy of Sciences, 579,* 183–191.

Dunn, A. J., et al. (1987). CRF-induced excessive grooming behavior in rats and mice. *Peptides, 8*(5), 841–844.

Dutia, R., et al. (2012). Beta-endorphin antagonizes the effects of alpha-MSH on food intake and body weight. *Endocrinology, 153*(9), 4246–4255.

Duvaux-Miret, O., & Capron, A. (1992). Proopiomelanocortin in the helminth schistosoma mansoni. Synthesis of beta-endorphin, ACTH, and alpha-MSH. Existence of POMC-related sequences. *Annals of the New York Academy of Sciences, 650*(1), 245–250.

Dwarkasing, J. T., et al. (2014). Hypothalamic food intake regulation in a cancer-cachectic mouse model. *Journal of Cachexia Sarcopenia and Muscle, 5*(2), 159–169.

Dyakonova, V. (2001). Role of opioid peptides in behavior of invertebrates. *Journal of Evolutionary Biochemistry and Physiology, 37*(4), 335–347.

Eberwine, J. H., et al. (1987). Complex transcriptional regulation by glucocorticoids and Corticotropin-releasing hormone of Proopiomelanocortin gene-expression in rat pituitary cultures. *DNA-a Journal of Molecular & Cellular Biology, 6*(5), 483–492.

Eddy, J. P., & Strahan, R. (1968). The role of the pineal complex in the pigmentary effector system of the lampreys, Mordacia mordax (Richardson) and Geotria australis gray. *General and Comparative Endocrinology, 11*(3), 528–534.

Edwards, A. V., & Jones, C. T. (1993). Autonomic control of adrenal function. *Journal of Anatomy, 183*(Pt 2), 291–307.

Egecioglu, E., et al. (2010). Ghrelin increases intake of rewarding food in rodents. *Addiction Biology, 15*(3), 304–311.

Eguibar, J. R., et al. (2017). Yawning reduces facial temperature in the high-yawning subline of Sprague-Dawley rats. *BMC Neuroscience, 18*(1), 3.

Elam, R., Bergmann, F., & Feuerstein, G. (1984). Simultaneous changes of catecholamines and of leu-enkephalin-like immunoreactivity in plasma and cerebrospinal-fluid of cats undergoing acute hemorrhage. *Brain Research, 303*(2), 313–317.

Elias, C. F., et al. (1998). Leptin activates hypothalamic CART neurons projecting to the spinal cord. *Neuron, 21*(6), 1375–1385.

Ellacott, K. L., & Cone, R. D. (2006). The role of the central melanocortin system in the regulation of food intake and energy homeostasis: Lessons from mouse models. *Philosophical Transactions of the Royal Society of London. Series B, Biological Sciences, 361*(1471), 1265–1274.

Endo, M., et al. (2007). Involvement of stomach ghrelin and hypothalamic neuropeptides in tumor necrosis factor-alpha-induced hypophagia in mice. *Regulatory Peptides, 140*(1–2), 94–100.

Endsin, M. J., et al. (2017). CRH peptide evolution occurred in three phases: Evidence from characterizing sea lamprey CRH system members. *General and Comparative Endocrinology, 240,* 162–173.

Engeland, W. C., & Arnhold, M. M. (2005). Neural circuitry in the regulation of adrenal corticosterone rhythmicity. *Endocrine, 28*(3), 325–332.

Enriori, P. J., et al. (2007). Diet-induced obesity causes severe but reversible leptin resistance in arcuate melanocortin neurons. *Cell Metabolism, 5*(3), 181–194.

Enriori, P. J., et al. (2016). Alpha-melanocyte stimulating hormone promotes muscle glucose uptake via melanocortin 5 receptors. *Molecular Metabolism, 5*(10), 807–822.

Epel, E. S., et al. (2000). Stress and body shape: Stress-induced cortisol secretion is consistently greater among women with central fat. *Psychosomatic Medicine, 62*(5), 623–632.

Epel, E., et al. (2001). Stress may add bite to appetite in women: A laboratory study of stress-induced cortisol and eating behavior. *Psychoneuroendocrinology, 26*(1), 37–49.

Ercil, N. E., Galici, R., & Kesterson, R. A. (2005). HS014, a selective melanocortin-4 (MC4) receptor antagonist, modulates the behavioral effects of morphine in mice. *Psychopharmacology, 180*(2), 279–285.

Ericson, M. D., et al. (2017). Bench-top to clinical therapies: A review of melanocortin ligands from 1954 to 2016. *Biochimica et Biophysica Acta, 1863*(10 Pt A), 2414–2435.

Ezeoke, C. C., & Morley, J. E. (2015). Pathophysiology of anorexia in the cancer cachexia syndrome. *Journal of Cachexia, Sarcopenia and Muscle, 6*(4), 287–302.

Fairburn, C. G. (1997). Bulimia outcome. *American Journal of Psychiatry, 154*(12), 1791–1791.

Fani, L., et al. (2014). The melanocortin-4 receptor as target for obesity treatment: A systematic review of emerging pharmacological therapeutic options. *International Journal of Obesity, 38*(2), 163–169.

Fardin, V., Oliveras, J. L., & Besson, J. M. (1984). A reinvestigation of the analgesic effects induced by stimulation of the periaqueductal gray matter in the rat. I. The production of behavioral side effects together with analgesia. *Brain Research, 306*(1–2), 105–123.

Feig, P. U., et al. (2011). Effects of an 11 beta-hydroxysteroid dehydrogenase type 1 inhibitor, MK-0916, in patients with type 2 diabetes mellitus and metabolic syndrome. *Diabetes Obesity & Metabolism, 13*(6), 498–504.

Fekete, E. M., & Zorrilla, E. P. (2007). Physiology, pharmacology, and therapeutic relevance of urocortins in mammals: Ancient CRF paralogs. *Frontiers in Neuroendocrinology, 28*(1), 1–27.

Fentress, J. C. (1988). Expressive contexts, fine-structure, and central mediation of rodent grooming. *Annals of the New York Academy of Sciences, 525*, 18–26.

Fernandez, M., et al. (2003). Sucrose attenuates a negative electroencephalographic response to an aversive stimulus for newborns. *Journal of Developmental and Behavioral Pediatrics, 24*(4), 261–266.

Ferrari, W. (1958). Behavioural changes in animals after intracisternal injection with adrenocorticotrophic hormone and melanocyte-stimulating hormone. *Nature, 181*(4613), 925–926.

Ferrari, W., Floris, E., & Paulesu, F. (1955). A peculiar impressive symptomatology induced in the dog by injections of ACTH injected into the cisterna magna. *Bollettino della Societa italiana di biologia sperimentale, 31*(7–8), 862–864.

Ferrari, W., Vargiu, L., & Gessa, G. L. (1963). Behavioral effects induced by intracisternally injected acth and msh. *Annals of the New York Academy of Sciences, 104*(1), 330.

Fletcher, M., & Kim, D. H. (2017). Age-dependent neuroendocrine signaling from sensory neurons modulates the effect of dietary restriction on longevity of caenorhabditis elegans. *PLoS Genetics, 13*(1), e1006544.

Ford, E. S. (2005). Risks for all-cause mortality, cardiovascular disease, and diabetes associated with the metabolic syndrome: A summary of the evidence. *Diabetes Care, 28*(7), 1769–1778.

Forman, L. J., & Bagasra, O. (1992). Demonstration by insitu hybridization of the proopiomelanocortin gene in the rat-heart. *Brain Research Bulletin, 28*(3), 441–445.

Forslin Aronsson, A., et al. (2007). Alpha-MSH rescues neurons from excitotoxic cell death. *Journal of Molecular Neuroscience, 33*(3), 239–251.

Foster, M. T., et al. (2009). Palatable foods, stress, and energy stores sculpt corticotropin-releasing factor, adrenocorticotropin, and corticosterone concentrations after restraint. *Endocrinology, 150*(5), 2325–2333.

Fothergill, E., et al. (2016). Persistent metabolic adaptation 6 years after "the biggest loser" competition. *Obesity, 24*(8), 1612–1619.

Fratta, W., et al. (1981). Reciprocal antagonism between ACTH1–24 and β-endorphin in rats. *Neuroscience Letters, 24*(1), 71–74.

Frederich, R. C., et al. (1995). Leptin levels reflect body lipid-content in mice – Evidence for diet-induced resistance to leptin action. *Nature Medicine, 1*(12), 1311–1314.

Fu, L.-Y., & van den Pol, A. N. (2008). Agouti-related peptide and MC3/4 receptor agonists both inhibit excitatory hypothalamic ventromedial nucleus neurons. *Journal of Neuroscience, 28*(21), 5433–5449.

Fujikawa, M., et al. (1995). Involvement of β-adrenoceptors in regulation of the yawning induced by neuropeptides, oxytocin and α-melanocyte-stimulating hormone, in rats. *Pharmacology Biochemistry and Behavior, 50*(3), 339–343.

Fujimoto-Ouchi, K., et al. (1995). Establishment and characterization of cachexia-inducing and -non-inducing clones of murine colon 26 carcinoma. *International Journal of Cancer, 61*(4), 522–528.

Fuxe, K., et al. (1987). Studies on the cellular-localization and distribution of glucocorticoid receptor and estrogen-receptor immunoreactivity in the central nervous-system of the rat and their relationship to the monoaminergic and peptidergic neurons of the brain. *Journal of Steroid Biochemistry and Molecular Biology, 27*(1–3), 159–170.

Gagner, J. P., & Drouin, J. (1985). Opposite regulation of pro-opiomelanocortin gene transcription by glucocorticoids and CRH. *Molecular and Cellular Endocrinology, 40*(1), 25–32.

Gagner, J. P., & Drouin, J. (1987). Tissue-specific regulation of pituitary proopiomelanocortin gene transcription by corticotropin-releasing hormone, 3′,5′-cyclic adenosine monophosphate, and glucocorticoids. *Molecular Endocrinology, 1*(10), 677–682.

Galimberti, D., et al. (1999). Alpha-MSH peptides inhibit production of nitric oxide and tumor necrosis factor-alpha by microglial cells activated with beta-amyloid and interferon gamma. *Biochemical and Biophysical Research Communications, 263*(1), 251–256.

Gallagher, J. P., et al. (2008). Synaptic physiology of central CRH system. *European Journal of Pharmacology, 583*(2–3), 215–225.

Gallopin, T., et al. (2006). Cortical sources of CRF, NKB, and CCK and their effects on pyramidal cells in the neocortex. *Cerebral Cortex, 16*(10), 1440–1452.

Gallup, A. C., & Eldakar, O. T. (2013). The thermoregulatory theory of yawning: What we know from over 5 years of research. *Frontiers in Neuroscience, 6*, 188.

Gamber, K. M., et al. (2012). Over-expression of leptin receptors in hypothalamic POMC neurons increases susceptibility to diet-induced obesity. *PLoS One, 7*(1), e30485.

Gao, Y., et al. (2014). Hormones and diet, but not body weight, control hypothalamic microglial activity. *Glia, 62*(1), 17–25.

Gautron, L., et al. (2005). Influence of feeding status on neuronal activity in the hypothalamus during lipopolysaccharide-induced anorexia in rats. *Neuroscience, 134*(3), 933–946.

Gautron, L., et al. (2012). Melanocortin-4 receptor expression in different classes of spinal and vagal primary afferent neurons in the mouse. *Journal of Comparative Neurology, 520*(17), 3933–3948.

Gelin, J., et al. (1991). Role of endogenous tumor necrosis factor alpha and interleukin 1 for experimental tumor growth and the development of cancer cachexia. *Cancer Research, 51*(1), 415–421.

Gessa, G. L., et al. (1967). Stretching and yawning movements after intracerebral injection of ACTH. *Revue Canadienne de Biologie, 26*(3), 229.

Getting, S. J., Flower, R. J., & Perretti, M. (1999a). Agonism at melanocortin receptor type 3 on macrophages inhibits neutrophil influx. *Inflammation Research, 48*(SUPPL. 2), S140–S141.

Getting, S. J., et al. (1999b). POMC gene-derived peptides activate melanocortin type 3 receptor on murine macrophages, suppress cytokine release, and inhibit neutrophil migration in acute experimental inflammation. *Journal of Immunology, 162*(12), 7446–7453.

Ghamari-Langroudi, M., et al. (2015). G-protein-independent coupling of MC4R to Kir7.1 in hypothalamic neurons. *Nature, 520*(7545), 94–98.

Gibson, E. L. (2006). Emotional influences on food choice: Sensory, physiological and psychological pathways. *Physiology & Behavior, 89*(1), 53–61.

Giles-Corti, B., & Donovan, R. J. (2002). Socioeconomic status differences in recreational physical activity levels and real and perceived access to a supportive physical environment. *Preventive Medicine, 35*(6), 601–611.

Gilhooly, C. H., et al. (2007). Food cravings and energy regulation: The characteristics of craved foods and their relationship with eating behaviors and weight change during 6 months of dietary energy restriction. *International Journal of Obesity, 31*(12), 1849–1858.

Girardet, C., & Butler, A. A. (2014). Neural melanocortin receptors in obesity and related metabolic disorders. *Biochimica et Biophysica Acta-Molecular Basis of Disease, 1842*(3), 482–494.

Girardet, C., et al. (2017). Melanocortin-3 receptors expressed in Nkx2.1(+ve) neurons are sufficient for controlling appetitive responses to hypocaloric conditioning. *Scientific Reports, 7*, 44444.

Gispen, W. H., et al. (1975). Induction of excessive grooming in rat by intraventricular application of peptides derived from ACTH – Structure-activity studies. *Life Sciences, 17*(4), 645–652.

Giuliani, D., et al. (2006). Both early and delayed treatment with melanocortin 4 receptor-stimulating melanocortins produces neuroprotection in cerebral ischemia. *Endocrinology, 147*(3), 1126–1135.

Giuliani, D., et al. (2007). Selective melanocortin MC4 receptor agonists reverse haemorrhagic shock and prevent multiple organ damage. *British Journal of Pharmacology, 150*(5), 595–603.

Glowa, J. R., et al. (1992). Effects of corticotropin releasing hormone on appetitive behaviors. *Peptides, 13*(3), 609–621.

Gluck, M. E., et al. (2004). Cortisol, hunger, and desire to binge eat following a cold stress test in obese women with binge eating disorder. *Psychosomatic Medicine, 66*(6), 876–881.

Goehler, L. E., et al. (1997). Vagal paraganglia bind biotinylated interleukin-1 receptor antagonist: A possible mechanism for immune-to-brain communication. *Brain Research Bulletin, 43*(3), 357–364.

Goehler, L. E., et al. (2000). Vagal immune-to-brain communication: A visceral chemosensory pathway. *Autonomic Neuroscience-Basic & Clinical, 85*(1–3), 49–59.

Gore, A. C., Attardi, B., & DeFranco, D. B. (2006). Glucocorticoid repression of the reproductive axis: Effects on GnRH and gonadotropin subunit mRNA levels. *Molecular and Cellular Endocrinology, 256*(1–2), 40–48.

Goto, M., et al. (2006). Ghrelin increases neuropeptide Y and agouti-related peptide gene expression in the arcuate nucleus in rat hypothalamic organotypic cultures. *Endocrinology, 147*(11), 5102–5109.

Gottlieb, S., & Ruvkun, G. (1994). Daf-2, Daf-16 and Daf-23 – Genetically interacting genes-controlling Dauer formation in Caenorhabditis-Elegans. *Genetics, 137*(1), 107–120.

Goyal, S. N., et al. (2006). Alpha-melanocyte stimulating hormone antagonizes antidepressant-like effect of neuropeptide Y in Porsolt's test in rats. *Pharmacology, Biochemistry, and Behavior, 85*(2), 369–377.

Greenfield, J. R., et al. (2009). Modulation of blood pressure by central melanocortinergic pathways. *The New England Journal of Medicine, 360*(1), 44–52.

Gropp, E., et al. (2005). Agouti-related peptide-expressing neurons are mandatory for feeding. *Nature Neuroscience, 8*(10), 1289–1291.

Gruenewald, T. L., Kemeny, M. E., & Aziz, N. (2006). Subjective social status moderates cortisol responses to social threat. *Brain Behavior and Immunity, 20*(4), 410–419.

Grueter, B. A., Rothwell, P. E., & Malenka, R. C. (2012). Integrating synaptic plasticity and striatal circuit function in addiction. *Current Opinion in Neurobiology, 22*(3), 545–551.

Grunfeld, C., et al. (1996). Endotoxin and cytokines induce expression of leptin, the ob gene product, in hamsters. *The Journal of Clinical Investigation, 97*(9), 2152–2157.

Guarini, S., et al. (1989). Reversal of hemorrhagic-shock in rats by cholinomimetic drugs. *British Journal of Pharmacology, 98*(1), 218–224.

Guarini, S., Bazzani, C., & Bertolini, A. (1997). Resuscitating effect of melanocortin peptides after prolonged respiratory arrest. *British Journal of Pharmacology, 121*(7), 1454–1460.

Guarini, S., et al. (2004). Adrenocorticotropin reverses hemorrhagic shock in anesthetized rats through the rapid activation of a vagal anti-inflammatory pathway. *Cardiovascular Research, 63*(2), 357–365.

Gyengesi, E., et al. (2010). Corticosterone regulates synaptic input organization of POMC and NPY/AgRP neurons in adult mice. *Endocrinology, 151*(11), 5395–5402.

Hagan, M. M., et al. (2002). A new animal model of binge eating: Key synergistic role of past caloric restriction and stress. *Physiology & Behavior, 77*(1), 45–54.

Hagan, M. M., et al. (2003). The role of palatable food and hunger as trigger factors in an animal model of stress induced binge eating. *International Journal of Eating Disorders, 34*(2), 183–197.

Hahn, T. M., et al. (1998). Coexpression of Agrp and NPY in fasting-activated hypothalamic neurons. *Nature Neuroscience, 1*(4), 271–272.

Haitina, T., et al. (2007). Further evidence for ancient role of ACTH peptides at melanocortin (MC) receptors;

pharmacology of dogfish and lamprey peptides at dogfish MC receptors. *Peptides, 28*(4), 798–805.

Hansson, G. K. (2005). Inflammation, atherosclerosis, and coronary artery disease. *The New England Journal of Medicine, 352*(16), 1685–1695.

Harrell, C. S., Gillespie, C. F., & Neigh, G. N. (2016). Energetic stress: The reciprocal relationship between energy availability and the stress response. *Physiology & Behavior, 166*, 43–55.

Hashimoto, H., et al. (2007). Parathyroid hormone-related protein induces cachectic syndromes without directly modulating the expression of hypothalamic feeding-regulating peptides. *Clinical Cancer Research, 13*(1), 292–298.

Haskell-Luevano, C., & Monck, E. K. (2001). Agouti-related protein functions as an inverse agonist at a constitutively active brain melanocortin-4 receptor. *Regulatory Peptides, 99*(1), 1–7.

Haskell-Luevano, C., et al. (1999). Characterization of the neuroanatomical distribution of agouti-related protein immunoreactivity in the rhesus monkey and the rat. *Endocrinology, 140*(3), 1408–1415.

Haynes, W. G., et al. (1999). Interactions between the melanocortin system and leptin in control of sympathetic nerve traffic. *Hypertension, 33*(1), 542–547.

Heinig, J. A., et al. (1995). The appearance of proopiomelanocortin early in vertebrate evolution: Cloning and sequencing of POMC from a lamprey pituitary cDNA library. *General and Comparative Endocrinology, 99*(2), 137–144.

Heinricher, M. M., et al. (2009). Descending control of nociception: Specificity, recruitment and plasticity. *Brain Research Reviews, 60*(1), 214–225.

Hellhammer, D. H., et al. (1997). Social hierarchy and adrenocortical stress reactivity in men. *Psychoneuroendocrinology, 22*(8), 643–650.

Herman, J. P. (2013). Neural control of chronic stress adaptation. *Frontiers in Behavioral Neuroscience, 7*, 61.

Herman, J. P., & Morrison, D. G. (1996). Immunoautoradiographic and in situ hybridization analysis of corticotropin-releasing hormone biosynthesis in the hypothalamic paraventricular nucleus. *Journal of Chemical Neuroanatomy, 11*(1), 49–56.

Hill, J. W., & Faulkner, L. D. (2017). The role of the Melanocortin system in metabolic disease: New developments and advances. *Neuroendocrinology, 104*(4), 330–346.

Hill, J. W., et al. (2010). Direct insulin and Leptin action on pro-opiomelanocortin neurons is required for normal glucose homeostasis and fertility. *Cell Metabolism, 11*(4), 286–297.

Himmelsbach, C. (1939). Studies of certain addiction characteristics of (a) Dihydromorphine (" Paramorphan"), (b) Dihydrodesoxymorphine-D (" Desomorphine"), (c) Dihydrodesoxycodeine-D (" Desocodeine"), and (d) Methyldihydromorphinone (" Metopon"). *Journal of Pharmacology and Experimental Therapeutics, 67*(2), 239–249.

Homberg, J. R., et al. (2002). Enhanced motivation to self-administer cocaine is predicted by self-grooming behaviour and relates to dopamine release in the rat medial prefrontal cortex and amygdala. *European Journal of Neuroscience, 15*(9), 1542–1550.

Hong, W., Kim, D.-W., & Anderson, D. J. (2014). Antagonistic control of social versus repetitive self-grooming behaviors by separable amygdala neuronal subsets. *Cell, 158*(6), 1348–1361.

Hosobuchi, Y., Adams, J. E., & Linchitz, R. (1977). Pain relief by electrical stimulation of the central gray matter in humans and its reversal by naloxone. *Science, 197*(4299), 183–186.

Hotamisligil, G., Shargill, N., & Spiegelman, B. (1993). Adipose expression of tumor necrosis factor-alpha: Direct role in obesity—Linked insulin resistance. *Science, 259*, 87–91 Zhang Y, Proenca R, Mafiei M, Barone M, Leopold L, Friedman JM: Positional cloning of the mouse obese gene and its human homologue. Nature, 1995. 372: p. 425-432.

Hu, X. X., Goldmuntz, E. A., & Brosnan, C. F. (1991). The effect of norepinephrine on endotoxin-mediated macrophage activation. *Journal of Neuroimmunology, 31*(1), 35–42.

Huang, Q. H., Hruby, V. J., & Tatro, J. B. (1999). Role of central melanocortins in endotoxin-induced anorexia. *The American Journal of Physiology, 276*(3 Pt 2), R864–R871.

Humbert, M., et al. (1995). Increased interleukin-1 and interleukin-6 serum concentrations in severe primary pulmonary hypertension. *American Journal of Respiratory and Critical Care Medicine, 151*(5), 1628–1631.

Hummel, A., & Zuhlke, H. (1994). Expression of 2 proopiomelanocortin messenger-RNAs in the islets of langerhans of neonatal rats. *Biological Chemistry Hoppe-Seyler, 375*(12), 811–815.

Hyun, M., et al. (2016). Fat metabolism regulates satiety behavior in C. elegans. *Scientific Reports, 6*, 24841.

Ichiyama, T., et al. (1999a). Alpha-melanocyte-stimulating hormone inhibits NF-kappa B activation and I kappa B alpha degradation in human glioma cells and in experimental brain inflammation. *Experimental Neurology, 157*(2), 359–365.

Ichiyama, T., et al. (1999b). Systemically administered α-melanocyte-stimulating peptides inhibit NF-κB activation in experimental brain inflammation. *Brain Research, 836*(1–2), 31–37.

Iqbal, J., et al. (2010). An intrinsic gut leptin-melanocortin pathway modulates intestinal microsomal triglyceride transfer protein and lipid absorption. *Journal of Lipid Research, 51*(7), 1929–1942.

Jacobowitz, D. M., & Odonohue, T. L. (1978). Alpha-melanocyte stimulating hormone – Immunohistochemical identification and mapping in neurons of rat-brain. *Proceedings of the National Academy of Sciences of the United States of America, 75*(12), 6300–6304.

Jaffe, S. B., Sobieszczyk, S., & Wardlaw, S. L. (1994). Effect of opioid antagonism on beta-endorphin processing and proopiomelanocortin-peptide release in the hypothalamus. *Brain Research, 648*(1), 24–31.

Jahng, J. W., et al. (2008). Dexamethasone reduces food intake, weight gain and the hypothalamic 5-HT concentration and increases plasma leptin in rats. *European Journal of Pharmacology, 581*(1–2), 64–70.

Jais, A., & Bruning, J. C. (2017). Hypothalamic inflammation in obesity and metabolic disease. *The Journal of Clinical Investigation, 127*(1), 24–32.

Jang, P. G., et al. (2010). NF-kappa B activation in hypothalamic pro-opiomelanocortin neurons is essential in illness- and leptin-induced anorexia. *Journal of Biological Chemistry, 285*(13), 9706–9715.

Jegou, S., Boutelet, I., & Vaudry, H. (2000). Melanocortin-3 receptor mRNA expression in pro-opiomelanocortin neurones of the rat arcuate nucleus. *Journal of Neuroendocrinology, 12*(6), 501–505.

Jiang, W., et al. (2007). Pyrrolidinones as potent functional antagonists of the human melanocortin-4 receptor. *Bioorganic & Medicinal Chemistry Letters, 17*(20), 5610–5613.

Jin, X.-C., et al. (2007). Glucocorticoid receptors in the basolateral nucleus of amygdala are required for postreactivation reconsolidation of auditory fear memory. *European Journal of Neuroscience, 25*(12), 3702–3712.

John, K., et al. (2016). The glucocorticoid receptor: Cause of or cure for obesity? *American Journal of Physiology-Endocrinology and Metabolism, 310*(4), E249–E257.

Jolles, J., Wiegant, V. M., & Gispen, W. H. (1978). Reduced behavioral effectiveness of acth-1-24 after a 2nd administration – Interaction with opiates. *Neuroscience Letters, 9*(2–3), 261–266.

Jones, S. J., & Frongillo, E. A. (2006). The modifying effects of food stamp program participation on the relation between food insecurity and weight change in women. *Journal of Nutrition, 136*(4), 1091–1094.

Joseph, S. A., Pilcher, W. H., & Bennettclarke, C. (1983). Immunocytochemical localization of acth perikarya in nucleus tractus solitarius – Evidence for a 2nd opiocortin neuronal system. *Neuroscience Letters, 38*(3), 221–225.

Jun, D.-J., et al. (2010). Melanocortins induce interleukin 6 gene expression and secretion through melanocortin receptors 2 and 5 in 3T3-L1 adipocytes. *Journal of Molecular Endocrinology, 44*(4), 225–236.

Kalange, A. S., et al. (2007). Central administration of selective melanocortin 4 receptor antagonist HS014 prevents morphine tolerance and withdrawal hyperalgesia. *Brain Research, 1181*, 10–20.

Kalueff, A. V., & Tuohimaa, P. (2004). Grooming analysis algorithm for neurobehavioural stress research. *Brain Research Protocols, 13*(3), 151–158.

Kalueff, A. V., & Tuohimaa, P. (2005). The grooming analysis algorithm discriminates between different levels of anxiety in rats: Potential utility for neurobehavioural stress research. *Journal of Neuroscience Methods, 143*(2), 169–177.

Kalueff, A. V., et al. (2007). Analyzing grooming microstructure in neurobehavioral experiments. *Nature Protocols, 2*(10), 2538–2544.

Kalueff, A. V., et al. (2016). Neurobiology of rodent self-grooming and its value for translational neuroscience. *Nature Reviews Neuroscience, 17*(1), 45–59.

Kalyuzhny, A. E., et al. (1996). Mu-opioid and delta-opioid receptors are expressed in brainstem antinociceptive circuits: Studies using immunocytochemistry and retrograde tract-tracing. *Journal of Neuroscience, 16*(20), 6490–6503.

Karami Kheirabad, M., et al. (2015). Expression of melanocortin-4 receptor mRNA in male rat hypothalamus during chronic stress. *International Journal of Molecular and Cellular Medicine, 4*(3), 182–187.

Kariagina, A., et al. (2004). Hypothalamic-pituitary cytokine network. *Endocrinology, 145*(1), 104–112.

Kas, M. J., et al. (2005). Differential regulation of agouti-related protein and neuropeptide Y in hypothalamic neurons following a stressful event. *Journal of Molecular Endocrinology, 35*(1), 159–164.

Katritch, V., Cherezov, V., & Stevens, R. C. (2013). Structure-function of the G protein–coupled receptor superfamily. *Annual Review of Pharmacology and Toxicology, 53*, 531–556.

Kawauchi, H., & Sower, S. A. (2006). The dawn and evolution of hormones in the adenohypophysis. *General and Comparative Endocrinology, 148*(1), 3–14.

Kelly, M. J., Loose, M. D., & Ronnekleiv, O. K. (1990). Opioids hyperpolarize beta-endorphin neurons via mu-receptor activation of a potassium conductance. *Neuroendocrinology, 52*(3), 268–275.

Kenyon, C., et al. (1993). A C-Elegans mutant that lives twice as long as wild-type. *Nature, 366*(6454), 461–464.

Kerman, I. A., Akil, H., & Watson, S. J. (2006). Rostral elements of sympatho-motor circuitry: A virally mediated transsynaptic tracing study. *The Journal of Neuroscience, 26*(13), 3423–3433.

Khachaturian, H., et al. (1986). Further characterization of the extra-arcuate alpha-melanocyte stimulating hormone-like material in hypothalamus: Biochemical and anatomical studies. *Neuropeptides, 7*(3), 291–313.

Kievit, P., et al. (2013). Chronic treatment with a melanocortin-4 receptor agonist causes weight loss, reduces insulin resistance, and improves cardiovascular function in diet-induced obese rhesus macaques. *Diabetes, 62*(2), 490–497.

King, C. M., & Hentges, S. T. (2011). Relative number and distribution of murine hypothalamic proopiomelanocortin neurons innervating distinct target sites. *PLoS One, 6*(10), e25864.

Kirschbaum, C., et al. (1995). Persistent high cortisol responses to repeated psychological stress in a subpopulation of healthy-men. *Psychosomatic Medicine, 57*(5), 468–474.

Kishi, T., et al. (2003). Expression of melanocortin 4 receptor mRNA in the central nervous system of the rat. *Journal of Comparative Neurology, 457*(3), 213–235.

Kissileff, H. R., et al. (2012). Leptin reverses declines in satiation in weight-reduced obese humans. *American Journal of Clinical Nutrition, 95*(2), 309–317.

Kivimaki, M., et al. (2006). Work stress, weight gain and weight loss: Evidence for bidirectional effects of job strain on body mass index in the Whitehall II study. *International Journal of Obesity, 30*(6), 982–987.

Kleen, J. K., et al. (2006). Chronic stress impairs spatial memory and motivation for reward without disrupting motor ability and motivation to explore. *Behavioral Neuroscience, 120*(4), 842–851.

Kleinridders, A., et al. (2009). MyD88 signaling in the CNS is required for development of fatty acid-induced Leptin resistance and diet-induced obesity. *Cell Metabolism, 10*(4), 249–259.

Klovins, J., et al. (2004). Cloning of two melanocortin (MC) receptors in spiny dogfish: MC3 receptor in cartilaginous fish shows high affinity to ACTH-derived peptides while it has lower preference to gamma-MSH. *European Journal of Biochemistry, 271*(21), 4320–4331.

Knight, Z. A., et al. (2010). Hyperleptinemia is required for the development of leptin resistance. *PLoS One, 5*(6), e11376.

Knuth, N. D., et al. (2014). Metabolic adaptation following massive weight loss is related to the degree of energy imbalance and changes in circulating leptin. *Obesity, 22*(12), 2563–2569.

Kobayashi, Y., et al. (2012). Melanocortin systems on pigment dispersion in fish chromatophores. *Frontiers in Endocrinology (Lausanne), 3*, 9.

Koch, M., et al. (2015). Hypothalamic POMC neurons promote cannabinoid-induced feeding. *Nature, 519*(7541), 45–U72.

Kojima, M., et al. (1999). Ghrelin is a growth-hormone-releasing acylated peptide from stomach. *Nature, 402*(6762), 656–660.

Konner, A. C., et al. (2007). Insulin action in AgRP-expressing neurons is required for suppression of hepatic glucose production. *Cell Metabolism, 5*(6), 438–449.

Koob, G. F. (2010). The role of CRF and CRF-related peptides in the dark side of addiction. *Brain Research, 1314*, 3–14.

Krahn, D. D., Gosnell, B. A., & Majchrzak, M. J. (1990). The anorectic effects of CRH and restraint stress decrease with repeated exposures. *Biological Psychiatry, 27*(10), 1094–1102.

Krashes, M. J., et al. (2011). Rapid, reversible activation of AgRP neurons drives feeding behavior in mice. *Journal of Clinical Investigation, 121*(4), 1424–1428.

Kreitzer, A. C., & Malenka, R. C. (2008). Striatal plasticity and basal ganglia circuit function. *Neuron, 60*(4), 543–554.

Krude, H., & Gruters, A. (2000). Implications of proopiomelanocortin (POMC) mutations in humans: The POMC deficiency syndrome. *Trends in Endocrinology and Metabolism, 11*(1), 15–22.

Kruk, M. R., et al. (1998). The hypothalamus: Cross-roads of endocrine and behavioural regulation in grooming and aggression. *Neuroscience & Biobehavioral Reviews, 23*(2), 163–177.

Kuhnen, P., et al. (2016). Proopiomelanocortin deficiency treated with a melanocortin-4 receptor agonist. *The New England Journal of Medicine, 375*(3), 240–246.

Kuo, J. J., Silva, A. A., & Hall, J. E. (2003). Hypothalamic melanocortin receptors and chronic regulation of arterial pressure and renal function. *Hypertension, 41*(3), 768–774.

la Fleur, S. E., et al. (2005). Choice of lard, but not total lard calories, damps adrenocorticotropin responses to restraint. *Endocrinology, 146*(5), 2193–2199.

Land, B. B., et al. (2014). Medial prefrontal D1 dopamine neurons control food intake. *Nature Neuroscience, 17*(2), 248–253.

Larhammar, D., et al. (2009). Early duplications of opioid receptor and peptide genes in vertebrate evolution. *Annals of the New York Academy of Sciences, 1163*, 451–453.

Laurent, H. K., Powers, S. I., & Granger, D. A. (2013). Refining the multisystem view of the stress response: Coordination among cortisol, alpha-amylase, and subjective stress in response to relationship conflict. *Physiology & Behavior, 119*, 52–60.

LeDoux, J. (2012). Rethinking the emotional brain. *Neuron, 73*(4), 653–676.

Lee, D. J., & Taylor, A. W. (2011). Following EAU recovery there is an associated MC5r-dependent APC induction of regulatory immunity in the spleen. *Investigative Ophthalmology & Visual Science, 52*(12), 8862–8867.

Lee, E. H. Y., et al. (1993). Hippocampal Crf, ne, and Nmda system interactions in memory processing in the rat. *Synapse, 14*(2), 144–153.

Lee, J. Y., et al. (2001). Saturated fatty acids, but not unsaturated fatty acids, induce the expression of cyclooxygenase-2 mediated through toll-like receptor 4. *The Journal of Biological Chemistry, 276*(20), 16683–16689.

Lee, M., et al. (2008). Effects of selective modulation of the central melanocortin-3-receptor on food intake and hypothalamic POMC expression. *Peptides, 29*(3), 440–447.

Leeson, R., Gulabivala, K., & Ng, Y. (2014). Definition of pain. Endodontics, 369 Merskey H Logic, truth and language in concepts of pain. *Qual Life Res, 1994*(Suppl 1), S69–S76.

Leone, S., Noera, G., & Bertolini, A. (2013). Melanocortins as innovative drugs for ischemic diseases and neurodegenerative disorders: Established data and perspectives. *Current Medicinal Chemistry, 20*(6), 735–750.

Leroy, J. L., et al. (2013). Cash and in-kind transfers lead to excess weight gain in a population of women with a high prevalence of overweight in rural Mexico. *Journal of Nutrition, 143*(3), 378–383.

Lesouhaitier, O., et al. (2009). Gram-negative bacterial sensors for eukaryotic signal molecules. *Sensors (Basel), 9*(9), 6967–6990.

Lewis, V. A., & Gebhart, G. F. (1977). Evaluation of the periaqueductal central gray (PAG) as a morphine-specific locus of action and examination of morphine-induced and stimulation-produced analgesia at coincident PAG loci. *Brain Research, 124*(2), 283–303.

Liang, L., et al. (2013). Evolution of melanocortin receptors in cartilaginous fish: Melanocortin receptors and the stress axis in elasmobranches. *General and Comparative Endocrinology, 181*, 4–9.

Libby, P. (2002). Inflammation in atherosclerosis. *Nature, 420*(6917), 868–874.

Lieberman, H. R., Wurtman, J. J., & Chew, B. (1986). Changes in mood after carbohydrate consumption among obese individuals. *American Journal of Clinical Nutrition, 44*(6), 772–778.

Lim, C. T., Grossman, A., & Khoo, B. (2000). Normal physiology of ACTH and GH release in the hypothalamus and anterior pituitary in man. In L. J. De Groot et al. (Eds.), *Endotext*. South Dartmouth, MA: MDText.com, Inc.

Lim, B. K., et al. (2012). Anhedonia requires MC4R-mediated synaptic adaptations in nucleus accumbens. *Nature, 487*(7406), 183–U64.

Lim, K., et al. (2016). Origin of aberrant blood pressure and sympathetic regulation in diet-induced obesity. *Hypertension, 68*(2), 491-+.

Lindberg, C., et al. (2005). Cytokine production by a human microglial cell line: Effects of beta-amyloid and alpha-melanocyte-stimulating hormone. *Neurotoxicity Research, 8*(3–4), 267–276.

Lindblom, J., et al. (2001). The MC4 receptor mediates alpha-MSH induced release of nucleus accumbens dopamine. *Neuroreport, 12*(10), 2155–2158.

Lipton, J. M., et al. (1991). Central administration of the peptide alpha-msh inhibits inflammation in the skin. *Peptides, 12*(4), 795–798.

Lisak, R. P., Nedelkoska, L., & Benjamins, J. A. (2016). The melanocortin ACTH 1-39 promotes protection of oligodendrocytes by astroglia. *Journal of the Neurological Sciences, 362*, 21–26.

Liu, H. Y., et al. (2003). Transgenic mice expressing green fluorescent protein under the control of the melanocortin-4 receptor promoter. *Journal of Neuroscience, 23*(18), 7143–7154.

Liu, J., et al. (2007). The melanocortinergic pathway is rapidly recruited by emotional stress and contributes to stress-induced anorexia and anxiety-like behavior. *Endocrinology, 148*(11), 5531–5540.

Liu, J., et al. (2013). Melanocortin-4 receptor in the medial amygdala regulates emotional stress-induced anxiety-like behaviour, anorexia and corticosterone secretion. *International Journal of Neuropsychopharmacology, 16*(1), 105–120.

Liu, Y., et al. (2016). Lipopolysaccharide rapidly and completely suppresses AgRP neuron-mediated food intake in male mice. *Endocrinology, 157*(6), 2380–2392.

Loh, K., et al. (2017). Insulin controls food intake and energy balance via NPY neurons. *Molecular Metabolism, 6*(6), 574–584.

Lopez-Soriano, J., et al. (1999). Leptin and tumor growth in rats. *International Journal of Cancer, 81*(5), 726–729.

Lu, X. Y., et al. (2003). Interaction between alpha-melanocyte-stimulating hormone and corticotropin-releasing hormone in the regulation of feeding and hypothalamo-pituitary-adrenal responses. *Journal of Neuroscience, 23*(21), 7863–7872.

Ludwig, J., et al. (2011). Neighborhoods, obesity, and diabetes – A randomized social experiment. *New England Journal of Medicine, 365*(16), 1509–1519.

Luisella, A., et al. (2007). Type 2 diabetes and metabolic syndrome are associated with increased expression of 11beta-hydroxysteroid dehydrogenase 1 in obese subjects. *International Journal of Obesity, 31*, S88–S88.

Luquet, S., et al. (2005). NPY/AgRP neurons are essential for feeding in adult mice but can be ablated in neonates. *Science, 310*(5748), 683–685.

Luscher, C., & Slesinger, P. A. (2010). Emerging roles for G protein-gated inwardly rectifying potassium (GIRK) channels in health and disease. *Nature Reviews. Neuroscience, 11*(5), 301–315.

Lute, B., et al. (2014). Biphasic effect of melanocortin agonists on metabolic rate and body temperature. *Cell Metabolism, 20*(2), 333–345.

Lutter, M., & Nestler, E. J. (2009). Homeostatic and hedonic signals interact in the regulation of food intake. *The Journal of Nutrition, 139*(3), 629–632.

Lutter, M., et al. (2008). The orexigenic hormone ghrelin defends against depressive symptoms of chronic stress. *Nature Neuroscience, 11*(7), 752–753.

Ma, X. M., Levy, A., & Lightman, S. L. (1997). Rapid changes in heteronuclear RNA for corticotrophin-releasing hormone and arginine vasopressin in response to acute stress. *Journal of Endocrinology, 152*(1), 81–89.

MacLean, P. S., et al. (2009). Regular exercise attenuates the metabolic drive to regain weight after long-term weight loss. *American Journal of Physiology-Regulatory Integrative and Comparative Physiology, 297*(3), R793–R802.

Magoul, R., et al. (1993). Tachykinergic afferents to the rat arcuate nucleus. A combined immunohistochemical and retrograde tracing study. *Peptides, 14*(2), 275–286.

Markison, S., et al. (2005). The regulation of feeding and metabolic rate and the prevention of murine cancer cachexia with a small-molecule melanocortin-4 receptor antagonist. *Endocrinology, 146*(6), 2766–2773.

Markowitz, C. E., et al. (1992). Effect of opioid receptor antagonism on proopiomelanocortin peptide levels and gene expression in the hypothalamus. *Molecular and Cellular Neurosciences, 3*(3), 184–190.

Marks, D. L., Ling, N., & Cone, R. D. (2001). Role of the central melanocortin system in cachexia. *Cancer Research, 61*(4), 1432–1438.

Marks, D. L., et al. (2006). The regulation of food intake by selective stimulation of the type 3 melanocortin receptor (MC3R). *Peptides, 27*(2), 259–264.

Marsh, D. J., et al. (1999). Response of melanocortin-4 receptor-deficient mice to anorectic and orexigenic peptides. *Nature Genetics, 21*(1), 119–122.

Martin, W. J., & MacIntyre, D. E. (2004). Melanocortin receptors and erectile function. *European Urology, 45*(6), 706–713.

Martin, C. K., et al. (2007). Effect of calorie restriction on resting metabolic rate and spontaneous physical activity. *Obesity (Silver Spring), 15*(12), 2964–2973.

Matthes, H. W., et al. (1996). Loss of morphine-induced analgesia, reward effect and withdrawal symptoms in mice lacking the mu-opioid-receptor gene. *Nature, 383*(6603), 819–823.

Mavrikaki, M., et al. (2016). Melanocortin-3 receptors in the limbic system mediate feeding-related motivational responses during weight loss. *Molecular Metabolism, 5*(7), 566–579.

Mayan, H., et al. (1996). Dietary sodium intake modulates pituitary Proopiomelanocortin mRNA abundance. *Hypertension, 28*(2), 244–249.

McEwen, B. S. (2007). Physiology and neurobiology of stress and adaptation: Central role of the brain. *Physiological Reviews, 87*(3), 873–904.

McEwen, B. S., De Kloet, E. R., & Rostene, W. (1986). Adrenal steroid receptors and actions in the nervous system. *Physiological Reviews, 66*(4), 1121–1188.

Meaney, M. J., Sapolsky, R. M., & McEwen, B. S. (1985). The development of the glucocorticoid receptor system in the rat limbic brain. II. An autoradiographic study. *Brain Research, 350*(1–2), 165–168.

Mechanick, J. I., et al. (1992). Proopiomelanocortin gene-expression in a distinct population of rat spleen and lung leukocytes. *Endocrinology, 131*(1), 518–525.

Melby, C. L., et al. (2017). Attenuating the biologic drive for weight regain following weight loss: Must what Goes down always go back up? *Nutrients, 9*(5), E468.

Mena, J. D., Selleck, R. A., & Baldo, B. A. (2013). Mu-opioid stimulation in rat prefrontal cortex engages hypothalamic orexin/hypocretin-containing neurons, and reveals dissociable roles of nucleus accumbens and hypothalamus in cortically driven feeding. *The Journal of Neuroscience, 33*(47), 18540–18552.

Mendez, I. A., et al. (2015). Involvement of endogenous enkephalins and beta-endorphin in feeding and diet-induced obesity. *Neuropsychopharmacology, 40*(9), 2103–2112.

Menzaghi, F., et al. (1993). Functional impairment of hypothalamic corticotropin-releasing factor neurons with immunotargeted toxins enhances food intake induced by neuropeptide Y. *Brain Research, 618*(1), 76–82.

Mercer, A. J., et al. (2014). Temporal changes in nutritional state affect hypothalamic POMC peptide levels independently of leptin in adult male mice. *American Journal of Physiology-Endocrinology and Metabolism, 306*(8), E904–E915.

Methlie, P., et al. (2013). Changes in adipose glucocorticoid metabolism before and after bariatric surgery assessed by direct hormone measurements. *Obesity, 21*(12), 2495–2503.

Michalaki, V., et al. (2004). Serum levels of IL-6 and TNF-alpha correlate with clinicopathological features and patient survival in patients with prostate cancer. *British Journal of Cancer, 90*(12), 2312–2316.

Michopoulos, V., Toufexis, D., & Wilson, M. E. (2012). Social stress interacts with diet history to promote emotional feeding in females. *Psychoneuroendocrinology, 37*(9), 1479–1490.

Miell, J. P., Englaro, P., & Blum, W. F. (1996). Dexamethasone induces an acute and sustained rise in circulating leptin levels in normal human subjects. *Hormone and Metabolic Research, 28*(12), 704–707.

Miklos, I. H., & Kovacs, K. J. (2002). GABAergic innervation of corticotropin-releasing hormone (CRH)-secreting parvocellular neurons and its plasticity as demonstrated by quantitative immunoelectron microscopy. *Neuroscience, 113*(3), 581–592.

Mizoguchi, K., et al. (2000). Chronic stress induces impairment of spatial working memory because of prefrontal dopaminergic dysfunction. *Journal of Neuroscience, 20*(4), 1568–1574.

Moberg, G. (2000). Biological response to stress: implications for animal welfare. In *The biology of animal stress: basic principles and implications for animal welfare* (pp. 1–21). London: CABI Publishing.

Moller, C. L., et al. (2011). Characterization of murine melanocortin receptors mediating adipocyte lipolysis and examination of signalling pathways involved. *Molecular and Cellular Endocrinology, 341*(1–2), 9–17.

Montero-Melendez, T. (2015). ACTH: The forgotten therapy. *Seminars in Immunology, 27*(3), 216–226.

Moore, C. J., et al. (2015). Antagonism of corticotrophin-releasing factor type 1 receptors attenuates caloric intake of free feeding subordinate female rhesus monkeys in a rich dietary environment. *Journal of Neuroendocrinology, 27*(1), 33–43.

Morgan, S. A., et al. (2014). 11 beta-HSD1 is the major regulator of the tissue-specific effects of circulating glucocorticoid excess. *Proceedings of the National Academy of Sciences of the United States of America, 111*(24), E2482–E2491.

Morgan, D. A., et al. (2015). Regulation of glucose tolerance and sympathetic activity by MC4R signaling in the lateral hypothalamus. *Diabetes, 64*(6), 1976–1987.

Morimoto, M., et al. (1996). Distribution of glucocorticoid receptor immunoreactivity and mRNA in the rat brain: An immunohistochemical and in situ hybridization study. *Neuroscience Research, 26*(3), 235–269.

Morton, N. M., et al. (2004). Novel adipose tissue-mediated resistance to diet-induced visceral obesity in 11 beta-hydroxysteroid dehydrogenase type 1-deficient mice. *Diabetes, 53*(4), 931–938.

Mostyn, A., et al. (2001). The role of leptin in the transition from fetus to neonate. *The Proceedings of the Nutrition Society, 60*(2), 187–194.

Mounien, L., et al. (2005). Expression of melanocortin MC3 and MC4 receptor mRNAs by neuropeptide Y neurons in the rat arcuate nucleus. *Neuroendocrinology, 82*(3–4), 164–170.

Mountjoy, K. G. (2010). Distribution and function of melanocortin receptors within the brain. In A. Catania (Ed.), *Melanocortins: Multiple actions and therapeutic potential* (pp. 29–48). New York: Springer.

Mountjoy, K. G., et al. (1992). The cloning of a family of genes that encode the melanocortin receptors. *Science, 257*(5074), 1248–1251.

Mul, J. D., et al. (2012). Melanocortin receptor 4 deficiency affects body weight regulation, grooming behavior, and substrate preference in the rat. *Obesity, 20*(3), 612–621.

Murphy, M. T., Richards, D. B., & Lipton, J. M. (1983). Anti-pyretic potency of centrally administered alpha-melanocyte stimulating hormone. *Science, 221*(4606), 192–193.

Myers, B., & Greenwood-Van Meerveld, B. (2007). Corticosteroid receptor-mediated mechanisms in the amygdala regulate anxiety and colonic sensitivity. *American Journal of Physiology-Gastrointestinal and Liver Physiology, 292*(6), G1622–G1629.

Myers, M. G., et al. (2010). Obesity and leptin resistance: Distinguishing cause from effect. *Trends in Endocrinology and Metabolism, 21*(11), 643–651.

Myers, B., McKlveen, J. M., & Herman, J. P. (2014). Glucocorticoid actions on synapses, circuits, and behavior: Implications for the energetics of stress. *Frontiers in Neuroendocrinology, 35*(2), 180–196.

Nakamura, Y., Walker, B. R., & Ikuta, T. (2016). Systematic review and meta-analysis reveals acutely elevated plasma cortisol following fasting but not less severe calorie restriction. *Stress, 19*(2), 151–157.

Naslund, E., et al. (2000). Associations of leptin, insulin resistance and thyroid function with long-term weight loss in dieting obese men. *Journal of Internal Medicine, 248*(4), 299–308.

Nathan, P. J., & Bullmore, E. T. (2009). From taste hedonics to motivational drive: Central mu-opioid receptors and binge-eating behaviour. *International Journal of Neuropsychopharmacology, 12*(7), 995–1008.

Nestler, E. J., & Hyman, S. E. (2010). Animal models of neuropsychiatric disorders. *Nature Neuroscience, 13*(10), 1161–1169.

Ni, X. P., et al. (2003). Genetic disruption of gamma-melanocyte-stimulating hormone signaling leads to salt-sensitive hypertension in the mouse. *The Journal of Clinical Investigation, 111*(8), 1251–1258.

Ni, X. P., et al. (2006). Modulation by dietary sodium intake of melanocortin 3 receptor mRNA and protein abundance in the rat kidney. *American Journal of Physiology - Regulatory Integrative and Comparative Physiology, 290*(3), R560–R567.

Nijenhuis, W. A. J., Oosterom, J., & Adan, R. A. H. (2001). AgRP(83-132) acts as an inverse agonist on the human-melanocortin-4 receptor. *Molecular Endocrinology, 15*(1), 164–171.

Nijenhuis, W. A. J., et al. (2003). Discovery and in vivo evaluation of new melanocortin-4 receptor-selective peptides. *Peptides, 24*(2), 271–280.

Nikolarakis, K. E., Almeida, O. F., & Herz, A. (1987). Feedback inhibition of opioid peptide release in the hypothalamus of the rat. *Neuroscience, 23*(1), 143–148.

Nilaver, G., et al. (1979). Adrenocorticotropin and beta-lipotropin in the hypothalamus - localization in the same arcuate neurons by sequential immunocytochemical procedures. *Journal of Cell Biology, 81*(1), 50–58.

Niu, Z. J., et al. (2012). Melanocortin 4 receptor antagonists attenuates morphine antinociceptive tolerance, astroglial activation and cytokines expression in the spinal cord of rat. *Neuroscience Letters, 529*(2), 112–117.

Norman, D., et al. (2003). ACTH and alpha-MSH inhibit leptin expression and secretion in 3T3-L1 adipocytes: Model for a central-peripheral melanocortin-leptin pathway. *Molecular and Cellular Endocrinology, 200*(1–2), 99–109.

Novoselova, T. V., et al. (2016). Loss of Mrap2 is associated with Sim1 deficiency and increased circulating cholesterol. *Journal of Endocrinology, 230*(1), 13–26.

Nyberg, S. T., et al. (2012). Job strain in relation to body mass index: Pooled analysis of 160 000 adults from 13 cohort studies. *Journal of Internal Medicine, 272*(1), 65–73.

Ochi, M., et al. (2008). Effect of chronic stress on gastric emptying and plasma ghrelin levels in rats. *Life Sciences, 82*(15–16), 862–868.

Odonohue, T. L., & Dorsa, D. M. (1982). The opiomelanotropinergic neuronal and endocrine systems. *Peptides, 3*(3), 353–395.

Ohata, H., & Shibasaki, T. (2011). Involvement of CRF2 receptor in the brain regions in restraint-induced anorexia. *Neuroreport, 22*(10), 494–498.

Okada, S., et al. (1998). Elevated serum interleukin-6 levels in patients with pancreatic cancer. *Japanese Journal of Clinical Oncology, 28*(1), 12–15.

Oliveira, J. F., et al. (2013). Chronic stress disrupts neural coherence between cortico-limbic structures. *Frontiers in Neural Circuits, 7*, 10.

Oliver, G., & Wardle, J. (1999). Perceived effects of stress on food choice. *Physiology & Behavior, 66*(3), 511–515.

Oliver, G., Wardle, J., & Gibson, E. L. (2000). Stress and food choice: A laboratory study. *Psychosomatic Medicine, 62*(6), 853–865.

Ollmann, M. M., et al. (1997). Antagonism of central melanocortin receptors in vitro and in vivo by Agouti-related protein. *Science, 278*(5335), 135–138.

Ortolani, D., et al. (2011). Effects of comfort food on food intake, anxiety-like behavior and the stress response in rats. *Physiology & Behavior, 103*(5), 487–492.

Ottaviani, E., Franchini, A., & Franceschi, C. (1997). Evolution of neuroendocrine thymus: Studies on POMC-derived peptides, cytokines and apoptosis in lower and higher vertebrates. *Journal of Neuroimmunology, 72*(1), 67–74.

Ottosson, M., et al. (2000). Effects of cortisol and growth hormone on lipolysis in human adipose tissue. *The Journal of Clinical Endocrinology and Metabolism, 85*(2), 799–803.

Pagano, R. L., et al. (2012). Motor cortex stimulation inhibits thalamic sensory neurons and enhances activity of PAG neurons: Possible pathways for antinociception. *Pain, 153*(12), 2359–2369.

Palkovits, M., Mezey, É., & Eskay, R. L. (1987). Pro-opiomelanocortin-derived peptides (ACTH/β-endorphin/α-MSH) in brainstem baroreceptor areas of the rat. *Brain Research, 436*(2), 323–338.

Pan, J. P., et al. (2004). The value of plasma levels of tumor necrosis factor-alpha and interleukin-6 in predicting the severity and prognosis in patients with congestive heart failure. *Journal of the Chinese Medical Association, 67*(5), 222–228.

Pan, X. C., et al. (2013). Melanocortin-4 receptor expression in the rostral ventromedial medulla involved in modulation of nociception in transgenic mice. *Journal of Huazhong University of Science and Technology-Medical Sciences, 33*(2), 195–198.

Parton, L. E., et al. (2007). Glucose sensing by POMC neurons regulates glucose homeostasis and is impaired in obesity. *Nature, 449*(7159), 228–2U7.

Parylak, S. L., Koob, G. F., & Zorrilla, E. P. (2011). The dark side of food addiction. *Physiology & Behavior, 104*(1), 149–156.

Pavlov, V. A., & Tracey, K. J. (2006). Controlling inflammation: The cholinergic anti-inflammatory pathway. *Biochemical Society Transactions, 34*, 1037–1040.

Peckett, A. J., Wright, D. C., & Riddell, M. C. (2011). The effects of glucocorticoids on adipose tissue lipid metabolism. *Metabolism-Clinical and Experimental, 60*(11), 1500–1510.

Pecoraro, N., et al. (2004). Chronic stress promotes palatable feeding, which reduces signs of stress: Feedforward and feedback effects of chronic stress. *Endocrinology, 145*(8), 3754–3762.

Pednekar, J., Mulgaonker, V., & Mascarenhas, J. (1993). Effect of intensity and duration of stress on male sexual behaviour. *Indian Journal of Experimental Biology, 31*(7), 638–640.

Peeke, P. M., & Chrousos, G. P. (1995). Hypercortisolism and obesity. *Annals of the New York Academy of Sciences, 771*, 665–676.

Pego, J. M., et al. (2010). Stress and the neuroendocrinology of anxiety disorders. *Current Topics in Behavioral Neurosciences, 2*, 97–117.

Pelletier, G. (1993). Regulation of proopiomelanocortin gene expression in rat brain and pituitary as studied by in situ hybridization. *Annals of the New York Academy of Sciences, 680*, 246–259.

Pennock, R. L., & Hentges, S. T. (2011). Differential expression and sensitivity of presynaptic and postsynaptic opioid receptors regulating hypothalamic proopiomelanocortin neurons. *The Journal of Neuroscience, 31*(1), 281–288.

Pennock, R. L., & Hentges, S. T. (2016). Desensitization-resistant and -sensitive GPCR-mediated inhibition of GABA release occurs by Ca2+−dependent and -independent mechanisms at a hypothalamic synapse. *Journal of Neurophysiology, 115*(5), 2376–2388.

Perello, M., Stuart, R. C., & Nillni, E. A. (2007). Differential effects of fasting and leptin on proopiomelanocortin peptides in the arcuate nucleus and in the nucleus of the solitary tract. *American Journal of Physiology-Endocrinology and Metabolism, 292*(5), E1348–E1357.

Perello, M., Stuart, R., & Nillni, E. A. (2008). Prothyrotropin-releasing hormone targets its processing products to different vesicles of the secretory pathway. *Journal of Biological Chemistry, 283*(29), 19936–19947.

Petervari, E., et al. (2011). Central alpha-MSH infusion in rats: Disparate anorexic vs. metabolic changes with aging. *Regulatory Peptides, 166*(1–3), 105–111.

Petrovich, G. D., et al. (2009). Central, but not basolateral, amygdala is critical for control of feeding by aversive learned cues. *Journal of Neuroscience, 29*(48), 15205–15212.

Pfaus, J. G., et al. (2004). Selective facilitation of sexual solicitation in the female rat by a melanocortin receptor agonist. *Proceedings of the National Academy of Sciences of the United States of America, 101*(27), 10201–10204.

Plotsky, P. M. (1986). Opioid inhibition of immunoreactive corticotropin-releasing factor secretion into the hypophysial-portal circulation of rats. *Regulatory Peptides, 16*(3–4), 235–242.

Pollard, T. M., et al. (1995). Effects of academic examination stress on eating behavior and blood lipid levels. *International Journal of Behavioral Medicine, 2*(4), 299–320.

Pradhan, A. D., et al. (2001). C-reactive protein, interleukin 6, and risk of developing type 2 diabetes mellitus. *JAMA, 286*(3), 327–334.

Pritchard, L. E., & White, A. (2007). Minireview: Neuropeptide processing and its impact on melanocortin pathways. *Endocrinology, 148*(9), 4201–4207.

Pritchard, L. E., Turnbull, A. V., & White, A. (2002). Pro-opiomelanocortin processing in the hypothalamus: Impact on melanocortin signalling and obesity. *Journal of Endocrinology, 172*(3), 411–421.

Prokop, J. W., et al. (2012). Leptin and leptin receptor: Analysis of a structure to function relationship in interaction and evolution from humans to fish. *Peptides, 38*(2), 326–336.

Prokop, P., Fancovicova, J., & Fedor, P. (2014). Parasites enhance self-grooming behaviour and information retention in humans. *Behavioural Processes, 107*, 42–46.

Quarta, C., et al. (2016). Renaissance of leptin for obesity therapy. *Diabetologia, 59*(5), 920–927.

Rauchhaus, M., et al. (2000). Plasma cytokine parameters and mortality in patients with chronic heart failure. *Circulation, 102*(25), 3060–3067.

Rebuffescrive, M., et al. (1992). Effect of chronic stress and exogenous glucocorticoids on regional fat distribution and metabolism. *Physiology & Behavior, 52*(3), 583–590.

Rees, J. L. (2000). The melanocortin 1 receptor (MC1R): More than just red hair. *Pigment Cell Research, 13*(3), 135–140.

Refojo, D., et al. (2011). Glutamatergic and dopaminergic neurons mediate anxiogenic and anxiolytic effects of CRHR1. *Science, 333*(6051), 1903–1907.

Regev, L., et al. (2012). Site-specific genetic manipulation of amygdala Corticotropin-releasing factor reveals its imperative role in mediating behavioral response to challenge. *Biological Psychiatry, 71*(4), 317–326.

Reinick, C. L., et al. (2012). Identification of an MRAP-independent Melanocortin-2 receptor: Functional expression of the cartilaginous fish, Callorhinchus milii, Melanocortin-2 receptor in CHO cells. *Endocrinology, 153*(10), 4757–4765.

Ren, P., et al. (1996). Control of C. elegans larval development by neuronal expression of a TGF-beta homolog. *Science, 274*(5291), 1389–1391.

Rene, F., et al. (1998). Melanocortin receptors and δ-opioid receptor mediate opposite signalling actions of POMC-derived peptides in CATH.A cells. *European Journal of Neuroscience, 10*(5), 1885–1894.

Renquist, B. J., et al. (2012). Melanocortin-3 receptor regulates the normal fasting response. *Proceedings of the National Academy of Sciences of the United States of America, 109*(23), E1489–E1498.

Retana-Marquez, S., et al. (2009). Naltrexone effects on male sexual behavior, corticosterone, and testosterone in stressed male rats. *Physiology & Behavior, 96*(2), 333–342.

Reul, J. M., & de Kloet, E. R. (1986). Anatomical resolution of two types of corticosterone receptor sites in rat brain with in vitro autoradiography and computerized image analysis. *Journal of Steroid Biochemistry, 24*(1), 269–272.

Reul, J., & Dekloet, E. R. (1985). 2 receptor systems for corticosterone in rat-brain – microdistribution and differential occupation. *Endocrinology, 117*(6), 2505–2511.

Reul, J., & Dekloet, E. R. (1986). Anatomical resolution of 2 types of corticosterone receptor-sites in rat-brain with invitro autoradiography and computerized image-analysis. *Journal of Steroid Biochemistry and Molecular Biology, 24*(1), 269–272.

Reul, J. M. H. M., & Holsboer, F. (2002). Corticotropin-releasing factor receptors 1 and 2 in anxiety and depression. *Current Opinion in Pharmacology, 2*(1), 23–33.

Reynolds, D. V. (1969). Surgery in the rat during electrical analgesia induced by focal brain stimulation. *Science, 164*(3878), 444–445.

Riddle, D. L., Swanson, M. M., & Albert, P. S. (1981). Interacting genes in nematode Dauer larva formation. *Nature, 290*(5808), 668–671.

Rinaman, L. (2010). Ascending projections from the caudal visceral nucleus of the solitary tract to brain regions involved in food intake and energy expenditure. *Brain Research, 1350*, 18–34.

Rivier, C., & Vale, W. (1984). Influence of corticotropin-releasing factor on reproductive functions in the rat. *Endocrinology, 114*(3), 914–921.

Roberts, L., et al. (2010). Plasma cytokine levels during acute HIV-1 infection predict HIV disease progression. *AIDS, 24*(6), 819–831.

Roberts, B. W., et al. (2014). Regulation of a putative corticosteroid, 17,21-dihydroxypregn-4-ene,3,20-one, in sea lamprey, Petromyzon marinus. *General and Comparative Endocrinology, 196*, 17–25.

Rodrigues, A. R., Almeida, H., & Gouveia, A. M. (2013). Alpha-MSH signalling via melanocortin 5 receptor promotes lipolysis and impairs re-esterification in adipocytes. *Biochimica et Biophysica Acta-Molecular and Cell Biology of Lipids, 1831*(7), 1267–1275.

Rodriguez, A. C. I., et al. (2015). Hypothalamic-pituitary-adrenal axis dysregulation and cortisol activity in obesity: A systematic review. *Psychoneuroendocrinology, 62*, 301–318.

Roeling, T. A. P., et al. (1993). Efferent connections of the hypothalamic grooming area in the rat. *Neuroscience, 56*(1), 199–225.

Romanowski, C. P., et al. (2010). Central deficiency of corticotropin-releasing hormone receptor type 1 (CRH-R1) abolishes effects of CRH on NREM but not on REM sleep in mice. *Sleep, 33*(4), 427–436.

Roozendaal, B. (2002). Stress and memory: Opposing effects of glucocorticoids on memory consolidation and memory retrieval. *Neurobiology of Learning and Memory, 78*(3), 578–595.

Roozendaal, B., PortilloMarquez, G., & McGaugh, J. L. (1996). Basolateral amygdala lesions block glucocorticoid-induced modulation of memory for spatial learning. *Behavioral Neuroscience, 110*(5), 1074–1083.

Roozendaal, B., et al. (2002). Involvement of stress-released corticotropin-releasing hormone in the basolateral amygdala in regulating memory consolidation. *Proceedings of the National Academy of Sciences of the United States of America, 99*(21), 13908–13913.

Roozendaal, B., McReynolds, J. R., & McGaugh, J. L. (2004). The basolateral amygdala interacts with the medial prefrontal cortex in regulating glucocorticoid effects on working memory impairment. *Journal of Neuroscience, 24*(6), 1385–1392.

Rosellirehfuss, L., et al. (1993). Identification of a receptor for gamma-melanotropin and other proopiomelanocortin peptides in the hypothalamus and limbic system. *Proceedings of the National Academy of Sciences of the United States of America, 90*(19), 8856–8860.

Rosenbaum, M., & Leibel, R. L. (2014). 20 years of leptin: Role of leptin in energy homeostasis in humans. *The Journal of Endocrinology, 223*(1), T83–T96.

Rosenstock, J., et al. (2010). The 11-beta-Hydroxysteroid dehydrogenase type 1 inhibitor INCB13739 improves hyperglycemia in patients with type 2 diabetes inadequately controlled by metformin monotherapy. *Diabetes Care, 33*(7), 1516–1522.

Royalty, J. E., et al. (2014). Investigation of safety, tolerability, pharmacokinetics, and pharmacodynamics of single and multiple doses of a long-acting alpha-MSH analog in healthy overweight and obese subjects. *Journal of Clinical Pharmacology, 54*(4), 394–404.

Rubinstein, M., et al. (1996). Absence of opioid stress-induced analgesia in mice lacking beta-endorphin by site-directed mutagenesis. *Proceedings of the National Academy of Sciences of the United States of America, 93*(9), 3995–4000.

Russo, S. J., et al. (2010). The addicted synapse: Mechanisms of synaptic and structural plasticity in nucleus accumbens. *Trends in Neurosciences, 33*(6), 267–276.

Ryan, K. K., et al. (2014). Loss of melanocortin-4 receptor function attenuates HPA responses to psychological stress. *Psychoneuroendocrinology, 42*, 98–105.

Rybkin, I. I., et al. (1997). Effect of restraint stress on food intake and body weight is determined by time of day. *The American Journal of Physiology, 273*(5 Pt 2), R1612–R1622.

Salzet, M., et al. (1997). Leech immunocytes contain proopiomelanocortin: Nitric oxide mediates hemolymph proopiomelanocortin processing. *The Journal of Immunology, 159*(11), 5400–5411.

Sanchez, M. M., et al. (2000). Distribution of corticosteroid receptors in the rhesus brain: Relative absence of glucocorticoid receptors in the hippocampal formation. *The Journal of Neuroscience, 20*(12), 4657–4668.

Sanchez, M. S., et al. (2001). Correlation of increased grooming behavior and motor activity with alterations in nigrostriatal and mesolimbic catecholamines after alpha-melanotropin and neuropeptide glutamine-isoleucine injection in the rat ventral tegmental area. *Cellular and Molecular Neurobiology, 21*(5), 523–533.

Sandman, C. A., & Kastin, A. J. (1981). Intraventricular Administration of Msh Induces Hyperalgesia in rats. *Peptides, 2*(2), 231–233.

Sarkar, S., Legradi, G., & Lechan, R. M. (2002). Intracerebroventricular administration of alpha-melanocyte stimulating hormone increases phosphorylation of CREB in TRH- and CRH-producing neurons of the hypothalamic paraventricular nucleus. *Brain Research, 945*(1), 50–59.

Sartin, J. L., et al. (2008). Central role of the melanocortin-4 receptors in appetite regulation after endotoxin. *Journal of Animal Science, 86*(10), 2557–2567.

Sato, I., et al. (2005). Insulin inhibits neuropeptide Y gene expression in the arcuate nucleus through GABAergic systems. *The Journal of Neuroscience, 25*(38), 8657–8664.

Scarinci, F. (2004). French women don't get fat: The secret of eating for pleasure. *Library Journal, 129*(18), 112-+.

Scarlett, J. M., et al. (2007). Regulation of central melanocortin signaling by interleukin-1 beta. *Endocrinology, 148*(9), 4217–4225.

Schackwitz, W. S., Inoue, T., & Thomas, J. H. (1996). Chemosensory neurons function in parallel to mediate a pheromone response in C. elegans. *Neuron, 17*(4), 719–728.

Schadt, J. C. (1989). Sympathetic and hemodynamic adjustments to hemorrhage – A possible role for endogenous opioid-peptides. *Resuscitation, 18*(2–3), 219–228.

Schaible, E. V., et al. (2013). Single administration of tripeptide alpha-MSH(11-13) attenuates brain damage by reduced inflammation and apoptosis after experimental traumatic brain injury in mice. *PLoS One, 8*(8), e71056.

Schiöth, H. B., et al. (2005). Evolutionary conservation of the structural, pharmacological, and genomic characteristics of the melanocortin receptor subtypes. *Peptides, 26*(10), 1886–1900.

Schreck, C. B. (2000). Accumulation and long-term effects of stress in fish. In G. P. Moberg & J. A. Mench (Eds.), *The biology of animal stress: Basic principles and implications for animal welfare* (pp. 147–158). Wallingford; New York: CABI Pub..

Schwabe, L., et al. (2013). Mineralocorticoid receptor blockade prevents stress-induced modulation of multiple memory Systems in the Human Brain. *Biological Psychiatry, 74*(11), 801–808.

Schwartz, N. B. (2000). Neuroendocrine regulation of reproductive cyclicity. In P. M. Conn & E. Marc (Eds.), *Freeman neuroendocrinology in physiology and m*. Humana Press. https://doi.org/10.1007/978-1-59259-707-9135–145.

Seevers, M. (1936). Opiate addiction in the monkey I. Methods of study. *Journal of Pharmacology and Experimental Therapeutics, 56*(2), 147–156.

Selkirk, J. V., et al. (2007). Identification of differential melanocortin 4 receptor agonist profiles at natively expressed receptors in rat cortical astrocytes and recombinantly expressed receptors in human embryonic kidney cells. *Neuropharmacology, 52*(2), 459–466.

Sellayah, D., Cagampang, F. R., & Cox, R. D. (2014). On the evolutionary origins of obesity: A new hypothesis. *Endocrinology, 155*(5), 1573–1588.

Seoane, L. M., et al. (2004). Ghrelin: From a GH-secretagogue to the regulation of food intake, sleep and anxiety. *Pediatric Endocrinology Reviews, 1*(Suppl 3), 432–437.

Serlachius, A., Hamer, M., & Wardle, J. (2007). Stress and weight change in university students in the United Kingdom. *Physiology & Behavior, 92*(4), 548–553.

Serova, L. I., et al. (2013). Intranasal infusion of melanocortin receptor four (MC4R) antagonist to rats ameliorates development of depression and anxiety related symptoms induced by single prolonged stress. *Behavioural Brain Research, 250*, 139–147.

Shah, S., et al. (2011). Efficacy and safety of the selective 11 beta-HSD-1 inhibitors MK-0736 and MK-0916 in overweight and obese patients with hypertension. *Journal of the American Society of Hypertension, 5*(3), 166–176.

Sharma, S., Fernandes, M. F., & Fulton, S. (2013). Adaptations in brain reward circuitry underlie palatable food cravings and anxiety induced by high-fat diet withdrawal. *International Journal of Obesity, 37*(9), 1183–1191.

Shepard, J. D., Barron, K. W., & Myers, D. A. (2000). Corticosterone delivery to the amygdala increases corticotropin-releasing factor mRNA in the central amygdaloid nucleus and anxiety-like behavior. *Brain Research, 861*(2), 288–295.

Shi, X., et al. (2013). Nuclear factor kappaB (NF-kappaB) suppresses food intake and energy expenditure in mice by directly activating the Pomc promoter. *Diabetologia, 56*(4), 925–936.

Shibata, M., et al. (2016). AgRP neuron-specific deletion of glucocorticoid receptor leads to increased energy expenditure and decreased body weight in female mice on a high-fat diet. *Endocrinology, 157*(4), 1457–1466.

Shin, A. C., et al. (2017). Insulin receptor signaling in POMC, but not AgRP, neurons controls adipose tissue insulin action. *Diabetes, 66*(6), 1560–1571.

Shinyama, H., et al. (2003). Regulation of melanocortin-4 receptor signaling: Agonist-mediated desensitization and internalization. *Endocrinology, 144*(4), 1301–1314.

Shoureshi, P., et al. (2007). Analyzing the evolution of beta-endorphin post-translational processing events: Studies on reptiles. *General and Comparative Endocrinology, 153*(1–3), 148–154.

Shutter, J. R., et al. (1997). Hypothalamic expression of ART, a novel gene related to agouti, is up-regulated in obese and diabetic mutant mice. *Genes & Development, 11*(5), 593–602.

Sinha, R. (2008). Chronic stress, drug use, and vulnerability to addiction. In Uhl, G. R. (Ed.), Addiction reviews 2008 (pp. 105–130).

Sirinathsinghji, D. J. S. (1987). Inhibitory influence of Corticotropin releasing-factor on components of sexual-behavior in the male-rat. *Brain Research, 407*(1), 185–190.

Skibicka, K. P., & Grill, H. J. (2009). Hypothalamic and hindbrain melanocortin receptors contribute to the feeding, thermogenic, and cardiovascular action of melanocortins. *Endocrinology, 150*(12), 5351–5361.

Skutella, T., et al. (1994a). Corticotropin-releasing hormone (CRH) antisense oligodeoxynucleotide induces anxiolytic effects in rat. *Neuroreport, 5*(16), 2181–2185.

Skutella, T., et al. (1994b). Corticotropin-releasing hormone (CRH) antisense oligodeoxynucleotide treatment attenuates social defeat-induced anxiety in rats. *Cellular and Molecular Neurobiology, 14*(5), 579–588.

Slominski, A., Ermak, G., & Mihm, M. (1996). ACTH receptor, CYP11A1, CYP17 and CYP21A2 genes are expressed in skin. *Journal of Clinical Endocrinology & Metabolism, 81*(7), 2746–2749.

Smart, J. L., et al. (2007). Central dysregulation of the hypothalamic-pituitary-adrenal axis in neuron-specific proopiomelanocortin-deficient mice. *Endocrinology, 148*(2), 647–659.

Smeets, T., et al. (2009). Stress selectively and lastingly promotes learning of context-related high arousing information. *Psychoneuroendocrinology, 34*(8), 1152–1161.

Smith, A. I., & Funder, J. W. (1988). Proopiomelanocortin processing in the pituitary, central nervous-system, and peripheral-tissues. *Endocrine Reviews, 9*(1), 159–179.

Smith, M. A., et al. (2007). Melanocortins and agouti-related protein modulate the excitability of two arcuate nucleus neuron populations by alteration of resting potassium conductances. *The Journal of Physiology, 578*(2), 425–438.

Sohn, J. W., et al. (2013). Melanocortin 4 receptors reciprocally regulate sympathetic and parasympathetic preganglionic neurons. *Cell, 152*(3), 612–619.

Sokolove, J., & Lepus, C. M. (2013). Role of inflammation in the pathogenesis of osteoarthritis: Latest findings and interpretations. *Therapeutic Advances in Musculoskeletal Disease, 5*(2), 77–94.

Spaccapelo, L., et al. (2013). Up-regulation of the canonical Wnt-3A and Sonic hedgehog signaling underlies melanocortin-induced neurogenesis after cerebral ischemia. *European Journal of Pharmacology, 707*(1–3), 78–86.

Speakman, J. R. (2008). Thrifty genes for obesity, an attractive but flawed idea, and an alternative perspective: The 'drifty gene' hypothesis. *International Journal of Obesity, 32*(11), 1611–1617.

Spranger, J., et al. (2003). Inflammatory cytokines and the risk to develop type 2 diabetes: Results of the prospective population-based European prospective investigation into cancer and nutrition (EPIC)-potsdam study. *Diabetes, 52*(3), 812–817.

Spruijt, B. M., et al. (1985). Comparison of structural requirements of alpha-MSH and ACTH for inducing excessive grooming and pigment dispersion. *Peptides, 6*(6), 1185–1189.

Srinivasan, S., et al. (2004). Constitutive activity of the melanocortin-4 receptor is maintained by its N-terminal domain and plays a role in energy homeostasis in humans. *The Journal of Clinical Investigation, 114*(8), 1158–1164.

Starowicz, K., & Przewlocka, B. (2003). The role of melanocortins and their receptors in inflammatory processes, nerve regeneration and nociception. *Life Sciences, 73*(7), 823–847.

Starowicz, K., et al. (2002). Modulation of melanocortin-induced changes in spinal nociception by mu-opioid receptor agonist and antagonist in neuropathic rats. *Neuroreport, 13*(18), 2447–2452.

Starowicz, K., et al. (2005). Inhibition of morphine tolerance by spinal melanocortin receptor blockade. *Pain, 117*(3), 401–411.

Starowicz, K., et al. (2009). Peripheral antinociceptive effects of MC4 receptor antagonists in a rat model of neuropathic pain – A biochemical and behavioral study. *Pharmacological Reports, 61*(6), 1086–1095.

Stefano, G. B., & Kream, R. (2008). Endogenous opiates, opioids, and immune function: Evolutionary brokerage of defensive behaviors. *Seminars in Cancer Biology, 18*(3), 190–198.

Stefano, G. B., Salzet-Raveillon, B., & Salzet, M. (1999). Mytilus edulis hemolymph contains proopiomelanocortin: LPS and morphine stimulate differential processing. *Molecular Brain Research, 63*(2), 340–350.

Stein, C., & Machelska, H. (2011). Modulation of peripheral sensory neurons by the immune system: Implications for pain therapy. *Pharmacological Reviews, 63*(4), 860–881.

Stengel, A., & Tache, Y. (2014). CRF and urocortin peptides as modulators of energy balance and feeding behavior during stress. *Frontiers in Neuroscience, 8*, 52.

Steptoe, A., Lipsey, Z., & Wardle, J. (1998). Stress, hassles and variations in alcohol consumption, food choice and physical exercise: A diary study. *British Journal of Health Psychology, 3*, 51–63.

Strassmann, G., et al. (1992). Evidence for the involvement of interleukin 6 in experimental cancer cachexia. *The Journal of Clinical Investigation, 89*(5), 1681–1684.

Sundstrom, G., Dreborg, S., & Larhammar, D. (2010). Concomitant duplications of opioid peptide and receptor genes before the origin of jawed vertebrates. *PLoS One, 5*(5), e10512.

Suzuki, H., et al. (2011). Similar changes of hypothalamic feeding-regulating peptides mRNAs and plasma leptin levels in PTHrP-, LIF-secreting tumors-induced cachectic rats and adjuvant arthritic rats. *International Journal of Cancer, 128*(9), 2215–2223.

Takahashi, A., et al. (1995). Isolation and characterization of melanotropins from lamprey pituitary-glands. *International Journal of Peptide and Protein Research, 46*(3–4), 197–204.

Takahashi, A., et al. (2001). Evolutionary significance of proopiomelanocortin in agnatha and chondrichthyes. *Comparative Biochemistry and Physiology Part B: Biochemistry and Molecular Biology, 129*(2–3), 283–289.

Takahashi, A., et al. (2006). Posttranslational processing of proopiomelanocortin family molecules in sea lamprey based on mass spectrometric and chemical analyses. *General and Comparative Endocrinology, 148*(1), 79–84.

Takahashi, A., et al. (2016). Characterization of melanocortin receptors from stingray Dasyatis akajei, a cartilaginous fish. *General and Comparative Endocrinology, 232*, 115–124.

Tao, Y. X. (2010). The Melanocortin-4 receptor: Physiology, pharmacology, and pathophysiology. *Endocrine Reviews, 31*(4), 506–543.

Tao, Y.-X., et al. (2010). Constitutive activity of neural melanocortin receptors. In Conn, P. M. (Ed.), Methods in enzymology, Constitutive activity in receptors and other proteins, Part A Vol 484: (pp. 267–279).

Tataranni, P. A., et al. (1996). Effects of glucocorticoids on energy metabolism and food intake in humans. *American Journal of Physiology-Endocrinology and Metabolism, 271*(2), E317–E325.

Taylor, A. W., & Lee, D. (2010). Applications of the role of alpha-MSH in ocular immune privilege. In A. Catania (Ed.), *Melanocortins: Multiple actions and therapeutic potential* (pp. 143–149). New York: Springer.

Taylor, A. W., & Namba, K. (2001). In vitro induction of CD25(+) CD4(+) regulatory T cells by the neuropeptide alpha-melanocyte stimulating hormone (alpha-MSH). *Immunology and Cell Biology, 79*(4), 358–367.

Taylor, A. W., Kitaichi, N., & Biros, D. (2006). Melanocortin 5 receptor and ocular immunity. *Cellular and Molecular Biology, 52*(2), 53–59.

Tedford, H. W., & Zamponi, G. W. (2006). Direct G protein modulation of Cav2 calcium channels. *Pharmacological Reviews, 58*(4), 837–862.

Telander, D. G. (2011). Inflammation and age-related macular degeneration (AMD). *Seminars in Ophthalmology, 26*(3), 192–197.

Tempel, D. L., & Leibowitz, S. F. (1994). Adrenal steroid receptors: Interactions with brain neuropeptide systems in relation to nutrient intake and metabolism. *Journal of Neuroendocrinology, 6*(5), 479–501.

Terenius, L., Gispen, W. H., & Dewied, D. (1975). Acth-like peptides and opiate receptors in rat-brain – Structure-activity studies. *European Journal of Pharmacology, 33*(2), 395–399.

Thaler, J. P., et al. (2012). Obesity is associated with hypothalamic injury in rodents and humans. *Journal of Clinical Investigation, 122*(1), 153–162.

Thistlethwaite, D., et al. (1975). Familial glucocorticoid deficiency – Studies of diagnosis and pathogenesis. *Archives of Disease in Childhood, 50*(4), 291–297.

Thornton, J. W. (2001). Evolution of vertebrate steroid receptors from an ancestral estrogen receptor by ligand exploitation and serial genome expansions. *Proceedings of the National Academy of Sciences, 98*(10), 5671–5676.

Ting, P. T., & Koo, J. Y. (2006). Use of etanercept in human immunodeficiency virus (HIV) and acquired immunodeficiency syndrome (AIDS) patients. *International Journal of Dermatology, 45*(6), 689–692.

Tiwari, A. (2010). INCB-13739, an 11beta-hydroxysteroid dehydrogenase type 1 inhibitor for the treatment of type 2 diabetes. *IDrugs, 13*(4), 266–275.

Tolle, V., & Low, M. J. (2008). In vivo evidence for inverse agonism of agouti-related peptide in the central nervous system of proopiomelanocortin-deficient mice. *Diabetes, 57*(1), 86–94.

Tong, Q., et al. (2008). Synaptic release of GABA by AgRP neurons is required for normal regulation of energy balance. *Nature Neuroscience, 11*(9), 998–1000.

Torres, S. J., & Nowson, C. A. (2007). Relationship between stress, eating behavior, and obesity. *Nutrition, 23*(11–12), 887–894.

Townsend, M. S., et al. (2001). Food insecurity is positively related to overweight in women. *Journal of Nutrition, 131*(6), 1738–1745.

Tracey, K. J. (2002). The inflammatory reflex. *Nature, 420*(6917), 853–859.

Tracey, K. J. (2007). Physiology and immunology of the cholinergic antiinflammatory pathway. *Journal of Clinical Investigation, 117*(2), 289–296.

Tran, J. A., et al. (2007). Design, synthesis, in vitro, and in vivo characterization of phenylpiperazines and pyridinylpiperazines as potent and selective antagonists of the melanocortin-4 receptor. *Journal of Medicinal Chemistry, 50*(25), 6356–6366.

Traslavina, G. A. A., & Franci, C. R. (2012). Divergent roles of the CRH receptors in the control of gonadotropin secretion induced by acute restraint stress at Proestrus. *Endocrinology, 153*(10), 4838–4848.

Tsatmali, M., et al. (2000). Skill POMC peptides: Their actions at the human MC-1 receptor and roles in the tanning response. *Pigment Cell Research, 13*, 125–129.

Tsou, K., et al. (1986). Immunocytochemical localization of pro-opiomelanocortin-derived peptides in the adult rat spinal cord. *Brain Research, 378*(1), 28–35.

Ulrich-Lai, Y. M., & Engeland, W. C. (2002). Adrenal splanchnic innervation modulates adrenal cortical responses to dehydration stress in rats. *Neuroendocrinology, 76*(2), 79–92.

Ulrich-Lai, Y. M., & Herman, J. P. (2009). Neural regulation of endocrine and autonomic stress responses. *Nature Reviews. Neuroscience, 10*(6), 397–409.

Ulrich-Lai, Y. M., & Ryan, K. K. (2014). Neuroendocrine circuits governing energy balance and stress regulation: Functional overlap and therapeutic implications. *Cell Metabolism, 19*(6), 910–925.

Ulrich-Lai, Y. M., et al. (2010). Pleasurable behaviors reduce stress via brain reward pathways. *Proceedings of the National Academy of Sciences of the United States of America, 107*(47), 20529–20534.

Ulrich-Lai, Y. M., et al. (2015). Stress exposure, food intake and emotional state. *Stress, 18*(4), 381–399.

Utter, A. C., et al. (1999). Effect of carbohydrate ingestion and hormonal responses on ratings of perceived exertion during prolonged cycling and running. *European Journal of Applied Physiology and Occupational Physiology, 80*(2), 92–99.

Valdearcos, M., et al. (2014). Microglia dictate the impact of saturated fat consumption on hypothalamic inflammation and neuronal function. *Cell Reports, 9*(6), 2124–2138.

Valentino, R. J., & Van Bockstaele, E. (2008). Convergent regulation of locus coeruleus activity as an adaptive response to stress. *European Journal of Pharmacology, 583*(2–3), 194–203.

Valentino, R. J., Foote, S. L., & Page, M. E. (1993). The locus-Coeruleus as a site for integrating Corticotropin-releasing factor and noradrenergic mediation of stress responses. *Corticotropin-Releasing Factor and Cytokines: Role in the Stress Response, 697*, 173–188.

Vallarino, M., et al. (1988). Alpha-melanocyte-stimulating hormone (alpha-MSH) in the brain of the cartilaginous fish. Immunohistochemical localization and biochemical characterization. *Peptides, 9*(4), 899–907.

Vallarino, M., et al. (1989). Proopiomelanocortin (POMC)-related peptides in the brain of the rainbow trout, Salmo gairdneri. *Peptides, 10*(6), 1223–1230.

Vallarino, M., d'Amora, M., & Dores, R. M. (2012). New insights into the neuroanatomical distribution and phylogeny of opioids and POMC-derived peptides in fish. *General and Comparative Endocrinology, 177*(3), 338–347.

Valles, A., et al. (2000). Single exposure to stressors causes long-lasting, stress-dependent reduction of food intake in rats. *American Journal of Physiology. Regulatory, Integrative and Comparative Physiology, 279*(3), R1138–R1144.

Valsamakis, G., et al. (2004). 11 beta-hydroxysteroid dehydrogenase type 1 activity in lean and obese males with type 2 diabetes mellitus. *Journal of Clinical Endocrinology & Metabolism, 89*(9), 4755–4761.

van Gaalen, M. M., et al. (2002). Effects of transgenic overproduction of CRH on anxiety-like behaviour. *European Journal of Neuroscience, 15*(12), 2007–2015.

van Wimersma Greidanus, T. B., et al. (1986). The influence of neurotensin, naloxone, and haloperidol on elements of excessive grooming behavior induced by ACTH. *Behavioral and Neural Biology, 46*(2), 137–144.

Varela, L., & Horvath, T. L. (2012). Leptin and insulin pathways in POMC and AgRP neurons that modulate energy balance and glucose homeostasis. *EMBO Reports, 13*(12), 1079–1086.

Vergoni, A. V., et al. (1989). Tolerance develops to the behavioral-effects of ACTH-(1-24) during continuous icv infusion in rats, and is associated with increased hypothalamic levels of beta-endorphin. *Neuropeptides, 14*(2), 93–98.

Vergoni, A. V., et al. (1998). Differential influence of a selective melanocortin MC4 receptor antagonist (HS014) on melanocortin-induced behavioral effects in rats. *European Journal of Pharmacology, 362*(2–3), 95–101.

Vergoni, A. V., et al. (1999a). Corticotropin-releasing factor (CRF) induced anorexia is not influenced by a melanocortin 4 receptor blockage. *Peptides, 20*(4), 509–513.

Vergoni, A. V., et al. (1999b). Selective melanocortin MC4 receptor blockage reduces immobilization stress-induced anorexia in rats. *European Journal of Pharmacology, 369*(1), 11–15.

Vogel, H., et al. (2016). GLP-1 and estrogen conjugate acts in the supramammillary nucleus to reduce food-reward and body weight. *Neuropharmacology, 110*(Pt A), 396–406.

Vogt, M. C., et al. (2014). Neonatal insulin action impairs hypothalamic Neurocircuit formation in response to maternal high-fat feeding. *Cell, 156*(3), 495–509.

Voisey, J., Kelly, G., & Van Daal, A. (2003). Agouti signal protein regulation in human melanoma cells. *Pigment Cell Research, 16*(1), 65–71.

Vowels, J. J., & Thomas, J. H. (1992). Genetic-analysis of chemosensory control of Dauer formation in Caenorhabditis-Elegans. *Genetics, 130*(1), 105–123.

Vrinten, D. H., et al. (2000). Antagonism of the melanocortin system reduces cold and mechanical allodynia in mononeuropathic rats. *Journal of Neuroscience, 20*(21), 8131–8137.

Vrinten, D. H., et al. (2001). Chronic blockade of melanocortin receptors alleviates allodynia in rats with neuropathic pain. *Anesthesia and Analgesia, 93*(6), 1572–1577.

Walker, J. M., et al. (1982). Non-opiate effects of dynorphin and des-Tyr-dynorphin. *Science, 218*(4577), 1136–1138.

Walker, B. R., et al. (1995). Carbenoxolone increases hepatic insulin sensitivity in man: A novel role for 11-oxosteroid reductase in enhancing glucocorticoid receptor activation. *The Journal of Clinical Endocrinology and Metabolism, 80*(11), 3155–3159.

Wallerius, S., et al. (2003). Rise in morning saliva cortisol is associated with abdominal obesity in men: A preliminary report. *Journal of Endocrinological Investigation, 26*(7), 616–619.

Wang, Y. J., et al. (2012). Effects of adrenal dysfunction and high-dose adrenocorticotropic hormone on NMDA-induced spasm seizures in young Wistar rats. *Epilepsy Research, 100*(1–2), 125–131.

Wardlaw, S. L. (2011). Hypothalamic proopiomelanocortin processing and the regulation of energy balance. *European Journal of Pharmacology, 660*(1), 213–219.

Wardlaw, S. L., McCarthy, K. C., & Conwell, I. M. (1998). Glucocorticoid regulation of hypothalamic proopiomelanocortin. *Neuroendocrinology, 67*(1), 51–57.

Watkins, L. R., et al. (1995). Blockade of interleukin-1 induced hyperthermia by subdiaphragmatic vagotomy – Evidence for vagal mediation of immune brain communication. *Neuroscience Letters, 183*(1–2), 27–31.

Watson, S. J., et al. (1978). Evidence for 2 separate opiate peptide neuronal systems. *Nature, 275*(5677), 226–228.

Wellen, K. E., & Hotamisligil, G. S. (2005). Inflammation, stress, and diabetes. *Journal of Clinical Investigation, 115*(5), 1111–1119.

Wendelaar Bonga, S. E. (1997). The stress response in fish. *Physiological Reviews, 77*(3), 591–625.

Wessells, H., et al. (2000). Effect of an alpha-melanocyte stimulating hormone analog on penile erection and sexual desire in men with organic erectile dysfunction. *Urology, 56*(4), 641–646.

Wessells, H., et al. (2003). MT-II induces penile erection via brain and spinal mechanisms. In R. D. Cone (Ed.), *Melanocortin system* (pp. 90–95).

Weyermann, P., et al. (2009). Orally available selective melanocortin-4 receptor antagonists stimulate food intake and reduce cancer-induced cachexia in mice. *PLoS One, 4*(3), e4774.

White, J. D. (1993). Neuropeptide Y: A central regulator of energy homeostasis. *Regulatory Peptides, 49*(2), 93–107.

White, B. D., Dean, R. G., & Martin, R. J. (1990). Adrenalectomy decreases neuropeptide Y mRNA levels in the arcuate nucleus. *Brain Research Bulletin, 25*(5), 711–715.

Wikberg, J. E. S., & Mutulis, F. (2008). Targeting melanocortin receptors: An approach to treat weight disorders and sexual dysfunction. *Nature Reviews Drug Discovery, 7*(4), 307–323.

Wilde, P. E., & Peterman, J. N. (2006). Individual weight change is associated with household food security status. *Journal of Nutrition, 136*(5), 1395–1400.

Wilkinson, C. W., & Dorsa, D. M. (1986). The effects of aging on molecular forms of beta- and gamma-endorphins in rat hypothalamus. *Neuroendocrinology, 43*(2), 124–131.

Willemse, T., et al. (1994). The effect of haloperidol and naloxone on excessive grooming behavior of cats. *European Neuropsychopharmacology, 4*(1), 39–45.

Wilson, B. D., et al. (1995). Structure and function of ASP, the human homolog of the mouse agouti gene. *Human Molecular Genetics, 4*(2), 223–230.

Wisse, B. E., et al. (2001). Reversal of cancer anorexia by blockade of central melanocortin receptors in rats. *Endocrinology, 142*(8), 3292–3301.

Wisse, B. E., Schwartz, M. W., & Cummings, D. E. (2003). Melanocortin signaling and anorexia in chronic disease states. *Melanocortin System, 994*, 275–281.

Wood, P., et al. (1978). Increase of hippocampal acetylcholine turnover rate and the stretching-yawning syndrome elicited by alpha-MSH and ACTH. *Life Sciences, 22*(8), 673–678.

Woods, C. P., et al. (2015). Tissue specific regulation of glucocorticoids in severe obesity and the response to significant weight loss following bariatric surgery (BARICORT). *Journal of Clinical Endocrinology & Metabolism, 100*(4), 1434–1444.

Wu, Q., & Palmiter, R. D. (2011). GABAergic signaling by AgRP neurons prevents anorexia via a melanocortin-independent mechanism. *European Journal of Pharmacology, 660*(1), 21–27.

Wyss-Coray, T. (2006). Inflammation in Alzheimer disease: Driving force, bystander or beneficial response? *Nature Medicine, 12*(9), 1005–1015.

Xu, H., et al. (2003). Chronic inflammation in fat plays a crucial role in the development of obesity-related insulin resistance. *The Journal of Clinical Investigation, 112*(12), 1821–1830.

Yamaguchi, N. (1992). Sympathoadrenal system in neuroendocrine control of glucose: Mechanisms involved in the liver, pancreas, and adrenal gland under hemorrhagic and hypoglycemic stress. *Canadian Journal of Physiology and Pharmacology, 70*(2), 167–206.

Yamano, Y., et al. (2004). Regulation of CRF, POMC and MC4R gene expression after electrical foot shock stress in the rat amygdala and hypothalamus. *Journal of Veterinary Medical Science, 66*(11), 1323–1327.

Yang, Y., et al. (2011). Hunger states switch a flip-flop memory circuit via a synaptic AMPK-dependent positive feedback loop. *Cell, 146*(6), 992–1003.

Yaswen, L., et al. (1999). Obesity in the mouse model of pro-opiomelanocortin deficiency responds to peripheral melanocortin. *Nature Medicine, 5*(9), 1066–1070.

Ye, D. W., et al. (2014). Motor cortex-periaqueductal gray-spinal cord neuronal circuitry may involve in modulation of nociception: A virally mediated transsynaptic tracing study in spinally transected transgenic mouse model. *PLoS One, 9*(2), 5.

Yi, C. X., et al. (2017). TNF alpha drives mitochondrial stress in POMC neurons in obesity. *Nature Communications, 8*, 15143.

Young, J. (1935). The photoreceptors of lampreys III. Control of color change by the pineal and pituitary. *The Journal of Experimental Biology, 12,* 258–270.

Young, E. A., et al. (1993). Altered ratios of beta-endorphin: Beta-lipotropin released from anterior lobe corticotropes with increased secretory drive. II. Repeated stress. *Journal of Neuroendocrinology, 5*(1), 121–126.

Zagon, I. S., & Mclaughlin, P. J. (1992). An opioid growth-factor regulates the replication of microorganisms. *Life Sciences, 50*(16), 1179–1187.

Zakrzewska, K. E., et al. (1999). Selective dependence of intracerebroventricular neuropeptide Y-elicited effects on central glucocorticoids. *Endocrinology, 140*(7), 3183–3187.

Zelissen, P. M. J., et al. (2005). Effect of three treatment schedules of recombinant methionyl human leptin on body weight in obese adults: A randomized, placebo-controlled trial. *Diabetes Obesity & Metabolism, 7*(6), 755–761.

Zellner, D. A., et al. (2006). Food selection changes under stress. *Physiology & Behavior, 87*(4), 789–793.

Zhan, C., et al. (2013). Acute and long-term suppression of feeding behavior by POMC neurons in the brainstem and hypothalamus. *Respectively. Journal of Neuroscience, 33*(8), 3624–3632.

Zhang, M., & Kelley, A. E. (2000). Enhanced intake of high-fat food following striatal mu-opioid stimulation: Microinjection mapping and Fos expression. *Neuroscience, 99*(2), 267–277.

Zhang, L., et al. (2011). Melanocortin-5 receptor and sebogenesis. *European Journal of Pharmacology, 660*(1), 202–206.

Zharkovsky, A., et al. (1993). Concurrent nimodipine attenuates the withdrawal signs and the increase of cerebral dihydropyridine binding after chronic morphine treatment in rats. *Naunyn-Schmiedeberg's Archives of Pharmacology, 347*(5), 483–486.

Zheng, H., et al. (2005a). Brain stem melanocortinergic modulation of meal size and identification of hypothalamic POMC projections. *American Journal of Physiology - Regulatory, Integrative and Comparative Physiology, 289*(1), R247–R258.

Zheng, S. X., Bosch, M. A., & Ronnekleiv, O. K. (2005b). mu-opioid receptor mRNA expression in identified hypothalamic neurons. *The Journal of Comparative Neurology, 487*(3), 332–344.

Zheng, J., et al. (2009). Effects of repeated restraint stress on gastric motility in rats. *American Journal of Physiology. Regulatory, Integrative and Comparative Physiology, 296*(5), R1358–R1365.

Zhou, Y., et al. (2013). Voluntary alcohol drinking enhances proopiomelanocortin gene expression in nucleus accumbens shell and hypothalamus of Sardinian alcohol-preferring rats. *Alcoholism, Clinical and Experimental Research, 37*(Suppl 1), E131–E140.

Zollner, C., & Stein, C. (2007). Opioids. *Handbook of Experimental Pharmacology, 177,* 31–63.

Obesity and the Growth Hormone Axis

12

Brooke Henry, Elizabeth A. Jensen, Edward O. List, and Darlene E. Berryman

12.1 Introduction to the growth hormone/insulin-like growth factor (GH/IGF-1) Axis

The GH/IGF-1 axis refers to the collective and coordinated actions of GH and IGF-1. Both hormones have a dramatic impact on adipose tissue (AT). GH promotes the release of stored energy by increasing lipolysis, decreasing lipogenesis, and influencing proliferation and differ-

B. Henry · E. O. List
The Diabetes Institute, Konneker Research Labs, Ohio University, Athens, OH, USA

Edison Biotechnology Institute, Konneker Research Labs, Ohio University, Athens, OH, USA
e-mail: bh997614@ohio.edu; list@ohio.edu

E. A. Jensen
Translational Biomedical Sciences, Graduate College, Ohio University, Athens, OH, USA

Edison Biotechnology Institute, Konneker Research Labs, Ohio University, Athens, OH, USA
e-mail: ej391209@ohio.edu

D. E. Berryman (✉)
The Diabetes Institute, Konneker Research Labs, Ohio University, Athens, OH, USA

Edison Biotechnology Institute, Konneker Research Labs, Ohio University, Athens, OH, USA

Department of Biomedical Sciences, Heritage College of Osteopathic Medicine, Ohio University, Athens, OH, USA
e-mail: berrymad@ohio.edu

entiation of the preadipocytes (Moller and Jorgensen 2009). In contrast, IGF-1 appears to be critical for hyperplasia and lipogenesis in both white and brown AT (Boucher et al. 2016). Thus, lipid metabolism and AT mass are drastically altered via this axis.

Human GH (hGH) is a 191 amino acid peptide hormone made up of four alpha helices and two disulfide bonds. The primary form of GH is 22 kDa; however, additional isoforms have been described with different molecular weights, including 20 kDa (which is the second most prominent form), 24 kDa, and 17.5 kDa forms. The vast majority of GH is produced, stored, and secreted by specialized cells in the anterior pituitary called somatotrophs. In humans, four additional closely related genes have been described in the GH gene cluster, including *CSL*, *CSA*, *GHV*, and *CSB*; however, production of these alternate forms of GH are limited to the placenta during pregnancy in females and are thought to be important for fetal growth and development (Chellakooty et al. 2004).

Secretion of GH occurs in a pulsatile fashion. GH synthesis and release are stimulated by GH-releasing hormone (GHRH) and inhibited by somatostatin (SST), which are both produced in the hypothalamus (Fig. 12.1a). Multiple feedback loops contribute to the regulation of GH release. For example, high serum GH levels act directly on the hypothalamus to inhibit GH release by inducing SST release and inhibiting GHRH secretion (Muller 1990; Muller et al.

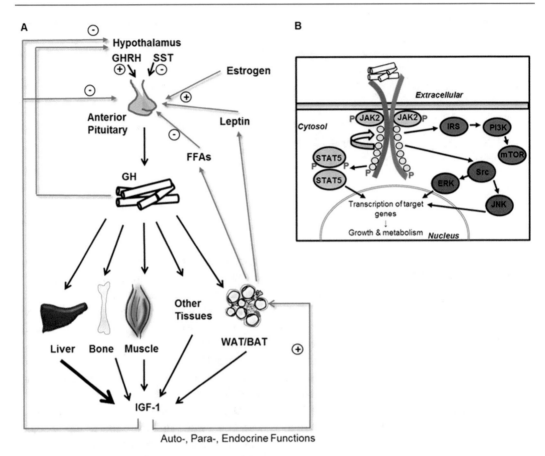

Fig. 12.1 (**a**) Regulation of pituitary GH. Growth hormone-releasing hormone (GHRH) and somatostatin (SST) are released from the hypothalamus and stimulate or inhibit GH secretion, respectively. The GH-GHR interaction results in production of IGF-1 by various target tissues. IGF-1 and GH act in a negative feedback manner to reduce GH secretion. GH secretion is altered by other factors such as leptin, free fatty acids (FFAs), and estrogen. (**b**) The intracellular signaling cascade of GHR in response to GH bind-ing. GH binding to the GHR induces a canonical pathway that includes GHR-JAK2-STAT5. STAT5 is phosphorylated and activated in response to GH binding and regulates gene transcription. Other signaling pathways include insulin receptor substrate (IRS) 1 and subsequent activation of mechanistic target of rapamycin (mTOR) and activation of extracellular signal-regulated kinase (ERK) and c-Jun N-terminal kinase (JNK) via Src. (Adapted with permission Berryman et al. (2015) and Kopchick and Andry (2000))

1999). Additionally, high serum IGF-1 levels act on the hypothalamus and the pituitary to decrease GH secretion (Muller 1990; Muller et al. 1999).

Once secreted, circulating GH binds to the GHR on various tissues (Fig. 12.1b). The GHR is a class 1 cytokine receptor that exists as a pre-formed homodimer (Brooks and Waters 2010). The receptor is expressed ubiquitously in tissues throughout the body, including but not limited to muscle, white and brown AT, liver, kidneys, heart, intestines, and lungs (Brown et al. 2005; Zou et al. 1997; Vikman et al. 1991). The extracellular GH-binding domain of the GHR can be cleaved by metalloproteases and/or alternatively spliced to produce a soluble GH-binding protein that is also found in circulation with ~50% of circulat-ing GH being bound to this GH-binding protein (Herrington et al. 2000). Upon GH binding to the predimerized receptor, a conformational change in the intracellular domain of GHR results in transphosphorylation and activation of associated tyrosine kinases (JAK2) (Brooks et al. 2014). As a result, tyrosines in the cytoplasmic domain of GHR are phosphorylated, which provide a docking site for signal transducer and activator of transcription 5 (STAT5) and result in the recruit-ment of additional intracellular signaling mole-cules (Brooks et al. 2014). Phosphorylation and activation of STAT5 allows for dissociation from the GHR, dimerization and translocation to the nucleus, where it regulates gene expression (Wells 1996). While the JAK-STAT pathway appears to predominate in most cells, additional pathways, such as mechanistic target of rapamycin (mTOR), extracellular signal-regulated kinase (ERK), and c-Jun N-terminal kinase (JNK) (Brooks and Waters 2010), are also activated in response to GH binding to GHR.

As stated previously, production of IGF-1 resulting from GH-GHR signaling is an important mediator of GH action as hinted in IGF-1's previously accepted name somatomedin or "to mediate the actions of GH." While IGF-1 is produced in many tissues and functions in a paracrine and autocrine manner, the vast majority (70–90%) of IGF-1 found in the bloodstream – referred to as "endocrine" IGF-1 – comes from the liver. While the majority of this chapter will focus on GH's action, it is important to note that the tight association between GH signaling and IGF-1 production often makes it difficult to discern the effects of GH versus that of IGF-1.

12.2 Role of GH in Nutrient Metabolism

GH plays a role in a variety of pathways related to energy metabolism and storage, counterbalancing the effect of insulin. In the postprandial state, GH secretion is suppressed, and insulin levels are increased, promoting glucose uptake and adipogenesis (Moller and Jorgensen 2009; Rabinowitz et al. 1965). In contrast, under fasting conditions, GH levels increase, and insulin levels are suppressed, increasing lipolysis and hepatic glucose output. Figure 12.2 further depicts the role of GH in major metabolic tissues.

As AT is the largest energy reservoir in the body, it is not surprising that GH has a significant influence on AT. For example, administration of human GH (hGH) stimulates lipolysis when injected into healthy humans or GH deficient (GHD) patients (Beck et al. 1957; Raben and Hollenberg 1959; Ikkos et al. 1959). The striking lipolytic effect of GH is thought to be mediated, in part, through the stimulation of hormone-sensitive lipase (HSL), a critical enzyme for

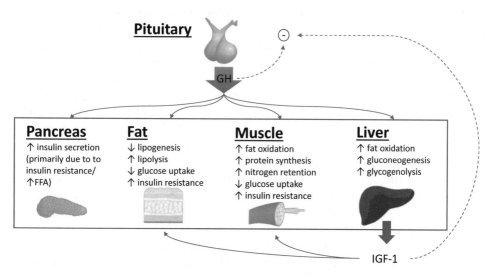

Fig. 12.2 Summary of metabolic effects of GH. GH promotes insulin secretion from the pancreas, which is primarily due to the increase in insulin resistance in other tissues as well as the increase in circulating free fatty acids. In fat, GH increases lipolysis and decreases lipogenesis, leading to increased free fatty acids in circulation and decreased adiposity. In the muscle, GH increases fat oxidation, protein synthesis, and nitrogen retention while decreasing glucose uptake. In the liver, GH increases fatty acid oxidation and glucose production while decreasing hepatic fat content. Endocrine IGF-1 is mainly produced by the liver and acts in a negative feedback loop to decrease GH secretion from the pituitary indicated by the dashed arrow. Chronic excess GH and increased IGF-1 induce insulin resistance and hyperglycemia contributing to the diabetogenic effect of GH

lipolysis (Florini et al. 1996; Franco et al. 2005; Yang et al. 2004; Yip and Goodman 1999; Richelsen et al. 2000). However, more recent evidence has implicated other molecules in GH's lipolytic action, including adipose triglyceride lipase, perilipin, and cell death-inducing DFFA-like effector A (Karastergiou et al. 2016). Because of GH's potent lipolytic effect, the total amount of AT is dramatically reduced under conditions of GH excess (Abrahamsen et al. 2004; Chihara et al. 2010). In addition to an increase in lipolysis, GH impairs lipogenesis, with many studies pointing to a decrease in lipoprotein lipase (LPL) protein activity (Richelsen et al. 2000; Murase et al. 1980; Asayama et al. 1984). Interestingly, both spontaneous and stimulated GH secretion are blunted with obesity (Rasmussen 2010), which could further exacerbate the excess in AT accumulation as will be discussed in more detail later in this chapter.

GH also impacts nutrient metabolism in the liver. GH stimulates hepatic glucose production through gluconeogenesis and glycogenolysis, but whether GH preferentially stimulates one pathway remains unclear as there is evidence to support both possibilities (Lindberg-Larsen et al. 2007; Sakharova et al. 2008; Ghanaat and Tayek 2005; Hoybye et al. 2008). In addition, GH has been shown to play a key role in hepatic lipid processing (Fan et al. 2009b; Cui et al. 2007). That is, a reduction in circulating levels of GH or defects in hepatic GH signaling results in hepatic steatosis that can be somewhat alleviated with GH treatment (Kredel and Siegmund 2014; Diniz et al. 2014; Nishizawa et al. 2012). More recently, GH has been shown to inhibit hepatic de novo lipogenesis (Cordoba-Chacon et al. 2015; Christ et al. 1999) and potentially promote ectopic fat deposition in extrahepatic tissues (Wang et al. 2007).

The skeletal muscle is another major target of GH. GH is well known to promote nitrogen retention and protein synthesis in this tissue (Jorgensen et al. 1990; Moller et al. 2009). In the fasting state, GH preserves positive nitrogen balance, maintains lean body mass, and increases resting energy expenditure (Mulligan et al. 1998). Unlike the role of GH in AT, GH upregulates LPL

mRNA expression and induces FFA uptake (Khalfallah et al. 2001; Oscarsson et al. 1999) and fat oxidation in skeletal muscle (Bastie et al. 2005; Kim et al. 2009; Ehrenborg and Krook 2009). Typically, GH secretion is inhibited by an excess of macronutrients; however, some evidence suggests that certain amino acids promote GH secretion (van Vught et al. 2008). Collectively, this suggests that there is a link between GH and amino acid metabolism and that the metabolic effects of GH are largely dependent on nutritional state.

As a chapter focused on GH and obesity, it is important to emphasize that GH is well known to be a diabetogenic hormone, inducing insulin resistance, hyperinsulinemia, and hyperglycemia (Fathallah et al. 2015). For example, classic studies show that fasting and postprandial perfusion of GH through the brachial artery inhibits muscle glucose uptake in an acute manner (Rabinowitz et al. 1965; Rabinowitz and Zierler 1963; Fineberg and Merimee 1974). There appears to be multiple ways in which GH contributes to insulin resistance. There is evidence that GH may directly block insulin-signaling intermediates, such as p85α, and suppress insulin receptor substrate 1 (IRS-1)-associated phosphatidylinositol 3-kinase (PI3K) activity, both key modulators of insulin signaling (del Rincon et al. 2007). However, this evidence is mainly limited to rodent models. The increase in glucose output by the liver as well as the increased lipolysis, which causes substrate competition between lipids and glucose, also contributes to GH's diabetogenic effect. Moreover, some evidence shows that GH may affect insulin secretion by influencing β-cell function (Zhang et al. 2004; Shin-Hye and Mi-Jung 2017). In fact, prolonged exposure to elevated levels of GH may lead to pancreatic β-cell failure, resulting in type 2 diabetes (Shadid and Jensen 2003). However, a study using β-cell growth hormone receptor knockout (βGHRKO) mice on a high-fat diet shows that GH is necessary for β-cell proliferation and glucose-stimulated insulin secretion (Wu et al. 2011). Thus, it is important to recognize that the affect of GH on β-cell function is complex.

12.3 Clinical Conditions and Mouse Lines with Alterations in the GH Axis

Extremes in the GH/IGF-1 axis in both humans and mice have allowed researchers to uncover many of GH's actions. Examples of these extremes include (1) elevated GH action found in humans with acromegaly/gigantism and bovine GH (bGH) transgenic mice, (2) decreased GH action found in GHD in humans and GHR antagonist (GHA) mice among other mouse lines, and (3) complete GH resistance found in humans with Laron syndrome (LS) and GHR gene-disrupted (GHR-/-) mice. While a more detailed summary will be provided below, Table 12.1 summarizes the general effects associated with these clinical conditions along with a comparable mouse line of excess, reduced, and absence of GH signaling. Note that these mouse lines share many features with the clinical conditions, making them a valuable resource to do more invasive experimental measures.

12.3.1 Acromegaly and Gigantism

Acromegaly is a rare disorder characterized by chronically elevated circulating GH levels and subsequently increased production of IGF-1. In adults and children, acromegaly can arise from various etiologies. The most common cause of acromegaly is a somatotropic pituitary adenoma; yet, other causes include ectopic tumors outside the pituitary gland that produce GH or GHRH (Melmed 2009, 2011; Katznelson et al. 2014). Adults with acromegaly can experience headaches and visual impairment associated with the tumor mass and symptoms related to elevated GH action, including frontal bossing, coarse facial features, and soft tissue swelling; as such, a common diagnostic sign for acromegaly is a change in ring, shoe, or hat size (Katznelson et al. 2011; Abreu et al. 2016). In children, prior to the closure of the epiphyseal growth plate, gigantism results (Kopchick et al. 2014). Since excess GH action disrupts glucose homeostasis, individuals with acromegaly are reported to have increased fasting glucose levels, insulin resistance, and

Table 12.1 Phenotypic summary of GH clinical conditions and comparable mouse line

| Variable | Excess GH action | | Reduced GH action | | Absent GH action | |
	Acromegaly/gigantism	bGH	GHD[a]	GHA[b]	Laron	GHR-/-
GH defect	Hypersecretion of GH	Transgenic for bovine GH	Many variations	Transgenic for GHR antagonist gene	Mutation of GHR	Disruption of Ghr gene
GH action	↑↑	↑↑	↓	↓	Absent	Absent
GH	↑↑	↑↑	↓	↑	↑	↑
IGF-1	↑↑	↑↑	↓	↓	↓↓	↓↓
Growth/body weight	↑↑[a]	↑↑	↓ ↔[a]	↓	↓↓	↓↓
Insulin sensitivity	↓	↓	↑	↓ ↔	↑↓[c]	↑
Lifespan	↓	↓	↔	↔	↔	↑

Adapted and reused with permission (Troike et al. 2017)
↑ increase, ↔ no change, ↓ decrease
[a]Depends on age of onset
[b]Comparable to congenital GHD
[c]Depends on Israeli or Ecuadorian cohort. Israeli cohort tends to have higher insulin levels (Laron et al. 2006)

type 2 diabetes (Abreu et al. 2016). Acromegaly results in a twofold increase in overall mortality with the most common causes of death being heart disease, such as cardiac hypertrophy and congestive heart failure (Abreu et al. 2016; Colao et al. 2004; Katznelson et al. 2011), and cancer, such as colon, breast, and thyroid cancer (Rokkas et al. 2008; Rogozinski et al. 2012; dos Santos et al. 2013; Jenkins 2006). Overall, if individuals with acromegaly do not receive treatment, their lifespan is projected to be decreased by an average of 10 years (Melmed 2009). Treatment resolves the elevated levels of GH and IGF-1, arrests tumor growth, manages the complications and pituitary function, and restores mortality rates to normal (Melmed et al. 2014). Typically to accomplish the above goals, an individual with acromegaly may undergo transsphenoidal surgical resection, radiation of the pituitary adenoma, or biochemical management through SST analogs or pegvisomant (Katznelson et al. 2014). After reduction of the tumor and suppression of GH action through surgery or medication, patients have reduced incidence of comorbidities and increased adiposity.

12.3.2 Growth Hormone Deficiency (GHD)

GHD is a disease in children and adults defined by low levels of GH and IGF-1, resulting from various abnormalities in the GH-IGF-1 axis (Mullis 2007). The pathogenesis of GHD differs between congenital and acquired deficiencies. Congenital GHD can broadly manifest from either the absence of the pituitary gland or genetic mutations in GH, GHRH, or pituitary transcription factors crucial to pituitary development (Mullis 2007). Meanwhile, acquired GHD arises from different and widely varying causes, including nonfunctioning pituitary adenoma, central nervous system (CNS) trauma, idiopathic hypopituitarism, and treatment of pituitary adenomas and CNS tumors that lead to hypopituitarism (Alatzoglou et al. 2014). The clinical presentation of GHD depends upon the age of onset. Children with congenital or acquired GHD exhibit stunted growth and increased fat mass;

they also have a higher risk of developing hyperlipidemia and dramatic hypoglycemic episodes (Smuel et al. 2015). The lack of growth tends to be more exaggerated for children with congenital GHD than children diagnosed with acquired GHD. Adults with GHD present with more generalized and nonspecific symptoms, including decreased muscle mass and energy, lower bone density, anxiety, depression, and increased adiposity along with an overall decline in quality of life (Molitch et al. 2011; Reed et al. 2013). Treatment goals for GHD also depend on the age of onset. Administration of recombinant hGH increases linear growth, reduces the risk of increased adiposity, and improves self-reported quality of life (Alatzoglou et al. 2014; Grimberg et al. 2016). However, long-term GH therapy in children has been correlated with a higher incidence of type 2 diabetes and is controversial on whether it promotes the development of malignancies or shortens lifespan (Alatzoglou et al. 2014). In adults, GH therapy reduces generalized symptoms yet increases the risk of developing type 2 diabetes. Due to these potential harmful effects of GH therapy, children and adults need to be carefully assessed and managed to ensure the best treatment outcome.

12.3.3 Laron Syndrome (LS)

GH insensitivity, more commonly known as LS, is a disorder characterized by high GH and low IGF-1 levels. Most commonly caused by an autosomal recessive mutation in the *GHR* gene, individuals with LS present similar to patients with congenital GHD with extremely stunted growth. This mutation confers the GHR protein nonfunctional; thus, the downstream effects of GH, including the production of IGF-1, are severely attenuated (Laron et al. 1968). Due to the loss of negative feedback from IGF-1, these individuals have elevated circulating GH levels (Laron et al. 1968). LS is characterized by severe growth retardation, obesity, and reduced lean body mass (Laron and Kauli 2016). The risk of developing insulin resistance and type 2 diabetes varies among the different LS cohorts in Israel, Turkey, and Ecuador. For instance, the cohorts in Israel and Turkey have higher levels of insulin and can develop type 2 diabetes later in life (Agladioglu et al. 2013; Laron and Kauli 2016). On the other hand, a cohort of LS patients from Ecuador appear to maintain insulin sensitivity and be protected from type 2 diabetes (Guevara-Aguirre et al. 2015). To date, the reason behind these differences in cohorts is not well understood and is being currently explored. Interestingly, all cohorts of LS appear to be protected from developing cancer (Shevah and Laron 2007; Steuerman et al. 2011; Laron and Kauli 2016; Guevara-Aguirre et al. 2011). Through treatment with recombinant IGF-1 (rIGF-1), LS individuals observe a resolve in the growth discrepancy and an increase in bone mineral density; however, severe adverse reactions, including increased adiposity and insulin resistance, can accompany these benefits (Laron 2015).

12.3.4 Mouse Lines

Many different animal models exist to explore the physiological and metabolic impact of GH action. This section will concentrate on mice with alterations in GH action for the following reasons: (1) mice are anatomically, physiologically, and genetically similar to humans with fairly short gestation periods and lifespans; and (2) mouse genomes are easy to manipulate, allowing for alterations in GH action at the whole-body and tissue-specific level. While many mouse lines are available with alterations in the GH/IGF-1 axis, this chapter will describe the most well-studied lines as well as those lines most in common to the aforementioned clinical conditions.

12.3.5 Increased GH Action

Bovine GH transgenic mice Bovine GH transgenic (bGH) mice have been genetically engineered to constitutively overexpress GH (Fig. 12.3). Similar to acromegaly, bGH mice have elevated plasma levels of GH and IGF-1 (Kopchick and Laron 1999). Overall, this mouse line experiences increased lean mass and

decreased fat mass, as will be discussed more in the next section. bGH mice also have similar comorbidities as acromegaly. For instance, bGH mice have disrupted glucose homeostasis, insulin resistance, and hyperinsulinemia (Kopchick and Laron 1999; Berryman et al. 2004; Palmer et al. 2009). bGH mice also develop cardiomegaly and hepatomegaly and have marked cardiac, vascular, and kidney damage (Berryman et al. 2006; Kopchick et al. 2014; Miquet et al. 2013). Additionally, bGH mice have a greater incidence of tumors and dramatically shorter lifespans, reduced by approximately 50% compared to their littermate controls (Bartke 2003; Kopchick et al. 2014). Due to these similarities, bGH mice provide an opportunity to more invasively study the condition of acromegaly.

12.3.6 Decreased GH Action

There are multiple mouse lines with decreased GH action that mimic the clinical states associated with reduced GH action. Examples include the GHA mouse line, which has decreased GH action and similar to congenital GHD, and the GHR-/- mouse line, which do not have a GHR and are similar to LS (Fig. 12.3). In addition, there are hypopituitarism mouse lines (Ames dwarf and Snell dwarf) that lack GH as well as other pituitary-derived hormones. More recently, two separate adult-inducible lines with reduced GH action have been described that mimic acquired GHD. In addition, several tissue-specific GHR knockouts have been generated to explore the influence of GH on individual cells, tissues, and organs (List et al. 2014; Romero et al. 2010; Lu et al. 2013; Wu et al. 2011; List et al. 2013; Fan et al. 2009a, 2014; List et al. 2015; Mavalli et al. 2010). These tissue-specific lines, while helpful in elucidating the effect of GH on a particular tissue, do not replicate a clinical condition and so will not be discussed in detail in this chapter.

Mice with hypopituitarism Ames and Snell dwarf mice were among the first mice described with a genetic mutation that resulted in GHD and overall hypopituitarism. Both mouse lines exhibit dwarfism, increased adiposity, improved glucose tolerance,

Fig. 12.3 Mice with altered GH action. From left to right: a wild-type mouse, a bGH mouse with increased GH action, a GHA mouse with decreased GH action, and a GHR-/- mouse with no GH action. (Figure adapted and reused with permission Berryman et al. (2011))

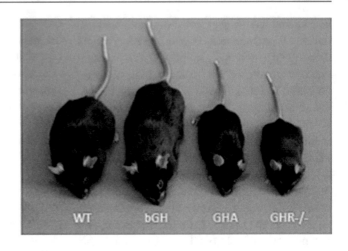

increased insulin sensitivity, and extended lifespans (Bartke & Westbrook, 2012; Hill et al. 2016). Similar to congenital GHD individuals, these mice have mutations in homeobox protein prophet of PIT1 (PROP1) for Ames mice and pituitary transcription factor (PIT1) for Snell mice that results in improper development of the pituitary gland (Bartke and Westbrook 2012; Flurkey et al. 2001; Andersen et al. 1995; Sornson et al. 1996; Junnila et al. 2013). Importantly, these dwarf lines experience deficiencies in three anterior pituitary hormones (prolactin, GH, and thyroid-stimulating hormone), making it difficult to determine the specific contribution of GH to their phenotypes (Bartke & Westbrook, 2012). Still, many of their features are shared with GHR-/- mice, suggesting that GH is critical to the observed phenotype (Bartke & Westbrook, 2012).

GHR antagonist transgenic mice GHA mice have a number of similarities with congenital GHD. GHA mice express a mutated bGH gene, in which the codon for glycine at position 119 is substituted for a larger amino acid (Chen et al. 1991). This single substitution results in the production of a protein that competes with endogenous GH for GHR binding and results in a marked reduction, but not elimination, of GH-induced intracellular signaling. GHA mice are intermediate in size between the global GHR knockout (GHR-/-) and wild-type (WT) mice (Fig. 12.3) (Kopchick et al. 2014). GHA mice have overall reduced levels of IGF-1 and lean mass but increased adiposity (Berryman et al.

2004). Yet, these mice maintain insulin sensitivity and are protected from developing type 2 diabetes and comorbidities associated with obesity, even when placed on a high-fat diet (HFD) (Yang et al. 2015), although there is evidence that their insulin sensitivity deteriorates with advancing age (Berryman et al. 2014). Strikingly, the GHA mice are the only mice that have decreased GH action and do not exhibit a significant increase in lifespan (Coschigano et al. 2003; Berryman et al. 2013).

GHR-/- mice The GHR-/- mice are analogous to LS patients (Kopchick and Laron 2011). Both are dwarf and resistant to GH action with very low IGF-1 and elevated GH levels (Zhou et al. 1997) (Fig. 12.3). GHR-/- mice have decreased lean mass, smaller organs, and increased total adiposity (Berryman et al. 2004; Berryman et al. 2010). Yet despite excess AT throughout life, these mice are also extremely insulin sensitive (Berryman et al. 2004; Lubbers et al. 2013). Interestingly, these mice are resistant to many age-associated complications compared to their littermate controls. In contrast to what is observed in bGH mice, GHR-/- mice are protected from neoplastic diseases (Ikeno et al. 2009). Furthermore, GHR-/- mice are extremely long-lived, living about a year longer than a typical mouse. This increased lifespan has been reproduced in different laboratories and under different experimental conditions, including alterations in sex, genetic background, and diet composition (Coschigano et al. 2003; Bonkowski et al. 2006; Brown-Borg et al. 2009).

Reduction of GH action in adulthood All the mouse lines reported thus far have the alteration in GH action throughout life. More recently, two mouse lines have been genetically engineered to have a decrease in GH action in adulthood; therefore, these lines more closely resemble adult-onset GHD. One of these mouse lines, the adult-onset-isolated GHD line (AOiGHD), employs the inducible Cre-Lox system and diptheria toxin to ablate somatotrophs (Luque et al. 2011). Circulating levels of GH and IGF-1 are decreased but still detectable in these mice after induction. Interestingly, even partial GH disruption, as shown in these mice, has a dramatic impact on metabolic function, resulting in improved insulin sensitivity. A second line, called aGHRKO for adult-onset GHR deletion, was developed in which the *Ghr* is temporally deleted using an ubiquitously expressed tamoxifen-inducible Cre-Lox system (Junnila et al. 2016). These mice have improved insulin sensitivity, increased adiposity, and, at least in the female mice, improved longevity (Junnila et al. 2016). However, it is important to note that there are limitations to this line as the *Ghr* gene is not disrupted in a similar fashion in all tissues, which could influence interpretation of the data generated with these mice.

12.4 Adipose Tissue and GH

Circulating GH levels have an inverse relationship with AT mass in both humans and mice. As expected and in accordance with significant changes in fat mass, GH influences adipokine secretion as well as other properties of AT. The action of GH on AT provides a unique perspective on the relationship between AT and health. That is, we typically associate excess fat mass with a decrease in insulin sensitivity and other obesity-related chronic conditions, whereas a reduced AT mass is usually considered favorable for metabolism and health. Mice with excess GH action, such as bGH mice, are lean but exhibit deleterious metabolic and lifespan consequences; conversely, mice with a reduction in GH action are obese yet lack the metabolic dysfunction

associated with obese states. These contradictory and counterintuitive phenotypes – unhealthy leanness and healthy obesity – allow us to examine not only the biological effects of GH on AT but also the properties of excess AT that contribute to overall metabolic dysfunction.

12.4.1 Body Composition

Excess GH reduces overall adiposity. For example, in individuals with acromegaly, AT mass is drastically reduced (Bengtsson et al. 1989; Moller and Jorgensen 2009; Berryman et al. 2004) while normalizing GH levels when treating acromegaly increases adiposity (Bengtsson et al. 1989; Gibney et al. 2007). Likewise, adult bGH mice are leaner than littermate controls, having notably less fat mass for most of their adult lives (Fig. 12.4) (Berryman et al. 2004). However, the longitudinal body composition data for bGH mice reveals several important observations regarding fat mass, more than can be feasibly observed with clinical populations. First, while they show resistance to midlife gains in AT, bGH mice have greater fat mass at younger ages (less than 3 months of age for males and 4 months of age in females), suggesting a role of the GH/IGF-I axis in establishing the number of adipocyte-competent cells within the tissue (Palmer et al. 2009). Second, male and female mice respond differently to excess GH, with females showing some delay and less exaggerated changes in body composition. This longitudinal study reveals the importance of both age and sex when studying the effects of GH on AT. Finally, bGH mice are also resistant to diet-induced obesity and exhibit distinct patterns of nutrient partitioning with high-fat feeding, in particular having preferential accumulation of lean tissue instead of AT (Olsson et al. 2005; Berryman et al. 2006).

Conversely, decreased GH action consistently increases adiposity. Children and adults with GHD have increased fat mass with greater truncal fat deposition (Boot et al. 1997; De Boer et al. 1992), which begins to reverse with the administration of hGH (Rodriguez-Arnao et al. 1999).

Similar to the decreased GH action in GHD adults and children, LS patients experience increased central and total adiposity (Laron & Kauli, 2016). Likewise, mice with decreased GH action have increased fat mass, although there is evidence that the trend may be age- and/or sex-dependent. For example, Ames and Snell mice have been reported to have increased adiposity (Bartke and Westbrook 2012; Hill et al. 2016). However, a cross-sectional study with Ames mice reports that while both female and male Ames mice have a higher percentage of body fat compared to WT controls at younger ages, the percentage of body fat is decreased relative to the WT controls at older ages (Heiman et al. 2003). Longitudinal body composition data are also available for GHR-/-, GHA, and aGHRKO mice; all three lines have increased fat mass throughout life, albeit the increase in AT is less dramatic for females (GHA and GHR-/- mice shown in Fig. 12.4) (Berryman et al. 2014; Berryman et al. 2010; Junnila et al. 2016). Further, when GHA and GHR-/- mice are fed a HFD, both lines are more susceptible to gaining additional fat mass when compared to WT mice and yet remain resil-ient to the detrimental effects of high-fat feeding on glucose homeostasis and insulin sensitivity (Robertson et al. 2006; Berryman et al. 2006; Yang et al. 2015).

12.4.2 Adipokines

As would be expected with such dramatic changes in adiposity, adipokine levels are altered in clinical states and mouse lines with altered GH-induced signaling. Overall, GH action is negatively correlated with leptin and adiponectin levels. For example, individuals with acromegaly have decreased levels of leptin and adiponectin, while treatment increases leptin levels (Lam et al. 2004; Silha et al. 2003; Sucunza et al. 2009; Reyes-Vidal et al. 2014). Other adipokines, such as vaspin, visfatin, and omentin, have also been shown to be increased in acromegaly and decreased upon or after treatment (Ciresi et al. 2015; Olarescu et al. 2015b). Notably, two visceral AT adipokines, vaspin and visfatin, have been proposed as biomarkers of visceral AT dysfunction with acromegaly (Ciresi et al. 2015).

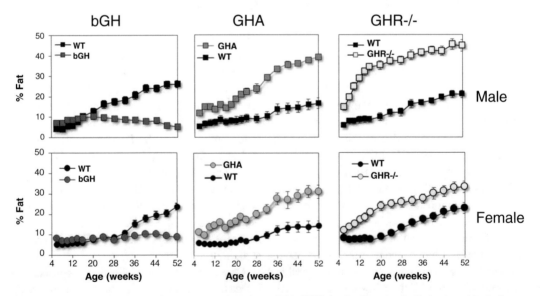

Fig. 12.4 Comparison of body fat percentage in mice with altered GH action. Male and female bGH mice have greater body fat percent than WT mice earlier in life, a trend that starts to reverse at 4 and 6 months of age, respectively (left). Fat percentage is greater in male and female GHA mice compared to controls throughout life. Male GHR-/- mice have markedly increased body fat percent compared to controls and appear to rapidly accumulate fat during the first 4 months of life. Increased percentage of fat is also observed in female GHR-/- mice compared to controls, albeit not as drastic. (Republished with permission Troike et al. (2017))

bGH mice, like the clinical correlate, also have decreased levels of leptin, adiponectin, and resistin (Lubbers et al. 2012).

While data from GHD patients varies based on age and likely etiology, adipokine levels in LS and mouse lines with reduced GH action are fairly consistent and show an overall opposite trend seen in acromegaly and bGH mice. For GHD, children before puberty have been shown to have no alterations in leptin, visfatin, resistin, and omentin levels (Ciresi et al. 2016). However, another study reports children with GHD treated with GH have increased leptin levels and reduced resistin and tumor necrosis factor-alpha (TNF-α) levels (Meazza et al. 2014). GH treatment tends to increase visfatin and decrease leptin and omentin (Ciresi et al. 2016). Children and adults with LS experience increased leptin levels; similarly,

total adiponectin and high molecular weight (HMW) adiponectin are elevated (Laron 2015). As for mouse lines with reduced GH action, Ames, Snell, GHA, GHR-/-, AOiGHD, and aGHRKO mice have either unchanged or elevated leptin levels, which is partly dependent on age, and have markedly elevated total and HMW adiponectin levels (Lubbers et al. 2012; Bartke and Westbrook 2012; Masternak et al. 2006. The exception to this observation is the AOiGHD mice, in which adiponectin levels are unchanged (Lubbers et al. 2012). Of note, when placed on a HFD, Ames mice appear to be protected from pro-inflammatory adipokines with no difference in adiponectin or leptin levels compared to Ames mice on a standard chow diet (Hill et al. 2016).

As summarized in Table 12.2, leptin and adiponectin show an interesting relationship with fat

Table 12.2 Summary of AT in mice with altered GH

Mouse line	GH signaling	% fat mass	Leptin	Adiponectin	Citations
bGH	↑	↑ early life (before 3 months) ↓ later life (after 4 to 6 months in males and females, respectively)	↓	↓	Palmer et al. (2009), Berryman et al. (2004), Wang et al. (2007)
AOiGHD	↓ in adulthood (starting at 10–12 weeks of age)	↑ (after induction)	↑	↔	Luque et al. (2011), Lubbers et al. (2013)
aGHRKO	↓ in adulthood (starting at 6 weeks of age)	↑ (after induction)	↔	↑	Junnila et al. (2016)
GHA	↓	↑ (increases dramatically with advancing age)	↑ (increases dramatically with advancing age)	↑	Berryman et al. (2004), Yakar et al. (2004), Lubbers et al. (2013), Berryman et al. (2014)
GHR-/-	Absent	↑	↔, ↑	↔,↑	Berryman et al. (2004), Berryman et al. (2006), Berryman et al. (2010), Li et al. (2003), Egecioglu et al. (2006), Lubbers et al. (2013)
Snell	↓	↑	↑	↑	Heiman et al. (2003), Wang et al. (2006), Wang et al. (2007)
Ames	↓	↑	↔, ↑	↑	(Bartke (2008), Flurkey et al. (2001), Alderman et al. (2009), Combs et al. (2003), Lubbers et al. (2013)

↑ increase, ↔ no change, ↓ decrease

mass and GH signaling. That is, adiponectin and leptin are both increased with a reduction in GH signaling and decreased with excess GH signaling. Although leptin levels are consistent with what would be expected based on fat mass, low levels of leptin rather than high levels of leptin are more commonly associated with improved insulin sensitivity, an improved metabolic profile, and increased longevity (Stenholm et al. 2011; Arai et al. 2011). Unlike leptin, adiponectin is considered a beneficial adipokine and is usually negatively associated with obesity and positively associated with insulin sensitivity and longevity (Stenholm et al. 2011; Arai et al. 2011). In these mice as well as the comparable clinical conditions, adiponectin is positively associated with fat mass. This unusual relationship has given significant attention to these adipokines and their role in promoting the GH-induced phenotype.

12.4.3 GH Alters AT in a Depot-Specific Manner

Data from both clinical studies and mice reveal that GH's effects are not uniform across all AT depots. In fact, even without considering clinical conditions of extreme GH action, visceral adiposity is a stronger indicator of 24-h endogenous GH secretion than total fat mass (Vikman et al. 1991), suggesting a close association between GH action and visceral depots in humans. Accordingly, while subcutaneous and visceral depots are decreased with acromegaly, the greatest reduction of white AT mass occurs in the visceral depot (Freda et al. 2008). In humans with GHD, GH treatment (0.013–0.026 mg/kg/day) reduces total body fat by 9.4%, with again the visceral AT being more impacted than subcutaneous AT (30% versus 13% reduction, respectively) (Bengtsson et al. 1993). In another study, after GHD adults were treated with GH, total body fat decreased with the largest decrease observed in the visceral depots as compared to subcutaneous (Johannsson et al. 1997). However, there is also clinical evidence for targeting subcutaneous depots. That is, male and female patients with LS have marked increases in subcutaneous and intra-abdominal AT and, in particular, have a larger percentage of fat distributed in the arms (subcutaneous) when compared to control patients (Laron et al. 2006). Clearly, data from humans has demonstrated a depot-specific role of GH's impact on fat mass; yet, the ability to directly compare depots from the same clinical sample is challenging.

A more detailed understanding of the depot differences is made possible by using mouse lines, where more invasive procedures and multiple depot sampling can be made with a single animal. Overall, a reduction in GH action causes a striking and specific enlargement of the subcutaneous depot (Fig. 12.5). This significant increase in the subcutaneous depot is illustrated for GHR-/- mice in Fig. 12.5c. The trend is not as clear with an excess of GH action since depot mass appears to be similarly reduced in all AT depots; however, molecular characterization of the AT depots reveals a more significant impact in subcutaneous depots as compared to others (Benencia et al. 2014). The targeted impact of GH on subcutaneous AT is readily apparent when comparing AT histology among depots. As shown in Fig. 12.5a and b, hematoxylin- and eosin-stained tissue sections from mouse lines with extremes in GH action show dramatic alterations in morphology and adipocyte size in subcutaneous AT; yet, the epididymal tissue is fairly uniform. Many other examples in the literature support depot-specific differences at the cellular or molecular level. Although not an exhaustive list, some recent examples of depot-specific differences are provided in Table 12.3.

12.4.4 GH as a Treatment Modality for Obesity? Pros vs Cons

With the escalating rates of obesity nationally and globally, methods for obesity management are desperately needed. For almost three decades, numerous studies have attempted to evaluate the efficacy of using GH for obesity management, many of which are summarized in Table 12.4. In part, the attention given to GH is related to its ability to reduce fat mass while preserving lean

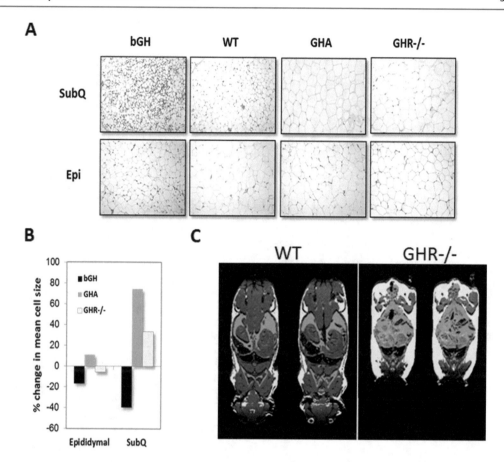

Fig. 12.5 Depot-specific differences due to extremes in GH. (**a**) Hematoxylin and eosin staining of subcutaneous (SubQ) and epididymal (Epi) AT. Tissue samples were obtained from 6-month-old GHR-/-, GHA, bGH, and control mice. (**b**) Quantification of adipocyte size from these mice. (**c**) Adiposity in GHR-/- mice. Regional body fat mass. In addition, human obesity, especially in the visceral depots, is associated with markedly suppressed spontaneous and stimulated GH secretion (Rasmussen 2010). Of note, circulating IGF-1 levels remain unaffected or only slightly lowered with obesity. Likewise, substantial weight loss has been shown to restore GH secretion patterns (Rasmussen et al. 1995). Thus, restoring GH levels appears to be a viable means to combat obesity. Overall, GH at therapeutic levels has shown a fairly consistent reduction in total and visceral AT in clinical trials (Table 12.4). Further, many of the studies that used calorie restriction showed the added benefit of attenuating the loss of lean body mass with GH treatment distribution of male WT mice (left) and male GHR-/- mice (right) using magnetic resonance imaging (MRI). The mouse is positioned with the anterior part at the bottom of the image. Subcutaneous AT is highlighted yellow and intra-abdominal blue. (Adapted with permission Berryman et al. (2011))

(Clemmons et al. 1987; Snyder et al. 1988; Snyder et al. 1990; Tagliaferri et al. 1998).

Despite the logic behind using GH therapeutically, its use is not without controversy. For example, a meta-analysis of 24 clinical studies on the use of GH therapy for obesity concludes that the effects on body composition and lipid profiles are very small and do not reduce weight sufficiently to be considered beneficial, even with very high doses (Mekala and Tritos 2008). It is important to note that there are many confounding factors in these studies that likely contribute to the different responses, including GH dose, inclusion/exclusion criteria, sample size, length of follow-up, and timing of glucose and

Table 12.3 Evidence for depot-specific differences in mice due to extremes in GH action

Reference	Model System	Research Focus	Findings
Benencia et al. (2014)	bGH mice	Immune cell infiltration in AT	↑ immune cell infiltration (macrophage, T cells) mainly in subcutaneous and mesenteric depots with little change in epididiymal AT ↑ in gene expression related to T-cell infiltration/activation in subcutaneous but not epididymal AT
Stout et al. (2014)	bGH, Snell, Ames, GHR-/-, and GH-injected mice	Cellular senescence in AT	bGH females: ↑ cellular senescence in all depots except periovarian GH-injected WT females: ↑ cellular senescence in subscapular and mesenteric depots GHR-/- and Snell females: ↓ cellular senescence in subcutaneous AT Ames: ↓ cellular senescence in paraovarian, mesenteric, and subcutaneous GHR-/-: ↓ cellular senescence all depots except mesenteric
Flint et al. (2006)	GHR-/- mice	Proliferation and differentiation of preadipocytes	Subcutaneous adipocytes proliferate, differentiate, and respond to hormones in a similar manner to controls, but perigonadal adipocytes do not
Kelder et al. (2007)	GHR-/- mice	CideA expression	↓ cell death-inducing DFF45-like effector-A (CideA) levels in subcutaneous AT but not retroperitoneal or epididymal
Lubbers et al. (2013)	bGH, GHA, GHR-/-, AoiGHD, and Ames mice	Adiponectin expression	Circulating adiponectin levels correlated strongly with subcutaneous fat mass Higher adiponectin levels in subcutaneous AT of GHR-/- mice
Stout et al. (2015)	GHR-/- mice	Depot whole-genome microarrays	Differences in gene expression related to metabolic function and inflammation among epididymal, subcutaneous, retroperitoneal AT were detected
Sackmann-Sala et al. (2013)	GHR-/- mice	Proteomic analysis of depot differences	Lower levels of Glut4 protein in subcutaneous AT of GHR-/- mice, no difference in epididymal AT Retroperitoneal depot particularly affected by GHR deletion and age
Hjortebjerg (2017)	bGH, GHR-/- mice	Expression of insulin, GH and IGF-1 receptors, and IGF-1	↑ IGF-1 RNA levels in subcutaneous, epididymal, and retroperitoneal depots in bGH mice, but ↓ in all depots in GHR-/- mice ↓ insulin receptor expression in retroperitoneal and mesenteric depots in bGH mice, but ↑ in subcutaneous, retroperitoneal, and mesenteric depots in GHR-/- mice
Olarescu et al. (2015a)	GHR-/-, bGH mice	AT-derived mesenchymal stem cell	Increased differentiation in cells isolated from subcutaneous AT vs epididymal Increased differentiation in cells isolated from subcutaneous GHR-/- than from bGH mice

insulin measurements relative to GH dose (Berryman et al. 2013). There are also adverse side effects that have been reported with GH therapy. For example, 20–40% of obese individuals that have received GH develop similar effects seen in acromegaly, including fluid reten-tion and carpal tunnel syndrome (Shadid and Jensen 2003). Of greater concern, obesity leads to an insulin-resistant state, which might be further exacerbated by the diabetogenic properties of GH. Although some studies show deterioration in glucose metabolism with GH treatment

Table 12.4 Select clinical studies in which GH was used to treat obesity

Citation	Duration	Study design	GH dose	Starting BMI kg/m²	Δ total body fat	ΔVisceral AT
Clemmons et al. (1987)	3 weeks	Open crossover + CR	0.1 mg/kg IBW	42.5% > IBW	↔	ND
Snyder et al. (1988)	10 weeks	Open + CR	0.1 mg/kg IBW	30–67% > IBW	↔	ND
Snyder et al. (1990)	5 weeks	Open crossover + CR	0.2 mg/kg IBW	30–65% > IBW	↔	ND
Skaggs and Crist (1991)	4 weeks	RPCT	0.08 mg/kg IBW	30–50	↓	ND
Richelsen et al. (1994)	5 weeks	RPCT crossover	0.03 mg/kg IBW	34.5	↓	↓
Johannsson et al. (1997)	9 months	RPCT	9.5 µg/kg	≈30	↓	↓
Tagliaferri et al. (1998)	4 weeks	RPCT + CR	0.3 mg/kg IBW	35.9	↔	ND
Kim et al. (1999)	12 weeks	RPCT + CR	8.6 µg/kg	≈29	↓	↓
Franco et al. (2005)	12 months	DB, RPCT	8 µg/kg	30	↓	↓
Bredella et al. (2012)	6 months	RPCT	Final dose 1.4 ± 0.1 mg	34.9	↓	↓

Symbols and abbreviations: ↑, increased; ↓, decreased; ↔, no difference; ND, not determined; DB, double blind; RPCT, randomized placebo-controlled trial; IBW, ideal body weight; CR, caloric restriction.

(Richelsen et al. 1994; Snyder et al. 1995, 1990; Bredella et al. 2012), many studies report improvement or no impact (Johannsson et al. 1997; Franco et al. 2005). For example, two studies that included obese individuals with diabetes show GH treatment not only reduced visceral AT mass and levels of LDL cholesterol but also improved insulin sensitivity (Nam et al. 2001; Ahn et al. 2006). Studies using diet-induced obese, diabetic mice would also suggest that high levels of GH are beneficial, improving body composition and glucose metabolism (List et al. 2009).

Another consideration is that changes in AT mass may not be the best clinical readout for GH's impact on obesity. As discussed previously in this chapter, GH causes a striking reduction in fat mass, but the quality of AT is suspect as increased cellular senescence, decreased adipogenesis, and altered immune cell infiltration occur in AT in mice exposed to excess GH. Overall, while GH has great potential for treating obesity, its cost, the concerns related to reported and suspected side effects, and the lack of larger and longer controlled trials make it unlikely that GH will be of therapeutic value for obesity in the near future.

Questions

1. What is the relationship between GH and IGF-1, and how do these hormones alter adipose tissue mass?
2. How does GH alter glucose metabolism? How would an increase or absence of GH signaling influence insulin sensitivity?
3. At least in mice, which depot appears to be most impacted by alterations in GH action?
4. How might excess GH in both mouse and man resemble a lipodystrophic state?
5. What are the pros and cons of using GH as a treatment modality for obesity?

References

Abrahamsen, B., Nielsen, T. L., Hangaard, J., Gregersen, G., Vahl, N., Korsholm, L., Hansen, T. B., Andersen, M., & Hagen, C. (2004). Dose-, IGF-I- and sex-dependent changes in lipid profile and body composition during GH replacement therapy in adult onset GH deficiency. *European Journal of Endocrinology, 150*(5), 671–679.

Abreu, A., Tovar, A. P., Castellanos, R., Valenzuela, A., Giraldo, C. M., Pinedo, A. C., Guerrero, D. P., Barrera, C. A., Franco, H. I., Ribeiro-Oliveira, A., Jr., Vilar, L.,

Jallad, R. S., Duarte, F. G., Gadelha, M., Boguszewski, C. L., Abucham, J., Naves, L. A., Musolino, N. R., de Faria, M. E., Rossato, C., & Bronstein, M. D. (2016). Challenges in the diagnosis and management of acromegaly: A focus on comorbidities. *Pituitary, 19*(4), 448–457. https://doi.org/10.1007/s11102-016-0725-2.

Agladioglu, S. Y., Cetinkaya, S., Savas Erdeve, S., Onder, A., Kendirci, H. N., Bas, V. N., & Aycan, Z. (2013). Diabetes mellitus with Laron syndrome: case report. *Journal of Pediatric Endocrinology & Metabolism: JPEM, 26*(9–10), 955–958. https://doi.org/10.1515/jpem-2012-0411.

Ahn, C. W., Kim, C. S., Nam, J. H., Kim, H. J., Nam, J. S., Park, J. S., Kang, E. S., Cha, B. S., Lim, S. K., Kim, K. R., Lee, H. C., & Huh, K. B. (2006). Effects of growth hormone on insulin resistance and atherosclerotic risk factors in obese type 2 diabetic patients with poor glycaemic control. *Clinical Endocrinology, 64*(4), 444–449.

Alatzoglou, K. S., Webb, E. A., Le Tissier, P., & Dattani, M. T. (2014). Isolated growth hormone deficiency (GHD) in childhood and adolescence: Recent advances. *Endocrine Reviews, 35*(3), 376–432. https://doi.org/10.1210/er.2013-1067.

Alderman, J. M., Flurkey, K., Brooks, N. L., Naik, S. B., Gutierrez, J. M., Srinivas, U., Ziara, K. B., Jing, L., Boysen, G., Bronson, R., Klebanov, S., Chen, X., Swenberg, J. A., Stridsberg, M., Parker, C. E., Harrison, D. E., & Combs, T. P. (2009). Neuroendocrine inhibition of glucose production and resistance to cancer in dwarf mice. *Experimental Gerontology, 44*(1–2), 26–33.

Andersen B, Pearse RV 2nd, Jenne K, Sornson M, Lin SC, Bartke A, Rosenfeld MG (1995) The Ames dwarf gene is required for Pit-1 gene activation. Developmental Biology 172 (2):495–503.

Arai, Y., Takayama, M., Abe, Y., & Hirose, N. (2011). Adipokines and aging. *Journal of Atherosclerosis and Thrombosis, 18*(7), 545–550. doi:JST.JSTAGE/jat/7039 [pii].

Asayama, K., Amemiya, S., Kusano, S., & Kato, K. (1984). Growth-hormone-induced changes in postheparin plasma lipoprotein lipase and hepatic triglyceride lipase activities. *Metabolism, 33*(2), 129–131.

Bartke, A. (2003). Can growth hormone (GH) accelerate aging? Evidence from GH-transgenic mice. *Neuroendocrinology, 78*(4), 210–216.

Bartke, A. (2008). Impact of reduced insulin-like growth factor-1/insulin signaling on aging in mammals: Novel findings. *Aging Cell, 7*(3), 285–290.

Bartke, A., & Westbrook, R. (2012). Metabolic characteristics of long-lived mice. *Frontiers in Genetics, 3*, 288. https://doi.org/10.3389/fgene.2012.00288.

Bastie, C. C., Nahle, Z., McLoughlin, T., Esser, K., Zhang, W., Unterman, T., & Abumrad, N. A. (2005). FoxO1 stimulates fatty acid uptake and oxidation in muscle cells through CD36-dependent and -independent mechanisms. *The Journal of Biological Chemistry, 280*(14), 14222–14229. https://doi.org/10.1074/jbc.M413625200.

Beck, J. C., McGarry, E. E., Dyrenfurth, I., & Venning, E. H. (1957). Metabolic effects of human and monkey growth hormone in man. *Science, 125*, 884.

Benencia, F., Harshman, S., Duran-Ortiz, S., Lubbers, E. R., List, E. O., Householder, L., Alnaeeli, M., Liang, X., Welch, L., Kopchick, J. J., & Berryman, D. E. (2014). Male bovine GH transgenic mice have decreased adiposity with an adipose depot-specific increase in immune cell populations. *Endocrinology:en20141794.* https://doi.org/10.1210/en.2014-1794.

Bengtsson, B. A., Brummer, R. J., Eden, S., & Bosaeus, I. (1989). Body composition in acromegaly. *Clinical Endocrinology, 30*(2), 121–130.

Bengtsson B. A., Edén S., Lönn L., Kvist H., Stokland A., Lindstedt G., Bosaeus I., Tölli J., Sjöström L., & Isaksson O.G. (1993). Treatment of adults with growth hormone (GH) deficiency with recombinant human GH. *The Journal of Clinical Endocrinology & Metabolism, 76*(2), 309–317.

Berryman, D., Householder, L., Lesende, V., List, E., & Kopchick, J. J. (2015). Living large: What mouse models reveal about growth hormone. In N. A. Berger (Ed.), *Murine models, energy, balance, and Cancer* (pp. 65–95). New York: Springer.

Berryman, D. E., Glad, C. A., List, E. O., & Johannsson, G. (2013). The GH/IGF-1 axis in obesity: Pathophysiology and therapeutic considerations. *Nature Reviews. Endocrinology, 9*(6), 346–356. https://doi.org/10.1038/nrendo.2013.64. nrendo.2013.64 [pii].

Berryman, D. E., List, E. O., Coschigano, K. T., Behar, K., Kim, J. K., & Kopchick, J. J. (2004). Comparing adiposity profiles in three mouse models with altered GH signaling. *Growth Hormone & IGF Research, 14*(4), 309–318.

Berryman, D. E., List, E. O., Kohn, D. T., Coschigano, K. T., Seeley, R. J., & Kopchick, J. J. (2006). Effect of growth hormone on susceptibility to diet-induced obesity. *Endocrinology, 147*(6), 2801–2808. doi:en.2006-0086 [pii]. https://doi.org/10.1210/en.2006-0086.

Berryman, D. E., List, E. O., Palmer, A. J., Chung, M. Y., Wright-Piekarski, J., Lubbers, E., O'Connor, P., Okada, S., & Kopchick, J. J. (2010). Two-year body composition analyses of long-lived GHR null mice. *The Journals of Gerontology. Series A, Biological Sciences and Medical Sciences, 65*(1), 31–40.

Berryman, D. E., List, E. O., Sackmann-Sala, L., Lubbers, E., Munn, R., & Kopchick, J. J. (2011). Growth hormone and adipose tissue: Beyond the adipocyte. *Growth Hormone & IGF Research, 21*(3), 113–123. https://doi.org/10.1016/j.ghir.2011.03.002. S1096-6374(11)00018-9 [pii].

Berryman, D. E., Lubbers, E. R., Magon, V., List, E. O., & Kopchick, J. J. (2014). A dwarf mouse model with decreased GH/IGF-1 activity that does not experience life-span extension: Potential impact of increased adiposity, leptin, and insulin with advancing age. *The Journals of Gerontology. Series A, Biological Sciences and Medical Sciences, 69*(2), 131–141. https://doi.org/10.1093/gerona/glt069. glt069 [pii].

Bonkowski, M. S., Rocha, J. S., Masternak, M. M., Al Regaiey, K. A., & Bartke, A. (2006). Targeted disruption of growth hormone receptor interferes with the beneficial actions of calorie restriction. *Proceedings of the National Academy of Sciences of the United States of America, 103*(20), 7901–7905.

Boot, A. M., Engels, M. A., Boerma, G. J., Krenning, E. P., & De Muinck Keizer-Schrama, S. M. (1997). Changes in bone mineral density, body composition, and lipid metabolism during growth hormone (GH) treatment in children with GH deficiency. *The Journal of Clinical Endocrinology and Metabolism, 82*(8), 2423–2428. https://doi.org/10.1210/jcem.82.8.4149.

Boucher, J., Softic, S., El Ouaamari, A., Krumpoch, M. T., Kleinridders, A., Kulkarni, R. N., O'Neill, B. T., & Kahn, C. R. (2016). Differential roles of insulin and IGF-1 receptors in adipose tissue development and function. *Diabetes, 65*(8), 2201–2213. https://doi.org/10.2337/db16-0212.

Bredella, M. A., Lin, E., Brick, D. J., Gerweck, A. V., Harrington, L. M., Torriani, M., Thomas, B. J., Schoenfeld, D. A., Breggia, A., Rosen, C. J., Hemphill, L. C., Wu, Z., Rifai, N., Utz, A. L., & Miller, K. K. (2012). Effects of GH in women with abdominal adiposity: A 6-month randomized, double-blind, placebo-controlled trial. *European Journal of Endocrinology, 166*(4), 601–611. https://doi.org/10.1530/EJE-11-1068. EJE-11-1068 [pii].

Brooks, A. J., Dai, W., O'Mara, M. L., Abankwa, D., Chhabra, Y., Pelekanos, R. A., Gardon, O., Tunny, K. A., Blucher, K. M., Morton, C. J., Parker, M. W., Sierecki, E., Gambin, Y., Gomez, G. A., Alexandrov, K., Wilson, I. A., Doxastakis, M., Mark, A. E., & Waters, M. J. (2014). Mechanism of activation of protein kinase JAK2 by the growth hormone receptor. *Science, 344*(6185), 1249783. https://doi.org/10.1126/science.1249783. 1249783 [pii]. science.1249783 [pii].

Brooks, N. L., Trent C. M., Raetzsch C. F., Flurkey K., Boysen G., et al. (2007). Low utilization of circulating glucose after food withdrawal in Snell dwarf mice. *The Journal of Biological Chemistry, 282*(48), 35069–35077.

Brooks, A. J., & Waters, M. J. (2010). The growth hormone receptor: Mechanism of activation and clinical implications. *Nature Reviews Endocrinology, 6*(9), 515–525. https://doi.org/10.1038/nrendo.2010.123.

Brown-Borg, H. M., Rakoczy, S. G., Sharma, S., & Bartke, A. (2009). Long-living growth hormone receptor knockout mice: Potential mechanisms of altered stress resistance. *Experimental Gerontology, 44*(1–2), 10–19.

Brown, R. J., Adams, J. J., Pelekanos, R. A., Wan, Y., McKinstry, W. J., Palethorpe, K., Seeber, R. M., Monks, T. A., Eidne, K. A., Parker, M. W., & Waters, M. J. (2005). Model for growth hormone receptor activation based on subunit rotation within a receptor dimer. *Nature Structural & Molecular Biology, 12*(9), 814–821.

Chellakooty, M., Vangsgaard, K., Larsen, T., Scheike, T., Falck-Larsen, J., Legarth, J., Andersson, A. M., Main, K. M., Skakkebaek, N. E., & Juul, A. (2004). A longitudinal study of intrauterine growth and the placental growth hormone (GH)-insulin-like growth factor I axis in maternal circulation: Association between placental GH and fetal growth. *The Journal of Clinical Endocrinology and Metabolism, 89*(1), 384–391. https://doi.org/10.1210/jc.2003-030282.

Chen, W. Y., Wight, D. C., Mehta, B. V., Wagner, T. E., & Kopchick, J. J. (1991). Glycine 119 of bovine growth hormone is critical for growth-promoting activity. *Molecular Endocrinology, 5*(12), 1845–1852.

Chihara, K., Fujieda, K., Shimatsu, A., Miki, T., & Tachibana, K. (2010). Dose-dependent changes in body composition during growth hormone (GH) treatment in Japanese patients with adult GH deficiency: A randomized, placebo-controlled trial. *Growth Hormone & IGF Research, 20*(3), 205–211.

Christ, E. R., Cummings, M. H., Albany, E., Umpleby, A. M., Lumb, P. J., Wierzbicki, A. S., Naoumova, R. P., Boroujerdi, M. A., Sonksen, P. H., & Russell-Jones, D. L. (1999). Effects of growth hormone (GH) replacement therapy on very low density lipoprotein apolipoprotein B100 kinetics in patients with adult GH deficiency: A stable isotope study. *The Journal of Clinical Endocrinology and Metabolism, 84*(1), 307–316. https://doi.org/10.1210/jcem.84.1.5365.

Ciresi, A., Amato, M. C., Pizzolanti, G., & Giordano, C. (2015). Serum visfatin levels in acromegaly: Correlation with disease activity and metabolic alterations. *Growth Hormone & IGF Research : Official Journal of the Growth Hormone Research Society and the International IGF Research Society, 25*(5), 240–246. https://doi.org/10.1016/j.ghir.2015.07.002.

Ciresi, A., Pizzolanti, G., Leotta, M., Guarnotta, V., Teresi, G., & Giordano, C. (2016). Resistin, visfatin, leptin and omentin are differently related to hormonal and metabolic parameters in growth hormone-deficient children. *Journal of Endocrinological Investigation.* https://doi.org/10.1007/s40618-016-0475-z.

Clemmons, D. R., Snyder, D. K., Williams, R., & Underwood, L. E. (1987). Growth hormone administration conserves lean body mass during dietary restriction in obese subjects. *The Journal of Clinical Endocrinology and Metabolism, 64*(5), 878–883.

Colao, A., Ferone, D., Marzullo, P., & Lombardi, G. (2004). Systemic complications of acromegaly: Epidemiology, pathogenesis, and management. *Endocrine Reviews, 25*(1), 102–152.

Combs, T. P., Berg, A. H., Rajala, M. W., Klebanov, S., Iyengar, P., Jimenez-Chillaron, J. C., Patti, M. E., Klein, S. L., Weinstein, R. S., & Scherer, P. E. (2003). Sexual differentiation, pregnancy, calorie restriction, and aging affect the adipocyte-specific secretory protein adiponectin. *Diabetes, 52*(2), 268–276.

Cordoba-Chacon, J., Majumdar, N., List, E. O., Diaz-Ruiz, A., Frank, S. J., Manzano, A., Bartrons, R., Puchowicz, M., Kopchick, J. J., & Kineman, R. D. (2015). Growth hormone inhibits hepatic De novo

Lipogenesis in adult mice. *Diabetes, 64*(9), 3093–3103. https://doi.org/10.2337/db15-0370.

Coschigano, K. T., Holland, A. N., Riders, M. E., List, E. O., Flyvbjerg, A., & Kopchick, J. J. (2003). Deletion, but not antagonism, of the mouse growth hormone receptor results in severely decreased body weights, insulin and IGF-1 levels and increased lifespan. *Endocrinology, 144*(9), 3799–3810.

Cui, Y., Hosui, A., Sun, R., Shen, K., Gavrilova, O., Chen, W., Cam, M. C., Gao, B., Robinson, G. W., & Hennighausen, L. (2007). Loss of signal transducer and activator of transcription 5 leads to hepatosteatosis and impaired liver regeneration. *Hepatology, 46*(2), 504–513. https://doi.org/10.1002/hep.21713.

De Boer, H., Blok, G. J., Voerman, H. J., De Vries, P. M., & van der Veen, E. A. (1992). Body composition in adult growth hormone-deficient men, assessed by anthropometry and bioimpedance analysis. *The Journal of Clinical Endocrinology and Metabolism, 75*(3), 833–837.

del Rincon, J. P., Iida, K., Gaylinn, B. D., McCurdy, C. E., Leitner, J. W., Barbour, L. A., Kopchick, J. J., Friedman, J. E., Draznin, B., & Thorner, M. O. (2007). Growth hormone regulation of p85alpha expression and phosphoinositide 3-kinase activity in adipose tissue: Mechanism for growth hormone-mediated insulin resistance. *Diabetes, 56*(6), 1638–1646. doi:db06-0299 [pii]. https://doi.org/10.2337/db06-0299.

Diniz, R. D., Souza, R. M., Salvatori, R., Franca, A., Gomes-Santos, E., Ferrao TO, Oliveira, C. R., Santana, J. A., Pereira, F. A., Barbosa, R. A., Souza, A. H., Pereira, R. M., Oliveira-Santos, A. A., Silva, A. M., Santana-Junior, F. J., Valenca, E. H., Campos, V. C., & Aguiar-Oliveira, M. H. (2014). Liver status in congenital, untreated, isolated GH deficiency. *Endocrine Connections, 3*(3), 132–137. https://doi.org/10.1530/EC-14-0078.

dos Santos, M. C., Nascimento, G. C., Nascimento, A. G., Carvalho, V. C., Lopes, M. H., Montenegro, R., Montenegro, R., Jr., Vilar, L., Albano, M. F., Alves, A. R., Parente, C. V., & dos Santos Faria, M. (2013). Thyroid cancer in patients with acromegaly: A case-control study. *Pituitary, 16*(1), 109–114. https://doi.org/10.1007/s11102-012-0383-y.

Egecioglu, E., Bjursell, M., Ljungberg, A., Dickson, S. L., Kopchick, J. J., Bergstrom, G., Svensson, L., Oscarsson, J., Tornell, J., & Bohlooly, Y. M. (2006). Growth hormone receptor deficiency results in blunted ghrelin feeding response, obesity, and hypolipidemia in mice. *American Journal of Physiology. Endocrinology and Metabolism, 290*(2), E317–E325.

Ehrenborg, E., & Krook, A. (2009). Regulation of skeletal muscle physiology and metabolism by peroxisome proliferator-activated receptor delta. *Pharmacol Rev, 61*(3), 373–393. https://doi.org/10.1124/pr.109.001560.

Fan, Y., Fang, X., Tajima, A., Geng, X., Ranganathan, S., Dong, H., Trucco, M., & Sperling, M. A. (2014). Evolution of hepatic steatosis to fibrosis and adenoma formation in liver-specific growth hormone receptor knockout mice. *Front Endocrinol (Lausanne), 5*, 218. https://doi.org/10.3389/fendo.2014.00218.

Fan, Y., Menon, R. K., Cohen, P., Hwang, D., Clemens, T., DiGirolamo, D. J., Kopchick, J. J., Le Roith, D., Trucco, M., & Sperling, M. A. (2009a). Liver-specific deletion of the growth hormone receptor reveals essential role of growth hormone signaling in hepatic lipid metabolism. *The Journal of Biological Chemistry, 284*(30), 19937–19944. https://doi.org/10.1074/jbc.M109.014308.

Fathallah, N., Slim, R., Larif, S., Hmouda, H., & Ben Salem, C. (2015). Drug-induced Hyperglycaemia and diabetes. *Drug Safety, 38*(12), 1153–1168. https://doi.org/10.1007/s40264-015-0339-z.

Fineberg, S. E., & Merimee, T. J. (1974). Acute metabolic effects of human growth hormone. *Diabetes, 23*(6), 499–504.

Flint, D. J., Binart, N., Boumard, S., Kopchick, J. J., & Kelly, P. (2006). Developmental aspects of adipose tissue in GH receptor and prolactin receptor gene disrupted mice: Site-specific effects upon proliferation, differentiation and hormone sensitivity. *The Journal of Endocrinology, 191*(1), 101–111.

Florini, J. R., Ewton, D. Z., & Coolican, S. A. (1996). Growth hormone and the insulin-like growth factor system in myogenesis. *Endocrine Reviews, 17*(5), 481–517.

Flurkey, K., Papaconstantinou, J., Miller, R. A., & Harrison, D. E. (2001). Lifespan extension and delayed immune and collagen aging in mutant mice with defects in growth hormone production. *Proceedings of the National Academy of Sciences of the United States of America, 98*(12), 6736–6741.

Franco, C., Brandberg, J., Lonn, L., Andersson, B., Bengtsson, B. A., & Johannsson, G. (2005). Growth hormone treatment reduces abdominal visceral fat in postmenopausal women with abdominal obesity: A 12-month placebo-controlled trial. *The Journal of Clinical Endocrinology and Metabolism, 90*(3), 1466–1474.

Freda, P. U., Shen, W., Heymsfield, S. B., Reyes-Vidal, C. M., Geer, E. B., Bruce, J. N., & Gallagher, D. (2008). Lower Visceral and Subcutaneous but Higher Intermuscular Adipose Tissue Depots in Patients with Growth Hormone and Insulin-Like Growth Factor I Excess Due to Acromegaly. *The Journal of Clinical Endocrinology and Metabolism, 93*(6), 2334–2343. http://doi.org/10.1210/jc.2007-2780.

Ghanaat, F., & Tayek, J. A. (2005). Growth hormone administration increases glucose production by preventing the expected decrease in glycogenolysis seen with fasting in healthy volunteers. *Metabolism, 54*(5), 604–609. https://doi.org/10.1016/j.metabol.2004.12.003.

Gibney, J., Wolthers, T., Burt, M. G., Leung, K. C., Umpleby, A. M., & Ho, K. K. (2007). Protein metabolism in acromegaly: Differential effects of short- and long-term treatment. *The Journal of Clinical*

Endocrinology and Metabolism, 92(4), 1479–1484. https://doi.org/10.1210/jc.2006-0664.

Grimberg, A., DiVall, S. A., Polychronakos, C., Allen, D. B., Cohen, L. E., Quintos, J. B., Rossi, W. C., Feudtner, C., & Murad, M. H. (2016). Guidelines for growth hormone and insulin-like growth factor-I treatment in children and adolescents: Growth hormone deficiency, idiopathic short stature, and primary insulin-like growth factor-I deficiency. *Hormone Research in Pædiatrics, 86*(6), 361–397. https://doi.org/10.1159/000452150.

Guevara-Aguirre, J., Balasubramanian, P., Guevara-Aguirre, M., Wei, M., Madia, F., Cheng, C. W., Hwang, D., Martin-Montalvo, A., Saavedra, J., Ingles, S., de Cabo, R., Cohen, P., & Longo, V. D. (2011). Growth hormone receptor deficiency is associated with a major reduction in pro-aging signaling, cancer, and diabetes in humans. *Science Translational Medicine, 3*(70), 70ra13.

Guevara-Aguirre, J., Rosenbloom, A. L., Balasubramanian, P., Teran, E., Guevara-Aguirre, M., Guevara, C., Procel, P., Alfaras, I., De Cabo, R., Di Biase, S., Narvaez, L., Saavedra, J., & Longo, V. D. (2015). GH receptor deficiency in Ecuadorian adults is associated with obesity and enhanced insulin sensitivity. *The Journal of Clinical Endocrinology and Metabolism, 100*(7), 2589–2596. https://doi.org/10.1210/jc.2015-1678.

Heiman, M. L., Tinsley, F. C., Mattison, J. A., Hauck, S., & Bartke, A. (2003). Body composition of prolactin-, growth hormone, and thyrotropin-deficient Ames dwarf mice. *Endocrine, 20*(1–2), 149–154.

Herrington, J., Smit, L. S., Schwartz, J., & Carter-Su, C. (2000). The role of STAT proteins in growth hormone signaling. *Oncogene, 19*(21), 2585–2597. https://doi.org/10.1038/sj.onc.1203526.

Hill, C. M., Fang, Y., Miquet, J. G., Sun, L. Y., Masternak, M. M., & Bartke, A. (2016). Long-lived hypopituitary Ames dwarf mice are resistant to the detrimental effects of high-fat diet on metabolic function and energy expenditure. *Aging Cell, 15*(3), 509–521. https://doi.org/10.1111/acel.12467.

Hjortebjerg, R., Berryman, D., Comisford, R., Frank, S., List, E., Bjerre, M., Frystyk, J., & Kopchick, J. (2017). Insulin, IGF-1, and GH receptors are altered in an adipose tissue depot-specific manner in male mice with modified GH action. *Endocrinology, 158*, 1406.

Hoybye, C., Chandramouli, V., Efendic, S., Hulting, A. L., Landau, B. R., Schumann, W. C., & Wajngot, A. (2008). Contribution of gluconeogenesis and glycogenolysis to hepatic glucose production in acromegaly before and after pituitary microsurgery. *Hormone and Metabolic Research = Hormon- und Stoffwechselforschung = Hormones et Métabolisme, 40*(7), 498–501. https://doi.org/10.1055/s-2008-1065322.

Ikeno, Y., Hubbard, G. B., Lee, S., Cortez, L. A., Lew, C. M., Webb, C. R., Berryman, D. E., List, E. O., Kopchick, J. J., & Bartke, A. (2009). Reduced incidence and delayed occurrence of fatal neoplastic diseases in growth hormone receptor/binding protein knockout mice. *The Journals of Gerontology. Series A, Biological Sciences and Medical Sciences, 64*(5), 522–529.

Ikkos, D., Luft, R., & Gemzell, C. A. (1959). The effect of human growth hormone in man. *Acta Endocrinologica, 32*, 341–361.

Jenkins, P. J. (2006). Cancers associated with acromegaly. *Neuroendocrinology, 83*(3–4), 218–223. https://doi.org/10.1159/000095531.

Johannsson, G., Marin, P., Lonn, L., Ottosson, M., Stenlof, K., Bjorntorp, P., Sjostrom, L., & Bengtsson, B. A. (1997). Growth hormone treatment of abdominally obese men reduces abdominal fat mass, improves glucose and lipoprotein metabolism, and reduces diastolic blood pressure. *The Journal of Clinical Endocrinology and Metabolism, 82*(3), 727–734.

Jorgensen, J. O., Moller, N., Lauritzen, T., Alberti, K. G., Orskov, H., & Christiansen, J. S. (1990). Evening versus morning injections of growth hormone (GH) in GH-deficient patients: Effects on 24-hour patterns of circulating hormones and metabolites. *The Journal of Clinical Endocrinology and Metabolism, 70*(1), 207–214. https://doi.org/10.1210/jcem-70-1-207.

Junnila, R. K., Duran-Ortiz, S., Suer, O., Sustarsic, E. G., Berryman, D. E., List, E. O., & Kopchick, J. J. (2016). Disruption of the GH receptor gene in adult mice increases maximal lifespan in females. *Endocrinology, 157*(12), 4502–4513. https://doi.org/10.1210/en.2016-1649.

Junnila, R. K., List, E. O., Berryman, D. E., Murrey, J. W., & Kopchick, J. J. (2013). The GH/IGF-1 axis in ageing and longevity. *Nature Reviews. Endocrinology, 9*(6), 366–376. https://doi.org/10.1038/nrendo.2013.67. nrendo.2013.67 [pii].

Karastergiou, K., Bredella, M. A., Lee, M. J., Smith, S. R., Fried, S. K., & Miller, K. K. (2016). Growth hormone receptor expression in human gluteal versus abdominal subcutaneous adipose tissue: Association with body shape. *Obesity, 24*(5), 1090–1096. https://doi.org/10.1002/oby.21460.

Katznelson, L., Atkinson, J. L., Cook, D. M., Ezzat, S. Z., Hamrahian, A. H., & Miller, K. K. (2011). American Association of Clinical Endocrinologists Medical Guidelines for clinical practice for the diagnosis and treatment of acromegaly--2011 update: Executive summary. *Endocrine Practice, 17*(4), 636–646.

Katznelson, L., Laws, E. R., Jr., Melmed, S., Molitch, M. E., Murad, M. H., Utz, A., & Wass, J. A. (2014). Acromegaly: An endocrine society clinical practice guideline. *The Journal of Clinical Endocrinology and Metabolism, 99*(11), 3933–3951. https://doi.org/10.1210/jc.2014-2700.

Kelder, B., Berryman, D. E., Clark, R., Li, A., List, E. O., & Kopchick, J. J. (2007). CIDE-A gene expression is decreased in white adipose tissue of growth hormone receptor/binding protein gene disrupted mice and with high-fat feeding of normal mice. *Growth Hormone & IGF Research, 17*(4), 346–351.

Khalfallah, Y., Sassolas, G., Borson-Chazot, F., Vega, N., & Vidal, H. (2001). Expression of insulin target genes in skeletal muscle and adipose tissue in adult patients with growth hormone deficiency: Effect of one year recombinant human growth hormone therapy. *The Journal of Endocrinology, 171*(2), 285–292.

Kim, D. S., Itoh, E., Iida, K., & Thorner, M. O. (2009). Growth hormone increases mRNA levels of PPARdelta and Foxo1 in skeletal muscle of growth hormone deficient lit/lit mice. *Endocrine Journal, 56*(1), 141–147.

Kim, K. R., Nam, S. Y., Song, Y. D., Lim, S. K., Lee, H. C., & Huh, K. B. (1999). Low-dose growth hormone treatment with diet restriction accelerates body fat loss, exerts anabolic effect and improves growth hormone secretory dysfunction in obese adults. *Hormone Research, 51*(2), 78–84.

Kopchick, J. J., & Andry, J. M. (2000). Growth hormone (GH), GH receptor, and signal transduction. *Molecular Genetics and Metabolism, 71*(1–2), 293–314.

Kopchick, J. J., & Laron, Z. (1999). Is the Laron mouse an accurate model of Laron syndrome? *Molecular Genetics and Metabolism, 68*, 232–236.

Kopchick, J. J., & Laron, Z. (Eds.). (2011). *Laron syndrome - from man to mouse*. Berlin: Springer.

Kopchick, J. J., List, E. O., Kelder, B., Gosney, E. S., & Berryman, D. E. (2014). Evaluation of growth hormone (GH) action in mice: Discovery of GH receptor antagonists and clinical indications. *Molecular and Cellular Endocrinology, 386*(1–2), 34–45. https://doi.org/10.1016/j.mce.2013.09.004. S0303-7207(13)00366-3 [pii].

Kredel, L. I., & Siegmund, B. (2014). Adipose-tissue and intestinal inflammation - visceral obesity and creeping fat. *Frontiers in Immunology, 5*, 462. https://doi.org/10.3389/fimmu.2014.00462.

Lam, K. S., Xu, A., Tan, K. C., Wong, L. C., Tiu, S. C., & Tam, S. (2004). Serum adiponectin is reduced in acromegaly and normalized after correction of growth hormone excess. *The Journal of Clinical Endocrinology and Metabolism, 89*(11), 5448–5453.

Laron, Z. (2015). Lessons from 50 years of study of Laron syndrome. *Endocrine Practice, 21*(12), 1395–1402. https://doi.org/10.4158/EP15939.RA.

Laron, Z., Ginsberg, S., Lilos, P., Arbiv, M., & Vaisman, N. (2006). Body composition in untreated adult patients with Laron syndrome (primary GH insensitivity). *Clinical Endocrinology, 65*(1), 114–117. doi:CEN2558 [pii]. https://doi.org/10.1111/j.1365-2265.2006.02558.x.

Laron, Z., & Kauli, R. (2016). Fifty seven years of follow-up of the Israeli cohort of Laron syndrome patients-from discovery to treatment. *Growth Hormone & IGF Research : Official Journal of the Growth Hormone Research Society and the International IGF Research Society, 28*, 53–56. https://doi.org/10.1016/j.ghir.2015.08.004.

Laron, Z., Pertzelan, A., & Karp, M. (1968). Pituitary dwarfism with high serum levels of growth hormone. *Israel Journal of Medical Sciences, 4*(4), 883–894.

Li, Y., Knapp, J. R., & Kopchick, J. J. (2003). Enlargement of interscapular brown adipose tissue in growth hormone antagonist transgenic and in growth hormone

receptor gene-disrupted dwarf mice. *Experimental Biology and Medicine (Maywood, N.J.), 228*(2), 207–215.

Lindberg-Larsen, R., Moller, N., Schmitz, O., Nielsen, S., Andersen, M., Orskov, H., & Jorgensen, J. O. (2007). The impact of pegvisomant treatment on substrate metabolism and insulin sensitivity in patients with acromegaly. *The Journal of Clinical Endocrinology and Metabolism, 92*(5), 1724–1728.

List, E. O., Berryman, D. E., Funk, K., Gosney, E. S., Jara, A., Kelder, B., Wang, X., Kutz, L., Troike, K., Lozier, N., Mikula, V., Lubbers, E. R., Zhang, H., Vesel, C., Junnila, R. K., Frank, S. J., Masternak, M. M., Bartke, A., & Kopchick, J. J. (2013). The role of GH in adipose tissue: Lessons from adipose-specific GH receptor gene-disrupted mice. *Molecular Endocrinology, 27*(3), 524–535. https://doi.org/10.1210/me.2012-1330. me.2012-1330 [pii].

List, E. O., Berryman, D. E., Funk, K., Jara, A., Kelder, B., Wang, F., Stout, M. B., Zhi, X., Sun, L., White, T. A., LeBrasseur, N. K., Pirtskhalava, T., Tchkonia, T., Jensen, E. A., Zhang, W., Masternak, M. M., Kirkland, J. L., Miller, R. A., Bartke, A., & Kopchick, J. J. (2014). Liver-specific GH receptor gene-disrupted (LiGHRKO) mice have decreased endocrine IGF-I, increased local IGF-I, and altered body size, body composition, and adipokine profiles. *Endocrinology, 155*(5), 1793–1805. https://doi.org/10.1210/en.2013-2086.

List, E. O., Berryman, D. E., Ikeno, Y., Hubbard, G. B., Funk, K., Comisford, R., Young, J. A., Stout, M. B., Tchkonia, T., Masternak, M. M., Bartke, A., Kirkland, J. L., Miller, R. A., & Kopchick, J. J. (2015). Removal of growth hormone receptor (GHR) in muscle of male mice replicates some of the health benefits seen in global GHR−/− mice. *Aging (Albany NY), 7*(7), 500–512.

List, E. O., Palmer, A. J., Berryman, D. E., Bower, B., Kelder, B., & Kopchick, J. J. (2009). Growth hormone improves body composition, fasting blood glucose, glucose tolerance and liver triacylglycerol in a mouse model of diet-induced obesity and type 2 diabetes. *Diabetologia, 52*(8), 1647–1655.

Lu, C., Kumar, P. A., Sun, J., Aggarwal, A., Fan, Y., Sperling, M. A., Lumeng, C. N., & Menon, R. K. (2013). Targeted deletion of growth hormone (GH) receptor in macrophage reveals novel Osteopontin-mediated effects of GH on glucose homeostasis and insulin sensitivity in diet-induced obesity. *The Journal of Biological Chemistry, 288*(22), 15725–15735. https://doi.org/10.1074/jbc.M113.460212. M113.460212 [pii].

Lubbers, E. R., List, E. O., Jara, A., Sackmann-Sala, L., Cordoba-Chacon, J., Gahete, M., Kineman, R. D., Boparai, R., Bartke, A., Kopchick, J., & Berryman, D. E. (2012). Adiponectin in mice with altered growth hormone action: Links to insulin sensitivity and longevity? *The Journal of Endocrinology*. https://doi.org/10.1530/JOE-12-0505.

Lubbers, E. R., List, E. O., Jara, A., Sackman-Sala, L., Cordoba-Chacon, J., Gahete, M. D., Kineman, R. D.,

Boparai, R., Bartke, A., Kopchick, J. J., & Berryman, D. E. (2013). Adiponectin in mice with altered GH action: Links to insulin sensitivity and longevity? *The Journal of Endocrinology, 216*(3), 363–374. https://doi.org/10.1530/JOE-12-0505. JOE-12-0505 [pii].

Luque, R. M., Lin, Q., Cordoba-Chacon, J., Subbaiah, P. V., Buch, T., Waisman, A., Vankelecom, H., & Kineman, R. D. (2011). Metabolic impact of adult-onset, isolated, growth hormone deficiency (AOiGHD) due to destruction of pituitary Somatotropes. *PLoS One, 6*(1), e15767.

Masternak, M. M., Al-Regaiey, K. A., Del Rosario Lim, M. M., Jimenez-Ortega, V., Panici, J. A., Bonkowski, M. S., Kopchick, J. J., Wang, Z., & Bartke, A. (2006). Caloric restriction and growth hormone receptor knockout: Effects on expression of genes involved in insulin action in the heart. *Experimental Gerontology, 41*(4), 417–429.

Mavalli, M. D., DiGirolamo, D. J., Fan, Y., Riddle, R. C., Campbell, K. S., van Groen, T., Frank, S. J., Sperling, M. A., Esser, K. A., Bamman, M. M., & Clemens, T. L. (2010). Distinct growth hormone receptor signaling modes regulate skeletal muscle development and insulin sensitivity in mice. *The Journal of Clinical Investigation, 120*(11), 4007–4020.

Meazza, C., Elsedfy, H. H., Pagani, S., Bozzola, E., El Kholy, M., & Bozzola, M. (2014). Metabolic parameters and adipokine profile in growth hormone deficient (GHD) children before and after 12-month GH treatment. *Hormone and Metabolic Research = Hormon- und Stoffwechselforschung = Hormones et Metabolisme, 46*(3), 219–223. https://doi.org/10.1055/s-0033-1358730.

Mekala, K. C., & Tritos, N. A. (2008). Effects of recombinant human growth hormone therapy in obesity in adults - a meta-analysis. *The Journal of Clinical Endocrinology and Metabolism.*

Melmed, S. (2009). Acromegaly pathogenesis and treatment. *The Journal of Clinical Investigation, 119*(11), 3189–3202. https://doi.org/10.1172/JCI39375.

Melmed, S. (2011). Pathogenesis of pituitary tumors. *Nature Reviews Endocrinology, 7*(5), 257–266. https://doi.org/10.1038/nrendo.2011.40.

Melmed, S., Kleinberg, D. L., Bonert, V., & Fleseriu, M. (2014). Acromegaly: Assessing the disorder and navigating therapeutic options for treatment. *Endocrine Practice, 20*(Suppl 1), 7–17. quiz 18–20. https://doi.org/10.4158/EP14430.RA.

Miquet, J. G., Freund, T., Martinez, C. S., Gonzalez, L., Diaz, M. E., Micucci, G. P., Zotta, E., Boparai, R. K., Bartke, A., Turyn, D., & Sotelo, A. I. (2013). Hepatocellular alterations and dysregulation of oncogenic pathways in the liver of transgenic mice overexpressing growth hormone. *Cell Cycle, 12*(7), 1042–1057. https://doi.org/10.4161/cc.24026. 24026 [pii].

Molitch, M. E., Clemmons, D. R., Malozowski, S., Merriam, G. R., & Vance, M. L. (2011). Evaluation and treatment of adult growth hormone deficiency: An Endocrine Society clinical practice guideline. *The Journal of Clinical Endocrinology and Metabolism, 96*(6), 1587–1609. https://doi.org/10.1210/jc.2011-0179.

Moller, N., & Jorgensen, J. O. (2009). Effects of growth hormone on glucose, lipid, and protein metabolism in human subjects. *Endocrine Reviews, 30*(2), 152–177.

Moller, N., Vendelbo, M. H., Kampmann, U., Christensen, B., Madsen, M., Norrelund, H., & Jorgensen, J. O. (2009). Growth hormone and protein metabolism. *Clinical Nutrition, 28*(6), 597–603. https://doi.org/10.1016/j.clnu.2009.08.015 S0261-5614(09)00177-0. [pii].

Muller, E. E. (1990). Clinical implications of growth hormone feedback mechanisms. *Hormone Research, 33*(Suppl 4), 90–96.

Muller, E. E., Locatelli, V., & Cocchi, D. (1999). Neuroendocrine control of growth hormone secretion. *Physiological Reviews, 79*(2), 511–607.

Mulligan, K., Tai, V. W., & Schambelan, M. (1998). Effects of chronic growth hormone treatment on energy intake and resting energy metabolism in patients with human immunodeficiency virus-associated wasting--a clinical research center study. *The Journal of Clinical Endocrinology and Metabolism, 83*(5), 1542–1547. https://doi.org/10.1210/jcem.83.5.4772.

Mullis, P. E. (2007). Genetics of growth hormone deficiency. *Endocrinology and Metabolism Clinics of North America, 36*(1), 17–36. https://doi.org/10.1016/j.ecl.2006.11.010.

Murase, T., Yamada, N., Ohsawa, N., Kosaka, K., Morita, S., & Yoshida, S. (1980). Decline of postheparin plasma lipoprotein lipase in acromegalic patients. *Metabolism, 29*(7), 666–672.

Nam, S. Y., Kim, K. R., Cha, B. S., Song, Y. D., Lim, S. K., Lee, H. C., & Huh, K. B. (2001). Low-dose growth hormone treatment combined with diet restriction decreases insulin resistance by reducing visceral fat and increasing muscle mass in obese type 2 diabetic patients. *International Journal of Obesity and Related Metabolic Disorders, 25*(8), 1101–1107.

Nishizawa, H., Iguchi, G., Murawaki, A., Fukuoka, H., Hayashi, Y., Kaji, H., Yamamoto, M., Suda, K., Takahashi, H., Seo, Y., Yano, Y., Kitazawa, R., Kitazawa, S., Koga, M., Okimura, Y., Chihara, K., & Takahashi, Y. (2012). Nonalcoholic fatty liver disease in adult hypopituitary patients with GH deficiency and the impact of GH replacement therapy. *European Journal of Endocrinology, 167*(1), 67–74. https://doi.org/10.1530/EJE-12-0252.

Olarescu, N. C., Berryman, D. E., Householder, L. A., Lubbers, E. R., List, E. O., Benencia, F., Kopchick, J. J., & Bollerslev, J. (2015a). GH action influences adipogenesis of mouse adipose tissue-derived mesenchymal stem cells. *The Journal of Endocrinology, 226*(1), 13–23. https://doi.org/10.1530/JOE-15-0012.

Olarescu, N. C., Heck, A., Godang, K., Ueland, T., & Bollerslev, J. (2015b). The metabolic risk in newly diagnosed patients with acromegaly is related to fat distribution and circulating Adipokines and improves after treatment. *Neuroendocrinology.* https://doi.org/10.1159/000371818.

Olsson, B., Bohlooly, Y. M., Fitzgerald, S. M., Frick, F., Ljungberg, A., Ahren, B., Tornell, J., Bergstrom, G., & Oscarsson, J. (2005). Bovine growth hormone transgenic mice are resistant to diet-induced obesity but develop hyperphagia, dyslipidemia, and diabetes on a high-fat diet. *Endocrinology, 146*(2), 920–930.

Oscarsson, J., Ottosson, M., Vikman-Adolfsson, K., Frick, F., Enerback, S., Lithell, H., & Eden, S. (1999). GH but not IGF-I or insulin increases lipoprotein lipase activity in muscle tissues of hypophysectomised rats. *The Journal of Endocrinology, 160*(2), 247–255.

Palmer, A. J., Chung, M. Y., List, E. O., Walker, J., Okada, S., Kopchick, J. J., & Berryman, D. E. (2009). Age-related changes in body composition of bovine growth hormone transgenic mice. *Endocrinology, 150*(3), 1353–1360.

Raben, M. S., & Hollenberg, C. H. (1959). Effect of growth hormone on plasma fatty acids. *The Journal of Clinical Investigation, 38*(3), 484–488. https://doi.org/10.1172/JCI103824.

Rabinowitz, D., Klassen, G. A., & Zierler, K. L. (1965). Effect of human growth hormone on muscle and adipose tissue metabolism in the forearm of man. *The Journal of Clinical Investigation, 44*, 51–61.

Rabinowitz, D., & Zierler, K. L. (1963). A metabolic regulating device based on the actions of human growth hormone and of insulin, singly and together, on the human forearm. *Nature, 199*, 913–915.

Rasmussen, M. H. (2010). Obesity, growth hormone and weight loss. *Molecular and Cellular Endocrinology, 316*(2), 147–153. https://doi.org/10.1016/j.mce.2009.08.017.

Rasmussen, M. H., Hvidberg, A., Juul, A., Main, K. M., Gotfredsen, A., Skakkebaek, N. E., Hilsted, J., & Skakkebae, N. E. (1995). Massive weight loss restores 24-hour growth hormone release profiles and serum insulin-like growth factor-I levels in obese subjects. *The Journal of Clinical Endocrinology and Metabolism, 80*(4), 1407–1415. https://doi.org/10.1210/jcem.80.4.7536210.

Reed, M. L., Merriam, G. R., & Kargi, A. Y. (2013). Adult growth hormone deficiency - benefits, side effects, and risks of growth hormone replacement. *Front Endocrinol (Lausanne), 4*, 64. https://doi.org/10.3389/fendo.2013.00064.

Reyes-Vidal, C., Fernandez, J. C., Bruce, J. N., Crisman, C., Conwell, I. M., Kostadinov, J., Geer, E. B., Post, K. D., & Freda, P. U. (2014). Prospective study of surgical treatment of acromegaly: Effects on ghrelin, weight, adiposity, and markers of CV risk. *The Journal of Clinical Endocrinology and Metabolism, 99*(11), 4124–4132. https://doi.org/10.1210/jc.2014-2259.

Richelsen, B., Pedersen, S. B., Borglum, J. D., Moller-Pedersen, T., Jorgensen, J., & Jorgensen, J. O. (1994). Growth hormone treatment of obese women for 5 wk: Effect on body composition and adipose tissue LPL activity. *The American Journal of Physiology, 266*(2 Pt 1), E211–E216.

Richelsen, B., Pedersen, S. B., Kristensen, K., Borglum, J. D., Norrelund, H., Christiansen, J. S., & Jorgensen,

J. O. (2000). Regulation of lipoprotein lipase and hormone-sensitive lipase activity and gene expression in adipose and muscle tissue by growth hormone treatment during weight loss in obese patients. *Metabolism, 49*(7), 906–911.

Robertson, K., Kopchick, J. J., & Liu, J. L. (2006). Growth hormone receptor gene deficiency causes delayed insulin responsiveness in skeletal muscles without affecting compensatory islet cell overgrowth in obese mice. *American Journal of Physiology. Endocrinology and Metabolism, 291*(3), E491–E498.

Rodriguez-Arnao, J., Jabbar, A., Fulcher, K., Besser, G. M., & Ross, R. J. (1999). Effects of growth hormone replacement on physical performance and body composition in GH deficient adults. *Clinical Endocrinology, 51*(1), 53–60.

Rogozinski, A., Furioso, A., Glikman, P., Junco, M., Laudi, R., Reyes, A., & Lowenstein, A. (2012). Thyroid nodules in acromegaly. *Arquivos Brasileiros de Endocrinologia e Metabologia, 56*(5), 300–304.

Rokkas, T., Pistiolas, D., Sechopoulos, P., Margantinis, G., & Koukoulis, G. (2008). Risk of colorectal neoplasm in patients with acromegaly: A meta-analysis. *World Journal of Gastroenterology, 14*(22), 3484–3489.

Romero, C. J., Ng, Y., Luque, R. M., Kineman, R. D., Koch, L., Bruning, J. C., & Radovick, S. (2010). Targeted deletion of somatotroph insulin-like growth factor-I signaling in a cell-specific knockout mouse model. *Molecular Endocrinology, 24*(5), 1077–1089. https://doi.org/10.1210/me.2009-0393.

Sackmann-Sala, L., Berryman, D. E., Lubbers, E. R., Zhang, H., Vesel, C. B., Troike, K. M., Gosney, E. S., List, E. O., & Kopchick, J. J. (2013). Age-related and depot-specific changes in white adipose tissue of growth hormone receptor-null mice. *The Journals of Gerontology. Series A, Biological Sciences and Medical Sciences*. https://doi.org/10.1093/gerona/glt110.

Sakharova, A. A., Horowitz, J. F., Surya, S., Goldenberg, N., Harber, M. P., Symons, K., & Barkan, A. (2008). Role of growth hormone in regulating lipolysis, proteolysis, and hepatic glucose production during fasting. *The Journal of Clinical Endocrinology and Metabolism, 93*(7), 2755–2759. doi:jc.2008-0079. [pii]. https://doi.org/10.1210/jc.2008-0079.

Shadid, S., & Jensen, M. D. (2003). Effects of growth hormone administration in human obesity. *Obesity Research, 11*(2), 170–175.

Shevah, O., & Laron, Z. (2007). Patients with congenital deficiency of IGF-I seem protected from the development of malignancies: A preliminary report. *Growth Hormone & IGF Research, 17*(1), 54–57.

Shin-Hye Kim, Mi-Jung Park, (2017) Effects of growth hormone on glucose metabolism and insulin resistance in human. Annals of Pediatric Endocrinology & Metabolism 22(3):145–152.

Silha, J. V., Krsek, M., Hana, V., Marek, J., Jezkova, J., Weiss, V., & Murphy, L. J. (2003). Perturbations in adiponectin, leptin and resistin levels in acromegaly: Lack of correlation with insulin resistance. *Clinical Endocrinology, 58*(6), 736–742.

Skaggs, S. R., & Crist, D. M. (1991). Exogenous human growth hormone reduces body fat in obese women. *Hormone Research, 35*(1), 19–24.

Smuel, K., Kauli, R., Lilos, P., & Laron, Z. (2015). Growth, development, puberty and adult height before and during treatment in children with congenital isolated growth hormone deficiency. *Growth Hormone & IGF Research : Official Journal of the Growth Hormone Research Society and the International IGF Research Society, 25*(4), 182–188. https://doi.org/10.1016/j.ghir.2015.05.001.

Snyder, D. K., Clemmons, D. R., & Underwood, L. E. (1988). Treatment of obese, diet-restricted subjects with growth hormone for 11 weeks: Effects on anabolism, lipolysis, and body composition. *The Journal of Clinical Endocrinology and Metabolism, 67*(1), 54–61.

Snyder, D. K., Underwood, L. E., & Clemmons, D. R. (1990). Anabolic effects of growth hormone in obese diet-restricted subjects are dose dependent. *The American Journal of Clinical Nutrition, 52*(3), 431–437.

Snyder, D. K., Underwood, L. E., & Clemmons, D. R. (1995). Persistent lipolytic effect of exogenous growth hormone during caloric restriction. *The American Journal of Medicine, 98*(2), 129–134.

Sornson, M. W., Wu, W., Dasen, J. S., Flynn, S. E., Norman, D. J., O'Connell, S. M., Gukovsky, I., Carriere, C., Ryan, A. K., Miller, A. P., Zuo, L., Gleiberman, A. S., Andersen, B., Beamer, W. G., & Rosenfeld, M. G. (1996). Pituitary lineage determination by the prophet of Pit-1 homeodomain factor defective in Ames dwarfism. *Nature, 384*(6607), 327–333. https://doi.org/10.1038/384327a0.

Stenholm, S., Metter, E. J., Roth, G. S., Ingram, D. K., Mattison, J. A., Taub, D. D., & Ferrucci, L. (2011). Relationship between plasma ghrelin, insulin, leptin, interleukin 6, adiponectin, testosterone and longevity in the Baltimore longitudinal study of aging. *Aging Clinical and Experimental Research, 23*(2), 153–158.

Steuerman, R., Shevah, O., & Laron, Z. (2011). Congenital IGF1 deficiency tends to confer protection against post-natal development of malignancies. *European Journal of Endocrinology, 164*(4), 485–489. https://doi.org/10.1530/EJE-10-0859.

Stout, M. B., Swindell, W. R., Zhi, X., Rohde, K., List, E. O., Berryman, D. E., Kopchick, J. J., Gesing, A., Fang, Y., & Masternak, M. M. (2015). Transcriptome profiling reveals divergent expression shifts in brown and white adipose tissue from long-lived GHRKO mice. *Oncotarget, 6*(29), 26702–26715. https://doi.org/10.18632/oncotarget.5760.

Stout, M. B., Tchkonia, T., Pirtskhalava, T., Palmer, A. K., List, E. O., Berryman, D. E., Lubbers, E. R., Escande, C., Spong, A., Masternak, M. M., Oberg, A. L., LeBrasseur, N. K., Miller, R. A., Kopchick, J. J., Bartke, A., & Kirkland, J. L. (2014). Growth hormone action predicts age-related white adipose tissue dysfunction and senescent cell burden in mice. *Aging (Albany NY), 6*(7), 575–586.

Sucunza, N., Barahona, M. J., Resmini, E., Fernandez-Real, J. M., Ricart, W., Farrerons, J., Rodriguez Espinosa, J., Marin, A. M., Puig, T., & Webb, S. M. (2009). A link between bone mineral density and serum adiponectin and visfatin levels in acromegaly. *The Journal of Clinical Endocrinology and Metabolism, 94*(10), 3889–3896.

Tagliaferri, M., Scacchi, M., Pincelli, A. I., Berselli, M. E., Silvestri, P., Montesano, A., Ortolani, S., Dubini, A., & Cavagnini, F. (1998). Metabolic effects of biosynthetic growth hormone treatment in severely energy-restricted obese women. *International Journal of Obesity and Related Metabolic Disorders, 22*(9), 836–841.

Troike, K. M., Henry, B. E., Jensen, E. A., Young, J. A., List, E. O., Kopchick, J. J., & Berryman, D. E. (2017). Impact of Growth Hormone on Regulation of Adipose Tissue. *Comprehensive Physiology, 7*(3), 819–840. https://doi.org/10.1002/cphy.c160027.

van Vught, A. J., Nieuwenhuizen, A. G., Brummer, R. J., & Westerterp-Plantenga, M. S. (2008). Effects of oral ingestion of amino acids and proteins on the somatotropic axis. *The Journal of Clinical Endocrinology and Metabolism, 93*(2), 584–590. https://doi.org/10.1210/jc.2007-1784.

Vikman, K., Carlsson, B., Billig, H., & Eden, S. (1991). Expression and regulation of growth hormone (GH) receptor messenger ribonucleic acid (mRNA) in rat adipose tissue, adipocytes, and adipocyte precursor cells: GH regulation of GH receptor mRNA. *Endocrinology, 129*(3), 1155–1161.

Wang, Z., Al-Regaiey, K. A., Masternak, M. M., & Bartke, A. (2006). Adipocytokines and lipid levels in Ames dwarf and calorie-restricted mice. *The Journals of Gerontology. Series A, Biological Sciences and Medical Sciences, 61*(4), 323–331.

Wang, Z., Masternak, M. M., Al-Regaiey, K. A., & Bartke, A. (2007). Adipocytokines and the regulation of lipid metabolism in growth hormone transgenic and calorie-restricted mice. *Endocrinology, 148*(6), 2845–2853. https://doi.org/10.1210/en.2006-1313.

Wells, J. A. (1996). Binding in the growth hormone receptor complex. *Proceedings of the National Academy of Sciences, 93*(1), 1–6.

Wu, Y., Liu, C., Sun, H., Vijayakumar, A., Giglou, P. R., Qiao, R., Oppenheimer, J., Yakar, S., & LeRoith, D. (2011). Growth hormone receptor regulates beta cell hyperplasia and glucose-stimulated insulin secretion in obese mice. *The Journal of Clinical Investigation, 121*(6), 2422–2426. https://doi.org/10.1172/JCI45027.

Yakar, S., Setser, J., Zhao, H., Stannard, B., Haluzik, M., Glatt, V., Bouxsein, M. L., Kopchick, J. J., & LeRoith, D. (2004). Inhibition of growth hormone action improves insulin sensitivity in liver IGF-1-deficient mice. *The Journal of Clinical Investigation, 113*(1), 96–105.

Yang, S., Mulder, H., Holm, C., & Eden, S. (2004). Effects of growth hormone on the function of beta-adrenoceptor subtypes in rat adipocytes. *Obesity*

Research, 12(2), 330–339. https://doi.org/10.1038/oby.2004.41.

Yang, T., Householder, L. A., Lubbers, E. R., List, E. O., Troike, K., Vesel, C., Duran-Ortiz, S., Kopchick, J. J., & Berryman, D. E. (2015). Growth hormone receptor antagonist transgenic mice are protected from hyperinsulinemia and glucose intolerance despite obesity when placed on a HF diet. *Endocrinology, 156*(2), 555–564. https://doi.org/10.1210/en.2014-1617.

Yip, R. G., & Goodman, H. M. (1999). Growth hormone and dexamethasone stimulate lipolysis and activate adenylyl cyclase in rat adipocytes by selectively shifting Gi alpha2 to lower density membrane fractions. *Endocrinology, 140*(3), 1219–1227. https://doi.org/10.1210/endo.140.3.6580.

Zhang, Q., Kohler, M., Yang, S. N., Zhang, F., Larsson, O., & Berggren, P. O. (2004). Growth hormone promotes ca(2+)-induced Ca2+ release in insulin-secreting cells by ryanodine receptor tyrosine phosphorylation. *Molecular Endocrinology, 18*(7), 1658–1669. https://doi.org/10.1210/me.2004-0044.

Zhou, Y., Xu, B. C., Maheshwari, H. G., He, L., Reed, M., Lozykowski, M., Okada, S., Cataldo, L., Coschigamo, K., Wagner, T. E., Baumann, G., & Kopchick, J. J. (1997). A mammalian model for Laron syndrome produced by targeted disruption of the mouse growth hormone receptor/binding protein gene (the Laron mouse). *Proceedings of the National Academy of Sciences of the United States of America, 94*(24), 13215–13220.

Zou, L., Menon, R. K., & Sperling, M. A. (1997). Induction of mRNAs for the growth hormone receptor gene during mouse 3T3-L1 preadipocyte differentiation. *Metabolism, 46*(1), 114–118.

Part V
Nutrition

Brain, Environment, Hormone-Based Appetite, Ingestive Behavior, and Body Weight

13

Kyle S. Burger, Grace E. Shearrer, and Jennifer R. Gilbert

13.1 Theoretical Underpinnings of Appetite Regulation

Multiple theories aim to explain the neurobehavioral underpinnings of appetite control and weight gain. These theories fall into two general categories: theories of homeostatic regulation and those of hedonic motivation. Homeostatic regulation theories involve both implicit and explicit regulation mechanisms such as the long-standing theory of implicit regulation which is the set point theory of obesity. Set point theory posits that body weight is maintained by primarily hormone-based feedback mechanisms to conserve a constant "body-inherent" weight, without conscious effort (Harris 1990). This theory of weight regulation arose from studies in which humans and rats are forced to lose or gain weight and eventually recover to a similar weight as before the involuntary change (for review, see Leibel 2008; Svetkey et al. 2008). However, this theory came under increasing scrutiny as subsequent evidence demonstrated that metabolic programming does not entirely protect against weight gain above the weight set point. For example, as energy intake increases, energy expenditure does not increase

to maintain energy balance (Westerterp 2010), resulting in excess caloric intake that raises weight beyond the set point, contributing to obesity. A more cogitative theory of explicit weight regulation is dietary restraint theory, which hypothesizes that some individuals follow an eating style that involves purposely reducing intake to prevent weight gain or lose weight (Herman and Mack 1975). Unlike the set point theory, restraint theory is founded on the notion that cognitive control, not physiological control, drives the majority of eating behavior. Further research expanded upon the consequences of dietary restraint, where restrained eating led to reduced sensitivity to satiety cues, further leading to dietary disinhibition and excessive intake in a cyclic fashion (Polivy and Herman 1985). Dietary restraint theory is actively being tested in eating behavior experiments through a variety of methods and samples, producing complex and nuanced results (for review, see Johnson et al. 2012).

13.1.1 Conflicting Theories of Hedonically Motivated Food Intake

Theories of hedonically motivated eating behavior center on how aberrant reward processing in the brain contributes to overeating pleasurable foods. Obesity has been characterized as a disease of both hyper- and hyporesponsivity to

K. S. Burger (✉) · G. E. Shearrer · J. R. Gilbert
Department of Nutrition, University of North Carolina at Chapel Hill, Chapel Hill, NC, USA
e-mail: kyle_burger@unc.edu

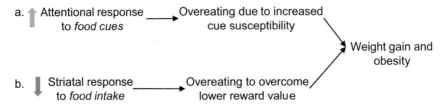

Fig. 13.1 Intersection of the (**a**) incentive sensitization and (**b**) anhedonia models of obesity

reward. The incentive sensitization model of obesity posits that intake of palatable food results in elevated response to food cues in reward-associated brain regions (Berridge 2012). This conditioning process alters in neurotransmitter signaling leading to further overconsumption (Berridge 2012). This theory is contrasted by an alternative theory of aberrant reward processing, the anhedonia model of obesity. The model suggests that individuals with decreased reward response to palatable food intake due to decreased dopamine D2 receptor density overconsume palatable foods to overcome reward deficit (Wang et al. 2004). In support, studies of dopamine signaling show that obese individuals have decreased dopamine synthesis (Wilcox et al. 2010) and receptor binding in the striatum (Volkow et al. 2008). While this original model proposed anhedonia as a predisposing factor for obesity, subsequent research demonstrated that decreased reward sensitivity to food consumption is much more likely a consequence and maintenance factor for overeating and obesity, but not an initial risk factor (Burger 2017; Johnson and Kenny 2010) (Fig. 13.1).

13.1.2 Dynamic Theory of Hedonically Motivated Food Intake

To synthesize across evidence of both hyper- and hyposensitivity to reward, Stice and Burger proposed a dynamic vulnerability model of obesity that suggests that brain response to palatable food intake changes over time to perpetuate overeating (Burger and Stice 2011). Similar to incentive sensitization, the dynamic vulnerability model suggests that individuals with elevated brain response in reward regions during palatable food consumption are at risk for overeating. Continued excess energy intake then leads to decreased dopamine signaling/receptor density, which results in food reward hyposensitivity, similar to the anhedonia model. In theory, individuals will then continue and/or escalate overeating to achieve the same level of pleasure that previously was received palatable food consumption (Fig. 13.2). Emerging models of overeating incorporating research from psychology and cognitive science to characterize how decision-making and habit formation may contribute to overeating complement the dynamic vulnerability model. Decision-making strategies, such as habit-based systems, can perpetuate overeating by preventing updated valuation of palatable food that may be less rewarding following repeated consumption (Gilbert and Burger 2016). While models of eating habit formation are new, they dovetail with existing models of obesity and may provide unique insight into the development of maladaptive eating behaviors.

With these theories as a foundation, we will next discuss in detail topics relating to the influence of the environment on overeating factors that increase hedonically motivated eating, such as brain response to food pictures, cues, and taste, brain-based contributions and consequences of weight gain, and the interaction of neural and endocrine systems on hedonically motivated eating. Consideration of hedonic, habitual, and environmental contributors to excess energy intake is important given that for about two-thirds of Americans (Flegal et al. 2012), homeostatic systems of energy balance regulation are not successful in maintaining a healthy body weight.

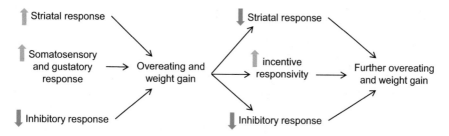

Fig. 13.2 Dynamic vulnerability model of weight gain

13.2 Environmental Contributors to Hedonically Motivated Overeating

Environmental cues of palatable food consumption contribute significantly to eating behavior by drawing attention to food stimuli and capitalizing on reward sensitivity to increase eating outside of hunger. Environmental factors such as an image of food can increase food intake, as the sight of palatable food increases the desire to eat, even in the absence of hunger (Cornell et al. 1989; Rolls et al. 1982). The effect of visual food cues on intake is especially powerful in the food cue-rich modern food environment. Consider this: how often during the day do you see pictures of food? From print media and television advertisements to social media postings online, to storefront posters and other advertisements, visual cues promoting consumption of palatable, energy-dense foods are ubiquitous. When the saturation of food cues is combined with the high availability of low-cost tasty food, the modern food environment presents very few barriers to excessive energy intake (Barthes 1997). In this section, we will discuss the neural and physiological underpinnings of environmental food cue's contribution to overeating.

13.2.1 The Impact of Food Volume Driving Intake

One facet of the food environment that increases intake is the increase in portion sizes of packaged and ready-made food. Since the 1970s, portion sizes of a number of foods, including fast-food and restaurant meals, frozen foods, and even recipes, have steadily increased (Fig. 13.3 Young and Nestle 2002). The characteristics of external food cues such as portion size are critical for influencing eating behavior, because food portion size serves as a normative cue that provides information about how much is acceptable to consume (Herman and Polivy 2008). As such, research demonstrates that when adults and children are presented with large portion sizes, they consume more kilocalories (Ello-Martin et al. 2005). Men and women served macaroni and cheese consumed on average 30% more calories when served the largest portion (500 g) compared to the smallest portion (100 g) (Rolls et al. 2002). Similar effects are observed in children. In one study, children consumed about 25% more kilocalories when presented with a very large but age-appropriate portion size (Fisher et al. 2003).The "portion size effect" has been replicated in a number of laboratory and naturalistic settings (Diliberti et al. 2004; Fisher and Kral 2008; Kling et al. 2016; Rolls et al. 2004a, b, 2006). The portion size effect also extends to ratings of food appeal. When compared to lean individuals, overweight individuals rated larger portion-sized food pictures are more desirable (Burger et al. 2011). Together, this research demonstrates that portion size can have a strong and uniform effect on overeating in both children and adults. Additionally, the portion size effect is resistant to behavioral interventions aimed at reducing intake in the face of large portions, as demonstrated by a randomized controlled trial

Fig. 13.3 Similar to the increases in portion size in the natural eating environment, the above shows a 100% increase in portion size of an amorphous food (cheesy pasta). These foods are typically used in feeding studies, as many individuals do not notice the larger amount of food

(Rolls et al. 2017). To reduce caloric intake, the best strategies often involve a "volumetric" approach, which involves replacing high energy density (high calories for volume) foods with low energy density foods (low calories for volume) (Rolls and Barnett 2000). Low energy density diets are associated with higher diet quality (Ledikwe et al. 2006), and interventions that increase consumption of low energy density foods are associated with significant weight loss (Ello-Martin et al. 2007; Ledikwe et al. 2007).

13.2.2 Impact of the Environment on Food Intake

Food advertisements are pervasive in the modern environment, particularly in developed countries. Ninety-eight percent of food advertisements promote items that are highly caloric and high in fat and sugar (Powell et al. 2007). Food advertisements for energy-dense, palatable foods serve as priming stimuli, which can subtly contribute to overeating by promoting intake of advertised foods. In children, the effect of food advertisements is striking – many studies show an increase in children's intake following exposure to food advertisements (Boyland and Whalen 2015; Halford et al. 2007; Harris et al. 2009). These effects are seen across types of media (Boyland et al. 2016) and are irrespective of the healthfulness of foods being advertised (Dovey et al. 2011).

In adults, food advertisements also increase intake, but the associations are less consistent (Vukmirovic 2015). Experimental studies in adults demonstrate both greater consumption of advertised foods following exposure (Koordeman et al. 2010) and a nonsignificant effect of advertisement exposure on intake (Anschutz et al. 2011; Martin et al. 2009a, b). A meta-analysis published in 2016 of 18 studies found a small-to-moderate effect of advertising on increasing food consumption, compared to controls, and that the results in children drove the effect (Boyland et al. 2016). These studies all examine acute response to food advertisements, so little is known about the effects of exposure to food advertisements on long-term eating habits. All things considered, given the sensitivity of children to food advertisements, researchers and policy makers have advocated for limits on food advertisements targeted at children (World Health Organization 2010).

13.2.3 Individual Differences in Environmental Response

An important caveat to the research presented thus far is that while environmental cues for food intake are omnipresent, about one-third of Americans remain at a healthy weight (Flegal et al. 2012). This may be due, in part, to individual differences in sensitivity to external food cues. Scales including the Power of Food Scale (Lowe et al. 2009) and the Dutch Eating Behavior

Questionnaire (Van Strien et al. 1986) measure individual responsiveness to external food cues. The Power of Food Scale was designed to measure hedonic hunger, or the tendency to develop an appetite from environment food cues (Lowe and Butryn 2007). The percent of individuals reporting high hedonic hunger, as measured by the Power of Food Scale, is higher among severely obese individuals compared to those at a healthy weight (Schultes et al. 2010). Also, high hedonic hunger is associated with greater motivation to consume food (Burger et al. 2016) and related to greater binge eating behavior (Burger et al. 2016; Lowe et al. 2016; Witt and Lowe 2014). Behavioral aspects of hedonic hunger are still being explored, but current evidence supports that hedonic hunger may explain some variability in susceptibility to external food cues. External eating, as measured by the Dutch Eating Behavior Questionnaire, is another construct developed to assess individual differences in sensitivity to environmental food cues. High external eating is associated with increased consumption of sugar-sweetened beverages (Elfhag et al. 2007) and cravings for high-fat foods (Burton et al. 2007). Critically both external eating and hedonic hunger are thought to interact with dietary restraint to impact long-term weight outcomes (Appelhans et al. 2011; Ely et al. 2015; Van Strien et al. 2009).

13.3 Aberrant Neural Response Associated with Obesity

Human neuroimaging techniques, such as functional magnetic resonance imaging (fMRI; Fig. 13.4) and positron emission tomography (PET), have provided valuable insight into the brain's rewa0rd circuitry implicated in eating behavior. Despite these advances, some of the most prominent theories of aberrant neural responses to food and obesity appear to be in conflict. Obesity is characterized by both a hyper- and hyporesponsivity of the regions within the reward circuitry (Berridge et al. 2010; Davis et al. 2004; Wang et al. 2001), presenting inconsistent patterns of response.

13.3.1 Food-Specific Aberrant Response in Obesity

Relative to healthy weight, obese individuals exhibit less activity in regions of the brain associated with reward in response to receipt of palatable food (Babbs et al. 2013; Frank et al. 2012; Green et al. 2011; Stice et al. 2008a, b). Further, data indicate that obese adults, relative to healthy weight, show both lower striatal dopamine D2 receptor availability (de Weijer et al. 2011; Volkow et al. 2008) and lower capacity of nigrostriatal neurons to synthesize dopamine (Wilcox et al. 2010). Animal work compliments these data where obese animals, relative to lean, show lower basal dopamine levels, lower dopamine D2 receptor availability, and less ex vivo dopamine release in response to electrical stimulation in the nucleus accumbens and dorsal striatum tissue (Fetissov et al. 2002; Huang et al. 2006; Thanos et al. 2008).

When presented with food cues and images, overweight and obese individuals show significantly higher activity in the striatum, insula, orbitofrontal cortex, and amygdala, compared to healthy weight counterparts (Bruce et al. 2010; Dimitropoulos et al. 2012; Frankort et al. 2012; Martin et al. 2009a, b; Ng et al. 2011; Nummenmaa et al. 2012; Rothemund et al. 2007; Eric Stice et al. 2008a, b; Stoeckel et al. 2008). These regions previously have been implicated in encoding the reward value of stimuli and consequently influencing goal-directed behavior (Beaver et al. 2006). Further, when exposed to visual food images or cues predicting palatable food receipt, obese individuals, relative to healthy weight, also consistently show greater activation (1) in brain regions associated with visual processing and attention (visual and anterior cingulate cortices), (2) in brain regions encoding stimulus salience (precuneus), (3) in the primary gustatory cortex (anterior insula, frontal operculum), (4) and in oral somatosensory regions (postcentral gyrus, Rolandic operculum; (Bruce et al. 2010; Dimitropoulos et al. 2012; Frankort et al. 2012; Martin et al. 2009a, b; Ng et al. 2011; Nummenmaa et al. 2012; Rothemund et al. 2007; Stice et al. 2008a, b; Stoeckel et al. 2008)).

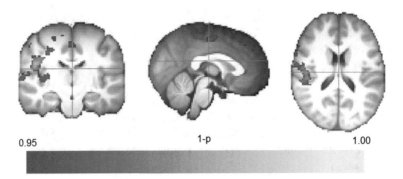

0.95   1-p   1.00

Fig. 13.4 Blood oxygen level-dependent response (BOLD) measures the release of oxygen to tissues. Neurons are a special tissue in that they do not store oxygen. Therefore, when they fire, neurons require an increase in blood-supplied oxygen. The change in the blood from oxygenated to deoxygenated is measured via MRI. Statistical analyses allow for modeling of the change from oxygenated to deoxygenated blood. Areas of high oxygen consumption are statistically "mapped" to areas in the brain as pictured here. The *red to yellow color* represents areas in the brain where the deoxygenation in the blood was significantly higher than the rest of the brain ($p > 0.05$ or $1-p > 0.95$). Measuring the turnover of oxygen in the blood has been shown to be a good proxy for neuron firing (Logothetis et al. 2001)

Together, this demonstrates that overweight and obese individuals have strong hyperresponsivity to food cues and images.

In sum, overweight/obese status is associated with elevated striatal, gustatory, and attentional responses to anticipatory food cues (e.g., food images and predictive cues) and decreased striatal response to food taste. As a result, these findings support both the hyper- and hypo-reward theories of obesity (Berridge et al. 2010; Davis et al. 2004; Wang et al. 2001). By nature of study design, the cross-sectional data collected in the majority of studies provide no ability to draw inferences regarding the temporal precedence of weight status and observed neural effects. Thus, results cannot distinguish between notions that the neural response patterns are an underlying cause of weight gain, or a consequence of habitual overeating and/or obesity.

13.3.2 Generalized Reward Sensitivity and Impulsivity

Obesity theory suggests that individuals with high levels of trait reward sensitivity display enhanced activity in brain regions implicated in food reward and therefore are at greater risk for hedonic overeating and weight gain (Davis et al. 2004). Similarly, data-driven hypotheses indicate that individuals at risk for substance abuse initially experience greater reward from substance use, increasing the risk for substance abuse (Davis and Claridge 1998; Dawe and Loxton 2004). Similar theories have also implicated a lack of cognitive control and/or impulsivity with overeating and obesity as well as substance abuse, problem gambling, and risky sexual behavior (Dawe and Loxton 2004; Verdejo-García et al. 2008). Despite mounting evidence that reward sensitivity and impulsivity are associated with obesity, it remains unclear whether these traits specifically relate to overeating, or whether they are simply generalized risk factors for any hedonic behavior.

Compared to other hedonic behaviors, such as smoking, or drinking alcohol, food consumption is necessary to sustain life, and the modern food environment inundates individuals with food cues, promoting intake. As a result, nearly all individuals in industrialized nations consume highly palatable foods, at least intermittently. The initiation of repeated intake of highly palatable foods precipitates weight gain and its associated adverse health consequences. Thus, examining risk factors for the transition from intermittent consumption of energy-dense foods to habitual intake of such foods may prove to be the most useful in understanding hedonic overeating and weight gain.

13.4 Neural Risk Factors, Brain-Based Prediction, and Consequences of Overeating and Weight Gain

With 35% of adult men and 40% of adult women in the USA classified as obese, the identification of risk factors and methods for predicting weight gain and obesity is a top priority (Flegal et al. 2016). Despite knowing that obesity is related to higher mortality rates (Abdelaal et al. 2017), the ability to lose weight and maintain weight loss remains elusive, with most people regaining their lost weight (Langeveld and DeVries 2015). The primary barrier to weight loss and maintenance of weight loss is not physiological but rather is behavioral (MacLean et al. 2015). This behavioral barrier implies that weight loss and weight gain are matter of the mind, not necessarily of the body. In line with this conjecture, preventing weight gain through behavioral means and predicting weight gain risk through neural mechanisms may be the best way to prevent and ameliorate obesity.

Along with behavioral barriers to weight loss, interindividual variability in susceptibility to obesity is an additional challenge for effective weight loss (MacLean et al. 2015). An estimated 40–70% of interindividual obesity risk is attributed to genetics (Elks et al. 2012). Research on genetic and heritable mental health disorders, such as substance abuse (Yu and McClellan 2016), have shown that genes can shape behaviors: protecting or predisposing an individual to weight gain risk factors.

13.4.1 Neural and Genetic Risk Factors of Overeating and Weight Gain

Studies in both humans and animals show differential neural response to palatable food across various weight statuses (Babbs et al. 2013; Contreras-Rodríguez et al. 2015; Cosgrove et al. 2015; Johnson and Kenny 2010). These studies provide a solid description of the differences in hedonically motivated food intake; however

they generally measure differences in already overweight and obese individuals. Thus, they are unable to predict weight gain or characterize risk factors related to obesity. Assessing normal weight adolescents at high risk for obesity (defined as dual parental overweight/obesity) compared to normal weight adolescents at low risk for obesity (defined as having two lean parents) allows researchers to evaluate the contribution of both genetic and environmental risk factors on neural response before the onset of obesity. Parental BMI is shown to be a reliable predictor of obesity in their offspring (Faith et al. 2006; Jacobson et al. 2006; Kumar et al. 2010; Rath et al. 2016). Additionally, children of obese parents exhibit increased preference for highly palatable foods (Birch and Fisher 1995; Wardle et al. 2001), which is further associated with increased obesity risk (Johnson and Wardle 2014; Salbe et al. 2002). Two studies used parental weight status as a proxy for obesity risk (Shearrer et al. 2017; Stice et al. 2011) and found that adolescents at risk for obesity, by virtue of parental overweight/obesity, compared to those at low risk for obesity showed increased BOLD response in areas related to gustatory (insula), somatosensory (postcentral gyrus), and reward processing (orbital frontal cortex, caudate) when given a palatable milkshake (Shearrer et al. 2017; Stice et al. 2011). In particular, adolescents at high risk for obesity, compared to those at low risk, presented increased BOLD response to high-sugar milkshakes compared to a tasteless solution in the caudate and central opercular cortex, areas related to taste and reward (Shearrer et al. 2017). However, in both studies, there was no difference between risk groups in BOLD response to a visual food stimuli (Shearrer et al. 2017; Stice et al. 2011). These results support the dynamic vulnerability model, which posits that initial hypersensitivity in striatal and gustatory regions to a palatable taste, but not food images, precedes and propagates overeating (Burger and Stice 2011; Stice et al. 2011). Habitual overeating and weight gain shift neural response from the palatable taste to food cues (such as images), resulting in the hyporesponsivity to palatable food tastes, and hyperresponsivity to food cues, seen in obesity (Burger and Stice 2011, 2012,

2013a, b; Burger 2017; Cosgrove et al. 2015; Demos et al. 2012; Geha et al. 2013; Stice et al. 2008a, b; Volkow et al. 2008; Wang et al. 2001; Yokum et al. 2012).

13.4.1.1 Genetic Influences on Brain Response to Food and Weight Regulation

To parse the contribution of genetics from that of the environment, researchers have examined genetic markers related to weight status. The ANKK1 TaqIA polymorphism has been significantly studied in relation to food reward and obesity (Barnard et al. 2009; Blum et al. 1996; Epstein et al. 2007; Felsted et al. 2010; Kirsch et al. 2006; Stice et al. 2008a, b; Sun et al. 2015). The TaqIA A1 allele is associated with reduced dopamine D2 receptor (D2R) density and dopamine signaling and with increased food intake and food reinforcement in both mice and humans (Barnard et al. 2009; Epstein et al. 2007; Thanos et al. 2008). The TaqIA A1 allele significantly moderated the relationship between striatal response to milkshake receipt and weight change over a 1-year follow-up, such that, among those with the A1 allele, low striatal response to milkshake receipt predicted future weight gain (Stice et al. 2008a, b). A separate study replicated this finding in an initially lean sample of 152 adolescents using a sensitive measure of percent body fat over a 3-year follow-up period (Stice et al. 2015). Similar effects have emerged for another genotype (no 7-repeat or longer alleles of the *DRD4* gene [*DRD4*-S]) that has been associated with elevated dopamine signaling (Stice et al. 2010a). Together, these studies imply that neural response to food receipt combined with genetic status may place individuals at risk for excess weight gain and identify aberrant dopamine signaling as a potential mechanism of brain-based obesity risk.

While dopamine-related alleles have received the majority of attention in the neuroimaging field, the fat mass and obesity-associated (FTO) gene was the first gene explicitly related to BMI and obesity (Loos and Yeo 2014). The effect of FTO on obesity risk has been shown to be substantial, with the odds of becoming obese increased from 18–27% per allele depending on the population (Loos and Yeo 2014). FTO encodes for an RNA demethylase enzyme broadly important for the control of protein expression (Jia et al. 2011), and while the exact mechanism as to how it increases obesity risk is under investigation, several studies have linked FTO to atypical neural processing of food stimuli, reward learning, and impulse control, as well as interacting with homeostatic hormones to increase risk (discussed further in Sect. 13.5) (de Groot et al. 2015; Heni et al. 2014b; Sevgi et al. 2015; Wiemerslage et al. 2016). Of note, the increased risk the FTO gene imparts appears to be related to eating behavior, rather than metabolic or endocrine, suggesting that the gene alters a neural mechanism (Sun et al. 2017). Combination of research on dopamine systems and FTO has revealed that the FTO gene interacts with the TaqIA A1 allele to regulate reward learning in humans (Sevgi et al. 2015). Those individuals with both the FTO and ANKK1 TaqIA polymorphism are shown to be insensitive to negative reward learning and exhibit reduced connectivity between the nucleus accumbens and medial prefrontal cortex and reduced connectivity between the ventral tegmental area (VTA) and substantia nigra in a "dose-by-gene response" (Sevgi et al. 2015). The valence-specific insensitivity further implicates the role of the D2 receptors (Mathar et al. 2017). Drops in dopamine when an outcome is worse than expected (negative prediction error) modulates D2R density (Frank et al. 2007). Similarly, insensitivity to negative reward learning with downregulated D2R was induced in rats with chronic overfeeding, resulting in obesity (Johnson and Kenny 2010). Therefore, excess caloric intake appears to downregulate D2R expression, and for individuals genetically predisposed to decreased D2R as in the TaqIA allele or with the TaqIA/FTO gene interaction, this effect appears to be magnified. The mechanism by which FTO influences the TaqIA polymorphism is being explored; however it has been suggested that FTO's encoding of a demethylase may alter the TaqIA polymorphism in an epigenetic manner (Hess et al. 2013).

13.4.2 Brain-Based Predication and Consequences of Weight Gain and Overeating

As mentioned in this section's introduction, finding neural correlates of weight gain may improve weight loss efficacy and prevent at-risk individuals from developing obesity. Prospective studies provide insight into preexisting risk factors for weight gain. In further support of the dynamic vulnerability model, greater response in the ventral striatum and medial prefrontal cortex/anterior cingulate during exposure to appetizing food images was shown to predict weight gain over a 6-month period (Demos et al. 2012). Further, individuals who showed greater activation in the orbitofrontal cortex (OFC) in response to a cue predicting palatable food intake gained more weight over 1 year compared to those who did not respond strongly (Yokum et al. 2011). Interestingly, the OFC is thought to play a prominent role in valuation and representation of specified rewards such as palatable food (Man et al. 2009). Together, this suggests that an increased valuation of food cues in areas of the brain associated with executive control is positively related to weight gain.

In addition, data indicates that reward-related anomalies predict weight gain, and eating patterns can influence how the brain then responds to food stimuli. For example, in a study, frequent ice cream consumption was related to a reduced striatal and prefrontal response during receipt of a palatable milkshake in lean adolescents (Burger and Stice 2012). Notably, these results were exclusive to ice cream consumption, as consumption of similar foods (e.g., chocolate sweets and baked goods) did not elicit significant reduced brain response, suggesting a marked specificity in this relationship. The specificity between the stimulus and the reduced neural response is not limited to milkshake but has also been seen with diet soda and artificial sweetener use (Green and Murphy 2012; Rudenga and Small 2012). Further, in a randomized controlled trial, participants assigned to consume a novel high-sugar beverage for 2 weeks exhibited a reduction in striatal response to the novel high-sugar beverage (Burger 2017). These results were not attenuated when controlling for BMI and are directly in line with the evidence that habitual consumption, not weight status per se, reduces striatal response. This study also showed how quickly the brain adapts to habitual consumption of an initially novel food. Again, these studies support the work of Johnson and Kenny in Sect. 13.4.1 and provide further evidence that habitual consumption of a food causes downregulation in D2R (Johnson and Kenny 2010). These studies in humans also coalesce with and inform the dynamic vulnerability hypothesis, showing that habitual consumption of food drives the shift from hyperresponsivity to food receipt to hyperresponsivity to a food cue through downregulation of D2R.

To further support the dynamic vulnerability model, several studies have looked at the impact of habitual excess caloric intake on food cue response. Increased caloric intake over a 2-week period measured by doubly labeled water was related to heightened parietal cortex, visual cortex, and insular response during cue-elicited anticipation of food intake (Burger and Stice 2013a, b). This suggests that overeating, in general, is associated with heightened response to food cues independent of weight status (Burger and Stice 2013a, b). Further, regular Coke™ consumers (>1/day) showed increased visual cortex and parietal response during exposure to Coke advertisements, relative to their nonconsuming counterparts, independent of BMI (Burger and Stice 2013a, b). These data are directly in line with research that suggests that the heightened striatal and attentional response to food cues is related to habitual eating behaviors and that weight gain is a secondary outcome from excessive caloric intake. In sum, it appears that eating behavior independent of weight status retrains the brain to focus on food cues, rather than the actual food reward. Animal models propose that this shift in reward learning is dependent on D2R density in the striatum and those with a genetic predisposition for low D2R appear to be a greater risk of future weight gain through unfavorable eating behaviors.

13.5   Neuroendocrine Systems and Ingestive Behavior in Humans

Understanding how hedonic desires integrate with or override the body's homeostatic signals is crucial for the study of eating behaviors and weight gain. While perceived hunger and eating behaviors are known barriers to weight loss, little is known of how hunger interacts with the hedonic desire to eat (Elfhag and Rossner 2005; MacLean et al. 2015). Historically, the desire to eat and hunger were considered to be the same thing. The glucostatic theory posited that hunger was a homeostatic signal for low blood sugar, and when blood sugar was returned to normal values, hunger was resolved (Mayer 1953). Research into hypoglycemia (low blood sugar) has shown that while accompanied by shaking, sweating, and weakness, feelings of hunger are often absent in hypoglycemia (Rogers and Brunstrom 2016). Indeed, hunger has more frequently been associated with gastric distention, circadian training (becoming hungry because it is lunch time), and/or food reward learning (eating out of celebration or consolation) (Rogers and Brunstrom 2016).

Although hypoglycemia does not appear to regulate the desire to eat, endocrine factors related to glucose balance may play a role. While the previous sections have outlined how excess caloric consumption can influence receptors and signaling in the brain, this section outlines how chronic overeating alters endocrine systems. Specifically, this part will discuss how overconsumption of high-sugar and fatty foods can result to insulin and leptin resistance and how insulin and leptin resistance connects the dynamic vulnerability model (Grosshans et al. 2012; Jastreboff et al. 2014; Kullmann et al. 2012; Simon et al. 2014). The dynamics between the fed and fasted state of the orexigenic hormone ghrelin also appears to influence perceived hunger and BOLD response (Karra et al. 2013). Finally, treatment with incretin hormones such as glucagon-like peptide 1 (GLP-1) analogues may resolve the abnormal BOLD response seen in obesity (ten Kulve et al. 2016).

13.5.1 Leptin

The fat-derived peptide hormone leptin plays a critical role as a long-term homeostatic feedback mechanism (Blundell et al. 2001). Leptin rises in proportion to body fat, which signals the adequacy of body fat stored in the central nervous system via the hypothalamus. This functions to maintain adequate body fat stored for reproduction and survival around a homeostatic set point (Blundell et al. 2001). Although one would expect high levels of leptin to reduce or inhibit feeding behaviors, leptin has been found to be elevated in obese subjects, and hyperleptinemia in normal weight subjects has not been shown to reduce food intake, suggesting obese individuals may experience "leptin resistance," or the inability to produce an adequate response to a given leptin stimulus (Kalra 2001).

Work in subjects with genetic leptin deficiency show increased BOLD response to high-calorie food images in the insula, parahippocampal gyrus, limbic lobe, parietal lobe, and precuneus, as well as high hunger ratings in response to food images (Baicy et al. 2007). Leptin replacement therapy suppressed BOLD response in the insula, parietal lobe, and temporal cortex and reduced hunger ratings in response to high-calorie foods while increasing BOLD response in the prefrontal cortex (Baicy et al. 2007). These findings posit that under normal physiological conditions, leptin works to reduce food intake and hunger, increasing brain regions associated with self-control and decreasing gustatory regions. Leptin deficiency has also been shown to increase liking of foods in fed and fasted states, and this liking is correlated with BOLD response in the nucleus accumbens (Farooqi et al. 2007). While in normal weight/leptin subjects, the correlation of BOLD response in the nucleus accumbens and ratings of liking were observed only in the fasted state (Farooqi et al. 2007). As such, individuals with leptin deficiency appear to consistently feel that they are in a fasted state.

Obesity can be considered a state of chronic elevated endogenous leptin circulation. In adolescents, elevated endogenous leptin, independent of body fat mass, was correlated with BOLD response in the putamen, caudate, thalamus, amygdala, posterior cingulate, and insula to images of high-calorie foods versus non-food objects (Grosshans et al. 2012; Jastreboff et al. 2014). Leptin was also shown to mediate a positive correlation between BOLD response in the ventral striatum to anticipation of a food reward and BMI (Simon et al. 2014). High levels of leptin appear to mimic the obese brain's response to food cues. However, it is unclear if high leptin levels decrease BOLD response to palatable food receipt. Additionally, further research is required to determine if the correlation between leptin and BOLD response is due to eating behaviors or body fat status.

13.5.2 Insulin

While best known for its action in glucose metabolism in the liver and muscle, insulin is critical within the CNS for glucose metabolism, satiety signaling, and, possibly, reward signaling (Kroemer and Small 2016; Murray et al. 2014). Insulin receptors are found within the hypothalamus in the arcuate nucleus, in the VTA, and within the striatum (Kroemer and Small 2016; Murray et al. 2014). Insulin concentrations peripherally and centrally and the activity of their receptors have been shown to influence dopamine metabolism, concentration, and receptor availability (Dunn et al. 2012; Kroemer and Small 2016; Potter et al. 1999). Insulin has been shown to decrease dopamine concentrations via upregulation of dopamine transport synthesis in the VTA (Figlewicz et al. 1994). Additionally, knockout of insulin receptors in the VTA leads to hyperphagia and increased fat mass (Brüning et al. 2000; Murray et al. 2014), suggesting that dopamine and insulin work synergistically to maintain food intake. The relationship between dopamine and insulin appears to extend beyond the central nervous system, as depletion of dopamine reduces peripheral insulin sensitivity, and insulin sensitivity is concomitantly related to D2/3 receptor availability in the striatum in healthy humans (Caravaggio et al. 2015). The connection between dopamine and peripheral

insulin is not linear. Work in animal models has shown peripheral insulin correlates with dopamine release in the nucleus accumbens (Potter et al. 1999). However, at high levels, peripheral insulin was correlated with suppressed dopamine release (Potter et al. 1999). This suggests that chronically elevated peripheral insulin, as in type 2 diabetes (T2D), is associated with a kind of insulin resistance at the level of dopamine release. Considering the previous research indicating the importance of dopamine signaling for negative reward learning (Coppin et al. 2014; Johnson and Kenny 2010; Mathar et al. 2017), suppression of dopamine response due to high insulin could alter reward learning behaviors and increase risk of overeating.

Further studies show that insulin has a profound effect on BOLD response in the occipital cortex, prefrontal cortex, and hypothalamus. Interestingly, insulin only stimulates brain glucose metabolism to a physiological set point; increasing insulin above fasting levels in the CNS does not increase glucose utilization in the brain. This is an important point for two reasons. First, in terms of imaging, this indicates that insulin administration, especially intranasal, does not artificially increase BOLD response due to increased brain metabolism. Second, this suggests that higher than fasting levels of insulin in the brain may signal for processes outside of simple glucose utilization. Uncoupling glucose metabolism from insulin may be particularly important in the CNS due to the sustained high demand for glucose to maintain neural function.

Insulin acts differentially in the CNS and peripherally; however both systems appear to regulate each other. Intranasal insulin administration allows researchers to artificially increase insulin in the CNS with minimal spill over to the periphery. Intranasal insulin in normal weight men, but not in overweight men, has been shown to suppress hepatic glucose production and stimulate peripheral glucose intake (Heni et al. 2017). Additionally, in normal weight subjects, intranasal insulin decreased cerebral blood flow in the orbitofrontal cortex; however this was not seen in overweight or obese subjects. This reduction in BOLD signaling with intranasal insulin in normal weight but not overweight or obese individuals may translate into eating behaviors. In healthy men and women, intranasal insulin reduced subsequent food intake in both fed and fasted conditions and increased feelings of satiety in women (Hallschmid et al. 2012; Jauch-Chara et al. 2012). This implies that CNS insulin action influences peripheral metabolism, signaling the fed state, and therefore a decreased need for hepatic glucose production. Conversely, the overweight men appear to exhibit insulin resistance in the CNS, which manifests as peripheral insulin resistance in the liver and a decreased BOLD response in areas of executive functioning. The fed signal is not relayed to the body, and hepatic glucose production is not suppressed. The increased gluconeogenesis may further increase fat accumulation in the liver, creating a peripheral insulin-resistant cycle (Seppälä-Lindroos et al. 2002).

Peripheral insulin also appears to affect nutrient utilization in the CNS, in particular in the ventral striatum and prefrontal cortex (Anthony et al. 2006). In normal, insulin-sensitive men, an increase in plasma insulin was related to increased metabolism (as measured through positron emission tomography (PET)) in the ventral striatum and prefrontal cortex and decreased metabolism in the right amygdala and hippocampus (Anthony et al. 2006). In insulin-resistant men, overall cerebral metabolism was reduced, and compared to the normal group, metabolism in the ventral striatum and prefrontal cortex was reduced (Anthony et al. 2006). The ventral striatum is made up of the nucleus accumbens and olfactory tubercle, and the nucleus accumbens in particular is critical for reward learning receiving dopaminergic projects from the VTA (Ubeda-Bañon et al. 2007). Further PET analysis investigating the ventral striatum and insulin sensitivity has shown increased D2R availability with decreasing peripheral insulin sensitivity in obese women, suggesting a decline in VTA projections of available dopamine to the nucleus accumbens for binding with increased insulin resistance (Dunn et al. 2012). Overall, these findings indicate that peripheral insulin sensitivity influences metabolism in the brain,

particularly in areas previously associated with cognition and reward learning (Anthony et al. 2006). This has bearing not only on ingestive but also cognitive disease states associated with T2D such as Alzheimer's disease (Haan 2006).

Previously, postprandial levels of peripheral insulin have been shown to predict food intake at the next meal, indicating that insulin can influence short-term appetite (Verdich et al. 2001). Several studies using fMRI have aimed to elucidate the influence of insulin on eating behaviors and satiety through the brain. Peripheral insulin following an oral glucose tolerance test (OGTT) was associated with decreased BOLD response in the prefrontal cortex and limbic system when looking at high-calorie food images in lean subjects (Heni et al. 2014a; Kroemer et al. 2013) and with a decrease in BOLD response in the orbitofrontal cortex (Heni et al. 2015). Moreover, cerebral blood flow (CBF) was negatively correlated to plasma insulin in the striatum during taste administration of a fructose or glucose drink (Page et al. 2013). In obese subjects, BOLD response to a food cue in the striatum and insula was shown to mediate the relationship between food craving and insulin resistance (Jastreboff et al. 2013). Peripheral insulin seems to act as a satiating signal decreasing activity in the brain to food cues (Balleine et al. 2007; Miller and Cohen 2001). In younger samples, fasting plasma insulin was positively correlated with hippocampal BOLD response when viewing food images (Wallner-Liebmann et al. 2010), and insulin resistance has been shown to be positively correlated with BOLD response in the dorsal prefrontal cortex, orbitofrontal cortex, and occipital cortex when anticipating a beverage receipt in overweight adolescents (Feldstein Ewing et al. 2016). This aligns with previous research of overweight and obese individuals; however it is unclear whether insulin is also a product of habitual overeating.

Since no area of the brain or the body acts independently, functional connectivity analysis aims to look at how the brain functions as a whole. Patients with T2D, a state of chronically elevated peripheral insulin and reduced brain insulin, exhibit decreased functional connectivity between the posterior cingulate cortex and the middle temporal gyrus (Chen et al. 2014; Xia et al. 2015), the middle occipital gyrus, and the precentral gyrus (Chen et al. 2014). In particular, the connection between the posterior cingulate cortex and middle temporal gyrus was negatively correlated with insulin resistance with the T2D group (Chen et al. 2014). The middle temporal gyrus is thought to be important for cognitive functioning and is a key region of impairment in Alzheimer's disease (Visser et al. 1999); thus, the breakdown of the connection between middle temporal gyrus and posterior cingulate cortex may be an important link between T2D and Alzheimer's disease. In a healthy population, functional connectivity was inversely related to insulin sensitivity in the putamen and the orbitofrontal cortex and was positively correlated to fasting plasma insulin (Kullmann et al. 2012). This suggests that normal levels of insulin may encourage reward learning; however chronic high plasma insulin, a precursor to T2D, may decrease connectivity in areas critical for cognition and predispose those with T2D to dementia-related diseases.

13.5.3 Ghrelin

Ghrelin is currently the only known orexigenic hormone and is critical for foraging behaviors, homeostatic maintenance, and reward learning (Pliquett et al. 2006). Despite its role in hunger and ability to stimulate feeding, ghrelin is reduced in obese individuals compared to lean humans (Tschöp et al. 2001). Under normal conditions, and in healthy humans, food intake rapidly suppresses plasma ghrelin. However, in the obese state, feeding does not suppress plasma ghrelin (English et al. 2002). This paradoxical response has led to increased research into ghrelin's role in healthy individuals and how it is disrupted in obesity.

In normal weight human studies, baseline plasma ghrelin levels positively correlate with BOLD response to food cues in the visual, limbic, and paralimbic areas (Kroemer et al. 2013; Malik et al. 2008). In particular, within the limbic system, BOLD response has been shown to be positively correlated to plasma ghrelin during administration of a milkshake taste but not a neutral taste (Sun et al. 2015). The BOLD activity in the amygdala and orbitofrontal cortex also correlates with self-reported feelings of hunger (Malik et al. 2008). Overall, these studies confirm that in normal weight humans, ghrelin is an orexigenic signal and is related to perceived hunger.

Additionally, ghrelin is also thought to be important for reward learning and cue salience. Ghrelin administration in normal weight men increased functional connectivity between the caudate and insula and between the amygdala and OFC cortex when the subjects were asked to memorize food images, but not when asked to memorize non-food objects (Kunath et al. 2016). Interestingly, ghrelin did not improve performance on working memory tasks or cognitive tasks, suggesting that ghrelin's effect is food specific (Kunath et al. 2016). Ghrelin in particular appears to increase attention to palatable food cues in the OFC and in the hippocampus (Goldstone et al. 2014). The dynamic fluctuations of ghrelin after feeding may also play a role in satiety. In normal weight subjects, larger postprandial reductions in ghrelin after feeding were associated with attenuated BOLD response in the midbrain, amygdala, pallidum, insula, hippocampus, and medial OFC to a milkshake taste (Sun et al. 2014). Therefore, a steep decrease in plasma ghrelin may be just as important as a satiety signal as high circulating ghrelin is a feeding signal.

As mentioned earlier, obese humans present lower circulating ghrelin even in the fasted state compared to normal weight (Tschöp et al. 2001). Obese humans exhibit different ghrelin fluctuations after feeding. Ghrelin is not suppressed in the postprandial state in subjects with obesity compared to normal weight (English et al. 2002). This lack of suppression in obese subjects may result in a diminished satiety response, or a maintained feeling of hunger despite overall lower circulating ghrelin. In overweight and obese subjects, compromised ghrelin suppression was associated with stronger intensity of food odors when individuals were hungry and with less reported satiation after a meal (Sun et al. 2016). Dysregulated circulating ghrelin and impaired ghrelin suppression may also have a genetic component. The FTO gene, discussed previously (Sect. 13.4.1.1), is associated with an increased risk of developing obesity (Loos and Yeo 2014). Homozygosity for the FTO gene is associated with depressed circulating ghrelin and attenuated postprandial appetite reduction compared to fat mass in BMI and fat mass-matched controls (Karra et al. 2013). The FTO gene was shown to mediate the response between circulating ghrelin and BOLD response in reward and appetite brain regions in normal weight subjects (Karra et al. 2013). Specifically, individuals with the FTO gene did not exhibit suppressed ghrelin postprandially, and postprandial ghrelin was related to increased BOLD response in the caudate to food images (Karra et al. 2013). This study indicates that an inability to suppress postprandial ghrelin may underlie the increased BOLD response to food cues seen in obesity and is in accordance with the dynamic vulnerability model. Further, this study implies that normal weight people with the FTO gene process food cues similarly to overweight and obese humans. This abnormal response to food cues may predispose those with the FTO gene to obesity via aberrant ghrelin signaling.

13.5.4 Glucagon-like Peptide-1

Incretins are peptide hormones that stimulate insulin secretion after meals, and one of the most researched incretion in relation to food intake and satiety is GLP-1 (Small and Bloom 2004). A majority of GLP-1 is synthesized in the L cells of the intestines (Small and Bloom 2004); however a smaller proportion is synthesized in the hindbrain (Sandoval and Sisley 2015). GLP-1 mRNA is found in the human brain (Sandoval and Sisley 2015), and GLP-1 receptors are found in the VTA, the nucleus accumbens (Merchenthaler et al. 1999), hypothalamus, medulla, and parietal cortex (Farr et al. 2016a). However GLP-1's ability to influence the brain directly is unclear due in large part to its very short half-life, only surviving in plasma for 1.5 to 5 min (Hui et al.

2002). Because of the quick metabolism of GLP-1, most research uses a more stable analogue, agonist, or antagonist to affect GLP-1 receptors. However, use of a synthetic, stable form casts doubt on the physiological effects of the endogenous hormone.

Exogenous GLP-1 has been shown to decrease food intake and increase satiety (Gallwitz 2012). One of the few studies to look at endogenous GLP-1 after an OGTT with a fMRI paradigm found an increase in GLP-1 was negatively correlated with BOLD response to a food cue in the OFC in both lean and obese subjects (Heni et al. 2015). This suggests that post prandial GLP-1 may modulate eating behaviors through the OFC, an area related to appraisal of food cues (Porubská et al. 2006). GLP-1 analogues have been shown to increase BOLD response to palatable milkshake in the caudate, putamen, insula, and OFC, as well as reduce food intake (van Bloemendaal et al. 2015). However in response to a food cue, treatment with a GLP-1 analogue resulted in decreased BOLD response in the putamen, OFC, insula, and parietal lobe (Farr et al. 2016a; van Bloemendaal et al. 2015). These results are similar to neural responses seen in normal weight populations and inverse to what is seen in overweight and obese subjects when given a palatable milkshake and when viewing food cues, indicating that GLP-1 may promote a "healthier" neural profile (Stice et al. 2008a, b, 2010a, b, 2015). When lean subjects were given a GLP-1 blocker, the BOLD response to the palatable milkshake was attenuated, bearing more resemblance to an obese individual's brain (ten Kulve et al. 2016; van Bloemendaal et al. 2015). This research suggests that GLP-1, at least in analogue form, may sensitize the brain to food intake, desensitize it to food cues, and promote satiety through reward learning networks effectively simulating a normal weight brain.

In the general population, GLP-1 analogues are used as a treatment for T2D (Tomkin 2014). In obese subjects with T2D, treatment with a GLP-1 analogue increased BOLD response in the caudate and insula in response to milkshake receipt compared to treatment with insulin (ten Kulve et al. 2016). The recovery of BOLD response to the milkshake receipt again suggests that GLP-1 is associated with increased feelings of satiety and normal responsivity in neural reward regions. As mentioned above, GLP-1 stimulates insulin secretion, and it has also been found to decrease leptin and increase gastric inhibitor peptide (GIP) (Farr et al. 2016b). After treatment with a GLP-1 analogue in T2D subjects, GIP was related to decreased BOLD response to a food image in the insula, and leptin was inversely correlated with BOLD response in the precuneus and dorsolateral prefrontal cortex and positively correlated with the parietal lobe and thalamus (Farr et al. 2016b). GLP-1 may not act directly on the CNS; rather it may influence peripheral hormones with longer half-lives to influence satiety. Finally, GLP-1 analogues have been found to be differentially effective, with some populations "responding" to treatment and others not responding. "Responders" to GLP-1 analogue treatment show higher connectivity in the hypothalamus to a food cue compared to "nonresponders" (Schlogl et al. 2013). While what makes one a "responder" or not is not understood, those who are "responders" may increase homeostatic signaling through the hypothalamus with GLP-1 treatment.

Questions with Correct Answer Highlighted

1. The dynamic vulnerability model suggests that individuals with decreased brain response in reward regions during palatable food consumption are at risk for overeating: True **False**

2. Portion size of food:
   (a) Has no effect on food intake in children or adults
   (b) Has increased in the environment
   (c) Has not changed in foods consumed in the home
   (d) All of the above

3. Obese relative to healthy weight adults show:
   (a) Both lower striatal dopamine D2 receptor availability and functioning
   (b) Both higher striatal dopamine D2 receptor availability and functioning
   (c) No difference in dopamine D2 receptors

   (d) None of the above

4. _____ does not appear to regulate the desire to eat.
   (a) Hypoglycemia
   (b) Glucose
   (c) Insulin
   (d) Ghrelin

5. People that report high dietary restraint are:
   (a) Show decreased acute food intake
   (b) More likely to be overweight
   (c) More likely to be underweight
   (d) Have an eating disorder

References

Abdelaal, M., le Roux, C. W., & Docherty, N. G. (2017). Morbidity and mortality associated with obesity. *Annals of Translational Medicine, 5*(7), 161. https://doi.org/10.21037/atm.2017.03.107.

Anschutz, D. J., Engels, R. C. M. E., van der Zwaluw, C. S., & Van Strien, T. (2011). Sex differences in young adults' snack food intake after food commercial exposure. *Appetite, 56*(2), 255–260.

Anthony, K., Reed, L. J., Dunn, J. T., Bingham, E., Hopkins, D., Marsden, P. K., & Amiel, S. A. (2006). Attenuation of insulin-evoked responses in brain networks controlling appetite and reward in insulin resistance the cerebral basis for impaired control of food intake in metabolic syndrome? *Diabetes, 55*(11), 2986–2992.

Appelhans, B. M., Woolf, K., Pagoto, S. L., Schneider, K. L., Whited, M. C., & Liebman, R. (2011). Inhibiting food reward: Delay discounting, food reward sensitivity, and palatable food intake in overweight and obese women. *Obesity, 19*(11), 2175–2182.

Babbs, R. K., Sun, X., Felsted, J., Chouinard-Decorte, F., Veldhuizen, M. G., & Small, D. (2013). Decreased caudate response to milkshake is associated with higher body mass index and greater impulsivity. *Physiology & Behavior.*

Baicy, K., London, E. D., Monterosso, J., Wong, M.-L., Delibasi, T., Sharma, A., & Licinio, J. (2007). Leptin replacement alters brain response to food cues in genetically leptin-deficient adults. *Proceedings of the National Academy of Sciences, 104*(46), 18276–18279. Retrieved from http://www.ncbi.nlm.nih.gov/pmc/articles/PMC2084333/pdf/zpq18276.pdf.

Balleine, B. W., Delgado, M. R., & Hikosaka, O. (2007). The role of the dorsal striatum in reward and decision-making. *Journal of Neuroscience, 27*(31), 8161–8165. https://doi.org/10.1523/JNEUROSCI.1554-07.2007.

Barnard, N. D., Noble, E. P., Ritchie, T., Cohen, J., Jenkins, D. J. A., Turner-McGrievy, G., et al. (2009). D2 dopamine receptor TaqlA polymorphism, body weight, and dietary intake in type 2 diabetes. *Nutrition, 25*(1), 58–65. https://doi.org/10.1016/j.nut.2008.07.012.

Barthes, R. (1997). Toward a psychosociology of contemporary food consumption. *Food and Culture: A Reader, 2,* 28–35.

Beaver, J. D., Lawrence, A. D., van Ditzhuijzen, J., Davis, M. H., Woods, A., & Calder, A. J. (2006). Individual differences in reward drive predict neural responses to images of food. *Journal of Neuroscience, 26*(19), 5160–5166. https://doi.org/10.1523/jneurosci.0350-06.2006.

Berridge, K. C. (2012). From prediction error to incentive salience: Mesolimbic computation of reward motivation. *The European Journal of Neuroscience, 35*(7), 1124–1143. https://doi.org/10.1111/j.1460-9568.2012.07990.x.

Berridge, K. C., Ho, C.-Y., Richard, J. M., & DiFeliceantonio, A. G. (2010). The tempted brain eats: Pleasure and desire circuits in obesity and eating disorders. *Brain Research, 1350,* 43–64.

Birch, L. L., & Fisher, J. A. (1995). Appetite and eating behavior in children. *Pediatric Clinics of North America, 42*(4), 931–953.

van Bloemendaal, L., Veltman, D. J., ten Kulve, J. S., Groot, P. F. C., Ruhé, H. G., Barkhof, F., et al. (2015). Brain reward-system activation in response to anticipation and consumption of palatable food is altered by glucagon-like peptide-1 receptor activation in humans. *Diabetes, Obesity and Metabolism, 17*(9), 878–886. https://doi.org/10.1111/dom.12506.

Blum, K., Braverman, E. R., Wood, R. C., Gill, J., Li, C., Chen, T. J., et al. (1996). Increased prevalence of the Taq I A1 allele of the dopamine receptor gene (DRD2) in obesity with comorbid substance use disorder: A preliminary report. *Pharmacogenetics.*

Blundell, J. E., Goodson, S., & Halford, J. C. (2001). Regulation of appetite: Role of leptin in signalling systems for drive and satiety. *International Journal of Obesity and Related Metabolic Disorders: Journal of the International Association for the Study of Obesity, 25*(Suppl 1), S29–S34. https://doi.org/10.1038/sj.ijo.0801693.

Boyland, E. J., & Whalen, R. (2015). Food advertising to children and its effects on diet: Review of recent prevalence and impact data. *Pediatric Diabetes, 16*(5), 331–337.

Boyland, E. J., Nolan, S., Kelly, B., Tudur-Smith, C., Jones, A., Halford, J. C. G., & Robinson, E. (2016). Advertising as a cue to consume: A systematic review and meta-analysis of the effects of acute exposure to unhealthy food and nonalcoholic beverage advertising on intake in children and adults. *The American Journal of Clinical Nutrition, ajcn120022.*

Bruce, A. S., Holsen, L. M., Chambers, R. J., Martin, L. E., Brooks, W. M., Zarcone, J. R., et al. (2010). Obese children show hyperactivation to food pictures in brain networks linked to motivation, reward and cognitive control. *International Journal of Obesity, 34*(10), 1494–1500. Retrieved from http://www.nature.com/ijo/journal/v34/n10/pdf/ijo201084a.pdf.

Brüning, J. C., Gautam, D., Burks, D. J., Gillette, J., Schubert, M., Orban, P. C., et al. (2000). Role of brain insulin receptor in control of body weight and reproduction. *Science (New York, N.Y.), 289*(5487), 2122–2125.

Burger, K. S. (2017). Frontostriatal and behavioral adaptations to daily sugar-sweetened beverage intake: A randomized controlled trial. *The American Journal of Clinical Nutrition, 105*(3), 555–563.

Burger, K. S., & Stice, E. (2011). Variability in reward responsivity and obesity: Evidence from brain imaging studies. *Current Drug Abuse Reviews, 4*(3), 182.

Burger, K. S., & Stice, E. (2012). Frequent ice cream consumption is associated with reduced striatal response to receipt of an ice cream-based milkshake. *The American Journal of Clinical Nutrition, 95*(4), 810–817. https://doi.org/10.3945/ajcn.111.027003.

Burger, K. S., & Stice, E. (2013a). Elevated energy intake is correlated with hyperresponsivity in attentional, gustatory, and reward brain regions while anticipating palatable food receipt. *The American Journal of Clinical Nutrition, 97*(6), 1188–1194.

Burger, K. S., & Stice, E. (2013b). Neural responsivity during soft drink intake, anticipation, and advertisement exposure in habitually consuming youth. *Obesity, 22*(2), 441–450. https://doi.org/10.1002/oby.20563.

Burger, K. S., Cornier, M. A., Ingebrigtsen, J., & Johnson, S. L. (2011). Assessing food appeal and desire to eat: The effects of portion size & energy density. *The International Journal of Behavioral Nutrition and Physical Activity, 8*(1), 101. https://doi.org/10.1186/1479-5868-8-101.

Burger, K. S., Sanders, A. J., & Gilbert, J. R. (2016). Hedonic hunger is related to increased neural and perceptual responses to cues of palatable food and motivation to consume: Evidence from 3 independent investigations. *The Journal of Nutrition, 146*(9), 1807–1812.

Burton, P., Smit, H. J., & Lightowler, H. J. (2007). The influence of restrained and external eating patterns on overeating. *Appetite, 49*(1), 191–197.

Caravaggio, F., Borlido, C., Hahn, M., Feng, Z., Fervaha, G., Gerretsen, P., et al. (2015). Reduced insulin sensitivity is related to less endogenous dopamine at d2/3 receptors in the ventral striatum of healthy nonobese humans. *The International Journal of Neuropsychopharmacology / Official Scientific Journal of the Collegium Internationale Neuropsychopharmacologicum (CINP), 18*(7), pyv014. https://doi.org/10.1093/ijnp/pyv014.

Chen, Y.-C., Jiao, Y., Cui, Y., Shang, S.-A., Ding, J., Feng, Y., et al. (2014). Aberrant brain functional connectivity related to insulin resistance in type 2 diabetes: A resting-state fMRI study. *Diabetes Care, 37*(6), 1689–1696. https://doi.org/10.2337/dc13-2127.

Contreras-Rodríguez, O., Martín-Pérez, C., Vilar-López, R., & Verdejo-Garcia, A. (2015). Ventral and dorsal striatum networks in obesity: Link to food craving and weight gain. *Biological Psychiatry.*

Coppin, G., Nolan-Poupart, S., Jones-Gotman, M., & Small, D. M. (2014). Working memory and reward association learning impairments in obesity. *Neuropsychologia, 65*, 146–155. https://doi.org/10.1016/j.neuropsychologia.2014.10.004.

Cornell, C. E., Rodin, J., & Weingarten, H. (1989). Stimulus-induced eating when satiated. *Physiology & Behavior, 45*(4), 695–704.

Cosgrove, K. P., Veldhuizen, M. G., Sandiego, C. M., Morris, E. D., & Small, D. M. (2015). Opposing relationships of BMI with BOLD and dopamine D2/3 receptor binding potential in the dorsal striatum. *Synapse (New York, N.Y.), 69*(4), 195–202. https://doi.org/10.1002/syn.21809.

Davis, C., & Claridge, G. (1998). The eating disorders as addiction: A psychobiological perspective. *Addictive Behaviors, 23*(4), 463–475.

Davis, C., Strachan, S., & Berkson, M. (2004). Sensitivity to reward: Implications for overeating and overweight. *Appetite, 42*(2), 131–138. https://doi.org/10.1016/j.appet.2003.07.004.

Dawe, S., & Loxton, N. J. (2004). The role of impulsivity in the development of substance use and eating disorders. *Neuroscience and Biobehavioral Reviews, 28*(3), 343–351. https://doi.org/10.1016/j.neubiorev.2004.03.007.

Demos, K. E., Heatherton, T. F., & Kelley, W. M. (2012). Individual differences in nucleus accumbens activity to food and sexual images predict weight gain and sexual behavior. *The Journal of Neuroscience, 32*(16), 5549–5552. https://doi.org/10.1523/JNEUROSCI.5958-11.2012.

Diliberti, N., Bordi, P. L., Conklin, M. T., Roe, L. S., & Rolls, B. J. (2004). Increased portion size leads to increased energy intake in a restaurant meal. *Obesity, 12*(3), 562–568.

Dimitropoulos, A., Tkach, J., Ho, A., & Kennedy, J. (2012). Greater corticolimbic activation to high-calorie food cues after eating in obese vs. normal-weight adults. *Appetite, 58*(1), 303–312. https://doi.org/10.1016/j.appet.2011.09.014.

Dovey, T. M., Taylor, L., Stow, R., Boyland, E. J., & Halford, J. C. G. (2011). Responsiveness to healthy television (TV) food advertisements/commercials is only evident in children under the age of seven with low food neophobia. *Appetite, 56*(2), 440–446.

Dunn, J. P., Kessler, R. M., Feurer, I. D., Volkow, N. D., Patterson, B. W., Ansari, M. S., et al. (2012). Relationship of dopamine type 2 receptor binding potential with fasting neuroendocrine hormones and insulin sensitivity in human obesity. *Diabetes Care, 35*(5), 1105–1111. https://doi.org/10.2337/dc11-2250.

Elfhag, K., & Rossner, S. (2005). Who succeeds in maintaining weight loss? A conceptual review of factors associated with weight loss maintenance and weight regain. *Obesity Reviews, 6*(1), 67–85. https://doi.org/10.1111/j.1467-789X.2005.00170.x.

Elfhag, K., Tynelius, P., & Rasmussen, F. (2007). Sugar-sweetened and artificially sweetened soft drinks in association to restrained, external and emotional eating. *Physiology & Behavior, 91*(2), 191–195.

Elks, C. E., den Hoed, M., Zhao, J. H., Sharp, S. J., Wareham, N. J., Loos, R. J. F., & Ong, K. K. (2012). Variability in the heritability of body mass index: A systematic review and meta-regression. *Frontiers in Endocrinology, 3*, 29. Retrieved from http://www.ncbi.nlm.nih.gov/pubmed/22645519.

Ello-Martin, J. A., Ledikwe, J. H., & Rolls, B. J. (2005). The influence of food portion size and energy density on energy intake: Implications for weight management. *The American Journal of Clinical Nutrition, 82*(1), 236S–241S.

Ello-Martin, J. A., Roe, L. S., Ledikwe, J. H., Beach, A. M., & Rolls, B. J. (2007). Dietary energy density in the treatment of obesity: A year-long trial comparing 2 weight-loss diets. *The American Journal of Clinical Nutrition, 85*(6), 1465–1477.

Ely, A. V., Howard, J., & Lowe, M. R. (2015). Delayed discounting and hedonic hunger in the prediction of lab-based eating behavior. *Eating Behaviors, 19*, 72–75.

English, P. J., Ghatei, M. A., Malik, I. A., Bloom, S. R., & Wilding, J. P. H. (2002). Food fails to suppress ghrelin levels in obese humans. *Journal of Clinical Endocrinology & Metabolism, 87*(6), 2984.

Epstein, L. H., Temple, J. L., Neaderhiser, B. J., Salis, R. J., Erbe, R. W., & Leddy, J. J. (2007). Food reinforcement, the dopamine D2 receptor genotype, and energy intake in obese and nonobese humans. *Behavioral Neuroscience, 121*(5), 877–886. https://doi.org/10.1037/0735-7044.121.5.877.

Faith, M. S., Berkowitz, R. I., Stallings, V. A., Kerns, J., Storey, M., & Stunkard, A. J. (2006). Eating in the absence of hunger: A genetic marker for childhood obesity in prepubertal boys? *Obesity, 14*(1), 131–138.

Farooqi, I. S., Bullmore, E., Keogh, J., Gillard, J., O'Rahilly, S., & Fletcher, P. C. (2007). Leptin regulates striatal regions and human eating behavior. *Science, 317*(5843), 1355. https://doi.org/10.1126/science.1144599.

Farr, O. M., Sofopoulos, M., Tsoukas, M. A., Dincer, F., Thakkar, B., Sahin-Efe, A., et al. (2016a). GLP-1 receptors exist in the parietal cortex, hypothalamus and medulla of human brains and the GLP-1 analogue liraglutide alters brain activity related to highly desirable food cues in individuals with diabetes: A crossover, randomised, placebo-controlled. *Diabetologia, 59*(5), 954–965. https://doi.org/10.1007/s00125-016-3874-y.

Farr, O. M., Tsoukas, M. A., Triantafyllou, G., Dincer, F., Filippaios, A., Ko, B.-J., & Mantzoros, C. S. (2016b). Short-term administration of the GLP-1 analog liraglutide decreases circulating leptin and increases GIP levels and these changes are associated with alterations in CNS responses to food cues: A randomized, placebo-controlled, crossover study. *Metabolism, 65*(7), 945–953. https://doi.org/10.1016/j.metabol.2016.03.009.

Feldstein Ewing, S. W., Claus, E. D., Hudson, K. A., Filbey, F. M., Yakes Jimenez, E., Lisdahl, K. M., & Kong, A. S. (2016). Overweight adolescents' brain response to sweetened beverages mirrors addiction pathways. *Brain Imaging and Behavior.* https://doi.org/10.1007/s11682-016-9564-z.

Felsted, J. A., Ren, X., Chouinard-Decorte, F., & Small, D. M. (2010). Genetically determined differences in brain response to a primary food reward. *Journal of Neuroscience, 30*(7), 2428–2432. https://doi.org/10.1523/jneurosci.5483-09.2010.

Fetissov, S. O., Meguid, M. M., Sato, T., & Zhang, L.-H. (2002). Expression of dopaminergic receptors in the hypothalamus of lean and obese Zucker rats and food intake. *American Journal of Physiology-Regulatory, Integrative and Comparative Physiology, 283*(4), R905–R910. Retrieved from http://ajpregu.physiology.org/content/ajpregu/283/4/R905.full.pdf.

Figlewicz, D. P., Szot, P., Chavez, M., Woods, S. C., & Veith, R. C. (1994). Intraventricular insulin increases dopamine transporter mRNA in rat VTA/substantia nigra. *Brain Research, 644*(2), 331–334.

Fisher, J. O., & Kral, T. V. E. (2008). Super-size me: Portion size effects on young children's eating. *Physiology & Behavior, 94*(1), 39–47.

Fisher, J. O., Rolls, B. J., & Birch, L. L. (2003). Children's bite size and intake of an entree are greater with large portions than with age-appropriate or self-selected portions. *The American Journal of Clinical Nutrition, 77*(5), 1164–1170.

Flegal, K. M., Carroll, M. D., Kit, B. K., & Ogden, C. L. (2012). Prevalence of obesity and trends in the distribution of body mass index among US adults, 1999-2010. *Jama-Journal of the American Medical Association, 307*(5), 491–497. https://doi.org/10.1001/jama.2012.39.

Flegal, K. M., Kruszon-Moran, D., Carroll, M. D., Fryar, C. D., & Ogden, C. L. (2016). Trends in obesity among adults in the United States, 2005 to 2014. *JAMA, 315*(21), 2284. https://doi.org/10.1001/jama.2016.6458.

Frank, M. J., Moustafa, A. A., Haughey, H. M., Curran, T., & Hutchison, K. E. (2007). Genetic triple dissociation reveals multiple roles for dopamine in reinforcement learning. *Proceedings of the National Academy of Sciences of the United States of America, 104*(41), 16311–16316. https://doi.org/10.1073/pnas.0706111104.

Frank, G. K. W., Reynolds, J. R., Shott, M. E., Jappe, L., Yang, T. T., Tregellas, J. R., & O'Reilly, R. C. (2012). Anorexia nervosa and obesity are associated with opposite brain reward response. *Neuropsychopharmacology, 37*(9), 2031–2046. Retrieved from http://www.ncbi.nlm.nih.gov/pmc/articles/PMC3398719/pdf/npp201251a.pdf.

Frankort, A., Roefs, A., Siep, N., Roebroeck, A., Havermans, R., & Jansen, A. (2012). Reward activity in satiated overweight women is decreased during unbiased viewing but increased when imagining taste: An event-related fMRI study. *International Journal of Obesity, 36*(5), 627–637. https://doi.org/10.1038/ijo.2011.213.

Gallwitz, B. (2012). Anorexigenic effects of GLP-1 and its analogues. In *Handbook of experimental pharmacology* (pp. 185–207). https://doi.org/10.1007/978-3-642-24716-3_8.

Geha, P. Y., Aschenbrenner, K., Felsted, J., O'Malley, S. S., & Small, D. M. (2013). Altered hypothalamic response to food in smokers. *The American Journal of Clinical Nutrition, 97*(1), 15–22. Retrieved from http://www.ncbi.nlm.nih.gov/pmc/articles/PMC3522134/pdf/ajcn97115.pdf.

Gilbert, J. R., & Burger, K. S. (2016). Neuroadaptive processes associated with palatable food intake: Present data and future directions. *Current Opinion in Behavioral Sciences, 9*, 91–96.

Goldstone, A. P., Prechtl, C. G., Scholtz, S., Miras, A. D., Chhina, N., Durighel, G., et al. (2014). Ghrelin mimics fasting to enhance human hedonic, orbitofrontal cortex, and hippocampal responses to food. *The American Journal of Clinical Nutrition, 99*(6), 1319–1330. https://doi.org/10.3945/ajcn.113.075291.

Green, E., & Murphy, C. (2012). Altered processing of sweet taste in the brain of diet soda drinkers. *Physiology & Behavior, 107*(4), 560–567.

Green, E., Jacobson, A., Haase, L., & Murphy, C. (2011). Reduced nucleus accumbens and caudate nucleus activation to a pleasant taste is associated with obesity in older adults. *Brain Research, 1386*, 109–117. Retrieved from http://www.ncbi.nlm.nih.gov/pmc/articles/PMC3086067/pdf/nihms284828.pdf.

de Groot, C., Felius, A., Trompet, S., de Craen, A. J. M., Blauw, G. J., van Buchem, M. A., et al. (2015). Association of the fat mass and obesity-associated gene risk allele, rs9939609A, and reward-related brain structures. *Obesity, 23*(10), 2118–2122. https://doi.org/10.1002/oby.21191.

Grosshans, M., Vollmert, C., Vollstädt-Klein, S., Tost, H., Leber, S., Bach, P., et al. (2012). Association of leptin with food cue-induced activation in human reward pathways. *Archives of General Psychiatry, 69*(5), 529–537. https://doi.org/10.1001/archgenpsychiatry.2011.1586.

Haan, M. N. (2006). Therapy insight: Type 2 diabetes mellitus and the risk of late-onset Alzheimer's disease. *Nature Clinical Practice. Neurology, 2*(3), 159–166. https://doi.org/10.1038/ncpneuro0124.

Halford, J. C. G., Boyland, E. J., Hughes, G., Oliveira, L. P., & Dovey, T. M. (2007). Beyond-brand effect of television (TV) food advertisements/commercials on caloric intake and food choice of 5–7-year-old children. *Appetite, 49*(1), 263–267.

Hallschmid, M., Higgs, S., Thienel, M., Ott, V., & Lehnert, H. (2012). Postprandial administration of intranasal insulin intensifies satiety and reduces intake of palatable snacks in women. *Diabetes, 61*(4), 782–789. https://doi.org/10.2337/db11-1390.

Harris, R. B. (1990). Role of set-point theory in regulation of body weight. *The FASEB Journal, 4*(15), 3310–3318.

Harris, J. L., Bargh, J. A., & Brownell, K. D. (2009). *Priming Effects of Television Food Advertising on Eating Behavior, 28*(4), 404–413. https://doi.org/10.1037/a0014399.

Heni, M., Kullmann, S., Ketterer, C., Guthoff, M., Bayer, M., Staiger, H., et al. (2014a). Differential effect of glucose ingestion on the neural processing of food stimuli in lean and overweight adults. *Human Brain Mapping, 35*(3), 918–928. https://doi.org/10.1002/hbm.22223.

Heni, M., Kullmann, S., Veit, R., Ketterer, C., Frank, S., Machicao, F., et al. (2014b). Variation in the obesity risk gene FTO determines the postprandial cerebral processing of food stimuli in the prefrontal cortex. *Molecular Metabolism, 3*(2), 109–113. https://doi.org/10.1016/j.molmet.2013.11.009.

Heni, M., Kullmann, S., Gallwitz, B., Häring, H.-U., Preissl, H., & Fritsche, A. (2015). Dissociation of GLP-1 and insulin association with food processing in the brain: GLP-1 sensitivity despite insulin resistance in obese humans. *Molecular Metabolism, 4*(12), 971–976. https://doi.org/10.1016/j.molmet.2015.09.007.

Heni, M., Wagner, R., Kullmann, S., Gancheva, S., Roden, M., Peter, A., et al. (2017). Hypothalamic and Striatal insulin action suppresses endogenous glucose production and may stimulate glucose uptake during Hyperinsulinemia in lean but not in overweight men. *Diabetes*, db161380. https://doi.org/10.2337/db16-1380.

Herman, C. P., & Mack, D. (1975). Restrained and unrestrained eating. *Journal of Personality, 43*(4), 647–660.

Herman, C. P., & Polivy, J. (2008). External cues in the control of food intake in humans: The sensory-normative distinction. *Physiology & Behavior, 94*(5), 722–728.

Hess, M. E., Hess, S., Meyer, K. D., Verhagen, W. L. A., Koch, L., Brönneke, H. S., et al. (2013). The fat mass and obesity associated gene (FTO) regulates activity of the dopaminergic midbrain circuitry. *Nature Publishing Group, 16*. https://doi.org/10.1038/nn.3449.

Huang, X.-F., Zavitsanou, K., Huang, X., Yu, Y., Wang, H., Chen, F., et al. (2006). Dopamine transporter and D2 receptor binding densities in mice prone or resistant to chronic high fat diet-induced obesity. *Behavioural Brain Research, 175*(2), 415–419.

Hui, H., Farilla, L., Merkel, P., & Perfetti, R. (2002). The short half-life of glucagon-like peptide-1 in plasma does not reflect its long-lasting beneficial effects. *European Journal of Endocrinology, 146*(6), 863–869.

Jacobson, P., Torgerson, J. S., Sjostrom, L., & Bouchard, C. (2006). Spouse resemblance in body mass index: Effects on adult obesity prevalence in the offspring generation. *American Journal of Epidemiology, 165*(1), 101–108. https://doi.org/10.1093/aje/kwj342.

Jastreboff, A. M., Sinha, R., Lacadie, C., Small, D. M., Sherwin, R. S., & Potenza, M. N. (2013). Neural correlates of stress- and food cue-induced food craving in obesity: Association with insulin levels. *Diabetes Care, 36*(2), 394–402. https://doi.org/10.2337/dc12-1112.

Jastreboff, A. M., Lacadie, C., Seo, D., Kubat, J., Van Name, M. A., Giannini, C., et al. (2014). Leptin is associated with exaggerated brain reward and emotion responses to food images in adolescent obe-

sity. *Diabetes Care, 37*(11), 3061–3068. https://doi.org/10.2337/dc14-0525.

Jauch-Chara, K., Friedrich, A., Rezmer, M., Melchert, U. H., Scholand-Engler, G., Hallschmid, H. M., & Oltmanns, K. M. (2012). Intranasal insulin suppresses food intake via enhancement of brain energy levels in humans. *Diabetes, 61*(9), 2261–2268. https://doi.org/10.2337/db12-0025.

Jia, G., Fu, Y., Zhao, X., Dai, Q., Zheng, G., Yang, Y., et al. (2011). N6-methyladenosine in nuclear RNA is a major substrate of the obesity-associated FTO. *Nature Chemical Biology, 7*(12), 885–887. https://doi.org/10.1038/nchembio.687.

Johnson, P. M., & Kenny, P. J. (2010). Dopamine D2 receptors in addiction-like reward dysfunction and compulsive eating in obese rats. *Nature Neuroscience, 13*(5), 635–641.

Johnson, F., & Wardle, J. (2014). Variety, palatability, and obesity. *Advances in Nutrition (Bethesda, Md.), 5*(6), 851–859. https://doi.org/10.3945/an.114.007120.

Johnson, F., Pratt, M., & Wardle, J. (2012). Dietary restraint and self-regulation in eating behavior. *International Journal of Obesity, 36*(5), 665–674.

Kalra, S. P. (2001). Circumventing leptin resistance for weight control. *Proceedings of the National Academy of Sciences of the United States of America, 98*(8), 4279–4281. https://doi.org/10.1073/pnas.091101498.

Karra, E., O'Daly, O. G., Choudhury, A. I., Yousseif, A., Millership, S., Neary, M. T., et al. (2013). A link between FTO, ghrelin, and impaired brain food-cue responsivity. *The Journal of Clinical Investigation, 123*(8), 3539–3551. https://doi.org/10.1172/JCI44403.

Kirsch, P., Reuter, M., Mier, D., Lonsdorf, T., Stark, R., Gallhofer, B., et al. (2006). Imaging gene-substance interactions: The effect of the DRD2 TaqIA polymorphism and the dopamine agonist bromocriptine on the brain activation during the anticipation of reward. *Neuroscience Letters, 405*(3), 196–201. https://doi.org/10.1016/j.neulet.2006.07.030.

Kling, S. M. R., Roe, L. S., Keller, K. L., & Rolls, B. J. (2016). Double trouble: Portion size and energy density combine to increase preschool children's lunch intake. *Physiology & Behavior, 162*, 18–26.

Koordeman, R., Anschutz, D. J., van Baaren, R. B., & Engels, R. C. M. E. (2010). Exposure to soda commercials affects sugar-sweetened soda consumption in young women. An observational experimental study. *Appetite, 54*(3), 619–622.

Kroemer, N. B., & Small, D. M. (2016). Fuel not fun: Reinterpreting attenuated brain responses to reward in obesity. *Physiology & Behavior, 162*, 37–45. https://doi.org/10.1016/j.physbeh.2016.04.020.

Kroemer, N. B., Krebs, L., Kobiella, A., Grimm, O., Vollstädt-Klein, S., Wolfensteller, U., et al. (2013). (still) longing for food: Insulin reactivity modulates response to food pictures. *Human Brain Mapping, 34*(10), 2367–2380. https://doi.org/10.1002/hbm.22071.

Kullmann, S., Heni, M., Veit, R., Ketterer, C., Schick, F., Häring, H.-U. H. H.-U., et al. (2012). The obese brain:

Association of body mass index and insulin sensitivity with resting state network functional connectivity. *Human Brain Mapping, 33*(5), 1052–1061. https://doi.org/10.1002/hbm.21268.

Kumar, S., Raju, M., & Gowda, N. (2010). Influence of parental obesity on school children. *The Indian Journal of Pediatrics, 77*(3), 255–258. https://doi.org/10.1007/s12098-010-0015-3.

Kunath, N., Müller, N. C. J., Tonon, M., Konrad, B. N., Pawlowski, M., Kopczak, A., et al. (2016). Ghrelin modulates encoding-related brain function without enhancing memory formation in humans. *NeuroImage, 142*, 465–473. https://doi.org/10.1016/j.neuroimage.2016.07.016.

Langeveld, M., & DeVries, J. H. (2015). The long-term effect of energy restricted diets for treating obesity. *Obesity, 23*(8), 1529–1538. https://doi.org/10.1002/oby.21146.

Ledikwe, J. H., Blanck, H. M., Khan, L. K., Serdula, M. K., Seymour, J. D., Tohill, B. C., & Rolls, B. J. (2006). Low-energy-density diets are associated with high diet quality in adults in the United States. *Journal of the American Dietetic Association, 106*(8), 1172–1180.

Ledikwe, J. H., Rolls, B. J., Smiciklas-Wright, H., Mitchell, D. C., Ard, J. D., Champagne, C., et al. (2007). Reductions in dietary energy density are associated with weight loss in overweight and obese participants in the PREMIER trial. *The American Journal of Clinical Nutrition, 85*(5), 1212–1221.

Leibel, R. L. (2008). Molecular physiology of weight regulation in mice and humans. *International Journal of Obesity, 32*, S98–S108.

Logothetis, N. K., Pauls, J., Augath, M., Trinath, T., & Oeltermann, A. (2001). Neurophysiological investigation of the basis of the fMRI signal. *Nature, 412* (6843), 150.

Loos, R. J. F., & Yeo, G. S. H. (2014). The bigger picture of FTO: The first GWAS-identified obesity gene. *Nature Reviews. Endocrinology, 10*(1), 51–61. https://doi.org/10.1038/nrendo.2013.227.

Lowe, M. R., & Butryn, M. L. (2007). Hedonic hunger: A new dimension of appetite? *Physiology & Behavior, 91*(4), 432–439. https://doi.org/10.1016/j.physbeh.2007.04.006.

Lowe, M. R., Butryn, M. L., Didie, E. R., Annunziato, R. A., Thomas, J. G., Crerand, C. E., et al. (2009). The power of food scale. A new measure of the psychological influence of the food environment. *Appetite, 53*(1), 114–118. https://doi.org/10.1016/j.appet.2009.05.016.

Lowe, M. R., Arigo, D., Butryn, M. L., Gilbert, J. R., Sarwer, D., & Stice, E. (2016). Hedonic hunger prospectively predicts onset and maintenance of loss of control eating among college women. *Health Psychology, 35*(3), 238.

MacLean, P. S., Wing, R. R., Davidson, T., Epstein, L., Goodpaster, B., Hall, K. D., et al. (2015). NIH working group report: Innovative research to improve maintenance of weight loss. *Obesity, 23*(1), 7–15. https://doi.org/10.1002/oby.20967.

Malik, S., McGlone, F., Bedrossian, D., & Dagher, A. (2008). Ghrelin modulates brain activity in areas that control appetitive behavior. *Cell Metabolism, 7*(5), 400–409. Retrieved from http://ac.els-cdn.com/S1550413108000788/1-s2.0-S1550413108000788-main.pdf?_tid=443beeaa-e1de-11e3-8bb1-00000aab0f6c&acdnat=1400783229_25281b85b5a58e9889dc4a0aa30cb36d.

Man, M. S., Clarke, H. F., & Roberts, A. C. (2009). The role of the orbitofrontal cortex and medial striatum in the regulation of prepotent responses to food rewards. *Cerebral Cortex, 19*(4), 899–906. Retrieved from http://cercor.oxfordjournals.org/content/19/4/899.full.pdf.

Martin, C. K., Coulon, S. M., Markward, N., Greenway, F. L., & Anton, S. D. (2009a). Association between energy intake and viewing television, distractibility, and memory for advertisements. *The American Journal of Clinical Nutrition, 89*(1), 37–44.

Martin, L. E., Holsen, L. M., Chambers, R. J., Bruce, A. S., Brooks, W. M., Zarcone, J. R., et al. (2009b). Neural mechanisms associated with food motivation in obese and healthy weight adults. *Obesity, 18*(2), 254–260. Retrieved from http://onlinelibrary.wiley.com/store/10.1038/oby.2009.220/asset/oby.2009.220.pdf?v=1&t=hvie1us5&s=9e7326e9ee6e5317496a95a030883ab36244d626.

Mathar, D., Neumann, J., Villringer, A., & Horstmann, A. (2017). Failing to learn from negative prediction errors: Obesity is associated with alterations in a fundamental neural learning mechanism. *Cortex.* https://doi.org/10.1016/j.cortex.2017.08.022.

Mayer, J. (1953). Glucostatic mechanism of regulation of food intake. *The New England Journal of Medicine, 249*(1), 13–16. https://doi.org/10.1056/NEJM195307022490104.

Merchenthaler, I., Lane, M., & Shughrue, P. (1999). Distribution of pre-pro-glucagon and glucagon-like peptide-1 receptor messenger RNAs in the rat central nervous system. *The Journal of Comparative Neurology, 403*(2), 261–280. https://doi.org/10.1002/(SICI)1096-9861(19990111)403:2<261::AID-CNE8>3.0.CO;2-5.

Miller, E. K., & Cohen, J. D. (2001). An integrative theory of prefrontal cortex function. *Annual Review of Neuroscience, 24*(1), 167–202. https://doi.org/10.1146/annurev.neuro.24.1.167.

Murray, S., Tulloch, A., Gold, M. S., & Avena, N. M. (2014). Hormonal and neural mechanisms of food reward, eating behaviour and obesity. *Nature Reviews. Endocrinology, 10*(9), 540–552. https://doi.org/10.1038/nrendo.2014.91.

ten Kulve, J. S., Veltman, D. J., van Bloemendaal, L., Groot, P. F. C., Ruhé, H. G., Barkhof, F., et al. (2016). Endogenous GLP1 and GLP1 analogue alter CNS responses to palatable food consumption. *Journal of Endocrinology, 229*(1), 1–12. https://doi.org/10.1530/JOE-15-0461.

Ng, J., Stice, E., Yokum, S., & Bohon, C. (2011). An fMRI study of obesity, food reward, and perceived caloric density. Does a low-fat label make food less appealing? *Appetite, 57*(1), 65–72. https://doi.org/10.1016/j.appet.2011.03.017.

Nummenmaa, L., Hirvonen, J., Hannukainen, J. C., Immonen, H., Lindroos, M. M., Salminen, P., & Nuutila, P. (2012). Dorsal striatum and its limbic connectivity mediate abnormal anticipatory reward processing in obesity. *PLoS One, 7*(2). https://doi.org/10.1371/journal.pone.0031089.

Page, K. A., Chan, O., Arora, J., Belfort-Deaguiar, R., Dzuira, J., Roehmholdt, B., et al. (2013). Effects of fructose vs glucose on regional cerebral blood flow in brain regions involved with appetite and reward pathways. *JAMA, 309*(1), 63–70. https://doi.org/10.1001/jama.2012.116975.

Pliquett, R. U., Führer, D., Falk, S., Zysset, S., von Cramon, D. Y., & Stumvoll, M. (2006). The effects of insulin on the central nervous system--focus on appetite regulation. *Hormone and Metabolic Research = Hormon- Und Stoffwechselforschung = Hormones et Métabolisme, 38*(7), 442–446. https://doi.org/10.1055/s-2006-947840.

Polivy, J., & Herman, C. P. (1985). Dieting and binging: A causal analysis. *American Psychologist, 40*(2), 193.

Porubská, K., Veit, R., Preissl, H., Fritsche, A., & Birbaumer, N. (2006). Subjective feeling of appetite modulates brain activity: An fMRI study. *NeuroImage, 32*(3), 1273–1280. https://doi.org/10.1016/j.neuroimage.2006.04.216.

Potter, G. M., Moshirfar, A., & Castonguay, T. W. (1999). Insulin affects dopamine overflow in the nucleus accumbens and the striatum. *Physiology & Behavior, 65*(4–5), 811–816.

Powell, L. M., Szczypka, G., Chaloupka, F. J., & Braunschweig, C. L. (2007). Nutritional content of television food advertisements seen by children and adolescents in the United States. *Pediatrics, 120*(3), 576–583. https://doi.org/10.1542/peds.2006-3595.

Rath, S. R., Marsh, J. A., Newnham, J. P., Zhu, K., Atkinson, H. C., Mountain, J., et al. (2016). Parental pre-pregnancy BMI is a dominant early-life risk factor influencing BMI of offspring in adulthood. *Obesity Science & Practice, 2*(1), 48–57. https://doi.org/10.1002/osp4.28.

Rogers, P. J., & Brunstrom, J. M. (2016). Appetite and energy balancing. *Physiology & Behavior.* https://doi.org/10.1016/j.physbeh.2016.03.038.

Rolls, B. J., & Barnett, R. A. (2000). *Volumetrics.* HarperCollins.

Rolls, B. J., Rowe, E. A., & Rolls, E. T. (1982). How flavour and appearance affect human feeding. *Proceedings of the Nutrition Society, 41*(2), 109–117.

Rolls, B. J., Morris, E. L., & Roe, L. S. (2002). Portion size of food affects energy intake in normal-weight and overweight men and women. *The American Journal of Clinical Nutrition, 76*(6), 1207–1213.

Rolls, B. J., Roe, L. S., Meengs, J. S., & Wall, D. E. (2004a). Increasing the portion size of a sandwich increases energy intake. *Journal of the American Dietetic Association, 104*(3), 367–372.

Rolls, B. J., Roe, L. S., Kral, T. V. E., Meengs, J. S., & Wall, D. E. (2004b). Increasing the portion size of a packaged snack increases energy intake in men and women. *Appetite, 42*(1), 63–69.

Rolls, B. J., Roe, L. S., & Meengs, J. S. (2006). Larger portion sizes lead to a sustained increase in energy intake over 2 days. *Journal of the American Dietetic Association, 106*(4), 543–549.

Rolls, B. J., Roe, L. S., James, B. L., & Sanchez, C. E. (2017). Does the incorporation of portion-control strategies in a behavioral program improve weight loss in a one-year randomized controlled trial? *International Journal of Obesity (2005), 41*(3), 434.

Rothemund, Y., Preuschhof, C., Bohner, G., Bauknecht, H.-C., Klingebiel, R., Flor, H., & Klapp, B. F. (2007). Differential activation of the dorsal striatum by high-calorie visual food stimuli in obese individuals. *NeuroImage, 37*(2), 410–421. https://doi.org/10.1016/j.neuroimage.2007.05.008.

Rudenga, K. J., & Small, D. M. (2012). Amygdala response to sucrose consumption is inversely related to artificial sweetener use. *Appetite, 58*(2), 504–507. https://doi.org/10.1016/j.appet.2011.12.001.

Salbe, A. D., Weyer, C., Harper, I., Lindsay, R. S., Ravussin, E., & Tataranni, P. A. (2002). Assessing risk factors for obesity between childhood and adolescence : II. *Energy Metabolism and Physical Activity, 110*(2), 307–314.

Sandoval, D., & Sisley, S. R. (2015). Brain GLP-1 and insulin sensitivity. *Molecular and Cellular Endocrinology.* https://doi.org/10.1016/j.mce.2015.02.017.

Schlogl, H., Kabisch, S., Horstmann, A., Lohmann, G., Muller, K., Lepsien, J., et al. (2013). Exenatide-induced reduction in energy intake is associated with increase in hypothalamic connectivity. *Diabetes Care, 36*(7), 1933–1940. https://doi.org/10.2337/dc12-1925.

Schultes, B., Ernst, B., Wilms, B., Thurnheer, M., & Hallschmid, M. (2010). Hedonic hunger is increased in severely obese patients and is reduced after gastric bypass surgery. *The American Journal of Clinical Nutrition, 92*, 277–283. https://doi.org/10.3945/ajcn.2009.29007.INTRODUCTION.

Seppälä-Lindroos, A., Vehkavaara, S., Häkkinen, A.-M., Goto, T., Westerbacka, J., Sovijärvi, A., et al. (2002). Fat accumulation in the liver is associated with defects in insulin suppression of glucose production and serum free fatty acids independent of obesity in normal men. *The Journal of Clinical Endocrinology & Metabolism, 87*(7), 3023–3028. https://doi.org/10.1210/jcem.87.7.8638.

Sevgi, M., Rigoux, L., Kuhn, A. B., Mauer, J., Schilbach, L., Hess, M. E., et al. (2015). An obesity-predisposing variant of the FTO gene regulates D2R-dependent reward learning. *Journal of Neuroscience, 35*(36), 12584–12592. https://doi.org/10.1523/JNEUROSCI.1589-15.2015.

Shearrer, G., Stice, E., & Burger, K. (2017). Adolescents with versus without parental obesity show greater striatal response to increased sugar, but not fat content of milkshakes. https://doi.org/10.17605/OSF.IO/7J4EH.

Simon, J. J., Skunde, M., Hamze Sinno, M., Brockmeyer, T., Herpertz, S. C., Bendszus, M., et al. (2014). Impaired cross-talk between Mesolimbic food reward processing and metabolic signaling predicts body mass index. *Frontiers in Behavioral Neuroscience, 8*, 359. https://doi.org/10.3389/fnbeh.2014.00359.

Small, C. J., & Bloom, S. R. (2004). Gut hormones as peripheral anti obesity targets. *Current Drug Targets. CNS and Neurological Disorders, 3*(5), 379–388.

Stice, E., Spoor, S., Bohon, C., Veldhuizen, M. G., & Small, D. M. (2008a). Relation of reward from food intake and anticipated food intake to obesity: A functional magnetic resonance imaging study. *Journal of Abnormal Psychology, 117*(4), 924–935. https://doi.org/10.1037/a0013600.

Stice, E., Spoor, S., Bohon, C., & Small, D. M. (2008b). Relation between obesity and blunted striatal response to food is moderated by TaqIA A1 allele. *Science, 322*(5900), 449–452. https://doi.org/10.1126/science.1161550.

Stice, E., Yokum, S., Bohon, C., Marti, N., & Smolen, A. (2010a). Reward circuitry responsivity to food predicts future increases in body mass: Moderating effects of DRD2 and DRD4. *NeuroImage, 50*(4), 1618–1625. https://doi.org/10.1016/j.neuroimage.2010.01.081.

Stice, E., Yokum, S., Blum, K., & Bohon, C. (2010b). Weight gain is associated with reduced Striatal response to palatable food. *Journal of Neuroscience, 30*(39), 13105–13109. https://doi.org/10.1523/jneurosci.2105-10.2010.

Stice, E., Yokum, S., Burger, K. S., Epstein, L. H., & Small, D. M. (2011). Youth at risk for obesity show greater activation of striatal and somatosensory regions to food. *The Journal of Neuroscience, 31*(12), 4360–4366. https://doi.org/10.1523/JNEUROSCI.6604-10.2011.

Stice, E., Burger, K. S., & Yokum, S. (2015). Reward region Responsivity predicts future weight gain and moderating effects of the TaqIA allele. *The Journal of Neuroscience, 35*(28), 10316–10324.

Stoeckel, L. E., Weller, R. E., Cook, E. W., III, Twieg, D. B., Knowlton, R. C., & Cox, J. E. (2008). Widespread reward-system activation in obese women in response to pictures of high-calorie foods. *NeuroImage, 41*(2), 636–647. https://doi.org/10.1016/j.neuroimage.2008.02.031.

Sun, X., Veldhuizen, M. G., Wray, A. E., de Araujo, I. E., Sherwin, R. S., Sinha, R., & Small, D. M. (2014). The neural signature of satiation is associated with ghrelin response and triglyceride metabolism. *Physiology & Behavior, 136*, 63–73. https://doi.org/10.1016/j.physbeh.2014.04.017.

Sun, X., Kroemer, N. B., Veldhuizen, M. G., Babbs, A. E., de Araujo, I. E., Gitelman, D. R., et al. (2015). Basolateral Amygdala response to food cues in the absence of hunger is associated with weight gain susceptibility. *The Journal of Neuroscience, 35*(20), 7964–7976.

Sun, X., Veldhuizen, M. G., Babbs, A. E., Sinha, R., & Small, D. M. (2016). Perceptual and brain response to odors is associated with body mass index and postprandial Total Ghrelin reactivity to a meal. *Chemical Senses, 41*(3), 233–248. https://doi.org/10.1093/chemse/bjv081.

Sun, X., Luquet, S., & Small, D. M. (2017). DRD2: Bridging the genome and Ingestive behavior. *Trends in Cognitive Sciences, 21*(5), 372–384. https://doi.org/10.1016/j.tics.2017.03.004.

Svetkey, L. P., Stevens, V. J., Brantley, P. J., Appel, L. J., Hollis, J. F., Loria, C. M., et al. (2008). Comparison of strategies for sustaining weight loss: The weight loss maintenance randomized controlled trial. *JAMA, 299*(10), 1139–1148.

Thanos, P. K., Michaelides, M., Piyis, Y. K., Wang, G.-J., & Volkow, N. D. (2008). Food restriction markedly increases dopamine d2 receptor (D2R) in a rat model of obesity as assessed with in-vivo mu PET imaging (C-11 raclopride) and in-vitro (H-3 spiperone) autoradiography. *Synapse, 62*(1), 50–61. https://doi.org/10.1002/syn.20468.

Tomkin, G. H. (2014). Treatment of type 2 diabetes, lifestyle, GLP1 agonists and DPP4 inhibitors. *World Journal of Diabetes, 5*(5), 636–650. https://doi.org/10.4239/wjd.v5.i5.636.

Tschöp, M., Weyer, C., Tataranni, P. A., Devanarayan, V., Ravussin, E., & Heiman, M. L. (2001). Circulating ghrelin levels are decreased in human obesity. *Diabetes, 50*(4), 707–709.

Ubeda-Bañon, I., Novejarque, A., Mohedano-Moriano, A., Pro-Sistiaga, P., de la Rosa-Prieto, C., Insausti, R., et al. (2007). Projections from the posterolateral olfactory amygdala to the ventral striatum: Neural basis for reinforcing properties of chemical stimuli. *BMC Neuroscience, 8*(1), 103. https://doi.org/10.1186/1471-2202-8-103.

Van Strien, T., Frijters, J. E. R., Bergers, G., Defares, P. B., Van Strien, T., Frijters, J. E. R., et al. (1986). The Dutch eating behavior questionnaire (DEBQ) for assessment of restrained, emotional, and external eating behavior. *International Journal of Eating Disorders, 5*(2), 295–315. https://doi.org/10.1002/1098-108X(198602)5:2<295::AID-EAT2260050209>3.0.CO;2-T.

Van Strien, T., Herman, C. P., & Verheijden, M. W. (2009). Eating style, overeating, and overweight in a representative Dutch sample. Does external eating play a role? *Appetite, 52*(2), 380–387.

Verdejo-García, A., Lawrence, A. J., & Clark, L. (2008). Impulsivity as a vulnerability marker for substance-use disorders: Review of findings from high-risk research, problem gamblers and genetic association studies. *Neuroscience & Biobehavioral Reviews, 32*(4), 777–810.

Verdich, C., Toubro, S., Buemann, B., Lysgård Madsen, J., Juul Holst, J., & Astrup, A. (2001). The role of postprandial releases of insulin and incretin hormones in meal-induced satiety—Effect of obesity and weight reduction. *International Journal of Obesity & Related Metabolic Disorders, 25*(8).

Visser, P. J., Scheltens, P., Verhey, F. R. J., Schmand, B., Launer, L. J., Jolles, J., & Jonker, C. (1999). Medial temporal lobe atrophy and memory dysfunction as predictors for dementia in subjects with mild cognitive impairment. *Journal of Neurology, 246*(6), 477–485. https://doi.org/10.1007/s004150050387.

Volkow, N. D., Wang, G.-J. J., Telang, F., Fowler, J. S., Thanos, P. K., Logan, J., et al. (2008). Low dopamine striatal D2 receptors are associated with prefrontal metabolism in obese subjects: Possible contributing factors. *NeuroImage, 42*(4), 1537–1543. https://doi.org/10.1016/j.neuroimage.2008.06.002.

Vukmirovic, M. (2015). The effects of food advertising on food-related behaviours and perceptions in adults: A review. *Food Research International, 75*, 13–19.

Wallner-Liebmann, S., Koschutnig, K., Reishofer, G., Sorantin, E., Blaschitz, B., Kruschitz, R., et al. (2010). Insulin and hippocampus activation in response to images of high-calorie food in normal weight and obese adolescents. *Obesity (Silver Spring, Md.), 18*(8), 1552–1557. https://doi.org/10.1038/oby.2010.26.

Wang, G.-J., Volkow, N. D., Logan, J., Pappas, N. R., Wong, C. T., Zhu, W., et al. (2001). Brain dopamine and obesity. *The Lancet, 357*(9253), 354–357. https://doi.org/10.1016/S0140-6736(00)03643-6.

Wang, G.-J., Volkow, N. D., Thanos, P. K., & Fowler, J. S. (2004). Similarity between obesity and drug addiction as assessed by neurofunctional imaging: A concept review. *Journal of Addictive Diseases, 23*(3), 39–53.

Wardle, J., Guthrie, C., Sanderson, S., Birch, L., & Plomin, R. (2001). Food and activity preferences in children of lean and obese parents. *International Journal of Obesity, 25*(7), 971–977. https://doi.org/10.1038/sj.ijo.0801661.

de Weijer, B. A., van de Giessen, E., van Amelsvoort, T. A., Boot, E., Braak, B., Janssen, I. M., et al. (2011). Lower striatal dopamine D2/3 receptor availability in obese compared with non-obese subjects. *EJNMMI Research, 1*(1), 1–5. https://doi.org/10.1186/2191-219X-1-37.

Westerterp, K. R. (2010). Physical activity, food intake, and body weight regulation: Insights from doubly labeled water studies. *Nutrition Reviews, 68*(3), 148–154.

Wiemerslage, L., Nilsson, E. K., Solstrand Dahlberg, L., Ence-Eriksson, F., Castillo, S., Larsen, A. L., et al. (2016). An obesity-associated risk allele within the FTO gene affects human brain activity for areas important for emotion, impulse control and reward in response to food images. *European Journal of Neuroscience, 43*(9), 1173–1180. https://doi.org/10.1111/ejn.13177.

Wilcox, C. E., Braskie, M. N., Kluth, J. T., & Jagust, W. J. (2010). Overeating behavior and Striatal dopamine with 6-[1 8 F]-Fluoro-L-m-tyrosine PET. *Journal of Obesity, 2010*.

Witt, A., & Lowe, M. R. (2014). Hedonic hunger and binge eating among women with eating disorders. *The International Journal of Eating Disorders, 47*(3), 273–280. https://doi.org/10.1002/eat.22171.

World Health Organization. (2010). Set of recommendations on the marketing of foods and non-alcoholic beverages to children.

Xia, W., Wang, S., Spaeth, A. M., Rao, H., Wang, P., Yang, Y., et al. (2015). Insulin resistance-associated Interhemispheric functional connectivity alterations in T2DM: A resting-state fMRI study. *BioMed Research International, 2015*, 719076. https://doi.org/10.1155/2015/719076.

Yokum, S., Ng, J., & Stice, E. (2011). Attentional Bias to food images associated with elevated weight and future weight gain: An fMRI study. *Obesity, 19*(9), 1775–1783. https://doi.org/10.1038/oby.2011.168.

Yokum, S., Ng, J., & Stice, E. (2012). Relation of regional gray and white matter volumes to current BMI and future increases in BMI: A prospective MRI study. *International Journal of Obesity (2005), 36*(5), 656–664. https://doi.org/10.1038/ijo.2011.175.

Young, L. R., & Nestle, M. (2002). The contribution of expanding portion sizes to the US obesity epidemic. *American Journal of Public Health, 92*(2), 246–249.

Yu, C., & McClellan, J. (2016). Genetics of substance use disorders. *Child and Adolescent Psychiatric Clinics of North America, 25*(3), 377–385. https://doi.org/10.1016/j.chc.2016.02.002.

Index

A

Acetyl-CoA carboxylase (ACC), 146
Acromegaly, 325–326
ACTH, *see* Adrenocorticotropic hormone
Activating transcription factor 4 (ATF4), 90–91
Activator protein-1 (AP1) complex, 61
Adipokines, 330–332
    anti-inflammatory adipokines
        adiponectin, 239–240
        apelin, 241–242
        BMP, 241
        FGF21, 240
        omentin-1, 241
        SFRP5, 240–241
        vaspin, 242
    neuroendocrine system, 242–243
    pro-inflammatory adipokines
        CCL2 and CCR5, 237–238
        chemerin, 238
        DPP-4, 238
        interleukin-6, 236–237
        interleukin-1β, 237
        leptin, 235–236
        RBP4, 238
        resistin, 239
        TNF-α, 236
Adiponectin, 239–240
Adipose tissue (AT)
    cellular and noncellular composition, WAT/BAT
        adipocytes, 209–210
        beige adipocytes, 211
        ECM, 210–211
        intra-depot adipocyte heterogeneity, 211–212
        lymph and vasculature, 210
        senescent cells, 210
        SVF, 210
    control of lipolysis, 213–214
    depots and depot differences, WAT/BAT
        definition, 207
        in humans and mice, 207, 209
        intra-abdominal depots, 207
        PCOS, 208
        subQ deposition, 207, 208

    ectopic fat deposition/lipodystrophy, 214–215
    endocrine function, 212–213
    growth hormone
        adipokines, 330–332
        body composition, 329–330
        depot-specific manner, 332, 333
        obesity, 332–335
    inflammation
        BAT, 231–232
        hyperlipidemia and hyperglycemia, 231
        immune responses, 232–234
        WAT, 231–232
    lipogenesis, 213–214
    remodeling, 215–217
    types, 205–206
Adrenocorticotropic hormone (ACTH), 39, 76–77, 126
Agouti-related protein (AgRP), 10, 34–36, 67, 90, 142, 272
Agouti signaling protein (ASIP), 272
Allan-Herndon-Dudley syndrome (AHDS), 257
AMP-activated protein kinase (AMPK)
    activation results, 146
    autophagy, 152
    BAT-mediated thermogenesis, 152
    CAMKKβ, 145
    glucose, 151
    ICV infusion, 151
    LKB1, 145
    mammals, 145
Anti-inflammatory adipokines
    adiponectin, 239–240
    apelin, 241–242
    BMP, 241
    FGF21, 240
    omentin-1, 241
    SFRP5, 240–241
    vaspin, 242
Apelin, 241–242
Apoptosis, 13
Appetite regulation
    hedonic motivation
        conflicting theories, 347–348
        dynamic theory, 348–349

Appetite regulation (*cont.*)
    environmental response, 350–351
      food intake environment, 350
      food volume driving intake, 349–350
    homeostatic regulation, 347
Arcuate nucleus (ARC)
    AgRP, 35–36
    α-MSH peptide, 36–37
    in vitro and in vivo experiments, 34
    low leptin signaling, 35
    median eminence, 34
    NPY, 35–36
    NTS, 35
    nutrient sensors, 141
    POMC, 34–37, 126–127
    populations, 34
    pro-NPY, 34
    proTRH, 34, 35
    subset of neurons, 34
ASIP, *see* Agouti signaling protein
Astrocyte-neuron lactate shuttle, 78
Astrocytic end feet, 78
AT, *see* Adipose tissue

B
Basic helix-loop-helix (bHLH), 65
Bed nucleus of the stria terminalis (BNST), 12
Behavioral switch hypothesis, 21
Beta-endorphin
    opioids, 275–277
    orexin-A, 128
    POMC, 277, 278
Blood-brain barrier (BBB), 84
Bone morphogenetic protein (BMP), 241
Bovine GH transgenic (bGH), 327, 328
Brain-derived neurotrophic factor (BDNF), 10
Brefeldin A (BFA), 85
Brown adipose tissue (BAT), 18
    cellular and noncellular composition
      adipocytes, 209–210
      beige adipocytes, 206, 211
      ECM, 210–211
      intra-depot adipocyte heterogeneity, 211–212
      lymph and vasculature, 210
      senescent cells, 210
      SVF, 210
    depots, 209
    endocrine, 212–213
    inflammation, 231–232
    UCP1, 205

C
Carboxypeptidase E (CPE), 272
Carnitine/acylcarnitine transporter (CACT), 261
Carnitine palmitoyltransferase I (CPT1), 146
CASTOR1, 144
CC chemokine ligand type 2 (CCL2), 237–238
CC chemokine receptor type 5 (CCR5), 237–238

Central nervous system (CNS)
    astrocytes, 78–79
    energy homeostasis, 82, 83
    glial cell, 78
    hypothalamus and energy homeostasis
      astrocytes, 81
      BBB, 84
      central inflammation, 80
      diet-induced, 80–81, 83–84
      fasting, 83
      fatty acid-induced inflammation, 81–82
      genetic and pharmacological studies, 84
      glucocorticoids, 82
      HFD feeding, 80–81
      JNK, 80, 82–83
      microglial depletion, 81
      peripheral inflammation, 80
      pro-inflammatory effects, 80
      tanycytes, 83
      TLR4, 83
      TNFα, 79–80
    microglia, 78–79
Chemerin, 238
Cholecystokinin (CCK), 41, 187–189
Chromatin immunoprecipitation (ChIP), 66
c-Jun N-terminal kinases (JNK), 88
Cocaine amphetamine-related transcript (CART) peptide, 10, 38
Colony-stimulating factor 1 receptor (CSF1R), 79
Corticotrophin-releasing hormone (CRH), 38–39, 128–130, 279
Corticotropin-releasing hormone/factor (CRH/F), 76–77
CREB-regulated transcription coactivator 2 (CRTC2), 89
CRISPR/Cas9 promoter analysis methods, 57

D
Diacylglycerol (DAG), 229
Diet-induced obesity (DIO), 36, 158–161
Dipeptidyl peptidase-4 (DPP-4), 238
Diseases of modern civilizations, 5
DNA binding, 57–58
Dorsal raphe (DR), 12
Dorsomedial nucleus (DMN), 35, 40

E
Early growth response protein 1 (Egr-1), 35
4E binding protein 1 (4E-BP1), 144
Ectopic fat deposition, 214–215
Endocrine function, 212–213
Endoplasmic reticulum (ER) stress
    ATF6, 88–89
    BFA, 85
    biological processes, 84
    energy balance
      AgRP, 90
      ATF4, 90–91
      canonical secretory pathway, 90
      diet-induced obesity, 92–93

factors, hormones and neuronal inputs, 89–90
IKK signaling, 93
leptin-responsive cells, 91–92
MTF1/2, 93
UPR, 91
XBP1, 91
function of, 85
intracellular expression profile, 84–85
IRE1, 87–88
neuropeptides, 93–95
NF-κB pathway, 87
PERK, 85–86
proteostatic signaling routes, 89
RSP, 116–117, 122
signaling components, 85
thapsigargin, 85
TUDCA, 85
tunicamycin, 85
uORF1, 86–87
uORF2, 86–87
UPR, 85, 87
XBP1, 87, 88
Energy balance, 8
ER-associated degradation (ERAD) pathways, 85
Extracellular matrix (ECM), 210–211

F
Fat mass and obesity-associated (FTO), 354
Feeding center, 10, 12
Fibroblast growth factor 21 (FGF21), 240
Follicle-stimulating hormone (FSH), 31
Forkhead box protein O1 (FOXO1), 43, 64

G
g-aminobutyric acid (GABA), 35
Gastrointestinal (GI) hormones
biosynthesis, 184
cholecystokinin, 187–189
closed-type cells, 184
features, 194
ghrelin, 185–188
GLP-1, 189–190
GPCRs, 185
G proteins, 185
obesity, 194–197
open-type cells, 184
oxyntomodulin, 192–193
pancreatic polypeptide, 193
PYY, 190–192
secretion, 184
somatostatin, 193
translational modifications, 184
GH-releasing hormone (GHRH), 321
Ghrelin, 185–188
Ghrelin O-acyltransferase (GOAT), 185
Gigantism, 325–326
Glucagon-like peptide 1 (GLP1), 41, 189–190, 360–361
Glucocorticoids (GCs), 39, 82, 280–281

Golgi complex (GC), 119
Gonadotropin-releasing hormone (GnRH), 56–57
G protein-coupled receptors (GPCRs), 184, 185
Granulocytes, 232–233
Growth hormone (GH), 31
Growth hormone deficiency (GHD), 326
Growth hormone/insulin-like growth factor (GH/IGF-1)
            axis
    acromegaly and gigantism, 325–326
    adipose tissue
        adipokines, 330–332
        body composition, 329–330
        depot-specific manner, 332, 333
        obesity, 332–335
    in adulthood, 329
    bGH mice, 327, 328
    GHD, 326
    GHR antagonist transgenic mice, 328
    GHRH secretion, 321, 322
    GHR-/-mice, 328
    human GH, 321
    hypopituitarism, 327–328
    Laron syndrome, 326–327
    mouse lines, 327
    nutrient metabolism, 323–324
    STAT5, 322, 323
Growth hormone secretagogue receptor type 1a (GHSR),
            187
Guanosine triphosphate (GTP), 185

H
High-fat diet (HFD) feeding, 80–81
Highly active antiretroviral therapy (HAART), 214
Histone acetyltransferases (HATs), 147
Hominoids, 5
HPA axis, *see* Hypothalamic-pituitary-adrenal axis
Human GH (hGH), 321
Human nutrition
    behavioral switch hypothesis, 21
    brain evolution and changes
        African human fossils, 5
        ancestral hominin diets, 6
        climate disadvantages, 5
        early anthropoid primates, 4
        features, 5
        food-processing procedures, 5
        great apes/hominids, 4
        modern technologies, 5
        primates group, 3–4
    environmental changes, 19–20
    evolutionary traits
        Barker hypothesis, 16
        childbearing women, 14
        early evolutionary hypotheses, 14
        genetically unknown foods hypothesis, 18
        genetic predisposition, 4, 14–15
        hunter-gatherer populations, 4, 14, 16
        industrial and agricultural revolution, 16–17
        multiple and intricate mechanisms, 18

Human nutrition (*cont.*)
  Neel's hypothesis, 15–16
  thrifty and drifty genotype hypotheses, 15
  thrifty phenotype hypothesis, 16–18
genetic tendency, 20
human migration, 19–20
lifestyle changes, 20
nutritional balance
  body weight set point, 10
  energy-dense diet, 13–14
  epigenetics, 9
  feeding/intake of calories, 9
  functional neuroimaging, 9
  genetics, 9
  hedonic obesity, 10–13
  hunger hormone ghrelin, 11, 13
  metabolic obesity, 9–10, 12
  neurons, 10
  optogenetic experiments, 12–13
  peptides, 10
  positive energy balance, 9
  PVN, 10
obesity
  classification, 8
  comorbidities/metabolic syndrome, 6
  energy balance, 8
  grain-based desserts, 7
  growth hormone deficiency, 8
  hyperinsulinemia, 8
  IGF-1, 8
  medical complications, 6
  metabolic syndrome, 8
  overweight people, 6–7
  PCOS, 8
  polyunsaturated fat intake, 6
  resting energy expenditure, 8
  saturated fats, 6
  self-destruction, 7
  taxpayers, 7
  treatments, 6–7
  visceral adiposity, 8
  WHR, 8
thermogenesis, 18–20
Hunger center, 39–40
Hypothalamic-pituitary-adrenal (HPA) axis
  CRH, 279
  PVN and functions, 38
  stress, 279
Hypothalamus
  afferent component, 30
  amino acids, 30
  ARC
    AgRP, 35–36
    α-MSH peptide, 36–37
    in vitro and in vivo experiments, 34
    low leptin signaling, 35
    median eminence, 34
    NPY, 35–36
    NTS, 35
    POMC, 34–37

    populations, 34
    pro-NPY, 34
    proTRH, 34, 35
    subset of neurons, 34
  brainstem, 31–33, 41
  DMN, 40
  DNA methylation, 62–63
  efferent component, 30
  energy availability
    fasting, 66–68
    POMC, 63–66
    up-and downregulation, 63
  energy homeostasis
    insulin, 43–44
    Janus tyrosine kinases, 42
    leptin, 41–42
    PI3K pathway, 43
    PTP1B, 43
    STAT3, 42–43
  feeding and energy expenditure, 30, 33
  free fatty acids, 30
  function of, 30
  gene expression localization, 56
  genetic/metabolic alterations, 34
  genomic transcriptional analysis, 58–59
  ghrelin, 30, 33
  glucose, 30
  histone H3 methylation, 62–63
  hypothalamic cell line, 56–57
  hypothalamic-feeding network, 30, 32–33
  hypothalamic-pituitary axis, 30–31
  in vivo methods, 57–58
  LHA, 39–40
  median eminence, 30
  melanocortin receptors, 34
  neuroendocrine and autonomic function, 30–32
  neuronal disruptions, 34
  peripheral hormones, 30
  promoters
    FANTOM site, 59–60
    MC4R gene, 61
    neural, 58–59
    POMC, 59–61
  PVN
    ACTH, 39
    α-MSH, 38
    CART, 38
    CRH, 38–39
    GC function, 39
    magnocellular divisions, 37–38
    oxytocin, 37
    parvocellular section, 37–38
    TRH, 37, 44–45
    vasopressin, 37
  tissue-specific transcription, 55
  transcription factors, 60–62
  VMN, 40–41
Hypothalamus-pituitary-thyroid (HPT) axis
  gene regulation, 258–259
  thyroid hormone levels

AHDS, 257
    MCT8 and MCT10, 257
    sulfation and glucuronidation, 257–258
    TRH, 255, 256
    TSH, 255–257
Hypothyroidism, 260, 261

I

Immature secretory granules (ISGs), 119–120
Inflammation
    acute inflammation, 75
    bacterial infection, 75
    basal homeostatic state, 75
    chronic inflammation, 75
    CNS (see (Central nervous system))
    hallmarks
        activators, 76
        HPA axis, 76–77
        IKKβ, 77
        IL-1, 76
        LPS, 76
        NF-κB activation, 77
        NLRP3 ablation, 78
        TLR activation, 77–78
        TNFα, 76
Ingestive behavior
    ghrelin, 359–360
    GLP-1, 360–361
    insulin, 357–359
    leptin, 356–357
Inhibitor of kappa B kinase (IKK) signaling, 93
Innate invariant natural killer T (iNKT) cells, 234
Inositol-requiring protein-1 (IRE1), 87–88
Insulin receptor gene (INSR) mutations, 229
Insulin receptor substrate (IRS), 227, 229
Insulin receptor substrate 1 (IRS1), 88
Insulin signaling, 227–228
Interleukin-1β, 237
Interleukin-6 (IL-6), 236–237
ISRIB, 86

L

Laron syndrome (LS), 326–327
Lateral hypothalamic area (LHA), 39–40
Lateral hypothalamus (LH), 10, 12–13
Leptin receptor (LepRb), 63–64
Lipodystrophy, 214–215
Lipogenesis, 213–214
Liver kinase B 1 (LKB1), 145–146
Long noncoding RNA (lncRNA), 66
Low-density lipoprotein receptor (LDLR), 262
Luteinizing hormone (LH), 31

M

Mechanistic target of rapamycin (mTOR)
    energy surplus and anabolic processes, 144
    MBH-specific overexpression, 150

    mTORC1, 144–145
    mTORC2 function, 150–151
    PI3K-dependent pathway, 149
    Rags, 144
    Rheb GTPase activity, 144
    S6 K1 KO mice, 149–150
Medial prefrontal cortex (mPFC), 12
Median eminence (ME), 34, 79
Melanin-concentrating hormone (MCH), 39–40
Melanocortin receptor 3 (MC3R), 34
Melanocortin receptor 4 (MC4R), 34, 61
Melanocortin receptors (MSRs), 273–275
Melanocortin system
    beta-endorphin
        opioids, 275–277
        POMC, 277, 278
    immune function, 282–284
    MCRs, 273–275
    pain, 284–285
    proopiomelanocortin, 271–273
    stress
        acute stress, 288–290
        behavioral responses, 285–287
        chronic stress, 290–293
        food insecurity and obesity, 287–288
        glucocorticoids, 280–281
        hippocampus and amygdala, 280
        HPA axis, 279, 280
        obesity, 293–294
        physiologic response, 281–282
        psychogenic and emotional stressors, 278
        SNS, 278–279
Melanocyte-concentrating hormone (MCH), 10
Metabolically healthy obese (MHO), 216
Metabolically obese but normal weight
        (MONW), 216, 217
Metabolic syndrome, 8
Mitofusin 1 and mitofusin 2 (MTF1/2), 93
Monocarboxylate transporters 8 and 10 (MCT8 and
        MCT10), 257
mTOR complex 1(mTORC1), 144–145
mTOR complex 2 (mTORC2), 144

N

Nescient helix-loop-helix 2 protein (NHLH2), 65–66
Neural risk factors
    BOLD response, 353
    brain-based predication and consequences, 355
    brain response, 354
Neuroendocrine system, 242–243
Neuropeptide Y (NPY), 10, 34–36, 67
Neutrophils, 233
NF-κB pathway, 87
NG2, 79
Noncoding RNAs, 66
Nucleus accumbens (NAc), 12
Nucleus of the solitary tract (NTS), 35, 127
Nutrient sensors
    AgRP, 142

Nutrient sensors (*cont.*)
  AMPK
    activation results, 146
    autophagy, 152
    BAT-mediated thermogenesis, 152
    CAMKKβ, 145
    glucose, 151
    ICV infusion, 151
    LKB1, 145
    mammals, 145
  ARC, 141
  energy homeostasis, 143
  histone deacetylases
    classes, 147
    expression pattern and regulation, 152
    FoxO1, 153–154
    NAD$^+$ dependence, 148–149
    NAD$^+$/NADH ratio, 147
    neuronal SIRT1 expression, 152–153
    N-terminal acetyltransferases, 146–147
    protein, 146
    SIRT1, 147–149
    sirtuins, 147
  hormonal and stress signals, 143
  JAK2-STAT3 pathway, 142–143
  mTOR
    energy surplus and anabolic processes, 144
    MBH-specific overexpression, 150
    mTORC1, 144–145
    mTORC2 function, 150–151
    PI3K-dependent pathway, 149
    Rags, 144
    Rheb GTPase activity, 144
    S6 K1 KO mice, 149–150
  OGT, 154
  PI3K-Akt pathway, 142
  prohormone theory (*see* (Prohormone theory))

O
Obesity
  adipokines (*see* (Adipokines))
  food-specific aberrant response, 351–352
  generalized reward sensitivity, 352
  GI hormones
    biosynthesis, 184
    cholecystokinin, 187–189, 195
    closed-type cells, 184
    features, 194
    ghrelin, 185–188
    GLP-1, 189–190
    GPCRs, 185
    G proteins, 185
    open-type cells, 184
    oxyntomodulin, 192–193, 196
    pancreatic polypeptide, 193, 196
    PYY, 190–192, 196
    RYGB, 194, 195
    secretion, 184
    somatostatin, 193
    translational modifications, 184

glucose homeostasis, 226
growth hormone, 332–335
impulsivity, 352
inflammation, adipose tissue
  BAT, 231–232
  hyperlipidemia and hyperglycemia, 231
  immune responses, 232–234
  WAT, 231–232
insulin resistance
  cellular and molecular stress, 228–229
  definition, 226
  glycogenolysis and gluconeogenesis, 230
  IRS, 229
  liver and muscle, 229, 230
insulin signaling, 227–228
melanocortin system, 293–294
thyroid hormone, 262–263
type 2 diabetes, 225
O-GlcNAcase (OGA), 154
O-GlcNAc transferase (OGT), 154
Oligodendrocytes (ODs), 79
Omentin-1, 241
Otsuka Long-Evans Tokushima Fatty (OLETF) rat, 242
Oxyntomodulin (OXM), 192–193, 196

P
Pancreatic polypeptide (PP), 193, 196
Paraventricular nucleus (PVN), 10
  ACTH, 39
  α-MSH, 38
  CART, 38
  CRH, 38–39
  GC function, 39
  magnocellular divisions, 37–38
  oxytocin, 37
  parvocellular section, 37–38
  TRH, 37, 44–45
  vasopressin, 37
PC1, *see* Prohormone convertase 1
PC2, *see* Prohormone convertase 2
PCs, *see* Prohormone convertases
Peptide tyrosine-tyrosine (PYY), 33, 190–192
Per-Arnt-Sim (PAS) kinases, 154
Periaqueductal gray matter (PAG), 115, 284–285
Periventricular nucleus (PeN), 62
4-phenyl butyric acid (4PBA), 85
Phosphatidylinositol-3-kinase (PI3K) pathway, 43
Phosphoinositide-dependent kinase 1 (PDK1), 43
Polycystic ovarian syndrome (PCOS), 8, 208
Polypeptide proopiomelanocortin (POMC), 271–273
POMC, *see* Pro-opiomelanocortin
POMC-processing enzymes PC2, 93
Portion size effect, 349–350
PPP1R15A, 86
PPP1R15B, 86
Pro-growth hormone-releasing hormone
    (pro-GHRH), 114
Prohormone convertase 1 (PC1), 95
Prohormone convertase 2 (PC2), 93–95
Prohormone convertases (PCs)

7B2, 114
cellular and extracellular compartments, 111–112
hyperglycemia inflammation, 115
mammalian hormones, 111
PC1 and PC2, 112–114
PCSK9, 112
posttranslational processing, 111
processing enzymes, 115–116
serine proteases, 111–113
SKI-1, 112
Prohormone theory
    amino acid sequences, 110–111
    carboxypeptidase E and D, 155
    CRH, 128–130
    pancreatic insulinoma, 110
    PCs
        7B2, 114
        cellular and extracellular compartments, 111–112
        hyperglycemia inflammation, 115
        mammalian hormones, 111
        PC1 and PC2, 112–114, 155
        PCSK9, 112
        posttranslational processing, 111
        processing enzymes, 115–116
        serine proteases, 111–113
        SKI-1, 112
    POMC
        ACTH, 126
        ARC, 126–127
        NTS, 127
        PC1 and PC2, 128
        posttranslational processing, 125–126
    posttranslational proteolysis, 154–155
    Pro-TRH
        cellular and biochemical, 118–119, 122
        C-terminal, 124
        15 and 10 kDa peptides, 118–119, 123
        Gln-His-Pro-Gly, 122–123
        Lysi-Arg/Arg-Arg, 118–119, 123
        MC4 receptor, 125
        N-terminal flanking peptide, 123, 124
        PC1 and PC2, 124
        STZ-induced animals, 125
        UCP1, 125
    pulse-labeling protocol, 110
    RSP
        constitutive pathway secretion, 117
        endocrine and neuroendocrine cells, 116–117
        ER, 116–117, 122
        GC, 119
        intracellular sorting mechanism, 121
        ISGs, 119–120
        N-terminal pro-segment, 122
        rat pro-TRH, 116, 118
        SGs, 119
        sorting signals, 118–119, 121–122
        TGN, 118–119
    SIRT1
        ARC, 156
        collaborative studies, 160–161

DIO, 158–161
    genetic ablation, 158
    ghrelin, 156
    orexigenic vs. anorexigenic role, 158
    orexins, 157
    proCRH, 161–164
    short-term refeeding, 157
    SINKO, 156–157
    Sprague-Dawley rat model, 155–156
vasopressin, 110
Pro-inflammatory adipokines
    CCL2 and CCR5, 237–238
    chemerin, 238
    DPP-4, 238
    interleukin-1β, 237
    interleukin-6, 236–237
    leptin, 235–236
    RBP4, 238
    resistin, 239
    TNF-α, 236
Prolactin (PRL), 31
Promoter-bashing, 57
Pro-opiomelanocortin (POMC), 10, 34–37
    ACTH, 126
    ARC, 126–127
    expression levels, 60–61
    FoxO1, 64
    homeobox transcription factors, 65
    hypothalamic-specific genes, 60
    initial studies, 59
    Jak-Stat signaling pathway, 63
    LepRb, 63–64
    NHLH2 gene, 65–66
    NTS, 127
    PC1 and PC2, 128
    PCSK1, 65
    polymorphisms, 61
    posttranslational processing, 125–126
    Stat5, 63–64
    transgenic mice, 59
Pro-protein convertase subtilisin kexin 9
    (PCSK9), 112
Protein kinase RNA (PKR)-like ER kinase
    (PERK), 85–86
Protein-tyrosine phosphatase 1B (PTP1B), 43
Pro-thyrotropin-releasing hormone (pro-TRH)
    cellular and biochemical, 118–119, 122
    C-terminal, 124
    15 and 10 kDa peptides, 118–119, 123
    Gln-His-Pro-Gly, 122–123
    Lysi-Arg/Arg-Arg, 118–119, 123
    MC4 receptor, 125
    N-terminal flanking peptide, 123, 124
    PC1 and PC2, 124
    STZ-induced animals, 125
    UCP1, 125

Q
Quantitative PCR, 58

R
Rag proteins (Rag A-D), 144
Regulated secretory pathway (RSP)
    constitutive pathway secretion, 117
    endocrine and neuroendocrine cells, 116–117
    ER, 116–117, 122
    GC, 119
    intracellular sorting mechanism, 121
    ISGs, 119–120
    N-terminal pro-segment, 122
    rat pro-TRH, 116, 118
    SGs, 119
    sorting signals, 118–119, 121–122
    TGN, 118–119
Resting energy expenditure (REE), 8, 260
Retinol-Binding Protein 4 (RBP4), 238
Roux-en-Y gastric bypass (RYGB), 194, 195

S
Secreted Frizzled-Related Protein 5 (SFRP5), 240–241
Secretory granules (SGs), 119
Senescent cells, 210
Serine/threonine kinase 11, 145–146
Signal transducer and activator of transcription 3
        (STAT3), 42–43
Signal transducer and activator of transcription 5
        (STAT5), 63–64, 322, 323
SIRT1 knockout (SINKO) model, 156–157
Somatostatin (SST), 193
Specificity protein 1 (Sp1), 64
S6 K1 knockout (S6 K1 KO) mice, 149–150
Stress
    acute stress and body weight, 288–290
    behavioral responses, 285–287
    chronic stress and body weight
        anorexia, 290, 291
        cortisol habituation, 291
        CRH, 292–293
        glucocorticoids, 291, 292
    food insecurity and obesity, 287–288
    glucocorticoids, 280–281
    hippocampus and amygdala, 280
    HPA axis, 279, 280
    obesity, 293–294
    physiologic response, 281–282
    psychogenic and emotional stressors, 278
    SNS, 278–279
Stromal vascular fraction (SVF), 210
Subtilisin kexin isozyme 1 (SKI-1), 112
Suprachiasmatic nucleus (SCN), 62
Sympathetic nervous system (SNS), 278–279

T
Taurour-sodeoxycholic acid (TUDCA), 85
Thiazolidinedione (TZD), 234

Thyroid hormone
    body weight regulation
        direct and indirect effects, 261–262
        hypothyroidism, 260, 261
        obesity, 262–263
        REE, 260
    HPT axis (*see* (Hypothalamus-pituitary-thyroid axis))
    weight loss effects, 263–264
Thyroid-stimulating hormone (TSH), 31, 255
Thyrotropin-releasing hormone (TRH), 10, 37, 44–45,
        115, 255
Toll-like receptor 4 (TLR4), 83
TRH-Gly, 115
Tumor necrosis factor-α (TNF-α), 236
Type 2 diabetes (T2D), 225

U
UDP-N-acetylglucosamine (UDP-GlcNAc), 154
Unc-51-like kinase 1 (ULK1), 146
Uncoupling protein 1 (UCP1), 18, 205, 231, 261
Uncoupling protein 2 (UCP2), 156
Unfolded protein response (UPR), 91
Upstream open reading frames (uORF), 86–87

V
Ventral tegmental area (VTA), 354
Ventromedial hypothalamus (VMH), 10, 40–41
Ventromedial nucleus (VMN), 40–41
Very low-density lipoprotein (VLDL), 88
Visceral adipose tissue-derived serine protease inhibitor,
        242

W
Waist circumference to hip circumference (WHR), 8
Weather forecast model, 16
White adipose tissue (WAT), 18
    cellular and noncellular composition
        adipocytes, 209–210
        beige adipocytes, 206, 211
        ECM, 210–211
        intra-depot adipocyte heterogeneity, 211–212
        lymph and vasculature, 210
        senescent cells, 210
        SVF, 210
    depots, 206–208
    endocrine, 212–213
    inflammation, 231–232
    lipodystrophy, 214–215
    obesity, 215–216

X
X-box-binding protein-1 (XBP1), 87, 88, 91